Lecture Notes on Data Engineering and Communications Technologies

Volume 153

Series Editor

Fatos Xhafa, Technical University of Catalonia, Barcelona, Spain

T0186762

The aim of the book series is to present cutting edge engineering approaches to data technologies and communications. It will publish latest advances on the engineering task of building and deploying distributed, scalable and reliable data infrastructures and communication systems.

The series will have a prominent applied focus on data technologies and communications with aim to promote the bridging from fundamental research on data science and networking to data engineering and communications that lead to industry products, business knowledge and standardisation.

Indexed by SCOPUS, INSPEC, EI Compendex.

All books published in the series are submitted for consideration in Web of Science.

Ning Xiong · Maozhen Li · Kenli Li · Zheng Xiao ·
Longlong Liao · Lipo Wang
Editors

Advances in Natural Computation, Fuzzy Systems and Knowledge Discovery

Proceedings of the ICNC-FSKD 2022

Volume 1

 Springer

Editors
Ning Xiong
Division of Intelligent Future Technologies
Mälardalen University
Västerås, Västmanlands Län, Sweden

Kenli Li
School of Information Science
and Technology
Hunan University
Changsha, Hunan, China

Longlong Liao
College of Computer and Data Science
Fuzhou University
Fuzhou, Fujian, China

Maozhen Li
Department of Electronic and Computer
Engineering
Brunel University London
Uxbridge, Middlesex, UK

Zheng Xiao
School of Information Science
and Technology
Hunan University
Changsha, Hunan, China

Lipo Wang
School of Electrical and Electronic
Engineering
Nanyang Technological University
Singapore, Singapore

ISSN 2367-4512   ISSN 2367-4520 (electronic)
Lecture Notes on Data Engineering and Communications Technologies
ISBN 978-3-031-20737-2   ISBN 978-3-031-20738-9 (eBook)
https://doi.org/10.1007/978-3-031-20738-9

This Springer imprint is published by the registered company Springer Nature Switzerland AG
The registered company address is: Gewerbestrasse 11, 6330 Cham, Switzerland

Organizing Committee

General Chairs

Lipo Wang   Nanyang Technological University, Singapore
Kenli Li   Hunan University, China

Organizing General Chairs

Wenzhong Guo   Fuzhou University, China
Yuanlong Yu   Fuzhou University, China

Program Chairs

Maozhen Li   Brunel University London, UK
Ning Xiong   Mälardalen University, Sweden

Organizing Chair

Longlong Liao   Fuzhou University, China

Publication Chairs

Xing Chen   Fuzhou University, China
Mingjian Fu   Fuzhou University, China
Guobao Xiao   Minjiang University, China

Publicity Chairs

Wenxi Liu   Fuzhou University, China
Chunyan Xu   Nanjing University of Science and Technology, China
Shanshan Fan   Beijing Language and Culture University, China

Finance Chair

Zheng Xiao   Hunan University, China

Sponsorship Chair

Xinqi Liu   University of Hong Kong, Hong Kong, China

Program Committee

Shigeo Abe	Kobe University, Japan
Henry N. Adorna	University of the Philippines, The Philippincs
Davide Anguita	University of Genoa, Italy
Sabri Arik	Istanbul University, Turkey
Krassimir Atanassov	Bulgarian Academy of Sciences, Bulgaria
Sansanee Auephanwiriyakul	Chiang Mai University, Thailand
Philip Azariadis	University of the Aegean, Greece
Vladan Babovic	Singapore National University, Singapore
Thomas Bäck	Leiden Institute of Advanced Computer Science, Netherland
Emili Balaguer-Ballester	Bournemouth University, UK
Valentina Balas	Aurel Vlaicu University of Arad, Romania
Yaxin Bi	University of Ulster, UK
Federico Bizzarri	Politecnico di Milano, Italy

Tossapon Boongoen	Mae Fah Luang University, Thailand
Pierre Borne	Ecole Centrale de Lille, France
Hamid Bouchachia	Bournemouth University, UK
Ivo Bukovsky	Czech Technical University in Prague, Czech
Sujin Bureerat	KhonKaen University, Thailand
Godwin Caruana	Harvest Technology, Malta
Michele Ceccarelli	University of Sannio, Italy
Kit Yan Chan	Curtin University, Australia
Chen-Tung Chen	National United University, Taiwan
David Daqing Chen	London South Bank University, UK
Jianxia Chen	Washington University in St. Louis, USA
Syuan-Yi Chen	National Taiwan Normal University, Taiwan
Chi Tsun (Ben) Cheng	RMIT University, Australia
Jao Hong Cheng	National Yunlin University of Science and Technology, Taiwan
France Cheong	RMIT University, Australia
Jen-Shiun Chiang	Tamkang University, Taiwan
Panagiotis Chountas	University of Westminster, UK
Huey-Der Chu	Takming University of Science and Technology, Taiwan
Hung-Yuan Chung	National Central University, Taiwan
Alessandro Colombo	Politecnico di Milano, Italy
José Alfredo F. Costa	Universidade Federal do Rio Grande do Norte, Brazil
Keeley Crockett	Manchester Metropolitan University, UK
Zoltán Ernö Csajbók	University of Debrecen, Hungary
Darryl N. Davis	University of Hull, UK
Andre C. P. L. F. de Carvalho	University of Sao Paulo, Brazil
Marc de Kamps	University of Leed, UK
Mingcong Deng	Tokyo University of Agriculture and Technology, Japan
Minghua Deng	Peking University, China
Milena Djukanovic	University of Montenegro, Montenegro
Mustafa Dogan	Baskent University, Turkey
Prabu Dorairaj	Broadcom Inc, India
Giorgos Dounias	University of the Aegean, Greece
António Dourado	University of Coimbra, Portugal
Abdelali El Aroudi	Universitat Rovira i Virgili, Spain
Mohammed El Abd	The American University of Kuwait, Kuwait
Zuhal Erden	ATILIM University, Turkey
Geoffrey Falzon	STMicroelectronics (Malta) Ltd, Malta
Xiannian Fan	City University of New York, USA
Saeed Panahian Fard	Universiti Sains Malaysia, Malaysia
Elisabetta Fersini	University of Milan Bicocca, Italy
Zbigniew Galias	AGH University of Science and Technology, Poland

Peter Geczy	AIST, Japan
Damian Giaouris	Newcastle University, UK
Onofrio Gigliotta	University of Naples Federico II, Italy
David Glass	University of Ulster, UK
Antonio Gonzalez	University of Granada, Spain
Giuseppe Grassi	University of Salento, Italy
Perry Groot	Radboud University Nijmegen, The Netherlands
Yuzhu Guo	University of Sheffield, UK
Jianchao (Jack) Han	California State University, USA
Thomas Hanne	University of Applied Sciences Northwestern Switzerland, Switzerland
Pitoyo Hartono	Chukyo University, Japan
Enrique Herrera-Viedma	University of Granada, Spain
Mhand Hifi	Université de Picardie, France
Ladislav Hluchy	Institute of Informatics, Slovak Academy of Sciences, Slovakia
Sean Holden	University of Cambridge, UK
Jun Hong	University of the West of England, Bristol, UK
Tzung-Pei Hong	National University of Kaohsiung, Taiwan
Wei-Chiang Samuelson Hong	Oriental Institute of Technology, Taiwan
Wen-xing Hong	Xiamen University, China
Xia Hong	University of Reading, UK
He Hu	Renmin University of China, China
Min Huang	Northeast University, China
Natthakan IamOn	Mae Fah Luang University, Thailand
Abdullah M. Iliyasu	Tokyo Institute of Technology, Japan
Raimundas Jasinevicius	Kaunas University of Technology, Lithuania
Richard Jensen	Aberystwyth University, UK
Zhuhan Jiang	University of Western Sydney, Australia
Colin Johnson	University of Kent, UK
Vladimir Jotsov	State University for Library Studies and Information Technologies, Bulgaria
Mehmet Karakose	Firat University, Turkey
Yoshiki Kashimori	University of Electro-Communications, Japan
Radoslaw Katarzyniak	Wroclaw University of Technology, Poland
A. S. M. Kayes	La Trobe University, Australia
DaeEun Kim	Yonsei University, South Korea
Mario Koeppen	Kyushu Institute of Technology, Japan
Vladik Kreinovich	University of Texas at El Paso, USA
Paul Kwan	University of New England, Australia
Wai Lam	The Chinese University of Hong Kong, China
Jimmy Lauber	University of Valenciennes, France
Chen Li	ETH Zurich, Switzerland
Gang Li	Deakin University, Australia
Kang Li	Queen's University Belfast, UK

Preface

The 2022 18th International Conference on Natural Computation, Fuzzy Systems and Knowledge Discovery (ICNC-FSKD 2022) was held from July 30 to August 1, 2022, online.

ICNC-FSKD is a premier international forum for scientists and researchers to present the state of the art of machine learning, data mining, and intelligent methods inspired from nature, particularly biological, linguistic, and physical systems, with applications to computers, systems, control, communications, and more. This is an exciting interdisciplinary area in which a wide range of theory and methodologies are being investigated and developed to tackle complex and challenging problems. We are delighted to receive many submissions from around the globe. After a rigorous review process, the accepted papers are included in this proceedings.

We have been looking forward to holding the conference in beautiful Fuzhou. However, the recent COVID-19 clusters in China have prompted the organizing committee to move this year's conference to online only. It is a pity that we cannot meet you all physically this year. But the pandemic will eventually be overcome, and we look forward to seeing you again next year in the sunny summer!

We would like to sincerely thank all organizing committee members, program committee members, and reviewers for their hard work and valuable contribution. Without your help, this conference would not have been possible. Special thanks go to the main organizer of this year's conference, College of Computer and Data Science of Fuzhou University, as well as the co-organizer, Science and Technology on Communication Information Security Control Laboratory. We thank Springer for publishing the proceedings. In particular, we thank Series Editors, Prof. Fatos Xhafa and Dr. Thomas Ditzinger, for their kind support. We are very grateful to the keynote speakers for their authoritative speeches. We thank all authors and conference participants for using this platform to communicate their excellent work.

August 2022

Ning Xiong
Maozhen Li
Kenli Li
Zheng Xiao
Longlong Liao
Lipo Wang

Contents

Machine Learning and Data Science (19)

Deep Learning (34)

Multiple Layers Global Average Pooling Fusion

Silei Cao, Shun Long, Weiheng Zhu[(✉)], Fangting Liao, Zeduo Yuan, and Xinyi Guan

Jinan University, Guangzhou 510632, China
{slaycao1998,yuanzd,guanxy98}@stu2020.jnu.edu.cn, {tlongshun,
tzhuwh}@jnu.edu.cn, 771791045@qq.com, 412442830@qq.com,
liaoft@stu2021.jnu.edu.cn

Abstract. We propose for deep convolutional neural network (CNN) a simple but effective feature fusion technique called *multiple layers global average pooling fusion (MLGAPF)*. It adds a branch at each CNN layer or module which uses global average pooling to extract global features, and these features are then fused for classification. Empirical experiments show that this technique can effectively improve the accuracy of ResNet, GoogleNet, SqueezeNet, MobileNetv2 and others. On average, MLGAPF brought additional performance enhancement of 2.62% and 2.49% on CIFAR100 and Tiny-ImageNet respectively.

Keywords: Convolutional neural network · Feature fusion · Global average pooling

1 Introduction

The past decade has witnessed a rapid development of deep neural networks in various domains, particularly in the areas of image processing and computer vision. The adoption of deep convolutional neural network (CNN) as a backbone for image processing plays a crucial role in classification tasks such as object detection and semantic segmentation. Various CNNs have been proposed, for instance VGG [1], GoogleNet [2], ResNet [3], Inception [4], DenseNet [5], SqueezeNet [6], ShuffleNet [7], MobileNet [8] and some networks for medical image classification [9–13]. Their effectiveness has been well validated in a large variety of computer vision tasks. However, most of these established CNNs use only the deep features (close to the output, of high level semantic, but of low resolution) extracted from the last convolution layer, while ignoring shallow features (close to the input, of low level semantic, but of high resolution) that may also provide vital hints for classification. Despite some CNN classifiers adopt global average pooling to capture global information from features in search for better results in practice, there is a lack of systematic use of both deep and shallow features extracted. Inspired by the combinatory nature of biological neural network, we believe that the pyramidal features formed by a proper combination of deep and shallow features can yield further improvement in classification.

This paper presents for image classification a novel yet simple feature fusion technique called multiple layers global average pooling fusion (MLGAPF), where a branch

N. Xiong et al. (Eds.): ICNC-FSKD 2022, LNDECT 153, pp. 3–10, 2023.
https://doi.org/10.1007/978-3-031-20738-9_1

is added at each layer or module to extract global features via the use of global average pooling, and these global features are then concatenated for fusion. This proposed architecture provides two major advantages in practice. First, after global average pooling, all feature maps are equal in size, suggesting that no alignment is needed between deep features with shallow features. Irrelevant parameters and information can therefore be avoided. Compared to other feature fusion methods, this architecture can reduce complexity (number of parameters) and in turn training overhead (time). In addition, because of the shortcuts between each layer and module and its last counterparts, this architecture can effectively prevent gradient disappearance when the network scales up.

To evaluate MLGAPF, we have carried out an empirical experiment on two datasets in Cifar100 [14] and Tiny-ImageNet [15], and compared the results against those of established models in ResNet, GoogleNet, SqueezeNet, MobileNetv2 and their variants. Experimental results confirm its effectiveness in performance improvement.

2   Related Work

Fusion of features at different scales is an important means to improve performance for deep learning tasks. Shallow features provide rich and detailed information such as location, but their semantic are at relatively low level due to less convolution undergone. On the contrary, deep features provide richer semantic information but without perception of details due to low resolution. Fusion is therefore of vital importance to fully exploit the rich semantic of deep features and the fine-grained information of shallow features in search for high accuracy.

Feature fusion has been widely adopted in computer vision practice. In object detection, Adaptive Spatial Feature Fusion (ASFF) [16] technique is a new data-driven pyramid feature fusion strategy that can effectively improve the accuracy of small object detection. In semantic segmentation, there are two main feature fusion strategies, namely Feature Pyramid Network (FPN) [17] and High-Resolution Net (HRNet) [18]. FPN constructs a feature pyramid by first obtaining high-level semantic features via bottom-up up-sampling, then up-sampling them in a top-down manner before fusing them with the shallow features of corresponding resolution. HRNet provides multiple-resolution bottom-up pathways from the bottom up which enables features of different resolutions be fused promptly during the bottom-up process. Both FPN and HRNet have to align deep and shallow features, which cannot be solved via simple up-sampling. Optical flow based amendments have been proposed, Semantic Flow [19] is a top-down strategy for FPN and AlignSeg [20] is a bottom-up strategy for HRNet. But they all make the whole computer vision task more complex.

In object detection, many networks (for instance YOLOv3 [21] or RetinaNet [22]) use FPN for feature fusion, where concatenation is used to directly connect features of various layers. Given two input features x and y of dimension p and q respectively, the dimension of the output feature is $p + q$. Alternatively, a direct summation is adopted where these two feature vectors are combined into a complex vector z as shown in Eq. (1), where i is an imaginary unit.

$$z = x + iy \tag{1}$$

In semantic segmentation, both shallow and deep features are desirable. Generally saying, deep features preserve the overall semantic structure of the image but at the cost of a severe loss of detailed information, and shallow features retain detailed information but of only small and local semantic structure. Both fully convolutional networks (FCN) [23] and U-Net [24] adopt a fusion strategy where deep features are first bilinearly interpolated with shallow features of the same resolution, before fused via summation or dimensional concatenation along the channel. However, these solutions need interpolation for up-sampling in order to align features, leading to extra irrelevant parameters and computational complexity.

3   Multiple Layers Global Average Pooling Fusion

The architecture of our MLGAPF is illustrated in Fig. 1 which shows a branch is added at each layer or module to extract global features via the use of global average pooling. These global features are then concatenated for fusion.

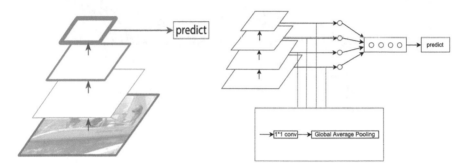

Fig. 1. A normal image classifier CNN (left) versus MLGAPF (right)

This section shows how our MLGAPF can be added to established models such as ResNet, SqueezeNet, MobileNetv2 [25], ShuffleNetv2 [26] and other models. These established CNN models are roughly divided into two major categories. One is the network without a convolution layer of kernel size 1*1 before the final classifier layer (the last layer in CNN, which is either a convolutional layer of kernel size 1*1 or a fully connected layer). CNNs of this category include SqueezeNet, ResNet, and GoogleNet, etc. The other is networks with a convolution layer of kernel size 1*1 before the classifier layer, such as MobileNetv2 and ShuffleNetv2, etc.

We take SqueezeNet as an example to show how MLGAPF is applied to the first category of CNN. As shown in Fig. 2, a 1*1 convolution layer is first added to not only the first convolution layer but also to each of the 8 fire modules in the original SqueezeNet. This enhances not only the nonlinearity of networks width but also the features of all 9 layers. Then, global average pooling is applied to extract global information of these enhanced features. Once accomplished, these global features are equal in size and ready for concatenation, i.e., no need for up-sampling for alignment. Once concatenated, these multi-layer global features are forwarded to a convolution layer of kernel size 1*1 (the green box on the right) for fusion. The channels are then re-mapped to the original one accordingly with a converted channel number, before the classifier layer yields the prediction.

In channel remapping, if there is a convolution layer of kernel size 1*1 before the classifier layer in the SqueezeNet, it can convert the channel number directly, otherwise (i.e. the original SqueezeNet does not have a layer of kernel size 1*1 before the classifier layer), a new convolutional layer of kernel size 1*1 for feature fusion must be added for conversion.

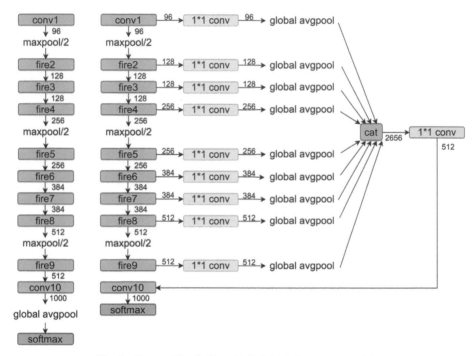

Fig. 2. SqueezeNet (left) and MLGAPF-SqueezeNet (right)

ShuffleNetv2 is used as an example to show how MLGAPF is applied to the second category of CNN. Similar to the prior case, Fig. 3 (right) shows that convolution layers of kernel size 1*1 (yellow boxes in Fig. 3) are added to both the first convolution layer and the three stage modules in the original ShuffleNetv2 to enhance not only the nonlinearity of network width but also the features of each layer and stage. Global average pooling is then applied to extract global information of the enhanced features. These enhanced global features are all equal in size and therefore ready for direct concatenation (the red cat box on right), i.e. without up-sampling for alignment. The concatenated global features are then directly fused by the convolution layer (of kernel size 1*1, the conv5 box in Fig. 3) provided in ShuffleNetv2, before forwarded to the classifier layer for final output. Unlike the prior case of Fig. 2, no additional convolution is needed. This direct feature fusion reduces the complexity caused by unnecessary parameters and extra computation.

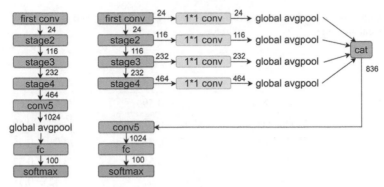

Fig. 3. ShuffleNetv2 (left) and MLGAPF-ShuffleNetv2 (right)

4 Experimental Results and Analysis

4.1 Experiment Setup

To evaluate the ability of MLGAPF, we carried out our experiments in two different settings. One used CIFAR-100 dataset, which consists of 60000 32 × 32 colored images in 100 classes, with 600 images per class, and the experiments were conducted on a NVIDIA GeForce RTX 2080Ti. The other setting used Tiny-ImageNet dataset which consists of 100000 64 × 64 colored images in 200 classes, with 500 images per class, and the experiments were conducted on a NVIDIA Tesla V100.

Our experiments followed the hyperparameter settings given in [27], with an initial value of 0.1 for the learning rate, the learning rate is divided by 5 at the 60th epoch, 120th epoch, and 160th epoch. The training last for 200 epochs with a batchsize of 128, a weight decay of 5e-4, and a Nesterov momentum of 0.9. Stochastic gradient descent is adopted as the optimizer in our experiments.

4.2 Ablation Experiments

The results of our ablation experiment are summarized in Table 1, which shows that the introduction of MLGAPF has brought performance improvement to all the deep learning models in our experiments. The most significant improvement was achieved on SqueezeNet, where the accuracy was raised from 70.37% to 74.68% on CIFAR100 and from 58.11% to 63.54% on Tiny-ImageNet, i.e. an improvement of 4.31% and 5.43% respectively. The modest improvements 0.87% was found in ShuffleNetv2 on CIFAR100, and similar achievements were found on SeResNet18 for Tiny-ImageNet (1.02%) and GoogleNet for CIFAR100 (1.18%). On average, MLGAPF brought additional performance enhancement of 2.62% on CIFAR100, and 2.49% on Tiny-ImageNet.

To demonstrate how MLGAPF makes the improvement, we plotted the accuracy against epoch on SqueezeNet, and the result is given in Fig. 4. It shows that the accuracy leaps after roughly 60, 120 and 160 epochs, where the learning rate was reduced. If the learning rate stays unchanged, the loss is likely to keep vibrating around its minimum. But once we lower the learning rate, the amplitude of vibration may drop to, leading its

Table 1 Comparison of the accuracy between the original network and the network with our MLGAPF enhancement

Dataset	Model	Accuracy (%)	
		Without MLGAPF	With MLGAPF
CIFAR100	SqueezeNet	70.37 ± 0.59	74.68 ± 0.61
	MobileNetv2	68.52 ± 0.52	72.08 ± 0.49
	ShuffleNetv2	70.25 ± 0.54	74.01 ± 0.58
	ResNet18	76.18 ± 0.55	78.08 ± 0.52
	SeResNet18	76.40 ± 0.37	77.42 ± 0.38
	GoogleNet	76.90 ± 0.42	78.08 ± 0.41
Tiny-ImageNet	SqueezeNet	58.11 ± 0.60	63.54 ± 0.66
	MobileNetv2	56.11 ± 0.51	59.67 ± 0.45
	ShuffleNetv2	60.35 ± 0.43	61.22 ± 0.42
	ResNet18	64.35 ± 0.49	66.12 ± 0.48
	SeResNet18	64.76 ± 0.39	66.35 ± 0.44
	GoogleNet	66.49 ± 0.55	68.18 ± 0.63

minimum, i.e. a higher accuracy. However, a more frequent reduction learning rate would keep the loss at its local minimum instead of reach its global optimum, as suggested by [27].

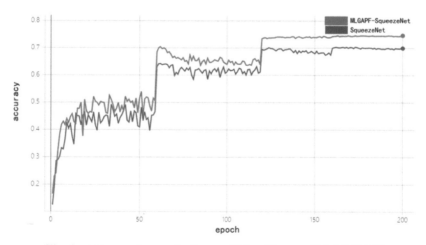

Fig. 4. Accuracy changes for SqueezeNet and SqueezeNet+MLGAPF

5 Conclusion

Fusion of features from multiple layers promises performance improvement in deep learning. It has been widely adopted in typical downstream deep learning tasks such as object detection and semantic segmentation. However, up-sampling or down-sampling is usually needed in these cases for feature alignment, which may bring in non-existing information whilst miss genuinely meaningful information. This paper presents a simple amendment to this problem, namely multiple layers global average pooling fusion (MLGAPF) for classification. A convolution layer of kernel size 1*1 and a global average pooling layer are added to each layer or module in a CNN. The resulting features from multiple layers or modules are then concatenated and sent to another convolution layer of kernel size 1*1 for fusion, right before the classifier layer makes its prediction. The innovation of this structure is that feature fusion can be performed without up-sampling for feature alignment and the global average pooling technique is used to fully extract the shallow features, which reduces the training parameters compared to other feature fusion methods and improves the accuracy of the original network. Empirical experiments show that MLGAPF can bring further accuracy improvements of 0.87–5.43% to various CNNs.

References

1. Simonyan, K., Zisserman, A.: Very deep convolutional networks for large-scale image recognition. In: 3rd International Conference on Learning Representations (ICLR 2015), pp. 1–14 (2019)
2. Szegedy, C., Liu, W., Jia, Y., Sermanet, P., Reed, S., Anguelov, D., Erhan, D., Vanhoucke, V., Rabinovich, A. (2015). Going deeper with convolutions. In: Proceedings of the IEEE Conference on Computer Vision and Pattern Recognition, pp. 1–9
3. He, K., Zhang, X., Ren, S., Sun, J.: Deep residual learning for image recognition. In: Proceedings of the IEEE Conference on Computer Vision and Pattern Recognition, pp. 770–778 (2016)
4. Szegedy, C., Ioffe, S., Vanhoucke, V., Alemi, A.A.: Inception-v4, inception-resnet and the impact of residual connections on learning. In: Thirty-first AAAI Conference on Artificial Intelligence (Feb 2017)
5. Huang, G., Liu, Z., Van Der Maaten, L., Weinberger, K.Q.: Densely connected convolutional networks. In: Proceedings of the IEEE Conference on Computer Vision and Pattern Recognition, pp. 4700–4708 (2017)
6. Iandola, F.N., Han, S., Moskewicz, M.W., Ashraf, K., Dally, W.J., Keutzer, K.: SqueezeNet: AlexNet-level accuracy with 50x fewer parameters and < 0.5 MB model size (2016). arXiv preprint arXiv:1602.07360
7. Zhang, X., Zhou, X., Lin, M., Sun, J.: Shufflenet: an extremely efficient convolutional neural network for mobile devices. In: Proceedings of the IEEE Conference on Computer Vision and Pattern Recognition, pp. 6848–6856 (2018)
8. Howard, A.G., Zhu, M., Chen, B., Kalenichenko, D., Wang, W., Weyand, T., Andreetto, M., Adam, H.: Mobilenets: efficient convolutional neural networks for mobile vision applications (2017). arXiv preprint arXiv:1704.04861
9. Badža, M.M., Barjaktarović, M.Č: Classification of brain tumors from MRI images using a convolutional neural network. Appl. Sci. **10**(6), 1999 (2020)

10. Ker, J., Wang, L., Rao, J., Lim, T.: Deep learning applications in medical image analysis. IEEE Access **6**, 9375–9389 (2017)

11. Yadav, S.S., Jadhav, S.M.: Deep convolutional neural network based medical image classification for disease diagnosis. J. Big Data **6**(1), 1–18 (2019). https://doi.org/10.1186/s40537-019-0276-2

12. Singh, S.P., Wang, L., Gupta, S., Goli, H., Padmanabhan, P., Gulyás, B.: 3D deep learning on medical images: a review. Sensors **20**(18), 5097 (2020)

13. Singh, S.P., Wang, L., Gupta, S., Gulyas, B., Padmanabhan, P.: Shallow 3D CNN for detecting acute brain hemorrhage from medical imaging sensors. IEEE Sens. J. **21**(13), 14290–14299 (2020)

14. Krizhevsky, A.: Learning multiple layers of features from tiny images. Master's thesis. University of Tront (2009)

15. Le, Y., Yang, X.: Tiny imagenet visual recognition challenge. CS 231N **7**(7), 3 (2015)

16. Liu, S., Huang, D., Wang, Y.: Learning spatial fusion for single-shot object detection (2019). arXiv preprint arXiv:1911.09516

17. Lin, T.Y., Dollár, P., Girshick, R., He, K., Hariharan, B., Belongie, S.: Feature pyramid networks for object detection. In: Proceedings of the IEEE Conference on Computer Vision and Pattern Recognition, pp. 2117–2125 (2017)

18. Wang, J., Sun, K., Cheng, T., Jiang, B., Deng, C., Zhao, Y., Liu, D., Mu, Y., Tan, M., Wang, X., Liu, W.: Deep high-resolution representation learning for visual recognition. IEEE Trans. Pattern Anal. Mach. Intell. **43**(10), 3349–3364

19. Li, X., You, A., Zhu, Z., Zhao, H., Yang, M., Yang, K., Tan, S., Tong, Y.: Semantic flow for fast and accurate scene parsing. In: European Conference on Computer Vision, pp. 775–793. Springer, Cham (Aug 2020)

20. Huang, Z., Wei, Y., Wang, X., Liu, W., Huang, T.S., Shi, H.: Alignseg: feature-aligned segmentation networks. IEEE Trans. Pattern Anal. Mach. Intell. **44**(1), 550–557 (2021)

21. Redmon, J., Farhadi, A.: Yolov3: an incremental improvement (2018). arXiv preprint arXiv: 1804.02767

22. Lin, T.Y., Goyal, P., Girshick, R., He, K., Dollár, P.: Focal loss for dense object detection. In: Proceedings of the IEEE International Conference on Computer Vision, pp. 2980–2988 (2017)

23. Long, J., Shelhamer, E., Darrell, T.: Fully convolutional networks for semantic segmentation. In: Proceedings of the IEEE Conference on Computer Vision and Pattern Recognition, pp. 3431–3440 (2015)

24. Ronneberger, O., Fischer, P., Brox, T.: U-net: convolutional networks for biomedical image segmentation. In: International Conference on Medical Image Computing and Computer-Assisted Intervention, pp. 234–241. Springer, Cham (Oct 2015)

25. Sandler, M., Howard, A., Zhu, M., Zhmoginov, A., Chen, L.C.: Mobilenetv2: inverted residuals and linear bottlenecks. In: Proceedings of the IEEE Conference on Computer Vision and Pattern Recognition, pp. 4510–4520 (2018)

26. Ma, N., Zhang, X., Zheng, H.T., Sun, J.: Shufflenet v2: practical guidelines for efficient cnn architecture design. In: Proceedings of the European Conference on Computer Vision (ECCV), pp. 116–131 (2018)

27. DeVries, T., Taylor, G.W.: Improved regularization of convolutional neural networks with cutout (2017). arXiv preprint arXiv:1708.04552

Han Dynasty Clothing Image Classification Model Based on KNN-Attention and CNN

Guan Ziwei[1](✉), Lv Zhao[1], and Teng Jinbao[2]

[1] Xi'an Polytechnic University, Xi'an Shaanxi 710000, China
Zicaijiangjiang@163.com, zhanggsumei@163.com, 19960713@xpu.edu.cn
[2] Xi'an University of Posts and Telecommunications, Xi'an Shaanxi 710121, China
210311029@stu.xpu.edu.cn

Abstract. Aiming at the problem that the traditional clothing image classification model cannot effectively extract the instance information of other samples in the training set, a clothing image classification model of Han Dynasty based on KNN-Attention and CNN was proposed. Firstly, KNN-Attention was used to extract the clothing image information of K instance samples similar to the original training samples. Secondly, CNN is used to further extract the local key features of clothing images. Finally, the output information of KNN-Attention and CNN is integrated, so as to achieve the purpose of effectively utilizing the training set instance information in the task of clothing image classification. The experimental results show that the proposed model is better than the traditional classification model and can effectively improve the classification effect of clothing images.

Keywords: Clothing image classification in Han Dynasty · KNN · Attention · CNN

1 Introduction

The research on clothing categories in Han Dynasty is an important part of the research on traditional clothing [1]. Classifying the clothing of Han Dynasty according to its categories based on the existing images not only saves the time spent in clothing retrieval, but also helps researchers to carry out the next scientific research work. The application of this in the museum system can help visitors to understand the clothing culture of the Han Dynasty more intuitively and conveniently, and further facilitate the spread of the clothing culture of the Han Dynasty.

Traditional clothing image classification algorithms are based on deep learning model [2]. Zhou [3] propose feature fusion Multi_MNet convolution neural network for clothing image classification to solve the problem of clothing image classification. This method is mainly composed of two branches of convolutional neural network. One branch extracts multi-scale features from the whole expression image through multix, which improves the abnormal network design, and the other branch extracts attention mechanism features from the whole expression image through MobileNetV3 small network. Before classification, multi-scale and attention mechanism features are aggregated. In addition,

N. Xiong et al. (Eds.): ICNC-FSKD 2022, LNDECT 153, pp. 11–17, 2023.
https://doi.org/10.1007/978-3-031-20738-9_2

in the training phase, the global average pool, convolution layer and Softmax classifier are used to classify the final features instead of the full connection layer, which speeds up the model training and alleviates the over fitting problem caused by too many parameters. Hong [4] propose research on clothing classification based on convolutional neural network, Two convolution neural network structures with different convolution layers and pooling layers are designed. Each model was trained and tested using the Fashion-MNIST dataset. The complex structure model can obtain higher classification accuracy, but it will increase the calculation cost. Although the above clothing image classification algorithm based on the traditional deep learning model has its own advantages, and the experimental results on the public data set show that it has a good classification effect, it does not take into account the instance information of other samples in the data set, so the classification effect needs to be improved.

In view of the above analysis, this paper proposes a Han Dynasty clothing image classification model based on KNN-Attention and CNN (KACNN), and takes the Han Dynasty clothing as an example. During model training, KNN-Attention is introduced to obtain the image information of example samples in the data set. At the same time, CNN is used to further extract the key features of the Han Dynasty clothing image, and further pool the operation. Finally, the output information of KNN-Attention is integrated with the output information of CNN, so that the image information extracted by the model is more comprehensive, and the purpose of improving the classification effect is achieved.

2 KACNN Model

The KACNN model proposed in this paper mainly includes CNN layer and KNN-Attention layer. Its overall architecture is shown in Fig. 1.

2.1 CNN Layer

As one of the most commonly used deep learning models, CNN performs well in the field of image processing [5]. CNN has strong feature extraction ability. By setting convolution kernels of different sizes, it can effectively extract the potential deep information of the image, and then compress the input feature map through the pooling layer to make the feature map smaller and simplify the network computing complexity. Finally, all features are connected by the full connection layer, and the output value is sent to the classifier. The image information input into the system is convoluted and pooled to obtain the output vector X° of CNN. The structure of CNN is shown in Fig. 2.

2.2 KNN-Attention Layer

KNN [6] is one of the traditional machine learning algorithms. Assume that the data sample of the whole training set is $X = \{X_1, X_2, \ldots X_n\}$, The label of the corresponding dataset sample is $Y = \{y_1, y_2, \ldots y_n\}$. Then the corresponding distance calculation formula is:

$$sim(X_i, X_j) = \sqrt{\sum_{t=1}^{n}(X_{it} - X_{jt})^2} \tag{1}$$

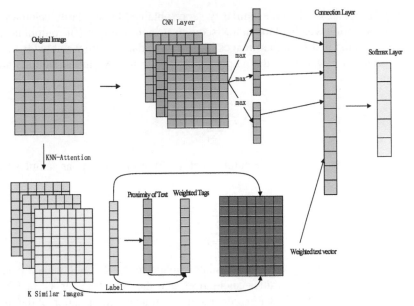

Fig. 1 Overall architecture

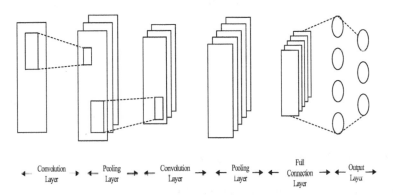

Fig. 2 CNN structure

Calculate the similarity between the predicted image and the whole image data set with the Euclidean distance calculation formula, and get the first K similar image sets closest to the whole data set: $\{X_1, X_2, \ldots X_k\}$, The label of the corresponding similar image set is: $\{y_1, y_2, \ldots y_k\}$. Use Eqs. (2) to (3) to calculate the similarity and get the first k similar images:

$$S_i = sim(X, X_i) \tag{2}$$

$$S = \{S_1, S_2, \ldots S_k\} \tag{3}$$

where $i \in \{1, 2, \ldots, k\}$, S_i is the similarity score obtained by similarity calculation. S refers to the attention weight of the first K sequences in the data set. The output tags and input vectors of the first K similar images in the image data set are weighted by S.

The formula for weighting the output labels of the first K similar samples in the image data set is:

$$y^* = \sum_{p=1}^{k} S_P * y_p \tag{4}$$

The formula for weighting the output vector of the first K similar samples in the image data set is:

$$X^* = \sum_{p=1}^{k} S_P * X_p \tag{5}$$

The weighted output tag y^* and the instance image vector X^* are obtained from the instance images in the dataset after calculation, and the output vector X° of CNN is fused with X^* to obtain X^θ, Better express the most authentic information of the image, and then integrate the fused X^θ input to the Softmax layer for classification.

In the training phase, Adam [7] optimizer is used to update the weight and redefine the cross entropy loss function as:

$$J(w, b) = -\frac{1}{m} \sum_{i=1}^{m} \{y_i \ln(y_i^*) + (1 - y_i) \ln(1 - y_i^*)\} \tag{6}$$

In Eq. (6) y_i refers to the real tag value, y_i^* refers to the weighted category label value, and m refers to the total number of samples for each training.

3   Experimental Design and Analysis

3.1   Experimental Environment and Data Set

The experiment was carried out on ubuntu 18.04 system, with Intel (R) Xeon (R) gold 5218 CPU, tensorflow 2.1.0 deep learning framework, python 3.6 programming language and cuda 10.1 for accelerated calculation. The experimental data set is the Han Dynasty clothing data set in the public deep fashion [8] and fashion ai [9] data sets. 30000 Han Dynasty clothing images obtained by manual screening are used as training sets, verification sets and test sets in the proportion of 5:3:1, and then input into the model for learning.

3.2   Evaluation Index

Accuracy rate (Acc), accuracy rate (Pre), recall rate (Rec) and F1 were used as evaluation indicators.

3.3 Comparative Experiment

To verify the performance of KACNN model, compare it with the following benchmark model under the same conditions:

(1) AlexNet [10] is composed of eight network layers, including the first five convolution layers and the last three full connection layers. After each convolution layer and full connection layer, ReLU nonlinear activation function is used, and then input into the classifier for classification.

(2) CNN-Attention [11] is composed of CNN and attention. The model first uses CNN to extract image information, and then uses attention to further focus the attention of the model on the pixels that have a great impact on the classification results. After being weighted by attention, it is input into the classifier for classification.

(3) KNN [12] is a traditional machine learning method. The K value is set to 8. The distance between the predicted sample and other samples is calculated through the Euclidean distance calculation formula, and the classification results of this sample are voted by the first 8 samples.

(4) Agrawal [13] proposed two novel CNN architectures: the number of filters in one architecture remains unchanged with the increase of network depth, while the number of filters in the other architecture decreases with the increase of depth. These two architectures all use convolution cores of size 8.

The experimental results of KACNN model and the above benchmark model on the public data set are shown in Table 1.

Table 1. Comparison of experimental results (unit: %)

Model	Acc	Pre	Rec	F1
AlexNet [12]	89.86	89.92	89.86	88.82
CNN-Attention [13]	91.23	91.96	91.07	91.23
KNN [14]	79.98	80.34	79.96	79.96
Agrawal [15]	91.27	90.72	90.86	90.37
KACNN	**93.29**	**93.98**	**93.67**	**93.86**

It can be seen from the table that KACNN model has the best classification effect and the best performance in the Han Dynasty clothing data set. Compared with the AlexNet model, the performance of KACNN model is improved by 3.43%, compared with the CNN-Attention model, the performance is improved by 2.06%, and compared with the model proposed by Agrawal, the performance is improved by 2.02%. AlexNet model, CNN-Attention model and Agrawal's model can more effectively extract the deep information of the current sample through its improved CNN, but it does not effectively use the instance information of other samples in the training set data, so the extracted features are not comprehensive, while KACNN model can not only extract the information of the current sample, On this basis, by introducing KNN-Attention, the information of the example samples in the data set is also fully used, so the extracted features are more comprehensive and more representative of the most real image information, so the performance effect on the Han Dynasty clothing data set is also better. Compared with KNN model, the performance of KACNN model is improved by 13.4%, because KNN model only focuses on the information of other samples similar to the current sample when classifying, and does not take into account the potential deep information of the current sample. On this basis, the proposed KACNN model introduces attention to make it more advantageous to extract the information of similar samples. In addition, it also introduces CNN to make up for the lack of extracting the information of the current sample, Therefore, the classification effect is also better.

4 Conclusion

In order to solve the problem that the traditional clothing image classification model can not effectively extract the instance information of other samples in the training set, a clothing image classification model of Han Dynasty based on KNN-Attention and CNN is proposed. By introducing KNN-Attention, the model has more advantages in extracting other similar sample information from the data set, and effectively solves the problem of incomplete feature extraction. By comparing with other models, the advantages of KACNN model are further verified, which can effectively improve the effect of clothing image classification in the Han Dynasty.

Acknowledgments. Fund projects: Shaanxi Art and Science Planning Project (NO. SYZ2021002).

References

1. Xiang, Y.H., Kong, J.C.: Exploring the design of Han clothing. In: 2020 Joint International Conference on Digital Arts, Media and Technology with ECTI Northern Section Conference on Electrical, Electronics, Computer and Telecommunications Engineering (ECTI DAMT & NCON), vol. 13(4), pp. 270–275. IEEE (2020)
2. Zhang, X., Deng, Z.: An improved method of clothing image classification based on CNN. Int. J. Advan. Netw. Appl. **12**(6), 4742–4745 (2021)
3. Zhou, H.L., Peng, Z.F., Tao, R., et al.: Feature fusion Multi_XMNet convolution neural network for clothing image classification. J. Donghua Univ. (English Edition) **18**(9), 426–430 (2021)
4. Hong, Y., Lv, C.: Research on clothing classification based on convolutional neural network. IOP Conf. Ser.: Mater. Sci. Eng. **768**(7), 136–143 (2020)
5. Tian, C., Xu, Y., Li, Z., et al.: Attention-guided CNN for image denoising. Neural Netw. **124**(6), 117–129 (2020)
6. Xing, W., Bei, Y.: Medical health big data classification based on KNN classification algorithm. IEEE Access **8**(5), 28808–28819 (2019)
7. Fang, W., Zhang, F., Sheng, V.S., et al.: A method for improving CNN-based image recognition using DCGAN. Comput., Mater. Continua **57**(1), 167–178 (2018)

8. Liu, Z.W., Luo, P., Qiu, S., et al.: Large-scale fashion (deepfashion) database [DB/OL]. The Chinese University of Hong Kong [25 April 2020]. http://mmlab.ie.cuhk.edu.hk/projects/DeepFashion.html

9. Fashion AI competition questions and data [DB/OL]. Alibaba Tianchi [25 April 2020]. https://tianchi.aliyun.com/competition/entrance/231648/introduction

10. Tang, H., Li, M., Chan, M.D., et al.: DC-AL GAN: pseudo progression and true tumor progression of glioblastoma multiform image classification based on DCGAN and AlexNet. Medical Phys. **47**(3), 1139–1150 (2020)

11. Guo, H., Zheng, K., Fan, X., et al.: Visual attention consistency under image transforms for multi-label image classification. Proc. IEEE/CVF Conf. Comput. Vision Pattern Recognition **38**(7), 729–739 (2019)

12. Larijani, M.R., Asli-Ardeh, E.A., Kozegar, E., et al.: Evaluation of image processing technique in identifying rice blast disease in field conditions based on KNN algorithm improvement by K-means. Food Sci. Nutrition **7**(12), 3922–3930 (2019)

13. Agrawal, A., Mittal, N.: Using CNN for facial expression recognition: a study of the effects of kernel size and number of filters on accuracy. Vis. Comput. **36**(2), 405–412 (2020)

Self-Attention SSD Network Detection Method of X-ray Security Images

Hong Zhang[1,2], Baoyang Liu[1,2(✉)] ⓘD, and Yue Gao[1,2]

[1] Xi'an University of Posts and Telecommunications, Xi'an, China
zhangh@unm.edu, {baoyang,gaoyue}@stu.xupt.edu.cn
[2] Automatic Sorting Technology Research Center, State Post Bureau of the People's Republic of China, Xi'an, China

Abstract. For the manual omission of X-ray security images inspection, SSD network extracts image features with a large number of convolutional layers, and some object feature information is missed in the deep feature layer after multiple convolutions and pooling. The self-attention mechanism can obtain images global features, and reduce the missing of image feature information. This paper proposes two new target detection algorithm BoT-SSD and ResT-SSD based on self-attention neural network, they apply BoTNet and ResT as the backbone network to extract more important features. The multi-scale network structure of SSD is combined with FPN and CBAM, which can use underlying and high-level feature and strengthen target feature information. Moreover, the self-attention modules are used to enhance the global under-standing of the image by high-level features layers. The X-ray images detection results show that the targets with complex background can identified more accurately. Compared with the original SSD network, the proposed BoT-SSD and ResT-SSD networks improve MAP value by 5.63% and 7.52% respectively.

Keywords: X-ray security inspection · Self-attention · SSD · BoTNet · ResT

1 Introduction

At present, the application of neural network in identification of dangerous objects is an important research area. In the field of computer vision, the most typical method is Convolutional Neural Network (CNN) [1], which effectively reduces the number of training parameters by extracting features through Convolutional kernel. The target detection network based on convolutional neural network is mainly divided into Two-stage target detection and One-stage target detection. Two-stage algorithms such as Faster RCNN [2]. The One-stage algorithms such as Yolo [3] and Single shot multibox detector (SSD) [4]. The SSD algorithm uses multiple feature maps of different scales to detect the target, which achieves a better balance in detection accuracy and speed.

Because the convolution operation lacks the global understanding of the image, the correlation of global features cannot be established. The self-attention [5] mechanism of Transformer [6] model is not limited by local interactions, which can both mine long-range dependencies and calculate in parallel. However, pure Transformer architecture has a huge amount of computation in the field of vision, so CNN combined with self-attention mechanism has became an important research direction recently.

2 Single Shot Multibox Detector Network Algorithm

SSD detection algorithm improved the backbone VGG16 [7] by replacing full-connection layer with Conv5. Its feature extraction network is composed of four multi-scale convolution modules, and its structure is shown in Fig. 1.

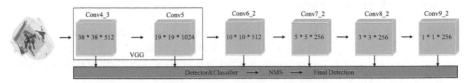

Fig. 1. The network structure of SSD

Conv4_3 and Conv5 convolutional layers in the backbone, Conv6_2, Conv7_2, Conv8_2 and Conv9_2 in the deep layer are taken as the effective characteristic layers, and the size of each layer is shown in the Fig. 1. SSD sets the prior boxes of different sizes and quantities on the feature maps of different scales to speed up the training. The feature maps of the six effective layers are respectively used for classification prediction and regression prediction.

In regression prediction, 4 parameters are usually needed to determine a prediction box, so the number of prior boxes in each layer are required to be multiplied by 4 to predict the change of each prior box on each grid point of the feature layer. In classification prediction, the number of prior boxes in each layer are multiplied by the classification number to predict the corresponding category of each prediction at each grid point. Finally, score sorting and non-maximum suppression (NMS) are needed to filter each prediction box, so as to obtain the final prediction result. Because SSD model use multi-scale structure to detect object, so the feature information of mutual occlusion targets in X-ray security images is likely to be lost in the high-level feature layer, which makes SSD network ineffective for X-ray security images inspection.

3 Single Shot Multibox Detector with Bottleneck Transformers

3.1 Self-Attention

In computer vision, the self-attention mechanism generates a new feature map by calculating the weight of attention between input features, where each position has information about any other features in the same image, so they can obtain the dependence between some long-distance interval features in space. The formula of standard self-attention is shown below:

$$\text{Attention}(Q, K, V) = \text{Softmax}\left(\frac{QK^T}{\sqrt{d_k}}\right)V \tag{1}$$

Firstly, the Q, K and V matrices are obtained through input. Second, the score which represents the degree of association of each feature information with other features needs to be calculated. Then it needs to be divided by the scaling factor and normalized by the softmax function in order to ensure that the gradient is more stable. Finally, each V vector is multiplied by softmax score to reduce the attention of irrelevant features while keeping the current feature attention unchanged. Multi-head self-attention (MSA) utilizes multiple self-attention modules to form multiple subspaces, enabling the model to focus on different aspects of information.

3.2 Bottleneck Transformers

ResNet [8] is made up of the series connection of residual blocks, Bottleneck Transformers (BoTNet) [9] replaces convolution with MSA in the last three blocks of Res-Net, which are often referred to as Transformer blocks (BoT). The structure of resid-ual block and BoT block is shown as follows (Fig. 2):

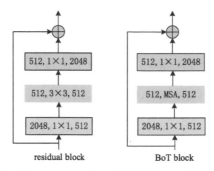

Fig. 2. The structure of residual block and BoT block

BoTNet uses convolution to extract feature information effectively from the input image. Meanwhile, in order to avoid the loss of feature information of small targets caused by too deep layers, it uses global self-attention to process and calculate the feature information obtained by convolution. The structure diagram of its self-attention realization is shown in Fig. 3:

It forms q, k and v matrix by 1*1 convolution, and the position codes in the graph are randomly initialized, then continuously learned during training. The calculation formula is as follows:

$$z = \text{softmax}\left(qr^T + qk^T\right)v \tag{2}$$

Firstly, the position score is obtained by qr^T, and then using qk^T to get the content score. The self-attention score is obtained by adding them together, after normalization by softmax function, the output is obtained by multiplying v.

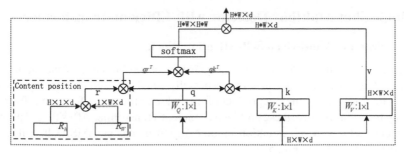

Fig. 3. The structure of BoT block diagram

3.3 Single Shot Multibox Detector with Bottleneck Transformer

BoTNet can obtain the relationship between global features due to its self-attention mechanism, so it can extract image features more effectively. Combined with the multi-scale network structure of SSD which can achieve better recognition accuracy. Here, we proposed BoT-SSD network. Firstly, BoTNet is used as the backbone net-work of SSD network to extract features, and then convolutional block attention module (CBAM) [10] module which contains Channel Attention Module and Spatial Attention Module is used to further process the extracted bottom features to strengthen the weight of target feature information. Finally, self-attention mechanism is added to the top feature layer to obtain the relevance of images global features. It can effectively avoid the loss of feature information caused by the small volume, complex background information and mutual occlusion of targets caused by too deep layers. At the same time, feature pyramid networks (FPN) [11] is added to the last layer, and up-sampling is carried out by bilinear interpolation. The up-sampling results are fused with the same size feature graph generated from the bot-tom up, and the high resolution of the bottom feature and the high semantic information of the top feature are utilized at the same time. The network structure diagram is shown below.

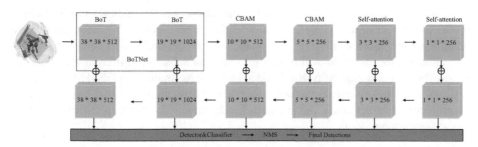

Fig. 4. The structure diagram of BoT-SSD

4 Single Shot Multibox Detector with Efficient Transformer

4.1 Efficient and Multi-Head Self-Attention Block

Each header in traditional MSA is just responsible for a submodule of the input, which may affect the capability of the network, especially when the channel dimension in each subset is too low, so that the dot product of q and k can no longer form an information matching function. To solve these problems, Efficient Transformers (ResT) [12] proposes an efficient multi-head self-attention module which names as EMSA. Its calculation mechanism is as follows (Fig. 5):

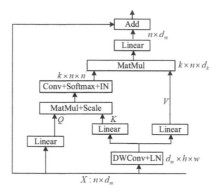

Fig. 5. Self-attention of EMSA structure diagram

Q, K and V are obtained by using depth-wise convolution (DWConv), Layer normalization (LN) [13] and Linear layers. The calculation formula in the figure is as follows:

$$\text{EMSA}(Q, K, V) = \text{IN}\left(\text{softmax}\left(\text{Conv}\left(\frac{QK^T}{\sqrt{d_k}}\right)\right)\right)V \tag{3}$$

Conv is a standard 1×1 convolution operation, which simulates the interaction between different heads. Therefore, the attention function of each header can depend on all keys and queries. However, this would weaken the multi-head attention's ability to combine information from different presentation subsets in different locations. To restore this diversity capability, ResT added Instance Normalization (IN) [14] after using softmax. EMSA can compensate for the short length limitation of input tokens per header by projecting interactions in the attention heads dimension, that improve efficiency of the self-attention module.

4.2 Efficient Transformer

The standard Transformer module consists of MSA, feed-forward network (FFN) and residual connections are used for each layer, then using LN before MSA and FFN. The calculation formula for each standard Transformer block is as follows:

$$y = x' + \text{FFN}\left(\text{LN}\left(x'\right)\right) \tag{4}$$

$$x' = x + \text{MSA}(\text{LN}(x)) \tag{5}$$

FFN is used for feature transformation and nonlinear processing. It consists of two linear layers with nonlinear activation. Efficient Transformers (ResT) replaces traditional MSA with EMSA, the x' of each ResT block is shown in formula (6).

$$x' = x + \text{EMSA}(\text{LN}(x)) \tag{6}$$

Finally, the output of ResT block can be obtained by combining with formula (4). The Transformer block structure diagram is shown in the Fig. 6.

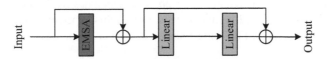

Fig. 6. The structure of Efficient Transformer block

ResT constructs four phases after uses Stem block of ResNet to extract the underlying features of images. Each stage contains patch embedding, Transformer blocks and position encoding modules with EMSA. ResT proposes an effective spatial attention module (pixel-attention) to encode positions. Its formula is shown as follows:

$$x' = x * \sigma(\text{DWConv}(x)) \tag{7}$$

The weight of the pixel direction is obtained by DWConv, then scaled by σ (sigmoid function), the position encoding x' is obtained by multiplying the input.

4.3 Single Shot Multibox Detector with Efficient Transformer

We also take Rest as the backbone network of SSD. Considering the size of extracted feature layer is compatible with SSD feature layer, Stage2 and Stage3 are used as effective low-level features, and bilinear interpolation is used for up-sampling to be compatible with SSD feature size. The remaining modules of ResT-SSD are the same as those of BoT-SSD which shown in Fig. 4, its backbone structure is shown in Fig. 7.

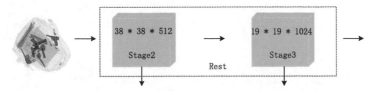

Fig. 7. The structure diagram of ResT-SSD

5   Experimental Results and Comparative Analysis

5.1   Experimental Operating Environment

The computer processor used in this experiment is the Intel Core i7-11800H model, the graphics card is GTX3060, the programming is pytorch 1.7.1 and python 3.7.12.

5.2   Result and Analysis

In this paper, 4000 X-ray security images relabeled on the public data set Sixray are randomly divided into training sets and verification sets in a ratio of 9:1. The objects in the dataset included four categories of objects, which are Knife, Gun, Wrench, and Plier. The indicators involved in model evaluation in this experiment include: Aver-age Precision (AP), MAP and FPS. Table 1 shows the AP comparison.

Table 1. AP comparison of five networks

Network/Categories	Gun	Knife	Wrench	Pliers
SSD	97.31	80.65	84.46	90.63
BoT-SSD	98.16	90.49	90.75	96.17
ResT-SSD	**99.29**	**93.14**	**94.35**	**97.09**
Yolov4	97.88	86.50	91.61	89.30
Faster RCNN	97.34	80.72	82.18	92.05

It can be seen from Table 1 that BoT-SSD and ResT-SSD have different degrees of improvement for the four types of targets, and ResT-SSD has a better recognition effect in X-ray security screening recognition for four kinds of targets.

We choose two representative X-ray security images with complex background information and small volume such as (a), mutual occlusion of target objects in (b). Then apply five networks to detect four kinds of targets, the results are shown in Fig. 8. From left to right are the detection results of SSD, Faster RCNN, Yolov4, BoT-SSD, ResT-SSD.

Fig. 8. Comparison of X-ray images detection results

Comparing the five methods, only Rest-SSD and BoT-SSD do not show any omission, the other three networks all miss some objects information.

Table 2 shows the comparison of target detection performance indicators. BoT-SSD slightly reduces FPS, but it still has a faster recognition speed compared with Yolov4 and Faster RCNN, and the MAP is second only to ResT-SSD. Although the ResT contains EMSA block to add some calculations and leads to a lower FPS, but the FPS is still close to the detection speed of Yolov4 and much faster than Faster RCNN. And the results show that the proposed Rest-SSD can achieve the highest MAP in X-ray security images.

Table 2. Comparison of MAP and FPS

Network	Backbone	MAP (%)	FPS
SSD	VGG	88.26	75
BoT-SSD	BoTNet	93.89	37
ResT-SSD	**ResT**	**95.78**	**28**
Yolov4	VGG	91.32	34
Faster RCNN	VGG	88.07	19

6 Conclusion

In this paper, BoT-SSD and ResT-SSD network models are proposed, which not only apply the self-attention mechanism to improve the model's learning between global features, but also add FPN feature fusion and CBAM blocks. New models improve the weight of the target object, and strengthen the feature reuse ability of the model. The experimental results of X-ray security inspection data set show that, comparing with SSD, Faster RCNN and Yolov4, proposed BoT SSD and ResT-SSD networks can improve the target detection performance. BoT-SSD gets FPS second only to SSD and the map is second only to ResT-SSD. Although ResT-SSD has reduced some FPS due to the use of EMSA modules which adds some calculations, but it is close to the detection speed of Yolo4 and obtain the highest MAP. It improves the MAP by 7.52% compared to SSD and 4.46% higher than Yolov4. ResT-SSD achieves the highest recognition accuracy while maintaining fast enough detection, so as to effectively prevent the omission in X-ray security inspection. In the future, we will optimize the self-attention mechanism to extract more effective features and improve its computational efficiency to achieve higher accuracy and speed. This research is supported in part by Shaanxi Provincial Natural Science Foundation of China (Grant No.2021 SF-478).

References

1. Gu, J., Wang, Z., Kuen, J., et al.: Recent advances in convolutional neural networks. Pattern Recogn. (77), 354–377 (2018)
2. Ren, S., He, K., Girshick, R., et al.: Faster RCNN: towards real-time object detection with region proposal networks. In: Twenty-ninth Conference on Neural Information Processing Systems (2015)
3. Tzou, T.-L., et al.: Detect safety net on the construction site based on YOLO-v4. In: Innovative Computing 2022, pp. 33–42. Springer, Singapore (2022)
4. Liu, W., Anguelov, D., Erhan, D., et al.: Ssd: single shot multibox detector. In: European Conference on Computer Vision, pp. 21–37. Springer, Cham (2016)
5. Shaw, P., Uszkoreit, J., Vaswani, A.: Self-attention with relative position representations. In: The North American Chapter of the Association for Computational Linguistics (2018)
6. Parmar, N., Vaswani, A., Uszkoreit, J., et al.: Image transformer. In: International Conference on Machine Learning (2018)
7. Kaiyan, L., Haoping, Z., Chang, L., et al.: An apple grading method based on improved VGG16 network. In: Smart Communications, Intelligent Algorithms and Interactive Methods 2022, pp. 75–85. Springer, Singapore (2022)
8. Zhu, H., Ma, M., Ma, W., et al.: A spatial-channel progressive fusion ResNet for remote sensing classification. Inform. Fusion **70**, 72–87 (2021)
9. Srinivas, A., Lin, T.Y., Parmar, N., et al.: Bottleneck transformers for visual recognition. In: Proceedings of the IEEE/CVF Conference on Computer Vision and Pattern Recognition (2021)
10. Woo, S., Park, J., Lee, J.Y., et al.: Cbam: convolutional block attention module. In: Proceedings of the European Conference on Computer Vision (2018)
11. Lin, T.-Y., et al.: Feature pyramid networks for object detection. In: Proceedings of the IEEE Conference on Computer Vision and Pattern Recognition (2017)
12. Zhang, Q., Yang, Y.B.: Rest: an efficient transformer for visual recognition. In: Thirty-fifth Conference on Neural Information Processing Systems (2021)
13. Xiong, R., Yang, Y., He, D., et al.: On layer normalization in the transformer architecture. In: International Conference on Machine Learning (2020)
14. Wang, J., Wen, C., Fu, Y., et al.: Neural pose transfer by spatially adaptive instance normalization. In: Proceedings of the IEEE/CVF Conference on Computer Vision and Pattern Recognition (2020)

Cross Architecture Function Similarity Detection with Binary Lifting and Neural Metric Learning

Zhenzhou Tian[1,2,3(✉)], Chen Li[1], and Sihao Qiu[1]

[1] Xi'an University of Posts and Telecommunications, Xi'an 710121, Shaanxi, China
tianzhenzhou@xupt.edu.cn, lichen@stu.xupt.edu.cn, brubbish@stu.xupt.edu.cn
[2] Shaanxi Key Laboratory of Network Data Analysis and Intelligent Processing,
Xi'an 710121, Shaanxi, China
[3] Xi'an Key Laboratory of Big Data and Intelligent Computing, Xi'an 710121,
Shaanxi, China

Abstract. Binary code similarity detection has extensive and important applications in IoT device security, yet which suffers the challenges from the differentiated underlying architectures of the diverse IoT devices. To this end, this paper presents XFSim (Cross-architecture Function-level binary code Similarity detection), through binary lifting and neural similarity metric learning. Firstly, to make the detection method architecture agnostic, the binaries to be analyzed are lifted to an intermediate code called LLVM-IR and normalized for an uniform representation, so as to alleviate the discrepancies between the raw assemblies of different instruction set architectures (ISAs). Secondly, we utilize FastText, a widely used word embedding algorithm, that learns on the functions' normalized intermediate codes to obtain high quality token embeddings. Then, an efficient CNN-based model is utilized to encode the semantics of each function into numerical vectors, meanwhile the siamese neural network structure is resorted to supervise the whole model training, with the goal of minimizing the contrastive loss. Finally, the similarity of two binary code snippets can measured by the cosine similarity of their encoded vectors. The experiments conducted on a public dataset show that, the strategy of lifting and normalizing the assemblies to uniform representations greatly alleviates the semantic-gaps between different ISAs, and XFSim outperforms two existing cross-architecture binary code similarity detectors.

Keywords: Binary code similarity detection · Instruction set architecture · Neural network · Binary lifting

ⓒ The Author(s), under exclusive license to Springer Nature Switzerland AG 2023
N. Xiong et al. (Eds.): ICNC-FSKD 2022, LNDECT 153, pp. 27–34, 2023.
https://doi.org/10.1007/978-3-031-20738-9_4

1   Introduction

Binary code similarity detection is the process of comparing two or more binary codes with different granularity, and then identifying their similarities and differences. There are two major challenges inherent to similarity detection for software code of IoT devices. One is the missing source code problem brought by closed source of commercial software, and the other is the cross-architecture problem brought by the diversity of underlying architectures of IoT devices. In practice, the majority of IoT devices use simple hardware solutions, and it has become the norm for them to rewrite and reuse a large number of third-party libraries in software. These vulnerable IoT devices are difficult to patch and automatically upgrade, which exacerbates the seriousness of this security problem. According to a large-scale case study of IoT devices, the percentage of IoT devices containing only known vulnerabilities has reached as high as 80.4%. Tencent Keen Lab [1] IoT security white paper further points out that Nday security risks caused by rewriting and reusing third-party libraries account for more than 90% of the total number of findings.

In 2015, Pewny et al. published MULTI-MH, the first binary code detection method for cross-architecture problems, which retrieves and detects functions by the semantics of the input and output, and simply provides the function-level code compiled by one instruction set to find function-level fragments similar to it in the function-level code of another instruction set [2]. In 2016, Lageman et al. trained neural networks by having the trained networks to determine whether two compiled function-level codes originate from the same source code [3]. As a result, deep learning has also started to be widely used in the field of binary code similarity detection research. In 2019, Luo et al. proposed the GeneDiff method, which first introduced the intermediate code VEX-IR to eliminate the differences between binary codes of different instruction architectures with the help of intermediate codes [4].

For the current situation of IoT devices, this paper proposes a binary code similarity detection method based on binary lifting. The main contributions of this method are reflected in the following three aspects.

- Intermediate code as a unified code representation. Using LLVM-IR [5] to lift binary code to function-level intermediate code. It avoids direct analysis of binary code and shields against the differences caused by different architectures.
- Siamese network model is introduced to learn the features of different source codes. The model maximizes the representations of dissimilar labels and minimizes the representations of similar labels by metric learning under supervised learning. The weights can be shared among neural networks because the intermediate code shields architectural differences.
- Experiments are conducted on the publicly available dataset Trex [6]. The experimental results show that the accuracy of the XFSim system proposed in this paper is better than other methods.

2   Proposed Method

In this section, we present the details of the proposed method. The main process of this study is shown in Fig. 1. The main process contains four parts. The first part is code preprocessing. Extracting instruction of binary code with using disassembly tools. Batch extraction of binary code based on different instruction set architectures and different optimization options in the Trex data set, to get the corresponding assembly information. The second part is binary lifting. Using the assembly information obtained for binary lifting and standardization, we divide function-level instructions based on the start and end addresses of functions in assembly instructions, and then perform binary lifting, which converts function-level instructions into a uniform intermediate code representation and standardization. The third part is to construct the sample pairs. The intermediate code is embedded to obtain the feature vector representation of the intermediate code. The similarity of the intermediate code feature vector is measured using cosine distance. Then the sample pair is constructed according to the result of the measure, which contains similar samples and dissimilar samples. The fourth part is model training. The constructed sample pairs are fed into the siamese neural network to train the model, and the contrastive loss function will evaluate the similarity of the feature vectors.

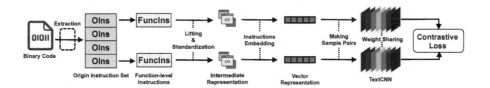

Fig. 1. Architecture of XFSim system.

2.1   Code Preprocessing

Binary code usually implies important information left during the compilation process. Extracting this important information requires disassembly of the binary code. This paper uses the Python interface provided by the IDA Pro disassembly tool [7] to disassemble the binary code in the Trex dataset (this paper considers the parsing results of IDA Pro disassembly to be accurate) and disassemble the high-level information extracted to correlate, organize and store in an orderly manner. Use the triple $N(f_{name}, f_{start}, f_{end})$ to uniquely represent the function obtained by disassembly, where f_{name} is the function name, f_{start} is the start address of the function, and f_{end} is the end address of the function.

2.2   Binary Lifting and Standardization

The same source code, compiled on different instruction set architectures with different optimization options, is bound to have more significant differences in the binary code. Therefore, the first problem in binary code similarity study is how to avoid the interference of different optimization options and different instruction set architectures on the compilation results.

This paper introduces LLVM-IR as an intermediate representation of function-level binary code. Use RetDec tool to disassemble binary code into LLVM-IR. The process of disassembling binary code into LLVM-IR is not limited to any specific target architecture, operating system or binary file format. Figure 2 gives the assembly code based on different instruction set architectures compiled from the same function of C program language in binutils-2.32 after O0 optimization, and the LLVM-IR code corresponding to the assembly code. As can be seen from the figure, the main operators of the LLVM-IR code are basically the same compared to the assembly code, masking the code differences between different instruction set architectures.

The LLVM-IR code obtained by using RetDec [8] disassembly cannot be directly word embedded yet. The LLVM-IR code still contains a large number of simple operators (e.g. "()", ",", etc.) and low-frequency vocabulary based on immediate data. Among them, the main purpose of operators is to highlight the code logic and improve code readability and low-frequency vocabulary based on immediate data also has randomness. Both of them have less impact on the semantics and functionality of the program, but significantly expand the size of the corpus during word embedding, which affects the effect of word embedding. In summary, this paper further standardizes the LLVM-IR code by removing the simple operators, replacing the immediate data in the intermediate code with "IMM", and extracting the token.

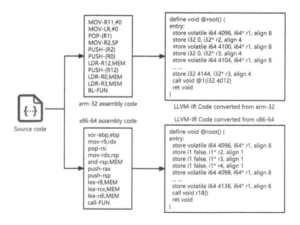

Fig. 2. Assembly code and its LLVM-IR in different ISA from a same source code.

2.3 Sample Pairs Construction

In this paper, we argue that the same source code compiled with different optimization options on different instruction set architectures produces two function-level binary codes that are similar. Specifically, for similar function-level binary codes f_n and f_m, there exists a discriminant $\pi(f_n, f_m) = 1$, and for dissimilar function-level binary codes f_x and f_y, there exists a discriminant $\pi(f_x, f_y) = 0$. In order to construct sample pairs, a mapping g needs to be found. The mapping g is able to map any function-level binary code f to a vector μ_f, denoted as $\mu_f = g(f)$. The purpose of finding function-level binary code mapping vectors is to consider $\pi(f_n, f_m) = 1$ if $Sim(\mu_{f_n}, \mu_{f_m}) \rightarrow 1$ exists and consider $\pi(f_x, f_y) = 0$ if $Sim(\mu_{f_x}, \mu_{f_y}) \rightarrow 0$ exists. Sim denotes the degree of similarity of function-level binary codes, and the degree of similarity between vectors is expressed using the distance between vectors. The function similarity is defined as follows.

$$Sim(f_x, f_y) = cos(\mu_{f_x}, \mu_{f_y}) \tag{1}$$

We consider LLVM-IR code as a special kind of text information, treat LLVM-IR instructions and operation objects as words, one LLVM-IR instruction as a sentence, and all sequences of LLVM-IR instructions in a function as paragraphs. Then use FastText [9] as a mapping g for word embedding of intermediate code. FastText's shallow network is able to greatly reduce the training time while ensuring the accuracy of the model.

2.4 Siamese Architecture-Based Similarity Detection Model

The embedding vectors of intermediate code tokens are stitched into a word matrix, and several tokens are convolved using multiple convolution kernels of different sizes to extract the key information in each line of intermediate code and better capture the local relevance of the token.

We use two identical TextCNN [10] embedding networks to form a siamese architecture [11], where both neural networks use the same hyperparameters and share parameters to feed sample pairs into the siamese network. The sample pair is constructed in such a way that $\langle f(I_p), f(I_q), y \rangle$. $f(I_p)$ denotes function-level binary code based on architecture p, y indicates whether the function-level binary code based on architecture p is similar to that based on architecture q. The loss function uses the contrastive loss, which is calculated as follows.

$$\mathbf{L} = y \times \|f(I_p) - f(I_q)\|_2^2 + (1 - y) \times \max\{0, m - \|f(I_p) - f(I_q)\|_2\}^2 \tag{2}$$

Contrastive loss [14] is introduced so that when the function-level binary code based on architecture p is similar to the binary code based on architecture q, the neural network is penalized for the distance between the two embedding vectors being too far. If the binary codes are not similar, the neural network will be penalized for the distance between the two embedding vectors being too close. For the binary codes that are not similar, a boundary m is set, and when the embedding vector distance is m, the neural network is considered to have made a separation of the function-level binary codes that are not similar.

3   Experimental Evaluation

3.1   Design and Preparation

This paper proposes an binary lifting-based method for function-level binary code similarity detection, which aims to remedy two shortcomings in the current study. Experiments are conducted to evaluate the overall accuracy of the XFSim system so as to answer the following two questions.

– Question1: Can the detection system perform effective cross-architecture function-level binary code similarity detection?
– Question2: Does the detection improve compared to the available method after boosting with LLVM-IR command?

We use the binutils-2.30 and coreutils-8.30 projects contained in the Trex dataset to obtain a library of function-level binary code samples for both x86-64 and arm-64 instruction set architectures with four compilation optimization options (O0, O1, O2, O3), and divide the library into a training set without intersection and a test set by 8:2. The statistics of binary function code classification and quantity are shown in Table 1. Among them, O0 means no optimization; O1 means only partial optimization; O2 means more register and instruction-level optimization than O1 configuration, but the compilation process needs more time and resources; O3 means more vectorization algorithm than O2 configuration.

Through training, the hyperparameters of the XFSim system are determined experimentally as follows: the dimension of the LLVM-IR code semantic embedding vector is 100, the window size is 5, the minimum word frequency is 5, and the number of iterative updates is 5. The TextCNN filter sizes are 2, 3, and 4, the activation function is ReLU, maximum pooling is used, and the number of iterative updates is 10.

The target task of function-level binary code similarity detection is to determine whether the given two cross-architecture function-level binary codes are similar, which is a binary classification problem, and the accuracy rate is used as the evaluation index for the experiments.

3.2   Experimental Results

The test sets for the experiments are divided into single instruction architecture and multi-instruction architecture, with x86-64 test set and arm-64 test set for single instruction architecture and x86-64 and arm-64 crossover test set for multi-instruction architecture. The accuracy of different models under the test sets is shown in Table 2. The comparison of the test data of Gemini [12], FuncSim [13] and XFSim systems shows that the accuracy of XFSim system is higher than other systems in all tests.

Comparing the results for single-instruction architectures with those for multi-instruction architectures, the differences are not significant, indicating that LLVM-IR can effectively mask the differences between different instruction set architectures. Specifically, the difference in accuracy between x86-64 test set,

Table 1. Number of binary function code

ISA	Optimization	Training set	Test set	Total
x86-64	O0	26387	6597	32984
	O1			
	O2			
	O3			
arm-64	O0	26387	6597	32984
	O1			
	O2			
	O3			

arm-64 test set and crossover test set does not exceed 5%. Therefore, the XFSim system is able to perform cross-architecture detection, answering Question 1.

The XFSim system introduces the LLVM-IR intermediate code representation for prediction using a TextCNN-based siamese network architecture. It can be seen from the data that the accuracy of the XFSim system is better than that of Gemini and FuncSim in x86-64 test set, arm-64 test set and crossover test set. Therefore, the binary lifting using LLVM-IR allows the network to more fully capture the semantic information in the function-level binary code, answering Question 2.

Table 2. Accuracy of models on different test set

Model	x86-64	arm-64	x86-64 and arm-64
Gemini	77.6	79.8	76.3
FuncSim	84.7	87.6	86.4
XFSim	**85.4**	**89.3**	**88.2**

4   Conclusion

With the rapid spread of digitalization and the growing efficiency and scale of software development, the act of code reuse or rewriting is increasingly becoming the norm. Cross-architecture binary code detection will play a bigger role in IoT device security. The binary code similarity detection technique based on binary lifting proposed in this paper utilizes the unified intermediate code LLVM-IR as a way to mask the differences in binary codes caused by the instruction set architecture. Meanwhile, this paper introduces TextCNN based on siamese architecture and uses contrastive loss to make similar function-level binary code feature vectors, which are gradually close in the same vector space, and dissimilar

similar function-level binary code feature vectors, which are gradually far away in the same vector space. Experimental results show that the XFSim prototype system is more accurate than existing solutions.

Acknowledgment. This work was supported in part by the Natural Science Basic Research Program of Shaanxi (2022JM-342, 2018JQ6078) the Science and Technology of Xi'an (2019218114GXRC017CG018-GXYD17.16), National Natural Science Foundation of China (61702414), the Key Research and Development Program of Shaanxi (2019ZDLGY07-08), and Special Funds for Construction of Key Disciplines in Universities in Shaanxi.

References

1. Keen Lab. https://keenlab.tencent.com/. Last accessed 2022/06/10
2. Pewny, J., Garmany, B., Gawlik, R., Rossow, C., Holz, T.: Cross-architecture bug search in binary executables. In: 2015 IEEE Symposium on Security and Privacy, pp. 709–724 (2015)
3. Lageman, N., Kilmer, E.D., Walls, R.J., McDaniel, P.D.: BINDNN: resilient function matching using deep learning. In: International Conference on Security and Privacy in Communication Systems, pp. 517–537 (2016)
4. Luo, Z., Wang, B., Tang, Y., Xie, W.: Semantic-based representation binary clone detection for cross-architectures in the internet of things. Appl. Sci. **9**(16), 3283 (2019)
5. Grech, N., Georgiou, K., Pallister, J., Kerrison, S., Morse, J., Eder, K.: Static analysis of energy consumption for LLVM IR programs. In: Proceedings of the 18th International Workshop on Software and Compilers for Embedded Systems, pp. 12–21 (2015)
6. Pei, K., Xuan, Z., Yang, J., Jana, S., Ray, B.: Trex: Learning execution semantics from micro-traces for binary similarity. arXiv:2012.08680 (2020)
7. Chris, E.: The IDA Pro Book, 2nd edn. William Pollock, San Francisco (2008)
8. Křoustek, J., Matula, P., Zemek, P.: Retdec: an open-source machine-code decompiler (2017)
9. Joulin, A., Grave, E., Bojanowski, P., Douze, M., Jégou, H., Mikolov, T.: Fasttext. zip: Compressing text classification models. arXiv:1612.03651 (2016)
10. Chen, Y.: Convolutional neural network for sentence classification. Master's thesis, University of Waterloo (2015)
11. Chopra, S., Hadsell, R., LeCun, Y.: Learning a similarity metric discriminatively, with application to face verification. In: 2005 IEEE Computer Society Conference on Computer Vision and Pattern Recognition, pp. 539–546 (2005)
12. Xu, X., Liu, C., Feng, Q., Yin, H., Song, L., Song, D.: Neural network-based graph embedding for cross-platform binary code similarity detection. In: Proceedings of the 2017 ACM SIGSAC Conference on Computer and Communications Security, pp. 363–376 (2017)
13. Fang, L., Wei, Q., Wu, Z., Du, J., Zhang, X.: Neural network-based binary function similarity detection. Comput. Sci. **48**(10), 286–293 (2021)
14. Wang, F., Liu, H.: Understanding the behaviour of contrastive loss. In: Proceedings of the IEEE/CVF Conference on Computer Vision and Pattern Recognition, pp. 2495–2504 (2021)

Pain Expression Recognition Based on Dual-Channel Convolutional Neural Network

Xuebin Xu, Meng Lei$^{(\boxtimes)}$, Dehua Liu, and Muyu Wang

School of Computer Science and Technology, Xi'an University of Posts and
Telecommunications, Xi'an Shaanxi 710121, China
xuxuebin@xupt.edu.cn, {leimeng,liudehua,
wmy961017}@stu.xupt.edu.cn, 1534961911@qq.com

Abstract. As an important branch in the field of artificial intelligence, deep learning has penetrated into all walks of life in society in recent years, especially in the recognition of human features has made great progress. Humans have a certain ability to feel pain, but there are some people who can't put it into words. At present, the recognition of pain expression mainly relies on traditional manual testing methods. This testing method is time-consuming and cumbersome, and subject to subjective influence by professionals, which may lead to biased test results. Therefore, a convenient, efficient and objective pain expression recognition system is needed as an auxiliary tool to help professionals make judgments. The pain expression recognition method proposed in this paper is based on dual-channel convolutional neural network. Firstly, the pain data set is preprocessed. For the dual-channel input data, facial expression features are extracted by using convolution neural networks with different parameters. Then the feature maps of the classification network are fused, and finally softmax is used for classification. The accuracy of the improved convolutional neural network model on COPE data set is 97.5%.

Keywords: Dual-channel convolutional neural network · Pain expression recognition · Deep learning

1 Introduction

In today's information age, computer technology has been widely used in all walks of life and is constantly changing people's production and lifestyles. Similarly, computer vision, deep learning and other fields have achieved remarkable results, and facial feature recognition technology has become more and more mature [1]. The recognition accuracy rate far exceeds traditional manual recognition methods, and very good results have been achieved. With the deepening of deep learning in feature recognition, there is a need for more complete and mature technology applications, which also puts forward higher requirements for developers.

A person's facial expression is an external form of expression that reflects a person's psychological activities and emotions, and it can intuitively express a person's mood. Pain is an individual's perceptual sensation, which is generally divided into chronic pain

N. Xiong et al. (Eds.): ICNC-FSKD 2022, LNDECT 153, pp. 35–42, 2023.
https://doi.org/10.1007/978-3-031-20738-9_5

and acute pain. It will be manifested when there is injury, illness, surgery or other health problems. It is accompanied by potential or real tissue damage, which is one of the common symptoms in medical clinics [2]. In hospitals, doctors usually communicate with patients to make decisions. But this method is not foolproof, especially when patients cannot express their pain (such as newborns, dementia patients, etc.), which is almost unreliable and ineffective [3]. Therefore, the development of a pain recognition system can facilitate medical staff to assess the pain level of these people and provide timely and effective treatment methods for different situations. It has important clinical significance for the medical diagnosis and care of special groups of people who cannot express their pain through natural language. It also has a wide range of medical fields, and its application prospects are widely favored by researchers at home and abroad, becoming a research hotspot.

Facial expression recognition is an artificial intelligence technology, which is a process of automatically extracting facial expression features and classifying emotions through intelligent terminals such as computers. Facial expression recognition is widely used in medical, security and other fields [4]. Similarly, pain is also expressed through the feature changes of facial expressions. The research on facial expression recognition based on computer vision technology has made great progress. Through the real-time monitoring and analysis of the change process of patients' facial expressions, we can not only timely understand the patients' physical status, but also detect the patients' facial expressions, so as to avoid the pain caused by medical staff in the treatment process. It provides a theoretical basis for the prediction and treatment of the disease. Therefore, using facial feature changes to recognize pain is a feasible solution.

Deep learning is a new direction of machine learning neighborhood extension. It is mainly to build a neural network that simulates the neural processing mechanism of the human brain [5]. The more network layers, the deeper the network. The number of network layers is related to the function's fitting ability. When the neural network contains multiple hidden layers, it greatly improves the neural network's ability to fit complex functions and has a more powerful learning ability. In addition, deep learning uses autonomous learning algorithms to obtain features from data, and it can automatically extract hidden features of data classification from a large amount of high-quality data, thereby improving the accuracy of classification and recognition [6].

The realization of painful facial expression recognition mainly goes through three steps: data pre-processing, feature extraction, and feature classification. The specific implementation process is shown in the Fig. 1:

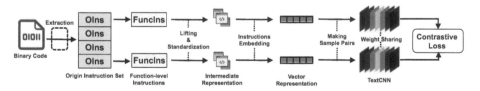

Fig. 1. The process of pain expression recognition

Input facial expression data: Collect some high-quality and certain amount of pain expression data, which can be video data or image data. Preprocessing: Since the collected data is complex and irregular, and facial expression data is the information we focus on, data preprocessing is required, such as unified processing of image size and resolution. Feature extraction: The extracted data is the core and essential feature of the image data. It has good robustness to pain expressions of the same category, and has essential differences in pain expressions of different categories. Feature classification: Choose an appropriate feature classification method based on the extracted features, map the pain expression features to the pain expression category, and classify similar features into the same expression category.

Tang [7] proposed an expression recognition algorithm combining CNN and SVM, and used hinge loss to optimize the loss of the network, which achieved good recognition results on FER2013 data set. Hamester et al. [8] proposed a facial expression recognition model based on dual channel convolutional neural network. Jung et al. [9] proposed to use two different types of neural networks to extract the temporal representation features and temporal geometric features of images respectively, and then integrate the two features to improve the recognition effect.

In order to improve the network performance and enhance the expression ability of network features, this paper proposes a convolution neural network model based on dual-channels. First, the data set is enhanced, and the data are obtained through the low-level network to obtain local features, and the high-level network to obtain global features. Then the features obtained by the two networks are fused, and finally the pain expression classification is directly output by softmax.

2   Related Work

2.1   The Dataset

The COPE [10] database, the first publicly available taxonomy of infant pain expressions, includes 230 color photographs of 26 healthy white newborns (13 boys and 13 girls) born between 18 h and 3 D ages. The photos in the database can also be divided into two categories, namely 82 photos with pain expressions and 148 photos without pain expressions, each of which is 120 by 100 pixels. In this paper, the original samples of the database were expanded and 1407 color photos were obtained, including 1176 photos for training and 231 photos for testing. There were 847 pain-free photos and 329 pain photos in the training set, which were divided into the training set and the verification set, accounting for 8:2. This is a relatively small still image database with limited pain information. The data example is shown in Fig. 2.

2.2   Data Normalization Processing

Data normalization is to process the input picture data so that the processed data is in the range of [0, 1] or [−1, 1]. After the data is normalized, the convergence speed of the model is greatly improved, the accuracy of the model is provided, the characteristics between different dimensions are numerically comparable, the accuracy of the classifier is improved, and the gradient explosion of the model can be prevented.

Vertical flipping   rotate   contrast ratio   Brightness

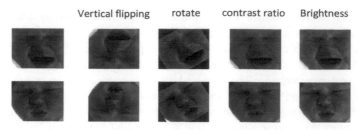

Fig. 2. Pain expression dataset

Data Norma in addition, the commonly used normalization method also has the Min-Max Normalization method. Min–Max Normalization, also known as deviation standardization, is a linear transformation of the original data to map its value between [0, 1]. The formula is as follows:

$$X^* = \frac{x - \min}{\max - \min}.$$ (1)

where max is the maximum value in the data and min is the minimum value in the data. In this method, when new values appear in the data, Max and min may change and need to be redefined.

In addition to the maximum-minimum standardization method of data normalization, there is also the Z-Score standardization method that is more commonly used. Its mathematical definition is:

$$X^* = \frac{x - \mu}{\sigma}$$ (2)

where μ is the mean value of the image data, σ is the variance of the image data, so this method is based on the mean value and variance of the original image data for standardization, the result obtained is the data distributed near 0 and the variance is 1. This method is suitable for input When the maximum and minimum data are unknown.

2.3 Data Preprocessing

In this paper, random horizontal and vertical up-and-down flipping, brightness, contrast adjustment and other methods are used to enhance data. Each time after data enhancement, the image data is different, which greatly improves the number of samples and makes the trained model more powerful in generalization [11].

3 The Proposed Method

3.1 Convolutional Neural Network

Nowadays, the use of convolutional neural networks is extremely common, and various new neural networks that appear now are also evolved on this basis [12]. Convolutional neural network is a multi-layer perceptron that contains multiple hidden layers. The hidden layer includes multiple convolutional layers, pooling layers and fully connected layers. Compared with general artificial neural networks, convolutional neural networks have the advantages of fewer parameters, less calculations, and strong feature extraction capabilities.

The network structure of convolutional neural network is shown in Fig. 3. Generally, in order to obtain more features, multiple layers of convolutional layers and pooling layers will be alternated design, arrange a pooling layer after each convolutional layer. The more layers of the convolutional neural network, the more complex the extracted features. The neurons in the input layer are used to obtain the input data, and then the neurons in each layer are a function containing weights and biases. After the data is obtained from the previous layer, a weighted sum is performed, and the calculated result is passed to the next then, through the fully connected layer, the weighted summation of the features after the front convolution is carried out, and then it can be mapped to the present label.

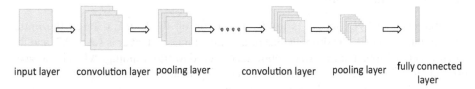

input layer   convolution layer   pooling layer   convolution layer   pooling layer   fully connected layer

Fig. 3. Convolutional neural network graph structure

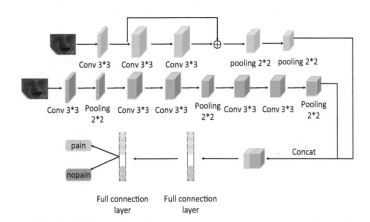

Fig. 4. Structure diagram of dual-channel convolutional neural network

3.2 Dual-Channel Convolutional Neural Network

Figure 4 shows the basic model of the network used in this paper, which is composed of two networks, named net1 and net2 respectively. Net1 contains a convolutional layer, a residual module, and 2 pooling layers. The residual module superimposes the input data according to the conventional weights and passes the activation function. After the weights are superimposed again, the superimposed result and the input data pass through the activation function. The addition of the residual module can capture more comprehensive information. And prevent the appearance of gradient explosion. Net2 contains 5 convolutional layers and 3 pooling layers, in which the pooling layer uses a mixture of average pooling and maximum pooling, which can retain more diverse feature information. The convolution layer and the full connection layer use the ReLu activation function, and the batch normalization technique is used after the third and fifth convolutions in the net2 network. Then the final characteristic graphs of the two networks are connected for planarization, and finally the task of classification is completed through two full connection layers.

4 Experimental Results and Analysis

4.1 Experimental Results

The dual-channel CNN model is used for training on the COPE dataset, the initial learning rate is 0.0001, the dropout size of the fully connected layer is set to 0.7, and the epoch is set to 1500, and the Adam optimizer is used to start training. As training goes through about 700 iterations, the model has converged, and then the accuracy fluctuates as the training progresses. The accuracy on the training set is stable at 98.2% and the accuracy on the validation set is stable at 97.5%. Visualize the accuracy and loss rate of the model during the training process, as shown in Figs. 5 and 6. It can be seen that the results obtained by this model are still very impressive and relatively accurate.

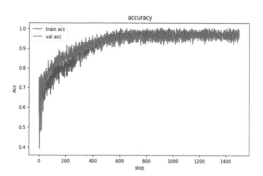

Fig. 5. Accuracy of dual-channel convolution neural network model

4.2 Comparative Test

Compared with the traditional feature extraction and classification methods of painful expression images, convolutional neural network has a very good performance in facial expression feature extraction. In this experiment, VGGNet, ResNet [13], MobileNet and dual-channel convolutional neural network (D-CNN) are used for comparison test. The

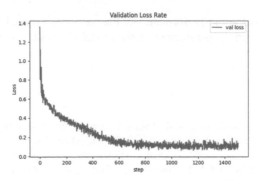

Fig. 6. Loss rate of two channel convolutional neural network model

comparison results are shown in Table 1. From Table 1, we can clearly see that the accuracy of the network model used in this experiment is higher than the other three models, and the recognition accuracy is as high as 97.5%.

Table 1. Compare the recognition accuracy of deep learning methods

Network model	Recognition accuracy (%)
D-CNN	97.5
MobileNet	93.6
ResNet50	96.1
VGGNet	79.36

5 Conclusion

The pain expression recognition method based on dual-channel convolution neural network is adopted in this paper. The neural network is trained by using COPE dataset, and the images are enhanced by flipping up and down, changing brightness and contrast. During model training, the learning rate is adjusted according to the training to improve the model fitting ability and save the best model. Finally, 97.5% accuracy is achieved on the verification set.

Pain expression recognition has practical application scenarios in many fields. Improving the accuracy of pain expression recognition is of great significance to people's life. Although the model used in this paper has achieved good recognition results, due to the limited data samples, which are limited to the expression data of dozens of infants, its universality cannot be extended to the facial expression data of the whole human, so the method in this paper needs more data to test. Secondly, during training, it is necessary to provide a computer with very powerful computing power. The training parameters reach the level of millions, and the training process is long. Therefore, on the basis of not reducing the performance of the model, the model should be properly optimized and the training parameters of the model should be reduced to reduce the amount of calculation.

References

1. Abdullah, S.M.S., Abdulazeez, A.M.: Facial expression recognition based on deep learning convolution neural network: a review. J. Soft Comput. Data Min. **2**, 53–65 (2021)
2. Bendaouia, L., Salhi, H., Si, M.K., et al.: FPGA-implementation of a bio-inspired medical hearing aid based DWT-OLA. In: International Conference on Audio, Language and Image Processing, pp. 806–811(2015)
3. Werner, P., et al.: Pain recognition and intensity rating based on comparative learning. In: 2012 19th IEEE International Conference on Image Processing, pp. 2313–2316 (2012)
4. Singh, S., Fatma, N.: Facial expression recognition with convolutional neural networks. In: 2020 10th Annual Computing and Communication Workshop and Conference (CCWC), pp. 0324–0328 (2020)
5. Wei, H., Zhi, Z.: A survey of facial expression recognition based on deep learning. In Proceedings of the 2020 15th IEEE Conference on Industrial Electronics and Applications (ICIEA), 9–13 Nov 2020, pp. 90–94. Kristiansand, Norway (2020)
6. Guo, J., et al.: Research on facial expression recognition technology based on convolutional-neural-network structure. Int. J. Softw. Innov. **6**, 103–116(2018)
7. Tang, Y.: Deep learning using support vector machines. In: Workshop on Representational Learning, ICML (2013)
8. Hamester, D., Barros, P., Wermter, S.: Face expression recognition with a 2-channel convolutional neural network. In: International Joint Conference on Neural Networks, IEEE, pp. 1–8 (2015)
9. Jung, H., Lee, S., Yim, J., et al.: Joint fine-tuning in deep neural networks for facial expression recognition. In: Proceedings of the IEEE International Conference on Computer Vision, pp. 2983–2991 (2015)
10. Brahnam, S., Chuang, C.F., Sexton, R.S., et al.: Machine assessment of neonatal facial expressions of acute pain. Decis SupportSyst. **43**(4), 1242–1254 (2007)
11. Barsoum, E., Zhang, C., Ferrer, C.C., Zhang, Z.: Training deep networks for facial expression recognition with crowd-sourced label distribution. In: Proceedings of the 18th ACM International Conference on Multimodal Interaction (ICMI'16), pp. 279–283. Association for Computing Machinery, New York, NY, USA (2016)
12. Liu, X., Deng, Z., Yang, Y.: Recent progress in semantic image segmentation. Artif. Intell. Rev. **52**(2), 1089–1106 (2018). https://doi.org/10.1007/s10462-018-9641-3
13. He, K., et al.: Deep residual learning for image recognition. In: 2016 IEEE Conference on Computer Vision and Pattern Recognition (CVPR), pp. 770–778 (2016)

A Noval Air Quality Index Prediction Scheme Based on Long Short-Term Memory Technology

Lijiao Ding, Jinze Sun, Tingda Shen, and Changqiang Jing[✉]

School of Information Science and Engineering Linyi University, Linyi, Shandong 276000, China

`2020120108@lyu.edu.cn`, `210854002047@lyu.edu.cn`, `2020120103@lyu.edu.cn`, `jingchangqiang@lyu.edu.cn`

Abstract. People have been paying more and more attention to their surroundings in recent years, and the problem of air quality, which affects a significant portion of our surroundings, warrants our attention. Ground-level ozone, particulate pollutants, carbon monoxide, sulfur dioxide, and nitrogen dioxide make up the five components of the Air Quality Index, which is a key index for describing the state of the air. Predicting Air Quality Index values using deep learning is crucial for addressing the issue of air pollution. This study compares the prediction accuracy of Long Short-Term Memory and Convolutional Neural Networks on time-series class data, and the results show that Long Short-Term Memory is significantly better than Convolutional Neural Networks in predicting Air Quality Index values. The study uses the actual Air Quality Index data of a specific city for 720 consecutive hours to predict Air Quality Index values using deep learning. When creating the model based on the visualization, a visual presentation is created to demonstrate the data, prediction model, and metrics for calculating the Air Quality Index. The visual display and model building of Air Quality Index indexes are beneficial to giving us a deeper understanding of air pollution problems, and play an important role in setting further policies to improve air quality and prevent respiratory diseases, etc.

Keywords: Air pollution · Air quality index · Long short-term memory

1 Introduction

1.1 Research Background and Significance

Environmental contamination has been a more severe issue in recent years. The current air pollutants impair people's daily life since they are present everywhere. Many air pollutants exist, such as PM2.5 (suspended particulate matter with an aerodynamic equivalent diameter less than or equal to 2.5) and PM10 (particles

N. Xiong et al. (Eds.): ICNC-FSKD 2022, LNDECT 153, pp. 43–52, 2023.
https://doi.org/10.1007/978-3-031-20738-9_6

with a particle size of less than 10), and the AQI is the main factor in determining the quality of the air (Air Quality Index). These invisible pollutants may readily enter the body via the respiratory system and cause respiratory conditions like asthma as well as a number of cardiovascular problems [1–3]. Therefore, it is beneficial for human life to find these pollutants.

The processing, analysis, and prediction of the data using LSTM (Long Short-Term Memory) and running tests on the dataset across a range of time periods are the essential steps in this study. Modeling and predicting many types of data using AI techniques is already standard practice [4–6]. The results of the final analysis show that the theories put forward in this study are viable and satisfy the necessary analytical objectives. Practically describing air quality using neural networks helps us decide on further strategies to enhance air quality.

1.2   Domestic and International Development Status

There are currently a number of mature domestic and international technologies for the detection of air pollutants. These are primarily divided into two categories: one is based on the machine to detect the content of air pollutants in the atmosphere separately, such as the Ishizu flag, through the dynamic dilution instrument to dilute the standard gas to the su code tank, and then after the gas bag transfer into the sample portable gas chromatogra- phy mass spectrometry detection, to achieve determination method of benzene in air [7], and the US has built a 2.1 g "jellyfish vehicle" to detect air pollution [8]. The other is based on machine learning algorithms and statistical models to predict these values, Wang Jing et al. proposed a hierarchical aggressive spatio-temporal model in a Bayesian framework to solve the problem of multi-site simultaneous PM2.5 prediction [9], Lin Chengyong predicted PM2.5 concentration based on the traditional network model based on air pollutants and the network model based on stepwise regression [10]. With more research being done on the atmospheric environment, machine learning models used to calculate the amount of air pollutants are becoming more accurate. However, due to the variety and complexity of air pollutants, we are unable to obtain the precise amounts and data of all emitted pollutants . Additionally, local policies are im- plemented to have some positive or negative effects on air pollution, which also affects the accuracy of the predicted values of air pollutant content.Since AQI is used to monitor the state of the air quality because air pollutants are more broadly categorized, the present research on AQI prediction is focused on in- creasing the accuracy of prediction, which is the top priority of research both domestically and internationally.

2   Visual Presentation of Relevant Data

2.1   Introduction to the Use of Visualization Techniques

Chinese integrate Python and to create PyEcharts [11], which makes using data to build graphs in Python relatively straightforward.

Matplotlib [12], a third-party library for Python data visualization, PyEcharts has not been used for a long time. PyEcharts has more beautiful colors and styles. The specific choice is based on the application scenario of the data.

2.2 Visualization of Data

The data set is Luggehalte data. The data includes values of PM2.5, PM10, NO$_2$ (nitrogen dioxide), SO$_2$ (sulfur dioxide), O$_3$ (Ozone), CO (carbon monoxide). The data format in the original table does not match the data format we expect to save. The data is then stored in a new table, as shown in Table 1. There are few missing values in the table and can be deleted directly.

Table 1. Partial data sets

Time	PM2.5	PM10	O$_3$	NO$_2$	SO$_2$	CO
20190601-0	31	94	36	91	10	0.667
20190601-1	32	107	32	88	8	0.556
20190601-2	30	96	44	72	10	0.571
20190601-3	31	91	49	72	9	0.567

The necessary data from each hour of every day in a certain city in June 2019 is chosen for this experiment, and the data is then displayed.Utilize the PyEcharts approach, as shown in Figs. 1, 2, 3, 4, 5 and 6.

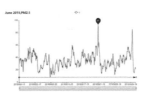

Fig. 1. PM2.5.

3   Data Prediction

3.1   Data Watch

The linear relationship between the scatter plot between the features and the observed features is shown in Fig. 7.

The first step in this design is based on the value of these 6 pollutants to the AQI. This design calculates the 1-h average index, calculates the IAQI (air

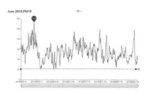

Fig. 2. PM10.

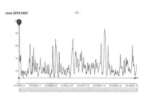

Fig. 3. NO$_2$.

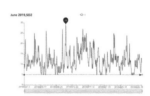

Fig. 4. SO$_2$.

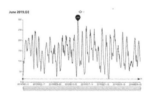

Fig. 5. O$_3$.

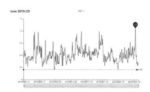

Fig. 6. CO.

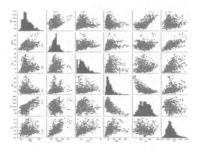

Fig. 7. Linear relationship.

quality sub-indicator) of each pollutant, and obtains the corresponding pollutant concentration value according to Table 2 (see the official website for details), and uses the calculation formula.

$$IAQI_P = \frac{IAQI_{Hi} - IAQI_{Lo}}{BP_{Hi} - BP_{Lo}} (C_P - BP_{Lo}) + IAQI_{Lo}. \tag{1}$$

Among them, $IAQI_P$-AQI of pollutant project P; C_P-Mass concentration value of pollutant item P; BP_{Hi}-Table 2 with the high values of similar pollutant concentration limits; BP_{Lo}-Table 2 with the low value of the concentration limits of similar pollutants; $IAQI_{Hi}$-Table 2 with the corresponding air quality sub-indices; $IAQI_{Lo}$-Table 2 with the corresponding air quality sub-indices.

Then select the largest value from the IAQI for each pollutant as the AQI.

Table 2. Pollutant project concentration limits (partial tables)

Air quality Sub-index (IAQI)	SO$_2$ 24 h on average (μg/m^3)	NO$_2$ 24 h on average (μg/m^3)	CO 24 h on average (mg/m^3)	O$_3$ 24 h on average (μg/m^3)	PM2.5 24 h on average (μg/m^3)
0	0	0	0	0	0
50	50	40	2	160	35
100	150	80	4	200	75
150	475	180	14	300	115

From the above table, the algorithm above is used to calculate the AQI values for the relevant period, and PyEcharts is used to depict the AQI values, as shown in Fig. 8.

3.2   Model Selection and Creation

Model Introduction and Selection The idea of neural network is to achieve intelligence by imitating the brain. Alahi A et al. pointed out the LSTM model, which can learn general human movements and predict their future trajectories [13]. In AQI forecasting, Zhang Lei et al. proposed a method for determining the number of nodes of air quality index prediction of BP neural network

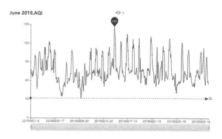

Fig. 8. AQI.

based on time series wavelet decomposition input, and modeled and predicted according to the recent 5a daily average air quality index data in Xuzhou area. In this paper [14], the best model suitable for the dataset is selected by comparing two models, CNN(Convolutional Neural Networks) [15] LSTM both of which can process sequence data. Compared with ordinary neural network, CNN has excellent performance in image processing, CNN also has convolution, which reduces the dimension of the feature vector output from the convolution layer and alleviates the over-fitting phenomenon.

LSTM has forget gate, input gate and output gate. The forget gate controls the degree to which the information from the previous storage unit is forgotten; The input gate controls the amount that the calculated new state is updated to the storage unit; The output current of the output gate depends on the current storage unit. Its calculation process is shown in Fig. 9.

Among them,C_{t-1}-The previous stage memory cell state;h_{t-1}-Hidden output of the previous stage; it-Percentage of learning at the current stage; $\delta, \tan h$-activation function

The formula is as follows, Oblivion Gate:

$$f_t = \delta \left(W_f \cdot [h_{t-1}, x_t] + b_f \right) \tag{2}$$

Input Gate:

$$i_t = \delta \left(W_i \cdot [h_{t-1}, x_t] + b_i \right) \tag{3}$$

$$\check{C}_t = \tan h \left(W_c \cdot [h_{t-1}, x_t] + b_c \right) \tag{4}$$

$$C_t = f_t * C_{t-1} + i_t * \check{C}_t \tag{5}$$

Output Gate:

$$O_t = \delta \left(W_o \cdot [h_{t-1}, x_t] + b_o \right) \tag{6}$$

$$h_t = O_t \cdot \tan h \left(C_t \right) \tag{7}$$

Compared to CNNs, there is a certain amount of memory capacity. The prediction of the selected AQI value in this experiment has a large correlation between the before and after values in a shorter period of time, so LSTM is chosen to create a network model.

Create Network Model Divide the data into training set and test set in the proportion of 8:2 to create a network model using LSTM. In the neural network, since the parameters and weights of the neural network are random each time, the results obtained from each run are different, and we set a random seed according to this characteristic to ensure the same results for each run and to facilitate comparison between models. Model creation using LSTM is done using Sequential model in Keras is the simplest linear, head to tail structural order. Due to the small amount of data, so LSTM used two layers, each layer used Dropout, dropping 20% of neurons, the third layer is fully connected layer units=1, because the final output is 1 dimensional.

In the compilation of the model, the adam optimizer is used, which has the advantages of fast convergence, easy to adjust the reference, etc. The loss function used is the mean square error. The mean square error is the most common regression loss function, this problem belongs to the regression problem, the mean square error performs better in the regression problem, the details of the model are shown in Fig. 10.

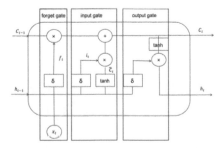

Fig. 9. LSTM calculation flow chart.

Use the training set in your experiment to train the model, select 50 times to train, and then test with the test set. During the 50 training sessions, the Loss value kept decreasing, in a way indicating that the model kept evolving, and finally decreased to 0.0088. Since the normalization of the data was performed in the preprocessing of the data, after the final training was completed, the data was normalized. After creating the model using LSTM, the accuracy of the model is measured by the R^2(coefficient of the degree of fit), and the closer to 1

Model: "sequential_14"

Layer (type)	Output Shape	Param #
lstm_17 (LSTM)	(None, 50, 32)	4352
dropout_19 (Dropout)	(None, 50, 32)	0
lstm_18 (LSTM)	(None, 50, 32)	8320
dropout_20 (Dropout)	(None, 50, 32)	0
dense_21 (Dense)	(None, 50, 1)	33

Total params: 12,705
Trainable params: 12,705
Non-trainable params: 0

Fig. 10. LSTM details.

indicates the more accurate prediction, which is calculated as follows:

$$R^2 = 1 - \frac{\sum_{i=1}^{m}\left(y_{\text{test}}^{(i)} - \hat{y}_{\text{test}}^{(i)}\right)^2}{\sum_{i=1}^{m}\left(y_{\text{test}}^{(i)} - \hat{y}_{\text{test}}^{(i)}\right)^2} = 1 - \frac{\text{MSE}\left(\hat{y}_{\text{test}}\ y_{\text{test}}\right)}{\text{Var}\left(y_{\text{test}}\right)} \tag{8}$$

The value calculated using the LSTM model is 0.558903396305362. The CNN model also divides the data into training and testing sets in an 8:2 ratio. The calculated value after creating the model using CNNR^2 is 0.24494932931597702, and the predicted and true values of the LSTM and CNN model results are shown in the Figs. 11 and 12.

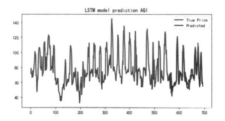

Fig. 11. LSTM.

Based on the R^2 calculated from the above two models, the accuracy of the prediction value of LSTM can be obtained 0.31395407 higher than CNN, and LSTM is more suitable for model construction of this dataset in this experiment.

4 Summary

In this paper, we first perform basic data visualization of the data in the dataset, then calculate the value of AQI, want to create a suitable AQI model by model prediction, and select the LSTM model for prediction by comparing the model

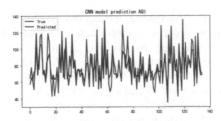

Fig. 12. CNN.

theory and experimental results. The introduction of the data source, loading of the data, pre-processing of the data, data observation, construction of the model, model validation, model training, and final data display are all included in the modeling process. This experiment applies LSTM to AQI prediction to establish a prediction model, which is beneficial for us to grasp the air condition, and lays a data foundation for the establishment and implementation of future policies.

Acknowledgment. This research was partly supported by 61901206 National Natural Science Foundation of China, and partly supported by 2019JZZY010134 Shandong Provincial Key Research and Development Program (Major Science and Technological Innovation Project).

References

1. Mi, Y.-Q., Wang, S.-S., Yu, H.-H., Zhang, J.-P., Wang, X., Luo, S., Li, W.: A cross-sectional study on the relationship between cardiovascular and cerebrovascular disease mortality and air quality. Chin. General Med. **21**(29), 3545–3550 (2018)
2. Dockery, D.W., Pope, C.A.: An association between air pollution and mortality in six US cities. New Engl. J. Med. **329**(24), 1753–1759 (1993)
3. Brook, R.D., Rajagopalan, S., Pope, C.A., Brook, J.R., Kaufman, J.D.: Particulate matter air pollution and cardiovascular disease: an update to the scientific statement from the American heart association. Circulation **121**(21), 2331–2378 (2010)
4. Liu, Y.-Y., Liu, Y.-S., Zheng, J.-W., Chai, F.-X., Li, M., Mu, J.: Research on urban waterlogging prediction method based on BP neural network and numerical model. J. Water Conservancy **53**(3), 12 (2022)
5. Zecchin, C., Facchinetti, A., Sparacino, G., De Nicolao, G., Cobelli, C.: A new neural network approach for short-term glucose prediction using continuous glucose monitoring time-series and meal information. In: Conference proceedings: ... Annual International Conference of the IEEE Engineering in Medicine and Biology Society. IEEE Engineering in Medicine and Biology Society. Conference, pp. 5653–5656 (2011)
6. Wang, Y., Wang, L., Yang, F., Di, W., Chang, Q.: Advantages of direct input-to-output connections in neural networks: the Elman network for stock index forecasting. Inf. Sci. **547**, 1066–1079 (2021)

7. Shi, J.-Q.: Application of portable gas chromatography-mass spectrometry in the detection of ambient air pollution incidents. Chem. Eng. **34**(8), 3 (2020)
8. China Science and Technology Network: U.S. creates 2.1 gram "jellyfish flying machine" to detect air pollution. Sci. Technol. Inf. **3**, I0020–I0020 (2014)
9. Zhang, Y.-W., Hu, J.-Y., Wang, R.: Pm2.5 prediction model based on neural network. J. Jiangsu Normal Univ. Nat. Sci. Ed. **33**(1), 3 (2015)
10. Wang, M., Zhou, B., Guo, Y., He, J.-Q.: Spatial prediction of urban pm2.5 concentration based on BP artificial neural network. Environ. Pollut. Prevent. **35**(9), 5 (2013)
11. Zhang, Y.-Y.: Data visualization based on Pyecharts. Comput. Knowl. Technol. Acad. Edn. **18**(2), 4 (2022)
12. Barrett, P., Hunter, J., Miller, J.T., Hsu, J.C., Greenfield, P.: matplotlib—a portable python plotting package. Astron. Data Anal. Softw. Syst. **XIV** (2005)
13. Alahi, A., Goel, K., Ramanathan, V., Robicquet, A., Savarese, S.: Social LSTM: human trajectory prediction in crowded spaces. In: 2016 IEEE Conference on Computer Vision and Pattern Recognition (CVPR) (2016)
14. Zhang, L., Fang, Z., Ma, T.-F.: Research on AQI prediction by BP neural network with time series wavelet decomposition. J. Xuzhou Eng. College Nat. Sci. Edn. **35**(1), 8 (2020)
15. Kamnitsas, K., Ledig, C., Newcombe, V.F.J., Simpson, J.P., Kane, A.D., Menon, D.K., Rueckert, D., Glocker, B.: Efficient multi-scale 3D CNN with fully connected CRF for accurate brain lesion segmentation. Med. Image Anal. **36**, 61 (2016)

Code Summarization Through Learning Linearized AST Paths with Transformer

Zhenzhou Tian[(✉)], Cuiping Zhang, and Binhui Tian

Xi'an University of Posts and Telecommunications, Xi'an 710121, China
tianzhenzhou@xupt.edu.cn, zhangcuiping@stu.xupt.edu.cn,
mx_info@stu.xupt.edu.cn

Abstract. The lack of code comments is common in software projects. This work proposes TFSum, which generates from a function's source code a readable summary to describe its functionality, with a Transformer based model trained on sequences linearized from the function's abstract syntax tree (AST). To ensure the quality of the generated summaries, TFSum firstly parses from the function's source code an AST of semantic richness, as the raw representation of the function; but linearizes and pre-processes it into a set of normalized token sequences, for efficient and effective following semantic representation learning. On this basis, an encoder-decoder based generative model that adopts the Transformer architecture is designed and trained to automatically generate code comments. The experimental evaluations conducted on a public dataset show the superiority of TFSum over state-of-the-art neural code summarization methods, with the BLEU score reaching 46.84.

Keywords: Code summarization · AST · Attention mechanism

1 Introduction

High-quality code comments can help software developers and maintainers understand source code faster and greatly reduce developers' work. However, the software industry generally suffers from a lack of high-quality code comments. Studies show that developers spend more than half of their time on program understanding [1]. In the software development process, developers manually write high-quality annotations time-consuming and laborious. In recent years, automatic code comments generation methods based on deep learning have received increasing attention from researchers.

Early work used methods based on specific templates and source code based information retrieval to generate comments. Template-based annotation generation methods [2–5] rely heavily on the structure of a particular programming language to generate templates, which need to be updated as the code structure changes, making this method narrowly applicable. The comment generation methods [6,7] based on information retrieval mainly searches for similar

© The Author(s), under exclusive license to Springer Nature Switzerland AG 2023
N. Xiong et al. (Eds.): ICNC-FSKD 2022, LNDECT 153, pp. 53–60, 2023.
https://doi.org/10.1007/978-3-031-20738-9_7

code fragments and reuses the already existing comments. This method relies on the ability to search for similar code fragments in the code base, and if similar code does not exist in the code base, it is difficult to generate accurate comments. In recent years, the use of deep neural network models to automatically generate code comments has attracted increasing attention from researchers. Many the methods use traditional Recurrent Neural Network (RNN)-based neural network models to generate comments in a sequence-to-sequence fashion [8]. However, when dealing with long sequences, such methods are unable to learn the correlation between tokens in the sequence, and thus lose more semantic information.

We propose to use three different representations of a complete java method to generate code comments for a given source code. Specifically, we first normalize the source code and extract Abstract Syntax Tree(AST). We propose the Covering AST Linear Path Extraction algorithm (LCPE) to traverse the AST and obtain the linearized AST. In addition, the function name of a java method contains information about the intent and function of the code, so we extract both the function name of the java method and the method called by this method. We use the source code tokens, the linearized AST and the function name as input to the encoder. To obtain the long-term dependencies between tokens, we design a neural network model incorporating attention to generate comments as well. In summary, the paper has the following main contributions:

- A novel approach to code summary generation is proposed, which automatically generates code summaries by learning the syntactic and semantic features of the source code.
- To improve the model training efficiency, a new AST linearization method is proposed, which traverses the AST and converts it into a normalized set of token sequences. Meanwhile, function names are used as additional features for improving the quality of summary generation.
- The experimental evaluations conducted on a public dataset show the superiority of TFSum over state-of-the-art neural code summarization methods, with the BLEU score reaching 46.84.

2   Proposed Method

In this section, we describe in detail the proposed approach. The overview of TFSum is shown in Fig. 1, which includes data processing, model training and prediction. It includes two main steps: code processing and AST linearizing, model training to generate comments.

2.1   Code Processing and AST Linearizing

Code Tokenization. The source code contains a wide variety of method names and identifiers. These words occur less frequently, and when building the lexicon of the training set, these words need to be removed due to time and memory constraints. However, the culling of the lexicon leads to the loss of semantic

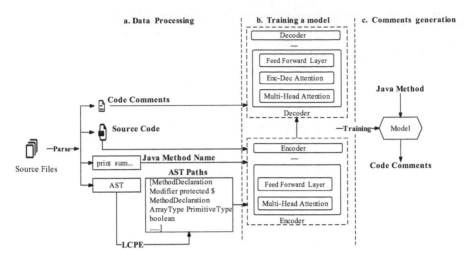

Fig. 1. Overall framework of TFSum.

information; meanwhile, when the model is trained and applied, words are not in the lexicon, which leads to the Out-Of-Vocabulary (OOV) problem and greatly reduces the quality of the comments generated by the model. In order to reduce the proportion of censored words and capture more semantic information of the source code, the identifiers and method names of the humped naming are first disassembled. For example, the method name *"createNewLaunchConfiguration"* will be split into *"create"*, *"New"*, *"Launch"* and *"Configuration"*. In this way, the remote words are split into common words with clearer meaning, and the proportion of OOV words is reduced, so that the model learns more semantic information about the words and generates more accurate comments when generating comments. The token sequence with length l_c is represented as: $\mathbf{X}_1 = (\mathbf{x}_1, \mathbf{x}_2, ..., \mathbf{x}_{l_c})$, where $\mathbf{x}_i$ is each token of the source code sequence.

AST Linearizing. Java code is a highly structured language, and if the source code is simply considered as pure textual information, the structural information will be ignored. We convert the source code into the corresponding AST by using the existing syntax analysis tools, and for each AST, we partition the traversal by the granularity of the AST path. Too large a granularity of AST separation, such as a complete AST, results in a longer sequence. Too small a granularity results in an AST sequence that carries very little structural information. A path-based traversal separation of AST, which contains both node and path information in the traversed sequence, contains rich syntactic information.

We propose an LCPE traversal method to obtain a linear sequence of tokens of AST. The detailed traversal process is shown in Algorithm 1.

– Starting from the root node, the entire AST tree is traversed using depth-first search. During the traversal, the traversed nodes are recorded on a stack and the traversed nodes are popped up in the backtracking.

Algorithm 1 Leaf node Covering AST linear Path Extraction

Input:
    A: AST of a java method.
Output:
    P: AST path set.
 1: $nodeStack \leftarrow A$
 2: $pathStack \leftarrow A$
 3: $pathSeq \leftarrow \varnothing$ ▷ pathSeq is the sequence obtained after traversing the AST
 4: **while** $nodeStack$ not null **do**
 5:    $curNode \leftarrow nodeStack.pop$
 6:    $curPath \leftarrow pathStack.pop$
 7:    **if** $curNode$ has no children **then**
 8:       $pathSeq \leftarrow curPath$ ▷ Record a full path to pathSeq
 9:    **else**
10:       **for** child in curNode's children **do**
11:          $nodeStack \leftarrow child$
12:          $pathStack \leftarrow curPath + child$ ▷ Add child nodes to curPath
13:       **end for**
14:    **end if**
15: **end while**
16: return $pathSeq$

- When the traversal reaches the leaf nodes, each path of the tree from the root node to the leaf nodes is recorded in the array prepared in advance. After traversing the AST tree, the path matrix of the AST tree is obtained.

AST contains the structure information and variable information of the source code. LCPE traversal keeps the structure and content information of the AST intact, while traversal makes the different source codes generate unique structural representations compared to the prior-order traversal. Then we linearize the obtained path matrix. This results in the sequence of the LCPE traversal of the AST. For a LCPE sequence of length l_a is represented as: $\mathbf{X}_2 = (\mathbf{x}_1, \mathbf{x}_2, ..., \mathbf{x}_{l_a})$, where $\mathbf{x}_i$ is each token of the AST sequence.

Method Name Extraction. The method name of a java method contains information about the intent and function of the code, so we extract both the method name of the java method and the method called by this method. We traversal of each node of the AST, extracting the node when it is a method call type. Then, we obtain the linear sequence of function name: $\mathbf{X}_3 = (\mathbf{x}_1, \mathbf{x}_2, ..., \mathbf{x}_{l_n})$, where $\mathbf{x}_i$ is each method name token of the java method.

We concatenate the three parts of the input, so the input is represented as:

$$\mathbf{I}_1 = (\mathbf{X}_1 + \mathbf{X}_2 + \mathbf{X}_3) \tag{1}$$

$$\mathbf{I}_2 = (\mathbf{X}_1 + \mathbf{X}_2) \tag{2}$$

(2) does not include java method name.

2.2   Transformer-Based Comment Generation Model

Using Transformer as the basic architecture, a bidirectional encoder and an autoregressive decoder are designed. Where each layer of the encoder is a Transformer structure. The encoder mainly contains Multi-Head Attention Layer and Feed Forward Layer. In the encoding phase, the embedding vector of the stitched token sequence is input to the Multi-Head Attention Layer together with the position encoding vector. The Multi-Head Attention mechanism is able to learn both sequential and non-sequential syntactic structures within a sentence, and capture the association between tokens in a sentence. In the Feed Forward layer, the vector obtained by attention is further extracted by amplifying the vector. The final encoder outputs a vector for each token.

The decoder mainly consists of Masked Multi-Head Attention Layer, Enc-Dec Attention Layer, and Feed Forward Layer, and in the decoding stage, at time t, the starting symbol $\langle sos \rangle$ is input to the autoregressive decoder, and in predicting the token at time t token, the decoder learns the relationship between the tokens generated at the previous $t-1$ moments. The output generated by the encoder is input to the second layer of the decoder, the Enc-Dec Attention Layer, which is used to learn the contextual information of the token to be generated at moment t. The relationship between the current word of the decoder and the output word of the encoder is learned. The token vector output in the Feed Forward Layer, where the token vector is mapped into a vector of vocabulary size in the layer, and in the softmax layer, is transformed into a probability in the vocabulary.

3   Experimental Evaluation

3.1   Dataset

We use the dataset collected by Hu et al. [9], which contains Java projects with more than 20 stars on GitHub from 2015 to 2016. The dataset consists of java methods extracted from the projects and the corresponding comments. In this dataset, we randomly chose 8714 samples as the validation set and test set, and the other 69,708 samples as the training set. The average length of source code and comments is 99.94 and 8.86 tokens, respectively.

3.2   Implementation Details and Experimental Setup

The vocabulary sizes of the code in the Java dataset is set to 30000, and the maximum length of code and comments is 150 and 50 tokens respectively. The number of layers for both the encoder and decoder was set to 6 layers. The number of heads of the multi-headed attention mechanism is set to 8. We use the Adam optimizer with an initial learning rate of 0.001.

We compare our TFSum approach with six methods reported in Ahmad et al. [10] and their proposed FULL model. In addition, for all baseline methods, we refer to Ahmad et al. 's hyperparameter setting.

3.3   Experimental Results and Analysis

The experiments use the evaluation metrics BLEU and METEOR, which are commonly used in the field of automatic code annotation generation. BLEU, proposed by Papineni et al. [11]. It uses n-gram matching rules and can score the similarity between reference and candidate utterances, and the greater the similarity, the higher the score. METEOR was proposed by Denkowski [12] et al. It uses knowledge sources such as WordNet to introduce synonym matching while considering the word morphology of the words.

The result shows that the TFSum model outperforms the baselines. Compared with the RNN-based sequence models [9,13–17], the performance of TFSum is obviously improved. The reason for this phenomenon is that the RNN-based models may not be able to learn the interconnections between tokens in long sequences. In order to make up the deficiency of the RNN-based sequence models, Ahmad et al. proposed Full Model which based Transformer architecture. They use the source code sequence to generate the comments. Unlike the baseline approach, we design the encoder and decoder based on the Transformer architecture. In the encoding stage, the source code sequence of hump decomposition, LCPE traversal sequence and function name of the code are stitched together as the input of the encoder, and in the decoding stage, the decoder uses the self-attention mechanism, and the enc-dec attention mechanism, and the decoder finally outputs the probability of the annotated token that will be output through the optimization algorithm to continuously optimize the model.

Table 1. Comparison analysis results of TFSum model with other models.

Approaches	BLEU (%)	METEOR (%)
CODE-NN	27.60	12.61
Tree2Seq	37.88	22.55
RL + Hybrid2Seq	38.22	22.75
DeepCom	39.75	23.06
API + CODE	41.31	23.73
Dual model	42.39	25.77
Full model	44.58	26.42
TFSum	45.59	28.4
TFSum (pre)	46.84	30.63

As shown in Table 1, on the Java dataset, Full Model [12] only inputs the source code sequences into the model, so it can only learn the lexical information of the code. The TFSum model inputs spliced sequences into the model, effectively learning the lexical and semantic information of Java methods. Compared to the Full Model, the BLEU score and METEOR score improved by 2.26 and 4.2, respectively. Moreover, compared with the RNN-based models, the transformer-based TFSum model has three important attention mechanisms, which solve the problem that the traditional RNN sequence-based model can not capture the close dependencies between source code tokens. The TFSum model can efficiently learn the correlation between tokens in a sequence, while capturing both syntactic and semantic information in Java methods. The results show that the TFSum model can capture richer and more comprehensive information about the source code.

In addition, we did an ablation study to investigate the benefits of function names on the source code summarization task. TFSum takes I_2 as input and TFSum(Pre) takes I_1 as input. Table 2 demonstrates that learning the function names of java methods is effective as we can see, it improves the quality of model generated comments compared to when it is not excluded. Table 1 demonstrates that the names of the methods in the source code provide a concise description of the source code. These methods can describe the intent of the code and functional information. We provide an example in Fig. 2 to demonstrate the usefulness of our proposed method.

```
private static int findTarget(int[] array, int target) {
    for (int i = 0; i < array.length; i++)
        if (array[i] == target) return target;
    return -1;
}


TFSum: Find the number from an array.
TFSum(Pre): Find the target number from an array.
Human Written: Find the target number from an array.
```

Fig. 2. An example of code comments generated by different inputs.

4   Conclusion

We design the encoder and decoder based on Transformer architecture and use source code sequences, LCPE traversal sequences and function name sequences as input to the encoder to learn syntactic and semantic information of java methods. The attention mechanism of the model solves the long-term dependency problem between source code tokens, learns the relationship between tokens, and captures richer semantics. The experiment show that the method generates higher quality code summary compared with existing summary generation methods, which better meet the needs of practical work.

Acknowledgment. This work was supported in part by the Natural Science Basic Research Program of Shaanxi (2022JM-342, 2018JQ6078), the Science and Technology of Xi'an (2019218114GXRC017CG018-GXYD17.16), and the Key Research and Development Program of Shaanxi (2019ZDLGY07-08).

References

1. Xia, X., Bao, L., Lo, D., Xing, Z., Hassan, A.E., Li, S.: Measuring program comprehension: a large-scale field study with professionals. IEEE Trans. Softw. Eng. **44**(10), 951–976 (2017)
2. Sridhara, G., Hill, E., Muppaneni, D., Pollock, L., Vijay-Shanker, K.: PrTowards automatically generating summary comments for java methods. In: Proceedings of the IEEE/ACM International Conference on Automated Software Engineering, pp. 43–52 (2010)
3. McBurney, P. W., McMillan, C.: Automatic documentation generation via source code summarization of method context. In: Proceedings of the 22nd International Conference on Program Comprehension, pp. 279–290 (2014)
4. McBurney, P.W., McMillan, C.: Automatic source code summarization of context for java methods. IEEE Trans. Softw. Eng. **42**(2), 103–119 (2015)
5. Sridhara, G., Pollock, L., Vijay-Shanker, K.: Generating parameter comments and integrating with method summaries. In: 2011 IEEE 19th International Conference on Program Comprehension, pp. 71–80. IEEE (2011)
6. Eddy, B. P., Robinson, J. A., Kraft, N. A., Carver, J. C.: Evaluating source code summarization techniques: Replication and expansion. In: 2013 21st International Conference on Program Comprehension (ICPC), pp. 13–22. IEEE (2013)
7. Haiduc, S., Aponte, J., Moreno, L., Marcus, A.: On the use of automated text summarization techniques for summarizing source code. In: 2010 17th Working Conference on Reverse Engineering, pp. 35–44. IEEE (2010)
8. Sutskever, I., Vinyals, O., Le, Q. V.: Sequence to sequence learning with neural networks. Adv. Neural Inf. Process. Syst. **27** (2014)
9. Hu, X., Li, G., Xia, X., Lo, D., Lu, S., Jin, Z.: Summarizing source code with transferred API knowledge (2018)
10. Ahmad, W. U., Chakraborty, S., Ray, B., Chang, K. W.: A transformer-based approach for source code summarization. arXiv:2005.00653 (2020)
11. Papineni, K., Roukos, S., Ward, T., Zhu, W. J.: Bleu: a method for automatic evaluation of machine translation. In: Proceedings of the 40th Annual Meeting of the Association for Computational Linguistics, pp. 311–318 (2002)
12. Denkowski, M., Lavie, A.: Meteor universal: Language specific translation evaluation for any target language. In: Proceedings of the Ninth Workshop on Statistical Machine Translation, pp. 376–380 (2014)
13. Iyer, S., Konstas, I., Cheung, A., Zettlemoyer, L.: Summarizing source code using a neural attention model. In: Proceedings of the 54th Annual Meeting of the Association for Computational Linguistics (Volume 1: Long Papers), pp. 2073–2083 (2016)
14. Eriguchi, A., Hashimoto, K., Tsuruoka, Y.: Tree-to-sequence attentional neural machine translation. arXiv:1603.06075 (2016)
15. Wan, Y., Zhao, Z., Yang, M., Xu, G., Ying, H., Wu, J., Yu, P. S.: Improving automatic source code summarization via deep reinforcement learning. In: Proceedings of the 33rd ACM/IEEE International Conference on Automated Software Engineering, pp. 397–407 (2018)
16. Hu, X., Li, G., Xia, X., Lo, D., Jin, Z.: Deep code comment generation. In: 2018 IEEE/ACM 26th International Conference on Program Comprehension (ICPC), pp. 200–20010. IEEE (2018)
17. Wei, B., Li, G., Xia, X., Fu, Z., Jin, Z.: Code generation as a dual task of code summarization. Adv. Neural Inf. Process. Syst. **32** (2019)

Function Level Cross-Modal Code Similarity Detection with Jointly Trained Deep Encoders

Zhenzhou Tian[✉] and Lumeng Wang

Xi'an University of Posts and Telecommunications, Xi'an 710121, China
tianzhenzhou@xupt.edu.cn, wanglumeng@stu.xupt.edu.cn

Abstract. Binary source code matching, which matches a source code snippet to its similar binaries or vice versa, facilitates security-critical tasks such as software plagiarism detection and vulnerability confirmation; while the huge structural and syntactical gaps between the two different kind modalities, makes it challenging to calculating their similarity in an accurate manner. To this end, this work presents XMSim (Cross-Modal function Similarity detector), which compares directly between the source codes and the binary assemblies. XMSim resorts to an Transformer-based encoder and the TextCNN model respectively, to extract semantic vectors from the normalized source code tokens and the assembly instructions. To guarantee the effectiveness the semantic encoders, a large dataset consisting of 92,000 source-binary function pairs is constructed, on which the siamese neural network structure is adopted to get both encoders jointly trained. As the experimental evaluations show, XMSim can effectively identify similar or non-similar relationships between the source codes and the binaries, with the detection accuracy reaches 86.7%. Also, XMSim show good resilience against the disturbances from different compilers and optimization levels.

Keywords: Code similarity · Cross-modal · Transformer · Siamese network

1 Introduction

While source code is flexible and informative, but it is difficult to obtain and lacks a lot of information about program operation in real scene. Executable binary code fragments are generated by compiling the source code, once the source code is modified, some of the information in the binary code fragment will be changed as well. We can focus on the binary source code matching [1] and apply it to software plagiarism detection, vulnerability confirmation and other fields.

Traditional research methods are mainly based on attribute counting [2], mainly by extracting a variety of basic attributes such as number of variables, strings, integers and other code attributes features [3]. Then they use matching

N. Xiong et al. (Eds.): ICNC-FSKD 2022, LNDECT 153, pp. 61–68, 2023.
https://doi.org/10.1007/978-3-031-20738-9_8

algorithms to complete binary source code matching, such as BinPro [4] and B2SFinder [5]. Although these methods implement the problem, selecting suitable features require expert experience and these literal feature information has great limitations. Therefore, the accuracy of code similarity detection is poor.

We build a cross-modal network framework to accomplish binary source code matching. We use fine-grained and precise function sequences to describe code information. Source code and binary code need to feed into TextCNN and Transformer respectively. The siamese neural network is adopted to get both encoders jointly trained to obtain two vectors. The representations of source code and binary code are different. It is not possible to measure the similarity directly. The two encoded feature vectors need to be processed so that they can in the same vector space. Finally, binary source code matching is achieved through metric learning [6]. In summary, this paper has the following major contributions:

- We have constructed a dataset, which consisting of 92,000 pairs of source-binary codes for similarity analysis. It can be used for cross-modal code similarity detection.
- We focus on binary source code matching. We propose a function level cross-modal code similarity detection framework called XMSim (cross-modal functional similarity detector). The source and binary code are encoded by Transformer and TextCNN to extract the latent semantic features of code. Compared with several classical models, XMsim has good performance.
- Considering the difference sequence length between source and target code, different encoders are used separately. Mean-pooling is adopted as the context feature extraction method. Experimental results show that our model has robustness.

2   Proposed Method

In this section, we present details of cross-modal code similarity detection. XMsim is at the function level. Coarse-grained information such as function call cannot be applied to our task. We mainly use sequence information of function for code matching. The overall framework of our methods is shown in Fig. 1. Both source and binary code have a code sequence information input. We will describe three points in terms of dataset construction, word embedding and cross-modal retrieval based on representation learning.

2.1   Dataset Construction

There are few publicly available datasets on binary source code matching task. We need to construct a source code binary code similarity detection dataset. Sorted out numerous C project source files and binaries. For the source code, the functions of the C and CPP files in the project files were constructed. For binary code, an interactive disassembler IDA Pro [7] disassembles binary code

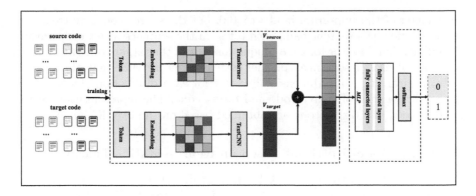

Fig. 1. The overall framework of XMSim.

into assembly code. We choose gcc-x64-O0 to build dataset. We focus on code similarity detection at the function-level. If a source function and a target function have the same project name and function name, it is considered similar, otherwise it is not similar. A total of 92,000 pairs of source-target codes were organized for cross-modal code similarity detection.

The binary information generated by compiling with different compilers or with different optimization options varies considerably. We hope that our model is robust enough to ignore compilation differences and still have good results in similarity analysis. We disassembled target code on 6 combinations of different compilers(gcc,clang,icc) and different optimization options(O0,O3). We construct 40,000 pairs of source-target code for training in each of the six compilation conditions.

2.2 Word Embedding

The source and binary codes are textual information, which cannot be directly fed into the neural network. We need word embedding.

The vectors generated by one-hot encoding are too sparse and the technique is computationally difficult. In contrast, word2vec [8] can learn dependencies and lexical semantics in an independent and unsupervised manner, which based on contextual information in the current source or target code corpus. It can compute a distributed word vector for each word in a given corpus context. Word2vec can be fully adapted to the task of word embedding in source and target code.

We use the skip-gram model to iteratively infer other words within the context window by the central word learning a d-dimensional vector for each unique token. When we use Word2Vec for word embedding, if the vector dimension is too high, it will lead to sparse representation space and too much invalid data, increasing the time and space cost of neural network model training. If the dimensionality is low, it may lead to similar representation vectors for different tokens. It's difficult to distinguish them accurately. We set the vector dimension d parameter of the skip-gram model to 100. For the source code, we treat each source function as a sentence and each token in the sentence as a word, learning a 100-dimensional vector for each unique token. For the target code, we consider each basic block as a sentence, consider each abstract instruction in the sentence as a word, learning a 100-dimensional vector for each instruction. Thus, we can get two dictionaries of source code and target code. The representation vectors of source code and target code are obtained based on the two dictionaries.

2.3 Cross-Modal Retrieval Based on Representation Learning

The source code and the binary code different in structure and representation. The similarity between them cannot be calculated directly. One of the research focuses of this paper is to mine the hidden semantic association between two codes and achieve cross-modal retrieval. Mature cross-modal retrieval can be broadly classified into two categories, hash-based methods and representation learning-based methods. In this paper, the cross-modal retrieval method based on representation learning, which is highly efficient to train and easy to implement.

The pseudo-siamese network is used for joint learning. The pseudo-siamese network [9] is trained to extract feature vectors of source code and binary code using TextCNN and Transformer [10] respectively, the extracted features are mapped to a common subspace [11] to achieve cross-modal [12] retrieval.

Source Code Feature Extraction Transformer proposed by Google consists only of self-attention and feed forward neural network. The structure of the Transformer encoder is shown in Fig. 2. It proposes to use only the attention mechanism. We use Transformer for source code presentation learning, using its self-attention mechanism to learn about dependencies and long-term dependencies within features. The encoder performs a nonlinear transformation of the input source code token features. Through a self-attentive mechanism, capture the internal correlation of features and learn long-term dependencies. The transformer encoder receives the input token sequence, which is recorded as:

$$\mathbf{S} = (\mathbf{s}_1^T, \mathbf{s}_2^T, ... \mathbf{s}_l^T)^T, \quad \mathbf{S} \in \mathbb{R}^{l * D_{model}} \tag{1}$$

where any s_i represents a certain token vector of the code, l is the length of the token sequence of the code and D_{model} is the dimensionality of the input token.

For each token vector of the input, the self-attention function in the encoder describes it as a set of query vectors and a set of key-value pair vectors. Mapping both of them to the output vector and calculates the attention score based on the query vector and the corresponding key vector. The output vector can be expressed as a weighted sum of the attention score and the corresponding value vector. Both the query vector and the key vector are mapped by s_i and the dimension is D_k. The dimension of the value vector is D_v.

In this paper, the dot product attention function used calculates the dot product of the query matrix $\mathbf{Q}$ and the key matrix $\mathbf{K}$ composed of each vector.

In order to avoid the problem that the input value is too large to lead to the softmax function enters the saturated region, we use $\sqrt{D_k}$ to scale. The attention score is calculated by the softmax function and weighted and summed with the value matrix $\mathbf{V}$ to obtain the attention output. The attention output is:

$$\mathbf{A} = \mathbf{softmax}(\mathbf{Q}\mathbf{K}^T/\sqrt{\mathbf{D_k}})\mathbf{V} \tag{2}$$

Self-attention is a feature extraction operation. For each token feature vector, the correlation between them is taken into account. Attention is calculated and then fused into a new feature vector.

The second sub-layer of transformer encoder is the feed forward neural network, which is composed of two full connection layers. In the middle of the two full connection layers, ReLU activation function is used for nonlinear transformation. To ensure the network converges quickly, we add residual connections and layer normalization after each sub-layer.

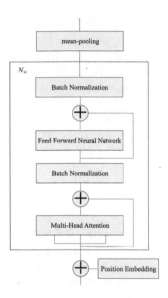

Fig. 2. Transformer encoder structure diagram.

Target Code Feature Extraction TextCNN is powerful for character-level target code sequences. Extract information from tokens using multiple convolutional kernels of different sizes to capture the local relevance of code sequence information. On the basis of word embedding, each target code sequence information is converted into the original feature matrix T.

$$\mathbf{T} = (\mathbf{t}_1, \mathbf{t}_2, ...\mathbf{t}_l)^T, \quad \mathbf{T} \in \mathbb{R}^{l*d} \tag{3}$$

where l is the length of the token sequence and d is the dimension of the token embedding. $\mathbf{t}_i \in \mathbb{R}^d$ is the corresponding embedding of the token in the target code sequence.

Then feature extraction is performed using a convolutional kernel. The token sequence expressed by word embedding is one-dimensional data, while TextCNN uses one-dimensional convolution. In the convolution layer, m convolution filters with structure $n \times d$ are used to convolve the original feature matrix T to obtain the feature matrix $\tau \in \mathbb{R}^{(l-n+1)*m}$. Where n denotes the size of the convolution kernel. To extract different views of the feature pattern, T is convolved using different convolution kernels of sizes 3, 5 and 7, respectively. The sizes of feature maps obtained by convolution kernels of different heights are different. Use aggregation functions to make feature vectors in the same size. The maximum value is extracted by 1D-maxpooling. Finally, eigenvectors are spliced together.

3   Experimental Evaluation

3.1   Implementation Details and Experimental Settings

We implement the cross-modal [13] code similarity detection model XMSim. We use the IDA Pro decompiler tool to parse the binary files to obtain the assembly instructions. The word embedding and the representation learning module are implemented by python and the PyTorch framework. We randomly assign the entire dataset to training set, validation set and testing set in the ratio of 70, 15 and 15%. The training set has a batch size of 128, Adam optimizer training for 100 epochs, learning rate of 0.001. Both of the source code and the target code, the length of the character-level sequence is 200 and the dimension of the word embedding layer is 100. We evaluate our model on the testing set in terms of accuracy, precision, recall and F1 as metrics.

3.2   Experimental Results

XMsim Performance Experiment   To evaluate the performance of the XMSim, the performance of our model is compared with other representation learning methods in terms of accuracy, precision, recall and f1-score. In Table 1, the left and right of '+' are the representation learning methods used for source code and binary code, respectively. It can be seen that XMSim outperforms other models in all four metrics. Both TextCNN and Transformer have powerful contextual semantic feature processing capabilities. The range of source code tag lengths is large. We use transformer for source code feature extraction. Setting the multi-head attention to 4, which can solve the long sequence dependency problem and extract rich source code features. The average length of the target token is small. The target code information is extracted by TextCNN. When processing short sequence code information, the convolution kernel of TextCNN is set to a reasonable size, which can focus on the global features while paying attention to the local features. We set the convolution kernel as 3, 5, and 7 to achieve the best result.

Table 1. Comparison with other methods.

Model	Accuracy (%)	Precision (%)	Recall (%)	F1 (%)
LSTM+LSTM	82.3	82.6	82	82
TextCNN+TextCNN	84.7	84.9	84.4	84.6
TextCNN+Transfomer	85.9	87	84.4	85.6
XMsim	**86.7**	**87.3**	**85.8**	**86.6**

XMSim Robustness Experiments We selected three compilers (gcc, clang, icc) and two levels of optimization options (O0,O3) to construct six datasets to explore the performance of our model under different compilers or different optimization options. Each dataset contains 28,000 source-binary pairs for training, 7000 pairs for validation, and 7000 pairs for testing. We use different encoders depending on the length of the source and target code tokens. Compared with the max-pooling which tends to extract local features, we choose the mean-pooling which can better represent the semantic features of sequence context. So the impact of different compilers and optimization options on the model performance is not significant. According to the results in Table 2, The accuracy of XMsim in different compilation environments is higher than 86.6%.

Table 2. The effect of the model under different compilation conditions.

Compiler optimization leve	Clang		gcc		icc	
	O0 (%)	O3 (%)	O0 (%)	O3 (%)	O0 (%)	O3 (%)
Accuracy	86.7	86.6	86.5	86.4	86.5	86.2
Precision	87.3	86.9	86.8	86.7	84.8	87
Recall	85.8	86.4	86.2	86	88.8	84.9
F1	86.6	86.7	86.5	86.3	86.8	86

4   Conclusion

In this paper, we propose a code similarity detection model that can be used for cross-modal code similarity detection. To evaluate the performance of the model, we constructed a dataset consisting of 92,000 source-target code pairs. XMSim performs well on the dataset. For the features of source and target code, the model use TextCNN and simplified Transformer for feature extraction of the code, respectively. The training speed is improved while maintaining the accuracy of the model. The model can focus on the global features while paying attention to the local features of the code sequence. According to the results, XMsim can be widely applied to code compiled from different compilers and optimization options. We will attempt to extract structural features of source and binary codes and fuse them with sequence features of source and target codes, so that the fused vectors can more fully reflect the code features, as a way to improve the accuracy of similarity detection.

Acknowledgment. This work was supported in part by the Natural Science Basic Research Program of Shaanxi (2022JM-342, 2018JQ6078) the Science and Technology of Xi'an (2019218114GXRC017CG018-GXYD17.16), National Natural Science Foundation of China (61702414), the Key Research and Development Program of Shaanxi (2019ZDLGY07-08), and Special Funds for Construction of Key Disciplines in Universities in Shaanxi.

References

1. Salvador, M.: Efficient Plagiarism Detection far Software Modeling assignments. Comput. Sci. Educ. 2–3 (2020)
2. Berghel, H.L., Sallach, D.L.: Measurements of program similarity in identical task environments. Sigplan Notices **19**, 65–76 (1984)
3. Sam, G.: A tool that detects plagiarism in pascal programs. ACM SI GCSE Bull. **13**, 15–20 (1981)
4. Miyani, D., Huang, Z., Lie, D.: Binpro: a tool for binary source code provenance (2017)
5. Yuan, Z., Feng, M., Li, F., et al.: B2SFinder: detecting open-source software reuse in COTS software. In: 34th IEEE/ACM International Conference on Automated Software Engineering (ASE), pp. 1038–1049 (2019)
6. Chopra, S., Hadsell, R., LeCun, Y.: Learning a similarity metric discriminatively, with application to face verification. In: 2005 IEEE Computer Society Conference on Computer Vision and Pattern Recognition (CVPR'05). June 20–25, San Diego, CA, USA. IEEE, pp. 539–546 (2005)
7. Justin, F., Dan K.: Reverse engineering code with IDA pro (2008)
8. Tu, N., Thu, H., Nguyen, V.A.: Language model combined with Word2Vec for product's aspect based extraction. ICIC Express Lett. 1033–1040 (2020)
9. Bromley, J., Bentz, J.W., Bottou, L., et al.: Signature verification using a "Siamese" time delay neural network. In: 1994 Series in Machine Perception and Artificial Intelligence, pp. 25–44 (1994)
10. Vaswani, A., Shazeer, N., Parmar, N., et al.: Attention Is All You Need. Advances in Neural Information Processing Systems, pp. 5998–6008 (2017)
11. Hzheng, Y.: Methodologies for cross-domain data fusion: an overview. IEEE Trans. Big Data, 16–34 (2015)
12. Zeng, D., Yu, Y., Oyama, K.: Audio-visual embedding for cross-modal music video retrieval through supervised deep CCA. In: Proceedings of 2018 IEEE International Symposium on Multimedia (ISM). Piscataway. IEEE Press, pp. 143–150 (2018)
13. Wei, J., Xu, X., Yang, Y., et al.: Universal weighting metric learning for cross-modal matching. In: Proceedings of the IEEE/CVF Conference on Computer Vision and Pattern Recognition, pp. 13005–13014 (2020)

Ease Solidity Smart Contract Compilation through Version Pragma Identification

Zhenzhou Tian$^{(\boxtimes)}$ and Ruikang He

Xi'an University of Posts and Telecommunications, Xi'an 710121, China
tianzhenzhou@xupt.edu.cn, heruikang@stu.xupt.edu.cn

Abstract. Solidity, the most popular language for developing smart contracts that run on blockchains, has experienced a fast version changing since its first release. Yet, the fact that different Solidity versions are generally incompatible, makes it torturous for developers to correctly pick the Solidity version to compile a smart contract missing version-indicative information. This work presents SolCom to ease the compilation of a Solidity smart contract. It infers from the source code of a smart contract its version pragma, with which the developers can accurately determine the right-version Solidity compiler to get the source code compiled rather than aimless attempts. SolCom processes the source code into normalized tokens, and resorts to a carefully-designed attention augmented convolutional neural network model to capture version indicative features. As the experimental evaluations show, SolCom can identify the version pragma with a rather high accuracy of 96.30%.

Keywords: Smart contract · Version pragma · Solidity · Neural networks

1 Introduction

In the current years, with the development of blockchain, smart contract technology has also developed rapidly [1,2]. Ether is the first blockchain to provide a sound framework for smart contract development. Among the many languages for developing smart contracts, Solidity stands out and is now the language of choice for developing smart contracts. Since the release of Solidity, it has evolved to its eighth major version. The language already has a complete syntax structure and development specifications, but the biggest problem with the language is that the compilers are not compatible between versions, which causes great problems for developers to compile contracts. Solidity officially recommends that you mark the version you are using to develop the contract on the first line of the contract, as shown in Listing1, to make it easier for developers to choose the right compiler to compile with, but there are still various reasons why this markup is missing. Influenced by the work of Tian et al. [3]. We hope that by putting various versions of the contract code into a neural network model for training, we can obtain a neural network model that can recognize unknown versions of the smart contract.

© The Author(s), under exclusive license to Springer Nature Switzerland AG 2023
N. Xiong et al. (Eds.): ICNC-FSKD 2022, LNDECT 153, pp. 69–75, 2023.
https://doi.org/10.1007/978-3-031-20738-9_9

```
1     pragma solidity 0.5.12;
2     contract Migrations {
3         address public owner;
4         uint public lastCompletedMigration;
5         modifier restricted() {
6             if (msg.sender == owner) {
7                 -;
8             }
9         }
10    }
```

Listing 1.1. A Solidity program example

In the domain of natural language processing, text classification is an important part of information filtering, and literature retrieval [4,5]. Yahui Chen's proposed Convolutional neural network for sentence classification (TextCNN) [6] model has achieved excellent results in sentence classification. Pengfei Liu et al. present the Recurrent Neural Network for Text Classification (TextRNN) [7] pointing out the use of a multiple-task learning framework for combined learning across multiple related tasks to enhance the classification of text. Siwei Lai et al. then proposed the Recurrent Convolutional Neural Networks for Text Classification (TextRCNN) [8] model to make up for the shortcomings of TextCNN and TextRNN, which can maximize the extraction of contextual information and automatically decide which feature occupies a more important role, thus improving the classification effect. Peng Zhou et al. present the Attention-Based Bidirectional Long Short-Term Memory Networks (TextRNN_Att) [9] model, which can collect important semantical information at any position in a sentence and then perform classification. Armand Joulin et al. propose FastText [10], which is comparable to other classifiers, but the model is many orders of magnitude faster to train and evaluate. Rie Johnson et al. then proposed Deep Pyramid Convolutional Neural Networks (DPCNN) [11] model, can extract dependencies of long distance text, and is good at handling classification of long text. While the transformer [12] model proposed by Ashish Vaswani et al. We form a new model for contract classification by adding the TextRCNN model to Attention and do comparison experiments using the above model. In summarization, this article has the following major contributions:

- We propose the SolCom model to determine the correct compiler for Solidity.
- We constructed the dataset for Solidity version identification.

2   The Approach

The general diagram of the methodological framework present in this paper is shown in Figure 1. Next we first describe the pre-processing of the data, after which we will proceed to the detailed introduction of the SolCom model.

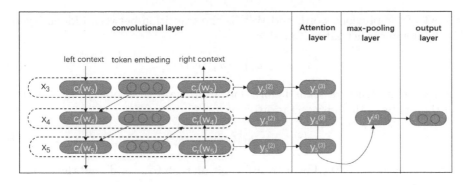

Fig. 1. SolCom general framework.

2.1 Preprocessing

All of our contract data is sourced from Ehterscan. We will discard the versions with less data during the subsequent processing. First we do the data de-duplication. In the second step we use Antlr parser to parse the contract code to get the token. The third step normalizes the token sequence to get the valid part of the token. At this point the data is ready. We selected three minor versions v0.4.24,v0.5.17 and v0.6.12 for training and classification. The data were constructed according to the proportion of training set, validation set, and test set of 8:1:1, and the constructed data are shown in Table 1.

Table 1. The constructed dataset.

Versions	Train	Dev	Test
v0.4.24	13713	1714	1715
v0.5.17	8628	1078	1079
v0.6.12	12993	1624	1625

2.2 Token Embedding

The model combines the context of each word to express a word as follows.

$$c_l(w_i) = f(W^{(l)}c_l(w_{i-1}) + W^{(sl)}e(w_{i-1})) \tag{1}$$

$$c_r(w_i) = f(W^{(r)}c_r(w_{i+1}) + W^{(sr)}e(w_{i+1})) \tag{2}$$

where $c_l(w_i)$ is the left context of the word w_i, and similarly $c_r(w_i)$ is the right context. $e(w_{i-1})$ is the word embedding of the word (w_{i-1}), $c_l(w_{i-1}$ is the left context of the previous word, and $W^{(l)}$ is the transition between hidden layers. f is the nonlinear activation function.

$$x_i = [c_l(w_i); e(w_i); c_r(w_i)] \tag{3}$$

The recurrent structure scans before and after the text to obtain all c_l and c_r and then obtains the representation x_i of w_i.

$$y_i^{(2)} = tanh(W^{(2)}x_i + b^{(2)}) \tag{4}$$

After getting x_i apply the linear transformation tanh to x_i and send the outcome to the next layer.

2.3   Version Pragma Identification Model

Attention mechanisms have been proven successful for many tasks, for example, machine translation, speech identification, answering questions, etc [13–15]. In this subsection, we add the attention mechanism after the convolutional layer learning of TextRCNN to enhance the accuracy of this model. where w^T is the trained parameter vector.

$$\alpha = softmax(w^T y_i^{(2)}) \tag{5}$$

$$y_i^{(3)} = \alpha^T y_i^{(2)} \tag{6}$$

When all the word representations have been computed, the word pooling is performed with the maximum pooling layer.

$$y_i^{(4)} = \max_i^n y_i^{(3)} \tag{7}$$

The pooling layer transforms text of different lengths into fixed-length vectors and also captures key information in the text. The last layer is the output layer of the common model, as follows.

$$y^{(5)} = W^{(4)}y^{(4)} + b^{(4)} \tag{8}$$

Finally, for $y(5)$ put into softmax function and convert the result into probability.

$$M = \frac{\exp(y_k^{(5)})}{\sum_{k=1}^n \exp(y_k^{(5)})} \tag{9}$$

3   Experimental Evaluation

The next paper describes some parameter settings and experimental results during the experiment.

3.1   Implementation and Experimental Settings

The experimental environment is Ubantu18.04.3, GeForce RTX 3090, and this experiment is mainly implemented in python, using pytorch as the framework to implement the network model. The length of the token sequence varies from a few hundred to a few thousand. Due to the limited experimental environment, it is not possible to cover the entire token sequence for each contract.So we set pad_size, which is the length of each sentence processed, to 1024 and batch_size to 64, so as to maximize the use of storage space. The learning rate is 0.001 and the number of training turns are 100 epoch. If the effect of more than 1000batch is not improved, the training is ended early and the model with the highest accuracy is selected as the final model.

3.2   Evaluation Results

We put the dataset into SolCom and get the prediction results after a period of training. After that, we conducted a comparison test and concluded that SolCom was the best, and the results are shown in Table 2, where the precision, recall and f1-score are macro-averaged. After training we randomly selected 50 contracts among the contracts for prediction and compiled the contracts based on the prediction results. All contracts can be compiled to prove that our model is valid.

Table 2. Comparison of experimental results.

Model	Precision (%)	Recall (%)	F1-score (%)	Accuracy ()%
SolCom	**95.79**	**96.21**	**95.98**	**96.30**
TextCNN	94.93	95.25	95.08	95.48
TextRNN	93.60	93.84	93.69	94.17
TextRNN_Att	94.20	94.50	94.34	94.77
TextRCNN	92.82	92.14	92.44	93.02
FastText	94.52	94.47	94.49	94.93
DPCNN	93.81	93.70	93.75	94.28
Transformer	90.64	89.90	90.21	90.96

4   Conclusion

In this paper, we extract and serialize Solidity contracts into SolCom for version classification and conduct comparative experiments to prove that the classification of our proposed model is the optimal, with an accuracy of 96.30%. The code version of the unknown contract can be accurately identified. And it shows that SolCom outperforms other models when dealing with code sequences. However, the results are not satisfactory when classifying some small adjacent versions. In future work, we hope to go even further to extract hidden features in the code for more fine-grained classification, thus making the version classification task more accurate.

Acknowledgment. This work was supported in part by the Natural Science Basic Research Program of Shaanxi (2022JM-342, 2018JQ6078) the Science and Technology of Xi'an (2019218114GXRC017CG018-GXYD17.16), National Natural Science Foundation of China (61702414), the Key Research and Development Program of Shaanxi (2019ZDLGY07-08), and Special Funds for Construction of Key Disciplines in Universities in Shaanxi.

References

1. Wohrer, M., Zdun, U.: Smart contracts: security patterns in the ethereum ecosystem and solidity. In: International Workshop on Blockchain Oriented Software Engineering (IWBOSE), pp. 2–8 (2018). https://doi.org/10.1109/IWBOSE.2018.8327565
2. Zou W., et al.: Smart contract development: challenges and opportunities. In: IEEE Transactions on Software Engineering, vol. 47, no. 10, pp. 2084–2106 (2021). https://doi.org/10.1109/TSE.2019.2942301
3. Tian, Z., Tian, J., Wang, Z., Chen, Y., Xia, H., Chen, L.: Landscape estimation of solidity version usage on Ethereum via version identification. Int. J. Intell. Syst. **37**(1), 450–477 (2022)
4. Aggarwal, C.C., Zhai, C.: A survey of text classification algorithms. In: Mining Text Data, pp. 163–222. Springer, Boston, MA (2012)
5. Collobert, R., Weston, J.: A unified architecture for natural language processing: deep neural networks with multitask learning. In: Proceedings of the 25th International Conference on Machine learning. pp. 160–167 (2008)
6. Chen, Y.: Convolutional Neural Network for Sentence Classification. MS thesis. University of Waterloo (2015)
7. Liu, P., Qiu, X., Huang, X.: Recurrent neural network for text classification with multi-task learning (2016). arXiv preprint arXiv:1605.05101
8. Lai, S., Xu, L., Liu, K., Zhao, J.: Recurrent convolutional neural networks for text classification. In: Twenty-ninth AAAI Conference on Artificial Intelligence (2015)
9. Zhou, P., Shi, W., Tian, J., Qi, Z., Li, B., Hao, H., Xu, B.: Attention-based bidirectional long short-term memory networks for relation classification. In: Proceedings of the 54th Annual Meeting of the Association for Computational Linguistics, pp. 207–212 (2016)
10. Joulin, A., Grave, E., Bojanowski, P., Mikolov, T.: Bag of tricks for efficient text classification (2016). arXiv preprint arXiv:1607.01759
11. Johnson, R., Zhang, T.: Deep pyramid convolutional neural networks for text categorization. In: Proceedings of the 55th Annual Meeting of the Association for Computational Linguistics, pp. 562–570 (2017)
12. Vaswani, A., Shazeer, N., Parmar, N., Uszkoreit, J., Jones, L., Gomez, A. N., Polosukhin, I.: Attention is all you need. Adv. Neur. Inform. Process. Syst. **30** (2017)
13. Hermann, K.M., Kocisky, T., Grefenstette, E., Espeholt, L., Kay, W., Suleyman, M., Blunsom, P.: Teaching machines to read and comprehend. Adv. Neur. Inform. Process. Syst. **28** (2015)

14. Bahdanau, D., Cho, K., Bengio, Y.: Neural machine translation by jointly learning to align and translate (2014). arXiv preprint arXiv:1409.0473
15. Chorowski, J.K., Bahdanau, D., Serdyuk, D., Cho, K., Bengio, Y.: Attention-based models for speech recognition. Adv. Neur. Inform. Process. Syst. **28** (2015)

Towards Robust Similarity Detection of Smart Contracts with Masked Language Modelling

Zhenzhou Tian$^{(\boxtimes)}$ and Xianqun Ke

Xi'an University of Posts and Telecommunications, Xi'an 710121, China
tianzhenzhou@xupt.edu.cn, xianqunke@stu.xupt.edu.cn

Abstract. Smart contracts are programs that run on blockchains. The whole smart contract ecosystem tends to be highly homogeneous, due to the immutable nature of contracts once deployed, as well as the copy-paste practice in developing smart contracts. Thus, similarity detection between smart contracts is of great value, which facilitates the quality assurance of the whole ecosystem, by providing a way to identify and track clones among the smart contracts. To this end, this work presents SoliSim, which encodes smart contracts into informative semantic vectors for effective and efficient similarity detection. The smart contract encoding procedure is enforced with masked language modelling on the Solidity programming language, which pre-trains a bert-like model by feeding in normalized token sequences extracted from the smart contracts' abstract syntax trees (ASTs); while the similarity detection procedure is enforced via simply calculating a score on the encoded numerical vectors. As the experimental results show, the pre-trained strategy adopted by SoliSim is capable of capturing the contextual and semantic information of smart contracts' code. The similarity scores calculated with SoliSim on pairs of real cloned contracts all exceed 96%, while the values between non-clone pairs are all below 50%.

Keywords: Smart contract · Similarity detection · Pre-trained model

1 Introduction

Recent years, with the widespread use of blockchain technology and the efficiency and reliability of smart contracts, its number has exploded in various fields, such as healthcare, business process management, and digital identity verification [1]. A smart contract is a computer protocol that disseminates, validates or enforces contracts in an informational way. And it is an important innovation of Ethereum [2], which monitors implementation of the agreements reached through digital forms, while the correctness of its execution is ensured by the consensus protocol of Ethereum. The core of blockchain is decentralization, and all its transactions are open, unmodifiable. Therefore, in peer-to-peer transactions, credit guarantee becomes a top priority. Smart contracts fill this

N. Xiong et al. (Eds.): ICNC-FSKD 2022, LNDECT 153, pp. 76–84, 2023.
https://doi.org/10.1007/978-3-031-20738-9_10

gap perfectly, acting as a fair, impartial and open role and regulating all transaction rules [3]. Because of the tamper-evident nature of the blockchain, once a
smart contract is deployed on the blockchain, its code cannot be modified, the
only way to modify a smart contract is to redeploy a new instance. Resulting
in a large number of cloned contracts in blockchain, so it is crucial to conduct
security audits and contract similarity checks on the smart contract code before
the code is released [4].

This naturally leads to a large number of similar smart contracts with potential security risks, given the prevalence of flaws in deployed smart contracts.
According to previous papers [5–7] code cloning will cause these problems: code
cloning increases the size of the source code; cloned code generates inconsistencies
during updates, making the software difficult to maintain. And there are many
previous studies on smart contracts and underlying blockchain bug features and
smart contract bug detection [8,9], which show that there are almost 81% of
code clones in smart contracts [10,11].In view of this, detecting the similarity
of smart contracts can allow developers to avoid using compromised contracts,
improve the efficiency of development and ensure the security of contracts. By
comparing contracts with known vulnerabilities, it helps to identify threats and
improve the security of Ethereum [2]. To address the above issues, in this paper,
we present a similarity detection model on the contract source code. We use
Bert [12], a deep pre-training model with excellent feature learning capability,
to learn code embeddings of source code sequences as well as semantic information. In summary, the main contributions of this paper are as follows:

- We use the ANTLR syntax parsing tool to generate the abstract syntax tree
  of the contract and use it to extract and normalize the code blocks.
- We propose a framework that extracts the representations of codes by pre-
  training the model on the source code and calculates the similarity of contracts based on the representations of contract codes.

2   Our Method

2.1   Contract Code Preprocessing

To prepare the smart contract source code used for our approach and evaluation,
First, we collect data using Etherscan, which is Ethereum's block browser and
analytics platform. More specifically, HTML pages describing smart contract
transaction information contain source code information, and we scrape and
parse these HTML pages to obtain our source code data. By the time we conducted our training and evaluation experiments, 134,449 verified smart contract
source codes had been collected.

We use the ANTLR tool to parse each smart contract and obtain an abstract
syntax tree (AST) for each contract. Since the AST captures program structure information, we build code embeddings based on the AST.Specifically, the
abstract syntax tree for each contract is traversed using a depth-first traversal method, during which the source code is divided into several different code

blocks depending on the type of node(e.g, branch and loop statement, simple statement). The code blocks are standardized (by removing comments, extra spaces, and normalized strings and constants) to eliminate non-essential differences between contracts, and we refer to the previous paper [13] for the above data preprocessing process. The obtained code blocks are fed into the pre-training model to learn the semantic information and code representations of the source code. We will describe the details of the code block processing process using specific Solidity code in the following.

```
1    pragma solidity ^0.5.17;
2    contract Twitter is ERC20{
3        //A record of deciamls for signing
4        uint8 public constant decimals = 18;
5        string public constant name = "Twitter";
6    }
```

Listing 1.1. An Example Solidity Smart Contract

We parsed the code of Listing 1 by ANTLR parsing tool, and used deep traversal to access the generated AST parse tree and extract all the terminal symbol when processing the data, and divided the source code into multiple code blocks according to the type information of the nodes, and normalized the extracted code blocks to eliminate some semantically irrelevant information. We use StringLiteral, DecimalNumber, HexNumber and HexLiteral to unify strings and constants respectively, and remove comments.

```
1    pragma solidity ^0.5.17;
2    contract Twitter is ERC20{}
3    uint8 public constant decimals = DecimalNumber;
4    string public constant name = StringLiteral;
```

Listing 1.2. Illustration of the Normalized Sample Program

2.2   Pre-training with MLM for Token Embedding

To train a deep semantic representation model, we pretrained the normalized data by Masked Language Modelling (MLM), first introduced in BERT [12], which aims to pre-train a deep bidirectional representation by jointly adjusting the left and right contexts in all Transformer layers. x_i is tagged as a token, and a block $X = x_1, x_2, ..., x_n$ is a sequence of consecutive tokens, for a given block X, we first randomly select 15% of tokens to be replaced, for the selected tokens, 80% of the tokens are replaced with [MASK] (a special token), 10% of the tokens are replaced with other tokens in the word list, and the rest 10% of the token remains unchanged. Then, together with the positional encoding, they are fed into the Transformer structure as input. By adding a softmax layer to the last hidden layer of the Transformer, these masked or replaced tokens are predicted, and the loss of the predicted output is measured by the cross entropy loss.

$$X = [CLS]x_1x_2...x_n[SEP] \tag{1}$$

Where X denotes the input of the model, x_i is each token of the code block, where $[CLS]$ is a special token that identifies the beginning of the sequence, and $[SEP]$ is used as the end flag of the code block.

$$H_0 = Embedding(X) \tag{2}$$

For the input X, the sum of the three embeddings (word embedding, positional embedding, and token type embedding) is first obtained by the Embedding layer as the input to the pre-trained model.

$$H_i = Transformer(H_{i-1}), i \in (1...L) \tag{3}$$

For the i-th hidden layer H_i, the output of the previous layer H_{i-1} is fed to the Transformer model to obtain the hidden vector of the i-th layer, and L denotes the number of Transformer blocks.

$$L_{mlm} = - \sum_{t_i \in m(I)} p(t_i) logp(t_i) \tag{4}$$

$$p(t_i) = \frac{exp(w_i H(S_i))}{\sum_{k=1}^{K} exp(w_i H(S_i))} \tag{5}$$

The loss of the final training model uses the cross-entropy loss function, where $m(I)$ denotes the set of masked tokens, H_i is the hidden state vector at the position corresponding to t_i on the last hidden layer of the Transformer and K is the token list length.

Figure 1 shows the pre-training process of the MLM-based model. The [CLS] in the token sequence is used to identify the beginning of a block of statements, and the token in red is a masked or substituted token; for example, "uint8" is masked with [MASK], and "18" is substituted with the "name" in the token list.

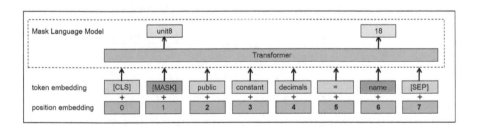

Fig. 1. MLM-based learning of code block token embedding.

2.3 Statement Block Embedding

Based on the previous Bert [12] pre-training model, we can extract the features and vector representation of a token or a code block from the Bert [12] embeddings. First a code block x is transformed into the standard input form of the pre-trained model and fed into the model, where $H_i(x)$ represents the output of the hidden layer of the model, and the vector of each code block is obtained by summing the last four hidden layers of the model, and the embedding vector of the code block X is shown below.

$$Embedding(x) = \sum_{i=L-4}^{L} H_i(x) \tag{6}$$

2.4   Contract Encoding

Each contract C extracts n code blocks X through the preprocessing stage, and each code block X_i is transformed into the standard input of the model to get the vector representation E_i in the pretrained model, and finally the vector representation of the contract can be obtained by summing n code blocks.

$$Embedding(C) = \sum_{i=0}^{n} E(i) \tag{7}$$

2.5   Contract Similarity Detection

Code similarity detection has been a long researched topic, Gao et al [13] encode smart contracts as vector representations by Fasttext and word2vec on the basis of source code, unlike their work, we use the Bert [12] pre-trained model to extract contract features and use Euclidean distance to compute similarity of two vectors, which provides a promising basis for subsequent contract cloning and bug detection. The similarity of two contracts is defined as follows: given two vectors e_1 and e_2 with the same dimension, we compute their Euclidean distance and set a threshold value to determine whether the two contracts are cloned pairs by comparing similarity score with the threshold value.

$$Similarity_{score}(e_1, e_2) = 1 - \frac{Euclidean(e_1, e_2)}{||e_1|| + ||e_2||} \tag{8}$$

3   Experimental Evaluation

Our approach is based on the Bert [12] pre-training model MLM for contract embedding representation to calculate the contract similarity and the degree of cloning between contracts. In this part, we evaluate the efficiency of the model and the correctness of the similarity approach by evaluating the embedding effect of contracts and using known contract cloning pairs.

3.1   Effectiveness Validation of Pre-trained Token Embeddings

We extract the top 100 high-frequency tokens from the collected dataset and extract the vector representations of the corresponding tokens by pre-training the model. To show the effectiveness of the embedding vectors, we draw 2D images by the T-SNE algorithm to demonstrate the embedding effect. The semantically similar tokens are close together in the vector space.

As shown in Figure 2, for example, special operators and contract names are each distributed in the same region of the space, such as "sub", "div", "mul" and "-", "/", "*" are distributed in the same region, while the names of more common functions like "balances", " transfer", "allowances" are clustered together, which shows that the pre-trained model is able to extract the semantic features and spatial distribution of each token as well as the embedded vector very well.

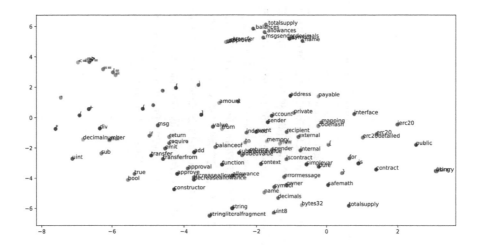

Fig. 2. Code embeddings of tokens.

3.2 Evaluation of SoliSim in Calculating Contract Similarity

According to the 10 known contract clone pairs provided in the study of Jiang [14] et al. We extracted vector representations of 10 smart contract pairs according to the pre-training model and calculated their similarity scores separately, and the experimental results showed that the similarity scores of each clone contract pair were above 96%. In addition, We construct 10 non-clone pairs by randomly selecting contracts from different types of contracts such as game contracts, ICO and AirDrop contracts, and ERC-20 contracts , and experimental results show that the similarity scores of non-cloned pairs are lower than 48.6 %. From the experimental results show, it is clear that the pre-trained model we used to extract the contract vector and to calculate the contract similarity is feasible. Table 1 gives the cloned contract pair names and similarity score results.

Table 1. Similarity scores calculated with SoliSim on ground-truth clone pairs.

Source contract	Target contract	Score (%)
0x613d0e9b91af3d0057fe376194580dd0048e91d4	0x613d0e9b91af3d0057fe376194580dd0048e91d4	98.8
0x00b9034425e357bf61b4abeb22299ec4a62c725b	0x001b6e5c7322899355eb65486e8cbb7dbbf19127	98.2
0x045c60df00aa2b7ff1753ee81b57bfe5e290e732	0x066b2063d4bd8cade177c55de659884c40bf2b8f	97.9
0x40e9a10fecd0350305d5e919e3963248fd6f845d	0x003516572f212dec785c3bcc97e6f354851eeb47	99.2
0x0035743c08768ad6558b49d751e0215762057754	0x008f81cbd97a3f59291aa0fed45a42491f10cfd2	98.2
0x23221fe28dadf788c7c59d0367bafef3b1607344	0x00446022 9a42542772f21ee82b8772cc6f2a502b	97.3
0x007b749fd9c28455f03a57c005f4249693550e51	0x015a8a0163ad54c86012ff57d3558b6271a2d2bd	96.4
0x00cf36853aa4024fb5bf5cc377dfd85844b411a0	0x0067 4045bb7c17f0aa1cde34780d6c51af548728	96.6
0x00eb84e7ca4ad6dbbef240056bc904003029a4cd	0x1b5d56bfe749e492ae226cf9aa23c1426f828b7b	96.0
0x09f55c2d116a5833d41ba9208216d11a7cdba4b3	0x0160ab3faf146f346b2cea49a7049d786aa1aafb	98.4

4   Conclusion

In this paper, we preprocess our collected data with ANTLR tool and we perform representation learning and similarity metric analysis on the contract source code by using Bert [12] pretrained language model. Based on the pre-trained model, we can obtain semantic features and vector representations of the code blocks or the whole contract, which can be useful for downstream task clone detection and similarity metric analysis. According to the experimental results, the framework model can well capture the internal and contextual information between utterances, while the detection rate of clone contract pairs is over 96%. This also confirms the existence of a large number of clones of contracts in the blockchain platform. In our future work, we will carry out vulnerability detection of contracts based on the above research, as well as cluster analysis of different types of contracts and versions of each contract type to analyze their iterations and the diversity phenomenon of the ethereum ecosystem.

Acknowledgment. This work was supported in part by the Natural Science Basic Research Program of Shaanxi (2022JM-342, 2018JQ6078) the Science and Technology of Xi'an (2019218114GXRC017CG018-GXYD17.16), National Natural Science Foundation of China (61702414), the Key Research and Development Program of Shaanxi (2019ZDLGY07-08), and Special Funds for Construction of Key Disciplines in Universities in Shaanxi.

References

1. Huh, S., Cho, S.,Kim, S.: Managing IoT devices using blockchain platform. In: 2017 19th International Conference on Advanced Communication Technology (ICACT), pp. 464-467. IEEE (2017)
2. Buterin, V.: A next-generation smart contract and decentralized application platform. White Paper **3**(37), 2–1 (2014)
3. He, N., Wu, L., Wang, H., Guo, Y., Jiang, X.: Characterizing code clones in the ethereum smart contract ecosystem. In: International Conference on Financial Cryptography and Data Security, pp. 654-67 (2020)
4. Liu, H., Liu, C., Zhao, W., Jiang, Y., Sun, J.: S-gram: towards semantic-aware security auditing for ethereum smart contracts. In: 2018 33rd IEEE/ACM International Conference on Automated Software Engineering (ASE), pp. 814-819. IEEE (2018)
5. Thummalapenta, S., Cerulo, L., Aversano, L., Di Penta, M.: An empirical study on the maintenance of source code clones. Emp. Softw. Eng. **15**(1), 1–34 (2010)
6. Kalra, S., Goel, S., Dhawan, M., Sharma, S.: Zeus: analyzing safety of smart contracts. In: Ndss, pp. 1–12 (2018)
7. Wohrer, M., Zdun, U.: Smart contracts: security patterns in the ethereum ecosystem and solidity. In: 2018 International Workshop on Blockchain Oriented Software Engineering (IWBOSE), pp. 2-8. IEEE (2018)
8. Bhargavan, K., Delignat-Lavaud, A., Fournet, C., Gollamudi, A., Gonthier, G., Kobeissi, N., ... Zanella-Béguelin, S.: Formal verification of smart contracts: Short paper. In: Proceedings of the 2016 ACM Workshop on Programming Languages and Analysis for Security, pp. 91–96 (2016)

9. Tikhomirov, S., Voskresenskaya, E., Ivanitskiy, I., Takhaviev, R., Marchenko, E., Alexandrov, Y.: Smartcheck: Static analysis of ethereum smart contracts. In: Proceedings of the 1st International Workshop on Emerging Trends in Software Engineering for Blockchain, pp. 9–16 (2018)

10. Kondo, M., Oliva, G.A., Jiang, Z.M.J., Hassan, A.E., Mizuno, O.: Code cloning in smart contracts: a case study on verified contracts from the Ethereum blockchain platform. Emp. Softw. Eng. **25**(6), 4617–4675 (2020)

11. Jiang, L., Misherghi, G., Su, Z.,Glondu, S.: Deckard: Scalable and accurate tree-based detection of code clones. In: 29th International Conference on Software Engineering (ICSE'07), pp. 96–105. IEEE (2007)

12. Devlin, J., Chang, M. W., Lee, K., Toutanova, K.: Bert: Pre-training of Deep Bidirectional Transformers for Language Understanding (2018)

13. Gao, Z., Jayasundara, V., Jiang, L., Xia, X., Lo, D., Grundy, J.: Smartembed: A tool for clone and bug detection in smart contracts through structural code embedding. In: 2019 IEEE International Conference on Software Maintenance and Evolution (ICSME), pp. 394-397. IEEE (2019)

14. Kondo, M., Oliva, G.A., Jiang, Z.M.J., Hassan, A.E., Mizuno, O.: Code cloning in smart contracts: a case study on verified contracts from the Ethereum blockchain platform. Emp. Softw. Eng. **25**(6), 4617–4675 (2020)

DeSG: Towards Generating Valid Solidity Smart Contracts with Deep Learning

Zhenzhou Tian[(⊠)] and Fanfan Wang

School of Computer Science and Technology, Xi'an University of Posts and
Telecommunications, Xi'an 710121, Shaanxi, China
tianzhenzhou@xupt.edu.cn, wangfanfan@stu.xupt.edu.cn

Abstract. Solidity being a young yet the most widely used programming language to develop smart contracts, which are programs that run
on the blockchains, is frequently exposed of bugs and under continues
improvement. To this end, this work presents DeSG to facilitate the fuzz
testing of the fast-evolving Solidity compiler, by automatically generating massive solidity smart contracts of diversity in a deep learning based
manner. An encoder-decoder model is designed to ensure the production of high-quality and valid smart contracts, by equipping with the
powerful representation learning ability from the Transformer, as well
as three carefully-designed generation strategies that well-match the features of the Solidity language. The experimental evaluations conducted
show that, DeSG can effectively generate syntactically valid and highly
compilable smart contracts. The impacts of different encoding neural
networks and generation strategies to DeSG are also evaluated, with the
best performance reaching 91.2% validity score.

Keywords: Solidity · Smart contract · Code generation · Deep
learning

1 Introduction

Smart contracts have been widely used with the development of blockchain technology, which is a good solution to the integrity problems in traditional transactions [1]. However, it is not easy to guarantee the development of a smart
contract without security issues, mainly because the language for developing
smart contracts is still in the development stage. According to the data available on Etherscan, more than 99% of smart contracts are implemented through
Solidity [2]. The Solidity language has gone through 8 major release iterations
and is rapidly evolving and widely used in smart contract development.

Solidity also has many potential compiler bugs during its rapid development,
and the most popular Solidity 0.4 version has the most bugs (28) announced, and
many known Solidity compiler bugs are reported in the official documentation.

© The Author(s), under exclusive license to Springer Nature Switzerland AG 2023
N. Xiong et al. (Eds.): ICNC-FSKD 2022, LNDECT 153, pp. 85–92, 2023.
https://doi.org/10.1007/978-3-031-20738-9_11

Therefore, it is crucial to avoid potential compiler errors and to design well during the development of smart contracts [3]. However, in reality, a large number of smart contracts are compiled in such a way that they trigger potential compiler errors, which can be exploited intentionally to gain benefit from them [4]. In order to improve this language and fix potential compiler errors, a lot of research has been conducted in smart contract compiler error detection [5–7]. But too clumsy, these require a lot of work to be done manually, specifying specific error patterns and rules for the tool. The more popular deep learning techniques can now do a lot of work instead of humans, and are widely used in traditional code compiler error detection [8–11]. However, it is less applied in smart contracts, so we propose to apply deep learning to smart contracts.

In this paper, we propose an approach to automatically generate smart contracts to test compiler security performance using an encoder-decoder model. Specifically, we train a neural network to learn the syntactic properties of Solidity and the association between upper and lower program statements, and then combine it with a designed generation strategy with syntactic fuzzification to automatically generate different syntactically valid and highly executable smart contracts based on data samples of smart contracts, which are then compiled to detect compiler bugs. In summary, our main contributions are as follows:

- We propose a deep learning-based method for automatically generating valid smart contracts to test compiler bugs. By training an encoder-decoder model and combining it with a generation policy, DeSG will automatically generate syntax-valid smart contracts to test the security of the compiler.
- We propose three different generation strategies to make DeSG generation contracts more valid and executable, and to solve the problem of generation boundary character mismatch that exists in traditional generation strategies.
- We performed different model effect evaluations, and the preprocessing of training data and the design of the generation strategy matching the Solidity language resulted in the maximum DeSG performance improvement and the maximum validity of the generated contracts could reach 91.2%.

2   Overview of the Approach

We propose the method DeSG is to automatically generate smart contracts to test Solidity compiler bugs, making the rapidly evolving Solidity compiler more secure. The workflow of DeSG is shown in Fig. 1. The first phase of our work feeds the training data into the Transformer model and performs the training. We pre-processed the training data, including the selection of sequence lengths and the removal of noise data from the data. The second phase uses the trained model and seed contracts for new smart contract generation. We first use the proposed LocalFun algorithm to identify the function bodies in the seed contracts, then use the model to generate new smart contracts in combination with the generation strategy, compile the newly generated smart contracts, record the smart contracts with the error type of internal compiler error, and analyze the reasons for the other errors and propose improvement solutions to further improve the validity. Therefore, the key metric to measure the effectiveness of our current work is the syntactic validity of the generated contracts. Validity is the ratio of the number of smart contracts with valid model-generated syntax and triggered compiler bugs to the total number of model-generated smart contracts. The work mentioned above is described in detail in the following sections.

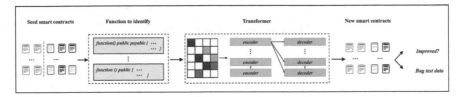

Fig. 1. The basic workflow of DeSG.

2.1 Preprocessing

Preprocessing is mainly done to optimize the training data features, including the selection of the length of the model training data and the removal of noise in the data. Sequence length selection: using a fixed-length sequence as input to predict the next character, the model works best when a fixed length of 50 characters is selected. The experimental phase compares the model effects of different length sequences. Noise Removal: Solidity is a contract oriented high level programming language for implementing smart contracts, allowing compilation instructions, import instructions and definitions of contracts, functions, interfaces, libraries, structures, enumerations and constants. According to the language characteristics of smart contracts, we need to remove invalid characters before generating training data. The noise we deal with consists of three parts: comments in the contract, blank characters, and invalid statements.

- Comments: The comments appearing in the contract are divided into two cases: multi-line comments and single-line comments, and according to the format of both types of comments, they are processed for direct deletion.
- Blank characters: The blank characters in the contract are unified, and multiple blank characters are replaced by a single space.
- Invalid statements: Invalid statements are handled by direct deletion, mainly referring to descriptive statements in the contract, including allow compile instructions and import instruction statements.

2.2 The Transformed-based Smart Contract Generation Model

We proposed DeSG is implemented on the encoder-decoder model. Initially, this model was proposed to be applied in the field of machine translation, and it has good performance results [12]. This provided the necessary experience for the implementation of our method, and we compared from several feature models and chose the transformer model that is more suitable for our method. In our method, the model is used in two phases, the training phase and the generation phase. In the training phase, we divide the input s into multiple fixed length d sequences $\mathbf{x}_i$ as the input of the encoder, and the feature vector of the input sequence is obtained through the attention mechanism to predict the corresponding next character $\mathbf{y}_i$ and obtain the output sequence of the decoder. Where $s[k : l]$ is the subsequence of s between indexes k and l.

$$x_i = s[\,i * d : (i+1) * d\,] \tag{1}$$

$$y_i = s[\,(i+1) * d + 1\,] \tag{2}$$

In the generation stage, we modify the output rule of the decoder and set a threshold value of probability, so that any character with output probability greater than the threshold value may become the output sequence, which effectively improves the diversity of the generated contract.

2.3   Generation Strategies

In order to generate contract statements based on prefix sequences with high validity and executability, we have carefully designed three generation strategies to generate and replace executable statements within function bodies, which well match the characteristics of the Solidity language. We also propose the function body identification algorithm LocalFun, which can accurately identify the function body and return the function body length. It makes the insertion position of DeSG more accurate. Algorithm 1 gives the pseudocode of LocalFun.

- STG_1: Insert two new lines in a function body, if a boundary character is introduced in the second sentence generated, then delete the second sentence and reinsert a new line.
- STG_2: Inserts a new line in each of the two different function bodies.
- STG_3: Replace two new lines in a function body.

3   Experimental Evaluation

3.1   Dataset Construction

The distribution of Solidity versions is found to be quite unbalanced according to the data on Ethernet, with Solidity 0.4 being the most popular in smart contract development [13]. So we are using the current most popular version 0.4 as the data selection object, and in order to reduce the errors caused by incompatible version features, we use the most widely available minor version 0.4.24 as the main data and other 5 minor versions as the auxiliary data, and construct about 5000 sample data as the training data set for our experiments, and use the version 0.4.24 data among them as the seed contract.

Algorithm 1 Function Body Identification for Smart Contracts

Input:
    A: Smart contracts after preprocessing.
Output:
    $FunIndex$: Representation of function body index.
    $FunLength$: Representation of function body length.
  1: $RNFunIndex \leftarrow \varnothing$
  2: $FunIndex \leftarrow getFun(A)$ ▷ the index string 'function' return the index value
  3: $RNFunIndex \leftarrow RemoveNoise(FunIndex)$ ▷ remove the non-keyword
  4: **for** j in $RNFunIndex$ **do**
  5:    **if** $A.find('\{'\, , \, j\,) \, ! = -1$ **then**
  6:      $startFunIndex \leftarrow A.find('\{'\, , \, j\,)$
  7:      $endFunIndex \leftarrow FunEndIndex(j)$ ▷ boundary character pairing principle

  8:      $FunLength \leftarrow endFunIndex - startFunIndex$
  9:    **end if**
10: **end for**
11: **return** $FunIndex$, $FunLength$

3.2 Experimental Settings

We use the Transformer model, with the number of encoder and decoder layers N set to 6 and the generation dimension size d set to 512 [14]. We conduct experiments on a server configured with RTX3090 GPU, and the data is randomly divided into training set, validation set and test set in the ratio of 8:1:1. The Adam optimizer was selected according to the experimental requirements, the size of the training batch was set to 300, the initial learning rate was 0.001, 100 epochs of training were performed, and the best performing parametric model during training was used for the experiments.

3.3 Experimental Results

The most important metric for evaluating the performance of DeSG is the validity of the generated contracts, only syntactically correct smart contracts can potentially trigger compiler bugs during the compilation process, so we conducted experiments to improve the validity of DeSG generated contracts from three aspects: model, training data, and generation strategy.

Model Performance Comparison. We have selected several representative encoder-decoder models for comparison, and the comparison results are shown in Table 1. Specifically, the models in the table were trained using the same data, and 2000 sample contracts were generated using the trained models under three strategies, and the average validity of the generated contracts was compared by compilation. Based on the experimental results, it is concluded that the Transformer model with strong representation learning capability is more suitable for our work, which on the one hand can input the training sequence as a whole and reduce the performance degradation due to long-term dependencies. On the other hand, both multi-headed attention and positional embedding provide information about the relationships between different words [14].

Table 1. DeSG validity under different models.

Model	STG-1 (%)	STG-2 (%)	STG-3 (%)
Bi-GRU	79.4	83.1	81.3
Bi-LSTM	82.3	85.2	83.7
Transformer	87.4	91.2	89.0

Training Sequence Length Effect Analysis. In order to make the training sequences summarize the contextual features well, we compared the validity of DeSG under different length training sequences, and Fig. 2 shows the comparison effect. Specifically, we generated training data with lengths of 40, 50, 60, and 70 characters, respectively, and used the STG_2 strategy to generate 3000 smart contracts for validity comparison. Through the analysis of the results, we found that the model corresponding to the sequence length of 40 characters performed slightly worse due to the lack of length, which could not summarize the contextual features well and easily resulted in the same sequence corresponding to the wrong label, thus leading to a decrease in validity. The validity of the other three sequence length correspondence models converge gradually as the number of samples increases, but the sequence correspondence models with lengths of 60 and 70 characters may have overfitting phenomena. Therefore, we choose a length of 50 characters as the fixed length of the training sequence.

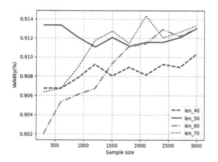

Fig. 2. Different length sequence models generate contract validity.

Generation Strategies Effect Comparison. Generation strategies are a core part of DeSG, and we compare three production strategies designed in conjunction with the features of the Solidity language with the traditional fuzzy test strategy of randomly selecting positions for insertion and replacement [15]. Figure 3 shows the validity of the two generation strategies in generating 2000 sample contracts under different scenarios. The experimental results show that DeSG achieves an effective performance of 91.2% under our designed generation strategies, and the STG_1 scheme improves the validity by 22.6% over the traditional generation strategies. Specifically, our proposed three generation strategies can perfectly apply to both object-oriented and procedural programming languages, compensate for the shortcomings of the traditional strategies that only apply to procedural-oriented programming languages, and solve the problem of compilation errors caused by the mismatch of generation boundary characters in the traditional strategies.

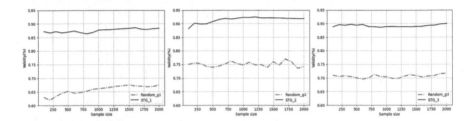

Fig. 3. Different generation strategies to generate contract validity.

4 Conclusion

In this paper, we propose DeSG, a method for automatically generating highly valid smart contracts, which is based on the Transformer model for character-level sequence prediction, using seed contracts for synthesis of utterances, and generating syntactically valid smart contracts for compiler bugs testing. The main metric for evaluating the performance of DeSG is the validity of the generated contracts. For this purpose, we perform training data preprocessing and propose three well-designed generation strategies, which result in a 91.2% validity of DeSG generated contracts. For the analysis of contracts with generation syntax errors, most of them are compilation errors caused by keywords containing other versions of features in the generated statements, which are unavoidable. A small part of them are compilation errors due to model prediction errors, because the character-level sequences lead to too much training data. In future work, we can try to use the output of lexical analyzer for serialization, and the amount of training data will be greatly reduced, then the validity will be improved. In the future, we plan to measure and improve the diversity of generative contracts, and implement a high-quality and automated method for testing bugs in Solidity compiler. We will also publish the bugs found in the tests to promote the rapid development of Solidity compiler and improve the security.

Acknowledgment. This work was supported in part by the Natural Science Basic Research Program of Shaanxi (2022JM-342, 2018JQ6078) the Science and Technology of Xi'an (2019218114GXRC017CG018-GXYD17.16), National Natural Science Foundation of China (61702414), the Key Research and Development Program of Shaanxi (2019ZDLGY07-08), and Special Funds for Construction of Key Disciplines in Universities in Shaanxi.

References

1. Wang, S., Ouyang, L., Yuan, Y., Ni, X., Han, X., Wang, F.Y.: Blockchain-enabled smart contracts: architecture, applications, and future trends. IEEE Trans. Syst. Man Cybern. Syst. **49**(11), 2266–2277 (2019)
2. Wang, S., Yuan, Y., Wang, X., Li, J., Qin, R., Wang, F. Y.: An overview of smart contract: architecture, applications, and future trends. In: IEEE Intelligent Vehicles Symposium, pp. 108-113. IEEE (2018)
3. Chen, J., Xia, X., Lo, D., Grundy, J., Luo, X., Chen, T.: Defining smart contract defects on ethereum. IEEE Trans. Softw. Eng. 327–345 (2020)
4. Solidity, List of known bugs. https://docs.soliditylang.org/en/v0.8.6/bugs.html
5. Tsankov, P., Dan, A., Drachsler-Cohen, D., Gervais, A., Buenzli, F., Vechev, M.: Securify: Practical security analysis of smart contracts. In: Proceedings ACM SIGSAC Conference on Computer and Communications Security, pp. 67–82 (2018)
6. Tikhomirov, S., Voskresenskaya, E., Ivanitskiy, I., Takhaviev, R., Marchenko, E., Alexandrov, Y.: Smartcheck: static analysis of ethereum smart contracts. In: Proceedings of the 1st International Workshop on Emerging Trends in Software Engineering for Blockchain, pp. 9–16 (2018)
7. Gao, Z., Jiang, L., Xia, X., Lo, D., Grundy, J.: Checking smart contracts with structural code embedding. IEEE Trans. Softw. Eng. (2020)
8. Liu, X., Li, X., Prajapati, R., Wu, D.: Deepfuzz: Automatic generation of syntax valid c programs for fuzz testing. In: Proceedings of the AAAI Conference on Artificial Intelligence, pp. 1044–1051 (2019)
9. Li, J., He, P., Zhu, J., Lyu, M. R.: Software defect prediction via convolutional neural network. In: 2017 IEEE International Conference on Software Quality, Reliability and Security (QRS), pp. 318–328. IEEE (2017)
10. Godefroid, P., Peleg, H., Singh, R.: Learn & fuzz: machine learning for input fuzzing. In: 2017 32nd IEEE/ACM International Conference on Automated Software Engineering (ASE), pp. 50–59. IEEE (2017)
11. Wang, S., Liu, T., Tan, L.: Automatically learning semantic features for defect prediction. In: 2016 IEEE/ACM 38th International Conference on Software Engineering (ICSE), pp. 297–308. IEEE (2016)
12. Chung, J., Gulcehre, C., Cho, K., Bengio, Y.: Empirical evaluation of gated recurrent neural networks on sequence modeling (2014). arXiv:1412.3555
13. Tian, Z., Tian, J., Wang, Z., Chen, Y., Xia, H., Chen, L.: Landscape estimation of solidity version usage on Ethereum via version identification. Int. J. Intel. Syst. **37**(1), 450–477 (2022)
14. Vaswani, A., Shazeer, N., Parmar, N., Uszkoreit, J., Polosukhin, I., et al.: Attention is all you need. Adv. Neural Inf. Process. Syst. **30** (2017)
15. Ganesh, V., Leek, T., Rinard, M.: Taint-based directed whitebox fuzzing. In: IEEE 31st International Conference on Software Engineering, pp. 474–484. IEEE (2009)

Combining AST Segmentation and Deep Semantic Extraction for Function Level Vulnerability Detection

Zhenzhou Tian[✉], Binhui Tian, and Jiajun Lv

Xi'an University of Posts and Telecommunications, Xi'an 710121, China
tianzhenzhou@xupt.edu.cn, mx_info@stu.xupt.edu.cn, lvjj@stu.xupt.edu.cn

Abstract. The explosive growth of software vulnerabilities poses a serious threat to computer system security and has become one of the urgent problems of the day. Yet, most existing vulnerability detection methods generally fail to capture the deep semantic features of code fragments, leading to the problem of high false negative rate easily. To this end, this paper proposes TrFVD (abstract syntax Tree based Function Vulnerability Detector), which mines deep semantics implied in source code fragments for accurate function level vulnerability detection. To ease the capture of fine-grained subtle semantic features, TrFVD converts the AST of a function into sequentially ordered sub-trees by splitting it in accordance with statements. The semantics of each sub-tree is then extracted with the Tree-LSTM, and a Text-RNN based model is utilized to summarize them up into a dense numerical vector to get the function represented. The experimental evaluations conducted on two C program vulnerability datasets show the effectiveness of TrFVD, which achieves 98.44% and 98.32% accuracy respectively. The averagely 12% more performance promotion against other vulnerability detection methods also indicates the superiority of TrFVD in capturing deeper subtle yet significant code semantics.

Keywords: Vulnerability detection · Abstract syntax tree · Deep learning · Deep semantic extraction

1 Introduction

Software vulnerabilities are code flaws caused by the negligence of software developers or limitations of programming languages during the software development process. With the rapid development of computer systems, software vulnerabilities have become a major threat affecting computer security. Once these vulnerabilities are exploited by attackers, their existence can lead to serious consequences. For example, the Apache Struts vulnerability in 2017 led to the compromise of 143 million consumers' financial data, causing incalculable financial losses [1,2]. However, due to the complexity of software, software vulnerabilities are inevitable. Given that it is impossible to prevent vulnerabilities, it is essential to detect and patch them as early as possible.

Existing vulnerability detection techniques include rule-based vulnerability mining techniques and deep learning-based source code vulnerability mining techniques. The traditional rule-based approach detects vulnerabilities by subjecting the source code to lexical analysis, which requires vulnerability experts to spend a lot of time defining vulnerability types in advance. Not only is the detection efficiency extremely low, but it is also prone to high false positive rate.

In recent years, as deep learning techniques have been widely used in the field of natural language processing, and have achieved good results [3,4]. Because of the powerful feature learning ability of deep learning models, they are also gradually used in vulnerability mining. Li et al. [5] proposed VulDeePecker to initiate the research of vulnerability detection using deep learning. They first sliced the program code based on the API calls in the program, then treated the sliced code fragments as plain text, mapped them into vectors using word embedding techniques, and finally input them into BiLSTM to determine whether the program has vulnerabilities. Duan et al. [6] proposed VulSniper, where they first model the source code using code property graph, then encode it into feature tensor using rule-based approach, and finally construct a fully connected network with attention mechanism to determine whether the program is vulnerable.

In order to solve the problem that existing vulnerability detection methods have high false positive rate when detecting vulnerabilities with unclear vulnerability characteristics. In this paper, we propose a fine-grained program vulnerability mining approach based on deep semantic extraction. First, we determine the granularity of vulnerability detection, we choose function rather than the whole program as the smallest vulnerability detection unit, and we use the ANTLR [7] tool to extract functions from the Software Assurance Reference Dataset(SARD) [8] to construct our dataset. Second, we use pycparser [9] tool to convert all functions into ASTs, which can be used as abstract intermediate code representation of function. In addition, we design a tree traversal algorithm based on preorder depth-first to traverse the AST into a series of sub-statement trees, and then We extract the lexical and semantic features of each sub-tree using tree-based long short-term memory network. The obtained vector sequence is fed into the sequence-based model Text-RNN [10] to capture its contextual information. Finally, the obtained vectors are input to the classification layer for function vulnerability detection. The main contributions of this work are as follows:

- We propose a fine-grained program vulnerability detection approach called TrFVD, which combines AST segmentation and deep semantic extraction, to enable the capturing of subtle yet significant code semantics from a function.
- An AST segmentation strategy is designed, which slices an AST into sequentially ordered sub-trees in accordance with code statements, to ease the extraction of subtle vulnerability-indicative features.
- Two vulnerability dataset targeting C programs are constructed, on which the experiments conducted show the superiority of TrFVD against other function-level vulnerability detection methods.

2   Our Approach

This section details our fine-grained program vulnerability mining method, the overall architecture of which is shown in Fig. 1. First, the function code fragment is parsed into an AST. second, the AST is split into a series of sub-statement trees, and the sub-statement trees are encoded. Finally, the obtained vector sequence is fed into the classifier model for training.

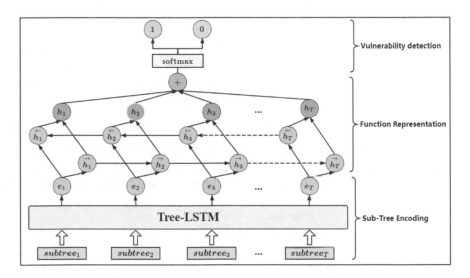

Fig. 1. The overview of TrFVD for function level vulnerability detection.

2.1   Funcion Parsing and AST Splitting

Parsing Function into AST. Abstract syntax trees(AST) are trees designed to represent the abstract syntactic structure of source code and have been widely used in the software development process. Recent research has shown that neural network models based on AST can adequately characterize information in source code by capturing lexical and syntactic knowledge [11–13]. We convert functions into abstract syntax trees using existing syntax parser tool pycparser, which is a complete C parser written in pure Python. It parses C code into ASTs and can be used as a front-end to C compilers or parsing tools. Figure 2 shows an example of a function and its corresponding AST.

Splitting AST into Sub-trees. For the AST of function fragment, we designed an algorithm to splitting it into a series of sub-statement trees. First we specify the granularity of the algorithm to slice the tree, and we set the minimum slice granularity to one statement or one statement block, as shown in Fig. 2, which slices local variable statement into a sub-tree, or conditional statement block

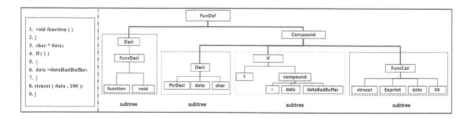

Fig. 2. An example of a C function's AST.

or function call statement block. After determining the granularity of the slice, we traverse each sub-tree of the AST in a preorder depth-first manner, which ensures that the sub-trees are added in order and the sequential nature of the statements can be guaranteed. The set of statement trees extracted using this algorithm.

$$TreeSet = \{subtree_1, subtree_2, ..., subtree_T\} \tag{1}$$

where T is the number of sub-statement trees for the specified cut, and the algorithm is described in Algorithm 1. After extracting features from all sub-statement trees by Tree-LSTM, the vector representations of all statement trees are obtained.

$$V_{TreeSet} = [e_1, e_2, ..., e_T] \tag{2}$$

where e_T represents the vector representation of the T_{th} sub-tree.

Algorithm 1 Split the AST into statement-segmented sub-trees

Input:
     AST: the abstract syntax tree of a function.
     subtreeSet: the sub-tree set of the input abstract syntax tree
Output:
     subtreeSet
1: **for** each *child* in *AST* **do**
2:     $name \leftarrow child.getName$
3:     **if** name in [Statement, If, For, While, Dowhile] **then**
4:         $subtreeSet \leftarrow child$
5:     **else if** name is Compound **then**
6:         SplittingTree(child,subtreeSet)
7:     **end if**
8: **end for**
9: return *subtreeSet*

2.2 Sub-Tree Encoding with Tree-LSTM

Given a sub-tree, we use tree-based neural network Tree-LSTM [14] to learn the vector representation of the sub-tree. Due to the presence of various special syntactic symbols in the AST, we extract all the token sequences from the preprocessed function code as the training corpus by preorder traversal of the AST. Word2vec [15] is used to learn the vector representation of each token, and the embeddings of the trained tokens are used as the initialization input for Tree-LSTM. Because the leaf nodes of each sub-tree represent lexical information, and also the tree-based structure reflects the syntactic information of the statement block, the representation of each sub-tree can adequately reflect the local information of this statement block.

2.3 Function Vulnerability Detection

Function Encoding Based on the sequence of sub-tree vectors, we use Text-RNN to track the sequential nature of statement. Text-RNN consists of a forward LSTM and a backward LSTM, where the forward LSTM can influence the backward information based on the forward information, and the backward LSTM can influence the forward information by the backward information. Therefore, this structure can capture the contextual information between statements very well. For a series of statement tree T, the transformation equation is as follows:

$$i_T = \sigma(W_{xi}e_T + W_{hi}h_{T-1} + W_{ci}c_{T-1}) \tag{3}$$

$$f_T = \sigma(W_{xf}e_T + W_{hf}h_{T-1} + W_{cf}c_{T-1}) \tag{4}$$

$$g_T = tanh(W_{xc}e_T + W_{hc}h_{T-1} + W_{cc}c_{T-1}) \tag{5}$$

$$c_T = i_T g_T + f_T c_{T-1} \tag{6}$$

$$o_T = \sigma(W_{xo}e_T + W_{ho}h_{T-1} + W_{co}c_{T-1}) \tag{7}$$

$$h_T = o_T tanh(c_T) \tag{8}$$

where i_T is the input gate, o_T is the output gate, c_T is the memory cell, h_T is the hidden state, f_T is the forget gate. This allows Text-RNN to capture the contextual information of the statement block.

Vulnerability Detection After obtaining the vector representation of the entire function, it is fed into the softmax classifier for training. Then, with the trained classifier, we use it to determine whether a function contains vulnerabilities or not.

3 Experimental Evaluation

3.1 Dataset Construction

To evaluate the performance of the model, we experimented on the C subset of the SARD, since pycparser tool can only convert C code to its corresponding AST. SARD is a vulnerability dataset maintained by the National Institute of Standards and Technology, which contains vulnerability data in different code forms, different languages, and different vulnerability types. Each source file in SARD contains one bad function and multiple good functions, with the bad functions containing specific vulnerabilities and the good functions fix the vulnerabilities in the bad functions in different ways.

As our goal is to uncover the existence of vulnerabilities in the code, we choose to collect specific types of vulnerability data in SARD. Specifically, we collected vulnerability data in the form of buffer errors and resource management errors in C source code in SARD for our experiments. These two types of vulnerabilities are two of the more common types of vulnerabilities in C/C++, and there is enough data in SARD for these two types of vulnerabilities to support model training. Our method uses function as the unit of vulnerability detection, so the source code files in the above dataset are extracted from the good and bad functions in the source files, and the good functions are labeled as negative samples with '0', representing that the function does not have vulnerabilities; the bad functions are labeled as positive samples with '1', representing the existence of vulnerabilities in the function. We construct SARD buffer error dataset and SARD resource management error dataset by function. The specific number of source files and the number of bad and good functions are shown in Table 1.

Table 1. Statistical information about dataset constructed.

Dataset	Source file	Bad function	Good function
Buffer error	4264	4242	6669
Resource management error	3116	3033	6550

3.2 Experimental Results

This subsection presents the results of our experiments and we used F1, Accuracy, Recall, Precison as our classifier evaluation metrics.

The dataset we conducted our experiments on was constructed entirely by ourselves, so we compared our proposed approach with the token-based approach on different models. Token sequences are equivalent to processing code by natural language processing, reflecting the natural order of the source code and to some extent reflecting the programming logic embodied in the source code. For the token-based approach, we use lexical analysis tools to extract the token sequences of functions and then use different sequence models for function-level vulnerability detection.

As shown in Table 2 and Table 3 respectively the experimental results on the SARD buffer error dataset and the SARD resource management error dataset, our approach can accurately detect the presence of vulnerabilities. It exhibits much better performance against the token-sequence based methods, with the scores of F1, Accuracy and Recall promoted by 12% above on average. The superiority performance indicates that the proposed approach is better at capturing deeper semantics implied in functions' source code.

Table 2. Experimental results on the SARD buffer error dataset.

Models	F1 (%)	Accuracy (%)	Recall (%)	Precision (%)
DPCNN	87.18	85.60	84.70	93.91
TextCNN	87.58	85.98	85.08	94.32
TextRNN	87.18	85.60	84.69	93.85
Transformer	85.58	83.87	83.46	92.77
TrFVD	98.47	98.44	98.48	98.54

Table 3. Experimental results on the SARD resource management error dataset.

Models	F1 (%)	Accuracy (%)	Recall (%)	Precision (%)
DPCNN	86.23	85.39	84.44	92.98
TextCNN	87.28	85.95	85.48	93.22
TextRNN	86.28	84.67	87.29	93.15
Transformer	85.87	83.27	82.46	92.53
TrFVD	98.38	98.32	98.12	98.24

4 Conclusion

To address the problem that existing vulnerability detection methods are prone to high false positive rate when detecting vulnerabilities, in this paper, we propose a fine-grained program vulnerability mining method based on deep semantic extraction, which extracts the contextual information of functions by splitting the AST into sub-statement trees, then encoding the sub-trees using Tree-LSTM, and finally using a sequence model, so the method can not only capture the fine-grained lexical and syntactic features of function code fragments, but also characterize their structural information. We use this method to test on the C vulnerability dataset, and the experimental results show that the method has better capability for fine-grained vulnerability detection, and our proposed method is expected to be extended to more languages for vulnerability detection.

Acknowledgment. This work was supported in part by the Natural Science Basic Research Program of Shaanxi (2022JM-342, 2018JQ6078), the Science and Technology of Xi'an (2019218124GXRC017CG018-GXYD17.16), and the Key Research and Development Program of Shaanxi (2019ZDLGY07-08).

References

1. Chen, X., Li, C., Wang, D., et al.: Android HIV: a study of repackaging malware for evading machine-learning detection. IEEE Trans. Inf. Forensics Sec. 987–1001 (2019)
2. Zhu, T., Li, G., Zhou, W., et al.: Differentially private data publishing and analysis: a survey. IEEE Trans. Knowl. Data Eng. **29**(8), 1619–1638 (2017)

3. Duan, X., Wu, J.Z., Luo, T.Y., Yang, M.T., Wu, Y.J.: Vulnerability mining method based on code property graph and attention BiLSTM. Ruan Jian Xue Bao/J. Softw. **31**(11), 3404–3420 (in Chinese) (2020)
4. Russell, R., Kim, L., Hamilton, L., Lazovich, T., Harer, J., Ozdemir, O., McConley, M.: Automated vulnerability detection in source code using deep representation learning. In: 2018 17th IEEE International Conference on Machine Learning and Applications (ICMLA), pp. 757–762. IEEE (2018)
5. Li, Z., Zou, D., Xu, S., Ou, X., Jin, H., Wang, S., Zhong, Y.: Vuldeepecker: a deep learning-based system for vulnerability detection. In: Proceedings of the 25th Annual Network and Distributed System Security Symposium, San Diego, CA, USA, pp. 1–15 (2018)
6. Duan, X., Wu, J., Ji, S., Rui, Z., Luo, T., Yang, M., Wu, Y.: VulSniper: focus your attention to shoot fine-grained vulnerabilities. In: IJCAI, pp. 4665–4671 (2019)
7. ANTLR Homepage, https://www.antlr.org/
8. SARD Homepage, https://samate.nist.gov/SARD/
9. pycparser Homepage, https://pypi.org/project/pycparser/
10. Zhou, P., Qi, Z., Zheng, S., Xu, J., Bao, H., Xu, B.: Text classification improved by integrating bidirectional LSTM with two-dimensional max pooling. In: COLING (2016)
11. Zhang, J., Wang, X., Zhang, H., Sun, H., Wang, K., Liu, X.: A novel neural source code representation based on abstract syntax tree. In: 2019 IEEE/ACM 41st International Conference on Software Engineering (ICSE), pp. 783–794 (2019)
12. Baxter, I. D., Yahin, A., Moura, L., Sant'Anna, M., Bier, L.: Clone detection using abstract syntax trees. In: Proceedings. International Conference on Software Maintenance (Cat. No. 98CB36272), pp. 368–377 (1998)
13. Yang, S., Cheng, L., Zeng, Y., Lang, Z., Zhu, H., Shi, Z.: Asteria: deep learning-based AST-Encoding for cross-platform binary code similarity detection. In: 2021 51st Annual IEEE/IFIP International Conference on Dependable Systems and Networks (DSN), pp. 224–236. IEEE (2021)
14. Tai, K. S., Socher, R., Manning, C. D.: Improved semantic representations from tree-structured long short-term memory networks. In: Proceedings ACL (2015)
15. Mikolov, T., Sutskever, I., Chen, K., Corrado, G. S., Dean, J.: Distributed representations of words and phrases and their compositionality. Adv. Neural Inf. Process. Syst. **26** (2013)

A Novel Variational-Mode-Decomposition-Based Long Short-Term Memory for Foreign Exchange Prediction

Shyer Bin Tan and Lipo Wang$^{(\boxtimes)}$ ⓘD

School of Electrical and Electronic Engineering, Nanyang Technological University,
Singapore 639798, Singapore
STAN137@e.ntu.edu.sg, elpwang@ntu.edu.sg

Abstract. The global foreign exchange (forex) market has a daily volume on the order of $5 trillion and is the largest financial market in the world. The importance of this market has attracted many research efforts. Numerous techniques have been developed, including technical analysis, where researchers attempt to predict forex rates based on past forex data. In this paper, we combine a signal processing technique Variational Mode Decomposition (VMD) and the Long Short-Term Memory Neural Network (LSTM) to predict forex and show significant improvements over other predicting models.

Keywords: Variational mode decomposition (VMD) · Foreign exchange (forex) · Long short-term memory neural network (LSTM)

1 Introduction

The daily trading volume for foreign exchange (forex) market was estimated at an immense amount of $5.1 trillion worldwide [1]. The forex market is operating continuously, 24 h from Sunday to Friday. Many factors affect the forex rate and massive transaction volume leads to the high liquidity of the asset. In the recent years, lots of techniques have been developed to study the forex market, including technical analysis, where researchers attempt to predict future forex rates based on past forex data.

Forecasting forex rate is a form of financial time series prediction, which is a part of time series analysis. Time series analysis is a well-developed technique and is applied in different research area such as electroencephalography (EEG) analysis [2], nonlinear dynamics [3], traffic [4] and finance [5]. In previous studies on time series analysis, the major models used are linear model such as Autoregressive Moving Average (ARMA) [6] and Autoregressive Integrated Moving Average (ARIMA) [7]. Nonlinear models usually perform better in forecasting actual time series data, such as financial data [8], compared to linear models. Therefore, nonlinear models such as Artificial Neural Network (ANN) [9] and Support Vector Regression (SVR) [10] are studied extensively in financial time series forecasting.

Liu et al., [11] studied the currency pairs Euro against United States Dollar (EUR/USD), British Pound Sterling against United Stated Dollar (GBP/USD) and United

© The Author(s), under exclusive license to Springer Nature Switzerland AG 2023
N. Xiong et al. (Eds.): ICNC-FSKD 2022, LNDECT 153, pp. 101–108, 2023.
https://doi.org/10.1007/978-3-031-20738-9_13

States Dollar against Japanese Yen (USD/JPY). Nagpure [12] used various forecasting models, namely ANN, Support Vector Regressor (SVR) and LSTM to predict the forex rates of 11 currency pairs, which includes EUR/USD, GBP/USD and USD/JPY.

However, the forex data are too liquid and noisy. Signal decomposition methods can be used to reduce the complexity of a signal. This paper aims to study forex prediction and trading strategies using LSTM and signal decomposition. Our results will be compared to the prediction capabilities of LSTM model [11, 12]. The hourly close prices of each currency will be used as input data to the neural networks. The comparison of the neural network models will be done on the average and standard deviation of the performance measures. All the neural network models in this paper are implemented with Python, using an open-source neural-network library Keras.

2   Methods

2.1   Convolutional Neural Network

CNN is a special type of ANN, which is among the most popular type of deep NN [13]. The rapid research progression of CNN can be partly attributed to the advancement Computer Vision in recent years [14]. This is because CNN can be utilized effectively to perform image classification tasks, due to its ability to operate on data in a two-dimensional (2D) structure. Besides image recognition [15], CNN is also used in other fields such as speech recognition [16] and semantic segmentation [17]. While other ANN architectures focus on processing the temporal information of input data, CNN preserves spatial information of input data.

Figure 1 shows the typical structure of a CNN. It consists of one or more convolution layers, pooling layers, fully connected layers, and finally an output layer. The convolutional layer aims to extract high-level feature of the 2D data. For example, in image recognition, this can be an edge. The convolutional layer is followed by pooling layer, which reduces the dimension of the data to save computational power. The fully connected layer is added for the CNN to learn non-linear combinations of the features of input data.

2.2   Recurrent Neural Network

In a typical Feedforward neural network, all input and output data and are trained independently. However, for tasks such as Natural Language Processing (NLP), simple ANN would not be a good idea because the outcome of a prediction instance in a sequence depends on the results before it. This is when Recurrent Neural Network (RNN) comes into play. RNN is a type of ANN that has outcome that depends on the previous computation results. This behavior renders RNN to store the information of previous data as a form of memory. A neuron in hidden layers of RNN computes the output not just based on input data from previous layers, but also based on the previous prediction results of this neuron in a sequence of data.

As shown in Fig. 1, the process for computing the output data of a neuron is described as follows. At any time step t, x_t denotes the input data at that time step, just like the

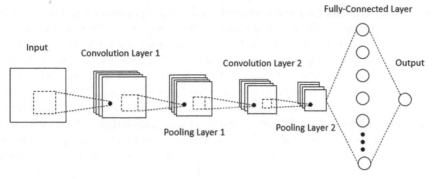

Fig. 1. Architecture of a typical CNN.

input to a normal ANN, while s_t is the input data based on previous prediction result, it can be regarded as the memory of the neuron. s_t can be computed from the formula: $s_t = f(u * x_t + w * s_t - 1)$, where f is the activation function. Finally, the output of a neuron at step t can be calculated as $o_t = \text{Softmax}(v * s_t)$. The parameters u, v and w above are the weights associated with the inputs and output (Fig. 2).

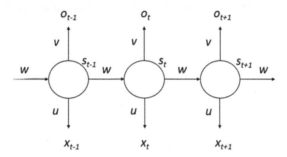

Fig. 2. Computation involved in a neuron of an RNN.

The ability for RNN to capture sequence dependencies greatly helps it to excel in the area of NLP [18]. Apart from that, RNN is also used in various time series analysis such as wind speed forecasting [19], stock price forecasting [20] and call centers call volume forecasting [21]. However, due to vanishing gradient problem, the gradient for optimizing might be very small in some training cases, preventing the weights from changing its value [22]. However, LSTM, a variation of RNN introduced by Hochreiter and Schmidhuber in 1997 [23] has been developed to solve this issue, LSTM retains long-term dependencies of data better than usual RNN [24]. Besides NLP, LSTM has also been applied extensively in many fields such as time series prediction [25] and emotional analysis [26].

2.3 Signal Mode Decomposition

Signal decomposition is a technique used to decompose a signal into different principal modes. The main decomposition methods include wavelet decomposition [27],

Fourier decomposition [28] and empirical mode decomposition [29]. The Empirical Mode decomposition (EMD) method proposed by Huang et al. [29] is a signal decomposition algorithm that recursively finds local minima and maxima in a signal. It was widely used in the decomposition and ensemble method together with neural network models to forecast time series data, such as financial time series data [30], crude oil price [31], and nuclear energy consumption [32]. In the decomposition and ensemble method, the original time series data are first decomposed into different modes, called Intrinsic Mode Function (IMF). Each IMF are then treated as separated data and used to train the NN model separately. The final predicted values are then ensembled together to form the final predicted values. The rationale for using such method is that the decomposition method reduces complexity in time series data [33].

However, EMD has some issues such as its sensitivity to noise and the lack of mathematical foundations. These issues may cause the predicted outcome by neural network to be inaccurate. VMD is a novel non-recursive signal decomposition model proposed by Zosso in 2015 [34]. It can be used to decompose time series data into subseries with specific frequency. This decomposition method is more robust in noisy data and is able to decompose data more accurately [35]. Recently, VMD has been used in many areas such as biomedical signal processing [36], fault detection [37] and financial time series analysis [38–46].

The theory of decomposition procedure of VMD method is described as below.

Step 1. The values $\{\hat{u}_k^1\}$, $\{\hat{\omega}_k^1\}$, λ^1 and n are initialized to 0.

Step 2. Update the values u_k, ω_k, and λ updated according to the equations follow:

$$\hat{u}_k^{n+1} = \frac{\hat{f}(\omega) - \sum_{i \neq k} \hat{u}_i(\omega) + \frac{\hat{\lambda}(\omega)}{2}}{1 + 2\alpha(\omega - \omega_k)^2}$$

$$\omega_k^{n+1} = \frac{\int_0^\infty \omega \left| \hat{u}_k^{n+1}(\omega) \right|^2 d\omega}{\int_0^\infty \left| \hat{u}_k^{n+1}(\omega) \right|^2 d\omega}$$

$$\lambda^{n+1} = \lambda^n + \tau \left(f(t) - \sum_{k=1}^{K} u_k^{n+1} \right)$$

Step 3. Iterate step 2 until the convergence criteria function

$$\sum_{k=1}^{K} \left\| u_k^{n+1} - u_k^n \right\|_2^2 / \left\| u_k^n \right\|_2^2 < \varepsilon$$

is satisfied, where ϵ is the accuracy required.

In the above,

f(t) is the original signal yet to be decomposed,
$\{u_k\} = \{u_1, u_2, \ldots, u_K\}$ stands for the set of IMFs,
$\{\omega_k\} = \{\omega_1, \omega_2, \ldots, \omega_K\}$ represents the center frequencies of respective kth IMFs,

$\lambda(t)$ is Lagrange multiplication factor for tightening restrain,
α is the balance parameter of data fidelity constraint, and
$\hat{f}(\omega), \hat{\lambda}(\omega), \hat{u}_i(\omega), \hat{u}_k^{n+1}$ are the Fourier transforms of $f(\omega), \lambda(\omega), u_i(\omega), u_k^{n+1}$ respectively.

Variational Mode Decomposition (VMD) [34] is useful for decomposing the time series. This paper combines VMD with LSTM to predict forex data.

3   Results

3.1   Data

The forex data is obtained from http://www.forextester.com/data/datasources, the same set of data as in [11]. Three currency pairs were chosen to train the neural network models, where the data are selected from 1 January 2001 to 31 December 2015, a span of 15 years. After the raw forex data were obtained, the data were processed as described follow. First, any record before 2001 and after 2015 are removed. Next, only one record for each hour window is preserved. After that, replicated data and error in data such as extreme values are removed, and empty in data are filled in respectively. Finally, the data will be rescaled to a standard range of [0, 1] interval using the min-max normalization technique. The formula to scale the data is given by $x = (x - x_{min})/(x_{max} - x_{min})$, where x_{min} and x_{max} denotes the minimum and maximum value in a set of data respectively. The data are saved in a Comma-Separated Values (CSV) file before feeding into neural networks for predictions (Fig. 3 and Table 1).

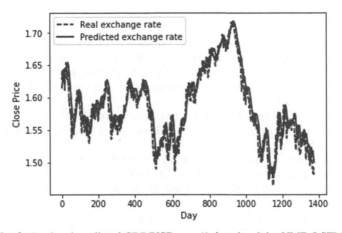

Fig. 3. Real and predicted GBP/USD rate (1 day ahead) by VMD-LSTM.

4   Conclusions

By comparing the prediction results given by CNN, LSTM and VMD-LSTM models, following conclusions can be made. The results obtained by our VMD-LSTM model is

Table 1. Statistic measures of forecast results for GBP/USD by VMD-LSTM.

Horizon (daysahead)	MSE		MAPE		R^2	
	Average	Standard deviation	Average	Standard deviation	Average	Standard deviation
1	0.0003	0.0006	0.007	0.0001	0.982	0.0002
2	0.0003	0.0012	0.008	0.0004	0.980	0.0010
3	0.0004	0.0007	0.008	0.0003	0.979	0.0008
4	0.0004	0.0003	0.009	0.0008	0.977	0.0004
5	0.0004	0.0002	0.009	0.0005	0.974	0.0008
6	0.0005	0.0010	0.009	0.0010	0.972	0.0003
7	0.0005	0.0009	0.010	0.0012	0.970	0.0006
8	0.0006	0.0014	0.010	0.0009	0.969	0.0006
9	0.0006	0.0014	0.010	0.0002	0.966	0.0010
10	0.0006	0.0011	0.011	0.0001	0.963	0.0015

generally better than results by plain LSTM model. Therefore, it can be concluded that the parameters for constructing VMD technique is well chosen, where it successfully separates the original forex data into less noisy subseries data.

References

1. Segal, T.: Forex market: who trades currency and why. Investopedia, 24 Oct 2019. Available: https://www.investopedia.com/articles/forex/11/who-trades-forex-and-why.asp
2. Gao, Z.K., Cai, Q., Yang, Y.X., Dong, N., Zhang, S.S.: Visibility graph from adaptive optimal kernel time-frequency representation for classification of epileptiform EEG. Int. J. Neural. Syst. **27**, 175005 (2017)
3. Gao, Z.K., Fang, P.C., Ding, M.S., Jin, N.D.: Multivariate weighted complex network analysis for characterizing nonlinear dynamic behavior in two-phase flow. Exp. Therm. Fluid. Sci. **60**, 157–164 (2015)
4. Hu, W., Yan, L., Liu, K., Wang, H.: A short-term traffic flow forecasting method based on the hybrid pso-svr. Neural Proc. Lett. **43**, 155–172 (2016)
5. Galeshchuk, S.: Neural networks performance in exchange rate prediction. Neurocomputing **172**, 446–452 (2016)
6. Box, G.E.P. Jenkins, G.M., Reinsel, G.C.: Linear Stationary Models, pp. 47–91 (2008)
7. Box, G.E.P., Jenkins, G.M., Reinsel, G.C.: Linear Nonstationary Models, pp. 93–136 (2008)
8. Chiarella, C., Peat, M., Stevenson, M.: Detecting and modelling nonlinearity in flexible exchange rate time series. Asia Pac. J. Manag. **11**, 159–186 (1994)
9. Ferreira, T.A.E., Vasconcelos, G.C., Adeodato P.J.L.: A new intelligent system methodology for time series forecasting with artificial neural networks. Neural Proc. Lett. **28**, 113–129 (2008)
10. Sermpinis, G., Stasinakis, C., Theofilatos, K., Karathanasopoulos, A.: Modeling, forecasting and trading the EUR exchange rates with hybrid rolling genetic algorithms support vector regression forecast combinations. Eur. J. Op. Res. **247**, 831–846 (2015)

11. Liu, C., Hou, W., Liu, D.: Foreign exchange rates forecasting with convolutional neural network. Neural Proc. Lett. **46**(3), 1095–1119 (2017). https://doi.org/10.1007/s11063-017-9629-z
12. Nagpure, A.R.: Prediction of multi-currency exchange rates using deep learning. Int. J. Innovative Technol. Exploring Eng. (IJITEE) **8**, 2278–3075 (2019)
13. Dertat, A.: Applied Deep Learning—Part 4: Convolutional Neural Networks. Medium, 9 Nov 2017. Available: https://towardsdatascience.com/applied-deep-learning-part-4-convolutional-neural-networks-584bc134c1e2
14. Saha, S.: A Comprehensive Guide to Convolutional Neural Networks—the FLI5 Way. Medium, 16 Dec 2018. Available: https://towardsdatascience.com/a-comprehensive-guide-to-convolutional-neural-networks-the-eli5-way-3bd2b1164a53
15. Zhang, Y., Zhao, D., Sun, J., Zou, G., Li, W.: Adaptive convolutional neural network and its application in face recognition. Neural Proc. Lett. **43**, 389–399 (2016)
16. Abdel-Hamid, O., Mohamed, A., Jiang, H., Penn, G.: Applying convolutional neural networks concepts to hybrid nn-hmm model for speech recognition. In: 2012 IEEE International Conference on Acoustics, Speech and Signal Processing (ICASSP), pp. 4277–4280 (2012)
17. Shelhamer, E., Long, J., Darrell, T.: Fully convolutional networks for semantic segmentation. IEEE Trans. Pattern Anal. Mach. Intell. **39**, 640–651 (2017)
18. Graves, A., Mohamed, A., Hinton, G.: Speech recognition with deep recurrent neural networks. In: IEEE International Conference on Acoustics, Speech and Signal Processing, pp. 6645–6649 (2013)
19. Cao, Q., Ewing, B.T., Thompson, M.A.: Forecasting wind speed with recurrent neural networks. Eur. J. Oper. Res **221**, 148–154 (2012)
20. Hsieh, T.J., Hsiao, H.F., Yeh, W.C.: Forecasting stock markets using wavelet transforms and recurrent neural networks: an integrated system based on artificial bee colony algorithm. Appl. Soft. Comput. **11**, 2510–2525 (2011)
21. Jalal, M.E., Hosseini, M., Karlsson, S.: Forecasting incoming call volumes in call centers with recurrent neural networks. J. Bus. Res **69**, 4811–4814 (2016)
22. Bengio, Y., et al.: Learning long-term dependencies with gradient descent is difficult. IEEE Trans. Neural Netw. **5**, 157–165 (1994)
23. Hochreiter, S., Schmidhuber, J.: Long short-term memory. Neural Comput. **9**, 1735–1780 (1997)
24. Mohan, A.: Recurrent neural network and long term dependencies. INFOLKS, 14 July 2019. Available: https://www.infolks.info/blog/recurrent-neural-network/
25. Wang, H., Yang, Z., Yu, Q., Hong, T., Lin, X.: Online reliability time series prediction via convolutional neural network and long short term memory for service-oriented systems. Knowl.-Based Syst. **159**, 132–147 (2018)
26. Wöllmer, M., Kaiser, M., Eyben, F., Schuller, B., Rigoll, G.: LSTM-modeling of continuous emotions in an audiovisual affect recognition framework. Image Vis. Comput. **31**, 153–163 (2013)
27. Nunez, J., Otazu, X., Fors, O., Prades, A., Pala, V., Arbiol, R.: Multiresolution-based image fusion with additive wavelet decomposition. IEEE Trans. Geosci. Remote Sens. **37**, 1204–1211 (1999)
28. Huang, N.E., Shen, Z., Long, S.R.: A new view of nonlinear water waves: the Hilbert spectrum. Annu. Rev. Fluid Mech. **31**, 417–457 (1999)
29. Huang, N.E., Shen, Z., Long, S.R., Wu, M.C., Shih, H.H., Zheng, Q., Yen, N.-C., Tung, C.C., Liu, H.H.: The empirical mode decomposition and the Hilbert spectrum for nonlinear and non-stationary time series analysis. In: Proceedings of the Royal Society A: Mathematical, Physical and Engineering Sciences, vol. 454, no. 1971, pp. 903–995 (March 1998)
30. Plakandaras, V., Papadimitriou, T., Gogas, P.: Forecasting daily and monthly exchange rates with machine learning techniques. J. Forecast. **34**, 560–573 (2015)

31. Yu, L., Wang, Z., Tang, L.: A decomposition—ensemble model with data-characteristic-driven reconstruction for crude oil price forecasting. Appl. Energ. **156**, 251–267 (2015)
32. Tang, L., Yu, L., Wang, S., Li, J., Wang, S.: A novel hybrid ensemble learning paradigm for nuclear energy consumption forecasting. Appl. Energ. **93**, 432–443 (2012)
33. Isham, M.F., Leong, M.S., Lim, M.H., Ahmad, Z.A.: Variational mode decomposition: mode determination method for rotating machinery diagnosis. J. Vibroengineering **20**, 2604–2621 (2018)
34. Dragomiretskiy, K., Zosso, D.: Variational mode decomposition. IEEE Trans. Signal Proc. **62**, 531–544 (2014)
35. Upadhyay, A., Pachori, R.B.: Instantaneous voiced/non-voiced detection in speech signals based on variational mode decomposition. J. Franklin Inst. B **352**, 2679–2707 (2015)
36. Lahmiri, S.: Comparative study of ECG signal denoising by wavelet thresholding in empirical and variational mode decomposition domains. Healthc. Technol. Lett. **1**, 104–109 (2014)
37. Wang, Y., Markert, R., Xiang, J., Zheng, W.: Research on variational mode decomposition and its application in detecting rubimpact fault of the rotor system. Mech. Syst. Signal Proc. **60**, 243–251 (2015)
38. Lahmiri, S.: A variational mode decompoisition approach for analysis and forecasting of economic and financial time series. Expert Syst. Appl. **55**, 268–273 (2016)
39. Duchi, J., Hazan, E., Singer, Y.: Adaptive subgradient methods for online learning and stochastic optimization. J. Mach. Learn. Res. **12**, 2121–2159 (2011)
40. Yu, L., Wang, S., Lai, K.K.: Forecasting crude oil price with an EMD-based neural network ensemble learning paradigm. Energ. Econ. **30**, 2623–2635 (2008)
41. Zhu, M., Wang, L.P.: Intelligent trading using support vector regression and multilayer perceptrons optimized with genetic algorithms. In: 2010 International Joint Conference on Neural Networks (IJCNN 2010) (2010)
42. Edwin, S., Wang, L.P.: Bitcoin price prediction using ensembles of neural networks. In: 2017 13th International Conference on Natural Computation, Fuzzy Systems and Knowledge Discovery (ICNC-FSKD 2017)
43. Teo, K.K., Wang, L.P., Lin, Z.P.: Wavelet packet multi-layer perceptron for chaotic time series prediction: effects of weight initialization. In: International Conference on Computational Science, pp. 310–317 (2001)
44. Gupta, S., Wang, L.P.: Stock forecasting with feedforward neural networks and gradual data sub-sampling. Aust. J. Intell. Inform. Proc. Syst. **11**, 14–17 (2010)
45. Wang, Y., Wang, L.P., Yang, F., Di, W., Chang, Q.: Advantages of direct input-to-output connections in neural networks: the Elman network for stock index forecasting. Inform. Sci. **547**, 1066–1079 (8 Feb 2021)
46. Wang, L.P., Teo, K.K., Lin, Z.P.: Predicting time series with wavelet packet neural networks. In: 2001 IEEE International Joint Conference on Neural Networks (IJCNN 2001), pp. 1593–1597 (2001)

Multi-modal Scene Recognition Based on Global Self-attention Mechanism

Xiang Li[1], Ning Sun[2(✉)], Jixin Liu[2], Lei Chai[2], and Haian Sun[2]

[1] College of Communication and Information Engineering, Nanjing University of Posts and Telecommunications, Nanjing 210003, China
[2] Engineering Research Center of Wideband Wireless Communication Technology, Ministry of Education, Nanjing University of Posts and Telecommunications, Nanjing 210003, China
{sunning,liujixin,chailei,haian}@njupt.edu.cn

Abstract. With the rapid development of deep neural network and the emergence of multi-modal acquisition devices, multi-modal scene recognition based on deep neural network has been known as a research hotspot. In view of the characteristics of various objects and complex spatial layout in scene images, and the complementarity of multi-modal data, this paper proposes an end-to-end trainable network model based on global self-attention mechanism for multi-modal scene recognition. This model, which is named MSR-Trans, is mainly consisted of two transformer-based branches for extracting feature from RGB image and depth data, respectively. Then, a fusion layer is used to fuse these two features for final scene recognition. To further explore the relationship between multi-modal information, the lateral connections are added on some layers between the two branches. And, a dropout layer is embedded in transformer block for preventing the model from overfitting. Extensive experiments are conducted to test the performance of the proposed method on SUN RGB-D and NYUD2 datasets, and the recognition accuracies of multi-modal scene recognition can be achieved at 69.0% and 74.1%, respectively.

Keywords: Scene recognition · Multi-modal · Transformer · RGB-D

1 Introduction

In recent years, machine vision-based products are blooming and change the way of people's daily life, such as unmanned driving, smart robot and intelligent monitoring systems. Among them, scene recognition achieves the recognition and understanding of the environment, which is one of the key technologies of machine vision.

With the impressive success of deep learning, especially convolutional neural networks (CNN), in the machine vision task. The performance of image recognition has been significantly improved with the help of large-scale image databases (such as ImageNet [1]). In order to support the scene recognition research based on large deep neural network (DNN), several large-scale scene image databases, such as Places database and Places2 [2] database, are recently released. With the help of these large databases, the scene

© The Author(s), under exclusive license to Springer Nature Switzerland AG 2023
N. Xiong et al. (Eds.): ICNC-FSKD 2022, LNDECT 153, pp. 109–121, 2023.
https://doi.org/10.1007/978-3-031-20738-9_14

recognition method based on DNN model [3] takes the accuracy of scene recognition to a new level.

Although large-scale CNN models trained with massive images can effectively improve the performance of scene recognition, there is still a huge gap between scene recognition performance obtained by relying solely the RGB image and human cognitive ability. Recently, with the rapid development of depth sensors, scene recognition methods using multi-modal information including RGB images and depth data have attracted the attention of many researchers. For the task of scene recognition, many research results show clear and strong complementarity between RGB image and depth data (RGB-D). Scene recognition methods based on RGB-D multi-modal data have obvious advantages over methods using unimodal data [4]. On the other hand, the scene image has the characteristic of large intra-class variations and small inter-class dissimilarity, which makes the scene recognition method based on overall modeling often ineffective. And, the attention mechanism can focus on useful information and suppress useless information during the learning, which is suitable for scene recognition.

Recently, transformer [5], as a DNN architecture based on self-attention mechanism, has been widely used in natural language understanding, machine vision and other fields. Therefore, this paper proposes an end-to-end trainable DNN model named MSR-Trans based on global self-attention architecture for multi-modal scene recognition. The model mainly consists of two branches based on transformer for extracting features from RGB images and depth data, respectively. Then, the fusion layer is used to fuse these two features for final scene recognition. To further explore the relationship between multimodal information, lateral connections are added at some layers between the two branches. In addition, a dropout layer is embedded in the transformer structure to alleviate the overfitting of the model. Extensive experiments are carried out on SUN RGB-D and NYUD2 datasets to verify the effectiveness of the proposed method.

2 Related Work

In this section, we review recent deep learning-based RGB-D scene recognition methods. In earlier works [6, 7], two CNNs, which usually pre-trained on larger scene image database, were used to respectively trained on RGB and depth modalities to obtain their respective feature representations. Then, the learned features of two modalities were fused and fed to SVM-based classifier for final recognition. In order to address the limitation of lack of depth data in training, Song et al. [8] proposed a smaller network trained in a two-step process. In the first step, the network was pre-trained with depth image (HHA) patches using weak supervision. Then, the network was fine-tuned with complete HHA. The results show that it can effectively learn more discriminative depth features, which are more complementary to RGB features, so as to provide higher gains in RGB-D models.

Intuitively, one scene image is usually composed of diverse objects with various spatial. Song et al. [9] proposed a novel DNN model to model the objects and their spatial relations in RGB-D multi-modal scene recognition. There are two types of object-based representation of the scene image. One is COOR, which represents co-occurring frequency of the object and their relation. Another is SOOR, which sequentially describes image contents with local captions. Finally, the SOOR is encoded as features using a sequence encoding model and fed into the classifier.

In order to focus on the discriminative parts of the scene image during the learning, attention mechanism is introduced in DNN model. Xiong et al. [10] proposed the modality separation networks to explicitly two kinds of features from RGB-D data simultaneously. One is the modal-consistent, another is modal-specific features. In follow-up work, Xiong et al. [11] designed a differentiable local feature selection module to adaptively select important local features for RGB-D scene recognition.

Furthermore, Du et al. [12] proposed a DNN with encode-decode architecture to improve the performance of RGB-D scene recognition with the help of cross-modal translation. The main idea of their method is that the modality transformation enhances the representation ability of the encoding network as it forces the one modality data to transfer information towards its complementary modality.

As mention above, the most obvious difference between the proposed method and the related work is that the proposed MSR-Trans is a transformer-based DNN model. As far as our knowledge, it is the first work of using transformer architecture for RGB-D scene recognition. In the following, we introduce the details of the proposed method and show its effectiveness experimentally.

3   The Transformer-Based Deep Neural Network for RGB-d Scene Recognition

The proposed MSR-Trans model is mainly consisted of three parts: RGB image global self-attention encoding branch (RGSE-B), depth data global self-attention encoding branch (DGSE-B), and feature fusion and recognition network (FR-Net). The architecture of MSR-Trans model is shown in Fig. 1.

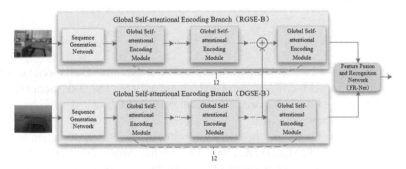

Fig. 1. The architecture of MSR-Trans model

3.1   Global Self-Attention Encoding Branch

The global self-attention encoding (GSE) branch is the backbone of the MSR-Trans, which is mainly followed the implemented of vanilla VIT [13] model. Firstly, the input

2D image is divided into image patches with size $P \times P$, and a series of image patches $X_p \in \mathbb{R}^{N \times (P^2 \cdot C)}$ is obtained. The number of image patches is $N = HW/P^2$, which H is the width of the image and W is the height of the image, N is also the length of input sequence of GSE branch. In the proposed method, $H = W = 224$, $P = 16$, and $N = 196$.

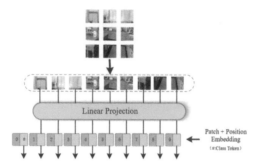

Fig. 2. The diagram of generating embedding from the input image

Then, each patch is projected into a vector with fixed size D. The output of this projection is called as the patch embedding. And, a position embedding E_{pos} is added to the patch embedding to retain the position information of image blocks, allowing the model to understand the image structure. Similar to other transformer model, a learnable class token is added to the patch embedding. As shown in Fig. 2. After these processing, the input RGB image and depth data are convert into embedding Z_{rgb} and Z_{depth}, respectively. Taking Z_{rgb} as an example, the expression corresponding to the above steps is shown in Eq. (1), where $X_{class} \in \mathbb{R}^{(P^2 \cdot C) \times D}$, $X_p^i \in \mathbb{R}^{P^2 \cdot C}(i = 1, 2, ..., n)$, $E_{pos} \in \mathbb{R}^{(N+1) \times D}$.

$$Z_{rgb} = \left[X_{class}; X_p^1 E; X_p^2 E; \dots; X_p^N E \right] + E_{pos} \tag{1}$$

After that, these embeddings are fed into multiple consecutive GSE modules to learn the dependencies between different parts of the input patches. The GSE module consists of alternating layers of multi-headed self-attention and MLP block. The architecture of the GSE module is shown in Fig. 3.

Self-attention usually adopts the query-key-value mode. For input sequence X, each input vector corresponds to Query, Key and Value vectors, where $Q = XW^Q$, $K = XW^K$ and $V = XW^V$. For a given Query vector Q, k key vectors K are matched by inner product calculation. Then k weights are normalized by softmax. So, for the Query vector, the output is the weighted average of the Value vector V corresponding to the k Key vectors. For N Query sequences, their attention output can be calculated by matrix, as shown in Eq. (2).

$$Attention(Q, K, V) = softmax\left(\frac{QK^T}{\sqrt{d_q}} \right) V \tag{2}$$

Here, $\sqrt{d_q}$ is the scaling factor to avoid the variance effect caused of the dot product. The multi-headed self-attention applies h self-attention mechanisms to the input sequence and the sequence is divided into h vectors with the size of $N \times d$, where $D = hd$. The outputs obtained by the h self-attention mechanisms are concatenated into a feature matrix by columns, which is the final output, and the size is still $N \times D$. The process is as follows.

$$MSAQ, K, V = concat(sa_1, \dots, sa_h)W^O \tag{3}$$

where, $sa_i = Attention\left(QW_i^Q, KW_i^K, VW_i^V\right)$, $W_i^Q \in \mathbb{R}^{d_{model} \times d_k}$, $W_i^K \in \mathbb{R}^{d_{model} \times d_k}$, $W_i^V \in \mathbb{R}^{d_{model} \times d_v}$, $W^O \in \mathbb{R}^{hd_v \times d_{model}}$. In this paper, set h to 12, $d_v = d_k = d_{model}/h$ = 64. Multi-headed self-attention mechanisms can be used to form multiple subspaces, which makes the model focus on the information of different locations and help the model to understand the local and global dependencies in the image.

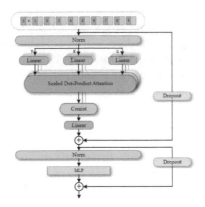

Fig. 3. The architecture of global self-attentional encoding module

In order to enhance the transmission of cross-modal information and improve the stability of model training, two improvements have been made to the vanilla VIT model. One is lateral connections between two GSE branches. There are several way to achieve lateral connection, such as concatenation, multiplication and addition, etc. In the MSR-Trans, we use the way of addition to connect the output of the N th GSE module in the DGSE-B and RGSE-B to be as the input of the $(N + 1)$ th GSE module in the RGSE-B. Secondly, a dropout mechanism is added to two the residual connections in the GSE module, the detail can be seen in Fig. 3. The effectiveness of these two improvements is verified by the following experiments.

3.2 Feature Fusion and Recognition Network

The feature fusion recognition network consists of a fusion layer and softmax activation function. Two features F_{rgb} and F_{dept} are the output of RGSE-B and DGSE-B, respectively. But they cannot be used to as the input of the fusion layer, because only class token is used for final recognition. Therefore, the first data of each dimension in F_{rgb} and F_{depth} is extracted to obtain F_{rgb}^1 and F_{depth}^1. And, F_{rgb}^1 and F_{dept}^1 are concatenated to fed fusion layer. The fusion layer is a fully connected layer, its output F_f has the same dimension as F_{rgb}^1. Finally, the fused feature F_f is fed to softmax layer to predict the scene category.

4 Experiments

4.1 Databases

SUN RGB-D database is currently the largest RGB-D scene image database, which contains 10355 RGB-D image pairs. These images are captured from different acquisition devices such as Asus Xtion, Microsoft Kinect v1 and so on. Following the standard experimental protocol reported in [14], 19 major scene categories are selected for experiments. As per standard splits, there are in total 4845 images for training and 4659 for testing.

NYU Depth v2 (NYUD2) is a relatively small RGB-D database consisting of 27 indoor categories. Following the standard split in [15], all the 27 well-presented categories are grouped into ten including nine most common categories and the other category representing the rest. Following the standard split, there are in total 795 images for training and 654 for testing, respectively.

4.2 Implementation Details

In the following experiments, the depth data are firstly converted into HHA-encoded images for better representation. And, all images are resized to 224×224. In the proposed MSR-Trans model, the number of GSE modules in RGSE-B and DGSE-B are all 12. The stochastic gradient descent (SGD) is used as the training optimizer. The batch size of the training data is 16, the number of iterations is 300 epochs, and the initial learning rate is 0.0003. During the training, the cosine function is used to reduce the learning rate. And the loss function is cross-entropy. All experiments were developed with the PyTorch framework and ran on an image processing workstation with an Intel Xeon 2.40 GHz 8 core CPU, 128 GB memory, and 2 NVIDIA 3080Ti GPUs.

4.3 Ablation Study

We conducted two aspects of experiments of ablation study to evaluate the impact of the settings of the MSR-Trans model on its final performance. The experiments in this section were all based on the SUN RGB-D databases.

(1) **Dropout effectiveness study**. In the MSR-Trans model, dropout is added to alleviate the problem of overfitting. The test accuracy and train loss of the MSR-Trans model with and without dropout are shown in Figs. 4 and 5. In Fig. 4, The abscissa represents the number of iterations and the ordinate represents the accuracy rate. In Fig. 5, The abscissa represents the number of iterations and the ordinate represents

the loss. From the left parts of Figs. 4 and 5, the test accuracy curves begin to drop significantly at 10k iterations, which indicates that overfitting has occurred. Correspondingly, the loss curve decreases rapidly at the beginning, and almost drops to 0 at 30k iterations, indicating that the training of the MSR-Trans model without dropout can no longer learn new information. From the right parts of Figs. 5 and 6, the test accuracy curve maintains an upward trend until 60k iterations, after which there is only a slight decrease. Although the loss curve fluctuated significantly after adding dropout to the model, it still maintained a downward trend at 100k iterations. This proves that adding dropout reduces the complex co-adaptation relationship between neurons, forcing the model to learn better stable features.

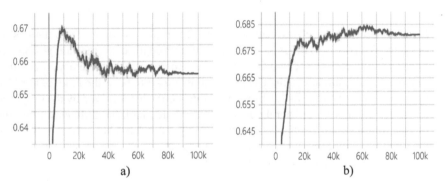

Fig. 4. **a** The test accuracy curves of the proposed model with dropout. **b** The test accuracy curves of the proposed model without dropout

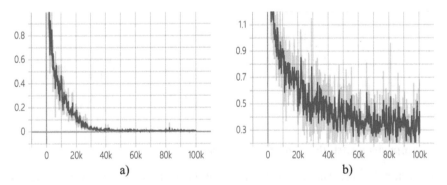

Fig. 5. **a** The loss curve of the proposed model with dropout. **b** The loss curve of the proposed model without dropout

Table 1 lists the recognition accuracies achieved by the proposed method with or without dropout on the SUN RGB-D database. It shows that the scene recognition accuracy is significantly improved by nearly 1.3% after adding dropout to the MSR-Trans model.

It clearly demonstrates that dropout can alleviate overfitting problem suffered in the training of the proposed method.

Table 1. Recognition accuracies achieved by the proposed method with or without dropout

Model	Dropout	Acc (%)
MSR-Trans	×	67.20
	√	68.45

(2) **Lateral connection effectiveness study**. To further explore the relationship between the two modalities of RGB image and depth data, the lateral connection is added between the two GSA branches. In the proposed method, the output of the 10th GSE module in the DGSE-B and RGSE-B to be as the input of the 11th GSE module in the RGSE-B. The recognition accuracies achieved by the proposed method with or without lateral connection are listed in Table 2. It shows that the recognition accuracy is effectively improved by using lateral connections between the two GSA branches. It shows that the using of lateral connections is helpful to explore the relationship between the two modalities.

Table 2. Recognition accuracies achieved by the proposed method with or without lateral connection

Model	Lateral connection (layer)	Acc (%)
MSR-Trans	×	68.45
	√(10)	68.96

4.4   Comparison with the State-Of-The-Arts

In this section, we compare the performance of the proposed method with the recently related FER methods on SUN RGB-D and NYUD2 databases.

(1) **Results on SUN RGB-D database**. The experimental results are listed in Table 3. It shows that the recognition accuracies obtained by the proposed method are obviously higher than those of other state-of-the-art multi-modal scene recognition methods, whether it is based on a single modality, or multi-modal. In the case of two modalities, the recognition accuracy of the proposed method is nearly 10% higher than other methods. The results clearly demonstrate the effectiveness of the

proposed method, and also illustrate that combining the multi-modal information and global self-attention mechanism can enable the model to focus on discriminative locations, which is helpful for the model to understand the local and global dependencies in the input data. Furthermore, the results also show that two transformer-based branches can better explore the complementary information between the two modalities, and improve the performance of the proposed method.

Table 3. Comparison between the proposed method and the related methods on the SUN RGB-D database

Method	Acc (%)		
	RGB	Depth	RGB-D
Caglayan et al. [16]	58.5	50.1	60.7
TRecgNet [12]	49.8	46.8	56.1
TRecgNet Aug [12]	50.6	47.9	56.7
Song et al. [9]	50.5	40.1	55.5
Song et al. [17]	41.5	40.1	52.3
PIA-SRN [18]	–	–	51.5
Song et al. [19]	–	–	52.4
Gong et al. [20]	–	–	52.6
Xiong et al. [21]	–	–	55.9
Song et al. [8]	–	–	53.8
Xiong et al. [11]	–	–	57.3
DF2Net [22]	–	–	54.6
MSN [10]	–	–	56.2
Du et al. [23]	–	–	53.3
Proposed method	**67.8**	**55.7**	**69.0**

Bold value indicates the highest accuracy in the comparison

The confusion matrix of the proposed method on SUN RGB-D database is shown in Fig. 6. The proposed method has a relatively high accuracy in categories as bathroom, bedroom, and restroom etc., but it is prone to confusion between classroom with conference rooms, discussion room and lecture theatre. This shows that the proposed method is still difficult to distinguish for which ambiguous categories.

(2) **Results on NYUD2 database**. Similarly, the proposed method is compared with other related methods on the NYUD2 multimodal scene database, and the experimental results are shown in Table 4. It shows that whether using only a single modality or two modalities of RGB image and depth data, the scene recognition accuracies obtained by the proposed method are much higher than the results of other methods. This once again proves the effectiveness of the proposed method. In the case of two modalities, the recognition accuracies of the proposed method are

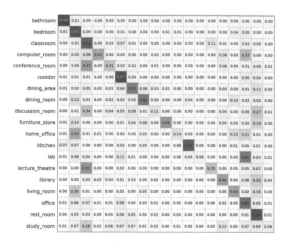

Fig. 6. Confusion matrix of the proposed method on SUN RGB-D database

generally more than 5% higher than that of other methods. And, the results based on two modalities are higher than those on single modality, which indicates that the depth data and RGB image are complementary and fusing multi-modal information can improve the performance of scene recognition. The confusion matrix of the proposed method on the NYUD2 database is shown in Fig. 7. Similar to the results on the SUN RGB-D database, the office category is prone to confusion with other semantically similar categories.

5   Conclusion

This paper proposes an end-to-end trainable transformer-based DNN model for RGB-D scene recognition, which called MSR-Trans. The proposed method is mainly consisted three parts: RGSE-B, DGSE-B, and FR-Net. Two branches based on transformer are designed for extracting features from RGB images and depth data, respectively. Then, the fusion layer is used to fuse these two features for final scene recognition. In order to enhance the transmission of cross-modal information and improve the stability of model training, lateral connections and dropout mechanism are added in the MSR-Trans. Ablation studies and comparison experiments are conducted on two benchmarks including SUN RGB-D and NYUD2. Results show the effectiveness of the proposed method.

Table 4. Comparison between the proposed method and the related methods on the NYUD2 database

Method	Acc (%)		
	RGB	Depth	RGB-D
Quattoni et al. [24]	53.5	51.1	67.8
TRecgNet [12]	63.8	56.7	66.5
TRecgNet Aug [12]	64.8	57.7	69.2
Song et al. [9]	64.2	62.3	67.4
Xiong et al. [11]	61.2	54.1	69.3
PIA-SRN [18]	–	–	65.4
Song et al. [19]	–	–	65.8
Wang et al. [6]	–	–	63.9
Gong et al. [20]	–	–	67.1
Song et al. [8]	–	–	67.5
Song et al. [25]	–	–	66.9
DF2Net [22]	–	–	65.4
MSN [10]	–	–	68.1
Proposed method	**72.3**	**64.6**	**74.1**

Bold value indicates the highest accuracy in the comparison

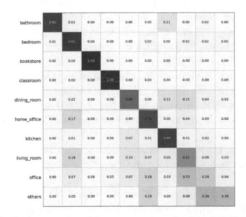

Fig. 7. Confusion matrix of the proposed method on NYUD2 database

Acknowledgement. This work was supported by the National Natural Science Foundation of China (Grant Nos. 61471206).

References

1. Deng, J., Dong, W., Socher, R., et al.: Imagenet: a large-scale hierarchical image database. In: 2009 IEEE Conference on Computer Vision and Pattern Recognition, pp. 248–255. IEEE (2009)
2. Zhou, B., Lapedriza, A., Khosla, A., et al.: Places: a 10 million image database for scene recognition. IEEE Trans. Pattern Anal. Mach. Intell. **40**(6), 1452–1464 (2017)
3. Patel, T.A., Dabhi, V.K., Prajapati, H.B.: Survey on scene classification techniques. In: 2020 6th International Conference on Advanced Computing and Communication Systems (ICACCS), pp. 452–458. IEEE (2020)
4. Zeng, D., Liao, M., Tavakolian, M., et al.: Deep learning for scene classification: a survey (2021). arXiv preprint arXiv:2101.10531
5. Vaswani, A., Shazeer, N., Parmar, N., et al.: Attention is all you need. Advan. Neural Inform. Proc. Syst. 30 (2017)
6. Wang, A., Cai, J., Lu, J., et al.: Modality and component aware feature fusion for rgb-d scene classification. In: Proceedings of the IEEE Conference on Computer Vision and Pattern Recognition, pp. 5995–6004 (2016)
7. Zhu, H., Weibel, J.B., Lu, S.: Discriminative multi-modal feature fusion for rgbd indoor scene recognition. In: Proceedings of the IEEE Conference on Computer Vision and Pattern Recognition, pp. 2969–2976 (2016)
8. Song, X., Jiang, S., Herranz, L., et al.: Learning effective RGB-D representations for scene recognition. IEEE Trans. Image Proc. **28**(2), 980–993 (2018)
9. Song, X., Jiang, S., Wang, B., et al.: Image representations with spatial object-to-object relations for RGB-D scene recognition. IEEE Trans. Image Proc. **29**, 525–537 (2019)
10. Xiong, Z., Yuan, Y., Wang, Q.: MSN: modality separation networks for RGB-D scene recognition. Neurocomputing **373**, 81–89 (2020)
11. Xiong, Z., Yuan, Y., Wang, Q.: ASK: adaptively selecting key local features for RGB-D scene recognition. IEEE Trans. Image Proc. **30**, 2722–2733 (2021)
12. Du, D., Wang, L., Wang, H., et al.: Translate-to-recognize networks for rgb-d scene recognition. In: Proceedings of the IEEE/CVF Conference on Computer Vision and Pattern Recognition, pp. 11836–11845 (2019)
13. Dosovitskiy, A., Beyer, L., Kolesnikov, A., et al.: An image is worth 16x16 words: transformers for image recognition at scale (2020) arXiv preprint arXiv:2010.11929
14. Song, S., Lichtenberg, S.P., Xiao, J., Sun rgb-d: a rgb-d scene understanding benchmark suite. In: Proceedings of the IEEE Conference on Computer Vision and Pattern Recognition, pp. 567–576 (2015)
15. Silberman, N., Hoiem, D., Kohli, P., et al.: Indoor segmentation and support inference from rgbd images. In: European Conference on Computer Vision, pp. 746–760. Springer, Berlin, Heidelberg (2012)
16. Caglayan, A., Imamoglu, N., Can, A.B., et al.: When CNNs meet random RNNs: towards multi-level analysis for RGB-D object and scene recognition, p. 103373. Computer Vision and Image Understanding (2022)
17. Song, X., Jiang, S., Herranz, L.: Combining models from multiple sources for RGB-D scene recognition. In: IJCAI, pp. 4523–4529 (2017)
18. Sun, N., Wang, L., Liu, J., et al.: Scene recognition based on privilege information and attention mechanism. J. Zhengzhou Univ. (Eng. Sci.) **42**(1), 42–49 (2021). In Chinese
19. Song, X., Herranz, L., Jiang, S.: Depth CNNs for RGB-D scene recognition: learning from scratch better than transferring from RGB-CNNs. In: Thirty-first AAAI Conference on Artificial Intelligence (2017)

20. Gong, W., Zhang, B., Li, X.: An efficient RGB-D scene recognition method based on multi-information fusion. IEEE Access **8**, 212351–212360 (2020)
21. Xiong, Z., Yuan, Y., Wang, Q.: RGB-D scene recognition via spatial-related multi-modal feature learning. IEEE Access **7**, 106739–106747 (2019)
22. Zeng, X., Peng, X., Qiao, Y.: Df2net: a dense-fine-finer network for detailed 3d face reconstruction. In: Proceedings of the IEEE/CVF International Conference on Computer Vision, pp. 2315–2324 (2019)
23. Du, D., Xu, X., Ren, T., et al.: Depth images could tell us more: Enhancing depth discriminability for rgb-d scene recognition. In: 2018 IEEE International Conference on Multimedia and Expo (ICME), pp. 1–6. IEEE (2018)
24. Quattoni, A., Torralba, A.: Recognizing indoor scenes. In: 2009 IEEE Conference on Computer Vision and Pattern Recognition, pp. 413–420. IEEE (2009)
25. Song, X., Chen, C., Jiang, S.: RGB-D scene recognition with object-to-object relation. In: Proceedings of the 25th ACM International Conference on Multimedia, pp. 600–608 (2017)

Sheep Herd Recognition and Classification Based on Deep Learning

Yeerjiang Halimu[1,3], Zhou Chao[1], Jun Sun[1(✉)], and Xiubin Zhang[2]

[1] School of Internet of Things Engineering, Jiangnan University, Wuxi 214122, China
yej@xjau.edu.cn, 6201910041@stu.jiangnan.edu.cn, junsun@jiangnan.edu.cn
[2] School of Electronic Information and Electrical Engineering, Shanghai Jiao Tong University, Shanghai 200240, China
zhangxb@sjtu.edu.cn
[3] School of Computer and Information Engineering, Xinjiang Agricultural University, Urumqi 830052, China

Abstract. Automatic identification of sheep herd is always an unsolved technical problem. This is because there are many kinds of sheep, and there are great differences in individual morphology and characteristics. Moreover, human's mastery of the individual characteristics of different breeds of sheep is only in the stage of experience. So far, people have not completely established the classification database of individual characteristics of sheep.This paper is a summary of the exploratory research on this technical problem. It provides a sheep recognition and classification algorithm based on deep learning. The algorithm adopts dual channel convolution neural network, and carries out reverse transmission according to the image characteristics in time to realize the adaptive adjustment of weight. Once the optimal or suboptimal weight is obtained, the iteration is ended, and the identified objects are located, counted and classified. The experimental results show that the algorithm can greatly reduce the calculation time and make the recognition and classification more accurate.

Keywords: Deep learning · Convolutional neural network · Sheep herd · Recognition · Classification

1 Introduction

When the scale of animal husbandry is increasing, herdsmen often are unable to do as well as one would wish in the management (tracking) of shepherds, cattle and horses. With the popularity and use of micro UAV, it will naturally cause people to obtain the information of herd distribution through the image sampling of UAV, and then realize the automatic tracking and analysis of herd,

N. Xiong et al. (Eds.): ICNC-FSKD 2022, LNDECT 153, pp. 122–129, 2023.
https://doi.org/10.1007/978-3-031-20738-9_15

which has become a logical thing. Naturally, it has become a technical problem that the majority of herdsmen urgently hope to solve.

As far as the current research situation is concerned, there is no more practical technical method on how to effectively identify sheep herd and their classification. In view of the objective limitations of many existing grazing management methods in the Qinghai Tibet Plateau due to factors such as family labor force, grassland scale and population variety number, article [1] introduced deep learning technology into animal husbandry image recognition, and performed translation, rotation, flip and zoom on the original image through data enhancement technology, so as to establish a common animal husbandry image data set in the Qinghai Tibet Plateau. The convolution neural network model designed in this paper has realized image recognition and statistics, and the recognition rate of animal husbandry image has reached 87.89%.

Although the recognition and classification of sheep herd can be realized by template matching, principal component analysis (PCA), support vector machine (SVM) and other algorithms, all these classical recognition algorithms involve the collection of huge sample data in the learning process, feature extraction, classification and even the establishment of learning database. Among them, a large number of acquisition quantity and operation time are not difficult to imagine [2]. However, as far as the available public literature is concerned, there is no practical example of applying PCA or SVM in the identification of livestock groups, because the experimental algorithm has shown that this kind of algorithm can-not adapt to the instantaneous and changeable characteristics of livestock distribution in the actual livestock identification, which makes it difficult to improve the recognition rate.

With the emergence of convolutional neural networks (CNN) algorithm, the recognition of various natural information such as image, sound, waveform and color can omit the previous "learning process" of information, and directly realize intelligent recognition and classified output of the information contained in the collected carrier. Referring to the idea of residual learning,article [3] designed identity residual block and convolution residual block, and used them to build a convolution neural network model with a network depth of 245 layers, and improved both identity residual block and convolution residual block. The improved constant residual block can make the network stack deeper without network degradation. At the same time, the loss function is optimized to reduce the occurrence of parameter update exceptions. Obviously, the network construction of the algorithm is more complex, and the amount of computation is large, which also has the practical problem in herd recognition.

Aiming at the technical difficulties in the process of sheep herd recognition and classification, this paper carries out the application research of convolution neural network algorithm. Its outstanding features are: the full connection layer has two outlets (the reverse transmission outlet and the coordinate positioning point outlet after calculating the residual); Starting from the coordinate positioning exit, return to the original image and form a second convolution channel. Therefore, in the process of recognition, the invalid pixels can be "skipped", which greatly reduces the amount of recognition operation, and finally obtains the good effect of fast recognition.

2 System Construction and Mathematical Model

Select the image local width v_a (number of pixels) according to the actual camera elevation H (m)

$$v_a = (k_1 H) \ldots (k_2 w) \tag{1}$$

where, k_1 and k_2 are proportional coefficients; $k_1 = \frac{W}{h}, W, h$ represents the viewing angle width (m) and the corresponding camera (image acquisition) height (m), and w is the local width (m) to be intercepted; $k_2 = \frac{v}{W}, v$ is the number of pixels corresponding to W.

Thus, a search two-dimensional space with a local size of $v_a \times v_a$ is formed. Construct a multifunctional convolutional neural network (see Fig. 1).

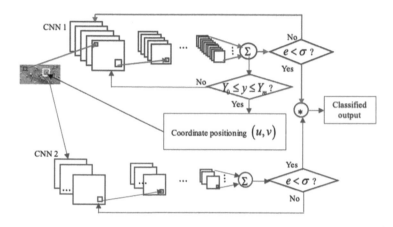

Fig. 1. Structure of multifunctional convolutional neural network

Wherein, the size of the convolution kernel is determined according to the size of the search local area, that is, the convolution kernel τ_j^i of the i-th channel in the j-th convolution layer is

$$\tau_j^i = [\tau_{kl}]_j^i = \begin{bmatrix} \tau_{1,1} & \tau_{1,2} & \cdots & \tau_{1,v_a} \\ \tau_{2,1} & \tau_{2,2} & \cdots & \tau_{2,v_a} \\ \vdots & \vdots & \vdots & \vdots \\ \tau_{v_a,1} & \tau_{v_a,2} & \cdots & \tau_{v_a,v_a} \end{bmatrix}_j^i \; ; \; k = 1, 2, \ldots, v_a; \; l = 1, 2, \ldots, v_a \tag{2}$$

The calculation process of convolution layer is

$$x_j^i = \sum_{i \in M_j} x_j^{i-1} \times \theta_j^i + b_j^i \tag{3}$$

where, x_j^i is the output of channel i in the j-th convolution layer, M_j is the characteristic diagram of the j-th convolution layer, b_j^i is the offset term, and θ_j^i is the kernel function element corresponding to the convolution layer, that is, the kernel function element matrix is

$$[\theta_j^i]_j^i = \begin{bmatrix} \tau_{v_a,v_a} & \tau_{v_a,v_a-1} & \cdots & \tau_{v_a,1} \\ \tau_{v_a-1,v_a} & \tau_{v_a-1,v_a-1} & \cdots & \tau_{v_a-1,1} \\ \vdots & \vdots & \vdots & \vdots \\ \tau_{1,v_a} & \tau_{1,v_a-1} & \cdots & \tau_{11} \end{bmatrix}_j^i \tag{4}$$

The active layer is configured after each convolution layer, that is

$$\hat{x}_j^i = f\left(x_j^i\right) \tag{5}$$

where, $f(\cdot)$ is the activation function and

$$\hat{x}_j^i = \text{sigmoid}\left(x_j^{i-1}\right) = \frac{1}{1 - e^{-x_j^{i-1}}} \tag{6}$$

Then, the next acquisition layer takes the maximum value through the 3×3 filter and the step size is 3 to pool, that is

$$\hat{y}_j^i = \max\left[\hat{x}_j^i\right]_{3\times3} \tag{7}$$

Further, the integrated feature vector is mapped to the corresponding tag space to form a full connection layer. The full connection layer has two exits: One is to calculate the residual error and reverse transmission; Second, find the coordinate positioning point (u, v).

In the process of calculating the residual e, according to whether it is less than the variance value σ of e: if $e < \sigma$, prepare for output; Otherwise, reverse transmission will continue to overlay the identically equal mapping layer on the basis of the shallow network, so that the network will not deteriorate with the increase of depth.

Once the transmission is reversed, a nonlinear element is used to approximate an objective function $h\left(x_j^i\right)$, and the objective function is divided into two parts: identity function x_j^i and residual function $h\left(x_j^i\right) - x_j^i$. Thus, the nonlinear element x_j^{i+1} can be used to approximate the residual function $h\left(x_j^i\right) - x_j^i$, and $x_j^{i+1} + x_j^i$ can be used to approximate the objective function $h\left(x_j^i\right)$.

Combining formulas (3) and (4), the output neuron of the approximated objective function after activation is [4]

$$\hat{x}_j^{i+1} = \text{sigmoid}\left[\sum_{i \in M_j} x_j^i \times \text{sigmoid}\left(\sum_{i \in M_j} x_j^{i-1} \times \theta_j^i + b_j^i\right) + b_j^{i+1} + x_j^i\right] \tag{8}$$

The coordinate positioning method of the second outlet of the full connection layer is: first, it is necessary to determine whether the total gray value y on the local area $v_a \times v_a$ meets the condition $Y_0 \leq y \leq Y_m$, where Y_0 and Y_m represent the base value and maximum value of y respectively, and Y_0 can take the average value of the gray value of the whole image. The search process is to "scan" the local area $v_a \times v_a$ with a step of 1 . Once $Y_0 \leq y \leq Y_m$ is found, take the central pixel of $v_a \times v_a$ as the positioning coordinate, otherwise "skip".

For the local area $v_a \times v_a$ of the location point, re-enter a "fine" convolution neural network system CNN2 to implement target reconfirmation.

Finally, the pixel matrix "dot product" operation is performed on the full connection layer of two "parallel" convolutional neural networks, that is

$$
\begin{aligned}
\mathbf{S} &= \mathbf{Z}_{CNN1} * \mathbf{Z}_{CNN2} \\
&= \begin{bmatrix} z_{1,1} & z_{1,2} & \cdots & z_{1,N} \\ z_{2,1} & z_{2,2} & \cdots & z_{2,N} \\ \vdots & \vdots & \vdots & \vdots \\ z_{N,1} & z_{N,2} & \cdots & z_{N,N} \end{bmatrix} * \begin{bmatrix} \dot{z}_{1,1} & \dot{z}_{1,2} & \cdots & \dot{z}_{1,N} \\ \dot{z}_{2,1} & \dot{z}_{2,2} & \cdots & \dot{z}_{2,N} \\ \vdots & \vdots & \vdots & \vdots \\ \dot{z}_{N,1} & \dot{z}_{N,2} & \cdots & \dot{z}_{N,N} \end{bmatrix} \\
&= [z_{ij} \times \dot{z}_{ij}]_{N \times N} = [s_l]_{N \times N}
\end{aligned}
\tag{9}
$$

Then output it to the classifier, and recognize and classify the identified object through Softmax function $y(s_l)$, that is

$$
y(s_l) = \frac{\exp(s_l)}{\sum_{l=1}^{n} [\exp(s_l)]}
\tag{10}
$$

where, s_l is the linear prediction of category l, and n is the number of categories; $0 < n < \left\lfloor \frac{v_m \times u_m}{v_a \times v_a} \right\rfloor$, v_m and u_m are the maximum number of pixels corresponding to the width and height of the picture of the collected image respectively [5].

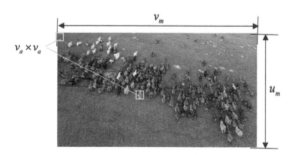

Fig. 2. An original top view of sheep herds

3   Experiment

In order to verify the correctness and feasibility of the above system construction and mathematical model, the sheep herd in Tianshan Mountain in Xinjiang region were tracked in real time by UAV. And the actual identification and classification of sheep herd distribution in the two environments.

3.1   Experiment 1

In the environment of scattered white rocks in the pasture and the presence of pure white and brown sheep in the sheep herd, the sheep herd shall be tracked and aerial photographed. Take the original top view of sheep herd (see Fig. 2), and construct convolutional neural networks CNN1 and CNN2.

Using the network structure in Fig. 1 and the above mathematical model, set the convolution kernel as the following matrix

$$\tau_1^i = [\tau_{kl}]_1^i = \begin{bmatrix} -1 & -1 & -1 & -1 & -1 \\ -1 & -1 & -1 & -1 & -1 \\ -1 & -1 & 8 & -1 & -1 \\ -1 & -1 & -1 & -1 & -1 \\ -1 & -1 & -1 & -1 & -1 \end{bmatrix} \tag{11}$$

Taking $b_j^1 = 1$, the full connection layer of identification and classification can be obtained through the operation process of CNN1 (see Fig. 3). Figure 3(a) shows the medium and shallow learning effect of CNN1; Fig. 3(b) shows the final effect after deep learning and residual error correction in reverse transmission.

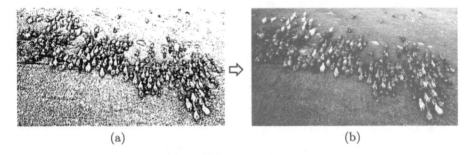

(a)   (b)

Fig. 3. Operation process of convolution neural network

According to the precise operation of pixel coordinate positioning (u, v) and CNN2, and then through the pixel matrix "dot product" operation at the output of CNN1 and CNN2, the total number of sheep herd is 241 (including 22 pure white sheep).

After practical verification, it is found that five white stones are misjudged as pure white sheep, but the actual situation is that the total number of sheep herd is 236 (including 17 pure white sheep).

Incidentally, with regard to this accuracy problem, we can intercept new images by re photographing the moving sheep herd, and then conduct multifunctional CNN operation again. After comparing the recognition results of the two images before and after, we can eliminate this kind of recognition error.

3.2 Experiment 2

Track and take aerial photos of sheep herd with similar colors distributed on the green one-color grassland. A top view of sheep herd is intercepted (see Fig. 4), and the network structure and mathematical model in Fig. 1 are also used to identify and classify them. The calculation results show that: first, the shadow will be automatically eliminated; then automatically distinguish sheep from shadows. Finally, the recognition results are accurately output: there are 302 sheep in the field of view.

(a) Original image   (b) Gray image   (c) Full connection diagram of convolutional neural network

Fig. 4. Recognition and classification process of sheep herds with shadow

3.3 Comparison with Other Algorithms

Using the above algorithm and through dozens of experiments, it is proved that the multi-functional CNN is compared with other algorithms in terms of operation speed and accuracy, as shown in Table 1.

Table 1. Comparison of operation functions.

Algorithm	Operation rate	Recognition and classification accuracy (%)	Do you need prior training?
Multifunctional CNN	<100 ms	>95	No need
PCA	<85 ms	>82	Need
SVM	<90 ms	>80	Need
Article [1]	(lack of data)	>87.89	No need
Article [3]	(lack of data)	>93.99	No need

It must be pointed out that both principal component analysis (PCA) and support vector machine (SVM) need to collect a sufficient number of training samples in advance to learn and build a training sample database before identifying the identified objects. It can be seen that PCA and SVM do not have the function of "autonomous learning" when the characteristics of objects are unknown. Therefore, in this regard, it can highlight the "inherent" advantages of the "multi-functional CNN" algorithm.

4   Conclusion

The outstanding advantage of the algorithm proposed in this paper is that it adopts dual channel convolutional neural network, and can carry out reverse transmission according to the image characteristics in real time, so as to realize the adaptive adjustment of weight. In the operation process, once the optimal or suboptimal weight is obtained, the iteration can be ended in real time, and the identified objects can be located, counted and classified. Therefore, it can greatly reduce the operation time of recognition and classification, and ensure the high accuracy of recognition and classification.

Although the sheep herd recognition and classification method based on deep learning provided in this paper can achieve high recognition and classification accuracy. In Experiment 1, the recognition accuracy is more than 98%; In Experiment 2, because the environment is relatively "simple" and the sheep's fur color is single, the recognition accuracy is close to 100%.

However, it must be pointed out that: First of all, there are great technical defects in classification only relying on gray images, that is, it is impossible to distinguish colors. Therefore, in order to realize the fine classification of sheep with different colors, it still needs to be realized according to the color classification algorithm; Secondly, it is difficult to distinguish the colored sheep, mainly because the proportion of different colors of wool in the whole-body picture of sheep cannot be determined, and there are fuzzy boundaries in the image of sheep. In the follow-up research, it is also necessary to add mathematical morphology and other algorithms to the sheep boundary algorithm to enhance the boundary definition, and then further subdivide the characteristics of sheep according to the operation of "algorithm layer" with different color proportion.

References

1. La, M., An, J.: Research on animal image recognition based on convolutional neural network. Comput. Eng. Softw. **41**(8), 43–45 (2020)
2. Zhang, X., Khan, M.: Intelligent Visual Perception Tutorial. China Water & Power Press, Beijing (2012)
3. Tian, Y.: Application research of image recognition based on deep learning. Master dissertation, China Jiliang University, Hangzhou (2019)
4. Wang, A.: Research and implementation of image recognition based on deep learning. Master thesis, University of Electronic Science and Technology of China, Chengdu (2018)
5. Xu, Q., Pan, G.: Sparse connect: regularising CNNs on fully connected layers. Electron. Lett. **53**(18), 1246–1248 (2017)

A Deep Learning-Based Innovative Points Extraction Method

Tao Yu[1], Rui Wang[1(✉)], Hongfei Zhan[1], Yingjun Lin[2], and Junhe Yu[1]

[1] Ningbo University, Ningbo 315000, China
{2111081243,wangrui,zhanhongfei,yujunhe}@nbu.edu.cn
[2] Zhongyin (Ningbo) Battery Co., Ltd., Ningbo 315040, China

Abstract. Most of the research on mining online reviews now focuses on the influence of reviews on consumers and the issue of sentiment analysis for analyzing consumer reviews, but few studies how to extract innovative ideas for products from review data. To this end, we propose a deep learning-based method to extract sentences with innovative ideas from a large amount of review data. First, we select a product review dataset from the Internet, and use a stacking integrated word embedding method to generate a rich semantic representation of review sentences, and then the resulting representation of each sentence will be feature extraction by a bidirectional gated recurrent unit (BiGRU) model combined with self-attention mechanism, and finally the extracted features are classified into innovative sentences through softmax. The method proposed in this paper can efficiently and accurately extract innovative sentences from class-imbalanced review data, and our proposed method can be applied in most information extraction studies.

Keywords: Information extraction · Deep learning · Word embedding · Text classification · Class imbalance problem

1 Introduction

Online shopping has become a mainstream consumption method for people at home and abroad, and with the severe situation of the new crown pneumonia epidemic, more and more people choose to shop online and like to make online reviews of the products they buy. Users' online reviews have the characteristics of fast dissemination, large amount of information and strong influence ability. Most of the products' features will exist to make users less satisfied, and users will leave some negative reviews about the product's shortcomings and innovative reviews about how the product should be improved. A large number of practical and theoretical studies have shown that the ideas and innovative knowledge provided by product users play a significant role in the improvement of enterprise products [1]. Therefore, extracting negative statements and innovative statements from many online reviews and making corresponding improvements and innovations to improve the market competitiveness of products are of great interest in the business and academic circles.

N. Xiong et al. (Eds.): ICNC-FSKD 2022, LNDECT 153, pp. 130–138, 2023.
https://doi.org/10.1007/978-3-031-20738-9_16

Users' online reviews are basically unstructured text data, and after collecting the online review text data to the need to use sentiment analysis and deep learning methods to extract negative and innovative reviews. Sentiment analysis methods can automatically analyze users' attitudes and opinions in online reviews of products. The traditional sentiment analysis method is to build a sentiment dictionary, cut the review text, and analyze the sentiment tendency at the level of words, phrases and sentences to extract the negative statements in the text. With the very good results of deep learning in several fields, the research on extracting negative reviews using deep learning models is gradually increasing. For extracting innovative statements, unlike negative statements, innovative statements are more complex and conceptual, plus the percentage of innovative statements in online reviews is very small, which is more difficult to extract, and in order to be able to extract them, they are often extracted by deep learning models plus loss functions. In this paper, we propose a new deep learning model to extract innovative statements from online reviews, and then introduce Gradient Harmonizing Mechanism (GHM) to correct the data in order to extract them accurately and effectively.

2 Related Research

The topic features in the review data are usually not obvious, and the information displayed is less, and the proportion of product innovation feature information extracted from it is too small, which will make the extraction difficult. Qi et al. [2] use the KANO method, a classical conjoint analysis model, it is innovatively used in the analysis of online reviews to mine useful information to improve and adjust the product. Qiao et al. [3] constructed a new Latent Dirichlet Allocation (LDA) model to identify and obtain the overall information of product defects, which helps relevant staff to do product quality analysis. Zhai et al. [4] achieved a breakthrough in named entity recognition of chemical patent information using ELMo for pre-trained word representation combined with BiLSTM-CRF model. Pasupa et al. [6] used deep learning combined with focal loss function to solve the problem of class imbalance in identifying abnormal blood cell morphology. Liu et al. [7] proposed to combine an integrated sampling technique incorporating oversampling and undersampling with the integration of support vector machines to improve prediction performance in unbalanced datasets. Li et al. [8] proposed a new gradient coordination mechanism in order to solve the problem of huge differences in the number of positive and negative examples of data, and huge differences in the number of easy and hard examples, which achieved better results than the state-of-the-art method focus loss. Beltagy et al. [9] proposed a SCIBERT model that has achieved significant results in addressing the lack of high-quality scientific data. Min et al. [10] proposed an integrated word embedding-based deep learning model to extract innovative sentences.

To sum up, the existing research on innovative sentence extraction has made some progress, which provides a good foundation for further research in this paper, but the quality of the extracted innovative sentences needs further research. A novel ensemble embedding method proposed in this paper can form a more accurate semantic and contextual representation, and then extract innovative sentence features through the bidirectional gated recurrent unit (BiGRU) model combined with the self-attention mechanism, in which a GHM loss function is used to solve its class imbalance problem. The final classified innovation reviews are of great significance to the upgrading of enterprise products.

3   Innovative Sentence Extraction Model

The Innovative Sentence extraction model consists of three parts: integrated word embedding, feature extraction and candidate word classification, and the model structure is shown in Fig. 1. First, the input text data is mapped into vector representation for each word by three integrated word embedding models; then these word vectors are passed through the feature extraction model; finally, the innovative words are identified by the softmax classifier.

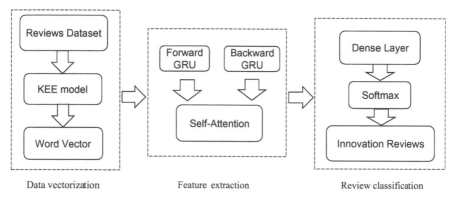

Fig. 1. Innovative sentence extraction model framework.

3.1   KNN-Based Ensemble Embedding (KEE)

Selection of Word Embedding Model. In order to identify the data in the text, it is necessary to use the technical method of word embedding to first convert the character data of the text into numerical data, and then convert these data into a numerical vector form that can be processed. There are many word embedding models designed now. Considering the wide application and the accuracy of generating word vectors, three word embedding models, Roberta, Skip-gram and ELMo, are selected. In order to integrate the advantages of the three word embedding models, this paper uses a stacking method to fuse the three models.

Word Embedding Ensemble Model. For Each Input Word is Part of the Sentence S, It Can Be Written as $S = [W_1, W_2 \ldots W_n]$, Any Word W_i Goes Through Three Word Embedding Models to Become $W_i^{Roberta}$, $W_i^{Skip-gram}$ and W_i^{ELMo}, and then the Three Word Embedding Models Are Used to Generate Feature Vectors for Stacking Embedding, the Method of Stacking Word Embedding is to Use Multiple Learning

Models as Basic Learners, and then Connect and Integrate Them Through a Meta-learner. Here, Three Word Embedding Models Are Used as the Base Learner, and the K-Nearest Neighbor (KNN) Algorithm is Used as the Meta-learner for Reprocessing. The Specific Model Fusion Structure is Shown in Fig. 2.

$$S = [W_1, W_2...W_n] \tag{1}$$

$$Y(W_i) = \begin{bmatrix} W_i^{Roberta} \\ W_i^{Skip-gram} \\ W_i^{ELMo} \end{bmatrix} \tag{2}$$

$$V_s = [Y(W_1), Y(W_2)...Y(W_n)] \tag{3}$$

$$KEE_s = KNN(V_s) \tag{4}$$

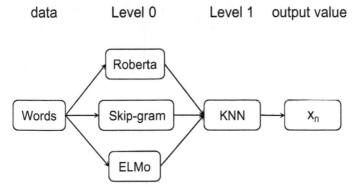

Fig. 2. KEE model structure diagram.

3.2 Feature Extraction Model

Extracting features by traditional BiGRU does not sufficiently consider the contextual connection of the text, and the extraction of features is not sufficient. The introduction of attention mechanism can capture global information and focus on important features. The model structure is shown in Fig. 3 and consists of a BiGRU layer and a self-attention layer.

BiGRU Layer. Gated recurrent network (GRU) and Long Short Term Memory (LSTM) have roughly the same functions, both of which have roughly the same function and are designed to solve the problem of gradient descent and gradient explosion in recurrent neural network (RNN), but the structure of GRU is simpler, so it is easier to train than LSTM and can save training time to a great extent. The structure of GRU is shown in Fig. 4.

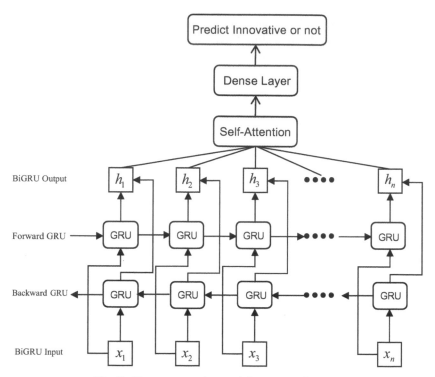

Fig. 3. Feature extraction model structure diagram.

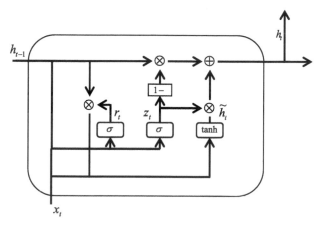

Fig. 4. GRU model structure diagram.

The information transfer in GRU is input sequentially from front to back, which leads to the GRU model can only learn the content before this moment, ignoring the influence of the following information at this moment. In order to fully learn the state information of the context, this paper adopts a bidirectional gated recurrent network (BiGRU), which can learn the contextual semantics of a word and fully mine the specific semantics of the word. BiGRU is composed of two GRUs with opposite directions, and the output information is jointly determined by the two GRUs. The structure of BiGRU is shown in Fig. 5.

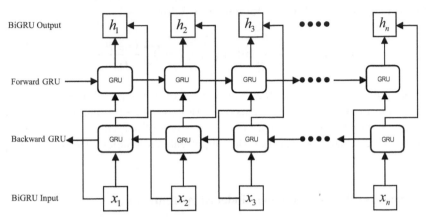

Fig. 5. Structure of BiGRU model.

Self-Attention Layer. Nowadays, the data volume of online product reviews is very large, and the effect of extracting information features by BiGRU model alone is poor, because the ability of BiGRU model to capture information decreases with the increase of textual information, then it is necessary to introduce a self-attention counting system, which makes the model pay more attention to key information and ignore unimportant information, i.e., not to lose important data and speed up the calculation.

The self-attention model is a Query-Key-Value (QKV) model, in which Q, K and V are derived from the input word vector matrix. Firstly, Q and K are calculated by similarity to get the weights, then the weights are normalized by the softmax function, and finally the normalized weights and the corresponding key values V are weighted and summed to get the results of the self-attention layer. The formula of self-attentiveness is shown as follows:

$$\text{Attention}(Q, K, V) = \text{softmax}\left(\frac{QK^T}{\sqrt{d_k}}\right)V \qquad (5)$$

where $Q \in R^{n \times d_k}$, $K \in R^{n \times d_k}$, $V \in R^{n \times d_k}$, and T represent transpose operations on the matrix, and dividing by $\sqrt{d_k}$ is to regulate the inner product so that it is not too large.

3.3 Loss Function

Class imbalance problem is often encountered in information extraction. In our problem, the proportion of innovative statements in the review data is too small, which will make the model difficult to train, and the classification effect will still be poor even if the model is trained. One of the mainstream methods to solve the class imbalance is to introduce a focal loss function, but the focal function will make the model focus too much on the particularly difficult to classify samples, which will be problematic and easily increase the risk of model overfitting. Therefore, this paper uses the GHM (Gradient Harmonizing Mechanism) loss function. GHM makes up for the defect of focal loss and avoids excessive focus on difficult classification problems. GHM is applied to classification problems to form GHM-C, and its formula is as follows.

$$L_{GHM-C} = \frac{1}{N} \sum_{i=1}^{N} \beta_i L_{CE}(p_i, p_i^*) = \sum_{i=1}^{N} \frac{L_{CE}(p_i, p_i^*)}{GD(g_i)} \tag{6}$$

Among them, N represents the total number of examples, β_i represents the loss weight of the ith sample, L_{CE} represents the cross entropy loss, g represents the gradient mean square, and the value of g represents the attribute of the sample, that is, the impact on the global gradient (easy/hard), GD(g) represents the gradient density with respect to g.

3.4 Innovative Points Extraction Model Algorithm

Our proposed innovation extraction model integrates several different word embeddings and then outputs the innovation statements at the sentence level by a text classification algorithm. For training, the GHM-C loss is calculated based on the target value to ensure that the trained model can be classified accurately rather than overfitting on one of the classes. Training algorithm diagram is shown in Fig. 6, and the prediction algorithm diagram.is shown in Fig. 7.

Algorithm 1: Innovation Extraction Training Algorithm

Data: Input: Review Sentence
Result: Output: Binary Classification of Innovation or Not (1 or 0)
for each sentence S in review R do
    for each word w_s ∈ sentence S do
        if word w_s is in contractions C then
            expand contraction on word w_s tokenize w_s
        end
    end
    compute feed-forward Stacking embedding K_v
    compute Self-Attention-based bidirectional GRU logits $G(K_v) = p_i$
    compute $L_{GHM-C} = \frac{1}{N} \sum_{i=1}^{N} \beta_i L_{CE}(p_i, p_i^*) = \sum_{i=1}^{N} \frac{L_{CE}(p_i, p_i^*)}{GD(g_i)}$
    backpropagate gradients to K_v
end

Fig. 6. Training algorithm diagram.

Algorithm 2: Innovation Extraction Prediction Algorithm

Data: Input: Review Sentence
Result: Output: Binary Classification of Innovation or Not (1 or 0)
for each sentence S in review R do

  for each word $w_s \in$ sentence S do

    if word w_s is in contractions C then

      expand contraction on word w_s tokenize w_s

    end

  end

  compute feed-forward Stacking embedding K_v

  compute Self-Attention-based bidirectional GRU logits G(K_v)= p_t

  take the argmax of logits argmax: p_t

end

Fig. 7. Prediction algorithm diagram.

4 Conclusion

We live in the era of network economy, making full use of product review data has attracted much attention, and extracting innovative sentences from it to improve products is especially valued by enterprises. In this paper, we propose an efficient deep learning extraction model. After data pre-processing, the KEE integrated word embedding model is used to vectorize the online review data, which is used as the input layer to train the BiGRU classification model with the self-attention mechanism, and then the GHM-C loss function is introduced to solve the serious class imbalance problem, and finally the innovative sentences in the user review data are extracted, which will be of great significance for related companies to carry out product innovation. This paper only proposes a model framework for innovative sentence extraction based on deep learning. Based on other people's research on innovative sentence extraction, the model proposed in this paper can theoretically effectively extract innovative sentences, and the accuracy effect will be better. Our work also has limitations. Although it can achieve better accuracy in theory, the specific situation needs to be experimentally verified. Our next work is to verify the effect of the model. Based on this work, how to improve the quality of innovative sentences and give corresponding solutions will become the focus of further research.

Acknowledgement. I would like to extend my gratitude to all those who have offered support in writing this thesis from National Key R&D Program of China (2019YFB1707101, 2019YFB1707103), the Zhejiang Provincial Public Welfare Technology Application Research Project (LGG20E050010, LGG18E050002) and the National Natural Science Foundation of China (71671097).

References

1. Chatterji, A.K., Fabrizio, K.: How do product users influence corporate invention? Organ. Sci. **23**(4), 971–987 (2012)

2. Qi, J., Zhang, Z., Jeon, S., Zhou, Y.: Mining customer requirements from online reviews: a product improvement perspective. Soc. Sci. Electron. Publishing. **53**(8), 951–963 (2016)
3. Qiao, Z., Zhang, X., Zhou, M., Wang, G.A., Fan, W.: A domain oriented LDA model for mining product defects from online customer reviews. In: Proceedings of the 50th Hawaii International Conference on System Sciences, pp. 1821–1830 (2017)
4. Zhai, Z., Nguyen, D.Q., Akhondi, S., et al.: Improving chemical named entity recognition in patents with contextualized word embeddings. In: Proceedings of the 18th BioNLP Workshop and Shared Task, pp. 328–338. Association for Computational Linguistics (2019)
5. Pasupa, K., Vatathanavaro, S., Tungjitnob, S.: Convolutional neural networks based focal loss for class imbalance problem: a case study of canine red blood cells morphology classification. Ambient Intelligence and Humanized Computing (2020). https://doi.org/10.1007/s12652-020-01773-x
6. Liu, Y., Yu, X.H., Xiang, J.H.J., et al.: Combining integrated sampling with SVM ensembles for learning from imbalanced datasets. Inf. Process. Manage. **47**(4), 617–631 (2011)
7. Li, B., Liu, Y., Wang, X.: Gradient harmonized single-stage detector. In: Proceedings of the AAAI Conference on Artificial Intelligence, pp. 33:8577–8584 (2019)
8. Beltagy, I., Lo, K., Cohan, A.: Scibert: a pretrained language model for scientific text. In: Proceedings of the 2019 Conference on Empirical Methods in Natural Language Processing and the 9th International Joint Conference on Natural Language Processing (EMNLP-IJCNLP), pp. 3615–3620 (2019)
9. Min, Z., Brandon, F., Ning, Z., Wang, W., Fan, W.: Mining product innovation ideas from online reviews. Inf. Proc. Manage. **58**(1) (2021)
10. Mikolov, T., Chen, K., Corrado, G., Jeffrey, D.: Efficient estimation of word representations in vector space. In: Proceedings of Workshop at ICLR (2013)

Deep Embedded Clustering with Random Projection Penalty

Kang Song[1,2(✉)], Wei Han[2], Chamara Kasun Liyanaarachchi Lekamalage[1], and Lihui Chen[1]

[1] Nanyang Technological University, Singapore, Singapore
song0149@e.ntu.edu.sg chamarakasun@ntu.edu.sg elhchen@ntu.edu.sg
[2] Shopee Pte, Singapore, Singapore
eric.song@shopee.com, wei.han@shopee.com

Abstract. Despite being studied for decades, current clustering approaches still face difficulties in handling high-dimensional datasets. For high-dimensional data, it is essential to get a good feature representation for the clustering algorithm to conduct on. Most of the clustering algorithms do not explicitly encourage the preservation of pairwise distance within the input data when learning feature representation. We proposed an improvement over the Deep Embedded Clustering (DEC) by including a penalty term in the loss function for the differences between the input data and the random projection of its corresponding feature embedding. The idea behind this penalty term is one of the properties of random projection, that is the pairwise distance is preserved between the low-dimensional manifold and the data space. In this way, the network encourages the learning towards preserving data similarities in the feature space. We named the proposed method as DEC-RPP. The experiments show significant improvements on clustering metrics over four datasets compared with the baseline DEC and another work IDEC that improves DEC by preserving local structure.

Keywords: Deep clustering · Random projection · Representation learning

1 Introduction

Clustering is a well-known technique to find patterns in data in an unsupervised manner based on similarity measures. Clustering on raw data directly is challenging and ineffective. It is unknown which input dimensions in the data points actually contribute to the patterns for clustering. This uneven and non-uniform nature of input dimensions in the data makes it difficult to group data points with similar patterns. Therefore it is essential to learn a low-dimensional feature manifold from high-dimensional data that generalise well, since key characteristics and similarity preservation are already in the nature of manifold.

© The Author(s), under exclusive license to Springer Nature Switzerland AG 2023
N. Xiong et al. (Eds.): ICNC-FSKD 2022, LNDECT 153, pp. 139–146, 2023.
https://doi.org/10.1007/978-3-031-20738-9_17

There was one influential work called Deep Embedded Clustering (DEC) [1] that can learn the feature representation and cluster assignment simultaneously in a self-supervised way and demonstrated significant improvement on clustering accuracy over traditional methods. However the feature representation it learns does not preserve pairwise distance in the data space, because the loss function only favors cluster assignment, and the manifold learned in the pre-training stage gets distorted without the constraints on local structure in data. That is why IDEC [11] was proposed to add such constraint during the training in order to preserve the local structure to some extent. However since the DEC at the pre-training stage already learned the low-dimensional manifold, keeping the same constraint in the loss function at the training stage will only reduce the undesirable distortion made in the feature space to some extent, but the approach it used to preserve local structure is homogeneous to the manifold learning at the pre-training stage, that's why the improvement of IDEC over DEC is marginal. The motivation to our proposed method came from the above analysis, it leads us to find another way to preserve pairwise distance in the data space other than the reconstruction approach. We found that performing random projection on a set of data points will only do rotation and scaling to the data points, thus their pairwise distance is preserved. Therefore we proposed an improvement over DEC by adding a random projection penalty term in the loss function to encourage the learned features to preserve similarities between data points. The experiment results showed significant improvements in evaluation metrics.

2   Literature Review

Clustering methods can be grouped into two categories: hierarchical and partitional. Hierarchical clustering methods iteratively update data and clusters using top-down (also known as divisive) or bottom-up (also know as agglomerative) approaches. Traditional hierarchical clustering methods are susceptible to outliers and noise, they are computationally inefficient, and the classification error during the clustering process cannot be corrected thus will cause accumulation of errors. Partitional clustering methods partition data into clusters that do not overlap with each other according to some optimisation targets. They are generally more efficient than hierarchical clustering methods in terms of computation due to the flat structure of the clusters. Those methods can be further grouped into three categories according to the type of targets they try to optimise: distance-based, density-based, and model-based.

Distance-based clustering methods use distance metric to evaluate the targets to be optimised for clustering. One of the most commonly used and simplest distance-based clustering method is k-means [2]. K-means is simple but it has a few drawbacks. It is compulsory to define the total number of clusters for a dataset, but in most of the cases there is no effective way of knowing the number of clusters beforehand. Defining a wrong number of clusters will result in bad clustering results. Another drawback of k-means is that it is susceptible to noise and outliers.

Density-based clustering methods consider the number of data points within a region as the main criterion for clustering. Two well-known density-based clustering methods are DBSCAN [3] and Mean Shift [4,5]. Comparing with k-means, neither DBSCAN nor Mean Shift requires the number of clusters to be specified, however they have specific parameters to be configured instead. Density-based clustering methods in general have better tolerance on noise and outliers without distorting the shape of clusters. They also perform better than k-means when the true cluster shape is irregular rather than spherical.

Model-based clustering methods attempt to find the mathematical model that best fits the dataset. A popular approach is to view clustering as modelling a mixture of probability distributions [6,7], with the assumption that each cluster of data is drawn from a particular probability distribution, and the whole dataset is a mixture of such distributions. One commonly used EM approach in clustering is Gaussian Mixture Model (GMM). It mixes a set of Gaussian functions with their own parameters to represent the joint distribution of the dataset.

Spectral clustering [9,10] reformulates the clustering problem as a graph partitioning problem. In recent years, the mainstream of research on clustering focuses on feature representation learning using deep neural networks. IDEC [11] is an improvement over DEC by combining the KL-divergence loss from the original DEC with a reconstruction loss from an autoencoder, such that the network can learn to produce feature embeddings that are not only to favor the clustering task (KL-divergence loss), but also can maintain the structural similarity between the input data and its corresponding feature representation (reconstruction loss). However in the original DEC, the autoencoder was already trained with reconstruction loss before the learning of the clustering task, keeping the reconstruction loss as a part of the total loss may be beneficial but the effectiveness is limited, this is reflected by the marginal improvement of IDEC over DEC in the experiment results. In a later work that did deep fuzzy clustering (GrDNFCS) [12], the authors improved IDEC by performing fuzzy clustering and used fuzzy membership to compute the KL-divergence loss with the prior distribution, and also added an additional graph-based regularisation term in the loss function to encourage the compactness within the clusters and separation between clusters. The experiment results show that our proposed solution has better evaluation metrics than DEC, IDEC, and GrDNFCS on four datasets.

3   The Proposed DEC-RPP

3.1   The Related Method DEC

The original DEC used Stacked Denoising Autoencoder (SDAE) [13] as the deep neural network for feature representation learning. The learnable parameters are the weights and bias of the fully-connected layers in the SDAE, and also the vectors that represent the cluster centers. There are two stages in the training of DEC: parameter initialisation on the learnable parameters, and parameter optimisation for clustering.

During the parameter initialisation stage, each of the Denoising Autoencoders (DAE) are trained separately on randomly and partially corrupted input using reconstruction loss, and the embeddings are fed to the next layer as input.

After all the DAEs are trained, the fully-connected layers and activation layers will stack up to form the encoder part of the SDAE, and the decoder part of the SDAE is formed by symmetry and tied weights. The weights trained on each fully-connected layers in the DAEs will be used as the initial weights for the SDAE, as they are generally better than randomly initialised. The SDAE will then perform training on the partially and randomly corrupted input using reconstruction loss. With the trained SDAE, the input data can be transformed into feature embedding vectors, and simple clustering methods like k-means can be used to perform a one-off clustering on the embedding vectors to get the initial set of cluster centers. Until then the parameter initialisation stage is done.

The parameter optimisation stage jointly optimises the fully-connected layers of the SDAE and the cluster centers towards the common objective function. The decoder part of the SDAE is discarded. The input data will pass through the SDAE encoder and produce feature embeddings. The soft cluster assignment is computed for every data point to indicate how likely they belong to each cluster. This is measured by the similarity between the feature embedding of the corresponding data point and all cluster centers. The Student's t-distribution [14] is used to calculate the probability of the soft assignment, as shown below:

$$q_{ij} = \frac{(1 + \|z_i - c_j\|^2/\alpha)^{-\frac{\alpha+1}{2}}}{\sum_{j'}(1 + \|z_i - c_{j'}\|^2/\alpha)^{-\frac{\alpha+1}{2}}} \tag{1}$$

where q_{ij} is the soft assignment probability of the ith data point belongs to the jth cluster, z_i is the feature embedding of the ith data point, c_j is the centre vector of the jth cluster, and α is set to 1 for all the experiments in the original DEC paper.

The soft assignments are predictions, the algorithm needs some guidance to correct the predictions for optimization. DEC uses an auxiliary target distribution computed as below:

$$p_{ij} = \frac{q_{ij}^2/f_j}{\sum_{j'} q_{ij'}^2/f_{j'}}, \quad \text{and} \quad f_j = \sum_i q_{ij} \tag{2}$$

where p_{ij} is the probability of the ith data point and jth cluster in the auxiliary target distribution, q_{ij} is the soft assignment probability for the ith data point and jth cluster, f_j is the frequency of the jth soft cluster.

The auxiliary target distribution was chosen with the assumption that the high confidence soft assignments are mostly correct when checking against the class labels, and the authors of DEC also verified this assumption with experiments. The square over soft assignment probabilities makes low confidence predictions even lower, which makes the high confidence predictions more dominating. The denominator serves as a normalization over all the clusters to prevent the feature space to be distorted by large clusters. So this auxiliary target distribution can be thought as the soft assignment distribution with noise reduced and normalized.

DEC uses KL divergence between the auxiliary target distribution and soft assignment distribution as the loss function for training, as shown below:

$$L = \mathbf{KL}(P\|Q) = \sum_i \sum_j p_{ij} \log \left(\frac{p_{ij}}{q_{ij}} \right)$$ (3)

Then backpropagation is performed to update the weights in the SDAE encoder and cluster centers.

DEC already initialized the SDAE weights and performed k-means clustering to initialise the cluster centers at the previous stage, so even at the initial state the clustering should be more or less reasonable. That's why DEC choose to conduct optimization using the KL divergence between the auxiliary target distribution and the soft assignment distribution. The idea is to strengthen the high confident predictions that are made based on a good initial state, and adjust the parameters to produce embeddings and cluster centre estimations that can further strengthen the predictions.

3.2   The DEC-RPP

As pointed by the authors in IDEC [11], by only focusing on the clustering loss, the DEC algorithm may distort the feature space, which in turn hurts clustering performance. IDEC's approach to preserve local structures in data is by keeping the autoencoder's reconstruction loss as part of the total loss in training, however the effect is limited as DEC already trained the autoencoder using reconstruction loss in its parameter initialisation stage. In this article we use random projection to preserve local structures in data.

Random projection is a technique usually used for dimensionality reduction. One important property of random projection is that it approximately preserves pairwise distance between any pair of data, stated in [15], according to Johnson-Lindenstrauss lemma [16]. Random projection matrix is an orthogonal matrix which will only do rotation and scaling when multiplied with another matrix, so that the pairwise distance is preserved. The authors in [17] used this property to learn embeddings through autoencoders that can preserve pairwise distance in data space. To our best knowledge, we are the first to introduce this technique in a clustering problem and our experimental results showed its effectiveness.

In DEC, there are three loss functions: the layer-wise pre-training of the DAEs use Mean Squared Error (MSE) to compute the reconstruction loss, the pre-training of the SDAE also uses MSE to compute the reconstruction loss, the training of the DEC uses KL divergence as the loss function. In our method, the random projection penalty term is added to all of those loss functions at their training stage, with the mathematical form:

$$\alpha\|ZA - X\|^2$$ (4)

where α is the adaptive weight, Z is the embedding vectors, A is the random projection matrix, X is the input vectors. This term will penalise the training loss if data points are dissimilar to the random projection of their corresponding feature embeddings. The random projection matrix will be re-initialised at the beginning of every epoch, in order to increase the generalisation and reduce the chance of overfitting. The random projection matrix by nature is orthogonal, and it is normalised to become orthonormal. The penalty term is weighted by an adaptive hyper-parameter α, such that the weight of the penalty term is small at the beginning of the training, and increase along the way. This is to let the main loss term (MSE or KL divergence) dominates and stabilise at the beginning, to prevent the learning goes wild due to the re-initialisation of the random projection matrix at every epoch. No scaling is needed for random projection that projects from low-dimension to high-dimension. However, if the random projection is from high-dimensional to low-dimensional, a scaling is required according to [15]:

$$\sqrt{d/k}\|Ax_1 - Ax_2\| \tag{5}$$

where $\sqrt{d/k}$ is the scaling factor, d is the dimension of the higher-dimensional space, and k is the dimension of the lower-dimensional space. In our method, we follow this scaling rule according to the dimensionality of the input layers and the embedding layers.

4   Experiment Results

4.1   Datasets

We ran experiments on four datasets to validate the effectiveness of our method on clustering:

- **MNIST**: the dataset [18] that contains 70,000 images of handwritten digits at 28×28 resolution. There are 10 classes in the dataset.
- **STL10**: the dataset [19] contains 13,000 labelled data for 10 classes, also contains 100,000 unlabelled images. Image resolution is 96×96.
- **USPS**: the dataset [20] contains 9,298 images of handwritten digits, similar to MNIST, with resolution of 16×16. There are 10 classes in the dataset.
- **REUTERS-10k**: the REUTERS dataset [21] contains 810,000 news stories in English labelled with categories. The REUTERS-10k dataset is sampled randomly from the REUTERS dataset with 10,000 data points.

We compared the proposed method against the original DEC [1] and its two variants IDEC [11] and GrDNFCS [12] on those four datasets on clustering accuracy (Table 1), normalised mutual information (Table 2), and adjusted rand index (Table 3). The proposed method significantly outperforms all the other methods in all metrics, except marginally worse than GrDNFCS on MNIST NMI.

Table 1. Comparison of clustering accuracy on benchmark datasets (in %).

	MNIST	REUTERS-10k	STL-10	USPS
K-means	53.48	54.17	74.21	66.79
DEC	86.53	70.74	89.44	72.78
IDEC	88.01	72.42	90.05	75.13
GrDNFCS	91.45	77.84	91.43	76.52
Ours	**95.00**	**83.96**	**91.95**	**88.54**

Table 2. Comparison of normalized mutual information (NMI) on benchmark datasets (in %).

	MNIST	REUTERS-10k	STL-10	USPS
K-means	49.99	34.69	72.34	62.56
DEC	83.69	45.31	82.01	73.52
IDEC	86.38	46.56	82.85	75.95
GrDNFCS	**90.74**	52.67	84.10	77.61
Ours	88.36	**64.80**	**84.80**	**79.92**

Table 3. Comparison of adjusted rand index (ARI) on benchmark datasets (in %).

	MNIST	REUTERS-10k	STL-10	USPS
K-means	36.67	27.39	52.97	54.50
DEC	80.29	42.46	78.26	66.22
IDEC	83.25	45.20	79.74	67.91
GrDNFCS	86.26	55.47	82.52	69.03
Ours	**89.37**	**65.67**	**83.31**	**80.46**

5 Conclusion

We proposed an improvement over DEC by adding a random projection penalty term in the loss functions to encourage the preservation of local structures in data during feature representation learning, and the experiment results show significant improvements in clustering accuracy, NMI, and ARI on four benchmark datasets. It also shows that a good feature representation should not only be low-dimensional and preserves key information, it should also preserve local structure or pairwise similarity as in the high-dimensional data space, this will make the distance metric easier to compute and thus more beneficial for the decision making of cluster assignments.

Some possible future work may include the application of this technique to CNN or transformer based feature extractors, multi-modality networks, or extension to fuzzy clustering and co-clustering.

References

1. Xie, J., Girshick, R., Farhadi, A.: Unsupervised deep embedding for clustering analysis. In: International Conference on Machine Learning, pp. 478–487 (2016)
2. MacQueen, J.: Some methods for classification and analysis of multivariate observations. In: Proceedings of the Fifth Berkeley Symposium on Mathematical Statistics and Probability, pp. 281–297 (1967)
3. Ester, M., et al.: A density-based algorithm for discovering clusters in large spatial databases with noise. Kdd, pp. 226–231 (1996)
4. Cheng, Y.: Mean shift, mode seeking, and clustering. IEEE Trans. Pattern Anal. Mach. Intel. **17**, 790–799 (1995)
5. Comaniciu, D., Meer, P.: Mean shift: a robust approach toward feature space analysis. IEEE Trans. Pattern Anal. Mach. Intel. **24**, 603–619 (2002)
6. Xu, R., Wunsch, D.: Clustering algorithms in biomedical research: a review. IEEE Rev. Biomed. Eng. **3**, 120–154 (2010)
7. Banfield, J., Raftery, A.: Model-based Gaussian and non-Gaussian clustering. Biometrics 803–821 (1993)
8. McLachlan, G., Krishnan, T.: The EM algorithm and extensions. Wiley (2007)
9. Ng, A., Jordan, M., Weiss, Y.: On spectral clustering: analysis and an algorithm. Adv. Neural Inf. Process. Syst. 849–856 (2002)
10. Von Luxburg, U.: A tutorial on spectral clustering. Stat. Comput. **17**, 395–416 (2007)
11. Guo, X., et al.: Improved deep embedded clustering with local structure preservation. In: IJCAI, pp. 1753–1759 (2017)
12. Feng, Q., et al.: Deep fuzzy clustering-a representation learning approach. IEEE Trans. Fuzzy Syst. **28**, 1420–1433 (2020)
13. Vincent, P., et al.: Stacked denoising autoencoders: learning useful representations in a deep network with a local denoising criterion. J. Mach. Learn. Res. **11** (2010)
14. Van der Maaten, L., Hinton, G.: Visualizing data using t-SNE. J. Mach. Learn. Res. **9** (2008)
15. Bingham, E., Mannila, H.: Random projection in dimensionality reduction: applications to image and text data. In: Proceedings of the Seventh ACM SIGKDD International Conference on Knowledge Discovery and Data Mining, pp. 245–250 (2001)
16. Johnson, W.: Extensions of Lipschitz mappings into a Hilbert space. Contemp. Math. **26**, 189–206 (1984)
17. Kasun, L., et al.: Distance preserving autoencoders for learning deep distance embedding. Rev. Trans. Neural Netw. Learn. Syst.
18. LeCun, Y., et al.: Gradient-based learning applied to document recognition. Proc. IEEE **86**, 2278–2324 (1998)
19. Coates, A., Ng, A., Lee, H.: An analysis of single-layer networks in unsupervised feature learning. In: Proceedings of the Fourteenth International Conference on Artificial Intelligence and Statistics, JMLR Workshop and Conference Proceedings, pp. 215–223 (2011)
20. Hull, J.: A database for handwritten text recognition research. IEEE Trans. Pattern Anal. Mach. Intel. **16**, 550–554 (1994)
21. Lewis, D., et al.: Rcv1: a new benchmark collection for text categorization research. J. Mach. Learn. Res. **5**, 361–397 (2004)

Multi-title Attention Mechanism to Generate High-Quality Images on AttnGAN

Pingan Qiao and Xiwang Gao[⊠]

Xi'an University of Posts and Telecommunications, Xi'an 710121, China
paqiao@xupt.edu.cn, gaoxiwang0805@163.com

Abstract. In the field of Text-to-image, text is essentially a constraint condition for the generated image, and the generation network guides to generate images that match the text according to the constraint conditions. However, if the image is generated only on the basis of a given text constraint condition, obviously, it can be imagined that the generated image without rich details, reducing the image visualization. With that in mind, we introduce Multi-title Attention Mechanism, regard the dataset as a prior condition, at first, select other titles in the dataset that are compatible with the given text according to given title, which is essentially the process of information retrieval, and then use the self-attention mechanism to integrate the embedding of multiple titles, the final text contains rich detail information, which guides the generation of high-quality images. In addition, in order to enable AttnGAN to generate clear image in the first stage, we introduce a mixed attention mechansim and an Residual Dense Block(RDB) model. The mixed attention mechanism includes: channel attention and pixel attention. Channel attention is mainly to guide what the image is generate, while pixel attention is responsible for where it is generated. Experiments on the CUB dataset show that the proposed approaches is significantly better than AttnGAN, and the Inception Score(IS) and R-precision of the evaluation index are improved by 4.12% and 10.43% respectively.

Keywords: Text-to-image · Muti-title attention · Mixed attention · Residual dense block

1 Introduction

The emergence of GAN [1] has made great progress in image generation tasks, and some interesting fields have emerged, such as: image restoration [2], image super definition [3], face synthesis [4], sketch coloring [5], etc. But in recent years, with the increasingly strict requirements for image generation, people want to achieve on-demand generation, so text-to-image has been researched.

Text-to-image is a combination of natural language and computer vision. Its includes two main tasks: mapping from text to vision and generating images that highly match text. T2I uses GAN as the basic framework, and has made a lot of contributions in image processing by integrating various technologies. T2I can design more attractive game

© The Author(s), under exclusive license to Springer Nature Switzerland AG 2023
N. Xiong et al. (Eds.): ICNC-FSKD 2022, LNDECT 153, pp. 147–156, 2023.
https://doi.org/10.1007/978-3-031-20738-9_18

renderings according to people's needs, reduce a lot of manpower and material resources, and greatly improve the work efficiency of game animation designers; according to the description of eyewitnesses, the computer will automatically extract the key words in the text information, and generate highly corresponding portraits of characters, which greatly improves the speed of solving cases; it can also automatically match illustrations and advertisements according to the needs of illustrators and advertising designers, reducing the pressure on workers.

At present, many scholars have also made great contributions to the field of text generation images, but there are still many difficulties and challenges in the field of T2I research. For example, since a single text provides less information, it will cause the generated image without enough detailed features, which reduces the quality and visualization of the image.

This paper makes the following key contributions:

- We integrate a multi-title attention network. According to a given single text, we retrieve the text compatible with the given text in the dataset, refine the title to get the best complement, and then fuse all compatible texts through the self-attention mechanism. The final text input into the generator has rich details.
- We embed a mixed attention mechanism, which extracts features in the two directions of channel attention and pixel attention, so that the image can generate a clearer image at the first stage.

2 Related Work

The method of text generation image, the commonly used method is to encode the entire text as a global sentence vector as a condition for GAN-based image generation [6–9]. Subsequently, Mansimov et al. established the align DRAW model, which extended the Deep Recurrent Attention Writer (DRAW) [10] to draw the details of the image in an iterative manner, while paying attention to the relevant words in the title. After DRAW, Nguyen et al. proposed an approximate Langevin method to generate images from subtitles [11], and Reed et al. used conditional PixelCNN [12] to synthesize images from text using a multi-scale model structure. Their follow-up work [12] also proved that GAN can generate better samples by combining other conditions. Zhang et al. [8, 13] stacked multiple GANs (StackGAN, StackGAN++) to synthesize text to image. The training process is divided into two stages. The first stage is based on the input text information, extracting the background, The rough information of color and contour is generated to generate a low-resolution (64 * 64) image. The second stage is to take the low-resolution image and text output in the first stage as input, and extract the details lost when the image is generated in the first stage. Generate high-resolution (256 * 256) images with rich details. However, they are all conditional on the global sentence vector and lack the fine-grained word-level information used to generate images.

In 2017, Zhang et al. proposed AttnGAN [14], which added the Attention attention mechanism to the generative confrontation network model. It not only uses the sentence features of the text as a constraint when the image is generated by the network, but also can extract the word level of the text. Fine-grained features are also input into the network as constraints. According to the weight of each word, the generator and discriminator are accurately optimized for the word part each time, and then generate more refined images, but sometimes generate unnatural local semantic details This is due to the inefficiency of convolutional neural networks in capturing high-level semantic information for pixel-level image synthesis.

In view of the fact that text-to-image is a popular crossing field, there is still much progress for improvement in the research and optimization of this problem. In summary, text-generated images have very good research value and application prospects.

3 Data

This paper uses the Caltech UCSD Birds dataset to establish an image library for text-generated images. The specific establishment standards are as Table 1:

(a) Quantity composition: a total of 200 categories, 11788 bird images, of which 6805 images are used for the training set and 4983 images are used for the test set.
(b) Resolution: 256×256
(c) Image text content: Each image provides 15 local area locations, 1 label box, and semantic level segmentation images.

Table 1. Statistic of datasets

Dataset	CUB	
	Train	Test
#samples	6805	4983
Caption/image	10	10

4 Methods

4.1 The Structure of the Network

The network architecture proposed in this paper is shown in Fig. 1. The network architecture mainly consists of three parts: multi-title fusion algorithm, using mixed attention to guide the generation of the first-stage image and edge enhancement deep network.

Firstly, according to a given single text, use a multi-title fusion algorithm to retrieve other texts in the dataset that are compatible with the given text. Through further refinement, the best complement of the given text is obtained, so that the text can be supplemented in many aspects, and then introduced Self-attention mechanism, fusion and embedding of mutually compatible texts, so that the text obtained by the generator contains sufficiently rich and detailed text features, provides more constraints for the generated image, and better guides the generator to generate the image.

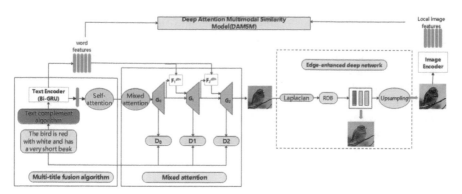

Fig. 1. The structure of the network

Secondly, after obtaining rich text semantics, it uses a hybrid attention mechanism to extract features from the text in two ways. Channel attention is to analyze what is generated by the image based on text information, and pixel attention is to analyze where the image is generated, so that sufficient text information can be obtained in the first stage of the image to generate a high-quality image. Finally, the edge enhancement network is introduced to refine the edges of the image, and generate images with clear edges, which improves the authenticity and visualization of the image.

4.2 Multi-title Fusion Algorithm

Text complement algorithm. After a text is given, the generator is not directly instructed to generate an image, but the text is supplemented first to obtain the best complement of the text, so that the final text has all aspects of information, and solves the problem of limited information. Improve the quality of the composite image, as shown in Fig. 2.

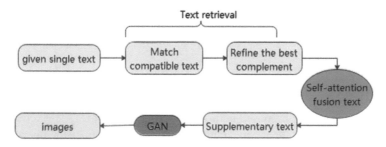

Fig. 2. The process of text complement

In the process of title matching, because an image may correspond to multiple texts in a common dataset, the title of each image and its training part should be taken as an item in the knowledge base. Then compatible texts are retrieved from the knowledge base and the best text complement is extracted. The process of text matching is actually an information retrieval problem, that is, retrieving relevant texts from training dataset.

The dataset is regarded as prior knowledge, namely knowledge base $\Omega = \{\omega_i\}$, where each item ω_i is composed of image I_i and its title $\{t_{ik}\}_{k=0}^{N^T}$. Given a titile t and a ω_i, RE2[15] to evaluate the score for this title and each item, the score is calculated as follows:

$$S(t, \omega_i) = \frac{1}{N^T} \sum_{k=0}^{N^T} S_{RE2}(t, t_{ik}) \qquad (1)$$

where, S_{RE2} is used to calculate the matching score of title t and t_{ik}.

Since the title $\{t_{ik}\}_{k=0}^{N^T} = 0$ of each item ω_i also describes the image I_i, these titles are compatible with each other. Construct a positive sample (t_i, w_{ic}) by randomly selecting a heading t_i from the knowledge base and the rest of the item as context ω_{ic}. Since titles of different categories may conflict with each other, negative samples $(t_{r(i)}, \omega_{ic})$ are constructed by $\omega_{r(i)}$ and titles $t_{r(i)}$. The training loss function has the following form:

$$L_1 = -\frac{1}{N^T} \sum_{i=0}^{N^T} \sigma(S(t_i, \omega_{ic})) + \sigma(S(t_{r(i)}, \omega_{ic})) \qquad (2)$$

where, σ is a sigmoid function. Given title t, K optimal candidate titles can be obtained from knowledge base Ω and represented by $\Omega_K(t)$. In order to improve semantic consistency and further eliminate conflicting titles, N^{test} titles whose cosine similarity is closer to title t are selected to achieve the purpose of refinement.

Multi-title attention generation network. The structure of Multi-title attention generation network as Fig. 3.

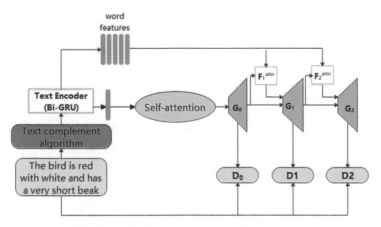

Fig. 3. Multi-title attention generation networks

Given a set of titles $T = \{t_j\}_j^{N_T} = 0$, use text editors f_{word}^{txt} and f_{cap}^{txt} as Bi-GRU to extract the word features and sentence features of t. In the first stage, the up-sampling module takes the total feature c of the self-attention mechanism as input to calculate the internal feature h1 and synthesize the image. In the next stage, the attention model F_i^{attn} takes the word feature and h_1 as input to obtain the attention feature of each title, and then uses the self-attention mechanism to calculate the total attention feature ω_1^j. The third stage is similar to the second stage.

The self-attention mechanism to fuse the embedding of the title. Given the hidden state h_i generated by the upsampling module, for each t_i, the corresponding embedding is calculated as follows:

$$c^j = F_{ca}\left(f_{cap}^{txt}\left(t_j\right)\right) \tag{3}$$

$$\omega_i^j = F_i^{attn}\left(h_i, f_{word}^{txt}\left(t_j\right)\right) \tag{4}$$

In order to extract the embedding of the title T, the self-attention module is used to fuse all the embeddings of the title for fusion:

$$c = f_{max}\left(f_{posw}\left(L_{MHA}\left(\left[c^0, ..., c^{N_T}\right]\right)\right)\right) \tag{5}$$

$$\omega_i = f_{max}\left(f_{posw}\left(L_{MHA}\left(\left[\omega_i^0, ..., \omega_i^{N_T}\right]\right)\right)\right) \tag{6}$$

Among them, the $f_{max}(x)$ function inputs a tensor, each element in the tensor is the maximum value in each column of x, f_{posw} is the position feedforward network, and L_{MHA} is the attention layer.

Mixed Attention. After adopting the self-attention mechanism in 3.2.2, the fusion of images and text is required, and the upsampling operation is used to guide the generation of images according to the constraints of the text. The upsampling part includes the convolutional layer, the residual layer, and the fully connected Layer, channel attention layer and pixel attention layer. This paper uses a mixed attention mechanism to extract features from different channels of text. The mixed attention mechanism model is shown in Fig. 4.

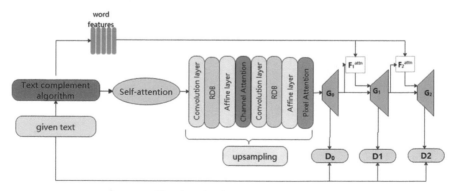

Fig. 4. Mixed attention model

In this stage, the text features are first extracted through the convolutional layer. Since the local convolutional layer cannot directly access the subsequent layers, it is difficult to make full use of the information of all layers inside. Therefore, the residual dense block (RDB) is introduced to make full use of the features of all layers. The output result of each RDB module, the output of an RDB module connected to it, and the information between all convolutional layers ensure the continuous storage and memory of the information. Each convolutional layer in the RDB can directly access all subsequent layers and transfer the information that needs to be saved. The densely connected layer also effectively alleviates the problem of gradient disappearance, strengthens feature propagation, encourages feature reuse, and reduces the amount of parameters. Then through the affine layer (that is, the fully connected layer), the final expected output is made, and the text and image are initially fused. Even through the first three layers of feature extraction and the fusion of graphics and text, the images and text information cannot be fully integrated. Therefore, the two attention mechanisms are deeply fused after the affine layer.

5 Loss

Based on the generative adversarial mechanism, the adversarial loss functions of generator and discriminator are used to constrain the min-maximum problem in the network to realize the optimal solution. The antagonistic loss function is expressed as:

$$l_{adv\_G}(X, Y) = -E_{X \sim p_{train}(X)}\big[\log D(G(X))\big] \tag{20}$$

$$l_{adv\_D}(X, Y) = E_{Y \sim p_{train}(Y)}\big[\log D(Y)\big] + E_{X \sim p_{train}(X)}\big[\log 1 - D(G(x))\big] \tag{21}$$

In this method, pixel loss, perceived loss and adversarial loss are combined by linear method to obtain the weighted sum loss function as the global loss function. To sum up, the objective function of the training network is expressed as:

$$l_G(X, Y) = l_{adv\_G}(X, Y) + \alpha l_{pixel} + \beta l_{feature} \tag{22}$$

$$l_D(X, Y) = -l_{adv\_D}(X, Y) + \beta l_{feature} \tag{23}$$

Then, β is the linear combination weight of the corresponding loss function and in the objective function.

6 Experiments

6.1 Evaluation Index

In this paper, Inception Score (IS) and R-Precision are used to evaluate the improved network performance of text.

Inception Score. IS calculated conditional distribution p (y | x) and marginal distribution p (y) KL divergence. If the model generation image real and diversity, then p (y | x) and p (y) between KL divergence will be bigger, the bigger the value IS also, means that generate the image quality IS higher, and the generated image belongs to a

Table 2. Comparison of IS values between different models

Model	Inception Score		
	Epoch0	Epoch50	Epoch100
Ground truth	11.63	–	–
AttnGAN	1.03	5.28	5.86
Ours	1.07	6.49	7.09

particular class. After implementation verification, as shown in Table 2, we observed that with nearly 100 training cycles, the IS values of AttnGAN and the model in this paper were displayed on the CUB data set, which indicated that the performance of the model in this paper was better than AttnGAN in image generation.

R-precision. Due to the IS reflect that the generated image matches the given text description one-sidely. So we also use R-Precision to evaluate the visual semantic similarity between the generated image and its corresponding text description. Table 3 shows the R precision values of AttnGAN and our model on the CUB dataset.

Table 3. Comparison of R-precision values between different models

Dataset	CUB		
Top-k	k=1	k=2	k=3
AttnGAN	57.58	58.48	60.21
Ours	63.38	64.32	66.23

As can be seen from Table 3, the score of ours is higher than AttnGAN. A higher score indicates a higher visual semantic similarity between the generated image and the input text.

6.2 Comparison of Experimental Results

Figure 5 shows a subjective visual comparison of AttnGAN and the generated images presented in this article. As can be seen from the figure, the images generated by AttnGAN (row 3, column 2) produce blurred images and lack of visual details. This phenomenon is caused by the fact that clear images are not generated in the first stage and high-resolution images cannot be optimized in the later generation stage. In addition, no clear background is generated in the figure (row 3, column 3 and row 3, column 5). This is because no complement of the given text is obtained in AttnGAN, and the image generated by the method in this paper (row 2, column 5), It can retrieve the best complement of a given text and extract various features of the text by introducing channel attention and pixel attention to ensure that it can fully learn the text features of multiple titles, so as to generate images with clearer background and better visual performance. Through

Input	a yellow bird with brown and white wings and a pointed bill.	this bird is blue and black in color,with a sharp black beak.	a small bird with a red belly,and a small bull and red wings.	a small bird with orange underbelly and black head and a small beak.
Ours				
AttnGAN				

Fig. 5. The generated results of AttnGAN and ours

experimental comparison, it is concluded that the images generated by our improved model are clearer and more realistic than those generated by AttnGAN.

7 Conclusion

In order to alleviate the single text provides limited information, we combine the muti-title fusion model, by retrieving dataset that is compatible with the given text, to select the best of texts, allows the generator to get enough rich text characteristic, more detail information to guide to generate high quality images; In addition, we extract text features at different levels by introducing channel attention and pixel attention. Channel attention mainly focuses on the features of the image corresponding to the text description that need to be paid attention to, and pixel attention made up for the inadequacy of channel attention capture image pixel space, positioning text description corresponding to the location of the image space, generate more detail image characteristics, make the image in the first stage can generate relatively clear images. Finally, experiments on CUB dataset verify that the quality of our model is better than AttnGAN.

Acknowledgment. This work was supported by the National Natural Science Foundation of China(Item NO:61105064),Scientific Research Project of Shaanxi Provincial Department of Education(Item NO:16JK1689),and the Key Laboratory of Network Data Analysis of Shaanxi Province. And I would like to reviewers for reading this paper, and thank my tutor and partners for their help during the process of my experiment. Especially thank the key laboratory for providing me with a good learning environment and experimental support condition.

References

1. Goodfellow, I., Pouget-Abadie, J., Mirza, M., Xu, B., Warde-Farley, D., Ozair, S., Courville, A., Bengio, Y.: Generative adversarial nets. In NIPS (2014)
2. Chen, Z., Nagy, J.G., Xi, Y., Yu, B.: Structured FISTA for Image Restoration. In EI (2020)
3. Tian, C., Zhuge, R., Wu, Z., Xu, Y.: Lightweight Image Super-Resolution with Enhanced CNN. In EI (2020)
4. Di, X., Patel, V.M.: Facial synthesis from visual attributes via sketch using multiscale generators. IEEE Trans Biometrics Behav Identity Sci. no. 1, pp. 55–67 (2020)
5. Jo, Y., Park, J.: SC-FEGAN—face editing generative adversarial network with user's sketch and color. ICCV, pp. 1745–1753 (2019)
6. Reed, S., Akata, Z., Yan, X., Logeswaran, L., Schiele, B., Lee, H.: Generative adversarial text-to-image synthesis. In ICML (2016)
7. Reed, S., Akata, Z., Mohan, S., Tenka, S., Schiele, B., Lee, H.: Learning what and where to draw. In NIPS (2016)
8. Zhang, H., Xu, T., Li, H., Zhang, S., Wang, X., Huang, X., Metaxas, D.N.: Stackgan: Text to photo-realistic image synthesis with stacked generative adversarial networks. In ICCV (2017)
9. Zhang, H., Xu, T., Li, H., Zhang, S., Wang, X., Huang, X., Metaxas, D.N.: Stackgan++: Realistic image synthesis with stacked generative adversarial networks. arXiv: 1710.10916 (2017)
10. Gregor, K., Danihelka, I., Graves, A., Rezende, D., Wierstra, D.: DRAW: A recurrent neural network for image generation. In ICML (2015)
11. Nguyen, A., Clune, J., Bengio, Y., Dosovitskiy, A., Yosinski, J.: Plug & play generative networks: conditional iterative generation of images in latent space. In CVPR (2017)
12. Van den Oord, A., Kalchbrenner, N., Espeholt, L., Vinyals, O., Graves, A.: Conditional image generation with pixelcnn decoders. In NIPS (2016)
13. Metaxas.: StackGAN++: Realistic image synthesis with stacked generative adversarial networks. arXiv: 1710.10916. (2017)
14. Xu, T., Zhang, P., Huang, Q., Zhang, H., Gan, Z., Huang, X., He, X.A.: AttnGAN: Fine-grained text to image generation with attentional generative adversarial networks. In CVPR, Salt Lake City, Utah (2018)
15. Xu, J., He, X., Li, H.: Deep learning for matching in search and recommendation[C]. The 41st International ACM SIGIR Conference on Research and Development in Information Retrieval. pp. 1365–1368 (2018)

Hyperspectral Remote Sensing Images Terrain Classification Based on LDA and 2D-CNN

Jing Liu[1][(✉)] [iD], Yang Li[1] [iD], Meiyi Wu[1] [iD], and Yi Liu[2] [iD]

[1] School of Electronic Engineering, Xi'an University of Posts and Telecommunications, Xi'an 710121, China
zyhalj1975@163.com, jingliu@xupt.edu.cn, {lyaa981030, 1193544209}@stu.xupt.edu.cn
[2] School of Electronic Engineering, Xidian University, Xi'an 710071, China
yiliu@xidian.edu.cn

Abstract. Hyperspectral remote sensing images (HRSIs) are characterized by high dimensionality and large data volume, but there is a large amount of redundant information between spectral dimensions, which not only affects the accuracy of terrain classification, but also increases the complexity of classification and recognition. In order to reduce the redundant information between spectral bands, reduce the computational complexity, and improve the efficiency of terrain classification of HRSIs, this paper firstly performs dimensionality reduction on HRSIs using linear discriminant analysis (LDA), projects the data to a lower dimensional feature subspace, extracts the most discriminative information, and then uses two-dimensional convolutional neural network (2D-CNN) for deep feature extraction and classification recognition. By classifying on three real HRSIs dataset, the experimental results show that the classification results of LDA-2D-CNN outperform those of the PCA-2D-CNN method using 2D-CNN combined with principal component analysis (PCA).

Keywords: Feature extraction · Convolutional neural network · Linear discriminant analysis · Hyperspectral remote sensing images

1 Introduction

Hyperspectral remote sensing images (HRSIs) have abundant spectral and spatial information of terrains, and these two kinds of information can be used to represent the targets and distribution states of terrains with different attributes, which can be used to classify terrains based on spectral and spatial information [1]. HRSIs are remote sensing images obtained by imaging spectrometer with many wavebands and large data volume, but the number of samples involved in training is usually smaller than the number of sample dimensions compared with the number of high-dimensional wavebands, resulting in the "Hughes" phenomenon [2], and the statistical parameters of samples will increase as the data volume increases [3]. The high correlation, high redundancy and strong noise among the bands in HRSIs lead to poor data separability and affect the accuracy of terrain classification [4]. Therefore, the spectral analysis of HRSIs is needed

© The Author(s), under exclusive license to Springer Nature Switzerland AG 2023
N. Xiong et al. (Eds.): ICNC-FSKD 2022, LNDECT 153, pp. 157–164, 2023.
https://doi.org/10.1007/978-3-031-20738-9_19

to reduce redundancy and noise to enhance data separability. For the above problems, feature extraction can be used to reduce dimensionality, extract effective features, reduce computational complexity, enhance data separability, improve classification efficiency, extract low-dimensional feature components of data while retaining effective information of high-dimensional data to the maximum extent, and classify and distinguish in the low-dimensional subspace [5].

In order to reduce the redundant information of HRSIs, reduce the computational complexity, and further extract deep features that are more conducive to classification, this paper proposes a HRSIs feature classification method combining linear discriminant analysis (LDA) and two-dimensional convolutional neural network (2D-CNN), denoted as LDA-2D-CNN. The LDA-2D-CNN method first uses LDA for feature extraction of HRSIs, projects the original data into a low-dimensional subspace to reduce the data dimensionality and computational complexity, then uses 2D-CNN to perform deep feature extraction on the reduced dimension data, and finally performs feature classification by softmax logistic regression layer.

2 LDA

LDA is widely used in face recognition, voice recognition, text retrieval, and has also achieved good results in terrain classification of HRSIs [6, 7]. LDA aims to find an optimal projection matrix through which the original data is projected into a low-dimensional subspace, enhancing the separability of data in the subspace [8, 9].

Suppose that the n-dimensional original space $\mathbf{X} = [\mathbf{x}_1, \mathbf{x}_2, \cdots, \mathbf{x}_u]$, there are u training samples, $\mathbf{x}_i \in \mathbf{R}^n$, $i = 1, 2, \cdots, u$, and the number of classes of the samples is C. The original data can be projected to the d-dimensional subspace $\mathbf{R}^d$ by the projection matrix A, $d < n$, and the projected sample set Y is obtained as

$$\mathbf{Y} = \mathbf{A}^\mathsf{T}\mathbf{X}, \tag{1}$$

the samples are projected into the low-dimensional subspace through the matrix A by maximizing the Fisher's criterion function, which is expressed as

$$\mathbf{J}(\mathbf{A}) = \max_{\mathbf{A}}\left\{ Tr\left[\left(\mathbf{A}^\mathsf{T}\mathbf{S}_w\mathbf{A}\right)^{-1}\left(\mathbf{A}^\mathsf{T}\mathbf{S}_b\mathbf{A}\right) \right] \right\}, \tag{2}$$

where $\mathbf{S}_b$ and $\mathbf{S}_w$ represent the between-class scatter matrix and the within-class scatter matrix of original data [10], respectively. In the process of feature extraction and classification using LDA, $\mathbf{S}_b$ and $\mathbf{S}_w$ are first calculated as

$$\mathbf{S}_b = \sum_{i=1}^{C} P_i(\mathbf{m}_i - \mathbf{m})(\mathbf{m}_i - \mathbf{m})^\mathsf{T}, \tag{3}$$

$$\mathbf{S}_w = \sum_{i=1}^{C} \frac{P_i}{u_i} \sum_{j=1}^{u_i} \left(\mathbf{x}_j^i - \mathbf{m}_i\right)\left(\mathbf{x}_j^i - \mathbf{m}_i\right)^\mathsf{T}, \tag{4}$$

where P_i and u_i represent the prior probability and the number of samples of class i, respectively. $\boldsymbol{m}_i$ denotes the mean of the samples of class i, $\mathbf{x}_j^i$ denotes the j-th sample of class i, and $\boldsymbol{m}$ denotes the overall mean vector of samples.

Therefore, when the dimension of the LDA subspace is d, the projection matrix A can be obtained by calculating the eigenvectors corresponding to the first d largest eigenvalues of $\mathbf{S}_w^{-1}\mathbf{S}_b$. Finally, the sample set Y projected to the subspace is obtained by Eq. (1).

3 LDA-2D-CNN Model

The proposed LDA-2D-CNN first uses LDA to extract the effective classification features in the original HRSIs to reduce the spectral dimensionality and computational complexity, and then the dimensionality-reduced data are used to extract the depth feature and complete the terrain classification of HRSIs using 2D-CNN.

After spectral feature extraction of HRSIs through LDA, the dimension is reduced to d-dimensional subspace, and the input image of 2D-CNN can be obtained. The neighborhood space pixel block $\boldsymbol{b}(k) \in \boldsymbol{R}^{r \times r \times d}$ of the k-th pixel is fed into 2D-CNN, where r is the size of the block, and d is the number of bands after spectral dimensionality reduction. The convolutional and pooling calculation in 2D-CNN are two-dimensional, according to Eq. (5), it can be derived that after two-dimensional convolution, the v-th feature map of the q-th layer is

$$\mathbf{x}_v^q = \sum_a f\left(\mathbf{w}_{v,a}^q * \mathbf{x}_a^{q-1} + b_v^q\right), \tag{5}$$

where $\mathbf{x}_a^{q-1} \in \mathbf{R}^{p \times p}$ is the a-th feature map extracted from layer $(q-1)$ in the network, with the size of $p \times p$, and connected to the v-th feature map of the convolutional layer q. $\mathbf{w}_{v,a}^q \in \mathbf{R}^{w \times w}$ is the convolutional kernel of $\mathbf{x}_a^{q-1}$, "$*$" represents the convolutional operation, b_v^q is the bias, and the size of $\mathbf{x}_v^q$ is $(p-w+1) \times (p-w+1)$. The pooling layer connected after the convolutional layer obtains invariance to rotation, translation and other operations by reducing the resolution of the convolution layer output feature map. In the experiment, the max-pooling is used to obtain the maximum value of the pooling area and reduce the resolution of the convolution layer output feature map.

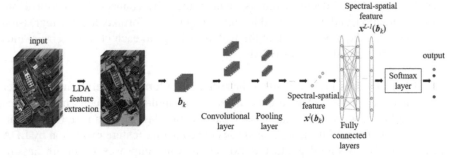

Fig. 1. LDA-2D-CNN model structure

After convolution and max-pooling, the extracted spectral-spatial feature $\mathbf{x}^l(\mathbf{b}_k)$ can be obtained. $\mathbf{x}^l(\mathbf{b}_k)$ is the output after passing the l-th convolutional layer and pooling layer, which can be considered as the result of deep feature extraction. Input the obtained feature into the fully connected layer, the output can be obtained as

$$\mathbf{x}^l(\mathbf{b}_k) = f\left\{\mathbf{W}^1 \cdot \mathbf{x}^l(\mathbf{b}_k) + \mathbf{b}^1\right\}, \tag{6}$$

where $\mathbf{W}^1$ and $\mathbf{b}^1$ are the weight matrix and bias of the feature passing through the first fully connected layer respectively. After further feature extraction of multiple fully connected layers, the input samples can be predicted through the softmax layer to complete the classification. LDA dimensionality reduction of HRSIs can project original data to low-dimensional subspace, so as to maximize the separability of data in low-dimensional subspace, extract the separability features of original HRSIs, reduce the computational complexity, and use 2D-CNN for depth feature extraction on the data after LDA dimensionality reduction, which can extract deeper separable features and improve the classification and recognition effect of HRSIs. The structure of LDA-2D-CNN model is shown in Fig. 1. and the steps of HRSIs terrain classification by LDA-2D-CNN are as follows.

- Step 1. According to the provided sample set X, the within-class scatter matrix $\mathbf{S}_w$ and between-class scatter matrix $\mathbf{S}_b$ of the dataset are calculated by Eqs. (3) and (4).
- Step 2. The optimal projection matrix A is obtained by Eq. (2), and the matrix A can make the samples have the largest between-class scatter matrix and the smallest within-class scatter matrix in the projected subspace. The original HRSIs data can be projected into the d-dimensional feature subspace through the projection matrix A and obtain the projected sample set Y, and $\mathbf{Y} = \mathbf{A}^\mathrm{T}\mathbf{X}$.
- Step 3. The target pixel in the image after LDA feature extraction is taken as the center point, and the sample set is obtained by taking blocks of the neighborhood space of the target pixel, each of which contains the spectral and spatial information of the target pixel.
- Step 4. The obtained sample set is fed into the 2D-CNN model proposed in Fig. 1. and pass through the network layers such as convolutional layer and pooling layer. The various parameters of the network are adjusted through several experiments to build out advanced features and perform deeper feature extraction.
- Step 5. The obtained depth feature is input to the fully connected layer for depth feature extraction, and the extracted feature is input to the softmax logistic regression layer to predict the probability that the feature belongs to each class of terrain, thus completing the classification of HRSIs.

The LDA-2D-CNN method reduces the redundant information between bands after feature extraction of original HRSIs data, increases the similarity between similar classes and separability between dissimilar classes, extracts the classification features containing more discriminative information, and uses the data after feature extraction by LDA as the input of 2D-CNN for depth feature extraction, which improves the terrain classification effect of HRSIs. In order to verify the feasibility of the designed LDA-2D-CNN, PCA-2D-CNN is used in this paper as a comparison method, which firstly performs dimensionality reduction and feature extraction on the original HRSIs spectral using PCA to retain the main information of the original data, and then takes the reduced data blocks as the sample set and uses 2D-CNN to further extract deep features to achieve terrain classification.

4 Network Structure and Parameter Setting

The network structure and parameters in CNN have a great influence on the experimental results, but there is no method to directly give the appropriate network structure and parameters for the required processing task from the theory. In this paper, we combine the results of repeated experiments to design a more suitable network structure and network parameters, and Table 1 shows the network structures of the proposed PCA-2D-CNN and LDA-2D-CNN models.

Table 1. PCA-2D-CNN and LDA-2D-CNN model structures

Layer name			$I1$	$C2$ $P3$	$C4$ $P5$	$F6$	$O7$
Convolutional kernel size	PU	PCA	$21 \times 21 \times 50$	3×3 2×2	3×3 2×2	–	1×9
		LDA	$21 \times 21 \times 6$				
	IP	PCA	$21 \times 21 \times 140$				1×16
		LDA	$21 \times 21 \times 15$				
	Bots	PCA	$21 \times 21 \times 40$				1×14
		LDA	$21 \times 21 \times 12$				

In Table 1, "IP", "PU" and "Bots" are abbreviations of "Indian Pines", "Pavia University" and "Botswana" dataset respectively. In addition, the input layer, convolutional layers, pooling layers, fully connected layers and output layer are represented as I, C, P, F, and O respectively in Table 1, and numbers 1 to 7 represent the position of the layer in the network. For example, $C4$ represents the convolutional layer, and this layer is the fourth layer in the whole network. When training the network, Adam is used as the optimizer, and the learning rate and batch size is set to 0.001 and 5, respectively, and the number of feature dimensions is selected according to the classification accuracy obtained after PCA and LDA dimensionality reduction in the experiment.

5 Experimental Results

In order to verify the feasibility of the method proposed above, three real HRSIs, namely Indian Pines, Botswana and Pavia University were used for the experiments. Indian Pines dataset includes 16 classes, it has 220 spectral bands. In addition, its image size is 145 × 145, with 21,025 pixels, of which 10,249 pixels are terrain pixels. The Pavia University dataset contains 9 classes of terrains, it has 103 spectral bands with an image size of 610 × 340, including 42,776 terrain pixels. The Botswana dataset has 145 spectral bands with 1476×256 image size and 14 terrain classes, of which 3248 pixels are terrain pixels.

In the experiment, for each dataset, we randomly selected 5, 10 and 20% samples of each class of terrain samples as training samples, and used these training samples to train the network, and finally compared the experimental results of PCA-2D-CNN and LDA-2D-CNN. In the experiment, the average overall accuracy (AOA) and standard deviation (SD) of the ten running results were used to evaluate the feasibility of the proposed method, and the optimal recognition results in each column were indicated using boldface. Tables 2, 3 and 4 show the AOAs and SDs for ten runs of classification for three HRSIs at 5, 10 and 20% training data.

Table 2. Pavia University dataset classification results

Ratio of training samples	5%	10%	20%
	AOA (%) ± SD (%)	AOA (%) ± SD (%)	AOA (%) ± SD (%)
PCA-2D-CNN	99.09 ± 0.23	99.41 ± 0.10	99.90 ± 0.05
LDA-2D-CNN	**99.34 ± 0.11**	**99.60 ± 0.06**	**99.91 ± 0.03**

From Table 2, it can be found that the proposed LDA-2D-CNN improves the AOAs by 0.25, 0.19 and 0.01 percentage points and has smaller SDs compared to PCA-2D-CNN at 5, 10 and 20% training data for the Pavia University dataset, respectively. It shows that the classification and recognition results are more stable. The feature extraction of HRSIs using LDA projects the original data to the low-dimensional subspace can make the samples more separable, which helps the classification and recognition of HRSIs. The LDA-2D-CNN method proposed in this paper can effectively reduce the redundant information between bands, retain more discriminative classification information, and help improve the classification results of HRSIs.

Table 3. Indian Pines dataset classification results

Ratio of training samples	5%	10%	20%
	AOA (%) ± SD (%)	AOA (%) ± SD (%)	AOA (%) ± SD (%)
PCA-2D-CNN	95.49 ± 0.39	98.01 ± 0.27	99.27 ± 0.11
LDA-2D-CNN	**96.17 ± 0.35**	**98.33 ± 0.12**	**99.33 ± 0.09**

From Table 3, we can see that the AOAs of the proposed LDA-2D-CNN improves by 0.68, 0.32 and 0.06% points and SDs are smaller compared with PCA-2D-CNN for the Indian Pines dataset at 5, 10 and 20% training data, respectively. It shows that the classification and recognition results are more stable and the average recognition rate increases more with fewer training samples, and the proposed LDA-2D-CNN method can better classify and extract effective features for small sample datasets than PCA-2D-CNN, so that the features can be better classified and recognized.

Table 4. Botswana dataset classification results

Ratio of training samples	5%	10%	20%
	AOA (%) ± SD (%)	AOA (%) ± SD (%)	AOA (%) ± SD (%)
PCA-2D-CNN	94.14 ± 1.51	98.72 ± 0.51	99.83 ± 0.14
LDA-2D-CNN	**96.48 ± 1.25**	**99.45 ± 0.33**	**99.95 ± 0.09**

From Table 4, we can see that the AOAs of the proposed LDA-2D-CNN improves by 2.34, 0.73 and 0.12% points compared with PCA-2D-CNN for 5, 10 and 20% training data, respectively, and the SDs are smaller. The most improvement in AOA is found in the case of 5% training samples, which indicates that the proposed LDA-2D-CNN can extract effective classification features and obtain better experimental results in the case of small-size samples.

From the AOAs and SDs of each data above, it can be found that the experimental results of 2D-CNN using LDA to reduce dimensionality and extract features have higher AOAs, smaller SDs and more stable experimental results than the method using PCA. And the average recognition rate improves more with fewer training samples, which indicates that the proposed LDA-2D-CNN method can classify small-size sample datasets better than the PCA-2D-CNN. The experimental results show that the proposed LDA-2D-CNN method can extract effective features and classify the features correctly to obtain better classification and recognition results. When using PCA for dimensionality reduction of high-dimensional data, most information of the original data is retained in the new subspace, while when using LDA for feature extraction of HRSIs, more separability features can be extracted by projecting the original data to the low-dimensional subspace, which makes the samples more separability in the low-dimensional subspace and helps to improve the terrain classification of HRSIs.

6 Conclusion

For HRSIs have high dimensionality and large data volume, but the strong inter-band correlation contains many redundant information which affects the terrain classification accuracy. In this paper, an LDA-2D-CNN method is proposed to reduce the correlation between bands, reduce redundant information and improve the efficiency of feature classification of HRSIs. First, HRSIs are dimensionally reduced by LDA, project the data to the feature subspace and extract the most discriminative information, and then 2D-CNN is used to extract deep feature and classify. By classifying on three real HRSIs dataset, it is shown that the classification results using LDA-2D-CNN method are better than PCA-2D-CNN method.

Acknowledgement. This work was funded by the National Natural Science Foundation of China (No. 61672405, No. 62077038), the Natural Science Foundation of Shaanxi Province of China (No. 2021JM-459, No. 2018JM4018).

References

1. Zeng, S., Wang, Z., Gao, C., et al.: Hyperspectral image classification with global-local discriminant analysis and spatial-spectral context. IEEE J. Sel. Top. Appl. Earth Observations Remote Sens. **11**(12), 5005–5018 (2018)
2. Donoho, D.L.: High-dimensional data analysis: the curses and blessings of dimensionality. AMS Math Challenges Lect **1**(2000), 1–32 (2000)
3. Yu, X., Wang, R., Liu, B., et al.: Salient feature extraction for hyperspectral image classification. Remote Sensing Letters **10**(6), 553–562 (2019)
4. Rasti, B., Scheunders, P., Ghamisi, P., et al.: Noise reduction in hyperspectral imagery: overview and application. Remote Sens. 10(3), 482 (2018)
5. Zhao, W., Du, S.: Spectral-spatial feature extraction for hyperspectral image classification: a dimension reduction and deep learning approach. IEEE Trans. Geosci. Remote Sens. **54**(8), 4544–4554 (2016)
6. Martinez, A.M., Kak, A.C.: PCA versus LDA. IEEE Trans. Pattern Anal. Mach. Intell. **23**(2), 228–233 (2001)
7. Tharwat, A., Gaber, T., Ibrahim, A., et al.: Linear discriminant analysis: a detailed tutorial. AI Commun. **30**(2), 169–190 (2017)
8. Li, W., Prasad, S., Fowler, J.E., et al.: Locality-preserving dimensionality reduction and classification for hyperspectral image analysis. IEEE Trans. Geosci. Remote Sens. **50**(4), 1185–1198 (2012)
9. Anowar, F., Sadaoui, S., Selim, B.: Conceptual and empirical comparison of dimensionality reduction algorithms (PCA, KPCA, LDA, MDS, SVD, LLE, ISOMAP, LE, ICA, t-SNE). Comput. Sci. Rev. **40**, 100378 (2021)
10. Liu, J., Chen, S., Tan, X.: A study on three linear discriminant analysis based methods in small sample size problem. Pattern Recogn. **41**(1), 102–116 (2008)

Design and Implementation of Vehicle Density Detection Method Based on Deep Learning

Jiale Yi$^{(\boxtimes)}$, Xia Zhang, Zhili Mao, Huimin Du, and Yu Ma

Xi'an University of Posts & Telecommunications, Xi'an 710121, China
yijiale990314@163.com, {zhangxia,fv}@xupt.edu.cn, mzl000@163.com

Abstract. With the development of society, more and more people rely on cars to travel and cause traffic congestion. Vehicle density detection can provide intelligent decision-making assistance for traffic video surveillance system and improve traffic efficiency. The camera view can cause the scale difference between vehicles in the same scene. To solve the multi-scale problem, a vehicle density detection method based on deep learning semantic segmentation is proposed. VGG-16 is used to extract vehicle features at the front end of the network, and the output feature image is 1/8 of the original image, which can improve the accuracy of the prediction density map. A dilated convolution module is designed at the back end to capture multi-scale features of the vehicle, enabling the network to capture more scale details and edge information. Finally, the network cascades the output with 1×1 convolution to get the prediction density map. And in the output characteristics of the last layer of the network, the quantity is predicted by adding the fully connected layer, and then the final predicted quantity is obtained by summing up the counting results of the network output density map. The performance of the network is tested on the TRANCOS dataset, and the test results show that the MAE of our network is 34.1% higher than that of the Hydra-3s network.

Keywords: Vehicle density detection · Dilated convolution · Density map

1 Introduction

In recent years, with the continuous development of technology, density detection methods continue to change. At present, the main density detection methods are divided into two kinds, one is the traditional density detection method. The traditional density detection methods are divided into detection-based methods and regression-based methods; the other is based on deep learning of density detection methods. Traditional methods usually achieve good results when dealing with low-density vehicle data, but are not applicable to high-density and severely artificially obscured scenarios. Therefore, this paper designs a new vehicle density detection network based on the deep learning density detection method for the problem that it is difficult to detect drastic changes in vehicle scale, which aims to deal with multi-scale problems and applies deep learning semantic segmentation to the vehicle density detection network. The front-end of this vehicle density detection network uses VGG-16 network to extract vehicle features, and then adds

© The Author(s), under exclusive license to Springer Nature Switzerland AG 2023
N. Xiong et al. (Eds.): ICNC-FSKD 2022, LNDECT 153, pp. 165–172, 2023.
https://doi.org/10.1007/978-3-031-20738-9_20

a new dilated convolution to make the receptive field larger and more complex, which can capture the multi-scale features of the vehicle, so that the network can capture more scale details and edge information, and adds a counting learning module at the end of the network to derive the final prediction number. Finally, it can improve the accuracy of vehicle density detection, and solve the problems of existing vehicle detection methods.

2 Related Work

2.1 VGG-16

The VGG-16 network [1] was proposed by the VGG group of Oxford University. VGG-16 all use 3×3 convolution kernel and 2×2 maximum pooling kernel to improve network performance by continuously deepening network structure. It uses the stacking of multiple small convolution kernels to replace the large convolution kernel, which makes more nonlinear layers to ensure that the network can learn more complex patterns and requires fewer parameters. The network structure is shown in Fig. 1. Show.

ConvNet Configuration					
A	A-LRN	B	C	D	E
11 weight layers	11 weight layers	13 weight layers	16 weight layers	16 weight layers	19 weight layers
input (224 × 224 RGB image)					
conv3-64	conv3-64 **LRN**	conv3-64 **conv3-64**	conv3-64 conv3-64	conv3-64 conv3-64	conv3-64 conv3-64
maxpool					
conv3-128	conv3-128	conv3-128 **conv3-128**	conv3-128 conv3-128	conv3-128 conv3-128	conv3-128 conv3-128
maxpool					
conv3-256 conv3-256	conv3-256 conv3-256	conv3-256 conv3-256	conv3-256 conv3-256 **conv1-256**	conv3-256 conv3-256 **conv3-256**	conv3-256 conv3-256 conv3-256 **conv3-256**
maxpool					
conv3-512 conv3-512	conv3-512 conv3-512	conv3-512 conv3-512	conv3-512 conv3-512 **conv1-512**	conv3-512 conv3-512 **conv3-512**	conv3-512 conv3-512 conv3-512 **conv3-512**
maxpool					
conv3-512 conv3-512	conv3-512 conv3-512	conv3-512 conv3-512	conv3-512 conv3-512 **conv1-512**	conv3-512 conv3-512 **conv3-512**	conv3-512 conv3-512 conv3-512 **conv3-512**
maxpool					
FC-4096					
FC-4096					
FC-1000					
soft-max					

Fig. 1. VGG-16 network [1]

2.2 Dilated Convolution

In order to solve the multi-scale problem, the CrowdNet [2] model uses the combination of deep and shallow convolution neural networks to deal with the scale problem. The MCNN [3] model also has a good effect on the multi-scale problem. Although the above two models have initially solved the multi-scale problem, because the size of the convolution kernel is fixed, it is equivalent to fixing the receptive field of a specific size

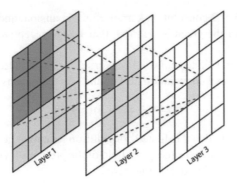

Fig. 2. Receptive field

(as shown in Fig. 2.), it is still not possible to extract much scale information for vehicle data with rapid scale changes.

Combined with the above two models, there are three problems: (1) The parameters of the up-sampling operation and the pooling layer operation cannot be learned, and the effective information of feature map cannot be used in the training process. (2) Too many pooling operations will lead to the loss of some details and spatial hierarchical information. (3) The information of small objects cannot be completely reconstructed, that is, for a network with three-layer pooling operation, theoretically, the information of objects smaller than the pixel size cannot be completely reconstructed. Therefore, this paper uses dilated convolution to solve the above problems. The dilated convolution [4] operation is shown in Fig. 3. Show.

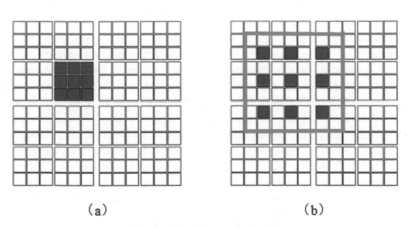

(a)   (b)

Fig. 3. Dilated convolution [4]

Figure 3. (a) is the dilated convolution with an expansion rate of 1. When the expansion rate is 1, the dilated convolution is an ordinary 3 × 3 convolution operation, and its receptive field is the red part of (a). Figure 3. (b), the expansion rate is 2, that is, zero padding is performed between each convolution kernel in (a), and only the red part is convoluted with the 3×3 convolution kernel after expansion, and the receptive field is enlarged to the part of the green box in (b), that is, the receptive field is expanded to 7×7.

The actual size of the dilated convolution kernel is shown in Eq. (1). In the equation, k is the size of the original convolution kernel and r is the expansion rate.

$$K = k + (k - 1)(r - 1) \tag{1}$$

The calculation method of the dilated convolution receptive field is shown in Eq. (2). In the equation, f is the size of the receptive field of the previous layer, s is the convolution step size, and k_s is the convolution kernel size.

$$F = (f - 1) \times s + k_s \tag{2}$$

As a result, it can be seen that dilated convolution [5] can solve the problem of reducing the spatial resolution in order to increase the receptive field in the convolution neural network. At the same time, the dilated convolution can set different dilated rates (that is the number of zeros filled in the convolution kernel) to change the receptive field to extract different scale information, which is very useful for the vehicle density detection task with multi-scale information processed in this paper.

3   Proposed Method

According to the introduction of the existing classical network and dilated convolution in the previous section, this section gives a detailed introduction to the structure of the vehicle density detection network designed in this paper. When designing the network structure to solve the vehicle density detection task, it is necessary to design two parts of the network: feature extraction and regression prediction density map. The feature extraction part of the designed vehicle density detection network structure is mainly to extract the vehicle features from the input data. When designing this module, it is necessary to ensure that the network is highly sensitive to vehicle features and can extract more vehicle information. Therefore, the network feature extraction part uses VGG-16 network [6], because the vehicle density detection does not need to classify the target, so the last three fully connected layers of the VGG-16 network are removed, and only the front convolution layers and pooling layers are used to extract the input image features. The VGG-16 uses three 3×3 convolution kernels instead of a 7×7 convolution kernel and two 3×3 convolution kernels instead of a 5×5 convolution kernel, so that several small convolution kernels are stacked instead of a large convolution kernel, which can not only increase the network depth to ensure learning efficiency and accuracy, but also make fewer parameters (the number of parameters of 7×7 convolution kernel is $49 \times R^2$, while the number of parameters for the stack of three 3×3 convolution kernels is $3 \times 9 \times R^2$, where R is the number of input channels).

The second part is the regression prediction density map part, which mainly reprocesses the output of the network feature extraction part, because the use of ordinary pooling and then up-sampling to restore the image resolution will lead to the loss of scale information. Therefore, the second part of the designed vehicle density detection network structure is composed of dilated convolution with an expansion rate of 2, so as to process the output of the feature extraction network part, which can retain more scale information and obtain a higher quality predicted density map. Finally, the prediction result is output by 1×1 ordinary convolution to obtain the predicted density map.

Due to the large variability in vehicle data, such as the size of vehicle (large buses and small cars), it will lead to miscalculation of super-large vehicles when regressing to the density map and calculating the density. Therefore, this paper adds a counting learning module on the basis of the designed vehicle density detection network (as shown in Fig. 4.). And in the output characteristics of the last layer of the network, the quantity is predicted by adding the fully connected layer, and then the final predicted quantity is obtained by summing up the counting results of the network output density map, as shown in Eq. (3).

$$C_{(i)} = G\left(N_{(i)}; \gamma\right) + N_{(i)} \tag{3}$$

where γ is the learnable parameter of the fully connected layer, and $N_{(i)}$ is the sum of each pixel in the predicted density map.

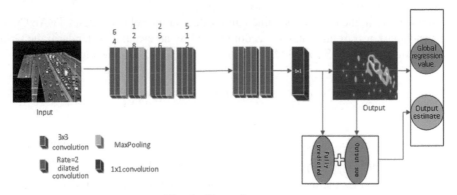

Fig. 4. Network structure

4 Experiment

4.1 Dataset

TRANCOS dataset [7] is a large-scale traffic statistical dataset, which contains 1244 annotated images of traffic scenes with different degrees of congestion captured by cameras. Each image is marked by the method of dot labeling, including the number and location of vehicles, and the location of streets, a total of 46,796 vehicles. In the experiment, 860 images were used as the training set, and the rest were the testing set. The image angle of the dataset is variable, so it has a good testability for the performance of the network.

4.2 Parameter Setting

In this paper, the platform CPU used for training and testing the network is Intel (R) Core (TM) i7–9700 @ 3.00GHz, and the graphics card is NVIDIA GeForce 2080. The whole experiment was carried out under the Ubuntu 16.04 operating system, and the network structure was designed under the framework of Pytorch deep learning. The experimental program is written in Python programming language, with an initial learning rate of $1e^{-6}$ and a training period of 400 epoch.

In this paper, when training the network, the Euclidean distance difference between the predicted density map and the real density map is used as the loss function, as shown in Eq. (4).

$$L(\theta) = \frac{1}{2N} \sum_{i=1}^{N} z(x_i; \theta) - z_{i2}^2 \tag{4}$$

where θ is the network parameter to be optimized, N is the number of training pictures, x_i is the input image, z_i represents the true value of density map corresponding to x_i, and $z(x_i; \theta)$ represents the density map estimated by the network. L is the loss between the estimated density map and the true density map.

In this section, the network is trained and tested on the vehicle dataset TRANCOS. The benchmark test of this dataset is accompanied by an evaluation standard: the Grid Average Mean absolute Error (GAME), which is calculated as shown in Eq. (5).

$$GAME(L) = \frac{1}{N} \sum_{n=1}^{N} \left(\sum_{l=1}^{4^L} \left| D_{I_n}^l - D_{I_n^{GT}}^l \right| \right) \tag{5}$$

where N is the number of images in the test set, $D_{I_n}^l$ is the estimated result in the l region of the input image n, and $D_{I_n^{GT}}^l$ is the true value in the l region of the input image n. GAME(L) subdivides the image using a grid of 4^L non-overlapping regions covering the entire image and calculates the error as the sum of the MAE of each of these regions. When $L = 0$, GAME is equivalent to MAE.

4.3 Results and Analysis

In order to directly show the effectiveness and advantages of the network proposed in this paper, the designed vehicle density detection network structure is compared with the other three classical network structures. The test accuracy of some networks is obtained by reproducing the original network, and the test accuracy of some networks comes directly from the relevant literature. The test results are shown in Table 1.

As can be seen from the data in Table 1, the designed vehicle density detection network structure can complete the vehicle density detection task and improve the accuracy compared with the classical network structure. The experimental results are shown in Fig. 5. It can be seen from the experimental results that the density map predicted by the designed vehicle density detection network is close to the real density map, and the gap between the predicted number and the true data can be calculated through the counting

Table 1. TRANCOS test results

Method	GAME 0	GAME 1	GAME 2	GAME 3
Fiaschi et al. [8]	17.77	20.14	23.65	25.99
Lempitsky et al. [9]	13.76	16.72	20.72	24.36
Hydra-3s [10]	10.99	13.75	16.69	19.32
Ours	**7.22**	**9.16**	**11.36**	**13.59**

learning module after the network. It is found that almost zero error detection can be achieved for the street scene data with relatively few vehicles, and for the street scene data with relatively dense vehicles, the error can be less than a few vehicles.

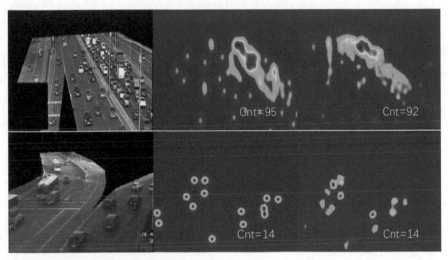

Fig. 5. Comparison of density maps. From left to right is the original image, true density map, and the prediction density map of the designed vehicle density detection network proposed in this paper.

The main reasons for the analysis are as follows: (1) Compared with the convolution layer structure with multiple columns of convolution kernels of different sizes, dilated convolution can better extract the scale information in the image and reduce the loss of fixed convolution kernel size when extracting scale information. (2) During feature extraction, using fewer pooling operations can retain more edge information and vehicle features. (3) A counting learning module is added in regression counting to make density detection more accurate.

References

1. Simonyan, K., et al.: Very deep convolutional networks for large-scale image recognition. arXiv preprint arXiv:1409.1556. (2014)
2. Boominathan, L., et al.: Crowdnet: a deep convolutional network for dense crowd counting. In: Proceedings of the 2016 ACM on Multimedia Conference, pages 640–644. ACM. (2016)
3. Zhang, Y., et al.: Single-image crowd counting via multi-column convolutional neural network. In Proceedings of the IEEE Conference on Computer Vision and Pattern Recognition, pages 589–597. (2016)
4. Yu, F., Koltun, V.: Multi-scale context aggregation by dilated convolutions. In: International Conference on Learning Representations, San Juan. ICLR. (2016)
5. hui, X.C., et al.: End-to-end dilated convolution network for document image semantic segmentation. J. Central South University. (2021)
6. Lei, G., et al.: Lung segmentation method with dilated convolution based on VGG-16 network. [J]. Comput. Assist. Surg. (Abingdon, England). (2019)
7. López-Sastre, R., Maldonado-Bascón, S., Guerrero-Gómez-Olmedo, R., et al.: Extremely overlapping vehicle counting. In: Iberian Conference on Pattern Recognition and Image Analysis, IbPRIA (2015)
8. Fiaschi, L., et al.: Learning to count with regression forest and structured labels. In: Proceedings of the 21st International Conference on Pattern Recognition, ICPR2012, pp. 2685–2688. (2012)
9. Lempitsky, V., Zisserman, A.: Learning to count objects in images. In: Advances in Neural Information Processing Systems, pages 1324–1332. (2010)
10. Onoro-Rubio, D., López-Sastre, R.J.: Towards perspective-free object counting with deep learning. In: European Conference on Computer Vision, pp 615–629. Springer. (2016)

Rotated Ship Detection with Improved YOLOv5X

Xuanhong Wang[1], Shuai Gao[1], Jingchen Zhou[1], and Yun Xiao[2(✉)]

[1] Xi'an University of Posts and Telecommunications, Xi'an 710061, China
wxh@xupt.edu.cn, {shuai_gao,zjc981208}@stu.xupt.edu.cn
[2] Northwest University, Xi'an 710127, China
yxiao@nwu.edu.cn

Abstract. Ship detection in optical remote sensing images is a vital yet challenging task. Now, more attention has been focused on increasing detection accuracy, while the detection speed is ignored. However, detection speed is as important as detection precision for ship detection. In this paper, we propose a new model, named ImYOLOv5X, which is based on YOLOv5X combined with a Squeeze-and-Excitation Module for fast and accurate rotated ship detection. Firstly, we incorporate a Squeeze-and-Excitation (SE) module into backbone of YOLOv5X, which enables the model to focus on detection objects, thus improving detection accuracy. Then we design an easy-to-insert module, containing a Convolution Set and Squeeze-and-Excitation Module (CS-SE), which can extract features and weigh the channels of features for prediction. Finally, we introduce the Gaussian Wasserstein Distance (GWD) loss as the regression loss of the model. The GWD loss resolves the boundary discontinuity and inconsistency in training and final detection metric. Extensive experiments on the HRSC2016 dataset show that our model can achieve highest detection accuracy and still maintain fastest detection speed compared with some other models, which proves the effectiveness of our model.

Keywords: Rotated ship detection · YOLOv5X · Attention mechanism

1 Introduction

Recently, with the increasing development of remote sensing detection technology, it has promoted the flourishing development of various applications in the field of remote sensing. Ship detection [1–3] in the optical remote sensing images always is an interesting and vital tasks. It can be used in military defense security, maritime rescue, environmental pollution, border management, vessel management, etc. For these applications, higher detection speed and better detection accuracy are essential. If rotated ship detection gets faster and more accurate, there will be a broad prospect in both military and civilian fields.

There are two main detection model for ship detection: two-stage detectors such as R2CNN [4], RRPN [5] and one-stage detectors such as RetinaNet-H [6], RetinaNet-H-GWD [7]. The previous two-stage detector can provide a higher detection accuracy than

N. Xiong et al. (Eds.): ICNC-FSKD 2022, LNDECT 153, pp. 173–181, 2023.
https://doi.org/10.1007/978-3-031-20738-9_21

the one-stage detector. While the detection speed of the two-stage detector is very low. And the one-stage detector has a faster detection speed. But the detection accuracy also decreases largely.

Recently, many works have exploited advanced one-stage detectors for faster and more accurate ship detection. Ming et al. [8] introduced a Representation Invariance Loss (RIL) to optimize the regression loss of bounding box for the rotated objects in the remote sensing images. Qian et al. [9] also designed the RSDET detector. It solves the problem of rotation sensitivity error caused by rotating frame regression. Yang et al. [7] introduced Gaussian Wasserstein Distance (GWD) loss into RetinaNet-H for accurate detection. For these one-stage detectors, in case of ensuring a certain level of detection accuracy, a faster detection speed is achieved. But for detection tasks requiring high detection accuracy, such as sea rescue and tracking of military enemy ships, the detection accuracy of one-stage detectors still needs to be increased.

To solve all the above challenges and problems, we propose an improved YOLOv5X combined with a Squeeze-and-Excitation Module (ImYOLOv5X). First, to meet the higher detection speed, we use the YOLOv5X model as the basic detection framework because of its high detection speed. Next, for adaptability of the model in complex inshore scenarios, we introduce the Squeeze-and-Excitation (SE) module into the backbone, called SE-DarkNetCSP. Then, in the backbone, a feature enhancement module including a Convolution Set and a Squeeze-and-Excitation Module (CS-SE) is added between the neck and predict module. Finally, as a regression loss function of the model, we use Gaussian Wasserstein Distance loss (GWD) as a good solution for the inconsistency of final test metrics and boundary discontinuity.

The main contributions of our paper are as follows:

A new model, named ImYOLOv5X is proposed for fast and accurate rotated ship detection. For ship detection under complex background, we propose SE-DarkNetCSP module, which introduces an attention mechanism in the backbone of YOLOv5X.

We also construct a novel and easy-to-insert feature enhancement module consisting of a Convolution Set and a Squeeze-and-Excitation Module, which obtains more information-rich features for prediction by further feature extraction and attention mechanism weighted mapping.

2   Proposed Method

In this section, we introduce the network architecture of ImYOLOv5X, which enables detection fast and accurate.

2.1   Network Architecture

The network architecture is divided into four modules: feature extraction module, feature fusion module, feature enhanced module and predict module. The network is shown in Fig. 1.

For the input of the model, the images are resized to an appropriate size. In this paper, the size of the images is $800 \times 800 \times 3$. In the feature extraction module, we propose the SE-DarkNetCSP backbone, which introduces an attention mechanism on the backbone

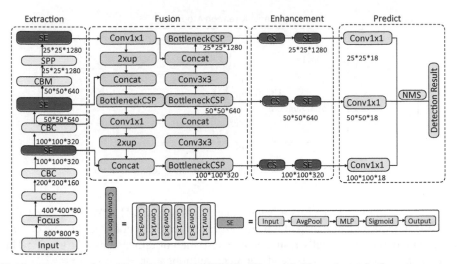

Fig. 1. The network architecture of the ImYOLOv5X model. The red part is the extra part we added. In the backbone, we added three Squeeze-and-Excitation modules. In the enhancement, we propose three convolution sets and SE modules.

of the model. Compared with the DarknetCSP, the SE-DarkNetCSP module can pay more attention to ship targets and increase detection accuracy.

In the feature fusion module, the ImYOLOv5X network predicts the bounding box at three different scales. Based on the structure of the feature pyramid network (FPN), By Path Aggregation Network (PANet), By bottom-up path enhancement, the information path between the lower and uppermost features is shortened, thus enhancing the accurate localization signal throughout the feature hierarchy. The feature enhanced module is the extra part we added. For the feature map after feature fusion, we construct a novel and easy-to-insert feature enhancement module including a Convolution Set and a Squeeze-and-Excitation Module, which obtains more information-rich features for prediction by further feature extraction and attention mechanism weighted mapping.

In the predict module, for the three branches of prediction, the last convolutional layer predicts the bounding box information. The size of the output feature map is $N \times M \times (3 \times (5 + 1))$. . Reflecting prediction information to the bounding box can generate the prediction box. Finally, by filtering the predicted boxes, we can get the prediction result via non-maximum suppression (NMS).

2.2 Squeeze-and-Excitation Module

Inspired by the application of the Squeeze-and-Excitation (SE) module [10] in other deep neural networks, we introduced the SE module into the feature extraction and enhancement module. SE module applies the attention mechanism in the channel dimension which explicitly models the relationships between the channels and channels in the feature map, which adaptively recalibrates the responses of channels and distinctly improves the detection results. The structure of SE is shown in Fig. 2.

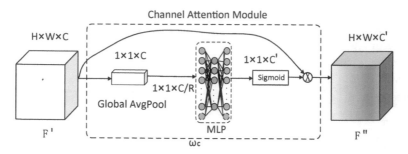

Fig. 2. The architecture of the squeeze-and-excitation module.

In Fig. 2, the input of SE is a feature map of dimensions H × W × C, denoted as F/; The output is a feature map of dimensions H × W × C/, denoted as F//; The channel attention module sequentially results in a feature map of dimensions 1 × 1 × C'; This function is denoted as ω_c; ⊗ presents element multiplication.

The input feature map adaptively recalibrates the channel- characteristic feature map responses. The result and input feature map perform element multiplication operations. The expression is shown in Eq. (1):

$$F'' = \omega_c(F') \otimes F' \tag{1}$$

The channel attention module consists of squeeze and excitation. For squeeze operation, using global average pooling generates channel- characteristic statistics. The input of the function is a feature map of dimensions $H \times W \times C$. The output of global average pooling is a feature map of dimensions $1 \times 1 \times C$.

Excitation operation adaptively recalibrate channel characteristic response by building the fully-connected network and sigmoid active function. To reduce the complexity of the FC, the number of input channels in the fully-connected network is compressed by a certain ratio of R. . The Sigmoid activation function could weigh the result of FC in the range of (0, 1). The above process is represented by Eq. (2):

$$\omega_c = \sigma\left(\text{MLP}(\text{AvgPool}(F'))\right) \tag{2}$$

2.3 Gaussian Wasserstein Distance Loss

The rotated bounding box is transformed into a two-dimensional Gaussian distribution, which can approximate the undifferentiated rotational IoU loss by Gaussian Wasserstein Distance (GWD) [7]. This distance loss can be learned effectively by gradient back-propagation.

First of all, the rotated bounding box $A = (x, y, h, w, \theta)$ is transformed into a two-dimensional Gaussian distribution $N = (m, B)$ by Eq. (3), where A represents rotated bounding box and (x, y, h, w, θ) represents the center point, height, width and angle of the rotated bounding box respectively.

$$
\begin{aligned}
B^{\frac{1}{2}} &= QMQ^{\top} \\
&= \begin{pmatrix} \cos\theta & -\sin\theta \\ \sin\theta & \cos\theta \end{pmatrix} \begin{pmatrix} \frac{w}{2} & 0 \\ 0 & \frac{h}{2} \end{pmatrix} \begin{pmatrix} \cos\theta & \sin\theta \\ -\sin\theta & \cos\theta \end{pmatrix} \\
&= \begin{pmatrix} \frac{w}{2}\cos^2\theta + \frac{h}{2}\sin^2\theta & \frac{w-h}{2}\cos\theta\sin\theta \\ \frac{w-h}{2}\cos\theta\sin\theta & \frac{w}{2}\sin^2\theta + \frac{h}{2}\cos^2\theta \end{pmatrix}
\end{aligned} \tag{3}
$$

where the rotation matrix is denoted as Q and the diagonal matrix of eigenvalues is denoted as M.

Next, computing the Euclidean distance of the 2-D Gaussian distribution evaluates the similarity between the ground truth and predict box form in detection literature in Eq. (4):

$$
d^2 = \|m_1 - m_2\|_2^2 + \mathrm{Tr}\left(B_1 + B_2 - 2\left(B_1^{\frac{1}{2}} B_2 B_1^{\frac{1}{2}} \right)^{\frac{1}{2}} \right) \tag{4}
$$

Then, for large errors, the GWD is very sensitive. Yang et al. [7] write the Eq. (5) by the mitation of IoU-based loss.

$$
\mathcal{L}_{\mathrm{reg}} = 1 - \frac{1}{\tau + f(d^2)}, \tau \geq 1 \tag{5}
$$

Finally, the total loss function is shown in Eq. (6):

$$
\alpha = \frac{\lambda_1}{N_{\mathrm{pos}}} \sum_{n=1}^{N} \mathcal{L}_{\mathrm{reg}}(b_n, gt_n) + \frac{\lambda_2}{N} \sum_{n=1}^{N} \mathcal{L}_{cls}(p_n, t_n\} \tag{6}
$$

3 Datasets and Implementation Details

3.1 HRSC2016 Dataset

The HRSC2016 dataset [11] is a public dataset for ship detection. All images of the entire dataset were collected from Google Earth. The entire dataset contains 1061 images and is divided into train, validation and test set including 436, 181 and 444 images, respectively. For data enhancement during training, we used the mosaic data enhancement and the images were resized to 800×800.

3.2 Implementation Details

In the experiments, we execute on Ubuntu 20.04 from a PC: Intel(R) Xeon(R) CPU E5–2650 v4 @ 2.20GHz RAM 62G NVIDIA TITAN RTX. The epoch is set at 120. All models are trained for 36,481 iterations and the learning rate is divided by 10 at 12,160 and 27,361 iterations. For the Gaussian Wasserstein Distance loss module is set to log and τ was 1.0. For Squeeze-and-Excitation (SE) module, the number of input channels in fully-connected network is compressed by a certain ratio of 8. For evaluation, we use mAP, FPS and FLOPs as the metrics.

4 Ablation Experiments

When the overall architecture is set to YOLOv5X, the loss function is set to the Gaussian Wasserstein Distance loss. The feature enhancement is set. And other conditions remain unchanged, the input for the feature fusion module is the $3\times, 4\times$ and $5\times$ downsample feature maps. The Squeeze-and-Excitation module [10] can extract features and weigh the channels of features. It can provide richer and more precise features for prediction. To enhance information richness after-downsampled feature for prediction, we introduce an SE module after these feature maps. And up to 5 SE modules can be introduced. Compared to introducing the SE modules in other ways, our model can maximize the detection effect while adding fewer parameters. We do the ablation experiments of the number of Squeeze-and-Excitation modules in darknet-csp.

After comprehensive consideration, the number of SE modules is increased sequentially after $5\times, 4\times, 3\times, 2\times, 1\times$ downsample feature map and up to five SE modules. From Fig. 3 and Table 1, it can be seen that the feature maps are the best structure when introduced into the SE module after $5\times$ $4\times$ and $3\times$ downsample and the structure is shown in Fig. 1, called SE-DarknetCSP. It can be seen that adding the SE module into YOLOv5X brings 3.37% mAP increment (85.19% versus 88.56%) with detection speed reduced little.

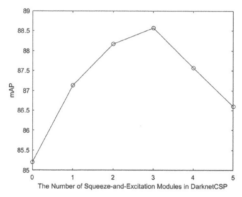

Fig. 3. Ship detection results were evaluated on the HRSC2016 dataset [11] by different numbers of SE modules in DarknetCSP.

5 Comparison Experiments

In this section, we compare the ImYOLOv5X network with the other state-of-the-art rotated detection method including R2CNN [4], Retinanet-H [6], Retinanet-H-GWD [7], RSDET [9]. As illustrated in Table 2, our method achieves an accuracy of 88.56% and a speed of 17.1 fps. ImYOLOv5X clearly outperforms R2CNN by 14.13% mAP (88.56% versus 74.43%), RetinaNet-H by 5.41% mAP (88.56% versus 83.15%), RetinaNet-H-GWD by 5.35% mAP (88.56% versus 83.21%) and RSDET by 5.33% mAP (88.56%

Table 1. Ship detection results were evaluated on the HRSC2016 dataset [11]. The squeeze-and-excitation module of the backbone in DarknetCSP is set as 1, 2, 3, 4 and 5. The entries with the best AP for the ship category are boldfaced.

Backbone	Loss	Recall	mAP50 (%)	Speed (fps)	FLOPs (G)
DarkNetCSP	GWD	93.61	85.19	**22.6**	**141.22**
DarkNetCSP + SE × 1 + CS-SE	GWD	94.32	87.12	17.8	233.46
DarkNetCSP + SE × 2 + CS-SE	GWD	94.85	88.16	17.3	233.46
DarkNetCSP + SE × 3 + CS-SE	GWD	**95.03**	**88.56**	17.1	233.46
DarkNetCSP + SE × 4 + CS-SE	GWD	94.68	87.56	16.6	233.48
DarkNetCSP + SE × 5 + CS-SE	GWD	94.19	86.59	16.5	233.48

versus 83.23%), while being much faster (17.1 fps). In addition, we show the visualized detection result in Fig. 4. For the result of other models except for ImYOLOv5X, many ships are missed and some bounding boxes are wrong. On the contrary, for the result of ImYOLOv5X model, little ships are missed and wrong. Furthermore, the bounding boxes of our ImYOLOv5X are most precise.

Compared with the other model, our model can achieves better detection accuracy and faster detection speed. These results can demonstrate the validity of ImYOLOv5X in this paper.

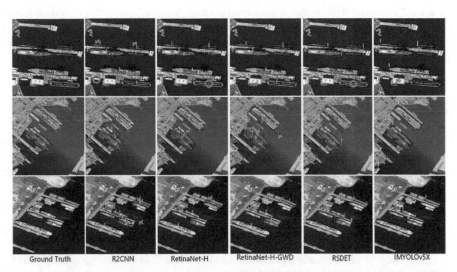

Fig. 4. The visualized detection result of Ground Truth, R2CNN [4], Retinanet-H [6], Retinanet-H-GWD [7], RSDET [9] and our ImYOLOv5X on the HRSC2016 dataset [11]. The missed targets are indicated by blue circles and the incorrectly marked targets are indicated by blue crosses.

Table 2. The detection accuracy and speed of different methods on the HRSC2016 dataset [11]. The best results for each metric are shown in bold.

Model	Backbone	mAP50 (%)	Speed (fps)	FLOPs (G)
R2CNN [4]	ResNet101	74.43	8.86	1179.87
RetinaNet-H [6]	ResNet101	83.15	14.67	1074.54
RetinaNet-H-GWD [7]	ResNet101	83.21	14.10	1074.54
RSDET [9]	ResNet101	83.23	13.98	1377.80
ImYOLOv5X (ours)	SE-DarknetCSP	**88.56**	**17.1**	**233.46**

6   Conclusions

For ship detection in the optical remote sensing images, we propose an improved YOLOv5X combined with a Squeeze-and-Excitation module for fast and accurate rotated ship detection. The results show the ImYOLOv5X achieves 88.56% mAP and 17.1fps on the HRSC2016 dataset. Extensive experiments show that our ImYOLOv5X can achieve the more accurate detection accuracy and the faster detection speed.

In the next work plan, we will further increase the detection accuracy while ensuring a certain detection speed. At the same time, the complexity of the model will be further reduced.

Acknowledgements. This work was partially supported by the NSFC under Grant (No.61972315), Shaanxi Province International Science and Technology Cooperation Program Project-Key Projects (No.2022KWZ-14), the National Key Laboratory of Science and Technology on Space Micrwave (No. 6142411412117).

References

1. Yao, Y., Jiang, Z., Zhang, H., et al.: Ship detection in optical remote sensing images based on deep convolutional neural networks. J. Appl. Remote Sens. 11(4), 042611 (2017)
2. Heiselberg, H.: A direct and fast methodology for ship recognition in sentinel-2 multispectral imagery. Remote Sensing 8(12), 1033 (2016)
3. Dong, C., Liu, J., Xu, F.: Ship detection in optical remote sensing images based on saliency and a rotation-invariant descriptor. Remote Sensing 10(3), 400 (2018)
4. Jiang, Y., Zhu, X., et al.: R2CNN: Rotational region CNN for orientation robust scene text detection. arXiv preprint arXiv:1706.09579 (2017)
5. Ma, J., Shao, W., et al.: Arbitrary-oriented scene text detection via rotation proposals. IEEE Trans. Multimedia 20(11), 3111–3122 (2018)
6. Yang, X., Yan, J., et al.: R3det: Refined single-stage detector with feature refinement for rotating object. In: Proceedings of the AAAI Conference on Artificial Intelligence 35, pp. 3163–3171 (2021)
7. Yang, X., Yan, et al.: Rethinking rotated object detection with Gaussian wasserstein distance loss. In: Proceedings of the 38th International Conference on Machine Learning, PMLR 139, pp. 11830–11841 (2021)

8. Ming, Q., Miao, L., et al.: Optimization for arbitrary-oriented object detection via representation invariance loss. IEEE Geosci. Remote Sens. Lett. **19**, 1–5 (2021)
9. Qian, W., Yang, X., et al.: Learning modulated loss for rotated object detection. In: Proceedings of the AAAI Conference on Artificial Intelligence, vol. 35, no. 3, pp. 2458–2466 (2021)
10. Hu, J., Shen, L., et al.: Squeeze-and-excitation networks. In: Proceedings of IEEE Conference on Computer Vision and Pattern Recognition, pp. 7132–7141 (2018)
11. Liu, Z., Yuan, L., et al.: A high resolution optical satellite image dataset for ship recognition and some new baselines. In: International Conference on Pattern Recognition Applications and Methods, SciTePress, vol. 2, pp. 324–331 (2017)

Application of End-To-End EfficientNetV2 in Diabetic Retinopathy Grading

Xuebin Xu, Dehua Liu[✉], Muyu Wang, and Meng Lei

School of Computer Science and Technology, Xi'an University of Posts & Telecommunications,
Xi'an Shaanxi 710121, China
xuxuebin@xupt.edu.cn, {liudehua,wmy961017,
leimeng}@stu.xupt.edu.cn

Abstract. Diabetic macular edema is a common cause of vision loss in patients with diabetic retinopathy, there are currently about 21 million diabetic macular edema patients worldwide. In this paper, we propose an automatic diagnosis method of diabetic retinopathy grading based on deep learning. This makes the disease diagnosis process more convenient and efficient. We preprocess the image including dark channel-based dehazing and color image histogram equalization to enhance the contrast and image brightness of the image. We propose a convolutional neural network called "RTNet" based on the EfficientNetV2 net-work. On the basis of the original squeeze-and-excite (SE) attention mechanism, two pooling layers are added, which can effectively reduce the size of the image, improve the calculation speed, effectively reduce the dimension and reduce the amount of calculation. Since the Stochastic Gradient Descent (SGD) optimizer learning rate adjustment strategy is limited by the pre-specified adjustment rules, we adjust its SGD optimizer to Adadelta optimizer, finding the parameters can be moved closer to the bottom of the slope, in order to speed up the convergence. The recognition accuracy rate reached 84.2%, which proved that RTNet can accurately identify diabetic retinopathy and has great application prospects.

Keywords: Diabetic retinopathy · Deep learning · Convolutional neural network (CNN)

1 Introduction

1.1 Research Background

With the rapid development of our country's social economy and improvement of people's living standards, more and more people had suffered from diabetes. There are about 116.4 million persons with diabetes in China, ranking first in the world. Now, the main methods to prevent diabetes-induced blindness are laser coagulation and vitrectomy, the cost is relatively high. Diabetic retinopathy (DR) classification and analysis of its severity can help to accurately formulate a DR treatment plan for the detection, diagnosis and treatment of early lesions. That shortens the treatment cycle and reduces the cost of treatment. In the early stage of DR, the patient has no obvious symptoms and no

N. Xiong et al. (Eds.): ICNC-FSKD 2022, LNDECT 153, pp. 182–190, 2023.
https://doi.org/10.1007/978-3-031-20738-9_22

obvious changes in vision. Therefore, patients often ignore the lesions, leading to delays in treatment and irreversible serious consequences. Venous dilatation, hard exudate, soft exudate, hemorrhagic spots, neovascularization and fibrosis eventually lead to retinal detachment. A recent study [1] found that if treated early in retinopathy, the rate of vision loss was slowed.

At present, the diagnosis of DR is mainly performed by doctors with rich clinical medical experience by analyzing color digital fundus images to diagnose micro aneurysms, small capillary sacs, retinal hemorrhages and vascular ruptures, and give the diagnosis results [2]. However, due to the complexity of the fundus images of diabetic retinopathy and the diversity and subtlety of DR lesions, the diagnosis process can be very long. According to statistics, clinician resources are very scarce, and the ratio between professional ophthalmologists and patients is as high as 3000. In addition, the differences in medical standards of regions and the limitation of patients' own financial ability will make many patients unable to receive timely diagnosis and treatment, which will lead to serious visual impairment.

With the advancement of science and technology, computer vision has been developed a lot, and many scholars have achieved some research results. Roychowdhury et al. used traditional machine learning methods, including support vector machines, k-nearest neighbor classification and other methods to perform a binary test of normal and lesions on image data. The final results were 100% sensitivity, 53.16% specificity and Area Under Curve (AUC) 0.904. Khalifa et al. [3] used the pre-trained weights of the four models AlexNet, Res-Net18, SqueezeNet and GoogleNet, finding AlexNet achieved the highest accuracy rate of 97.9%.

1.2   The Main Work

This paper proposes an automatic diagnosis method for diabetic retinopathy based on deep learning. We propose a model called "RTNet" by using public datasets on kaagle which is based on the SE attention mechanism fused on the basis of the EfficientNetV2 network. We adds two layers of pooling, and replace the SGD optimizer of the EfficientNetV2 network with the Adadelta optimizer. During training, end-to-end dark channel dehazing is performed on the images of the dataset first, and the contrast enhancement is performed on the dehazed images through histogram equalization of the images. Then training is used as input to learn the characteristics of different degrees of lesions. Network can classify retinal color images into five stages, namely normal DR, mild DR, moderate DR, severe DR and proliferative DR. The input images are trained and tested according to five types to attest the availability of the method, please see Fig. 1.

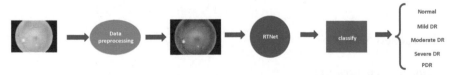

Fig. 1. RTNet main workflow

1.3 Dataset

The dataset we use is from public datasets on kaagle, which includes more than 80,000 color images and 35,126 images labeled by clinicians according to the severity of DR. Some examples of dataset from the kaggle DR dataset are shown below (see Fig. 2):

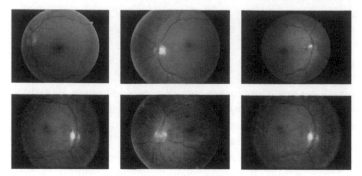

Fig. 2. Examples of various classes in the dataset

The dataset is characterized by a very uneven number of images in each category and most of the datas in the normal range and a few at the level of other lesions.

The distribution of images in the datasets is shown in the Fig. 3.

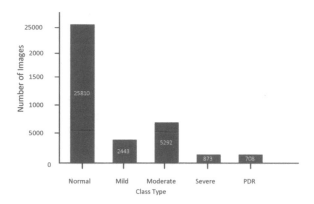

Fig. 3. The distribution map of various classes in the dataset

2 Proposed Work

2.1 Image Preprocessing

This article first preprocesses the dataset obtained from kaagle. Since the color fundus images in the dataset usually have different variations, in order to enable the network exclude misjudgments caused by differences other than lesions, actual training and learning are according to the location of diabetic retinal lesions to obtain better CNN training models and test results. The preprocessing methods used in this paper are mainly: removing the image background and uniformly resize, like using color image histogram algorithm to enhance image contrast, using an end-to-end dark channel dehazing process to make the image clearer and normalization to enhance the expression of the image. Specific steps are as follows.

Remove the background and resize uniformly. The fundus images in the dataset are circular retinal areas with a black background. We delete the background area and the resolution of the image is adjusted to 300×300.

End-to-end de-fogging processing. First calculate the atmospheric light component value and transmittance of the input image, and then calculate the dehazed image. Definition of Dark Channel: In a color image J(x) with three RGB channels, performs two minimum filtering on three RGB channels and a filter template $\Omega(x)$ of size n * n (usually n is 15), which is.

$$J^{dark}(x) = \min_{y \in \Omega(x)} \left(\min_{c \in r,g,b} J^c(y) \right) \tag{1}$$

After statistics on lots of outdoor fog-free images, it was found that the dark channel of outdoor fog-free images has attributed and the dark channel is close to 0.

Image Enhancement—Color Image Histogram Equalization. This paper adopts this method to enhance the global contrast of images especially in terms of brightness, which enhances the local contrast of the image without affecting the overall contrast, making each class in the dataset reach equilibrium.

Data augmentation. This paper uses mirroring, rotation, scale transformation, color dithering and other methods to expand the data set to solve the problem of data imbalance.

Adadelta optimizer [4] Since the SGD optimizer is used in the original EfficientNetV2 network, it is very difficult for the optimizer to choose an suitable initial learning rate. SGD introduces noise while randomly selecting gradients, resulting the weight update is not necessarily in the correct direction. The descending speed is slow, and it continues to oscillate on both sides of the gully and finally stays at a local optimum point. Therefore, this paper converts the SGD optimizer into an Adadelta optimizer. The algorithm iteration formula is as follows.

$$\Delta x_t = \frac{\eta}{\sqrt{\sum_{i=1}^{t} g_i^2}} * g_t \tag{2}$$

$$x_t = x_{t-1} \Delta x_t \tag{3}$$

where g_t represents the gradient value of the current iteration number. This solves the problem of the constant learning rate in the SGD optimizer.

2.2 Building Deep Learning CNN Model

At present, there are many excellent convolutional neural network constructions in the territory of deep [5] learning in computer vision. EfficientNet [6] was proposed by Google in 2019 stands out among many networks with its multi-dimensional mixed model scaling method, which simultaneously improve model performance in three dimensions: depth, width, and resolution and quantifies the relationship between the three dimensions.EfficientNetV2 network outperforms EfficientNet with faster training speed and better parameter efficiency.

There are some problems with EfficientNetV1, for example [7] using Depthwise convolutions in shallow networks is slow, etc. Based on these problems, EfficientNetV2 [8] designed a search space, such as Fused MBConv, applying training-aware NAS and scaling to jointly optimize model accuracy, training speed, and parameter size. The network training speed of EfficientNetV2 is 4 times faster than the previous one, and the number of parameters is 6.8 times smaller. The network structure is shown in the Table 1.

Table 1. EfficientNetV2 network structure

Stage	Operator	Stride	Channels	Layers
0	Conv3*3	2	24	1
1	Fused-MBConv1, k3*3	1	24	2
2	Fused-MBConv4, k3*3	2	48	4
3	Fused-MBConv4, k3*3	2	64	4
4	MBConv4, k3*3, SE0.25	2	128	6
5	MBConv6, k3*3, SE0.25	1	160	9
6	MBConv6, k3*3, SE0.25	2	256	15
7	Conv1*1 & Pooling & FC	–	1280	1

As shown above, EfficientNetV2-S is divided into Stage0 to Stage7, which are ordinary convolution module, fused MBConv module and MBConv module respectively.

Convolution module. Includes an ordinary 3*3 convolution + activation function (SiLU) + Batch Normalization (BN).

Fused-MBConv module. After image input, go through a 3*3 standard convolution, then use a Dropout layer of type Stochastic Depth on the output feature map. Use residuals to connect input and output when stride = 1 and the block's input image and the convolved output image has the same shape. If and only if the down sampling stage with stride = 2, directly output the feature map of the convolution output. As the picture shows (see Fig. 4):

When you need to increase the number of channels: after image input, go through a 3*3 standard convolution, then use 1*1 convolution to drop the number of channels, finally use a Dropout layer. Use residuals to connect input and output when stride = 1. If and only if the down sampling stage with stride = 2, directly output the feature map of the convolution output. As Fig. 5 shows.

MBConv module. In the basic module, image input is performed first, and the number of channels is increased through 1*1 convolution, then use depthwise convolution in

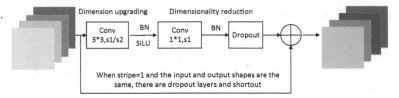

Fig. 4. Block diagram of fused MBConv module

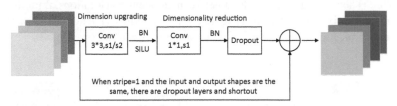

Fig. 5. Block diagram of Fused MBConv modules requiring increasing number of channels

high latitude space. Use the SE attention mechanism to optimize the feature map data, after that, the number of channels is reduced by 1*1 convolution and the first activation function is used. If the input feature map has the same shape as the output feature map, add a Dropout layer of Stochastic Depth type to the feature map after 1*1 convolutional dimension reduction to prevent overfitting and connect the input and output with the final residual (see Fig. 6).

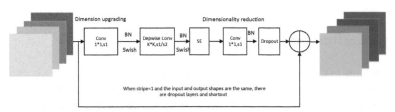

Fig. 6. Block diagram of MBConv module

Downsampling module (stride = 2).The Dropout layer and residual connection are not used, and the feature map is directly output after 1 * 1 convolutional dimension reduction, as Fig. 7 shows:

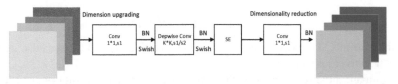

Fig. 7. Downsampling module

2.3 Attentional Mechanism

SE emerged to address the loss caused by the different importances of different channels of the texture map during convolutional pooling. The process of SEblock [9] is mainly divided into two parts: Squeeze and Excitation. It consists of a global average pooling and two fully connected layers. We use the RELU activation function. Based on the original SE attention mechanism of the original EfficientNetV2, this paper adds two pooling layers, an average pooling layer and a maximum pooling layer to the SE attention mechanism. The pooling layer is added between adjacent convolutions, finding the size of parameter matrix can be effectively reduced, which reduces the number of parameters in the final connection layer, contributing to prevent overfitting (see Fig. 8).

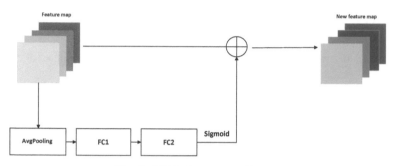

Fig. 8. SEblock structure diagram

3 Experimental Results

All experiments in this paper are carried out under Pytorch's deep learning framework. The laboratory currently has 16 RTX3080 GPU cluster platforms with 1024G memory and 100T hard disk storage, which can provide hardware support for experiments. The dataset is randomly divided into training sets and test sets with a ratio of 8:2, the pixel size of the input images is 300 × 300, the size of the batch-size is 64, when training. We use the AdaDelta optimizer and use SiLU as the loss function. The initial learning rate of the experiment is 0.0001, and a total of 300 epochs is trained, the final accuracy rate reaches 84.2% (see Fig. 9).

The formula for accuracy (Acc) is as follows

$$Acc = \frac{TP + TN}{TP + FP + FN + TN} \tag{4}$$

In this formula, TP, FP, FN, and TN are true positive, false positive, false negative, and true negative, respectively.

To compare the effectiveness of this paper, we use other convolutional neural networks to complete the DR diagnosis. The classification accuracy of each CNN is shown in the Table 2.

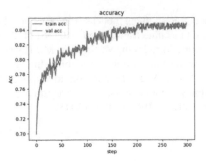

Fig. 9. Accuracy

Table 2. Conventional CNN comparison experiment

Network name	Accuracy
RTNet	84.2
Resnet50	77.8
DenseNet169	76.7
MoileNetV2	78.2

4   Conclusion

In this paper, "RTNet" is used to train Kaagle data set, and dark channel based defogging and color image histogram equalization and data enhancement are used to preprocess the image. Modify SE attention mechanism and SGD optimizer to adjust learning rate and improve data fitting ability. The final recognition accuracy is 84.2%. There are still many shortcomings in this paper, such as the lack of clinical data, and the experimental results in this paper are not enough to detect the location of lesions.

References

1. LNCS Homepage. http://www.springer.com/lncs. Accessed 21 Nov 2016
2. Shah, A.R., Gardner, T.W.: Diabetic retinopathy: research to clinical practice. Clin. Diab. Endocrinol. **3**(1), 1–7 (2017)
3. Mungloo-Dilmohamud, Z., Heenaye-Mamode Khan, M., Jhumka, K., Beedassy, B.N., Mungloo, N.Z., Peña-Reyes, C.: Balancing data through data augmentation improves the generality of transfer learning for diabetic retinopathy classification. Appl. Sci. **12**(11), 5363 (2022)
4. Qummar, S., Khan, F.G., Shah, S., Khan, A., Shamshirband, S., Rehman, Z.U., Khan, I.A., Jadoon, W.: A deep learning ensemble approach for diabetic retinopathy detection. IEEE Access 7, 150530–150539 (2019)
5. Kingma, D.P., Ba, J. Adam.: A method for stochastic optimization. arXiv preprint arXiv:1412. 6980. (2014)
6. Krizhevsky, A., Sutskever, I., Hinton, G.E.: Imagenet classification with deep convolutional neural networks. Adv. Neural Inf. Process. Syst. 25 (2012)

7. Tan, M., Le, Q.: Efficientnet: Rethinking model scaling for convolutional neural networks. In: International conference on machine learning, pp. 6105–6114. PMLR. (2019)
8. Tan, M., Chen, B., Pang, R., Vasudevan, V., Sandler, M., Howard, A., Le, Q., V.Mnasnet.: Platform-aware neural architecture search for mobile. In: Proceedings of the IEEE/CVF Conference on Computer Vision And Pattern Recognition, pp. 2820–2828. (2019)
9. Tan, M., Le, Q.: Efficientnetv2: smaller models and faster training. In: International conference on machine learning, pp. 10096–10106. PMLR. (2021)

Sternal Fracture Recognition Based on EfficientNetV2 Fusion Spatial and Channel Features

Xuebin Xu, Muyu Wang$^{(\boxtimes)}$, Dehua Liu, Meng Lei, and Xiaorui Cheng

School of Computer Science and Technology, Xi'an University of Posts & Telecommunications, Xi'an Shaanxi 710121, China
xuxuebin@xupt.edu.cn, {wmy961017,liudehua,leimeng, cxxxxxr}@stu.xupt.edu.cn

Abstract. Fractures are one of the most common injuries in medicine. A fracture is a partial or complete break in the continuity of bone structure. Fractures can occur for many reasons. For example, a certain part of the bone is fractured due to the direct action of violence on the part, resulting in a fracture of the injured part, which is often accompanied by varying degrees of soft tissue damage; fracture. The sternum is the most dense and complex part of human fractures, and it is very difficult to identify manually. The recognition of sternal fractures has great research significance in the field of fracture recognition, so this paper proposes an image recognition method based on a balanced improved convolutional neural network, and introduces the SA module combined with the SE module to design the model of the network MESCNet (fusion space and channel features), the network can help us identify sternal fractures very well. We used a dataset of 1227 sternum X-rays containing fractures from a dataset collected by the Radiology Department of Xi'an Red Cross Hospital. The images are not processed in any way and have different resolutions. We preprocessed the data and transformed it into a coloured database of sternal fractures. The recognition accuracy rate reached 78.12%. Therefore, it is proved that MESCNet can accurately identify sternum fractures and has broad application prospects.

Keywords: Sternal fracture · Image recognition · MESCNet

1 Introduction

There are 206 bones in the human body with different sizes, shapes and structures. As an important part of the human body, the health of bones has a very important impact on people's lives. Sternal fractures are relatively rare in thoracic trauma, but sternal fractures are easy to be combined with intrathoracic heart, great blood vessel damage, and sometimes even abdominal organ damage, with a fatality rate of 5 to 15%.

Sternal fracture is a very rare injury, occurring in only 0.5% of all fractures, and is estimated to be 3–8% in internal trauma patients [1]. Illness, sports injuries, and accidental injuries are common causes of cases. Doctors usually scan the chest using CT scans

© The Author(s), under exclusive license to Springer Nature Switzerland AG 2023
N. Xiong et al. (Eds.): ICNC-FSKD 2022, LNDECT 153, pp. 191–200, 2023.
https://doi.org/10.1007/978-3-031-20738-9_23

of the chest and lateral X-rays. Fractures of the sternum may aggravate other injuries [2]. CT and X-ray films are the easiest and fastest way to study bone diseases, especially the development of CT scanning technology has greatly improved the diagnosis and treatment of bone diseases [3]. In the field of orthopaedics and related medicine, there has been some use of deep learning to identify, segment, detect fractures in radiographs of different sites [4]. At present, deep learning methods have made significant progress in the application of pattern recognition, medicine, computer vision and many other fields. Kim and MacKinnon retrained the Inception v3 network using x-rays and obtained a classification model. Rajpurkar et al. employed MURA [5]. Using a large dataset of skeletal radiographs, a binary classification model was trained through a 169-layer DenseNet network [6].Because the sternum fracture recognition is difficult and the labor cost is high, so a neural network for sternum fracture recognition can greatly reduce the consumption of human resources. Therefore, this experiment is of great significance.

2   Related Work

We will introduce the relevant work in this section.

2.1   The Dataset

The application of deep learning in medical imaging screening and diagnosis systems has gradually become a trend. In deep learning training, medical data is the basis of training, and the sources of medical data are mainly obtained from public datasets and hospitals. It is difficult to obtain a large amount of sternal X-ray fracture data in a large-scale public data set of sternal X-ray fractures. The dataset used in this study was collected from the Department of Radiology, Xi'an Red Cross Hospital. It included sternum X-rays of different ages and genders. This dataset is divided into two types: fracture and normal data. The fracture data contains 1227 X-rays of sternum fractures, and the normal data contains 1341 X-rays of normal sternum. X-ray films are easily affected by light and have high identification difficulties, because they can be used as experimental data this time. The pictures are X-ray films in JPG format, and an example of the dataset is shown in Fig. 1:

2.2   Data Enhancement and Data Preprocessing

Data preprocessing can not only reduce the memory size, shorten the calculation process, but also effectively improve the accuracy. In the process of image processing, preprocessing is a key process that can effectively eliminate other types of interference. When taking X-ray images, due to differences in exposure, operating techniques, and machines used, the resulting images have great differences in brightness, contrast, etc. The contrast between the target part and the background is too small or the edge and background of the X-ray film are not clearly distinguished, so the fracture recognition cannot be directly performed on the original image. Sternal fracture recognition is that the database is a single-channel grayscale image, but this network can only recognize

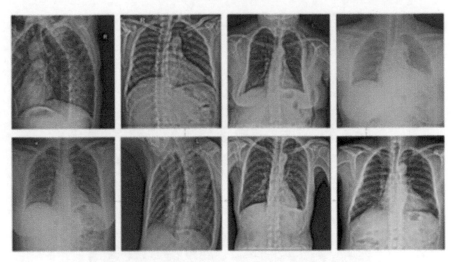

Fig. 1. Example graph of the sternal fracture recognition dataset.

color images, so we preprocess the data of the sternum X-ray database, and convert the single channel three times into RGB three-channel image format in JPG format.

The grayscale range of the original images in the fracture dataset used in this paper is large, which means that some images in the figure are brighter, while others are darker; on the contrary, the background of some images is close to gray, and the background of some images is pure black. The different resolutions of different radiographs can make it difficult to identify sternal fractures. The maximum resolution of the image in this experimental data set is 3712 * 4565, and the minimum image resolution is 465 * 512. Due to the large resolution span, this paper proposes to normalize the image, the width of the image is unified to 400 pixels, and the length is 400 pixels. Adaptive.

An important reason for overfitting in deep learning is that the training data set is too small. In this paper, data augmentation is used to expand the original data set, that is, the existing data is used for brightness, contrast, scaling, flipping, denoising, rotation and other processing processes, in order to increase the number of data sets and effectively prevent the occurrence of overfitting. The image enhancement effect of an image in the fracture dataset is shown in Fig. 2:

3 The Proposed Method

3.1 Convolutional Neural Network

For the oldest CNN convolutional neural network, the traditional neural network consists of an input layer, several convolutional layers, normalization layers, pooling layers, at least one fully connected layer and an output layer. A traditional convolutional neural network is shown in Fig. 3.

Input layer: It is used to input and process multi-dimensional data for the entire convolutional neural network. When processing images, the input layer refers to the

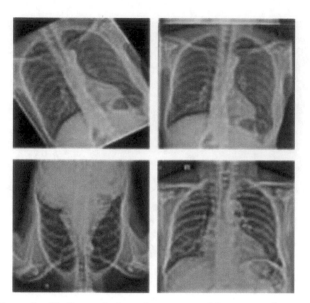

Fig. 2. Example of before and after enhancement of the sternum fracture dataset.

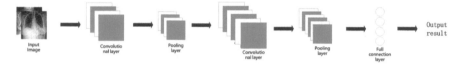

Fig. 3. Basic structure of convolutional neural network

input feature map of the input image, which is usually expressed as the size of the input image: W(input) × H(input) × C(input) are the width, width, and number of channels, respectively.

Convolutional layer: The convolutional layer can effectively extract feature information and is the core part of the entire convolutional neural network. The main function of this layer is to perform feature extraction on the input data, which is to extract and generate a new feature image from the feature image of the input image, also known as the feature extraction layer.

The formula of the feature map obtained by the convolutional layer is as follows:

$$Z^{m+1}(i,j) = \sum_{k=1}^{K_m} \sum_{x=1}^{f} \sum_{y=1}^{f} \left[Z_k^m(s_0 i + x, s_0 j + y)\omega_k^{m+1}(x, y1) \right] + b \qquad (1)$$

$$(i,j) \in \{0, 1, ...L_{m+1}\} \qquad (2)$$

$$L_{m+1} = \frac{L_m + 2p - f}{s_0} \qquad (3)$$

In the formula, b is the amount of deviation, Z^{m+1} represents the feature output of the convolution kernel of the $m + 1$th layer, also known as the feature map, $Z(i, j)$ corresponds to the pixels in the feature map, k represents the number of channels in the feature map, f, p, s_0 and respectively represent different parameters in the convolution layer, corresponding to the size of the convolution kernel, the number of filling layers and the step size of the convolution respectively [7]. When the input image size is large, the convolutional neural network only needs a small convolution kernel to extract features such as edges, Each neuron in a convolutional layer can only be connected to one region of the input body [8]. This greatly reduces the number of parameters of the convolutional neural network, which helps to reduce network training time and save storage space.

Pooling layer: It often used for downsampling. There are many forms of nonlinear pooling functions, such as average pooling, maximum pooling. The effectiveness of this mechanism is that after it finds a feature, the relative position of other features is more important than its precise position. The pooling layer not only continuously reduces the size of the data space, but it also greatly reduces the amount of computation, memory consumption and computation, which effectively prevents overfitting to a certain extent.

Normalization layer: Normalization in convolutional neural networks is to prevent gradient explosion and gradient disappearance. The most classic and widely used is Batch Norm. During the back-propagation process of the neural network, the gradient of each layer is multiplied by the gradient from the previous layer. If the gradient value of each layer is relatively small, if there are many gradients in the process of backpropagation, then the gradient will become smaller and smaller, which will cause the gradient to disappear.

Fully connected layer and output layer: It plays the role of "classification" on the entire convolutional neural network. The convolutional layer is connected to each neuron in the fully connected layer, so the number of parameters in the fully connected layer is the largest. The Softmax function is commonly used in the output layer of the convolutional neural network for classification tasks. The definition of Softmax function is given below:

$$\text{Softmax}(z_i) = \frac{e^{z_i}}{\sum_{d=1}^{D} e^{z_d}} \tag{4}$$

where D represents the number of output nodes, i represents the number of classification categories, and z_i is the output value of the ith node. Traditional neural networks are mainly composed of several convolutional layers, several pooling layers, and at least one fully connected layer. Figure 4 shows the structure of the handwritten digit convolutional neural network:

3.2 EfficientNetV2 Network Framework

EfficientNetV2 [9] is a new series of convolutional networks. It trains faster than the EfficientNet network and uses less parameters.In addition to the MBConv module proposed by the EfficientNet network, and also introduces the Fused-MBConv module. The diagram of the MBconv module is shown in Fig. 5:

EfficientNetV2 uses the Fused-MBConv module. Use a 3 x 3 convolution on the main branch to replace the original 1 x 1 convolution and a Depthwise Conv, followed

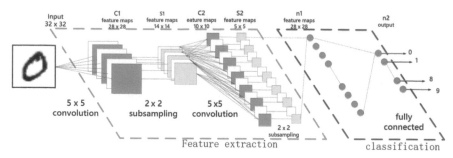

Fig. 4. Structure diagram of convolutional neural network for character recognition

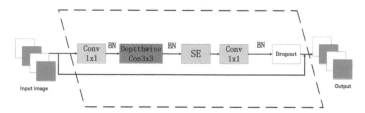

Fig. 5. MBconv module structure diagram

by BN and SILU activation functions, and a Dropout layer. The structure diagram of Fused-MBconv module is shown in Fig. 6:

Fig. 6. Fused-MBConvmodule structure diagram

3.3 Squeeze-and-Attention Network

SA-Net [10] is a simple and effective attention mechanism network structure model, which improves the performance of dense prediction, alleviates the local constraints of convolution kernels, and is specially responsible for pixel grouping. The spatial attention mechanism introduced by the SA module, which emphasizes the attention in groups of pixels of the same class at different spatial scales. The structure of SA-Net is shown in Fig. 7, and its main processing process is as follows.

Preprocessing: The original image is passed through the convolutional layer and the eating session layer to generate a feature map of size CxHxW. Squccze processing: Average pooling is used for operations that are not fully compressed. Note that channel

X is upsampled and output on the main channel to aggregate non-local features. The representation of the SA module is given below:

$$X_{out} = X_{att} * X_{res} + X_{att} \qquad (5)$$

$$X_{att} = Up\left(\sigma\left(\hat{X}_{att}\right)\right) \qquad (6)$$

$$\widehat{X}_{att} = F_{att}(APool(X_{in}); \theta_{att}, \Omega_{att}) \qquad (7)$$

where Up() is the upsampling function used to expand the output of the channel of interest. X_{att} represents the output of the attention convolution channel F_{att} (), which is parameterized by θ_{att} and the structure att of the attention convolution layer. The average pooling layer APool() is used to perform operations that are not fully compressed, and then upsample the output of the attention channel X_{att} to match the output of the main convolution channel X_{res}.

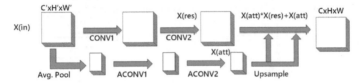

Fig. 7. Structure of the SA module

3.4 Image Classification of Sternal Fracture

The process of sternum fracture recognition algorithm proposed in this paper is shown in Fig. 8. The classification process of sternum fracture is as follows: After preprocessing the images of the sternal fracture data set, it is divided into training data and test data according to the ratio of 8 to 2, and the training data set is enhanced to achieve the purpose of expanding the data set. Preprocessed data is trained in MESCNet, and finally the divided test set data is input into the trained MESCNet for testing. Classification of sternal fracture images completed.

The basic network model used in this paper is EfficientNetV2, There are very few training parameters, and all the training parameters in the entire network are far less than those in EfficientNet, Reseet18, DenseNet121, etc. However, EfficientNetV2 only realizes the fusion of channel characteristics, while sanet adds SA module, which combines sanet with EfficientNetV2, so that the network can understand the relationship between channels and the importance of different channel characteristics. Drawing on the idea of SANet, this paper proposes a network of spatial features and channel features for Sternal Fracture Recognition (MESCNet), which can not only achieve spatial fusion, but also learn the relationship between channel features to further improve network performance. Design the MESCNet network model according to the position where the SA module is inserted. The insertion position of the SA module is shown in Fig. 9.

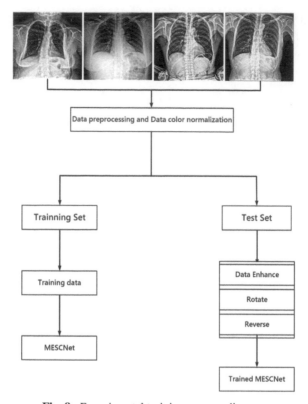

Fig. 8. Experimental training process diagram

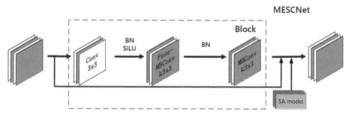

Fig. 9. The insertion location of the SA module in the EfficientNetV2 network model.

4 Experimental Results and Analysis

The standard recognition rate algorithm for sternal fracture image recognition algorithm is as follows:

$$R_{acc} = \frac{Z_a}{Z_s} \tag{8}$$

R_{acc} is the recognition accuracy, Z_a represents the number of correctly recognized images, and Z_s represents the total number of sternum fracture pictures in the model. This experiment uses Resnet18, DenseNet121, EfficientNet, EfficientNetV2, and MESCNet for comparative experiments. The comparison results of the sternum fracture experiment are shown in Table 1 below.

Table 1. Comparison of recognition performance of five algorithms.

Number	Algorithm	Learning efficiency	Recognition accuracy (%)
1	MESCNet	96.35	78.12
2	EfficientNetV2	95.14	76.79
3	EfficientNet	93.26	75.66
4	DenseNet121	91.54	74.56
5	Resnet18	90.16	67.02

According to the experimental results, MESCNet has higher classification accuracy than other sternal fracture recognition image methods, and the final experimental results are shown in Fig. 10.

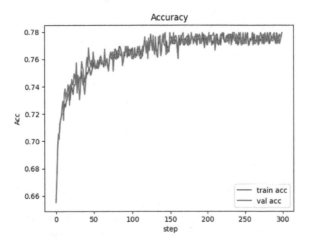

Fig. 10. Result of sternal fracture recognition training.

5   Conclusion

This paper proposes a new MESCNet-based method for sternal fracture recognition, which integrates image space and channel feature information. This paper proposes a new MESCNet-based method for sternal fracture recognition, which integrates image space and channel feature information. MESCNet is based on CNN models EfficientNetV2 and SANet. In efficientnetv2 model, SA module is used to extract spatial features, Se module extracts channel features, and finally fuses channel feature information with spatial feature information to realize image spatial features. Fusion with channel features. MESCNet is designed according to the position of inserting SA modules. The method uses a color normalization method and different data augmentation operations to process the sternal fracture identification dataset. Data preprocessing and data augmentation can further avoid overfitting during model training.Finally, the experimental results show that the accuracy of sternal fracture recognition reaches 78.12%. Our experimental results on the sternal fracture recognition data set of Xi'an Honghui Hospital show that this method can improve the verification accuracy of the experimental results, while meeting the requirements of practical applications.

References

1. Klei, D.S., de Jong, M.B., Öner, F.C., Leenen, L.P.H., van Wessem, K.J.P.: Current treatment and outcomes of traumatic sternal fractures—a systematic review. Int. Orthop. 43(6), 1455–1464 (2018). https://doi.org/10.1007/s00264-018-3945-4
2. Bentley, T.P., Ponnarasu, S., Journey, J.D.: Sternal Fracture. (2018)
3. Hallas, P., Ellingsen, T.: Errors in fracture diagnoses in the emergency department—characteristics of patients and diurnal variation. BMC Emerg. Med. 6(1), 4 (2006)
4. Kalmet, P.H.S., Sanduleanu, S., Primakov, S., et al.: Deep learning in fracture detection: a narrative review. Acta Orthop. 91(2), 215–220 (2020)
5. Rajpurkar, P., Irvin, J., Bagul, A., Ding, D., Duan, T., Mehta, H., Yang, B., Zhu, K., Laird, D., Ball, R.L., Langlotz, C., Shpanskaya, K., Lungren, M.P.: Large dataset for abnormality detection in musculoskeletal radiographs, 1–10 (2018). http://arxiv.org/abs/1712.06957v4 (Accessed September 15, 2018)
6. Huang, G., Liu, Z., Van Der Maaten, L., Weinberger, K.Q.: Densely connected convolutional networks. In: Proceedings 30th IEEE conference on computer vision and pattern recognition, CVPR 2017, pp. 2261–2269. 10.1 109/CVPR.2017.243 (2017)
7. Goodfellow, I., Bengio, Y., Courville, A.: Deep learning (Vol. 1): Cambridge: MIT press, 326–366 (2016)
8. O'Shea, K., Nash, R.: An introduction to convolutional neural networks. arXiv preprint arXiv: 1511.08458 (2015)
9. Tan, M., Le, Q.: Efficientnetv2: smaller models and faster training//International Conference on Machine Learning. PMLR, 10096–10106 (2021)
10. Zhong, Z., Lin, ZQ., Bidart, R., et al.: Squeeze-and-attention networks for semantic segmentation. In: Proceedings of the IEEE/CVF conference on computer vision and pattern recognition, 13065–13074 (2020)

Three-Segment Waybill Code Detection and Recognition Algorithm Based on Rotating Frame and YOLOv5

Jiandong Shen⬦ and Wei Song(✉)⬦

School of Automation, Xi'an University of Posts and Telecommunications, Xi'an 710121,
Shaanxi, China
sjd761107@xupt.edu.cn, 15353169568@stu.xupt.edu.cn

Abstract. Aiming at the inefficiency of the manual sorting method used in sorting centers at district and county levels, an algorithm is proposed for the detection and recognition of three-segment waybill codes based on rotating frames and YOLOv5. First of all, the detection of rotation angle is achieved by converting the regression problem into a classification problem through a Circular smooth label to solve the influence of angle periodicity on training. Secondly, the improved algorithm is used to detect the waybill Logo, according to the position relationship between the Logo and the three-segment code, and then combine the Logo position and rotation angle to locate the three-segment code, and use Radon transform to correct the tilt of the three-segment code to improve the three-segment code locating accuracy. Finally, the template matching algorithm is used to identify the three-segment codes. The experimental results show that the proposed algorithm achieves 93.6% detection accuracy of courier Logo and more than 99% positioning accuracy of three-segment code, which can effectively improve courier sorting efficiency.

Keywords: Target detection · Rotating frame · Three-segment waybill code detection · YOLOv5 · Circular smooth label

1   Introduction

In recent years, automated courier sorting has gradually gained popularity and replaced manual sorting, improving sorting efficiency and accuracy. Due to the late start of logistics in China, only large logistics sorting centers have adopted automated sorting systems, while courier sorting at the district and county levels is entirely based on human operations, which cannot meet the high speed and accuracy requirements of modern logistics distribution [1]. Therefore, there is an urgent need to study small automatic sorting systems for courier sorting centers at the district and county levels.

The key to the realization of automated sorting in the accurate and fast identification of waybill information, domestic courier companies mostly use waybill barcode information detection, but the barcode is susceptible to stain interference, and the use of barcode identification in district and county-level outlets requires communication with the courier company's backend database to obtain waybill information, which will be constrained by the courier company. In addition to the bar code, the face sheet can also be used to obtain express information by the three-segment code composed of numbers and letters. The advantage of three-segment code detection is that according to the encoding rules can be directly read out courier information, without the need to obtain waybill information by querying the courier company's background database, and the recognition rate is high [2]. However, the direct detection of three-segment code is susceptible to interference by other digital information in the waybill, resulting in insufficient detection accuracy. In contrast, the Logo in the waybill has obvious features and relatively fixed position, which can be easily detected, so the Logo can be detected first, and then located to the three-segment code according to the relative position relationship between the Logo and the three-segment code. Since the courier can be at any angle during the sorting process, it is necessary to obtain the rotation angle of the courier Logo while detecting its position.

Along with the development of neural network in target detection technology, it has been widely used in various application scenarios, such as the detection of steel surface defects based on deep residual network [3], which has achieved better detection results. You Only Look Once(YOLO) algorithm was proposed by Redmon [4] et al. xu [5] et al. compared the YOLO series of models and showed that YOLOv5 has greater advantages in model size and detection speed, and can meet the performance requirements of the actual sorting scenario of express delivery, but lacks angle prediction.

For the problem of target angle prediction, many scholars have conducted research. Jiang et al. proposed R^2CNN [6] algorithm, adding some small-scale anchor to improve the detection accuracy of small text; Ma et al. proposed RRPN [7] algorithm, using a new strategy of arbitrary rotated text region optimization to improve the detection performance of rotated text; Zhang et al. proposed CAD-Net [8], which acts on remote sensing images; Yang et al. proposed rotation detectors SCRDet [9], R^3Det [10]. However, most of the above algorithms are based on regression-based arbitrary orientation methods to predict angles, which ignore boundary discontinuities and suffer from the problem of periodicity of angular (PoA).

Combining the above problems and methods, the target detection algorithm based on rotating frame and YOLOv5 is proposed. First, Circular smooth label (CSL) angle classification method is introduced for accurate prediction of target angle. Then the improved algorithm is used to detect the waybill logo, derive the Logo position and rotation angle, locate the three-segment code according to the relative position of Logo and the three-segment code and the rotation angle, and perform tilt correction on the located three-segment code. Finally, identify the three-segment code that was located.

2   Courier Face Slip Detection Recognition Algorithm

2.1   Courier Bill Logo Rotation Detection

The YOLOv5 network currently contains four network models: YOLOv5s, YOLOv5m, YOLOv5l and YOLOv5x [11]. The YOLOv5s model, which has the simplest network structure and the fastest training and detection rate, is selected for improvement to complete the waybill Logo rotation detection.

The angle encoding method used is the five-parameter method with an angle range of 180°. As shown in Fig. 1, where x and y are the central coordinates of the rotating coordinate system, θ is the acute angle between the rotation coordinate system and the x-axis, the long side of the rotating frame is h, and the short side is w;which is designated as positive angle in counterclockwise direction, so the angle range is [−90°, 90°).

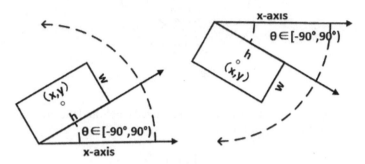

Fig. 1. Five-parameter method with an angle range fo 180°

The five-parameter method with an angle range of 180° has an PoA problem, as shown in Fig. 2 when the target is at the boundary value, the rotation angle of box ① is −90°, box ① rotates 1° clockwise in the upper right corner into box ② position; at this time the angle is 89°; while box ① rotates 1° counterclockwise to reach box ③, the rotation angle is −89°, although the actual error value is small in this rotation, the angle by definition produces a jump of one cycle.

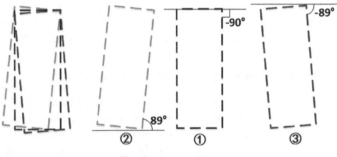

Fig. 2. PoA problem

Therefore, the problem of boundary discontinuity due to the angular periodicity problem is solved by converting the regression problem of angles into a classification problem by adding 180 angular classification channels and using a CSL.

CSL involves circular label codes with periodicity [12], and the assigned label values are smooth and have some tolerance, by setting the window function allows the model

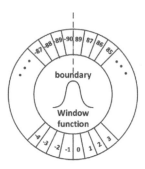

Fig. 3. CSL

to measure the angular distance between the predicted labels and the ground truth labels. The expression for CSL is as follows.

$$GSL(x) = \begin{cases} g(x), \theta - r \\ 0, \quad \text{otherwise} \end{cases} \tag{1}$$

where: g(x) is the window function, r is the radius of the window function, and θ is the angle of the current bounding box.

The Gaussian function is selected as the window function, the window radius is set to 6, and the final expression is shown below.

$$GSL(x) = \begin{cases} ae^{-(x-b)^2/2c^2}, \theta - 6 < x < \theta + 6 \\ 0, \quad \text{otherwise} \end{cases} \tag{2}$$

where: a, b and c are constant terms; x is the input angle information; and θ is the value of the angle to be rotated. Due to the setting of the window function, the closer the predicted value is to the true value within a certain range, the smaller the loss value is.

2.2 Courier Manifest Three-Segment Code Positioning

Through the center of the Logo and the rotation angle, combined with the relative position of the Logo and the three-segment code, you can locate the center point of the three-segment code. Then, according to the ratio of the length and width of the inspection frame of Logo to the three-segment code, locate the three-segment code. For the problem of angle deviation in the rotation detection of the courier Logo, which leads to errors in the positioning of the three-segment code, the Radon transform [13] based on the rotation projection is used to correct the tilt of the positioned three-segment code.

The functional expression for the two-dimensional Radon transformation is

$$R(\rho, \theta) = \iint_D f(x, y)\omega(\rho - x\cos\theta - y\sin\theta)dxdy \tag{3}$$

where: D is the whole image plane, and f (x, y) is the pixel gray value of point (x, y); characteristic function ω is Dirac function; as shown in Fig. 4, ρ is the distance from the

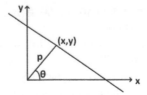

Fig. 4. Polar coordinate form of the straight line

straight line to the origin; θ is the angle between the x-axis and the vertical line from the origin to the line.

The angle against which the maximum projection value is selected is the tilt angle of the three-segment code by multi directional projection of the edge image of the three-segment code, and then the tilted three-segment code is rotated in the opposite direction to correct the three-segment code.

2.3   Courier Manifest Three Segment Code Identification

At present, the mainstream character recognition algorithms are feature-based method [14], template-based matching method [15] and deep learning-based method [16]. For the case of limited recognition target set and complete characters, the accuracy and efficiency of template matching are relatively high. Therefore, the template matching method is used to recognize three-segment codes. The images are first pre-processed with denoising, graying, binarization and character segmentation to improve the image quality and enhance the accuracy of the recognition algorithm.

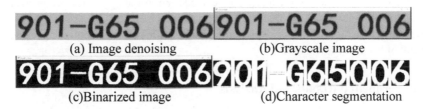

| (a) Image denoising | (b)Grayscale image |
| (c)Binarized image | (d)Character segmentation |

Fig. 5. Image pre-processing

After image pre-processing, the matched image is found by comparing the existing template with a single three-segment code character; the higher the degree of match between the two, the higher the likelihood of being identical.

3   Experimental Setup and Analysis of Results

3.1   Experimental Configuration and Data Set

The courier logo rotation detection training is implemented under the deep learning framework pytorch with the following hardware configuration: Windows 10 system, processor Intel(R) Xeon(R) CPU E5–2640 @2.60GHZ, memory capacity of 32.00GB, and graphics card model GTX1080Ti.

The main source of the dataset was taken by logistics companies, and the samples were distributed as follows: 2766, 2750, 2496, 2457 and 2318 for YTO, ZTO, STO, BEST and Yunda respectively. 12,787 courier face sheet images are included, of which 11,510 are used as the training set and 1277 as the test set. The image annotation tool rolabelimg was chosen for the dataset.

3.2 Courier Bill Logo Rotation Detection Model Training

Since there are differences in the Logo patterns of each company, the models are trained separately according to different courier companies. The courier logo detection model training loss function is shown in Fig. 6.

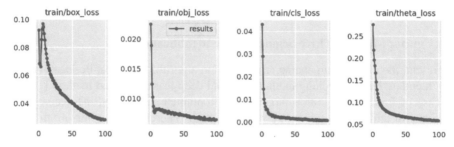

Fig. 6. Courier logo detection model training loss function

Evaluation of the model on a test set. As can be seen from Fig. 7, the average accuracy mean value obtained is 93.6% and the detection speed is 0.02–0.03s. The courier Logo rotation detection algorithm based on rotating frame with YOLOv5 achieves good results in terms of accuracy and speed.

3.3 Experimental Results and Analysis

Courier Face Sheet Logo Rotation Detection Experiment. The effect of courier Logo rotation detection is shown in Fig. 8, the algorithm has good detection effect for each company's courier Logo as well as its rotation angle.

Four commonly used rotating target detection networks were selected for comparison with this network under each test set: (1) R^3Det; (2) R^2CNN; (3) SCRDet; (4) RRPN; (5) YOLOv5s + CSL. The algorithm accuracy is evaluated by using the mean average precison(mAP) method. The calculation formula is

$$mAP = \frac{\sum_{i=1}^{k} AP_i}{k} \quad (4)$$

The mAP and the detection time are shown in Table 1.

The results show that the improved network based on YOLOv5s outperforms other network structures in terms of detection speed and also has high accuracy.

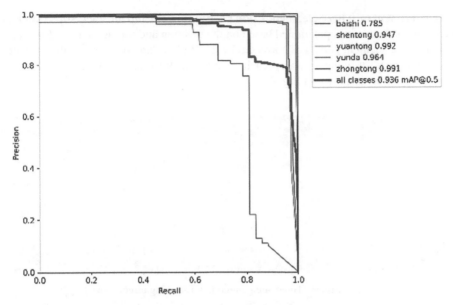

Fig. 7. Logo detection model training precision-recall curve

Fig. 8. Results of rotation detection of some courier logos

Table 1. Comparison of different algorithms.

Web	mAP/%	Time/s
R^{3}Det	90.8	0.18
R^{2}CNN	89.8	0.21
SCRDet	94.2	0.37
RRPN	93.3	0.42
YOLOv5s + CSL	93.6	0.03

Experiment on Three-Segment Code Positioning of Courier Face Slips. After the three-segment code was positioned based on the position and rotation angle of the logo, and the three-segment code was corrected by the Radon transform-based tilt correction algorithm, the positioning accuracy reached more than 99%, and the three-segment code positioning effect is shown in Fig. 9.

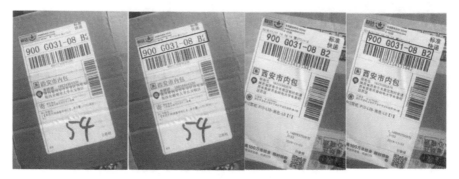

Fig. 9. Three-segment code positioning effect

Experiment on three-segment code recognition of courier face slips. Evaluation of three-segment code recognition algorithm performance on a test set.The recognition results are shown in Fig. 10, and it can be seen that the character recognition algorithm based on template matching can accurately recognize the three-segment code.

Fig. 10. Three-segment code recognition result

4 Concluding Remarks

For courier three-segment code positioning detection and recognition, a combination of YOLOv5 and rotating frame is used, and courier Logo is used to assist in detecting three-segment codes, and finally the three-segment codes are recognized. The experimental results show that the accuracy of logo detection by the courier three-segment code detection and recognition algorithm based on the rotating frame and YOLOv5 reaches 93.6%, and the accuracy of positioning and recognition of three-segment code after the logo is correctly detected reaches more than 99% effectively improving the accuracy and speed of three-segment code detection and recognition.

References

1. Jianfeng, Y.: Analysis of logistics express sorting methods in the context of e-commerce. Mod. Trade Ind. **42**(23), 25–26 (2021)
2. Weimin, Z., Gongmu, Z.: Design of dispatch information code based on automatic sorting application. Postal Res. **37**(01), 4–8 (2021)
3. Konovalenko, I., Maruschak, P., Brevus, V.: Steel surface defect detection using an ensemble of deep residual neural networks. J. Comput. Inf. Sci. Eng. **22**(1), 1–8 (2021)
4. Redmon, J., Divvala, S., Girshick, R., et al.: You Only Look Once: Unified, Real-Time Object Detection. IEEE (2016)
5. Xu, D.-G., Wang, L., Li, F.: A review of research on typical target detection algorithms for deep learning. Comput. Eng. Appl. **57**(08), 10–25 (2021)
6. Jiang, Y., Zhu, X., Wang, X., et al.: R2CNN: rotational region CNN for orientation robust scene text detection (2017)
7. Ma, J., Shao, W., Ye, H., et al.: Arbitrary-oriented scene text detection via rotation proposals. IEEE Trans. Multimedia PP(99):1–1 (2017)
8. Zhang, G., Lu, S., Zhang, W.: CAD-Net: a context-aware detection network for objects in remote sensing imagery. IEEE Trans. Geosci. Remote Sens. PP(99):1–10 (2019)
9. Yang, X., Yang, J., Yan, J., et al.: SCRDet: towards more robust detection for small, cluttered and rotated objects. In: 2019 IEEE/CVF international conference on computer vision (ICCV). IEEE (2019)
10. Yang, X., Liu, Q., Yan, J., et al.: R3Det: refined single-stage detector with feature refinement for rotating object (2019)
11. Zhao, L., Wang, X., Zhang, Y., Zhang, M.: Research on vehicle target detection technology based on YOLOv5s fusion SENet. J. Graphology 1–8 (2022)
12. Yang, X., Yan, J.: on the arbitrary-oriented object detection: classification based approaches revisited. Int. J. Comput. Vision **130**(5) (2022)
13. Lixia, G., Yanping, B.: Application of radon transform in tilted license plate image correction. J. Test. Technol. **23**(05), 452–456 (2009)
14. Zhang, F., Wang, X., Hao, X.: Intelligent vehicle character recognition based on edge features. Autom. Instrum. (06):11–14 (2020)
15. Ladislav, K., Elena, P.: comparative study of feature extraction and classification methods for recognition of characters taken from vehicle registration plates. Imaging Sci. J. **68**(1), 56–68 (2020)
16. XinSheng, Z., Yu, W.: Industrial character recognition based on improved CRNN in complex environments. Comput Ind 142 (2022)

A Method for Classification of Skin Cancer Based on VisionTransformer

Xuebin Xu, Haichao Fan[✉], Muyu Wang, Xiaorui Cheng, and Chen Chen

School of Computer Science and Technology, Xi'an University of Posts and Telecommunications, Xi'an Shaanxi 710121, China
xuxuebin@xupt.edu.cn, {fhc7911,wmy961017,cxxxxxr, 2103210107}@stu.xupt.edu.cn

Abstract. Human skin, long exposed to air and sunlight, is the body's first line of defense against natural hazards, which may be breached in a variety of ways, leading to cancer. According to the current level of clinical treatment, the death rate of skin cancer is still high, and it is also a kind of cancer with high treatment cost. Aiming at the current situation that the existing medical technology relies too much on the diagnosis of clinicians, we propose a method which could help greatly reduce time resuming and energy of clinical treatment utilizing artificial intelligence combined with neural network, this paper proposes a method of skin cancer image recognition and classification based on VisionTransformer (ViT) network classification model, which is greatly different from traditional Convolutional Neural Network (CNN) [1]. At the same time, SGD optimization method was used to adjust the learning rate, so as to realize the recognition and classification of skin cancer images. Experimental results show that the accuracy of this model on dataset reaches 85.3%, which is an important application of ViT model in the field of image classification.

Keywords: Skin cancer · Neural network · Vision transformer

1 Introduction

In recent years, skin canceration is a common malignant tumor [2]. The reason is that skin is the body's first defense mechanism, and human skin is directly exposed to air and sunlight, so it may cause skin cancer due to a variety of reasons (such as environmental, food and genetic factors). The malignant tumor on the skin is caused by the canceration of the skin. According to different types of tumor cells, it can be divided into epidermis, skin soft tissue, melanocyte, skin lymphatic reticular tissue and hematopoietic tissue. Observational studies of skin cancer degeneration are now primarily done through doctors' eyes. There are four steps: clinical screening, dermoscopy for analysis, biopsy, and histopathology. Classifying skin cancer using automated images is a daunting challenge.

Among many types of cancer, the high incidence of skin cancer makes it one of the most serious health problems in the world. With the continuous change of people's living environment, melanoma is increasing in skin cases. For melanoma patients, the survival rate is 96% in the early stage and only 14% in the advanced stage. This very significant difference highlights how early detection and effective treatment of skin cancer is a key means of reducing its incidence.

In fact, the incidence of skin cancer in China is pretty low, but it is one of the most common malignant tumors among white people. Data showed that in 1988, data from Shanghai Urban Cancer Institute showed that the incidence of malignant tumors in Shanghai urban area was 1.53/100,000 except for melanoma. Secondly, Boring et al. reported that in 1991, there were 600,000 new cases of skin canceration in the United States, excluding malignant melanoma. It is estimated that 40 to 50% of white Americans who survive to age 65 have at least experienced cutaneous cancer.

Skin cancer diagnosis, which initially focused on visual observations, has long been controversial because of differences between observers, even among experienced medical experts, and misdiagnosis. So, for this reason, neural network models in ARTIFICIAL intelligence have only recently been used to diagnose skin cancer. Basically, object recognition uses machine learning methods. In order to improve the ability of target recognition, we should collect larger data sets, learn powerful knowledge modules, and apply excellent techniques to reduce the problem of over-fitting. This technology analyzes images of skin lesions and is manipulated by machine learning software to provide objective data on melanoma to doctors. The AI was trained using thousands of images of skin and their corresponding melanin and hemoglobin levels. Not only can this reduce unnecessary biopsies, but it can also save a lot of money on medical expenses. It can provide doctors with objective information about the characteristics of lesions so that they can take less invasive treatments to help them rule out melanoma. Different from traditional CNN, which was widely used in the field, such as disaster management [3], sentiment analysis [4], medical image analysis [5]. However, we adopt a new method different from the conventional CNN, and even obtain better results.

2   Related Work

2.1   Dataset

The dataset used in this paper is ISIC 2018 Skin Lesions Classification dataset, which is publicly published by ISIC (International Skin Imaging Collaboration). It mainly contains two dermoscope image datasets, which are independent, namely, training set and test set. The training set contained 2239 images and the test set contained 118 images. The training set was divided into nine categories, including actinic keratosis, basal cell carcinoma, etc. The test set also includes dermoscope images of these diseases. The sample images of the dataset are shown in Fig. 1.

2.2   Data Preprocessing

Since the small-scale and deficiency of public datasets have impeded the training of neural networks for the computer-assisted diagnosis of pigmented skin lesions. We solved this problem by performing data Augmentation. In view of this diversity, we must adopt different methods and use specially trained neural networks to develop semi-automatic

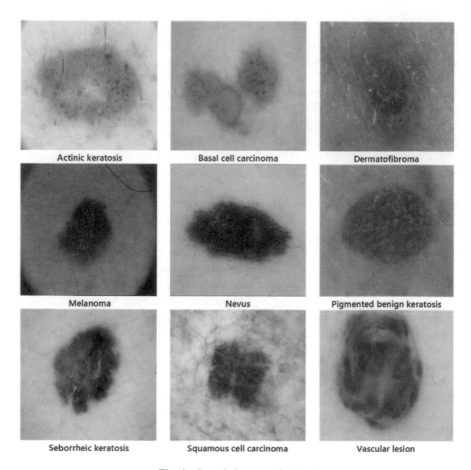

Fig. 1. Sample images of dataset

workflows. In order to improve the robustness of the network model and reduce over-fitting, it is necessary to adopt a data enhancement strategy to expand the training data set. The operation of data enhancement includes two contents: horizontal flip and rotation (90°, 180°), which can be shown in Fig. 2.

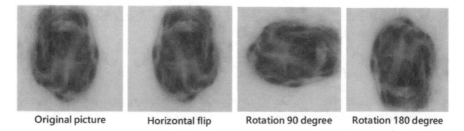

Fig. 2. Sample images of dataset augmentation

3 Method

3.1 Transformer

Transformer [6] is a model that uses the attention mechanism to increase the speed of model training. Transformer can be said to be a deep learning model completely based on the self-attention mechanism, because it is suitable for parallel computing, and the complexity of its own model leads to its higher accuracy and performance than the previously popular RNN recurrent neural network.

Starting in 2020, Transformer is starting to shine in the CV space: Image classification, object detection, Semantic segmentation, image generation and more. Nevertheless, because of the sudden prevalence of deep learning, CNN has become the main current model in CV field and achieved good results, at the same time, Transformer govern the NLP field. Exploration in CV with Transformer is precisely what the researchers want to utilize Transformer's experience in NLP field for reference in CV field. The core of transformer is the self-attention mechanism, which can effectively obtain remote indirect information. But training speed is slower than convolutional network.

ViT is a model that applies a transformer directly to image classification. The idea behind ViT is straightforward: divide the image directly into fixed-size patches, and then get the patch embeddings through linear transformations, which are similar to NLP words and word embeddings. Since the input of the Transformer is a series of token embeddings, it is possible to send patch embeddings of the image to the Transformer for feature extraction and classification. The structure of the ViT is shown as Fig. 3.

Patch Embedding. For the ViT, the original 2-d image was first converted into a series of 1-D patch embeddings. The input 2-D image is denoted as $x \in \mathbb{R}^{H \times W \times C}$, where H and W are the height and width of the image respectively, and C is the number of channels, which is 3 for RGB images. if the pattern is to be divided into patches of size $P \times P$ a sequence of Patches can be: $X_P \in \mathbb{R}^{N \times (P^2 \cdot C)}$, the image is segmented into $N = HW/P^2$ patches, which is also the length of sequence. Note that the patch is directly flattened into 1-D and its feature size is $P^2 \cdot C$. Patches is then mapped to the dimension of D size by a simple linear transformation, which is patch Embeddings: $X'_P \in \mathbb{R}^{N \times D}$.

Position Embedding. In addition, the model requires another special location embedding. Unlike CNNs, Transformer requires location embedding to encode the Token's location information. This is mainly because self-attention is constantly aligned. randomly arranging the order of to-ken in the sequence does not change the result. If the location information of the patch is not provided to the model, the puzzle needs to be learned through the patch semantics, which increases the cost of learning. ViT compared several different location embedding schemes and ultimately found that other types of location embeddings had similar effects if no location embedding was provided. This is mainly due to the fact that vit input is larger than pixel, so learning location information is much easier.

Transformer Encoder. The core operation of transformer is self-attention, which aggregates input information based on the current query by placing different weights on it. Attention has three main concepts: query, key and value, where key and value are paired. Specific structure can be shown in Fig. 4 for a given query vector $q \in \mathbb{R}^d$, k key

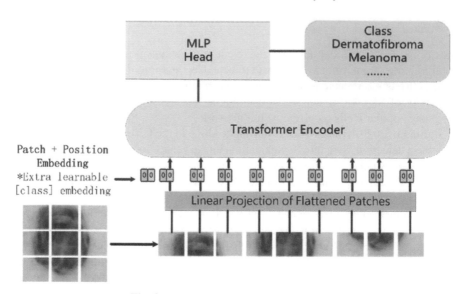

Fig. 3. The structure of vision transformer

vectors (dimension is also d, stacking up is matrix $K \in \mathbb{R}^{k \times d}$) are matched by inner product calculation. The inner product obtained is normalized by SoftMax to obtain K weights. For query, the output of attention is the weighted average of the value vectors corresponding to k key vectors (I.E. the matrix $V \in \mathbb{R}^{N \times d}$). For a series of N queries, we can calculate their attention output using the matrix:

$$Attention(Q, K, V) = Softmax\left(\frac{QK^T}{\sqrt{d_k}}\right) \tag{1}$$

$\sqrt{d_k}$ here is the scaling factor to avoid the variance effect of the dot product. The aforementioned Attention mechanism is called Scaled dot Product Attention. If Q, K, V are all derived from a linear transformation of a sequence ($X \in \mathbb{R}^{N \times D}$): $Q = XW_Q, K = XW_K, V = XW_V$ containing N vectors, then it becomes self-attention, and then there are N key pairs, then $k = N$. Self-attention is the most important part of Transformer. In fact, self-attention is the mutual attention between input vectors to learn new features. As mentioned above, we have obtained the patch sequence of the image, so the sequence output of the same size can be obtained after self-attention, but the features are changed.

3.2 Experiment

In our experiment, CPU was Xeon (R) E5–2673 V3, memory was 64G, GPU was Nvidia GeForce RTX 3070, and video memory was 32G. To facilitate the downloading and management of Python packages, the framework was PyTorch and the programming language Python were used. Conda was used for installation and configuration, which allows to switch randomly between multiple versions of Python without causing conflicts.

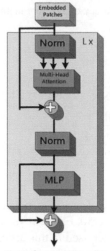

Transformer Encoder

Fig. 4. Structure of transformer encoder

In this experiment, we used ViT model as our training model. Batch size was set as 32, network image size was set as 224 * 224, network channel was set as 3, optimization function was SGD, momentum was 0.9, initial learning rate was 0.001, the number of training epochs is 300. The network training result graph is shown in Fig. 5.

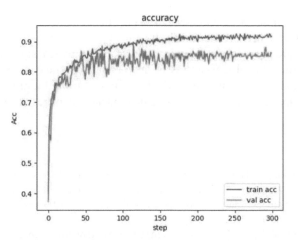

Fig. 5. Recognition result graph

After 200 epochs of network training, the training is completed and the convergence stable state is achieved after 200 epochs.

We have compared the accuracy of this network with other method, and the specific data is shown in the Table 1.

Table 1. The accuracy of different method on the dataset

Number	Method	Accuracy (%)
1	ViT	85.3
2	CNN-PA [8]	55.4
3	EfficientNet	74.2
4	HSV [10]	74.75
5	Inception-v3 [9]	80
6	ResNet50	83.4

By comparing the figures above, it can be seen that the convolutional neural network methods such as inceptionV3, resnet50, EfficientNet, etc. are used, and their effects are lower than 85.3% of the transformer used in this article. These effects benefit from the transformer's unique encoder-decoder structure.

After comparing the data in the table, we can get the good performance of Vision-Transformer. By setting certain parameter values, we not only achieve good results in the small size of the data set, but also reduce a lot of unnecessary operations. Greatly reduce the complexity of network as well as the training difficulty. The performance of our method is much higher than that of traditional convolutional networks.

4 Conclusion

Skin cancer is one of the most common cancers in human beings in recent years. Timely detection of this case and effective early treatment are very effective to help, and the mortality will be greatly reduced. Therefore, it is of great practical significance to use the advantages of artificial intelligence to assist doctors in non-contact automatic diagnosis. In this paper, transformer architecture, which is different from traditional convolutional neural network, applied to skin cancer classification and better results are obtained.

References

1. LeCun, Y., Bengio, Y., Hinton, G.: Deep learning. Nature **521**(7553), 436–444 (2015)
2. Rogers, H.W., Weinstock, M.A., Feldman, S.R., et al.: Incidence estimate of nonmelanoma skin cancer (keratinocyte carcinomas) in the US population, 2012. JAMA Dermatol. **151**(10), 1081–1086 (2015)
3. Yar, H., Hussain, T., Khan, Z.A., et al.: Vision sensor-based real-time fire detection in resource-constrained IoT environments. Comput. Intell. Neurosci. (2021)
4. Feldman, R.: Techniques and applications for sentiment analysis. Commun. ACM **56**(4), 82–89 (2013)
5. Muhammad, K., Ullah, H., Khan, Z.A., et al.: WEENet: an intelligent system for diagnosing COVID-19 and lung cancer in IoMT environments. Frontiers Oncol, 11 (2021)
6. Han, K., et al.: A survey on vision transformer. IEEE Trans. Pattern Anal. Mach. Intell. https://doi.org/10.1109/TPAMI.2022.3152247 (2022)

7. Dosovitskiy, A., Beyer, L., Kolesnikov, A., et al.: An image is worth 16 x 16 words: transformers for image recognition at scale. arXiv preprint arXiv:2010.11929 (2020)
8. Esteva, A., Kuprel, B., Novoa, R.A., et al.: Dermatologist-level classification of skin cancer with deep neural networks. Nature **542**(7639), 115–118 (2017)
9. Szegedy, C., Vanhoucke, V., Ioffe, S., et al.: Rethinking the inception architecture for computer vision. In: Proceedings of the IEEE Conference on Computer Vision and Pattern Recognition. 2016: 2818–2826 (2016)
10. Pham, T C., Tran, G S., Nghiem, T P., et al.: A comparative study for classification of skin cancer. In: 2019 International Conference on System Science and Engineering (ICSSE). IEEE, 2019: 267–272 (2019)

MOOC Courses Recommendation Algorithm Based on Attention Mechanism Enhanced Graph Convolution Network

Xiaoyin Wang and Xiaojun Guo[⊠]

Xi'an University of Posts and Telecommunications, Xi'an 710121, China
xywang@xupt.edu.cn, gxj@stu.xupt.edu.cn

Abstract. With the widespread popularity of massive online courses, facing the problem of information overload and users' difficulty in choosing suitable courses for themselves. Although the recommendation algorithm based on graph convolution network can extract the high-order correlation between users and items, but it ignores the influence of different neighbor nodes on the target node and cannot assign different weights to neighbor nodes. We propose a MOOC Courses Recommendation Algorithm Based on Attention Mechanism Enhanced Graph Convolution Network (AEGCN), apply it to MOOC courses recommendations areas, high-order association between nodes is extracted by graph convolution network; according to the importance of different neighbor nodes, the attention mechanism is used to assign different neighbor weights to the neighbor nodes of each node, so that the weights of nodes depend on the feature expression between nodes, and finally fuse the embedding representation of each layer to predict the user's preference. The recommendation results on two real datasets show that the performance indicators of this algorithm are better than other recommendation algorithms. Compared with the optimal recommendation algorithm, its Recall indicators are increased by 1.09% and 0.96% respectively, and it shows that the algorithm has better recommendation performance.

Keywords: Recommender system · Graph convolution network · Attention mechanism · Courses recommendation

1 Introduction

In recent years, with the rapid development of the Internet and the growing demand for educational resources, large-scale online course platforms such as MOOC have developed rapidly. However, in the face of large-scale MOOC platform resources, the problem of information overload is increasingly seriously, recommender system is used to solve the problem of information overload, it can find items that users may be interested in and recommend them to users based on their historical data. In the field of education [1], to solve the difficulty of users in finding relevant courses, by analyzing the historical behavior data of users on the MOOC platform, the recommendation system can

N. Xiong et al. (Eds.): ICNC-FSKD 2022, LNDECT 153, pp. 218–227, 2023.
https://doi.org/10.1007/978-3-031-20738-9_26

accurately recommend the required courses to students from a large number of course resources.

Today, deep learning models are widely used in the field of recommendation systems. Among all deep learning algorithms, Graph Convolution Network (GCN) is undoubtedly one of the most effective technologies [2], it can effectively capture high-order correlations between nodes through multi-layer graph convolution operations. However, it cannot aggregate the different influences of neighbor nodes for effective modeling, ignore the degree of influence of different neighbor nodes, and make the node features depend on the graph structure, reducing the recommendation accuracy. Aiming at the above problems, we propose a MOOC Courses Recommendation Algorithm Based on Attention Mechanism Enhanced Graph Convolution Network (AEGCN), which uses the graph convolution network to extract the high-level node features of users and courses, the attention mechanism is used to assign corresponding weights to different neighbor nodes to improve the generalization ability of the model. Finally, the node embedding representation obtained in different convolutional layers is subjected to mean aggregation operation as the final embedding of prediction. The proposed algorithm is compared with the benchmark algorithms commonly used in the recommendation field in two real public datasets, which proves the rationality of the proposed algorithm and its effectiveness on the MOOC dataset.

2 Related Work

2.1 Recommendation Algorithms for MOOC

Most research on MOOC recommendation algorithms mainly use collaborative filtering algorithms. Chen et al. [3] proposed a content-based recommendation algorithm, which compares course content with learner behavior to achieve Top-N recommendation. Esteban et al. [4] proposed a hybrid recommendation method for MOOC courses, using attribute information related to students and courses to recommend appropriate courses to students. However, the above methods cannot build a recommendation model with better performance, and the recommendation performance is poor.

2.2 Recommendation Algorithms Based on Graph Convolution Network

Different from collaborative filtering algorithms, graph convolution networks are widely used, because they can capture higher-order correlations of nodes. GC-MC [5] aggregates the features of neighbor nodes for each node through the graph convolution layer and obtains a set of vectors representing the node features, but it only focuses on the first-order neighbors of the node. NGCF [6] combines collaborative filtering algorithm with graph convolution network to refine the embedding propagation layer by mining high-order associations between users and items. He et al. [7] proposed a lightweight graph convolution network, namely LightGCN, which removes the nonlinear activation function and feature transformation matrix in traditional GCN and uses a simple weighted sum operation to aggregate nodes in each layer.

2.3 Attention Mechanism

In recent years, attention mechanism has been widely used in recommendation systems to extract important features of users and items and give them more weight. The DAML model [8] proposed by Liu et al. uses a dual attention mechanism to extract the correlation of potential features between users and comments. Velikovi et al. proposed a graph attention network (GAT) [9], which assigns weights to different neighborhood nodes through self-attention mechanism, avoiding the loss of important node information during the propagation process.

3 AEGCN Model

The model framework proposed in this paper is shown in Fig. 1. It consists of three parts: embedding layer, propagation layer and prediction layer. The embedding layer is responsible for encoding the nodes of users and courses in the bipartite graph, taking the initial embedding expression of users and courses as the input of the model; The propagation layer obtains the high-level embedding expression between nodes and adopts the attention mechanism to assign different neighbor weights to the neighbor nodes of each node; The prediction layer generates the final representation of users and courses by aggregating the embedding representations of each layer.

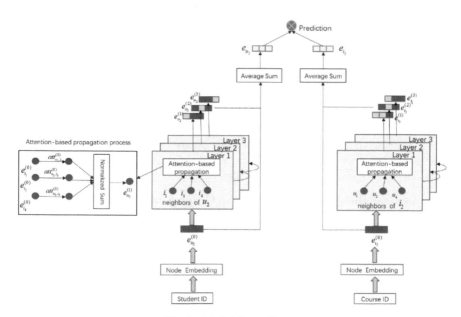

Fig. 1. Model frame diagram

3.1 Attention-Based Propagation Process

Firstly, use the historical interaction information of the user and the course to construct a bipartite graph of the interaction between the user and the course, encode the nodes in the bipartite graph and sort them according to a certain dimension, embedding generates the initialization embedding vectors of users and courses.

A standard graph convolution layer consists of feature transformation, neighborhood aggregation and nonlinear activation function, but a recent study shows that the two traditional designs of feature transformation and nonlinear activation function do not have a positive impact on collaborative filtering algorithms. In this paper, a propagation method like to LightGCN is adopted to extract the relationship between users and courses. Take user node u as an example, The calculation rule for one propagation in the graph convolutional layer is as follows:

$$e_u^{(1)} = \sum_{i \in N_u} \frac{1}{\sqrt{|N_u|}\sqrt{|N_i|}} e_i^{(0)} \tag{1}$$

where N_u are the neighbor nodes set of node u in the user-course bipartite graph, $e_u^{(1)}$ is the expression of the first hop embedding of node u, $e_i^{(0)}$ is the initial embedding expression of its neighbor node i, $1/\left(\sqrt{|N_u|}\sqrt{|N_i|}\right)$ is symmetric normalization.

Most GCN-based recommendation methods do not consider the degree of influence of different neighbor nodes on the target node when performing graph convolution operations, resulting in the loss of useful information for the target node and inaccurate recommendation results. Inspired by Literature [9] and Literature [10], the attention mechanism is introduced to calculate the embedding weight of different neighbor nodes to the target node, and softmax function is used to normalize, so that the target node can effectively make use of the information of the neighbor node. The calculation rules of the attention weight of the neighbor node i at layer l-1 is as follows:

$$att_{u,i}^{(l-1)} = \text{softmax}(g(w_1 e_u^{(l-1)}, w_2 e_i^{(l-1)})) = \frac{\exp((w_1 e_u^{(l-1)})^T w_2 e_i^{(l-1)})}{\sum\limits_{j \in N_u} \exp((w_1 e_u^{(l-1)})^T w_2 e_j^{(l-1)})} \tag{2}$$

where w_1 and w_2 is the weight parameter, $e_u^{(l-1)}$ and $e_i^{(l-1)}$ is the embedding expressions of the l-1th layer of node u and neighbor node i under the curriculum interaction bipartite graph, $att_{u,i}^{(l-1)}$ is the weight coefficient of the neighbor node i to the target node u on the l-1th convolutional layer.

After calculating the embedding weight of each neighbor node, modify the propagation rule of the node in the graph convolution network layer, and update the embedding representation of the node. The calculation rule of the l-layer propagation is as follows:

$$e_u^{(l)} = \sum_{i \in N_u} \frac{att_{u,i}^{(l-1)}}{\sqrt{|N_u|}\sqrt{|N_i|}} e_i^{(l-1)} \tag{3}$$

$$e_i^{(l)} = \sum_{u \in N_i} \frac{att_{i,u}^{(l-1)}}{\sqrt{|N_i|}\sqrt{|N_u|}} e_u^{(l-1)} \tag{4}$$

where $e_u^{(l)}$ and $e_i^{(l)}$ is the embedding expressions of node u and node i at layer l under the bipartite graph of courses interaction, N_i and N_u are the set of neighbor nodes of node i and node u in the interactive bipartite graph.

3.2 Rating Prediction

After going through the above steps, we can obtain the hierarchical embedding representation of users and courses, it is necessary to combine these embedding representations to generate the final representations of users and courses, and set uniform weights for the embeddings of each layer by means of mean aggregation:

$$e_u = \frac{1}{L+1}\left(\sum_{l=0}^{L} e_u^{(l)}\right) \tag{5}$$

$$e_i = \frac{1}{L+1}\left(\sum_{l=0}^{L} e_i^{(l)}\right) \tag{6}$$

We inner product the final embeddings of users and courses to get the predicted scores for the courses by the users:

$$y_{ui} = (e_u)^T e_i \tag{7}$$

3.3 Model Optimization

We use BPR loss to optimize model parameters, it considers that observed items can better reflect the user's preference than unobserved items and should be given a higher prediction score. The loss function formula is as follows:

$$L_{BPR} = -\sum_{(u,i,j)\in T} \ln \sigma\left(y_{ui} - y_{uj}\right) + \lambda \left\|E^{(0)}\right\|^2 \tag{8}$$

where $T = \{(u,i,j)|(u,i)\in N^+,(u,j)\in N^-\}$, N^+ represents the observed interactions, N^- represents the unobserved interactions, λ is an adjustable regularization parameter, $E^{(0)}$ is the initial embedding expressions for users and items, this paper uses the Adam algorithm as the optimizer to optimize this loss function.

4 Experimental Results and Analysis

4.1 Experimental Datasets and Evaluation Metrics

We use real datasets from two different scenes, MOOC, Movielens-100k, two datasets differ in number of users, number of items, and sparsity. The relevant information statistics are shown in Table 1.

The MOOC dataset is a dataset provided by Xuetang Online, it contains 1302 courses, 82,535 users and 458,454 interactive data. The dataset includes related attributes such

Table 1. Dataset related information

Dataset	Users	Items	Interactions
MOOC	8109	682	131182
Movielens-100k	943	1682	100000

as user ID, course ID, course category, etc., preprocess the dataset to remove useless information, it only keeps user ID and course ID and deletes users with less than 10 interactions.

The Movielens-100k dataset is a dataset in the field of movie recommendation provided by GroupLens, it contains 943 users, 1682 movies, and about 100,000 explicit ratings data, to facilitate the training of the model, the explicit scoring data is changed to invisible data.

This algorithm belongs to the top-K recommendation type, it recommends a group of top-ranked items to the user. Therefore, Recall@K and NDCG@K commonly used in the top-K recommendation type is used to evaluate the performance of the algorithm. The value of K is 20.

4.2 Compare Models and Parameter Settings

To show the effectiveness of this model, several commonly used recommendation models in the recommendation field are selected for comparison. The descriptions of the comparison models are as follows:

NeuMF [11]: combining matrix factorization techniques with multilayer perceptrons, using multilayer perceptrons to capture nonlinear interaction features between users and items.

ENMF [12]: proposes a neural network recommendation model without sampling and improves the effectiveness of the neural network model through a learning strategy based on all data.

GC-MC: a graph encoder framework for matrix completion is proposed by generating implicit features for users and items through the encoder; at the same time, the matrix completion task is transformed into a link prediction problem in the bipartite graph of user-item interaction.

NGCF: combining graph convolution networks with collaborative filtering and applying bipartite graphs of user and item interactions to the embedding layer, a multi-layer graph convolution network is used to obtain high-order correlations between users and items.

LightGCN: removes redundant nonlinear activation functions and feature transformation matrices and simplifies the structure of GCN.

To ensure the fairness and feasibility of the experiment, the experimental parameters are set, the dataset is divided into a ratio of 8:1:1, the learning rate is set to 0.001, the batch-size is set to 2048, and the dimension of the embedding vector is 64. The data sparsity of different datasets is different, which may affect the experimental results, so the settings of the number of embedding layers for the two datasets are adjusted. The remaining parameter settings in the comparison model remain unchanged.

4.3 Experimental Results and Analysis

The comparison results of the model proposed in this paper and other models on the two datasets are shown in Table 2, with the best performance of the model on the dataset in bold.

Table 2. Model performance

Model	MOOC		Movielens-100k	
	Recall (%)	NDCG (%)	Recall (%)	NDCG (%)
NeuMF	44.06	33.9	34.21	28.51
ENMF	44.34	34.3	35.45	29.86
GC-MC	45.77	34.76	36.87	30.62
NGCF	45.93	35.02	37.57	31.86
LightGCN	46.7	35.79	38.21	32.43
AEGCN	**47.21**	**36.14**	**38.58**	**32.7**
Improve	1.09	0.98	0.96	0.83

NeuMF model has the lowest metrics on both datasets. The ENMF model applies the neural network model to the matrix factorization model. Compared with the NeuMF model, the recommendation performance of ENMF model is improved, but two models are essentially a kind of matrix factorization model, and it is difficult to capture the characteristics of users and items on sparse datasets. Therefore, the improvement effect is not obvious.

The GC-MC model completes matrix completion by combining GCN and encoder. Compared with the NeuMF and ENMF models, its recommendation performance is better, and the effect is significantly improved, which proves the effectiveness of adding first-order neighborhoods.

The NGCF model extracts the high-level effective information of the project through the graph convolution network. Compared with the ENMF model, the recommendation result of the NGCF model is improved by 4.2%, indicating that when the dataset is sparse, GCN can effectively alleviate the data sparse problem, thereby improving the recommendation performance.

Compared with the NGCF model, the performance of the LightGCN model is significantly improved on the two datasets, indicating that the existence of nonlinear activation functions and feature transformation matrices in the traditional graph convolution network model will reduce the recommendation accuracy.

The recommendation performance of the AEGCN model proposed in this paper is better than other recommendation models on both datasets. Compared with the Light-GCN model, on the MOOC dataset, the Recall is increased by 1.09%, and the NDCG is increased by 0.98%; on the Movielens-100k dataset, the Recall is increased by 0.96%, and the NDCG is increased by 0.83%, which verifies the importance of obtaining different neighbor nodes through the attention mechanism, it demonstrates the effectiveness of the proposed model, improves the recommendation performance, and proves the recommendability on the MOOC dataset.

4.4 The Influence of Embedding Layers

To verify the influence of embedding layers on recommendation performance, the embedding layers are set to 1 to 4, and other parameter settings remained unchanged. The experimental results are shown in Table 3.

Table 3. Embedding layer adjustment comparison experiment

Dataset		MOOC		Movielens-100k	
Layer	Model	Recall (%)	NDCG (%)	Recall (%)	NDCG (%)
Layer 1	AEGCN	46.54	35.44	38.25	32.35
Layer 2	AEGCN	46.82	35.88	38.21	32.48
Layer 3	AEGCN	46.96	35.96	38.58	32.7
Layer 4	AEGCN	47.21	36.14	36.75	31.08

On the MOOC dataset, the recommendation performance of the AEGCN model improves with increasing number of graph convolution layers. When the number of embedding layers is 4, the recommendation performance of the model is best; on the Movielens-100k dataset, the recommendation performance of this model is improved as the number of graph convolution layers increases, when the number of embedding layers is 3, the model effect is optimal. The reason for this is because the sparsity of the different datasets is different. When the dataset is sparse, the higher the number of embedding layers, the higher-order neighborhood information can be obtained.

4.5 Ablation Experiment

To verify the influence of nonlinear activation function, feature transformation matrix and attention mechanism on model recommendation performance, two variants of the AEGCN model are designed, and ablation experiments are performed on two real datasets to prove the effectiveness of this model (see Fig. 2).

AEGCN-1: the nonlinear activation function, feature transformation matrix, and attention mechanism are retained. The variant model is like adding an attention mechanism to the NGCF model to obtain the weight of neighbor nodes.

AEGCN-2: the nonlinear activation function, feature transformation matrix, and attention mechanism are removed, and its variant model is similar to the LightGCN model.

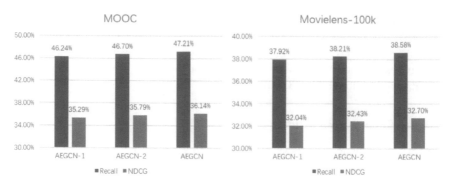

Fig. 2. Ablation experiments

It can be seen from Fig. 2, the recommendation performance of the AEGCN-1 model is lower than that of the AEGCN model, indicating that the existence of nonlinear activation functions and feature transformation matrices will increase the model burden and reduce the recommendation accuracy; the recommendation performance of the AEGCN model is better than that of AEGCN-2, indicating that it is effective to add attention mechanism to obtain the weights of neighbor nodes, which can improve the recommendation accuracy of the model.

5   Conclusion

Most recommendation algorithms based on graph convolution network ignore the influence of different neighbor nodes, a MOOC Courses Recommendation Algorithm Based on Attention Mechanism Enhanced Graph Convolution Network is proposed, and the algorithm is applied to the field of MOOC courses recommendation, to solve the problem that current users are difficult to choose related courses. It removes the nonlinear activation function and feature transformation matrix, obtains the relevant features of the node through the graph convolution network, and uses the attention mechanism to assign different neighbor weights to the neighbor nodes of each node, so that the weight of the node depends on the features between the nodes, enabling it to accurately capture user preferences. The proposed algorithm and the commonly used recommendation algorithm in the field of recommendation are experimentally verified on two real datasets, showing that the recommendation performance of the proposed algorithm is better than the comparison algorithm, and the recommendation reliability is higher.

Acknowledgments. This research was supported by the National Natural Science Foundation of China (Grant No. 61876138), the Key R&D Project of Shaanxi Province (Grant No. 2020GY-010), the Major Graduate Innovation Fund Project of Xi'an University of Posts & Telecommunications (Grant No. CXJJZL2021009).

References

1. Huang, Z., Liu, Q., Zhai, C., Yin, Y., Chen, E., Gao, W.: Exploring multi-objective exercise recommendations in online education systems. In: 28th ACM International Conference on Information and Knowledge Management, pp. 1261–1270. (2019)
2. Kipf, T.N., Welling, M.: Semi-supervised classification with graph convolutional networks. In: 5th International Conference on Learning Representations, pp. 1–14. (2017)
3. Chen, Y., Zhao, X., Gan, J., Ren, J.: Content-based top-n recommendation using heterogeneous relations. In: 27th Australasian Database Conference, pp. 308–320. (2016)
4. Esteban, A., Zafra, A., Romero, C.: Helping university students to choose elective courses by using a hybrid multi-criteria recommendation system with genetic optimization. Knowl. Based Syst. **194**, 105385 (2020)
5. Berg, R.V.D., Kipf, T.N., Welling, M.: Graph convolutional matrix completion. In: ACM SIGKDD: Deep Learning Day, pp. 145–155. (2018)
6. Wang, X., He, X., Wang, M., Feng, F., Chua, T.S.: Neural graph collaborative filtering. In: 42nd International ACM SIGIR Conference on Research and Development in Information Retrieval, pp. 165–174. (2019)
7. He, X., Deng, K., Wang, X., Li, Y., Zhang, Y., Wang, M.: Lightgcn: Simplifying and powering graph convolution network for recommendation. In: 43rd International ACM SIGIR Conference on Research and Development in Information Retrieval, pp. 639–648. (2020)
8. Liu, D., Li, J., Du, B., Chang, J., Gao, R.: Daml: Dual attention mutual learning between ratings and reviews for item recommendation. In: 25th ACM SIGKDD International Conference on Knowledge Discovery & Data Mining, pp. 344–352. (2019)
9. Veličković, P., Cucurull, G., Casanova, A., Romero, A., Lio, P., Bengio, Y.: Graph attention networks. arXiv preprint arXiv:1710.10903 (2017)
10. Song, W., Xiao, Z., Wang, Y., Charlin, L., Zhang, M., Tang, J.: Session-based social recommendation via dynamic graph attention networks. In: Twelfth ACM International Conference on Web Search and Data Mining, pp. 555–563. (2019)
11. He, X., Liao, L., Zhang, H., Nie, L., Hu, X., Chua, T.S.: Neural collaborative filtering. In: 26th International Conference on World Wide Web, pp. 173–182. (2017)
12. Chen, C., Zhang, M., Zhang, Y., Liu, Y., Ma, S.: Efficient neural matrix factorization without sampling for recommendation. ACM Transactions on Information Systems **38**(2), 1–28 (2020)

Network Traffic Anomaly Detection Based on Generative Adversarial Network and Transformer

Zhurong Wang$^{(\boxtimes)}$, Jing Zhou, and Xinhong Hei

School of Computer Science and Engineering, Xi'an University of Technology, Xi'an, China
{wangzhurong,heixinhong}@xaut.edu.cn, suiyuejinghao0626@163.com

Abstract. In order to solve the problem of low prediction accuracy of network traffic data and unbalanced sample data set, a network traffic anomaly detection method combining Generative Adversarial Network and Transformer is proposed. The model is built using Generative Adversarial Networks (GAN) and Vision Transformer. The original dataset is sampled, and the sampled imbalanced dataset is class-balanced through generative adversarial network, and the balanced dataset is input into the Vision Transformer for prediction. The data is normalized using the Max-Min criterion (Max-Min) method, and only N encoders are included in the Vision Transformer, each encoder consists of a multi-head self-attention layer and a feed-forward network layer. The CIC-IDS-2017 network intrusion detection data set is used for experimental evaluation. Through comparative experiments, it is proved that the proposed model can predict network traffic data with high accuracy, and also effectively solve the problem of network traffic data imbalance.

Keywords: Anomaly detection · Generative adversarial networks · Vision transformer · Class imbalance

1 Introduction

With the advent of the era of big data and the continuous maturity of artificial intelligence technology, intelligent digital products have gradually become an indispensable main force in people's pursuit of a convenient and fast life. According to the 48th "Statistical Report on China's Internet Development Status" recently published by the China Internet Network Information Center (CNNIC), As of June 2021, the number of Internet users in my country has reached 1.011 billion. It can be seen that my country has entered a new era of interconnection and big data.

In the Internet era, information transmission and communication are more portable, which inevitably leads to an increasingly complex network environment. According to the statistics of the National Information Security Vulnerability Sharing Platform (CNVD), there are 13,083 general-purpose security vulnerabilities, an increase of 18.2% year-on-year. Network intrusion seriously threatens the health of the network environment [1], and it becomes crucial to detect anomalies in network traffic.

© The Author(s), under exclusive license to Springer Nature Switzerland AG 2023
N. Xiong et al. (Eds.): ICNC-FSKD 2022, LNDECT 153, pp. 228–235, 2023.
https://doi.org/10.1007/978-3-031-20738-9_27

Nowadays, some researchers have done a lot of work on network traffic anomaly detection. Zhang et al. [2] proposed a network intrusion detection method based on auto-encoder network (AE) and long short-term memory neural network (LSTM). By building an auto-encoding network model, and then using the optimized LSTM model to train and predict network traffic. Hwang et al. [3] proposed D-PACK, an abnormal traffic detection mechanism composed of an unsupervised deep learning model and convolutional neural network (CNN), which can automatically analyze traffic patterns and filter abnormal traffic. Wei et al. [4] proposed a deep learning-based hierarchical spatiotemporal feature learning (HAST-HAD) anomaly detection method, using the spatial features of CNN learning data, the temporal features of RNN's long-short-term memory learning data, and the CNN and RNN. The combination effectively improves the accuracy of the model.

This paper takes network traffic anomaly detection as the research topic. In the continuous exploration of machine learning and deep learning, We proposed a GAN-Transformer model, and the model is implemented in Python language.

The structure of this paper is as follows. In Part 2, the proposed network traffic anomaly detection model is introduced in combination with the network traffic anomaly detection problem, and a generative adversarial network and a Vision Transformer are selected as the basic parts of the model. The third part is the introduction of experimental data and the construction of the GAN-Transformer neural network model. The fourth part is the evaluation index and experimental comparative analysis. The fifth part presents the research conclusions.

2 Anomaly Detection Model Based on Generative Adversarial Network and Transformer

2.1 Anomaly Detection Methods

Existing Anomaly Detection Methods are mainly divided into two categories, one is Ordinary Classification Methods, and the other is Deep Learning Methods based on neural networks. Common Classification Methods include K-Nearest Neighbor (KNN), Decision Tree and Support Vector Machines (SVM), etc. Deep learning methods based on neural networks include Convolutional Neural Network (CNN), Recurrent Neural Network (RNN), and Long Short-Term Memory (LSTM), etc.

2.2 Generative Adversarial Networks

Generative adversarial network (GAN) [5] uses antagonism training to make the samples generated by the generated network obey the true data distribution. The generative adversarial network is composed of a generator G and a discriminator D. The main task of G is to generate a fake sample close to the real data from random noise, and the discriminator tries to distinguish the real and fake samples as much as possible. GAN uses the generator and the discriminator to "game" each other, so that G learns the distribution of real samples. After the game is balanced, a random noise can be changed into a sample that is very close to the real data through the generator G. The objective optimization function of GAN As shown in formulas (1) and (2).

$$\max_{D} V(D, G) = E_{x \sim p_{data(x)}}[\log D(x)] + E_{z \sim p_{z(z)}}[\log(1 - D(G(z)))] \qquad (1)$$

$$\min_{G} V(D, G) = E_{z \sim p_{z(z)}}[\log(1 - D(G(z)))] \tag{2}$$

Formula (1) is used to train the discriminator D. Formula (2) is to keep the discriminator D unchanged and train the generator G.

In the above formula, $G(z)$ represents the output of G. $D(G(z))$ is the probability that D judges that the data synthesized by G is true. $D(x)$ is the probability that the discriminator will judge x as real data. The data x in formula (1) comes from real samples, so we hope D can recognize it as a real sample. Data z comes from random vectors, and $G(z)$ is the generated sample, so we hope D can recognize it as a false sample. Both items should be as large as possible, that is, the whole should be as large as possible. Formula (2) is to keep D unchanged and train G, because we want to confuse D. when training G, we hope that D will not recognize false samples as much as possible, that is, $D(G(Z))$ is as close to 1 as possible, that is, the whole is as small as possible.

2.3 Vision Transformer

Since the attention mechanism was proposed, it has received extensive attention from the academic community. In 2017, Google proposed the Transformer model in "Attention is your need" [6] and showed excellent performance.

Transformer is composed of N encoders and decoders. The encoder consists of a multi-head attention mechanism and a feedforward layer, and a layer is added after each multi-head attention mechanism and feedforward layer. Combined with the characteristics of anomaly detection, this paper selects the Vision Transformer [7]. Compared with the Transformer, it abandons the decoder part.

$$Attention(Q, K, V) = soft \max(\frac{QK^T}{\sqrt{d_k}})V \tag{3}$$

$$FFN(x) = \max(0, xW_1 + b_1)W_2 + b_2 \tag{4}$$

The QK dot product in formula (3) is a scalar, which reflects the similarity of two vectors: if the dot product is larger, it means that the two vectors are more similar. Formula (4) The fully connected layer is a two-layer neural network, which is linearly transformed first, then filtered by ReLU, and then linearly transformed.

2.4 Network Traffic Anomaly Detection Algorithm Based on GAN and Transformer

The algorithm steps are as follows: First, the network traffic data is preprocessed, and then the GAN is used to balance the preprocessed data. Then the Transformer model is trained, and the test set is put into the trained anomaly detection model for classification test. The flow of Anomaly Detection Algorithm Based on GAN and Transformer is shown in Fig. 1.

The Anomaly Detection Algorithm Based on GAN and Transformer is shown in Algorithm 1.

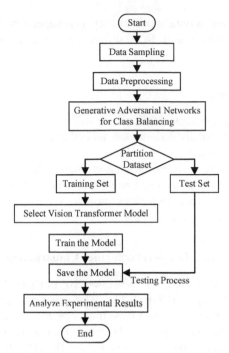

Fig. 1. Flow chart of anomaly detection model based on GAN and transformer

Algorithm 1 Anomaly Detection Algorithm Based on GAN and Transformer.

(1) Input raw data and sample.
(2) Cleaning and preprocessing the sampled data, including filling missing data, eliminating useless data, data standardization and normalization.
(3) Class Balancing Sampled Data with GAN.
(4) Divide training set and test set.
(5) Building an anomaly detection model based on Vision transformer.
(6) Use the training set to train the anomaly detection model and structural parameters based on the combination of GAN and Transformer, and obtain a trained anomaly detection model.
(7) Get the classification result by feeding the test set into the trained anomaly detection model.

3 Data Sample Introduction and Model Building

3.1 Data Sample Introduction

In this paper, the dataset uses the CICIDS2017 dataset [8]. This dataset collects network traffic data from Monday to Friday, with both normal and abnormal traffic. Abnormal traffic in network traffic includes malicious File Transfer Protocol(FTP), malicious Secure Shell (SSH), Denial of Service (Dos) and other attacks. According to statistics, there are 2,830,608 pieces of data in this dataset. This paper aims to detect and research the abnormal traffic in this dataset, so 10 different attack categories are selected from the entire dataset.

3.2 Data Preprocessing

The data preprocessing steps are as follows:

Step 1: The data sets are divided into test set and training set, of which 20% are test set and 80% are training set.

Step 2: Delete all outliers from the dataset, mainly Nan and Infinity. Converts a classification label to a number using a unique thermal encoding method.

Step 3: Use mean values for continuous variables to fill in missing values.

Step 4: Ultimately standardize all data.

3.3 Parameter Setting and GAN-Transformer Construction

The experiment was conducted on Windows 10 Intel(R) Core(TM) i5-11400H CPU@2.70GHz 2.69 GHz, 16.0 GB RAM. The generative adversarial network is implemented with Tensorflow and the Vision transformer is realized by pytorch.

The GAN-Transformer model proposed is divided into two parts. The first part firstly samples 11 categories of normal data and attack data. The GAN model optimizer selects Adam, the learning rate selects 0.0001, and the loss function Select crossentropy and choose 14,000 iterations. The second part is to input the balanced data into the Vision Transformer. The model parameters of visual transformer are as follows: the data volume of the training set of visual transformer is 88,000, the number of iterations is 20,000, the learning rate is 0.001, the number of hidden nodes is 128, the number of encoders is 6, the data volume of the test set is 22,000, the forgetting rate is set to 0.4, and the number of input features is 78.

This paper tests the proposed GAN-Transformer model and compares the prediction results of the SVM and Decision Tree Model and the model without class balance.

4   Test Study

4.1 Evaluation Indicators

The evaluation indexes of this experiment include Accuracy (A), Precision (P), Recall (R), and F1-score (F1). In the classification problem: TN (True Negative Prediction), n_{TN} (True negative prediction sample); TP (True Positive Prediction), n_{TP} (True positive prediction sample); FN (False Negative Prediction), and n_{FN} (False negative prediction sample); FP (False Positive Prediction), n_{FP} (False positive prediction sample).

$$A = \frac{n_{TP} + n_{TN}}{n_{TP} + n_{TN} + n_{FP} + n_{FN}} \tag{5}$$

$$P = \frac{n_{TP}}{n_{TP} + n_{FP}} \tag{6}$$

$$R = \frac{n_{TP}}{n_{TP} + n_{FN}} \qquad (7)$$

$$F_1 = \frac{2}{\frac{1}{P} + \frac{1}{R}} \qquad (8)$$

The definition of accuracy (A) in formula (5) represents the proportion of successful prediction samples in the total prediction samples.

The definition of precision (P) in formula (6) represents the proportion of predicted positive samples in the predicted positive samples.

The definition of recall (R) in formula (7) represents the proportion of the actual positive samples in the predicted successful positive samples.

The definition of F1-score(F1) in formula (8) represents the tradeoff between precision and recall.

4.2 Experiment Analysis

Experiments use SVM, Decision Tree and GAN-Transformer Model to test and compare the performance of the unbalanced dataset and the balanced dataset. The test results are shown in the Table 1, and Table 1 shows the accuracy rates obtained by the three models respectively. In terms of accuracy, neither a single SVM nor a Decision Tree Model can predict network traffic with high accuracy during training, and the GAN-Transformer Model has a stable test effect on the training set and test set. On the test set, the prediction accuracy of the Vision Transformer is 91% for the unbalanced dataset, and the prediction accuracy of the Vision Transformer is 97% for the dataset after using the Generative Adversarial Network balance. The GAN-Transformer Model has a 6% improvement in the accuracy of the prediction results of the unbalanced dataset, and has better prediction accuracy.

Table 1. Accuracy under different models.

	Unbalanced data set	Balanced data set
SVM	0.86020	0.92959
Decision tree	0.90046	0.94043
GAN-transformer	0.91019	0.97241

The experiment uses a Generative Adversarial Network to balance the sampled data set. Figure 2 shows the accuracy of the discriminator for real and fake samples during the training process of GAN. In this paper, the number of iterations is selected to be 14,000. It can be seen from the Fig. 2 that when the number of iterations is about 14,000, the recognition accuracy of the discriminator is relatively stable and can reach $99 \pm 0.5\%$.

The experiment is further analyzed by specific prediction error indicators. Under the precision, recall and F1-score metrics, the metrics of the unbalanced dataset are

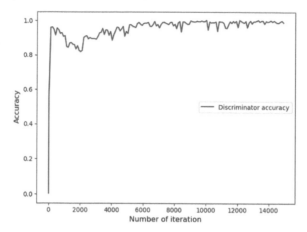

Fig. 2. Resolution accuracy

significantly inferior to the balanced dataset. After the data set is balanced, the long-tail effect can be effectively solved, and the prediction accuracy is higher. As shown in Table 2, the precision rate, recall rate and F1-score of 11 kinds of network traffic in the unbalanced dataset and the balanced dataset are compared.

Table 2. Comparison before and after balancing data sets.

Class	Sample number	Unbalanced data set			Balanced data set		
		Precision	Recall	F1-score	Precision	Recall	F1-score
Bengin	10,000	0.8020	0.8901	0.8437	0.9980	0.9950	0.9964
Dos-Hulk	10,000	0.9250	0.9029	0.9138	0.9750	0.9779	0.9764
Port-Scan	10,000	0.9745	0.9878	0.9811	0.9905	0.9909	0.9906
DDos	10,000	0.9445	0.9174	0.9307	0.9885	0.9724	0.9803
Dos-GoldenEye	10,000	0.9415	0.9396	0.9405	0.9720	0.9619	0.9669
FTP Patator	7938	0.9829	0.8997	0.9394	0.9914	0.9914	0.9914
SSH Patator	5897	0.9812	0.8859	0.9311	0.9913	0.9834	0.9873
Dos-slowloris	5796	0.8076	0.8901	0.8468	0.9181	0.9209	0.9194
Dos-Slowhttptest	5499	0.9198	0.9140	0.9168	0.9044	0.9242	0.9141
Bot	1966	0.8508	0.9043	0.8767	0.9909	0.9909	0.9909
Web-Attack	1507	0.7607	0.8875	0.8192	0.9755	0.9853	0.9803

Acknowledgements. This work is partially supported by Natural Science Foundation of China (No. U20B2050,61773313), Key Research and Development Program of Shaanxi Province

(2019TD-014), Science and technology research project of Shaanxi Province (2021JM-346), and the Key Laboratory of Shaanxi Provincial Department of Education (No.17JS100).

References

1. Buczak, A.L., Guven, E.: A survey of data mining and machine learning methods for cyber security intrusion detection. IEEE Commun. Surv. Tutorials **18**(2), 1153–1176 (2015)
2. Zhang, Y., Zhang, Y., Zhang, N.: A network intrusion detection method based on deep learning with higher accuracy. Procedia Comput. Sci. **17**(4), 50–54 (2020)
3. Hwang, R.H., Peng, M.C., Huang, C.W.: An unsupervised deep learning model for early network traffic anomaly detection. IEEE Access **8**(1), 30387–30399 (2020)
4. Wei, G., Wang, Z.: Adoption and realization of deep learning in network traffic anomaly detection device design. Soft. Comput. **25**(2), 1147–1158 (2020). https://doi.org/10.1007/s00 500-020-05210-1
5. Creswell, A., White, T., Dumoulin, V.: Generative adversarial networks: an overview. IEEE Signal Process. Mag. **35**(1), 53–65 (2018)
6. Vaswani, A., Shazeer, N., Parmar, N.: Attention is all you need. Adv. Neural Inf. Process. Syst. 6000–6010 (2017)
7. Arnab, A., Dehghani, M., Heigold, G.: A video vision transformer. In: Proceedings of the IEEE/CVF International Conference on Computer Vision, pp. 6836–6846 (2021)
8. Sharafaldin, I., Lashkari, A.H., Ghorbani, A.A.: Toward generating a new intrusion detection dataset and intrusion traffic characterization. ICISSp **1**(3), 108–116 (2018)

A Deep Learning-Based Early Patent Quality Recognition Model

Rongzhang Li[1], Hongfei Zhan[1(✉)], Yingjun Lin[2], Junhe Yu[1], and Rui Wang[1]

[1] Ningbo University, Ningbo 315000, China
{2111081046,zhanhongfei,yujunhe,wangrui}@nbu.edu.cn
[2] Zhongyin (Ningbo) Battery Co., Ltd., Ningbo 315040, China
yjlin0819@163.com

Abstract. Patent quality is important for the operation of the intellectual property market and the strategic layout of enterprises. However, the number of patent applications is increasing every year and only a small part of them are used. In this study, we propose a classification model based on deep learning to identify the quality of early patents. According to invention patents, utility model patents and design patents, the abstract, claims and technical efficiency phrases of each patent are taken as text features; take patent "reputation", patent protection scope and patent technology diffusion as digital features simultaneously. Finally, the combination of digital and text features for each category of patents is used for early patent quality classification prediction. Theoretically, this model combines patent text features and digital features more comprehensively as the evaluation of early patent quality, and can assist patent market layout personnel to quickly screen early valuable patents.

Keywords: Early patent · Deep learning · Quality classification model · Quality evaluation

1 Background

Transformation of scientific and technological achievements is very important for technological management. Although the number of patents granted is gradually increasing each year, only a small proportion is predicted to have a greater impact [1]. The object of intellectual property is the product of knowledge, and patents are an important component of intellectual property. Patent race theory suggests that patent quality is more important than patent quantity, with high quality patents generating greater economic and legal value. Patent quality measures the technical competitiveness and market potential of patents, and is positively correlated with the company's financial performance [2, 3]. The early identification of patent quality is therefore particularly important for the assessment of patent assets, the strategic layout of patents and patent litigation in an enterprise or industry. Existing scholarly research on patents, such as patent recommendation and patent trading, cannot be separated from patent quality analysis.

In September 2021, the State Council of the Central Committee of the Communist Party of China issued *The Outline for the Construction of a Strong Intellectual Property State* (2021–2035), which mentions the high-quality development of intellectual property: on the one hand, we should build an intellectual property market operation mechanism that stimulates innovation and development, improve the high-quality creation mechanism with enterprises as the main body and the market as the guide, and reform and improve the quality and value as the standard. On the other hand, we should build a humanistic and social environment that promotes the high-quality development of intellectual property. The evaluation method of patent quality is particularly important for the identification of early patent quality, because the different evaluation standards may result in large differences in evaluation results. So this paper analyzes the early patent quality identification from the dimensions of patent quality evaluation.

2 Related Studies

Discussion on research methods of patent quality analysis, some scholars have adopted the AHP method [4] relying on experts to qualitatively analyze the indicators affecting patent quality. Liu [5] built a patent competitiveness evaluation index system from the four dimensions of technical competitiveness, legal competitiveness, market competitiveness and social benefits. The weight of each secondary index is given by experts, and the final weight is calculated by principal component analysis as the evaluation standard. Li [6] constructed quantitative indicators, quality indicators and value indicators as a three-dimensional evaluation system for patents, and used a systematic analysis method to discuss the correlation between each dimension and patents. With the development of artificial intelligence, scholars continue to analyze the indicators affecting patent quality from the aspects of algorithm optimization and machine learning. For example, Huang [7] established a quantum genetic algorithm combined with hierarchical analysis theory from three aspects of patent technical value, legal value and economic value to make a comprehensive evaluation of patent quality. Fu [8] constructed patent quality characteristics from other indicators such as technical protection scope, regional protection scope and patent citation, and used these indicators to predict patent quality classification by machine learning. Some scholars analyse patent quality from the perspective of patent quality evaluation ontology, e.g. Lee [9] built a patent ontology by mining the relationship between the patented technology and the owner of that patent, and used fuzzy markup language (FML) to describe and evaluate patent quality.

In addition, some scholars analyze the value of patent by studying the diffusion of patent technology themes. Hang [10] constructed a patent citation network to reveal technology evolution paths and predict technology trends using a conceptual life cycle model as the basis for curve fitting. Yin [11] analyses high-value patents by evaluating technology diffusion paths and effects in terms of active diffusion and inactive diffusion dimensions. The high rate of diffusion of patented technology themes facilitates the transformation of technological achievements into the industry chain resources required by enterprises, and at the same time enables enterprises to grasp the pulse of market demand and enable them to respond quickly to the purpose of market demand. This reflects the possibility of measuring high quality patents by the diffusion of patent technology.

Although the above evaluation dimensions are comprehensive enough to reflect trends in quality changes in patents in terms of numerical characteristics, however, they fail to consider the detailed contextual information obtained from patent documents as the object of study. Some scholars have predicted the classification of patent quality from the perspective of patent textual features e.g. Wu [12] constructed 15 numerical features in terms of patent value, patent protection scope, and forward patent value; meanwhile, she extracted core words from patent abstract by means of word frequency statistics to produce co-occurrence networks, and combined patent numerical features with textual features based on graph convolutional network to automatically identify high-quality patents. Chung [13] takes patent abstract and claims as patent text features, and 11 factors affecting patent quality such as number of independent claims and number of dependent as digital features. Build Convolutional Neural Networks (CNN) and Bi-directional Long Short-Term Memory (BiLSTM) model to realize patent quality classification. However, the accuracy of patent information extraction needs to be improved. Therefore, this paper uses the patent digital features combined with the patent text features as the patent classification model, and extracts the patent text by labeling. The model also considers the semantics of context, which makes the model more explanatory.

3 Methodology

3.1 Patent Quality Evaluation Indicators

According to the concept and evaluation criteria of high-value patents proposed in China in 2021, the measurement method is for invention patents that satisfy any of the following conditions: (1) strategic emerging industries; (2) homologous patent rights overseas; (3) maintenance period of more than 10 years; (4) invention patents that achieve a high pledge financing amount; (5) winning the National Science and Technology Award or the China Patent Award. As the model proposed in this paper is only applicable to early patent quality identification, it is not practical for certain dimensions affecting patent quality, such as the number of patent transactions and the number of patent litigations. Combining the above dimensions of patent quality classification, the dimensions that are not highly relevant to patent quality are filtered out. In this paper, we choose to classify patent types into invention patents, utility patents and design patents, which are given weights respectively [5]. Patent text features are classified into patent abstract, claims and technical efficacy phrases. Patent digital features are divided into patent "reputation", patent protection scope and patent technology diffusion.

Patent "reputation" includes the number of inventors, the number of inventor patent transfers, the number of assignee, and the number of inventor countries. The number of inventors and assignees means the degree of effort in patent development. The number of patent transfers of inventors represents the reputation of inventors [14], and the number of inventors' countries has a strong correlation with the quality of patents [15].

The scope of patent protection includes the number of claims, patent families and IPCs as the dimensions of patent quality analysis. The number of claims indicates the scope of protection of the patented technology, the number of patent families represents the business scope of the patent, and the number of IPCs represents the classification of patents. The greater the number of IPCs, the stronger the applicability of the patent.

Patent technology diffusion includes the number of backward citation, the length of abstract, the length of claims, the total number of pages of patents, and the number of national economic classifications as the dimensions of patent quality analysis. Among them, the number of backward citation indicates the breadth of diffusion. The more the number, the wider the scope of patent diffusion and the higher the quality [14]. The length of the abstract, the length of the claims and the total number of pages of the patent can reflect the technical complexity of the patent to a certain extent [12]. The national economic classification number indicates the scope of applicability of the patent.

It is worth noting that the above dimensions of patent characteristics can be obtained for early patents. And these dimensions cover most of the quality related features of patents, and filter out other features with less impact, such as indirect quotation and co quotation between patents, which can more simply and comprehensively reflect the quality dimension of early patents. Its patent digital characteristics are shown in Table 1.

Table 1. Patent digital features

Category	Patent numerical indicators
Patent "reputation"	Number of inventors Number of inventor patent transfers Number of assignees Number of inventors' countries
Patent protection scope	Number of claims Number of patent families Number of IPCs
Patent technology diffusion	Number of backward citation Number of words in abstract Number of words in claim Total number of patent pages Number of national economic classifications

3.2 Quality Classification Model

3.2.1 Experimental Data Processing

The experimental data are from the global patent database incopat. Patents are classified into four grades A, B, C and D which based on their maintenance period of more than 20 years, 10–20 years, 10–5 years and less than 5 years [12]. Class A had the highest value. Divide the data set into training set 80% and test set 20% to process the data. As the data obtained cannot be directly used for experiments, the data needs to be pre-processed. This includes text processing of the data, cut words, delete stop words and extract verb roots and punctuation to delete special characters. The semi-structured data was transformed into structured data to facilitate the experiments.

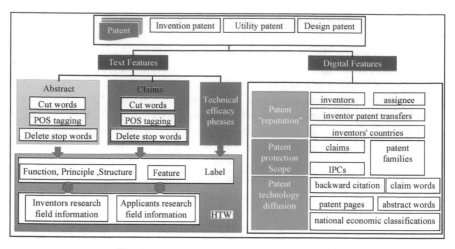

Fig. 1. Patent quality feature extraction model

On the one hand, the patent text is divided into patent abstract, claims and technical efficacy phrases. The patent abstract are classified according to function, structure and principle as classification labels, representing the technology that the patent enables, the composition of the patented design and the means by which the patent is implemented, respectively. Because the patent claim defines the scope of protection given by the patent or patent application in scientific terms, and the written expression forms a specification, "feature" is taken as the classification label of the patent claim in this paper. The technically effective phrases describe the function of the patent and include the advantages it has, and are composed of standardized phrases. On the other hand, the numerical features of the patent are extracted as a vector representation for the subsequent classification of the patent quality. The patent quality feature extraction model is shown in Fig. 1.

3.2.2 Structure of Quality Classification Model

Firstly, the patent abstract, claims and technical efficacy phrases embed sentences and phrases through Heterogeneous Topic model exploiting Word embedding to enhance word semantics (HTW) [16], as shown in Fig. 1. In contrast to term frequency–inverse document frequency, Word2vec and Doc2vec embedding method, the extraction of context-based word meanings and the association of words with the same meaning are enhanced when using HTW. In this model, the corresponding patent abstract, claims, and technical efficacy phrases are integrated with information about the patent inventor and the applicant's study field in this paper [14].Then, the text feature extraction model is used to extract text features, as shown in Fig. 2. Using Bi-directional Long Short-Term Memory (BiLSTM) can better capture long-distance dependencies in sentences, and better capture bidirectional semantic dependencies. The Attention mechanism (ATT) is added to improve the classification results to a certain extent. The purpose of adding Conditional Random Fields (CRF) is to make the learned sentences dependent before and after, so as to add some constraints to ensure the effectiveness of the final results.

Thus, word embedding is transformed into word vector through BiLSTM-ATT-CRF model.

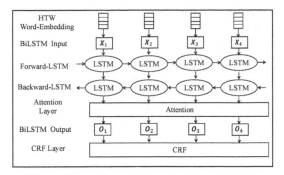

Fig. 2. Text feature extraction model

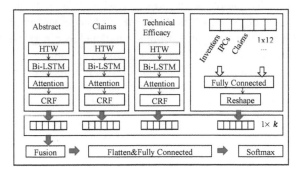

Fig. 3. Patent quality classification model

Secondly, the patent text features are combined with the patent digital features for classification. The patent text features are subjected to the text feature extraction model to form a $1 \times k$ dimensional vector respectively. At the same time, 12 digital feature dimensions related to patent quality are passed through the full connection layer. The purpose of the Reshape layer is to set the dimension of the output vector to $1 \times k$. Fusion layer is to fuse text feature matrix and patent digital feature matrix:

$$V^{fusion} = k_1 W_1 + k_2 W_2 + k_3 W_3 + k_4 W_4 \tag{1}$$

where W_1, W_2, W_3 and W_4 represents the patent abstract feature matrix, claim feature matrix, technical efficacy phrase feature matrix and patent numerical feature matrix. k_1, k_2, k_3 and k_4 are the corresponding weight coefficients, which adjust the influence degree of the abstract, claims, technical efficacy phrases and patent digital features respectively.

The patent text features and digital features through the Fusion layer are taken as the input of the Flatten and Fully Connected layer, the activation function is set to Rectified Linear Unit (ReLU), which is to make the input greater than 0 and directly return the value provided as the input; If the input is 0 or less, the return value is 0. The loss function is cross entropy. Through the classifier softmax, the quality classification results of patent are obtained. The patent quality classification model are presented in Fig. 3.

Thirdly, after the quality classification labels are obtained from the patent quality classification model, the accuracy of the classification is judged using accuracy, precision, recall, and F_1 measures:

$$ACC = \frac{TP_i + TN_i}{TP_i + TN_i + FP_i + FN_i} \tag{2}$$

$$P_i = \frac{TP_i}{TP_i + FP_i} \tag{3}$$

$$R_i = \frac{TP_i}{TP_i + FN_i} \tag{4}$$

$$F_{1i} = 2 \times \left(\frac{P_i \times R_i}{P_i + R_i} \right) \tag{5}$$

where ACC, P_i, R_i and F_{1i} represent accuracy, precision, recall, and F_1 measures. TP is the number of positive sample predicted as a positive, TN is the number of positive sample predicted as a negative, FP is the number of negative sample predicted as a positive, FN is the number of negative sample predicted as a negative [13]. The class i is the patent grade,namely A, B, C and D. Finally by training to obtain F_1 measures values are used for practical analysis. Theoretically, the patent quality recognition model based on deep learning is an improvement on the traditional combined patent text quality classification. This model more comprehensively combines other patent features as an index to measure patent quality.

4 Summary

Combined with the existing patent quality evaluation system, this paper propose an early patent quality identification model based on deep learning to predict the early patent quality. Firstly, the model according to the invention patents, utility patents and design patents, the patents of each category are combined with text features and digital features for early patent quality classification prediction. Secondly, a model based on HTW and BiLSTM-ATT-CRF is constructed to classify early patent quality. Finally, the process of processing experimental data is briefly introduced. Theoretically, this model combines patent text features and digital features more comprehensively as the evaluation of early patent quality. At the same time, the model can assist patent market layout personnel to quickly screen early valuable patents. In future studies, it will be used in model application and practical analysis and further combine patents and literature to do more accurate patent quality analysis and make a prediction of the patent technology market.

Acknowledgement. I would like to extend my gratitude to all those who have offered support in writing this thesis from National Key R&D Program of China (2019YFB1707101, 2019YFB1707103), the Zhejiang Provincial Public Welfare Technology Application Research Project (LGG20E050010, LGG18E050002) and the National Natural Science Foundation of China (71671097).

References

1. Mariani, M.S., Medo, M., Lafond, F.: Early identification of important patents: design and validation of citation network metrics. Technol. Forecast. Soc. Chang. **146**, 644–654 (2019)
2. Schankerman, M., Pakes, A.: Estimates of the value of patent rights in European countries during the post-1950 period. Soc. Sci. Electron. Publishing **96**(384), 1052–1076 (1986)
3. Agostini, L., Caviggioli, F., Filippini, R., Nosella, A.: Does patenting influence SME sales performance? A quantity and quality analysis of patents in Northern Italy. Eur. J. Innov. Manag. **18**(2), 238–252 (2015)
4. Gu, L., Yan, W.C., Han, X., Hong, C.: A research on the patent application quality evaluation index system. Scientific Res Manage **39**(S1), 130–136 (2018)
5. Liu, P.P., Lu, D.T., Su, W., Shi, B.X., Yu, D.H., Yen, L.: Patent competitiveness assessment system design and high-quality patent identification. Sci. Technol. Manage. Res. **41**(07), 110–115 (2021)
6. Li, Z.Y., Meng, F.S., Cao, X.: Research on patent three-dimensional evaluation system. Intell. Sci. **28**(10), 1569–1573 (2010)
7. Huang, X.F., Feng, L., Zhan, W.Q.: Comprehensive evaluation of patent value based on grey system-evidence theory. Intell. Explor. **02**, 35–42 (2022)
8. Fu, C.C., Chen, G.H., Yuan, Q.J.: Research on patent quality analysis and classification forecast based on machine learning-taking blockchain as an example. J. Mod. Inf. **41**(07), 110–120 (2021)
9. Lee, C.-S., Wang, M.-H., Hsiao, Y.-C., Tsai, B.-H.: Ontology-based GFML agent for patent technology requirement evaluation and recommendation. Soft. Comput. **23**(2), 537–556 (2017). https://doi.org/10.1007/s00500-017-2859-1
10. Huang, Y., Li, R., Zou, F., Jiang, L., Porter, A.L., Zhang, L.: Technology life cycle analysis: from the dynamic perspective of patent citation networks. Technol. Forecast. Soc. Chang. **181**, 121760 (2022)
11. Yin, C.H., Ren, S.Z., Jiang, Y.C.: Multi-dimensional evaluation on technology diffusion paths of high value patents in China. Forum Sci Technol China **02**, 125–132 (2022)
12. Wu, J., Gui, L., Liu, P.: Indicator and textual features-based patent evaluation with graph convolutional networks. J. Intell. **41**(01), 88–95 (2022)
13. Chung, P., Sohn, S.Y.: Early detection of valuable patents using a deep learning model: case of semiconductor industry. Technol. Forecast. Soc. Chang. **158**, 120146 (2020)
14. Du, W., Wang, Y., Xu, W., Ma, J.: A personalized recommendation system for high quality patent trading by leveraging hybrid patent analysis. Scientometrics **126**(12), 9369–9391 (2021). https://doi.org/10.1007/s11192-021-04180-x
15. Ferrucci, C., Lissoni, F.: Foreign inventors in Europe and the United States: diversity and patent quality. Res. Policy **48**(9), 103774 (2019)
16. Chen, J., Chen, J., Zhao, S., Zhang, Y., Tang, J.: Exploiting word embedding for heterogeneous topic model towards patent recommendation. Scientometrics **125**(3), 2091–2108 (2020). https://doi.org/10.1007/s11192-020-03666-4

Defect Detection of Exposure Lead Frame Based on Improved YOLOX

Wanyu Deng, Dunhai Wu[⊠], Jiahao Jie, and Wei Wang

Xi'an University of Posts and Telecommunications, Xi'an 710121, China
dengwanyu@xupt.edu.cn, weibin82@126.com, {wudunhai,
china_jjh_1998,wangwei}@stu.xupt.edu.cn

Abstract. In view of the unpredictable defects in the industrial lead frame production line, many factories invest a lot of manpower resources and material re-sources to participate in the inspection process, which will cause huge losses to the producers, and the human eyes have the problems such as easy fatigue and low efficiency. In this paper, for the research on defect detection in lead frame exposure production line, the Efficient Channel Attention module is added to YOLOX algorithm to optimize the backbone network to improve the feature extraction capability and detection efficiency of the algorithm, and Varifocal Loss is introduced to solve the problem of very few defect samples in industry. After a lot of experiments, it has been proved that the evaluation indicator mAP@0.5 of the model trained by the improved YOLOX algorithm is increased by 2.10%, which is lightweight and has a higher accuracy.

Keywords: Lead frame · Efficient channel attention · Varifocal loss

1 Introduction

As the foundation of the information industry, the semiconductor industry plays a vital role in the advancement of science and technology. As the basic material of semiconductor packaging, lead frame is widely used in the semiconductor industry. In the industry, there are some defects such as overflow and shortage of material in the exposure lead frame production line, so that the final yield of the product cannot be improved. Therefore, it is necessary to intersperse the inspection process in the production line. Problems such as easy fatigue and low efficiency require multiple manual detection links to improve yield of product and greatly increase the operating cost of the enterprises. Machine inspection can make up for the above shortcomings, and machine inspection has a huge contribution to saving costs and improving yield, so how to use machines to detect defects in exposure lead frame is imminent.

The traditional detection algorithm is divided into three steps: firstly, extract key areas from the image, then extract object features, and finally classify the features.

This work is supported by Science Research Plan of Shaanxi Provincial Department of Education under Grant No. 19JC036.

The region selection adopts the sliding window mechanism, and then uses algorithms such as Histogram of Oriented Gradient (HOG) and Scale-Invariant Feature Transform (SIFT) for feature extraction, and finally uses classifiers such as SVM and AdaBoost for classification. Based on this, Ding [1] et al. proposed the application of AdaBoost learning based on HOG and SVM in fabric defect detection, but this method is not effective for datasets with complex background. Deep learning can explore patterns in images and learn specific patterns of defects by itself. The research of this topic benefits from the development of object detection under deep learning. Object detection in the field of deep learning can be divided into two branches, one is the object detection method of two execution stages: Faster-R-CNN [2], FPN [3], Mask-R-CNN [4], R-FCN [5], etc. The main features of these algorithms are high detection accuracy; the second is the one-stage object detection algorithm: YOLOV series, YOLOX [6], SSD [7], etc. The main feature of these algorithms is high real-time performance. Based on this, Chengfei Li [8] et al. proposed a steel defect detection method based on YOLOv4, using Complete-IOU (CIOU) as the position loss and obtaining a priori frame through K-means++ to adapt to the steel data, which greatly improves the steel defect detection efficiency, but this method is not suitable for complex background and multi-color image detection; Li Wei et al. [9] proposed a method of wire rope defect detection based on Faster-R-CNN, using Hue-Saturation-Value (HSV) color segmentation algorithm to extract wire rope and Faster-R-CNN algorithm to identify defects, and effectively detect wire rope defects, but this method is not suitable for extracting lead frame defects and the real-time performance is not high.

Through the investigation of the above existing methods, since most of the defects in the exposure data set are overflow and shortage of materials, and the color characteristics of the defects in the data set in this paper are very similar to the background color characteristics and the background characteristics are very complicated, the idea of introducing Efficient Channel Attention [10] module is proposed to enhance local feature extraction; in order to save model construction time to adapt to the timeliness of industrial detection, the idea of transfer learning is introduced; finally, Binary Cross Entropy confidence loss is replaced by Varifocal Loss [11] to predict IOU Aware Classification Score(IACS) [12]. Extensive experiments have shown that the improved YOLOX algorithm has higher accuracy.

2 Related Work

2.1 YOLOX Algorithm

Compared this to the previously published YOLO series, YOLOX has made many effective improvements: the introduction of the Focus channel augmentation technology of YOLOV5, the Focus module slices the image before the image is input to the backbone, which can be understood as taking a value every other pixel, and finally an image is divided into four images. In other words, the number of input channels has become four times that of original number of channels. It can be understood that the original RGB 3-channel mode has become a 12-channel mode, which not only ensures that the image information is not lost, but also makes the feature extraction is more sufficient for subsequent convolution operation, the purpose is to provide more complete picture sampling information for subsequent backbone. In general, the calculation amount and parameters of Focus are four times that of ordinary convolution, but the picture information will not be lost during down sampling. Figure 1 shows the principle of the Focus module.

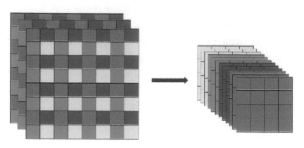

Fig. 1. Principle of focus

Using YOLOV4's feature extraction network Cross Stage Partial (CSP) Darknet architecture, the CSP module has two structures, CSP1_X and CSP2_X: CSP1_X has two input branches, one branch first performs Convolution, Batch Normalization and SiLu function, and then passes through several Residual structures, and then performs again Convolution, the other branch only performs Convolution, and finally combine the two branches. The CSP2_X module replaces the residual blocks of the CSP1_X module with a combination of several Convolution, Batch Normalization and SiLu function. The CSP1_X module is applied to the backbone part, and the introduction of the residual structure can increase the gradient value to ensure that gradient do not disappear as the network deepens and avoid network structure degradation. CSP2_X module is applied to the neck part, and the network is not as deep as CSP1_X, which improves the learning capacity of Convolutional Neural Network and decreases the amount of computation. Fig. 2 shows the CSP module analysis.

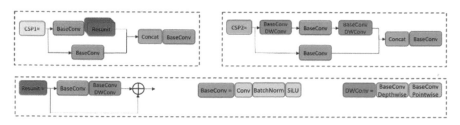

Fig. 2. CSP structural analysis

YOLOX also implements the classification and regression modules separately in the detection head, and integrates them in the final prediction. The decoupling detection head also speeds up the model convergence speed and improves the detection accuracy, but it also brings a little extra calculation cost; using the Anchor Free detector, which greatly reduces the heuristic adjustments and skills required to achieve good performance, and speeds up the detection speed; using the Sim Optimal Transport Assignment (Sim OTA) dynamic matching positive sample mechanism, the ground truth can be automatically calculated. Determine the candidate area of positive samples according to the number of positive samples; Mosaic data augmentation is used, and four images are randomly cropped and then spliced into one image for training, which makes the background of the image richer.

2.2 YOLOX Execution Process

There are three main parts to the YOLOX execution process: Backbone network, Feature Pyramid Networks (FPN) and YOLO head part. First, the size of the input image will be adjusted to 640 * 640 * 3, and then enter the Focus module to perform width and height compression and channel number expansion to 320 * 320 * 12, and Then the channel becomes 64 after Convolution, Batch Normalization and SiLU function operations. Finally, after four Res block body modules, three effective feature layers are extracted and put into FPN for enhanced feature extraction, and feature fusion is achieved through up sampling and down sampling. The three feature layers output by the FPN network will be put into YOLO Head to obtain the final prediction result. YOLO Head will implement classification and regression separately. Three prediction results can then be obtained from each feature layer: confidence, regression parameters and object judging the result, the drawn prediction box will be filtered by score and non-maximum suppression operation to get the final prediction box. In addition, the IOU loss of real box and predicted box can be calculated according to the regression parameter part, that is, the regression loss, and calculate whether the positive and negative samples feature points contain objects according to the object judgment, so as to calculate the cross-entropy loss as the object judgment part. The cross-entropy loss is calculated as the classification loss according to the type prediction result of the feature points extracted from the classification part.

3   The Proposed Method

The method proposed in this paper adds the Efficient Channel Attention [10] (ECA) module after the effective feature layer extracted by the backbone feature extraction network CSP Darknet, so as to enhance the local attention of CNN to improve the recognition ability of CNN on local features, and then put the enhanced feature layer into the FPN network module. Finally, the confidence prediction loss is changed to Varifocal Loss to predict the IACS, and the training focus is on high-quality samples to enhance the ability of detect of the algorithm. Figure 3 shows the improved network structure of YOLOX.

3.1   Efficient Channel Attention

The attention mechanism is a method that ignores irrelevant information and mainly concentrates on more important areas when the computing power of the algorithm is limited. The attention mechanism is an effective means to enhance the depth of CNN. The attention mechanism can be summarized into two parts according to differentiability, one is soft attention mechanism, the other is hard attention mechanism. The score of

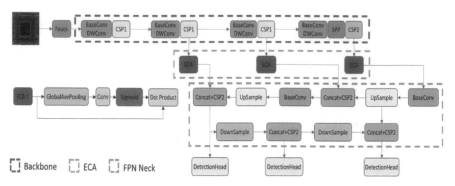

Fig. 3. The improved network structures

the attention degree of the soft attention mechanism for each area is between [0, 1], and the hard attention mechanism is that divides the area into the concerned area and the unconcerned area. According to the attention domain such as space, channel and time, it can be divided into the following parts: the spatial domain is to extract key information through information exchange in the image space domain, the channel domain is to learn features through channel interaction, and the time domain is to be added the temporal information, the mixed domain is to fuse attention of spatial information and channel information.

The ECA module introduced in this paper is based on channel attention. In view of the fact that the compression operation of the input feature map channel by the SE attention mechanism [13] will have a negative effect on the dependencies between the learning channels, the ECA module avoids dimensionality reduction and increases the Cross-channel interaction reduces the large computational burden caused by model complexity, and has high performance and efficiency in channel attention. Therefore, the ECA attention mechanism is introduced after the backbone feature extraction network extracts features. First, the feature map output by the backbone feature extraction network is executed to a global average pooling, and then a $1 * 1 * C$ feature map is obtained and then the convolution kernel is k. The one-dimensional convolution and the sigmoid activation function are operated to obtain the weights of each channel, and finally the input feature map and the weights are subjected to a dot product operation. Figure 4 shows the principle analysis of ECA.

The value of k in the ECA module depends on the coverage of the channel interaction information. In different network structures and convolution modules, although the optimal range can be adjusted manually, it will consume a lot of computing resources. The size k of kernel of one-dimensional convolution increases with the increase of channel dimension C. Therefore, there is a relationship between k and channel C, and channel C is generally a power of 2, so the simple linear mapping is extended into a nonlinear, it can be defined as follows:

$$C = \varphi(k) = 2^{(\gamma * k - b)} \qquad (1)$$

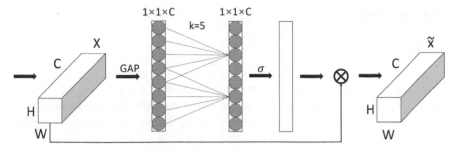

Fig. 4. ECA principle analysis

According to the above formula, the k value of the convolution kernel can be determined according to channel dimension C:

$$k = \psi(C) = \left| \frac{\log_2(C)}{\gamma} + \frac{b}{\gamma} \right|_{odd} \tag{2}$$

The above formula odd represents the nearest odd number. After the above operations, the value of k will be adaptive to the channel C. When the number of channels C becomes larger, the value of the convolution kernel k will become larger. When the number of channels C becomes smaller, the value of convolution kernel k also becomes smaller.

3.2 Varifocal Loss

For the loss function used to predict object and Classification, this paper replaces the binary crossover loss function in the YOLOX network with Varifocal Loss, and uses Varifocal Loss to predict confidence and IOU-aware classification scores. Varifocal Loss is an improved version of Focal Loss. The problem solved by Focal Loss [14] is the imbalance between foreground and background classes in dense object detector training. Varifocal Loss borrows the idea of sample weighting from Focal Loss, but Focal Loss treats positive sample and negative sample loss the same way, which causes the problem of asymmetric treatment, so Varifocal Loss is introduced, it can be defined as follows:

$$VFL(p, q) = \begin{cases} -q(q \log(p) + (1 - q) \log(1 - p)) \, q > 0 \\ -\alpha p^\gamma \log(1 - p) \, q = 0 \end{cases} \tag{3}$$

The IACS score is defined as p and the object IOU score is defined as q. When training defect samples, that is, q > 0, set q to generate the IOU between the Bounding Box and the Ground Truth Box, and when training without defect samples, that is, q = 0. After using q as the scaling loss, it makes up for the asymmetry of the Focal Loss of defect and without defect samples. And q only decreases the loss contribution of without defect samples, does not make corresponding operations on defect samples. After the above processing, if the Ground Truth IOU of a defect sample is relatively high, its contribution to the loss will also become larger, so the training focus will be placed on high-quality defect samples. This idea greatly fits the problem that the number of defect samples of the lead frame is quite different from that of the without defect samples.

4   Experiment

4.1   Data Set

This paper mainly studies the four defects of copper shortage (CS), copper overflow (CO), silver shortage (SS) and silver overflow (SO) on the lead frame semi-finished product in the lead frame exposure production line. Among the 2000 lead frame images, 500 images with defects were selected, and each image has one or more defects, and then the lead frame defects were marked by LabelImg software. Finally, lead frame images were divided into training set and test set by 4:1 ratio, including 400 training set and 100 validations set. Figure 5 shows the defect examples.

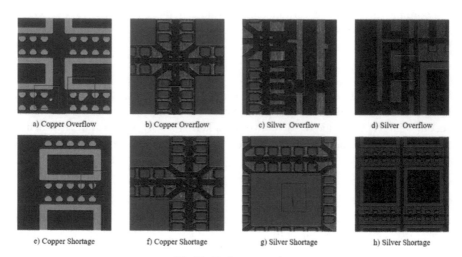

a) Copper Overflow   b) Copper Overflow   c) Silver Overflow   d) Silver Overflow

e) Copper Shortage   f) Copper Shortage   g) Silver Shortage   h) Silver Shortage

Fig. 5. Defect examples

4.2   Experimental Platform and Parameter Setting

The development environment for this experiment is as follows: Operating system: ubuntu16.04, GPU: NVIDIA GeForce RTX 3060, CPU: Intel(R) Core(TM) i7-9750H@2.60GHz, learning framework: PyTorch 1.8.0, Python 3.7, CUDA 11.1. In the experiment, the yolo-s pre-trained weight model is used, the batch size is 8, the input image resolution is 640 × 640 × 3, the learning rate is 0.0001, and training for 300 epochs.

4.3   Evaluation Indicators

This paper takes Precision (P), Recall (R), Average Precision (AP), Mean Average Precision (mAP) as the evaluation indicators of the model recognition accuracy, P represents the ratio of the accurate prediction to all results, and R represents the accurate prediction. The ratio of results to positive examples in all results, AP measures the accuracy of the model on each category, and mAP measures the accuracy of the model on all categories. And the confusion matrix is used to divide the predicted values into TP, FP, and FN. The positive sample whose prediction result is a positive class is TP, the negative sample whose prediction result is the positive class is FP, and the positive sample whose prediction result is the negative class is FN. According to these indicators, the final mAP@0.5 is calculated.

According to the different thresholds of the PR curve, different combinations of P and R are obtained. When the difference between defect and without defect samples is relatively large, the performance of the classifier can be reflected. AP@0.5 is the area enclosed by the PR curve and the coordinate axes when the threshold of the confusion matrix is 0.5. The P, R, AP@0.5 formulas can be defined as follows:

$$P = \frac{TP}{TP + FP} \tag{4}$$

$$R = \frac{TP}{TP + FN} \tag{5}$$

$$AP@0.5 = \frac{1}{n} \sum_{n}^{1} P_i = \frac{1}{n}(P_1 + P_2 + ... + P_n) \tag{6}$$

mAP@0.5 can reflect the trend of model accuracy with recall, and is the average of AP@0.5 of all categories, defined as follows:

$$mAP@0.5 = \frac{1}{nc} \sum_{n}^{1} P_i = \frac{1}{nc}(P_1 + P_2 + ... + P_n) \tag{7}$$

4.4 Results

In this study, the effects of different improved parts on the model performance are tested through ablation experiments. This paper is divided into four schemes: using the original YOLOX algorithm for model prediction, using the YOLOX algorithm with the added ECA module for model prediction, and using modified confidence loss. The YOLOX algorithm makes model prediction, and finally uses the YOLOX algorithm that adds the ECA module and modifies the confidence loss for model prediction. The original YOLOX model mAP@0.5 is 94.23%. After adding the ECA module, the mAP@0.5 is 95.62%. After modifying the confidence loss, the mAP@0.5 is 95.37%. Both adding the ECA module and modifying the confidence loss mAP@0.5 can reach 96.33%. Experiments show that the improved YOLOX has an increase of 2.10% compared to the original algorithm mAP@0.5. Table 1 shows the results of the experiments.

In this study, Faster-R-CNN, YOLOV3 and SSD were selected as model comparison tests. Under the same parameter settings, compared with Faster-R-CNN, SSD and YOLOV3, the proposed algorithm improves mAP@0.5 by 21.01%, 22.20% and 3.53%, respectively. The SSD algorithm and the Faster-R-CNN algorithm have serious missed detection of small objects. The algorithm in this paper is good at detecting small objects, and its efficiency is better than the current mainstream algorithms. Table 2 shows the results of the experiments.

Figure 6 shows part of the detection effect of the lead frame. The improved YOLOX performs better in the detection of copper shortage, copper overflow, silver shortage, and silver overflow.

Table 1. Ablation experiment results

Algorithm	AP@0.5% (CO)	AP@0.5% (CS)	AP@0.5% (SO)	AP@0.5% (SS)	mAP@0.5/%
YOLOX	95.34	94.28	97.15	90.13	94.23
YOLOX + ECA	94.93	97.52	93.19	96.82	95.62
YOLOX + VFL	97.67	95.25	94.76	93.78	95.37
YOLOX + ECA + VFL	96.76	96.09	94.59	97.89	96.33

Table 2. Contrast experiment results

Algorithm	AP@0.5% (CO)	AP@0.5% (CS)	AP@0.5% (SO)	AP@0.5% (SS)	mAP@0.5/%
SSD	79.22	67.62	68.66	81.01	74.13
Faster-R-CNN	73.90	75.62	70.73	81.03	75.32
YOLOV3	97.00	92.50	90.70	90.80	92.80
OURS	96.76	96.09	94.59	97.89	96.33

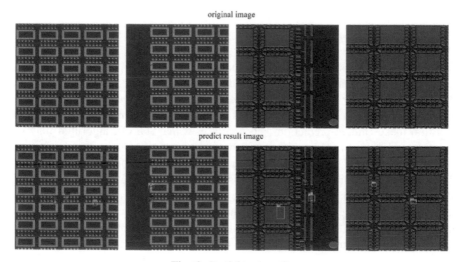

Fig. 6. Partial test results

5   Conclusion

This paper proposes an improved YOLOX-based exposure lead frame defect detection method, which mainly includes introducing ECA into YOLOX, modifying its confidence loss function to Varifocal Loss, and finally using the idea of transfer learning to meet the timeliness. The method effectively solves the problems of small defects and large defect density of the exposure lead frame, and has higher detection efficiency than mainstream object detection algorithms, and has certain practical significance for industrial defect detection.

References

1. Shumin, D., Zhoufeng, L., Chunlei, L.: AdaBoost learning for fabric defect detection based on HOG and SVM. In: International Conference on Multimedia Technology. pp. 2903–2906 (2011)
2. Ren, S., He, K., Girshick, R., Sun, J.: Faster R-CNN: towards real-time object detection with region proposal networks. IEEE Trans. Pattern Anal. Mach. Intell. **39**(6), 1137–1149 (2017)
3. Lin, T.Y., Dollár, P., Girshick, R., He, K., Hariharan, B., Belongie, S.: Feature pyramid networks for object detection. In: IEEE Conference on Computer Vision and Pattern Recognition (CVPR). pp. 2117–2125 (2017)
4. Karunakaran, V.: Deep learning based object detection using mask RCNN. In: International Conference on Communication and Electronics Systems (ICCES). pp. 1684–1690(2021)
5. Dai, J., Li, Y., He, K., Sun, J.: R-FCN: object detection via region-based fully convolutional networks. In: Advances in Neural Information Processing Systems, 29 (2016)
6. Ge, Z., Liu, S., Wang, F., Li, Z., & Sun, J.: YOLOX: Exceeding YOLO series in 2021. arXiv preprint arXiv: 2107.08430 (2021)
7. Liu, W., Anguelov, D., Erhan, D., Szegedy, C., Reed, S., Fu, C. Y., Berg, A.C.: SSD: Single shot multibox detector. In: European Conference on Computer Vision. Springer, Cham. pp. 21–37 (2016)
8. Li, C., Cai, J., Qiu, S., Liang, H.: Defects detection of steel based on YOLOv4. In: China Automation Congress (CAC). IEEE, pp. 5836–5839 (2021)
9. Li, W., Dong, T., Shi, H., Ye, L.: Defect detection algorithm of wire rope based on color segmentation and Faster RCNN. In: 2021 International Conference on Control, Automation and Information Sciences (ICCAIS). IEEE, pp. 656–661 (2021)
10. Wang, Q., Wu, B., Zhu, P., Li, P., Hu, Q.: ECA-Net: efficient channel attention for deep convolutional neural networks. In: 2020 IEEE/CVF Conference on Computer Vision and Pattern Recognition (CVPR) (2020)
11. Zhang, H., Wang, Y., Dayoub, F., Sunderhauf, N.: VarifocalNet: An IoU-aware dense object detector. In: 2021 IEEE/CVF Conference on Computer Vision and Pattern Recognition (CVPR). IEEE, pp. 8514–8523 (2021)
12. Wu, S., Li, X., Wang, X.: IoU-aware single-stage object detector for accurate localization. Image Vis. Comput. **97**, 103911 (2020)
13. Hu, J., Shen, L., Sun, G.: Squeeze-and-excitation networks. In: Proceedings of the IEEE Conference on Computer Vision and Pattern Recognition. pp. 7132–7141 (2018)
14. Lin, T.Y., Goyal, P., Girshick, R., He, K., Dollár, P.: Focal loss for dense object detection. In: Proceedings of the IEEE International Conference on Computer Vision. pp. 2980–2988 (2017)

Deep Neural Network for Infrared and Visible Image Fusion Based on Multi-scale Decomposition and Interactive Residual Coordinate Attention

Sha Zong, Zhihua Xie$^{(\boxtimes)}$, Qiang Li, and Guodong Liu

Key Lab of Optic-Electronic and Communication, Jiangxi Science and Technology Normal
University, Nanchang 220031, China
{1020101062,xiezhihua,1020101066,Liuguodong95}@jxstnu.edu.cn,
xie_zhihua68@aliyun.com

Abstract. To preserve diverse details, this paper proposes an infrared and visible image fusion model based on frequency-domain multiscale decomposition and interactive residual attention fusion strategy. Firstly, the source image is into three sub-bands of the high-frequency, low-frequency, and mid-frequency for powerful multiscale representation in the frequency domain. To further address the limitations of the straightforward fusion strategy, a learnable coordinate attention module in the fusion layer is incorporated to adaptively fuse representative information based on the characteristics of the corresponding feature map. Finally, the detail-preserving loss function and feature-enhancing loss function are introduced to train the entire fusion model for good detail retainability. Qualitative and quantitative comparisons demonstrate the rationality and validity of our model, which can consistently generate fusion images containing both highlight targets and legible details, outperforming the state-of-the-art methods.

Keywords: Image fusion · Multi-scale decomposition · Interactive residual fusion

1 Introduction

The purpose of infrared and visible image fusion (IVIF) is to obtain an improved image that contains both rich texture information and thermal radiation information of the target [1]. To be specific, Infrared images can avoid the effects of illumination and occlusion, but the low-texture detail characteristic of infrared images is detrimental to the visual understanding of the scene. In contrast, visible images are rich in texture information, while being susceptible to external conditions. In general, IVIF algorithms are mainly divided into traditional methods and deep learning-based methods.

Traditional approaches mainly include multi-scale transform-based fusion algorithms, sparse representation-based fusion algorithms, subspace-based fusion algorithms, and hybrid fusion methods. A multi-scale transform decomposition model based

© The Author(s), under exclusive license to Springer Nature Switzerland AG 2023
N. Xiong et al. (Eds.): ICNC-FSKD 2022, LNDECT 153, pp. 254–262, 2023.
https://doi.org/10.1007/978-3-031-20738-9_30

on target enhancement was proposed in [2]. This algorithm uses the Laplace transform to decompose the source image into high-frequency and low-frequency components. The improvement in the fusion performance of these algorithms depends on the decomposition method and fusion strategy. For deep learning-based fusion algorithms, [3] uses an encoder-decoder model as the backbone network, first exploiting the encoder for feature extraction, then performing the designed fusion rules, and finally reconstructing the fused image with a decoder. GANMcC [4] adopts a generative adversarial mechanism to transform image fusion into a multi-classification limited problem. However, the target edges in the fused images are blurred.

Recently, Zhao et al. proposed an efficient image fusion model called AUIF [5], which uses a learnable neural network to replace the iterative formulation in traditional optimization models to decompose the source image into high-frequency and low-frequency information. However, the AUIF discards the medium-frequency information during the scale decomposition. In addition, the addition fusion strategy of the AUIF also limits the quality of the fused images. To alleviate these problems, we propose a network based on frequency-domain multiscale decomposition and interactive residual coordinated attention fusion. Our contributions are as follows:

(1) A feature extraction network based on multi-scale decomposition in the frequency domain is developed to solve the problem of information loss in the multi-scale decomposition process.
(2) A dense network of residual gradients incorporating pyramidal segmentation attention (PSA) is designed to enhance the feature extraction from the medium-frequency feature information with complex components.
(3) A novel learnable interactive residual coordinate attention fusion network (IRCAFN) is proposed to substitute the traditional handcrafted fusion strategy. The inability to adaptively fuse feature maps is a weakness of many methods, which is overcome by IRCAFN, although many methods evade it by enhancing feature extraction capabilities.

The interactive attention module of IRCAFN considers the feature correlation between infrared and visible images while building spatial feature attention across channels, aiming to balance the distinguishing information between infrared and visible images in the fused images. Results show that the generated fused images retain richer detail information while focusing on the brightness of the infrared target.

2 Related Works

2.1 Multi-scale Decomposition

Multi-scale decomposition (MSD) is one of the classical operations of the IVIF algorithm. Representative multi-scale decomposition solutions are pyramid transform, bilateral filtering, etc. In contrast, the autoencoder (AE)-based approach is gradually gaining popularity and its structure consists of an encoder and a decoder. A Typical fusion model is the dense block-based DenseFuse [6]. The encoder corresponds to the MSD to extract significant features from the input image, while the decoder corresponds to the inverse

MSD to reconstruct the image based on the encoded features [7]. Additionally, the loss of feature information in the multi-scale decomposition process is a key issue. As a result, we construct a low-loss AE network based on multi-scale decomposition in the frequency domain.

2.2 Fusion Rules

Most existing fusion strategies are simple traditional fusion rules. The choice of fusion strategies remains limited, including addition, max, average, l1-normal, etc. Recently, a Few fusion strategies based on deep learning have been presented and implemented in IVIF tasks. RFN-Nest develops a learnable residual fusion network to fuse the multi-scale feature information proposed by the encoders [8]. Different feature information should be fused using corresponding fusion strategies. Besides, the information interaction between infrared and visible images should be emphasized when designing the fusion strategy. To cope with these issues, we purposed the learnable IRCAFN in high-frequency feature information fusion to retain more detailed information in the fused images. Meanwhile, simple fusion traditional strategies are applied to fuse features at low and medium frequencies to improve fusion efficiency.

3 Method

3.1 The Framework of the Proposed Method

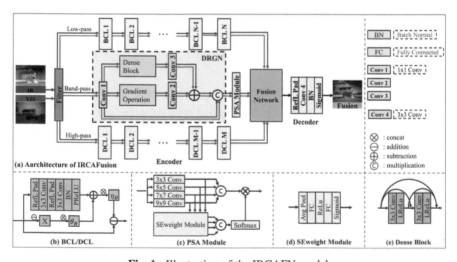

Fig. 1. Illustration of the IRCAFN model.

The proposed network consisted of the encoder, fusion network, and decoder. The detailed structure is shown in Fig. 1. First, the source image is decomposed in the frequency domain using different filters for multi-scale decomposition and performed encoding operations. For high-frequency and low-frequency feature information, we use the detail convolutional layer (DCL) and the base convolutional layer (BCL) in AUIF. BCL and DCL with the same structure and different parameters. Their structure is shown in Fig. 1(b). The reflection-padded structure is to prevent artifacts at the image edges [5]. For medium-frequency feature information, we employ a dense residual gradient network (DRGN) to extract features. An additional PSA module [9] is used to suppress redundant information.

This work proposes IRCAFN based on coordinate attention structure [10] to fuse high-frequency feature information, and its structure is shown in Fig. 2. Specifically, the coordinate attention weights of the two source images are weight interacted, then the fusion of high-frequency features is accomplished by using the residual connection structure. The traditional addition strategy is used to fuse the low-frequency and medium-frequency features of the image. The input of the decoder is the fused features and the output is the reconstructed fused image.

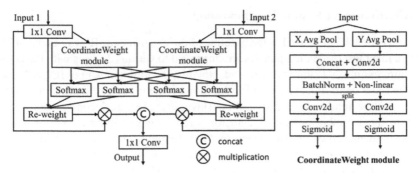

Fig. 2. The architecture of the IRCAFN

3.2 Two-Stage Training Strategy

During the first stage of training, the loss function of the codec network is represented using L_{total}, L_{total} defined as follows,

$$L_{total} = \left\| I_f - I_{in} \right\|_F^2 + \alpha\left[1 - SSIM\left(I_f, I_{in}\right)\right] \tag{1}$$

where I_{in} denote fusion image and the input image. α is the hyperparameter. $\|\cdot\|_F^2$ is the Frobenius norm, $SSIM(,)$ denote the structural similarity which quantifies the structural similarity between the output image and source image.

In the second stage, we employ a learnable IRCAFN combined with a traditional addition strategy as the final fusion strategy. We employ a mutually compensating loss function that uses the Frobenius norm to bound the luminance loss of the fused image to the visible and infrared images. The additional l1-norm loss is used to compensate for the resulting loss of visible detail texture loss. L_{IRCAFN} Defined as follows,

$$L_{IRCAFN} = \frac{1}{HW}\left[\beta\left\| I_f - I_{ir} \right\|_F^2 + \left\| I_f - I_{vi} \right\|_F^2 + \left\| I_f - I_{vi} \right\|_1\right] \tag{2}$$

where H and W denote the height and width of the source image. I_f, I_{ir}, I_{vi} denote the fusion image, infrared image, and visible image. β is the balanced hyperparameter, $\|\cdot\|_F^2$ is the Frobenius norm, $\|\cdot\|_1$ is the l1-norm.

4 Experimental Validation

All experiments were conducted on Intel Core i5-10400F CPU@2.90GHz and NVIDIA GeForce RTX3070 GPU. The FLIR [11] dataset is chosen as the training set for both stages. In the test phase, we used TNO [12]. Six common evaluation metrics are used to validate the performance of the proposed network: entropy (EN), standard deviation (SD), spatial frequency (SF), visual information fidelity (VIF), average gradient (AG), and the sum of the correlations of differences (SCD). More details of the metrics can be found in [1].

4.1 Effectiveness of Feature Decomposition

The results of AUIF were used as the baseline of the experiments. Rows 1 and 2 of Tab. 1 illustrate that most evaluation metrics are improved to various degrees with the addition of medium frequency feature information. Notably, the improvement of EN, SD, and SF indicates better definition and richer information of fusion images. Further, attempts are made to vary the number of the convolutional layer in the feature decomposition network. Results in Tab. 1 display the best effect fusion when numbers of the convolutional layer of the high-frequency, medium-frequency, and low-frequency feature extraction networks are 7, 10, and 12, respectively. Finally, a network combining PSA and DRGN is proposed as the medium-frequency feature extraction branch. The proposed network obtains the maximum improvement in all metrics, confirming the validity of the scale decomposition method.

Table 1. Frequency domain multi-scale decomposition network experiment. The xx-xx-xx indicates the number of the convolutional layer for the low-frequency, medium-frequency, and high-frequency feature extraction branches. The bold fonts are the best values in corresponding indexs.

Network structure	EN	SD	SF	VIF	AG	SCD
10–00-10(AUIF)	6.944	43.447	11.053	0.811	4.146	1.860
10–10-10	6.954	43.109	11.611	0.814	4.336	1.866
08–10-10	6.928	44.697	11.649	0.808	4.276	1.838
10–08-10	6.930	44.998	11.694	0.809	4.342	1.839
10–10-08	6.917	42.191	11.148	0.802	4.163	1.854
07–10-12	6.989	45.441	11.350	0.823	4.279	1.862
07-(DRGN + PSA)-12	**7.083**	**45.502**	**12.739**	**0.842**	**4.832**	**1.884**

Results of the quantitative ablation experiments are given in Tab. 2. All metrics are optimal when the medium frequency feature information is used with the DRGN and PSA modules. It demonstrates the feature extraction ability of DRGN and the ability of the PSA module to suppress the interference information.

Table 2. Ablation experiments of DRGN and PSA modules.

	DRGN	PSA	EN	SD	SF	VIF	AG	SCD
7-12-10			6.961	45.578	11.418	0.815	4.270	1.851
7-12-10		✓	7.013	45.462	11.528	0.832	4.336	1.870
DRGN	✓		6.923	44.202	11.840	0.804	4.377	1.852
DRGN	✓	✓	7.083	45.502	12.739	0.842	4.832	1.884

4.2 Fusion Strategy Experiments

The first group of experiments in Tab. 3 shows that the addition fusion strategy performs best when the same fusion strategy is used for different feature information. Suitable fusion strategies should be used for feature information at different scales, respectively. Compared to other groups of experiments, almost all evaluation metrics reach optimal values when the low and medium frequency feature information is using the addition fusion strategy and the high-frequency feature information is using the proposed IRCAFN fusion strategy. The results illustrate that the IRCAFN network is more suitable for critically high-frequency detailed feature information.

4.3 Comparison with Other Methods

Qualitative Comparison: We compare our method with influential methods in the last three years, including Densefuse [6], DIDFuse [13], NestFuse [14], Dual Branch [15], UNFusion [16], RFN-Nest [8], and AUIF [5]. The results of the fusion are shown in Fig. 3. To directly illustrate the comparison results, orange and blue boxes are used to subjectively marker and enlarge the selected area.

The proposed fusion method maintains the balance of meaningful information between different source images. Fusion images biased toward visible images lead to fusion results containing rich texture information but losing salient infrared information. For example, the fusion results of NestFuse and UNFusion in Fig. 3 all lose the cloud detail information. Conversely, the proposed fusion method retains the visible image detail information while highlighting the infrared image saliency target, keeping a good balance between them. The DIDFuse results appear to have an unnatural brightness distribution, and the RFN-Net results have a blurred situation. Although the results of DenseFuse, Dual Branch, and AUIF exhibit excellent performance, our method demonstrates a higher contrast ratio, which provides clearer details in the images. In addition,

Table 3. Comparison of fusion methods. The xx-xx-xx denotes the fusion strategy used for low, medium, and high-frequency feature information. The bold fonts are the best values in corresponding indexs.

Fusion strategy	EN	SD	SF	VIF	AG	SCD
avg-avg-avg	6.445	27.188	7.686	0.691	2.957	1.629
ll-ll-ll	6.877	40.348	9.635	0.819	3.704	1.735
max-max-max	6.139	23.621	7.218	0.778	2.690	1.178
add-add-add	7.083	45.502	12.739	**0.842**	4.832	1.884
avg-add-add	6.512	27.956	8.111	0.710	3.169	1.656
ll-add-add	6.935	40.788	9.959	0.836	3.883	1.752
max-add-add	6.313	26.153	8.096	0.839	3.042	1.210
add-avg-add	7.086	45.354	12.544	0.842	4.748	1.882
add-ll-add	7.085	45.342	12.536	0.842	4.744	1.882
add-max-add	7.084	45.315	12.515	0.842	4.737	1.882
add-add-avg	7.015	44.849	12.522	0.827	4.695	1.873
add-add-ll	7.015	44.849	12.522	0.827	4.695	1.873
add-add-max	7.014	44.826	12.518	0.827	4.694	1.873
Ours	**7.126**	**45.610**	**13.083**	0.836	**5.020**	**1.886**

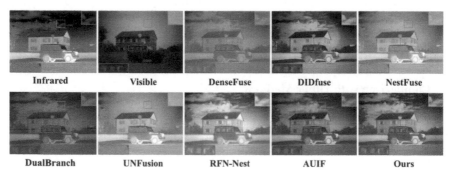

Fig. 3. Results of qualitative comparison experiments with other methods.

the luminance distribution of our results is more uniform with a wider distribution and better visual performance.

Quantitative Comparison: We tested the above-mentioned quantitatively using the same evaluation metrics. The test results are shown in Tab. 4. For the EN, SD, SF, AG, and SCD metrics, the results showed the best or second significant lead. As for the VIF metric, our results achieved a medium standing. Specifically, the larger EN and SCD show that our fusion images retain richer details. This means that the fusion images not only contain rich detail information but also have high contrast and the best visual quality. The best values are obtained in the metrics SF and AG, which represent images containing more detailed edge information and higher clarity. In combination, our approach is suitable for IVIF tasks in different scenes.

Table 4. Quantitative comparison with other methods. The bold fonts are the best values in corresponding indexs.

Methods	EN	SD	SF	VIF	AG	SCD
denseFuse	6.352	24.782	6.367	0.669	2.515	1.606
DIDFuse	7.006	**46.884**	*11.266*	0.832	*4.294*	1.784
NestFuse	*7.037*	41.751	10.105	*0.970*	3.892	1.721
DualBranch	6.547	29.551	6.597	0.729	2.645	1.679
UNFusion	7.022	41.954	10.268	**0.999**	3.907	1.696
RFN-Fusion	6.991	37.248	5.868	0.823	2.682	1.798
AUIF	6.944	43.447	11.053	0.811	4.146	*1.860*
Ours	**7.126**	*45.610*	**13.083**	0.836	**5.020**	**1.886**

5   Conclusions

To improve the image quality of the IVIF theme, this paper develops a new deep learning fusion model with complete frequency domain multi-scale decomposition. Especially, an additional feature extraction branch is introduced to address the useful feature extraction from the informative medium frequency region, which is based on the DRGN and PSA modules. Considering specific characteristics of the various spectrum, we fuse the features of different frequency bands independently. Crucially, an interactive residual coordinate attention fusion network is applied for high-frequency feature information fusion. Finally, the decoder is used to generate fusion images. With this low-loss multi-scale decomposition and adaptive fusion strategy, more valuable contents will be retained in the feature extraction stage and the fusion performance can also be enhanced. Both qualitative and quantitative experiments validate the superiority of our approach over state-of-the-art methods. In the future, we will design a fully adaptive fusion network based on deep learning, which can further improve the quality and effectiveness of fusion images.

Acknowledgements. This paper is supported by the National Nature Science Foundation of China (No.61861020).

References

1. Ma, J., Ma, Y., Li, C.: Infrared and visible image fusion methods and applications: a survey. Inf. Fusion **45**, 153–178 (2019)

2. Chen, J., Li, X., Luo, L., Mei, X., Ma, J.: Infrared and visible image fusion based on target-enhanced multiscale transform decomposition. Inf. Sci. **508**, 64–78 (2020)
3. Xu, H., Zhang, H., Ma, J.: Classification saliency-based rule for visible and infrared image fusion. IEEE Trans. Comput. Imaging **7**, 824–836 (2021)
4. Ma, J., Zhang, H., Shao, Z., Liang, P., Xu, H.: GANMcC: a generative adversarial network with multiclassification constraints for infrared and visible image fusion. IEEE Trans. Instrum. Meas. **70**, 1–14 (2021)
5. Zhao, Z., Xu, S., Zhang, J., Liang, C., Zhang, C., Liu, J.: Efficient and model-based infrared and visible image fusion via algorithm unrolling. IEEE Trans. Circuits Syst. Video Technol. **32**, 1186–1196 (2022)
6. Li, H., Wu, X.J.: DenseFuse: a fusion approach to infrared and visible images. IEEE Trans. Image Process. **28**(5), 2614–2623 (2018)
7. Long, Y., Jia, H., Zhong, Y., Jiang, Y., Jia, Y.: RXDNFuse: a aggregated residual dense network for infrared and visible image fusion. Inf. Fusion **69**, 128–141 (2021)
8. Li, H., Wu, X.-J., Kittler, J.: RFN-Nest: an end-to-end residual fusion network for infrared and visible images. Inf. Fusion **73**, 72–86 (2021)
9. Zhang, H., Zu, K., Lu, J., Zou, Y., Meng, D.: EPSANet: An efficient pyramid squeeze attention block on convolutional neural network. arXiv preprint arXiv:2105.14447 (2021)
10. Hou, Q., Zhou, D., Feng, J.: Coordinate attention for efficient mobile network design. In: Proceedings of the IEEE/CVF conference on computer vision and pattern recognition. pp. 13713–13722 (2021)
11. Xu, H., Ma, J., Le, Z., Jiang, J., Guo, X.: Fusiondn: A unified densely connected network for image fusion. In: Proceedings of the AAAI Conference on Artificial Intelligence. vol. 34, pp. 12484–12491(2020)
12. Toet, A.: "TNO Image Fusion Dataset," 4 2014. [Online]. Available: https://figshare.com/articles/dataset/TNO Image Fusion Dataset/1008029
13. Zhao, Z., Xu, S., Zhang, C., Liu, J., Li, P., Zhang, J.: DIDFuse: Deep image decomposition for infrared and visible image fusion. arXiv preprint arXiv:2003.09210 (2020)
14. Li, H., Wu, X.-J., Durrani, T.: NestFuse: an infrared and visible image fusion architecture based on nest connection and spatial/channel attention models. IEEE Trans. Instrum. Meas. **69**, 9645–9656 (2020)
15. Fu, Y., Wu, X.-J.: A dual-branch network for infrared and visible image fusion. In: 2020 25th International Conference on Pattern Recognition (ICPR), pp. 10675–10680 (2021)
16. Wang, Z., Wang, J., Wu, Y., Xu, J., Zhang, X.: UNFusion: a unified multi-scale densely connected network for infrared and visible image fusion. IEEE Trans. Circuits Syst. Video Technol. **32**, 3360–3374 (2022)

An Adaptive Model-Free Control Method for Metro Train Based on Deep Reinforcement Learning

Wenzhu Lai[1,3](✉), Dewang Chen[2,3](✉), Yunhu Huang[1,3], and Benzun Huang[1,3]

[1] College of Computer and Data Science, Fuzhou University, Fuzhou, China
200327042@fzu.edu.cn, N190310001@fzu.edu.cn, 200327163@fzu.edu.cn
[2] School of Transportation, Fujian University of Technology, Fuzhou, China
dwchen@fjut.edu.cn
[3] Key laboratory of Intelligent Metro of Universities in Fujian Province, Fuzhou University, Fuzhou 350108, China

Abstract. The current metro train control system has achieved automatic operation, but the degree of intelligence needs to be enhanced. To improve the intelligence of train driving, this paper adopts the proximal policy optimization (PPO) algorithm to study the intelligent train operation (ITO) of metro trains by drawing on the successful application of deep reinforcement learning in games. We propose an adaptive model-free control (MFAC) method for train speed profile tracking, named as intelligent train operation based on PPO (ITOP), and design reinforcement learning policies, actions, and rewards to ensure the accuracy of the train tracking speed profile, passenger comfort, and stopping accuracy. Simulation experiments are conducted using real railroad data from the Yizhuang Line of Beijing Metro (YLBS). The results show that the tracking curve generated by ITOP is highly coincident with the target curve with good parking accuracy and comfort, and responds positively to the changes of the target curve during the operation. This provides a new solution for the intelligent control of trains.

Keywords: Intelligent train operation · Model free adaptive control · Deep reinforcement learning

1 Introduction

A metro train is a servo system, and optimizing a single train operation entails two critical issues: speed profile optimization design and speed profile tracking control [1]. To achieve nearly equal performance between design and practice, the speed profile is designed offline and the speed profile tracking control is designed online. The construction of an excellent controller is the key to speed profile track.

N. Xiong et al. (Eds.): ICNC-FSKD 2022, LNDECT 153, pp. 263–273, 2023.
https://doi.org/10.1007/978-3-031-20738-9_31

Proportional-integral-derivative (PID) control is the speed profile tracking control technique that is most frequently employed in industrial process control [2]. PID controllers, which are frequently used in train operation control [3], regulate the system by regulating a linear mixture of proportional, integral, and differential errors. PID algorithms are widely used in industrial controllers because of its structural simplicity, high level of robustness, and user-friendliness. But for nonlinear systems with changing characteristics, it is not precise enough.

As a result, several academics have concentrated on adaptive control approaches for tracking adaptive train speed profiles. Ke et al. [4] proposed using a fuzzy PID gain approach to meet the recommended speed profile. Liu et al. [5] suggested a fuzzy control method-based high-speed train control system and built the system using the extensive expertise and knowledge. Gu [6] et al. suggested a new approach to energy-efficient train operation control that is based on geometric and topographical data from real-time traffic statistics. Gao et al. [7] presented two robust adaptive control methods that take into account actuator saturation as well as unknown system factors. Recently, Pu et al. [8] proposed a model-free adaptive speed controller based on neural network (NN) and PID algorithm, and verified the validity of the proposed algorithm in accurately tracking speed-distance trajectories by numerical tests and real-time implementation.

Reinforcement learning (RL) [9] has been used to deal with optimal control problems in many different fields, such as robot control [10], micro unmanned aircraft control [11], etc. To the best of our knowledge, few studies have used deep reinforcement learning to train speed profile tracking. If we look at metro train driving as a driving game, we can consider introducing deep reinforcement learning algorithms that have been successfully applied in the game field [12] and designing reinforcement learning strategies, actions, and rewards to ensure the accuracy of train tracking speed profile, passenger comfort, and stopping accuracy. This is the main driving force behind this paper. The rest of this paper is organized as follows. In Sect. 2, we define the necessary mathematical notation and performance metrics for metro train operation. Section 3 presents the design of the PID controller and the PPO-based ITOP algorithm. In Sect. 4, we construct an ITOP simulation platform and give three numerical examples of real data and compare the operational performance of the PID controller and ITOP. Section 5 draws conclusions.

2   Problem Formulation

Generally, train control can be described as an optimal control problem, with the goal of determining the best control strategy for traction and braking force during the trip. Therefore, we define Δt as the shortest time interval, and the train travel duration can be described as Eq. (1):

$$t_{i+1} = t_i + \Delta t, i = 0, \ldots, n \tag{1}$$

where $t = 0(s)$ and $\Delta t = 0.02(s)$ are initial run time, and the minimum time interval respectively.

2.1   Control Model of Train

The output force, the resistance created by the gradient of the railway, the resistance to motion, the curve resistance, and the resistance caused by interaction collisions among the vehicles all contribute to the train's motion. Its movement equation is determined by Newton's second law as Eq. (2)

$$M(1 + \mu)u = F - f_g - f_r - f_c - f_d \tag{2}$$

where M is the static mass of the train, μ is the rotation factor of the train, which is defined as $\mu = M_\mu/M$, and M_μ is the reduced mass of the train rotator. F is the traction or braking force output by the metro, $f_g = Mg\sin(\alpha(s))$ is the resistance due to gradient, where $\alpha(s)$ denotes the angle of the gradient of the railroad at different position s, $f_r = D_1 + D_2 v_i + D_3 v_i^2$ is the resistance to motion given by David's equation, where v_i is the current speed and D_1, D_2 and D_3 are the vehicle ratio coefficients measured by the running experiment. $f_c = 6.3M/[r(s) - 55]$ is the curve resistance, where $r(s)$ is the radius of the curve at position s; $f_d = \sum_{i-1}^{k-1} \left(\Delta \ddot{l}_i \sum_{j=i+1}^{k} m_j \right)$ is the interaction effect between vehicles, where $\Delta \ddot{l}_i$ is the second-order derivative of the distance between the center of the ith vehicle and the reference point.

The transfer functions for the acceleration and deceleration operations are provided by Eq. (3) due to the nonlinearity and time delays included in the train control model.

$$G_u(s) = \frac{u_0}{1 + T_\tau s} e^{-T_\sigma s} \tag{3}$$

where $G_u(s)$ denotes the actual acceleration and deceleration of the train; u_0 denotes the acceleration or deceleration performance gain; T_τ and T_σ denote the time delay and time constant of the acceleration or deceleration process respectively.

2.2   Indices of Model Evaluation

Metro control models are generally evaluated in terms of four aspects: safety, accuracy in tracking the reference speed profile, passenger comfort, and parking accuracy, they are defined as follows:

- **Safety**: where the corresponding speed limitations for the various sections between the two stations are V_1^{limit} , V_2^{limit} , V_3^{limit} and V_4^{limit}. That indicates that in order to maintain safety during the travel duration, the train's speed must be lower than the existing railroad section speed restriction. The safety evaluation index I_s is represented by Eq. (4) as follows

$$I_s = \begin{cases} 1 & v_i \leq V_i^{\text{limit}}(\forall i) \\ 0 & v_i > V_i^{limit}(\exists i) \end{cases} . \tag{4}$$

- **Trajectory Tracking Performance**: The maximum absolute speed error is used to measure the index of the model tracking the offline speed curve in this paper, The maximum absolute speed error is the maximum absolute value of the difference between the reference curve and the actual running speed of the train.

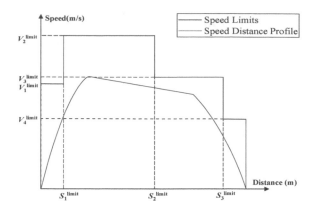

Fig. 1. Speed limit.

- *Passenger Comfort*: The magnitude of the change in the train joystick affects the comfort level of train operation. Comfort C_i can be measured by the rate of change in acceleration Δu_i, and the relationship between comfort and Δu_i, $C_i = \frac{1}{1+|\Delta u_i|}$. I_c denotes the comfort level, where $I_c = \frac{\sum_{i=1}^{n} C_i}{n}$.
- *Parking accuracy*: The parking error of the metro is generally required to be within ± 30 cm so that the screen door can be opened. Therefore, the parking accuracy index I_p can be defined as $I_p = |s_i - s_D|$, where s_D is the length of the segment between adjacent stations and s_i is the current running distance of the train.

3   ITOP Algorithm Design

RL is a machine learning paradigm that aims to learn to control systems in environments to maximize the numerical performance associated with long-term goals. Meanwhile, deep reinforcement learning is considered to be used for the control of continuous actions [13]. The following reasons motivate us to adapt deep reinforcement learning in train control tasks. First, during train control, behavior affects not only the immediate reward but also the reward of future states. It is what reinforcement learning excels at. Second, the use of deep reinforcement learning can improve the control strategies used in current ATO systems for discrete actions.

3.1   Principles of the PPO Algorithm

The ITOP algorithm is based on the PPO algorithm [14], which is a policy gradient (PG) algorithm and can be used to handle continuous action control and model-free problems. The PG algorithm adopts an online update strategy that resamples every parameter update, resulting in a learning rate that is not easy to determine. The PPO algorithm converts the online update strategy to an offline update strategy, i.e., it adopts the old and new Actor strategies, and the training data of the new Actor can be obtained from the old Actor. The action probability ratio $r_i(\theta)$ of the old and new strategies are used to denote the new strategy weights, which are expressed as Eq. (5):

$$r_i(\theta) = \frac{\pi_\theta\,(a_i \mid x_i)}{\pi_{\theta'}\,(a_i \mid x_i)}. \tag{5}$$

where θ' is the sampled neural network parameter. If the probability distributions obtained from two neural network parameters θ' and θ in the same state differ greatly, and in the case of insufficient sampling, it will lead to a large variance between them. Therefore, the PPO algorithm adds the CLIP function to the objective function basis to limit the parameters θ' and θ to differ too much, as shown in Eq. (6):

$$J_{PPO}(\theta) = E_{(x_i,a_i)\,\pi_\theta}\left[\min\left(r_i(\theta)A_t, \mathrm{clip}\left(r_i(\theta), 1 - \varepsilon, 1 + \varepsilon\right)A_t\right)\right] \tag{6}$$

$$A\left(x_i, a_i\right) = Q_\pi\left(x_i, a_i\right) - V_\pi\left(x_i\right) \tag{7}$$

where $A(x_i, a_i)$ is the estimate of the advantage function at moment i, where $Q_\pi(x_i, a_i)$ is the state action value function, denoting the expected reward obtained by the Agent following the policy π, after performing action a_i in state x_i until the end of the episode. Similarly, the state value function represents the expected reward obtained by the agent following the policy π from state x_i to the end of the episode.

3.2   Design Learning Strategy of PPO

Before adopting reinforcement learning methods, we construct our problem as a Markov Decision Process (MDP), which gives a mathematical foundation for decision making. The following are the fundamental parts of reinforcement learning: state, action, policy, and reward.

State x_i: The state contains a variety of factors that can influence the agent (train) to make the next decision. The agent receives the information provided by the status and analyzes the current situation to know how good or bad each action is. The current and historical train tracking curves have an important reference role as they directly affect the next train operations. In this case, the difference between the reference curve and the actual current running speed of the train, the acceleration, and the difference between the reference curve and the actual current running distance of the train is used as the train state, which can be described as Eq. (8).

$$x_i = [v_i', u_i, d_i'] \quad i = 0, \ldots, m. \tag{8}$$

Action a_i: Since the train moves forward all by the joystick and there is no steering wheel, only one action for deep reinforcement learning can be designed to simulate the joystick. According to the definition of passenger comfort, we can define the rate of acceleration change of the train at the moment as the output of the action $a_i = \Delta u_i$. and, the acceleration of the train at time t_i is $u_i = u_{i-1} + \Delta u_i$.

Policy π: The policy represents the probability of performing an action while performing a discrete action task. Considering ITOP is developed to handle continuous action control tasks, a policy is a statistic of a probability distribution in this study. It can be expressed as

$$\pi(a \mid x, \theta) = \mathcal{N}(\mu(x, \theta), \sigma(x, \theta)). \tag{9}$$

where θ is the weight.

Reward function $r(x_i, a_i)$: This function indicates the degree of influence on the train tracking target curve after executing the action. In reinforcement learning, each time the process from the initial state to the final state is called a round. Since the train operation control system is a highly complex, multi-objective nonlinear dynamical system, when the original target task is so complex that it is difficult to complete, the learning and training can be accelerated by "course-based" reinforcement learning [15]. i.e., an agent can start with simple and relevant tasks, and then gradually learn more complex tasks by increasing the difficulty of the tasks.

(1) Let the train learn to run the full distance within the set distance difference D. The train is given a certain positive reward λ_1 for each step taken. When the distance difference exceeds the D, the episode ends and starts again.

$$r(x_i, a_i) = \begin{cases} \lambda_1 & |d'|_i \le D \\ \text{End episode} & |d'|_i > D \end{cases} \tag{10}$$

(2) When the train can run the full distance under (1), set a speed error range V and a smaller D, if this condition is not met then give a negative reward λ_2, and vice versa give a positive reward λ_1, and at this point, if $|v'_i|$ is less than V give a positive reward λ_1. Otherwise, a negative reward λ_2 is given.

$$r(x_i, a_i) = \begin{cases} \lambda_1 & |d'_t| \le D \\ \lambda_2 & |d'_t| > D \\ 2\lambda_1 & |v'_i| \le V, d'_t \le D \\ 0 & |v'_i| > V, d'_t \le D \end{cases} \tag{11}$$

(3) Further minimize the range of D and V, and continue to use the reward method of (2).
(4) When the train largely learns to run close to the target curve, if the train stops within $\pm 30\,\text{cm}$ of the end point, then it is given a considerable bonus λ_3, otherwise, it is given a considerable negative bonus λ_4.

We give the algorithmic process of ITOP in Algorithm 1.

Algorithm 1 Detailed process of ITOP

1: Initialize:
   Initialize reply buffer capacity T.
   Randomly initialize Actor-New, Actor-Old networks and assign random weights θ, θ', and $\theta' = \theta$.
   Randomly initialize the Critic network and assign a random weight ϕ.
2: **for** $i = 1, N$ **do**
3:    **for** $j = 0, T$ **do**
4:       The environment information x_j is input to the actor-new network, and then an action a_j is sampled out by a normal distribution, which is then input to the environment to obtain the reward r_j and the next state $x_j + 1$(Reward the action a_j or end the episode according to the reward function $r(x_i a_i)$ of different stages), which is then stored in the experience buffer $[(x_j, a_j, r_j), \cdots]$, and then x_{j+1} is input to the actor-new network.
5:    **end for**
6:    Input x_t into the critic network to calculate and estimate the advantage function $V_\phi(x_t)$

$$A_t = \sum_{t > t'} \gamma^{t'-t} r_{t'} - V_\phi(x_t)$$

7:    Update the Critic network parameters by minimizing the loss function:

$$L_c = \frac{1}{N} \left(\sum_{t > t'} \gamma^{t'-t} r_{t'} - V_\phi(x_t) \right)^2.$$

8:    **for** $k = 1, M$ **do**
9:       The combination of all states x_i stored in the buffer is input into the Actor-New, Actor-Old network to obtain two state action probability rate distribution $\pi_\theta(a_k \mid x_k), \pi_{\theta'}(a_k \mid x_k)$, and calculate:

$$r_k(\theta) = \frac{\pi_\theta(a_k \mid x_k)}{\pi_{\theta'}(a_k \mid x_k)}$$

       Update the actor-new network weights θ by Eq. (6).
10:   **end for**
11:   Update the actor-old network with the actor-new network weights: $\theta' = \theta$.
12: **end for**

---

4 Simulations

We designed three numerical simulation experiments based on real-world data acquired by YLBS to test the effectiveness, flexibility, and robustness of ITOP. On December 30, 2010, YLBS began operations in Beijing. The YLBS runs for 23.3 km, beginning at Songjiazhuang station and finishing at Ciqu station. The train type used in YLBS is DKZ32 rolling stock with 6 vehicles, whose parameters are shown in Table 1. We use the ITOE proposed by [16] as the reference curve. In the training process of ITOP, the learning rate is set to

3×10^{-4}. The discount factor of the value function is 0.95, and the size of the mini-batch for the memory reply is 64. the values of $\lambda_1, \lambda_2, \lambda_3$ and λ_4, concerning the reward function are $0.1, -0.1, 100, -100$, respectively. In addition, ITOP has a total of three hidden layers, each with 128 units, and each hidden layer is followed by a Relu activation function.

Table 1. Parameters of DKZ32.

Value	Parameters
M (kg)	1.99×10^5
$m_i, i = 1, 6$ (kg)	3.3×10^4
$m_i, i = 3$ (kg)	2.8×10^4
$m_i, i = 2, 4, 5$ (kg)	3.5×10^4
$\Delta \ddot{l}_i, i = 1, 3, 6$ (mm)	$0.1 \sin(t)$
$\Delta \ddot{l}_i, i = 2, 4, 5$ (mm)	$0.15 \cos(t)$
Timeconstant(Braking)T'_τ	0.4
Timedelay(Braking)(s)T'_σ	0.8
Timeconstant(Accelerating)T_τ	0.4
Timedelay(Accelerating)(s)T_σ	1
(α, β, γ)	$(1.244, 1.45 \times 10^{-2}, 1.36 \times 10^{-4})$

In this section, we presents simulation results for three cases. In Case 1, we compare the operational performance of tracking the reference speed profile at Rongjing (RJ)—Wanyuanjie (WYJ) using the ITOP control algorithm and the PID control algorithm (with a planned trip time of 101 s). In Case 2, we test the flexibility of ITOP to track the reference curve at different trip times by varying the planned travel times (95 and 115 s) for the same rail segment. In Case 3, we test the operational performance of ITOP on the Songjiazhuang-XiaoCun railroad section with more complex speed limits and gradients to verify the robustness of ITOP. We give the speed limits and gradients for RJ-WYJ and SJZ-XC in Fig. 2.

Figure 3 shows the comparison of the operation performance of ITOP and PID controller in RJ-WYJ in Case 1, we can see that the ITOP controller can fit the reference curve well and the velocity error is significantly lower than that of the PID controller. the variance of the velocity vs. distance and position vs. time tracking errors of the ITOP controller is smaller than that of the classical PID controller, which implies that the suggested ITOP controller outperforms the existing controller under the same operating conditions. Note that each error is calculated under the same horizontal axis of the relevant curve, and for different running times and stopping positions, some errors are not equal to 0 at the end of the curve. in addition, both ITOP and PID controllers can guarantee that the speed is less than the speed limit, i.e., both ITOP and PID can guarantee safe driving of the train. In fact, the performance of ITOP in Case 2 and Case 3 is

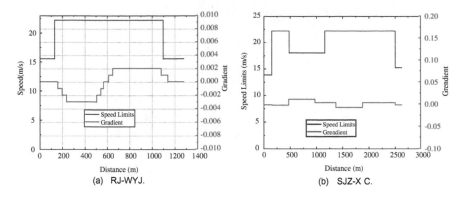

Fig. 2. Speed limit and slope of RJ-WYJ and SJZ-XC

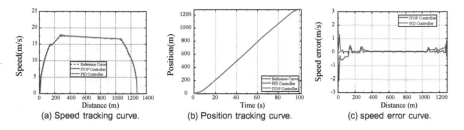

Fig. 3. Comparison of the operation performance of ITOP and PID controller in RJ-WYJ.

similar to that in Fig. 3. For space reasons, we do not give comparative graphs of the operational performance of Case 2 and Case 3. Table 2 shows the detailed data of the operational performance of the three cases.

As can be seen from Table 2, the ITOP was able to ensure accurate parking in all three cases, i.e., the parking error was less than 30 cm. The PID controller had a parking error of more than 30 cm in some cases. the I_v and I_c of the ITOP also outperformed the PID controller, which demonstrates the effectiveness of the ITOP. In Case 2, the trajectory of the reference curve changes due to the change of the planned trip time. While the mean values of I_v and I_p of ITOP are 57.3% and 50% lower than PID, respectively, and the mean value of comfort index I_c of ITOP is 13.7% higher than PID . This indicates that ITOP can achieve accurate tracking of the reference curve while satisfying the multi-objective constraint. In other words, ITOP has the flexibility to track the reference curves under different travel times under the same railroad section. ITOP has some flexibility. It can track the reference curve for the same railroad section with different travel times. We used the SJZ-XC railroad section with more complicated speed limit and gradient in Case 3. The results show that the I_v of ITOP is slightly higher at 1.56m/s, which is 19.59% lower than the PID controller. This verifies the robustness of ITOP and shows that ITOP can also adhere to the reference curve well in the complex railroad section.

Table 2. Comparison of performance in different cases

Case	Controller	t (s)	I_s	I_v	I_c	I_p
Case 1	ITOP	101	1	1.29	0.89	0.15
	PID	101s	1	2.24	0.77	0.20
Case 2	ITOP	95	1	1.31	0.82	0.22
	PID	95	1	3.20	0.69	0.47
	ITOP	115	1	0.53	0.93	0.09
	PID	115	1	1.11	0.82	0.21
Case 3	ITOP	190	1	1.56	0.81	0.26
	PID	190	1	1.94	0.71	0.39

5 Conclusion

In this paper, a PPO algorithm-based train speed profile tracking adaptive control method ITOP was proposed. By designing reinforcement learning state, action, policy, and reward, we divide the complex target task into several simple and easy to train steps, and gradually achieve the train control learning through rewards. At the same time, three numerical examples are given in the simulation platform using data from the real line YLBS. The results show that ITOP can achieve accurate tracking of the reference curve with certain flexibility and robustness while ensuring safety, passenger comfort, and stopping accuracy. However, tracking the reference curve offline, to some extent limits the handling of accident occurrences when the train is running between stations. In our future work, we will focus on how to achieve online control of trains without the need to calculate offline reference speed curves.

Acknowledgements. This work was supported by the National Natural Science Foundation of China (NSFC) under grants 61976055 and 61906043.

References

1. Cao, Y., Wang, Z.-C., Liu, F., Li, P., Xie, G.: Bio-inspired speed curve optimization and sliding mode tracking control for subway trains. IEEE Trans. Veh. Technol. **68**(7), 6331–6342 (2019)
2. Zhu, Y., Hou, Z.: Data-driven mfac for a class of discrete-time nonlinear systems with rbfnn. IEEE Trans. Neural Netw. Learn. Syst. **25**(5), 1013–1020 (2014)
3. Fu, P., Gao, S., Dong, H., Ning, B., Zhang, Q.: Speed tracking error and rate driven event-triggered pid control design method for automatic train operation system. In: 2018 Chinese Automation Congress (CAC), pp. 2889–2894 (2018)
4. Ke, B.-R., Lin, C.-L., Lai, C.-W.: Optimization of train-speed trajectory and control for mass rapid transit systems. Control Eng. Practice **19**(7), 675–687 (2011)
5. Liu, W.Y., Han, J.G., Lu, X.N.: A high speed railway control system based on the fuzzy control method. Exp. Syst. Appl. **40**(15), 6115–6124 (2013)

6. Qing, Gu., Tang, Tao, Cao, Fang, Song, Yong-duan: Energy-efficient train operation in urban rail transit using real-time traffic information. IEEE Trans. Intel. Transp. Syst. **15**(3), 1216–1233 (2014)
7. Gao, S., Dong, H., Chen, Y., Ning, B., Chen, G., Yang, X.: Approximation-based robust adaptive automatic train control: an approach for actuator saturation. IEEE Trans. Intel. Transp. Syst. **14**(4), 1733–1742 (2013)
8. Qian, P., Zhu, X., Zhang, R., Liu, J., Cai, D., Guanhua, F.: Speed profile tracking by an adaptive controller for subway train based on neural network and pid algorithm. IEEE Trans. Veh. Technol. **69**(10), 10656–10667 (2020)
9. Kaelbling, L.P., Littman, M.L., Moore, A.W.: Reinforcement learning: a survey. J. Artif. Intel. Res. **4**, 237–285 (1996)
10. Wen, Y., Si, J., Brandt, A., Gao, X., Huang, H.H.: Online reinforcement learning control for the personalization of a robotic knee prosthesis. IEEE Trans. Cybern. **50**(6), 2346–2356 (2020)
11. Xian, B., Zhang, X., Zhang, H., Gu, X.: Robust adaptive control for a small unmanned helicopter using reinforcement learning. IEEE Trans. Neural Netw. Learn. Syst. 1–9 (2021)
12. Mnih, V., Kavukcuoglu, K., Silver, D., Graves, A., Antonoglou, I., Wierstra, D., Riedmiller, M.: Playing atari with deep reinforcement learning (2013). arXiv:1312.5602
13. Lillicrap, T.P., Hunt, J.J., Pritzel, A., Heess, N., Erez, T., Tassa, Y., Silver, D., Wierstra, D.: Continuous control with deep reinforcement learning (2015). arXiv:1509.02971
14. Schulman, J., Wolski, F., Dhariwal, P., Radford, A., Klimov, O.: Proximal policy optimization algorithms (2017). arXiv:1707.06347
15. Florensa, C., Held, D., Wulfmeier, M., Zhang, M., Abbeel, P.: Reverse curriculum generation for reinforcement learning. In: Conference on Robot Learning, pp. 482–495. PMLR (2017)
16. Yin, J., Chen, D., Li, L.: Intelligent train operation algorithms for subway by expert system and reinforcement learning. IEEE Trans. Intel. Transp. Syst. **15**(6), 2561–2571 (2014)

Neighborhood Graph Convolutional Networks for Recommender Systems

Tingting Liu[1], Chenghao Wei[1,3], Baoyan Song[1], Ruonan Sun[2], Hongxin Yang[1], Ming Wan[1], Dong Li[1], and Xiaoguang Li[1(✉)]

[1] Liaoning University, No. 66 Chongshan Middle Road, Huanggu District, Liaoning Province, Shenyang, China
{liutingting,bysong,yanghongxin,wanming,dongli,xgli}@lnu.edu.cn,
weichenghao5@jd.com
[2] China Academy of Urban Planning and Design, No. 5 Chegongzhuang West Road, Haidian District, Beijing, China
sunruonan@caupd.com
[3] Beijing Jingdong Zhenshi Information Technology Co., LTD, No. 76 Zhichun Road, Haidian District, Beijing, China

Abstract. Recommendation system based on collaborative filtering has attracted much attention because it can make personalized recommendations according to the different needs of different users. However, a training recommender system based on collaborative filtering needs a large amount of user-item interaction information. In practical applications, the collaborative filtering recommender system has problems of cold start and sparsity due to the lack of information in the data. The researchers noticed that the knowledge graph contained a lot of attribute relations and structured information. In order to alleviate and solve the problems of cold start and sparsity in the existing recommendation system, the researchers tried to introduce the knowledge graph into collaborative filtering. However, the existing recommendation systems based on knowledge graphs usually only pay attention to the project information and use the project information to model user preferences, but seldom pay attention to the user information. In this paper, we discuss the necessity of using user information and propose a new convolutional neural network framework, which is an end-to-end framework, which explores the correlation attributes in the knowledge graph to capture the correlation, and then to model high-order user information and high-order project information. It is worth mentioning that we use 9 different ways to aggregate high-order user neighborhood information and high-order project neighborhood information based on 3 different aggregators. The proposed model is applied to two data sets about books and music, and the results show that the accuracy is significantly improved, which proves the superiority of our method.

This research has been supported by the National Natural Science Foundation of China (U1811261), the Science Research Fund of Liaoning Province Education Department (LJKQZ2021023, LJKZ0082, LJKZ0094, LQN202010), the Natural Science Foundation of Liaoning Province (2020-BS-082, 2021-BS-090), and the Natural Science Foundation of Liaoning Province of China in 2022 (No. 191), and the Major Science and Technology Plan of Liaoning Province of China in 2022 (No. 28).

Keywords: Recommender systems · Knowledge graph · Graph convolutional networks

1 Introduction

People will have access to more and more information in their daily life through various network platforms, including various information, news, movies, books and so on. A large amount of rich information will broaden the horizons of people, but it will also make people unable to find the information they usually want in a large amount of data. The recommendation system came into being. According to different people's different needs for different information, the recommendation system will try to make personalized recommendations.

Knowledge graph (KG), a famous structured knowledge base, is a hot topic recently. The knowledge graph is composed of a head entity, relation entity and tail entity. For example: < Ding Xia athlete. Project Volleyball >. It represents all kinds of relationships as the attributes of entities, and connects entities with common attributes, which shows a good effect in expressing the correlation between entities.

KGCN is an information network combining KG and GCN, which can automatically capture the advanced features of events and complete the recommended process. However, he did not consider the influence caused by the user's neighborhood in the knowledge graph in his recommendation. Recalling our previous example. The goal of KGCN is to judge the probability that user Ding Xia likes basketball under the current recommendation background. When calculating the probability, KGCN takes advantage of the domain information of the project, that is, aggregates some neighborhoods adjacent to basketball in the knowledge graph. In this way, Ding Xia can calculate the interaction probability as long as there have been interactions between her and the projects adjacent to basketball. This approach can be understood as refining project information and analyzing users' personal preferences and potential interests. However, there is a problem with this approach, that is, if Ding Xia never interacts with the sports adjacent to basketball (CBA, NBA, volleyball), we will still face the same problems of cold start and sparsity. In this case, we add the neighborhood information of Ding Xia in the knowledge graph, so as to expand the question of whether the neighborhood of Ding Xia and basketball interacted before into the question of whether the neighborhood of Ding Xia and basketball interacted. That is, we consider not only the user's personalized preference, but also the user's neighborhood's personalized preference, and then the user's neighborhood's personalized preference biased, selective and user preferences together to better analyze the user's potential interest.

In this paper, we attempt to obtain both high-level user and project neighborhood information and propose a new framework, NGCN. We try to use both user neighborhood information and project neighborhood information on the basis of the KG graph and use knowledge graph to capture the particularity of higher-order attributes, to obtain user preferences and neighborhood user preferences. Meanwhile, the neighborhood information of the project is used to enrich the project information.

Our contribution to this article can be summarized as follows:

- We propose a new end-to-end recommendation framework, NGCN, to solve the sparsity problem. By using a graph neural network, NGCN can more accurately represent user embedding and item embedding by considering the structural and semantic dependencies of users and items in KG.
- The sampling efficiency of the whole model is significantly improved by using the attribute linkage in KG and considering the preferences of users and users' neighbors.
- We experimented with two real-world recommendation scenarios. The results demonstrate that the NGCN exceeds the most advanced baseline.

2 Related Work

The traditional methods using KG include distance based model, translation based model, and Transse-based models include Transe [1], Transd [2], and Transr [3]. But these methods do more to map entities and relationships to low-dimensional vectors, and are more suitable for Kg completion than Kg-based prediction. To use KG for prediction, the algorithm must be designed using the entire KG structure. PER [4] and FMG regard KG as a heterogeneous information network, but the entity path must be designed manually, which is difficult to be applied in reality. In [5, 6], KG is used to build a new heterogeneous information network, and the correlation between data is measured through various source paths. DKN [7] is a framework that references deep learning, and they also use a new way to learn embedded representations [8]. These models all use knowledge embedding as a pre-trained embedding representation [9]. It is used to capture micro-behaviors and use micro-behaviors to enhance the recommendation effect. KGAT [10] applied the framework of graph-attention network to collaborative knowledge graphs to learn the embedding of users, projects and entities in an end-to-end manner. Ripplenet [11] is an in-memory network that propagates a user's potential preferences through links [12]. Modeling user representation through KG and hash is applied to the collaborative filtering matrix, which significantly improves the accuracy of recommendation [13]. Project attributes extracted were used as auxiliary information. After the introduction of graph neural network [14], KGCN [15] combines the interaction history of users and projects with KG, focuses on projects with relationships, and learns the embedding of users, objects and entities in an end-to-end manner. Then [16] is used in reinforcement learning, and the rich prior knowledge of KG is used to strengthen the state representation and improve the efficiency of recommendation to users. However, the model based on knowledge graph has never found the higher-order information work that focuses on users.

3 NGCN

3.1 Problem Definition

On the basis of KGCN, we continue to optimize the effect of recommendation under the background of a typical recommendation problem. We have two inputs, one is the interaction matrix Y; The other is the user and project knowledge graph G; So first we have a group of N users $U^{(1)} = (u_1, u_2 \ldots u_N)$ and A group of M items $V^{(1)} = (v_1, v_2 \ldots v_N)$,

To use, browse, excel at, or buy. With respect to knowledge graph, it is composed of entity-relationship-entity triples (h, r, t). Here h ∈ Z, r ∈ R, t ∈ Z, and the head of a triple respectively knowledge relation, tail, and R is the set of entities and relationships in knowledge map. Platoon league (platoon league, match. Competition. Players, Ding Xia) can represent the fact that Ding Xia participated in the Super League of Volleyball. When we get what we need input, the traditional KGCN when it is in an effort to predict whether the user U no interaction had a project has the potential to his previous interest, as we have stressed before, in fact he only consider the item neighbor information, the efficiency and accuracy of this way, he needs to be improved. In fact, Ding Xia has never touched basketball before, but if basketball exists as a neighbor of volleyball in the knowledge graph, an excellent result can be achieved quickly. However, if basketball does not exist as a neighbor of volleyball in the knowledge graph, it is difficult to achieve a good effect at this time. In this case, our model also finds Ding Xia's neighbors. Through the personalized preferences of neighbors, not only the accuracy of recommendation can be improved, but also the problem is optimized.

3.2 Framework Diagram

The framework consists of three parts. The first part is the embedding layer of users and projects. In the second part, neighbor N(u) and neighbor N(v) are obtained from the knowledge graph as the high-order neighborhood representation, which is used to refine the embedding layer. The third part is to aggregate the neighborhood information through the aggregator and make the final prediction.

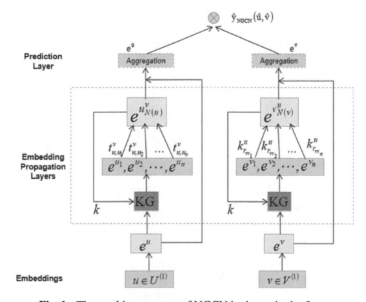

Fig. 1. The working process of NGCN is shown in the figure

According to the classic recommendation model collaborative filtering, we built two embedded vector scales U and V to store the embedded vector representation of us ser and item vector representation respectively. Where, the embedding vector $e_u \in R^d$ in U is used to describe a user u; similarly, the embedding vector in V is used to describe an item v, where d represents the embedding size. In this way, we can express all of U and V as embedded vectors:

$$U = (e_{u1}, e_{u2} \ldots e_{uN}), V = (e_{v1}, e_{v2} \ldots e_{vM}) \tag{1}$$

In contrast, in NGCN framework, we use knowledge graphs to capture the neighbors of User and Item to enrich user embedding and item embedding. This will make our recommendations more accurate and more efficient.

High-order embedding refines the embedding layer. NGCN is aimed to find the user in the knowledge map of neighborhood information and item of neighborhood information, and put these information embedded as high order to continue to refine embedded layer, rich user and item embedded, make recommendation result more accurate, NGCN as shown in Fig. 1 can have layers, we first describe the single-layer NGCN, first is to find the user and the item in the knowledge map of entities in the neighborhood. We can write it in terms of N(u) and N(v). In the previous Sect. 3.2.1, we obtained the embedded lookup table, we can obtain an initial e_u and e_v from the interaction matrix Y and this lookup table, and then we obtain N(u) and N(v) neighbors of entities U and V. In order to refine the original e_u and e_v with high-order neighborhood information, we need to find N(u) and N(v) representations and aggregate them into high-order neighborhood information. First by u and $v_i(v_i \in N(v))$ the relationship $r \in R$ between the to define the weight $K \in R$, according to the weight of k to find v each neighbor embedding said finally embedded into a new item neighborhood said And then through v and $u_j(u_j \in N(u))$ to define the relationship $r \in R$ between the weight $t \in R$, the same, according to the weight of K to find u every neighbor embedding said finally embedded into a new user neighborhood said formula is as follows:

$$v_{N(v)}^u = \sum_{e \in N(v)} k_{r_{v,e}}^u e \tag{2}$$

$$u_{N(u)}^v = \sum_{p \in N(v)} t_{r_{u,p}}^v p \tag{3}$$

The embedding vector representation of each entity obtained from V, and the embedding vector representation of each entity obtained from User embedding lookup table U, and the embedding vector representation of each entity obtained from Item embedding lookup table. k and t are the weights of each neighbor calculated. The formula is as follows:

$$k_{r_{v,e}}^u = \sum \frac{\exp\left(S_{r_{v,e}}^u\right)}{\sum_{e \in N(v)} \exp\left(S_{r_{v,e}}^u\right)} \tag{4}$$

$$t_{r_{u,p}}^v = \sum \frac{\exp\left(Z_{r_{u,p}}^v\right)}{\sum_{p \in N(u)} \exp\left(Z_{r_{u,p}}^v\right)} \tag{5}$$

where S is the score of User u ∈ U and each relation r ∈ R, and Z is the score of each relation r ∈ R of Item v ∈ V. Represents the relationship $r_{t,o}$ between entity t and entity o in the knowledge graph G. Now we can compute the fraction of u and r(v and r) using H: R × R → R.

$$s^u = H(u, r) \tag{6}$$

$$z^v = H(v, r) \tag{7}$$

We are looking for neighbors can choose fixed at a time when the user and item K neighbor, namely, K can be regulated (in our experiment using the K value is 2).

Predict layer. The ultimate goal of NGCN is to enrich our initial embedded representations of User and Item e_u and e_v through knowledge graphs, so the task of the predictive layer is to aggregate the higher-order neighborhood representations we obtained above into our e_u and e_v through aggregators. Two final embedded representations are formed. There are three aggregators we can choose from in this process.

$$agg_{sum}^{item} = \delta\left(W \cdot \left(v + v_{N(v)}^u + b\right)\right) \tag{8}$$

$$agg_{sums}^{user} = \delta\left(W \cdot \left(u + u_{N(u)}^v + b\right)\right) \tag{9}$$

where W and B are respectively transformation weights and deviations, is a nonlinear function, such. as ReLU the connection aggregator [17].

$$agg_{concat}^{item} = \delta\left(W \cdot concat\left(v, v_{N(v)}^u\right)\right) \tag{10}$$

$$agg_{concats}^{user} = \delta\left(W \cdot concat\left(u, v_{N(u)}^v\right)\right) \tag{11}$$

The neighbor aggregator [18] directly takes the neighbor representation of entities U and V as the output representation:

$$agg_{neighbor}^{item} = \delta\left(W \cdot \left(v_{N(v)}^u + b\right)\right) \tag{12}$$

$$agg_{neighbors}^{user} = \delta\left(W \cdot \left(u_{N(u)}^v + b\right)\right) \tag{13}$$

It is worth mentioning that we aggregate User and Item at the same time, looking for their higher-order neighborhood representations and using them to enrich e_u and e_v to improve the prediction effect. So in practice we can choose nine combinations to see more possibilities, namely select different aggregators when aggregating User information and Item information. At this point, we have the final representation sum of user u^ and item v^, and we can calculate the probability of the final representation interaction between user and item by a function f: R × R → R

$$\hat{y}_{u,v} = f\left(\hat{u}, \hat{v}\right) \tag{14}$$

where $\hat{y}_{u,v}$, represents the probability of candidate item V being clicked next time. The complete loss function is as follows:

$$\xi = \sum_{u \in U} \left(\sum_{v:y_{uv}=1} \zeta\left(y_{u,v}, \hat{y}_{u,v}\right) - \sum_{i=1}^{T^u} E_{v \sim p(v)} \zeta\left(y_{u,v}, \hat{y}_{u,v}\right) \right) + \lambda ||\Phi||^2 \quad (15)$$

where ζ is the cross entropy loss, P is the negative sampling distribution, and represents the negative sampling number of users. With a single NGCN layer, we have a neighbor representation of a single User and Item layer, which can be used to enrich the user and item representation. We can call this single-level neighborhood representation a first-order neighborhood representation, and likewise, first-order neighborhood representation naturally applies to higher-order neighborhood representation. Specifically, by repeating the entire NGCN single-layer process over the first-order neighborhood, we can get a second-order neighborhood representation, and so on, using a richer representation to apply higher-order neigh borhood information to our User and Item representations.

4   Experiment

4.1   Data Set

We used the following two data sets in the book and music recommendation experiments.

Table 1. Two data sets of basic statistics and parameter setting

Book-crossing				Last.FM
User-item interaction	#users		17,860	1872
	#items #interactions		14,967	3846
			139,746	42,346
KG	#Entities		77,903	9366
	#Relation Types #Triples		25	60
			151,500	15,518
	K		8	8
	d		64	16
hyper Parameter	H		1	1
	λ		2×10^{-5}	10^{-4}
	η		2×10^{-4}	5×10^{-4}
	batch size		256	128

We used [15] Microsoft Satori to build a knowledge graph for each data set. We first select a subset of triples with confidence greater than 0.9 from the entire KG. The basic statistics of the two datasets are shown in Table 1.

4.2 Comparison of Variants

WE used AUC and F1 to evaluate CTR. Since we are improving on the basis of KGCN, in order to facilitate the comparison with variants of KGCN. In the prediction layer of 3.2.3 we described 9 different aggregation approaches that we might use. We divided the nine variants into three groups, and the results are shown in Table 2.

Table 2. Comparison of results of 9 variants

Model	Book-crossing		Last.FM	
	AUC	F1	AUC	F1
NGCN-ss	**0.744**	0.680	**0.801**	0.733
NGCN-sc	0.743	0.678	0.800	0.733
NGCN-sn	0.734	**0.682**	0.784	**0.736**
NGCN-cc	0.742	0.678	0.798	0.731
NGCN-cs	0.743	0.678	0.800	0.731
NGCN-cn	0.734	**0.682**	0.783	0.736
NGCN-ns	0.736	0.659	0.758	0.680
NGCN-nc	0.735	0.659	0.758	0.680
NGCN-nn	0.733	0.667	0.750	0.682

We chose NGCN-SS, NGCN-CS, and NGCN-NS as the superior variants, and we tried to compare the results of these three variants with other baselines.

4.3 Experiments Settings

We compare the proposed NGCN with the following baseline, including LibFM [19], PER, CKE [12], Ripplenet [11] and KGCN [15].

We set the function H and f as the inner product, δ as the ReLU of the non-last-layer aggregator, and tanh as the last-layer. Table 1 provides additional hyper parameter Settings. The hyper parameters are determined by optimizing the AUC on the validation set. About RippleNet: Book-Crossing: $d = 4$, $H = 3$, $\lambda 1 = 10 \times 10 - 5$, $\lambda 2 = 0.01$, $\eta = 0.001$; Last. FM: $d = 16$, $H = 3$, $\lambda 1 = 10 \times 10 - 5$, $\lambda 2 = 0.02$, $\eta = 0.005$ Apply to the last FM. Other hyper parameters are the same as they were reported in the original paper or by default in their code.

4.4 Experiments Settings

The predicted results of CTR and the recommended results are shown in Table 3 and Fig. 2 (in order to more clearly explain the situation, we only used NGCN-SS to compare KGCN-SUM), and we observed the following:

Table 3. Comparison of baseline experiments

Model	Book-crossing		Last.FM	
	AUC	F1	AUC	F1
LibFM	0.685	0.639	0.778	0.710
PER	0.623	0.588	0.633	0.596
CKE	0.674	0.635	0.744	0.673
RippleNet	0.729	0.662	0.780	0.702
KGCN-sum	0.685	0.642	0.796	0.711
KGCN-concat	0.683	0.637	0.800	0.719
KGCN-neighbor	0.552	0.573	0.532	0.520
NGCN-ss	**0.744**	**0.680**	**0.801**	**0.733**
NGCN-cs	0.743	0.678	0.800	0.731
NGCN-ns	0.736	0.659	0.758	0.680

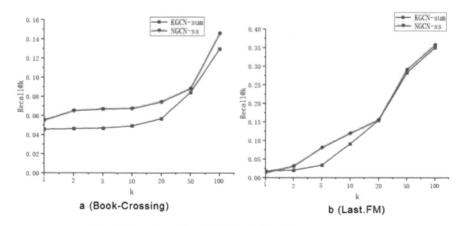

a (Book-Crossing)   b (Last.FM)

Fig. 2. Results of Recall@K in the Top-K recommendation

In general, we found that variants of NGCN had higher CTR and Top-K for both Book-Crossing and Last.FM than corresponding variants of KGCN. This is also in line with our assumption that NGCN improves the sample efficiency after adding higher-order information of the user's neighborhood.

We have found that the best effect on contrast is the use of the sum's aggregator, the worst is the use of the neighbors' aggregator, you can see, from the formula for 8 and 12 sum aggregator is combine neighborhood information and target, and neighbors' aggregators just bring neighborhood information directly, so the effect of the in our experiment, performance is very bad. This means that when we use neighborhood information, we only allow it to be used as an auxiliary information to enrich target information, to improve our recommendation level, which is also illustrated in our baseline experiment comparison in Table 3. This allows us to use this kind of neighborhood information later.

The performance of the Kg-free baseline (LIBFM) is better than the two Kg-aware baselines (PER and CKE), indicating that PER and CKE can't take full advantage of KG with manual design meta-paths and cross-class regularization.PER performed the worst of all baselines because the optimal meta-path is difficult to define. We can see from the results that NGCN are better than KGCN, for NGCN higher-order neighborhood of the reference to the basis of the user information, only exist project before the high order neighborhood information, based on further deepening the higher order neighborhood of the user information, further improve the richness of the sample.

5 Conclusion

NGCN extends the KGCN approach to users' higher-order neighborhood information by selectively and biasedly aggregating user and project neighborhood information. This not only learns about the user's personalization and potential interests. We pointed out two directions for future work. (1) It is a very good direction to design an algorithm to combine high-order neighborhood information of users and high-order neighborhood information of projects. (2) Here we enrich the information to fully improve the sample efficiency. We can combine this part as a state with reinforcement learning.

References

1. Bordes, A., et al.: Translating Embeddings for Modeling Multi-relational Data. Neural Information Processing Systems Curran Associates Inc (2013)
2. Ji, G., et al.: Knowledge graph embedding via dynamic mapping matrix. In: Meeting of the Association for Computational Linguistics and the International Joint Conference on Natural Language Processing (2015)
3. Lin, Y., Liu, Z., Sun, M., Liu, Y., Zhu, X.: Learning entity and relation embeddings for knowledge graph completion. In: Proceedings of AAAI, pp. 2181–2187 (2015)
4. Koren, Y.: Factorization meets the neighborhood: a multifaceted collaborative filtering model. In: Proceedings of the 14th ACM SIGKDD International Conference on Knowledge Discovery and Data Mining. Las Vegas, Nevada, USA (24–27 Aug 2008)
5. Yu, X., Ren, X., Sun, Y., Gu, Q., Han, J.: Personalized entity recommendation: a heterogeneous information network approach. In: Proceedings of the 7th ACM International Conference on Web Search and Data Mining (2014)
6. Palumbo, E., Rizzo, G., Troncy, R.: Entity2rec: learning user-item relatedness from knowledge graphs for top-n item recommendation. In: The Eleventh ACM Conference (2017)
7. Wang, H., et al.: DKN: deep knowledge-aware network for news recommendation. In: The 2018 World Wide Web Conference (2018)
8. Yang, D., et al.: Knowledge embedding towards the recommendation with sparse user-item interactions. In: ASONAM'19: International Conference on Advances in Social Networks Analysis and Mining (2019)
9. Meng, W., Yang, D., Xiao, Y.: Incorporating User Micro-behaviors and Item Knowledge into Multi-task Learning for Session-based Recommendation. SIGIR (2020)

10. Liu, Y., et al.: Contextualized Graph Attention Network for Recommendation with Item Knowledge Graph. IEEE Trans. Knowl Data Eng (2021)
11. Wang, H., et al.: RippleNet: Propagating User Preferences on the Knowledge Graph for Recommender Systems. ACM (2018)
12. Shi, S., Ma, W., Min, Z., Zhang, Y., Yu, X., Shan, H., Liu, Y., Ma, S.: Beyond user embedding matrix: Learning to hash for modeling large-scale users in recommendation, pp. 319–328 (2020)
13. Yang, D., Song, Z., Xue, L., Xiao, Y.: A knowledge-enhanced recommendation model with attribute-level co-attention. SIGIR (2020)
14. Kipf, T.N., Welling, M.: Semi-supervised classification with graph convolutional networks. ICLR (2016)
15. Wang, H., Zhao, M., Xie, X., Li, W., Guo, M.: Knowledge graph convolutional networks for recommender systems. In: World Wide Web Conference (2019)
16. Zhou, S., Dai, X., Chen, H., Zhang, W., Ren, K., Tang, R., He, X., Yu, Y.: Interactive recommender system via knowledge graph-enhanced reinforcement learning. In: SIGIR '20: The 43rd International ACM SIGIR Conference on Research and Development in Information Retrieval (2020)
17. Hamilton, W.L., Ying, R., Leskovec, J.: Inductive representation learning on large graphs. NIPS (2017)
18. Velikovi, P., Cucurull, G., Casanova, A., Romero, A., Liò, P., Bengio, Y.: Graph attention networks. ICLR (2017)
19. Rendle, S.: Factorization machines with libFM. ACM Trans. Intell. Syst. Technol. 3(3), 1–22 (2012)

Emotion Recognition from Multi-channel EEG via an Attention-Based CNN Model

Xuebin Xu, Xiaorui Cheng$^{(\boxtimes)}$, Chen Chen, Haichao Fan, and Muyu Wang

School of Computer Science and Technology, Xi'an University of Posts and
Telecommunications, Xi'an Shaanxi 710121, China
xuxuebin@xupt.edu.cn, {cxxxxxr,2103210107,fhc7911,
wmy961017}@stu.xupt.edu.cn

Abstract. People'S daily communication is inseparable from emotional interaction, and good results can be achieved by using EEG signals for emotion recognition. To this end, this paper proposes an EEG emotion recognition model that classifies emotions into three types: negative, neutral and positive. First, this paper preprocesses the raw EEG signal. By extracting the DE and PSD features in the EEG signal, and the information of five frequencies of Delta, Theta, Alpha, Beta and Gamma in each feature. The information from each 1S is then constructed into a brain map of size $10 \times 9 \times 9$ based on the location of the 62 electrodes. An emotion recognition model was then used to learn the features of this multichannel brain map and implement the classification. The model uses the residual idea of ResNet to ensure that the model learns enough feature information. Meanwhile, we design a channel attention mechanism module which obtains the relationship between channels in the brain map to further improve the recognition performance of the model. Finally, the proposed EEG emotion recognition model was evaluated using the SEED dataset. The experimental data show that the model achieves an average accuracy of 96.01%.

Keywords: EEG · Emotion recognition · Attentional mechanisms · CNN

1 Introduction

Emotion plays an important role in our daily life. Using emotion recognition technology to analyze and evaluate emotion has important application and research value, for example, it is helpful to explore the mechanism and attack the rule of mental diseases and psychological disorders and help doctors to make the most targeted treatment plan. Early studies on human emotions are mostly based on non-physiological signals, such as human facial expressions [1], voice signals [2] and body movements [3], which are difficult to induce a real emotional state. In contrast, physiological signals are governed by the human autonomic nervous system and endocrine system, which can relatively objectively respond to the emotional state of an individual, mainly including myoelectricity [4], blood pressure and brain electricity [5]. The features commonly used for emotion recognition using EEG signals are time domain [6], frequency domain [7], and null

© The Author(s), under exclusive license to Springer Nature Switzerland AG 2023
N. Xiong et al. (Eds.): ICNC-FSKD 2022, LNDECT 153, pp. 285–292, 2023.
https://doi.org/10.1007/978-3-031-20738-9_33

domain [8]. EEG signals can be divided into five basic waveforms by frequency, namely Delta, Theta, Alpha, Beta and Gamma, which are generally used for the extraction of emotion-related features except Delta waves. The differences between EEG signals and other physiological signals are mainly in weak amplitude, non-smoothness, non-linearity and frequency domain, therefore, the correct identification of the emotional information contained in EEG signals is of great significance to people's daily life.

In order to accurately identify emotions through EEG signals, we consider that EEG signals with different frequencies and electrode channels have different degrees of emotional responses, and thus propose an attention mechanism-based network approach, which is built by modifying RseNet18 and adding the proposed attention mechanism module. After validation, the algorithm in this paper can identify and classify different emotion types based on EEG signals. The feature information in the EEG signal is extracted by a preprocessing method and a brain map of size $10 \times 9 \times 9$ is constructed as the model input based on the location of 62 electrodes. The model was trained and tested on the SEED dataset to classify emotions into three major categories: positive, neutral and negative. The convolutional attention approach incorporates an attention mechanism module in the pre-activated residual units of the residual network to model the frequency of EEG signals and electrode channel information, respectively, thus enhancing the import of emotionally salient information.

The structure of this paper is shown below. Sect. 2 describes related work such as EEG signal data processing and feature extraction. In Sect. 3, we choose to analyze and improve the SE-ResNet network model. In Sect. 4, we analyze and compare the experimental results. Conclusions are given in the chapters.

2   Method

2.1   Pre-processing

Due to the complexity of EEG data, data pre-processing can remove noise from the data and at the same time can simplify the data, thus improving the training speed and recognition accuracy of the network model. And EEG data is a kind of data in sequence form, which cannot be directly extracted and recognized by CNN with features. Therefore, this paper first generates a long eigenvectors from the EEG signal using a series of feature extraction methods. The generated feature vector can roughly reflect the temporal variation of the EEG signal, which is considered to be closely related to the underlying emotional state. Two features are extracted from the EEG dataset, DE and PSD, PSD describes the power present in the signal and is the most commonly used frequency domain feature to achieve sentiment recognition [9], The complexity of the DE measurement signal [10] divides the signal into the five frequency bands mentioned above. Since the system of signal acquisition has 62 channels, the extracted information of both frequency bands is converted into a brain map, keeping the electrode positions constant to generate a 9×9 matrix. And for the other non-electrode counterparts, the positions are initialized using random values that satisfy a Gaussian distribution. Therefore, based on the two features of PSD and DE and the information corresponding to the five frequency bands, a brain map with a size of $10 \times 9 \times 9$ can be constructed, and the construction process is shown in Fig. 1.

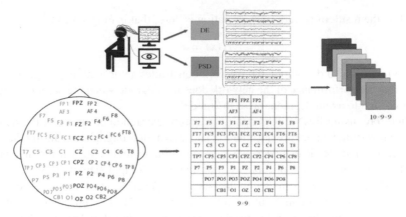

Fig. 1. Pre-processing process of constructing brain map

The constructed brain map data were then normalized to a distribution with mean 0 and variance 1 using Eq. (1) to avoid the presence of extreme values in the data that could negatively affect the recognition results.

$$x_{scale} = (x - x_{mean})/\sigma \tag{1}$$

where x_{mean} is the mean of the brain map features, σ is the standard deviation of each group of brain map features, x is each feature value, and x_{scale} is the normalized feature value.

2.2 CNN Model Based on Attention Mechanism

In this paper, a brain map of size $10 \times 9 \times 9$ is constructed from EEG signals as input by data preprocessing operation, so for the input brain map data, both spatial feature relationships and feature relationships between channels will largely affect the final recognition accuracy. Therefore, in order to be able to obtain sufficient important information from the input brain maps, a CNN model based on the attention mechanism is proposed in this paper. The model uses ResNet, a residual neural network, to extract features from the brain map and to learn the relationship between channels in the brain map through an attention mechanism module. Because the combination of ResNet and the attention mechanism module makes the CNN model pay more attention to the channel relationships in the feature map, so that the feature information that is more conducive to the final recognition can be obtained.

In response to the problems such as model degradation that occur in deep networks [11], Kai He et al. proposed ResNet in 2015, which is different from the traditional convolutional neural network in that the residual module adds a jump connection between the input and output of the two convolutional layers, constituting a layer back off mechanism.

The input goes through the convolution and activation layers to obtain the feature-transformed output F(x), followed by the final output by adding the elements of F(x) to the corresponding positions of the elements of the input.

$$F(x) = H(x) + x \tag{2}$$

When the result of feature transformation is zero, that is $F(x) = 0$, so:

$$H(x) = x \tag{3}$$

The input is also the output, the network constitutes a constant mapping, at least to ensure that the performance of the network does not degrade when the number of layers increases, and to avoid degradation of the model.

Therefore, using ResNet for brain map feature extraction ensures that the network model learns sufficient feature information and also avoids degradation during the training process, which leads to a decrease in recognition accuracy. Meanwhile, in order to make the network model applicable to brain maps with input size of $10 \times 9 \times 9$, this paper modifies the network model on the basis of ResNet18. The initial 7×7 convolution as well as the maximum pooling in the original ResNet18 were removed, and the step size of the convolution in the residual block was reduced to 1, except for the last residual block. The detailed structure is shown in Table 1.

Table 1. Network structure table

Layer name	Output size	Operation
Conv1_x	9×9	$\left\{ \begin{array}{l} 3 \times 3, 64 \\ 3 \times 3, 64 \end{array} \right\} \times 2$
Conv2_x	9×9	$\left\{ \begin{array}{l} 3 \times 3, 128 \\ 3 \times 3, 128 \end{array} \right\} \times 2$
Conv3_x	9×9	$\left\{ \begin{array}{l} 3 \times 3, 256 \\ 3 \times 3, 236 \end{array} \right\} \times 2$
Conv4_x	5×5	$\left\{ \begin{array}{l} 3 \times 3, 512 \\ 3 \times 3, 512 \end{array} \right\} \times 2$
Liner layer	1×1	Average pool, 3-d fc, softmax

For the input brain map with 10 channels, this paper designs a channel attention mechanism module for learning different EEG signal characteristics and the relationship between different frequencies, i.e., the relationship between channels in the brain map. The architecture is shown in Fig. 2. First, each channel feature of the input feature map is learned using three different pooling operations, respectively local average pooling and local maximum pooling, both with a pooling window of 2×2, a step size of 2, a fill of 1, and a global average pooling. Since the input feature map is $9 \times 9 \times C$, two feature maps of size $5 \times 5 \times C$ and one $1 \times 1 \times C$ feature map can be obtained. Next, two $5 \times 5 \times C$ feature maps are stitched together and converted to $1 \times 1 \times C$ using a convolution operation with a convolution kernel of 5×5. The resulting 2 feature maps of $1 \times 1 \times C$ are then weighted and summed over the corresponding channels. The final calculated

$1 \times 1 \times C$ feature map is the information representing the channels in the feature map, and after multiplying it with the corresponding channels of the input feature map, it is the complete process of the attention mechanism module. The calculation formula is Eq. (4).

$$\overline{X} = \left(\left(F_{Conv}\left(F_{avg}(X) \oplus F_{\max}(X)\right)\right) + F_{gap}(X)\right) \otimes X \tag{4}$$

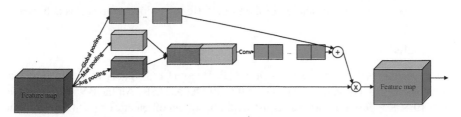

Fig. 2. Channel attention mechanism module architecture diagram

Finally, in this paper, the modified ResNet network model is connected to the channel attention mechanism module as shown in Fig. 3. The channel attention mechanism module is added after the residual block, and the output of the residual block is first multiplied by the result of the attention module and then added to the output of the previous residual block.

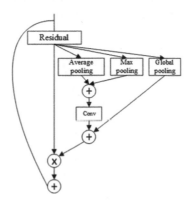

Fig. 3. CAModel and RseNet connection diagram

3   Experiment

3.1   SEED Dataset

This paper uses the publicly available EEG dataset from the BCMI Lab of Shanghai Jiaotong University. The dataset consists of 15 clips from 6 movies, each of which is about 4 min in length and contains three types of emotions: positive, neutral, and negative. Each experiment required 15 clips to be shown to the subjects, and the order of the clips was arranged so that clips of the same type were not viewed consecutively, with a prompt time of 5s before the clip was shown, 15 s to complete the questionnaire at the end of the viewing, and 15 s to take a break. There were 15 subjects, and each subject performed the experiment every approximately 1 week for a total of three times. This paper uses not the raw SEED data, but the extracted EEG signal features. These features were extracted from the raw EEG data as DE, PSD, DASM, RASM, and ASM. Based on these features fast identification of the three different emotion types was achieved.

3.2   Results

In the experiments this paper extracts DE and PSD from EEG signals, as well as information in the corresponding five frequency bands of Delta, Theta, Alpha, Beta and Gamma. And a $10 \times 9 \times 9$ brain map was constructed based on 62 electrode positions, and the information of every 1 s was used as an input data, and the SEED data was finally processed into a shape of $15{,}2730 \times 10 \times 9 \times 9$. Ultimately, the network model uses Adam as the optimizer, the initial learning rate is set to 0.001, and the cross-entropy loss is used as the loss function with 100 iterations. The accuracy variation curve of the algorithm in this paper on the training set is shown in Fig. 4. During the training process, the recognition accuracy gradually increases and changes steadily, and the final recognition accuracy on the test set is 96.01%. The effectiveness of the algorithm in this paper is also verified by comparing ResNet for adding the attention mechanism module, and SE-ResNet for adding the SE module. The comparison results are shown in Table 2. The recognition accuracy of both ResNet and SE-ResNet is lower than that of the algorithm in this paper.

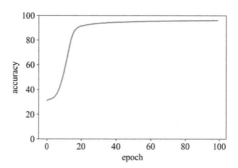

Fig. 4. Accuracy change curve of the algorithm in this paper during the training process

3.3   Comparison with Other Emotion Recognition Methods

Also to verify the superiority of the algorithm in this paper, we compared it with several EEG emotion recognition algorithms. Hwang et al. [10] proposed a CNN emotion recognition method with two convolutional layers and obtained an average accuracy of

Table 2. The results of the proposed algorithm were compared with ResNet18 without attention and with SE

Method	Accuracy (%)
ResNet18	85.24
SE-ResNet18	91.36
Proposed method	96.01

90.41% on the SEED dataset. Cimtay et al. [12] utilized the pre-trained CNN architecture of Inception ResnetV2 to perform sentiment classification on EEG data, and obtained an average accuracy of 78.34% in triple classification experiments on the SEED dataset. A new concept of electrode frequency distribution map (EFDM) based on short-time Fourier transform (STFT) was presented for the first time by Wang et al. [13]. A deep convolutional neural network based on residual blocks was then used for sentiment classification, and the final average accuracy obtained on the SEED dataset was 90.59%. Khare et al. [14] proposed to automatically extract and classify sentiment by using four different CNNs. The method uses a time-frequency representation to convert the filtered EEG signal into an image. The configurable CNN is the most effective, with an average accuracy of 93.01% obtained on the SEED dataset. The detailed results are shown in Table 3. The above results show that the algorithm in this paper is 3.0% higher than the highest one of 93.01%.

Table 3. Comparison results with other EEG emotion recognition methods

Method	Accuracy (%)
Hwang [10]	90.41
Cimtay [12]	78.34
Wang [13]	90.59
Khare [14]	93.01
Proposed method	96.01

4 Conclusion

In order to recognize emotions using EEG signals, a CNN-based emotion recognition method is designed in this paper. First, we preprocess the EEG signal to extract DE and PSD features, and obtain information in five frequency bands of these two features. A brain map with 10 channels is then constructed from this information and the 62 electrode locations as input to the network model. Finally, to recognize emotions, in this paper, we propose an attention-based mechanism of CNN for the recognition of emotions in multichannel EEG signals. In this paper, the proposed sentiment recognition method is evaluated using the SEED dataset. The experimental results showed that the method achieved 96.01% average accuracy. The effectiveness of the channel attention mechanism module and the superiority of the proposed emotion recognition method in this paper are also verified by experimental comparison.

Acknowledgement. This work was supported by the Postgraduate Innovation Fund of Xi'an University of Posts and Telecommunications (Program No. CXJJYL2021039).

References

1. Du, S., Martinez, A.M.: Compound facial expressions of emotion: from basic research to clinical applications. In: Dialogues in Clinical Neuroscience (2022)
2. Yadav, S.P., Zaidi, S., Mishra, A., et al.: Survey on machine learning in speech emotion recognition and vision systems using a recurrent neural network (RNN). Arch. Comput. Methods Eng. **29**(3), 1753–1770 (2022)
3. Ahmed, F., Bari, A.S.M.H., Gavrilova, M.L.: Emotion recognition from body movement. IEEE Access **8**, 11761–11781 (2019)
4. Kehri, V., Ingle, R., Patil, S., et al.: Analysis of facial EMG signal for emotion recognition using wavelet packet transform and SVM. Machine intelligence and signal analysis, pp. 247–257. Springer, Singapore (2019)
5. Song, T., Zheng, W., Song, P., et al.: EEG emotion recognition using dynamical graph convolutional neural networks. IEEE Trans. Affect. Comput. **11**(3), 532–541 (2018)
6. Almanza-Conejo, O., Almanza-Ojeda, D.L., Contreras-Hernandez, J.L., et al.: Emotion recognition using time-frequency distribution and GLCM features from EEG signals. In: Mexican Conference on Pattern Recognition, pp. 201–211. Springer, Cham (2022)
7. Gao, Q., Yang, Y., Kang, Q., et al.: EEG-based emotion recognition with feature fusion networks. Int. J. Mach. Learn. Cybern. **13**(2), 421–429 (2022)
8. Jafarifarmand, A., Badamchizadeh, M.A., Khanmohammadi, S., et al.: A new self-regulated neuro-fuzzy framework for classification of EEG signals in motor imagery BCI. IEEE Trans. Fuzzy Syst. **26**(3), 1485–1497 (2017)
9. Sarma, P., Barma, S.: Emotion recognition by distinguishing appropriate EEG segments based on random matrix theory. Biomed. Signal Proc. Control **7**, 102991 (2021)
10. Hwang, S., Hong, K., Son, G., Byun, H.: Learning CNN features from DE features for EEG-based emotion recognition. Pattern Anal. Appl. **23**(3), 1323–1335 (2019). https://doi.org/10.1007/s10044-019-00860-w
11. Saini, S.S., Rawat, P.: Deep residual network for image recognition. In: 2022 IEEE International Conference on Distributed Computing and Electrical Circuits and Electronics (ICDCECE), pp. 1–4. IEEE (2022)
12. Cimtay, Y., Ekmekcioglu, E.: Investigating the use of pre-trained convolutional neural network on cross-subject and cross-dataset EEG emotion recognition. Sensors **20**(7), 2034 (2020)
13. Wang, F., Wu, S., Zhang, W., et al.: Emotion recognition with convolutional neural network and EEG-based EFDMs. Neuropsychologia **146**, 107506 (2020)
14. Khare, S.K., Bajaj, V.: Time–frequency representation and convolutional neural network-based emotion recognition. IEEE Trans. Neural Netw. Learn. Syst. **32**(7), 2901–2909 (2020)

Multi-feature Short-Term Power Load Prediction Method Based on Bidirectional LSTM Network

Xiaodong Wang$^{(\boxtimes)}$, Jing Liu, Xiaoguang Huang, Linyu Zhang, and Yingbao Cui

State Grid Information and Telecommunication Group, Beijing, China
{dong2,liujing,huangxiaoguang,zhanglinyu,
cuiyingbao}@sgitg.sgcc.com.cn

Abstract. Power system load prediction is an important task of power enterprises, and the analysis method and prediction accuracy of regional grid load characteristics are a key factor in building smart grids and improving the consumption capacity of distributed power generation. Aiming at the problem of reducing the accuracy of the prediction model due to the large number of factors affecting the load forecast, the degree of influence is different, and the strong correlation between multiple influencing factors is caused. In this paper, a multi-feature short-term load prediction method based on deep learning is proposed. The core is to organically combine the improved bidirectional long-short-term memory (BiLSTM) model with the multi-feature data mining method, extract the high-dimensional features of the input vector by using deep convolution through the Inception structure, optimize the weight distribution of the output vector based on the Attention attention mechanism, and construct the Inception-BiLSTM-Attention model. Based on Inception-BiLSTM-Attention, a multi-chain fusion model is constructed, and the feature learning of multiple time period dimensions is carried out, and the short-term power load prediction with high accuracy is realized. This study can provide a reference for regional power system optimization decisions.

Keywords: Neural networks · Power load forecasting · Deep convolution

1 Introduction

With the increasing demand for electricity, the traditional power grid in the centralized distribution, manual monitoring and recovery, two-way communication and other aspects of the beginning of the challenge, the emergence of smart grid to solve the above difficulties to provide technical possibilities, help monitor the power production, transmission and consumption, and balance the relationship between the three. However, due to the influence of climate, economy, environment and other uncertainties, the power load fluctuates greatly, and it is difficult to simply predict it. In order to ensure the safe operation of the power system, reduce power consumption, meet market demand, balance the supply and demand relationship of power load, and maximize economic benefits, it is necessary to accurately predict the power load [1].

© The Author(s), under exclusive license to Springer Nature Switzerland AG 2023
N. Xiong et al. (Eds.): ICNC-FSKD 2022, LNDECT 153, pp. 293–303, 2023.
https://doi.org/10.1007/978-3-031-20738-9_34

The existing power load forecasting methods are mainly divided into four categories: traditional forecasting, modern forecasting, hybrid forecasting and combined forecasting [2]. Among them, traditional forecasting methods, including time series analysis, regression analysis, and other statistical methods, can better handle simple linear problems and estimate future power loads, but are not effective at dealing with nonlinear problems.

As a result, modern prediction techniques based on nonlinear mapping are gradually used to predict nonlinear problems [3], mainly including fuzzy logic, gray systems, artificial neural networks, and support vector regression. However, these methods still have inherent limitations such as computational complexity, poor generalization ability and over fitting, which bring new challenges to accurate prediction of power load. Aiming at the above deficiencies, a hybrid prediction model is proposed by optimizing the algorithm of particle swarms, Bayesians and other parameters and preprocessing the data [4]. In order to further improve and optimize the forecasting model, to overcome the inherent defects of a single forecasting model in traditional, modern and hybrid forecasting methods, more than two different forecasting models are combined in a specific weighted method to obtain a combined forecasting method, and the improvement of the combined forecasting method to the prediction requires not only multiple forecasting models, but also the weighting of each model, does not emphasize the importance of data preprocessing, and usually combines existing mature forecasting models. There are problems such as poor generalization and insufficient robustness.

At present, with the development of convolutional neural networks, deep learning has surpassed traditional machine learning algorithms in the fields of natural language processing [5], image recognition [6], and video tracking [7]. In the field of electric power, due to the strong feature extraction ability of convolutional neural networks, more and more scholars apply deep learning to power equipment fault diagnosis, power system planning, power load prediction and other fields [8–10]. Hochreiter, Schmidhuber et al. [11] proposed long short-term memory (LSTM), which adds a unit state to the RNN to store the long-term state, laying the foundation for deep learning in time series problems. Subsequently, a large number of LSTM-based network models were proposed and applied to fields such as power load prediction.

For example, in the literature [5], a model based on LSTM and LGBM power load prediction is proposed, and good prediction results are obtained. Literature [12] Based on LSTM, a sequence-to-sequence model structure of LSTM is proposed. Improved the accuracy of power load prediction. However, the time series semantic information extracted in these methods is not closely related, resulting in the prediction accuracy not being able to meet the intelligent requirements of power grid dispatching; In the literature [13], a method combining CNN and LSTM is proposed to predict short-term power load, and the high-level characteristics of power load are extracted by CNN, which improves the accuracy when the LSTM input is too long. However, the level of convolution in this method is not enough, the dimensional information of the extracted features is lacking, and the one-way LSTM ignores the reverse characteristics of the data, so that the prediction accuracy is still insufficient; Although increasing the number of convolutional layers can enhance the ability of feature extraction, as the convolutional depth increases, it may cause gradient disappearance and network difficulty training. Aiming at these problems, this paper proposes a multi-feature short-term power load prediction method based on two-way long-term memory network, which uses deep convolution to extract the high-dimensional characteristics of input vectors through Inception structure, and optimizes the weight distribution of output vectors based on The Attention attention mechanism to achieve high-precision short-term power load prediction.

2 Related Work

2.1 Data Preparation

The data sources in this paper are the power load value (sampling every 15 min, 96 h per day) and meteorological factor data (daily maximum temperature, daily minimum temperature, daily average temperature, daily relative humidity and daily rainfall) in a certain region from January 1, 2012 to January 10, 2015, a total of 1106 raw data records. Each data is a 101-dimensional vector of 96 load values and 5 weather data.

2.2 Data Pre-Processing

Missing value processing. Missing data can cause the model to fail to train or severely affect the prediction accuracy of the model. Usually, the methods for dealing with missing values are fixed value method, filling with central, median or mean, interpolation method, decision tree, similar sample value method, and direct culling method.

Normalization treatment. In power load prediction, different impact factors have different dimensions, the difference in their numerical size may be very large, the data of different dimensions may be directly trained in the model without processing, which may affect the training speed of the model, and may also affect the prediction accuracy of the model, so the normalized processing of load data sample set is the basic work of load prediction. This paper uses min-max normalization to normalize the data to [0, 1] to eliminate the difference in the dimension between different impact factors:

$$X_i^* = \frac{X_i - X_{min}}{X_{max} - X_{min}} \tag{1}$$

In the formula, X_i is the original measured value of each sample point of i, X_i^* is the normalized value, X_{min} is the minimum measured value, and X_{max} is the maximum value.

2.3 Schematic Diagram of the Research Method Flow

In this paper, a multi-feature short-term power load prediction method based on bidirectional long-term memory network is proposed, namely the Inception-BiLSTM-Attention model. The steps of the method are as follows: First, the data is read and pre-processed, the data is cleaned to achieve missing filling and outlier processing, and the data is standardized and integrated; Secondly, the model is constructed and optimized, the Bi-LSTM model is constructed, the high-latitude features of the input vector are extracted by CNN, the data feature relationship is learned by multi-layer LSTM, and the parameter weights are assigned by using the Attention attention mechanism to obtain the optimal parameter matrix. Finally, the accuracy evaluation is carried out and the generalization verification is carried out. The overall flow is shown in Fig. 1.

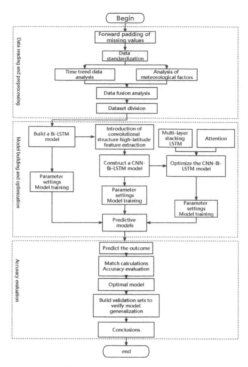

Fig. 1. Method flowchart

3 Model Structure

3.1 Inception Structure

The Inception module is an important achievement in the development of convolutional neural networks from shallow to deep. Before the Inception module, convolutional neural networks simply stacked convolutional layers directly to increase the hierarchical scale of convolutional neural networks for better performance, but this would make the network's parameter volume explode, making training more difficult and overfitting. In this context, in 2014, the GoogLeNet team proposed the Inception net V1 structure, the core of which is the use of convolutional kernels of different sizes in the same layer of network, which improves the model perception and reduces the amount of parameters and calculations. The Inception structure used in this article is shown in Fig. 2:

Compared with the original Inceptionet V1 structure, the following improvements have been made: (1) Using the BN layer, normalize the output of each layer to a normal distribution of $N(0, 1)$, which will help with training because the next layer does not have to learn the offset in the input data and can focus on how to better combine features. (2) Use 2 3×3 convolutions instead of 5×5 convolutions, so that while obtaining the same field of view, it also has fewer parameters and indirectly increases the depth of the network.

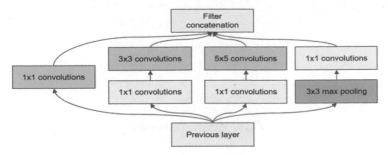

Fig. 2. Inception structure

3.2 Bi-lSTM Network Structure

Bi-LSTM of two-way long-term memory neural network is an optimization improvement of traditional one-way LSTM, which fully considers positive and negative two-way information on the basis of making full use of one-way LSTM to avoid the generation of long-distance dependence, which is conducive to further improving the accuracy of model prediction. Bi-LSTM structure as shown in Fig. 3 where $X_1, X_2 \ldots X_i$ represents the corresponding input data at each moment, $A_1, A_2 \ldots A_i, B_1, B_2 \ldots B_i$ represents the corresponding forward and backward iteration of LSTM hidden states, $Y_1, Y_2 \ldots Y_i$ indicates the corresponding output data.

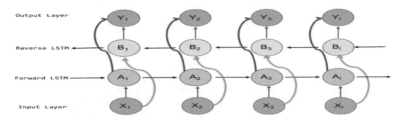

Fig. 3. Two-way long and short time memory network structure

3.3 Attention Attention Mechanism

Attention mechanism is an important concept in neural networks, and has been widely used in many fields such as object detection, image classification, instance segmentation, and text processing. The core idea is to achieve a reasonable distribution of attention to the target information by assigning different weight parameters to the input feature vectors, reducing or ignoring irrelevant information, and amplifying the important information required. At the same time, it can be achieved without increasing the amount of model computation. Optimize model learning and make better choices. The structure of the Attention mechanism is shown in the Fig. 4:

In the figure, $x1 \ldots x_i$ is the input eigenvalue; $h1 \ldots h_i$ is the status value of the hidden layer corresponding to the input eigenvalue; $a1 \ldots a_i$ is the weight value of the current

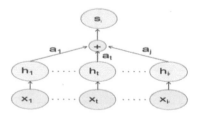

Fig. 4. Attention schematic

input corresponding to the hidden layer state of the historical input; s1 is the hidden layer state value output by the last node.

3.4 Inception-BiLSTM-Attention Model Structure

Because the Inception module enables multi-scale feature extraction of signals, Bi LSTM is able to fully extract time features from time series data. This article combines the two to improve the effect of model extraction features, and builds an Inception-Bi LSTM model as shown in Fig. 5

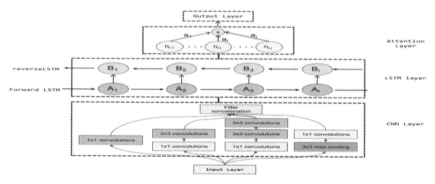

Fig. 5. Inception-BiLSTM-Attention model structure

The prediction model in this paper is mainly divided into input layer, CNN layer, LSTM layer, Attention layer and output layer.

(1) Input layer. After the regional power load dataset is preprocessed, it is imported into the model through the input layer.

(2) CNN layer. The CNN layer mainly extracts the eigenvectors entered by the input layer, mines and screens out the important eigenvectors, and inputs them to the LSTM layer. The CNN layer uses the Inception structure to reduce the amount of parameters while achieving high-latitude feature extraction of deep convolutional networks.

(3) LSTM layer. The feature vector extracted by the CNN layer is input to the LSTM layer, and the LSTM layer in this paper uses a two-layer stacked bidirectional LSTM

network to achieve positive and negative bidirectional data feature extraction and learn the characteristics of regional power load.

(4) Attention layer. The output vector input of the LSTM layer is used as the input of the Atom layer, and according to the weight distribution principle, different weight parameters are assigned to different characteristics in the vector to obtain a better weight parameter matrix.

(5) Output layer. The output vector of the Attention layer is input to the output layer, and the output layer obtains the power load prediction through the fully connected layer.

4   Experimental Verification

4.1   Experimental Data Preparation

The purpose of this experiment is to predict the daily load at various times of the future day by using the previous daily load changes and weather conditions. In this paper, the daily load prediction is carried out in weekly time spans, and in order to construct a dataset to meet the daily load forecasting requirements, the original data set is processed as follows:

(1) The date type is processed, and the factor of forming the holiday variable is assigned a value of 0 for the working day and the holiday position is 1. Then the dataset shape is (1106, 102).

(2) Divide the training set, according to the first 80% of the data as the training set for a total of 128 weeks, 896 days. The remaining 30 weeks total 210 days as a test set. i.e. the training set shape is (896, 102) and the test set is (210, 102)

4.2   Experimental Model Construction

The following figure shows the random continuous power load trend chart of 21 in a certain region, according to the trend chart, it can be seen that the power load is cyclical in weeks, but this cycle feature has a relatively long time span, the LSTM model has limited learning ability for long-term dependence, which is sensitive to the time period information near the prediction point, but does not have the ability to capture the characteristics of the long period. Based on this, based on the Inception-BiLSTM-Attention model, this paper constructs a multi-chain fusion load prediction model, and extracts time series features from multiple time periods at the same time, and the structure is shown in the following Fig. 6:

The model of multi-chain fusion is characterized by the input sequence of the model to be built for the periodic characteristics of the power load data. Construct the input sequence of the model from three dimensions: adjacent time period, short period, and long period. This paper takes the load change of one consecutive week as a data block, takes the seventh day of the data block as the prediction output, the data of the 6th day of the data block as the input sequence of the adjacent time period, the data of the 5th and 6th days of the data block as the short-term input sequence, and the data of the 1st-6th day of the data block as the long-term data input sequence.

Based on tensorflow Keras framework, a multi-chain fusion load prediction model is built, and the Adam optimizer is used to use the MSE loss function to set dropout for each convolution operation. The number of neurons in each LSTM layer hidden layer is set to 115, and the number of iterations is 100. After the channel attention of the enhanced model from the attention layer, after the global average pooling and dimensionality reduction, a 7-dimensional vector is output by Softmax.

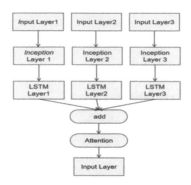

Fig. 6. Structure diagram of a multi-chain fusion model

4.3 Model Evaluation Indicators

In order to quantitatively analyze and compare the estimated performance of the model, the coefficient of determination (R2) root mean square error (RMSE) and mean absolute percent error (MAPE) were used as evaluation indicators. In these three evaluation indicators, the value range of R2 is (0, 1), and the closer R2 is to 1, the closer the estimated power load value is to the actual measured value; RMSE has different dimensions on different datasets, and the smaller the value of the root mean square error in the same dataset, the better the model. MAPE is to calculate the ratio of the difference between the predicted value and the true value to the true value, and then divide it by the number of samples to get a percentage. Smaller MAPE values represent better model effect. The formula for each evaluation indicator is as follows. In addition, the quantitative description of the error of the prediction result is also carried out by mean absolute error (MAE) and mean-square error (MSE).

$$R^2 = 1 - \frac{\sum_{i=1}^{m}(S_i - E_i)^2}{\sum_{i=1}^{m}\left(S_i - \overline{E_i}\right)^2} \tag{2}$$

$$RMSE = \sqrt{\frac{1}{m}\sum_{i=1}^{m}(S_i - E_i)^2} \tag{3}$$

$$MAPE = \frac{1}{M}\sum_{i=1}^{M}\left|\frac{x_i - \dot{x_i}}{x_i}\right| \times 100\% \tag{4}$$

where S_i represents the actual value of the i sample; E_i represents the estimation of the i sample; $\overline{E}$ represents the average of the estimated values; en represents the average of the actual values; m represents the sample size of the estimated model.

4.4 Analysis of Experimental Results

In order to visualize the prediction effect of the multi-chain fusion model, this paper constructs both the traditional LSTM model and the Inception-BiLSTM-Attention model as a comparison model. The following diagram shows the test results for the three models. As can be seen from the Fig. 7, the Inception-BiLSTM-Attention model and multi-chain fusion model constructed in this paper have better prediction effect than the traditional LSTM model. It is closer to the real trend line, and the prediction results of the multi-chain fusion model are less error than the Inception-BiLSTM-Attention model. It shows that the feature learning ability of multi-chain fusion model is stronger.

Table 1 shows the results of the comparison of the quantifications of the three models. As can be seen from the table, the R2 value of the traditional LSTM model is 96.51%, indicating that the traditional LSTM model can basically reflect the change of power load trend, but its MAE value, MSE value, RMSE value and MAPE value are 213.78, 80426.86, 283.59 and 2.8%, respectively, indicating that the feature learning ability of the traditional LSTM model is insufficient and the prediction error is large. Inception-BiLSTM-Attention model due to the two-way long short-term memory (BiLSTM) model and multi-feature data mining method for organic combination, through the Inception structure using deep convolution to extract the high-dimensional features of the input vector, based on the Attention attention mechanism to optimize the weight distribution of the output vector, so better feature learning and mining ability, the prediction effect compared with the traditional LSTM has a great improvement, Its MAE value is 97.44, MSE value is 1,469,128, RMSE value is 121.21, MAPE value is 1.40%, and R2 value is 99.36%; Due to the further consideration of the multi-dimensional periodic characteristics, the prediction results of the multi-chain fusion model have been further reduced, with maE values of 61.34, MSE values of 4944.92, RMSE values of 70.32, MAPE values of 0.89%, and R2 values of 99.78%; This also indicates the effectiveness of the method in this paper.

5 Conclusion

In this paper, the Inception-BiLSTM-Attention model and the multi-chain fusion model based on the model are constructed, the power load prediction is carried out, the real load data, meteorological data and date load data of a certain region are selected for training, and the performance of the algorithm is compared by the traditional LSTM algorithm, which verifies the effectiveness of the Inception-BiLSTM-Attention model and the multi-chain fusion model on power load prediction. The algorithm embodies the following characteristics and advantages in the theoretical analysis and experimental analysis in the article:

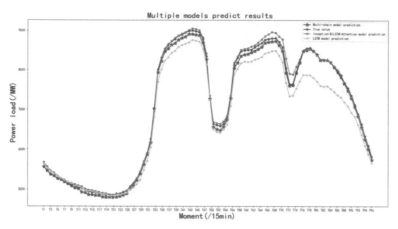

Fig. 7. Comparison of model results

Table 1. Table comparison of evaluation indicators table.

Model name	MAE	MSE	RMSE	MAPE (%)	R2 (%)
Multi-chain fusion model	61.34	4944.92	70.32	0.89	99.78
Inception-BiLSTM-Attention	97.44	1,469,128	121.21	1.40	99.36
LSTM model	213.78	80426.86	283.59	2.8	96.51

(1) Convolution operation based on Inception structure can extract high latitude characteristics, which can better learn the inherent characteristics of power load data.

(2) Data mining based on Bi-LSTM can achieve positive and negative feature extraction.

(3) The Attention attention mechanism can be used to optimize the configuration of weights and improve the output of the model.

(4) The input sequence of adjacent time periods, short periods and long periods can make up for the long-term dependence problem, improve the long-term characteristic learning ability of the model, and improve the prediction effect of the model.

References

1. Li, H.Z., Guo, S., Li, C.J., et al.: A hybrid annual power load forecasting model based on generalized regression neural network with fruit fly optimization algorithm. Knowl.-Based Syst. **37**(1), 378–387 (2013)
2. Niu, D.X., Chen, Z.Y., Xing, M., et al.: Combined optimum gray neural network model of the seasonal powerload forecasting with the double trends. Chin. J. Manage. Sci. (2001)

3. Niu, D., Bo, Z., Ming, M., et al.: Grey model of power load forecasting based on particle swarm optimization. In: World Congress on Intelligent Control and Automation. IEEE (2006)
4. Zhang, Y., Jiang, S.B., Zheng, L.I.: A gray RBF model improved by genetic algorithm for short-term load forecasting. Electr. Measur. Instrum. (2014)
5. Xi, X.F., Zhou, G.D.: A survey on deep learning for natural language processing. Acta Automatica Sin. (2016)
6. Garcia, G.R., Michau, G., Ducoffe, M., et al.: Temporal signals to images: Monitoring the condition of industrial assets with deep learning image processing algorithms. Proc. Inst. Mech. Eng. Part O J. Risk Reliab. (2021). 1748006X2199444
7. Zhai, M., Roshtkhari, M.J., Mori, G.: Deep learning of appearance models for online object tracking (2016)
8. Xianglong, L.I., Longfei, M.A., Xiangyang, Z.H.A.O., et al.: Multi-time scale electric heating load forecasting based on long short-term memory network. Proc. CSU-EPSA **33**(4), 71–75 (2021). (in Chinese)
9. Yongzhi, W.A.N.G., Bo, L.I.U., Yu, L.I.: A power load data prediction method based on LSTM neural network model. Res. Explor. Lab. **39**(5), 41–45 (2020). (in Chinese)
10. Bing, Z.H.A.O., Zengping, W.A.N.G., Weijia, J.I., et al.: A short-term power load forecasting method based on attention mechanism of CNN-GRU. Power Syst. Technol. **43**(12), 4370–4376 (2019). (in Chinese)
11. Pei, Y., Zhenglin, L., Qinghui, Z., et al.: Load forecasting of refrigerated display cabinet based on CEEMD-IPSO-LSTM combined model. De Gruyter Open Access (1) (2021)
12. Chuanjun, P.A.N.G., Jianming, Y.U., Changyou, F.E.N.G., et al.: Clustering modeling and characteristic analysis of power load based on long-short-term memory. Autom. Electr. Power Syst. **44**(23), 57–63 (2020). (in Chinese)
13. Marino, D.L., Amarasinghe, K., Manic, M.: Building energy load forecasting using deep neural net-works. In: IECON 2016—42nd Annual Conference of the IEEE Industrial Electronics Society, pp. 7046–7051. Florence, (2016)

Natural Computation Application (29)

Prediction of Yak Weight Based on BP Neural Network Optimized by Genetic Algorithm

Jie He[1], Yu-an Zhang[1(✉)], Dan Li[1], Zhanqi Chen[1], Weifang Song[2], and Rende Song[3]

[1] Department of Computer Technology and Application, Qinghai University, Xining 810016, China
{Y200854000183,2011990029,Y200854000190, Y200854000181}@qhu.edu.cn
[2] Menyuan Animal Disease Prevention and Control Center, Menyuan 810300, China
[3] Animal Disease Prevention and Control Center of Yushu Tibetan Autonomous Prefecture, Yushu 815099, China
songrende66@163.com

Abstract. Yak weight is an important physiological indicator of plateau yak. The body weight of yak is required for breeding, supplementary feeding, epidemic prevention, and slaughtering. However, due to the large size and strong wild nature of yak, it is difficult to pull it to the weighbridge, resulting in Common weighing methods being time-consuming and labor-intensive and having certain errors. To better predict the weight of yak, in this study, a BP neural network estimation model based on genetic algorithm optimization is proposed to measure the weight of yak. The results showed that there is a significant positive correlation between yak weight and height, body oblique length, chest circumference, and tube circumference. After optimization by genetic algorithm, the root mean square error of yak weight estimation decreased from 0.090 to 0.048, and the error between the predicted value of yak weight and the actual value decreased from 15.4 to 3.1%. Therefore, the algorithm can accurately predict the body weight of yak, and the research results can provide a reference for the estimation of yak body weight in the future.

Keywords: Body weight estimation · Genetic algorithm · BP neural network · Eigenvalue selection · Plateau yak · Data analysis

1 Introduction

The Qinghai-Tibet Plateau Yak is one of the rare animals endemic to China. Yaks live mainly in the mountainous regions of southern and northern Qinghai Province at altitudes above 3,000 m [1]. Yaks are not only an indispensable part of the local livestock economy but also an important source of livelihood and economy in the pastoral areas of the Tibetan plateau. The traditional method of measuring yak weight is to weigh the yak using a scale, but this method has disadvantages such as low efficiency, troublesome operation, and time consumption. With the advent of technology, the use of computer algorithms

N. Xiong et al. (Eds.): ICNC-FSKD 2022, LNDECT 153, pp. 307–316, 2023.
https://doi.org/10.1007/978-3-031-20738-9_35

supplemented by underlying equipment can be used to accurately estimate yak weight, thus saving a great deal of manpower and time.

However, the body weight of highland yaks does not present an accurate one- or multivariate linear relationship with its various physical data, which can present the problem of large errors in the accuracy of estimation. [2–4] BP neural networks have been extensively utilized for function approximation as well as numerical model prediction. However, it has the disadvantage of a low convergence rate, which leads to the appearance of local outliers. Also, the topology design and the selection of initial parameters can have a great impact on the performance of BP neural networks [5–8].

On this basis, this paper proposes a BP neural network yak weight estimation model optimized by a Genetic Algorithm for the following points.

(1) Improve the neural network structure and determine parameter selection as a means of improving the performance and learning effect of the neural network [9, 10].
(2) The identification of weight parameters that are closely related to yak weight using the correlation coefficient selection method.
(3) The optimized neural network is compared with the traditional BP neural network, multiple linear regression, and random forest to verify that the Genetic Algorithm-optimized BP neural network had better performance in terms of accuracy in terms of both the determination coefficient and root mean square error.
(4) Using the optimized neural network for weight estimation, it is confirmed that the optimized neural network has better prediction accuracy, precision, and practicality, thus providing theoretical and data support for the accurate estimation of yak weight.

2 BP Neural Network Optimized by Genetic Algorithm

The topology of BP (Back Propagation) network is shown in Fig. 1(a). The computational process is roughly divided into two stages: one is a forward propagation process in which results are obtained in the order of input, hidden layer, and output. In each iteration, the state of the neurons in each layer only affects the next layer. [11] The second is the error back propagation process [12], where the output layer may not be accurate every time, and once the calculated value deviates from the expected value, the error will be transmitted to the input layer again through the hidden layer and then processed by adjusting the weights to minimize the error [13]. In this study, the body height, body oblique length, chest circumference, and tube circumference of the yak are used as input nodes and the weight of the yak is used as the output node. A three-layer BP neural network model is developed, in which the hidden layer had m nodes.

A Genetic Algorithm is a reliable optimization method to overcome the disadvantages of simple BP neural networks where the parameters are difficult to determine and easily fall into partial optimization, and thus the long convergence time of simple BP neural networks is overcome [14], thus speeding up the computation of the problem. The whole process of the Genetic Algorithm optimized BP Neural Network algorithm (hereafter referred to as GA-BP) is illustrated in Fig. 1(b).

The following are the specific implementation steps:
Input: Characteristic parameters that determine the weight of the yak
Output: Yak weight

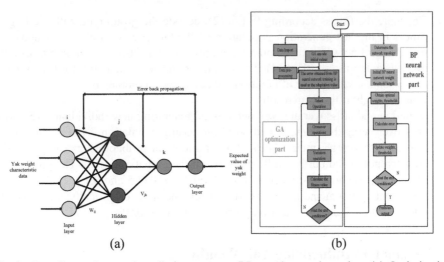

Fig. 1. Overall neural network prediction structure: BP neural network topology (**a**); Optimization process ——optimized by genetic algorithm (**b**)

(1) The neural network is encoded with real numbers to initialize the populations. Each initial population contains all the weights and threshold values associated with the BP network. Calculate the chromosome length L:

$$L = lm + \ln + m + n \tag{1}$$

where, l, m, n denote the number of nodes in the input, hidden, and output layers, respectively;

(2) The fitness of each individual is determined using the error from neural network training, as well as the individual with the highest fitness is directly replicated to the following generation. Fitness calculation formula:

$$f(x) = \frac{1}{2} \sum_{i=1}^{n} (m_i - c_i)^2 \tag{2}$$

where: n is the number of data sets; m_i is the actual weight measurement of the yak; c_i is the calculated value of the neural network;

(3) Roulette method selection according to Eqs. (3) and (4).

$$f_i = \frac{k}{F_i} \tag{3}$$

$$p_i = \frac{f_i}{\sum_{j=1}^{N} f_i} \tag{4}$$

where: F_i denotes the fitness of an individual, p_i denotes the probability of selection, k is the coefficient, and the population size is represented by N;

(4) Generate crossover and mutation operators in new generation populations;

(5) Calculate the fitness according to Eq. (2), decode the group type with the largest fitness in the last generation population of the Genetic Algorithm, and update it to BP neural network part to obtain the optimal connection weight and threshold, otherwise, loop steps (3) to (4). Then return to the neural network for training and exit when the set number of iterations or accuracy is completed;

(6) Return weight estimation results.

Mutation and crossover are two methods of generating new individuals [15], with crossover being the primary method of increasing population diversity and mutation as a secondary method [16]. Larger mutation probability can improve the convergence speed but it is difficult to reach the global optimum. The probability of variation is chosen between empirical values of 0.005 and 0.1 [17]. To achieve the global optimum and maximize the convergence rate, empirically, the crossover rate is set to 0.8 and the mutation probability is set to 0.01.

3   Parameters Influencing Yak Weight

The appropriate and proper selection of feature parameters is crucial for the model training and the prediction of the final results. In this sub-section, two methods are used to determine the feature parameters of the neural network, the Person correlation coefficient method, and the Spearman method. The formula is given in (5)

$$r = \frac{\sum (X - \bar{X})(Y - \bar{Y})}{\sqrt{\sum (X - \bar{X})^2 (Y - \bar{Y})^2}} \tag{5}$$

where: $\bar{X}$ is the mean of the independent variable X and $\bar{Y}$ is the mean of the dependent variable Y. The Pearson correlation coefficient and Spearman correlation coefficient are shown in Fig. 2.

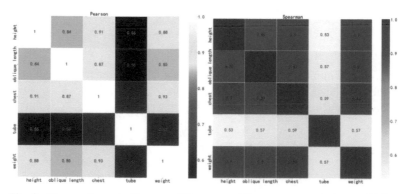

Fig. 2. Thermodynamic diagram of Pearson and Spearman correlation coefficient

Considering the two feature selection methods together, it can be concluded that yak weight data are very strongly correlated with height, body oblique length, and chest circumference data (0.8–1.0), and weight and tube circumference data are moderately correlated (0.4–0.6), so in the next experiments, height, body oblique length, chest circumference, and tube circumference were selected as features for weight estimation.

4 Results and Analysis

4.1 Experimental Evaluation Metrics

RMSE is the square root of the mean square error which is used to measure the difference between the observed and authentic values. If the actual value is $X = \{X_1, X_2, ..., X_n\}$, and the model's predicted value is $Y = \{Y_1, Y_2, ..., Y_n\}$, then the RMSE can be expressed as

$$\text{RMSE} = \sqrt{\frac{1}{N} \sum_{i=1}^{N} (X_i - Y_i)^2} \tag{6}$$

It is generally considered that a smaller value of RMSE is better, indicating that the calculated value is closer to the true value.

The formula for determination coefficient R^2 is

$$R^2 = 1 - \frac{\text{SSE}}{\text{SST}} \tag{7}$$

It is generally accepted that the range of R^2 values is 0 to 1. The closer to 1, the better the fit.

4.2 Parameter Setting

The experimental environment is Intel(R) Core(TM) i5-9400CPU@2.90GHz Processor, 16GB RAM Windows 10 operating system, and using Python programming language for algorithm simulation experiments. The yak dataset provided by the Yushu Livestock Veterinary Station was used and 600 data were selected for the experiment.

This paper's algorithm is validated using the yak dataset from the Yushu Livestock Veterinary Station. Data pre-processing was first performed to fill and clean the yak data to avoid the influence of erroneous data on the training model, and then 80% are chosen as the training set and the remaining 20% as the test data.

As the rounds of different iterations can have an impact on the accuracy of the neural network predictions. Figure 3 below visualizes the root mean square error of the two neural networks at different rounds of iterations.

From Fig. 3, it is concluded that when the round of iterations is 4000, the Genetic Algorithm-optimized BP neural network has the smallest root mean square error. If the round of training is too much more than 4000 iterations, the gradient descent may cross the minimum point, instead of leading to a trend of the error falling before rising, reducing the prediction accuracy. After normalizing the original data and inputting it into the model for calculation, the initial data for this experiment is indicated in Table 1 below:

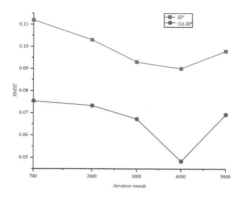

Fig. 3. Variation of root mean square error of BP and GA-BP under different iteration rounds

Table 1. Data setting parameters of GA-BP neural network

Data name	Value
Population size	100
Crossover probability	0.8
Variation probability	0.01
Iterations rounds	4000
Input parameter dimension	4

4.3 Algorithm Validation

Comparison of Model Predictions. The experimental results of the multiple linear regression model, random forest model, BP neural network model, and GA-BP model are presented in Table 2.

Table 2. Determination coefficient and root mean square error of different prediction methods

Algorithm name	R^2	RMSE
Multiple linear regression	0.853	19.700
Random forest	0.983	7.960
Traditional BP neural network	0.807	0.090
GA-BP	0.942	0.048

Through Table 2, comparing several prediction models, we can see that the determination coefficient of the GA-BP neural network prediction model is 0.942, which is slightly lower than the 0.983 predicted by random forest, and its root mean square error is 0.048, which is a hundred times and ten times more refined than the multiple linear regression and random forest algorithms respectively, and compared with the traditional BP neural network, it has improved 0.135 in the index of coefficient of determination and 0.042 in the index of root mean square error. Therefore, the Genetic Algorithm optimized BP neural network can accurately predict the weight of yaks.

GA-BP Neural Network Prediction Error of Yak Weight. Traditional BP neural network and GA-BP neural network were trained on six hundred yak data, then an estimation model is built and tested on 120 samples, and the results were plotted, as shown in Figs. 4 and 5 below:

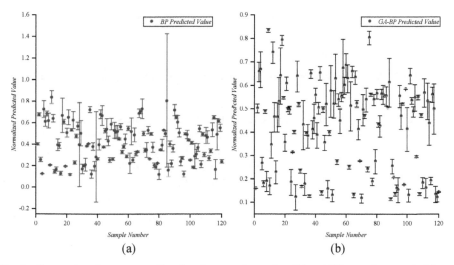

(a)   (b)

Fig. 4. Comparison between predicted and true values of traditional BP neural network (a) and GA-BP neural network (b)

Combined observation of Figs. 4 and 5, it is simple to discover that the BP prediction model has multiple samples, the error bars span a large interval, the results fluctuate greatly, and there are often abnormal spikes, with errors between ($-0.5 \sim 0.7$), and the prediction is poor; while the GA-BP model has sample error bars spanning a small interval, without abnormally large values, and the errors are stable between ($-0.15 \sim 0.10$). As well as the predicted values of the GA-BP are very close to the true value and have better accuracy than BP, so the Genetic Algorithm optimized BP neural network can better meet the requirements in the case of higher accuracy requirements.

Estimation of Yak Weight by GA-BP Neural Network. To verify the validity and accuracy of the optimized model, 10 yak samples were randomly selected for weight calculation and the results are shown in Table 3. For ease of writing, h in the table represents body height, bol represents body oblique length, cc represents chest circumference, and tc represents tube circumference.

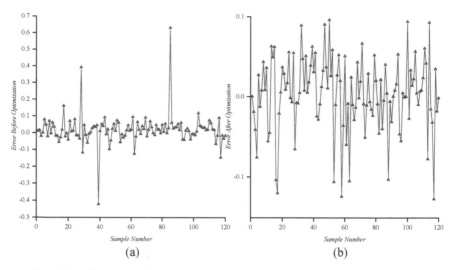

Fig. 5. Prediction error between traditional BP (a) and GA-BP neural network (b)

The average weight of the yak is 121.3 kg, the average mass of the yak estimated by the traditional BP neural network model is 136.6 kg and the average mass of the yak estimated by the GA-BP neural network model is 125 kg. The GA-BP neural network model is closest to the actual average mass of the yak. Also, the error of the GA-BP model in estimating the yak mass is reduced from the original 15.4–3.1%. As a result, the GA-BP neural network weight estimation model has better estimation accuracy than the individual BP networks and is closest to the actual weight. In practical applications, it has thus been demonstrated to perform better than a single neural network.

5 Discussion and Outlook

In this study, we proposed a Genetic Algorithm optimized method for yak weight prediction. The correlation between body size factors and body weight was obtained by Pearson and Spearman. A Genetic Algorithm optimized BP neural network yak weight prediction model was developed using Python algorithm. The results showed that the prediction results obtained by the method had less error and higher accuracy than the actual results.

At present, research on predicting yak weight by correlating body size parameters with yak weight is still in its infancy, and yak weight determination is tedious and time-consuming. The measurement methods still need to be improved to reduce the manual effort and make yak weight estimation work towards a smarter, faster, and more accurate goal.

Table 3. Weight estimation results and errors of two BP neural networks

Sample	h (cm)	bol (cm)	cc (cm)	tc (cm)	Actual weight (kg)	BP estimated value (kg)	BP relative error (%)	GA estimated value (kg)	GA relative error (%)
1	106	107	149	17	154.0	170.8	10.9	161.7	5.0
2	106	107	148	17	154.0	168.6	9.5	159.2	3.4
3	91	96	101	15	55.0	70.1	27.5	55.9	1.6
4	105	110	146	17	147.0	164.6	12.0	155.5	5.8
5	107	108	150	17	156.0	174.7	12.0	165.5	6.1
6	105	110	144	17	147.0	160.4	9.1	150.6	2.5
7	106	105	134	17	130.0	138.8	6.8	125.0	−3.9
8	104	103	131	17	117.0	129.8	11.0	116.3	−0.6
9	104	103	130	17	110.0	127.9	16.3	114.2	3.8
10	88	90	95	15	43.0	59.9	39.3	46.1	7.2
Mean error	–	–	–	–	–	–	15.4	–	3.1
Average weight	–	–	–	–	121.3	136.6		125	

Acknowledgement. This work is supported by Science and Technology Project in Qinghai Province (No: 2020-QY-218); Ministry of Finance and Ministry of Agriculture and Rural Affairs of the People's Republic of China: National Modern Agricultural Industry Technology System Grant (CARS-37); and received the support of the "High-end Innovative Talents Thousand Talents Program" in Qinghai Province.

References

1. Qinghai, Y.: Tibetan autonomous prefecture people's government. Overview of Yushu. http://www.yushuzhou.gov.cn/html/2/7.html
2. Man, J., Yang, W.: Based on multiple collinearity processing methods. Math. Theor. Appl. **30**(2), 105–109 (2010)
3. Zhang, G., Liu, X., Lu, S., et al.: Occurrence of typical antibiotics in nansi lake's inflowing rivers and antibiotic source contribution to nansi lake based on principal component analysis-multiple linear regression model. Chemosphere **242**, 125269 (2020)
4. Jiang, Q., Huang, R., Huang, Y., et al.: Application of BP neural network based on genetic algorithm optimization in the evaluation of power grid investment risk. IEEE Access **7**, 154827–154835 (2019)
5. Gallant, P.J., Aitken, G.: Genetic algorithm design of complexity-controlled time-series predictors. In: 2003 IEEE 13th Workshop on Neural Networks for Signal Processing, 2003. NNSP'03. IEEE (2003)

6. Tian, L., Noore, A.: Evolutionary neural network modeling for software cumulative failure time prediction. Reliab. Eng. Syst. Saf. **87**(1), 45–51 (2005)
7. Zhang, J.Y., Pan, G.Y.: Comparison and application of multiple regression and BP neural network prediction model. J. Kunming Univ. Sci. Technol. (Nat. Sci. Ed.) **38**(6), 61–67 (2013)
8. Feng, H., Song, Q., Ma, S., et al.: A new adaptive sliding mode controller based on the RBF neural network for an electro-hydraulic servo system. ISA Trans. (2022)
9. Singh, P., Chaudhury, S., Panigrahi, B.K.: Hybrid MPSO-CNN: multi-level particle swarm optimized hyperparameters of convolutional neural network. Swarm Evol. Comput. **63**, 100863 (2021)
10. Dang, Y., Liu, L., Li, Z.: Optimization of the controlling recipe in quasi-single crystalline silicon growth using artificial neural network and genetic algorithm. J. Cryst. Growth **522**, 195–203 (2019)
11. Hu, C., Zhao, F.: Improved methods of BP neural network algorithm and its limitation. In: 2010 International Forum on Information Technology and Applications, vol. 1. IEEE (2010)
12. Li, J., Cheng, J., Shi, J., et al.: Brief introduction of back propagation (BP) neural network algorithm and its improvement. In: Advances in Computer Science and Information Engineering. Springer, Berlin, Heidelberg, pp. 553–558 (2012)
13. Lin, J., Minghai, X.: Construction of performance evaluation system for intelligent simulation of university research under big data. In: International Conference on Forthcoming Networks and Sustainability in the IoT Era. Springer, Cham (2022)
14. Yao, D., Duan, Y., Li, M., et al.: Hybrid identification method of coupled viscoplastic-damage constitutive parameters based on BP neural network and genetic algorithm. Eng. Fract. Mech. **257**, 108027 (2021)
15. Zhao, Z., Xu, Q., Jia, M.: Improved shuffled frog leaping algorithm-based BP neural network and its application in bearing early fault diagnosis. Neural Comput. Appl. **27**(2), 375–385 (2015). https://doi.org/10.1007/s00521-015-1850-y
16. Jia, Z., Lu, X., Yang, J., et al.: Research on job-shop scheduling problem based on genetic algorithm. Int. J. Prod. Res. **49**(12), 3585–3604 (2011)
17. Ludwig, O., Nunes, U., Rui, A., et al.: Applications of information theory, genetic algorithms, and neural models to predict oil flow. Commun. Nonlinear Sci. Numer. Simul. **14**(7), 2870–2885 (2009)

Hybrid Sweep Algorithm and Modified Ant System with Threshold for Travelling Salesman Problem

Petcharat Rungwachira[(⊠)] [iD] and Arit Thammano[iD]

Computational Intelligence Laboratory, Faculty of Information Technology, King Mongkut's Institute of Technology Ladkrabang, Bangkok, Thailand
59606004@kmitl.ac.th, arit@it.kmitl.ac.th

Abstract. Travelling salesman problem is a special case of the vehicle routing problem. The objective of the travelling salesman problem is to find the shortest path for visiting every city without repeating city. Among metaheuristic algorithms, Ant System has been the most popular algorithm for solving the travelling salesman problem. However, Ant System has a disadvantage of often falling into local optimal solutions. This research proposed a modified Ant System with a modified pheromone density updating to reduce the rate of convergence. A threshold is also used to create a greater variety of routes. Moreover, the proposed algorithm used a Sweep Algorithm to generate initial population so that the next city to visit is close to each other. To prevent the ants from getting trapped at a local optimum, three types of local search, swap, insert, and reverse, are used to veer away the paths towards the higher pheromone density at a local optimum. The tested results of the proposed algorithm were compared to those of other three algorithms: GA-PSO-ACO, Hybrid VNS, and HAACO. On 13 out of 15 small- and medium-sized datasets, the proposed method outperformed or performed as well as the others.

Keywords: Travelling salesman problem · Ant system · Sweep algorithm · Combinatorial optimization

1 Introduction

Current lifestyle has changed greatly. The internet has played a part in making our lifestyle different from the past. Nowadays, people shop online instead of going to the store. This is especially true during Corona virus outbreak all over the world. As a result, online sales and the accompanying transportation logistics are on the rise. Transportation routes are one of the most important factors in transportation logistics. Efficient routing reduces agency cost. In addition, on-time delivery increases customer satisfaction. Travelling Salesman Problem (TSP) is a combinatorial optimization problem of finding routes to serve all customers for the shortest distance. Although there have been many researchers doing research on TSP, as it is a very complex problem, its solution remains to be improved by a more efficient algorithm. Metaheuristic, an algorithm that can provide a good result within a reasonable time, such as Ant System (AS), Genetic

© The Author(s), under exclusive license to Springer Nature Switzerland AG 2023
N. Xiong et al. (Eds.): ICNC-FSKD 2022, LNDECT 153, pp. 317–326, 2023.
https://doi.org/10.1007/978-3-031-20738-9_36

Algorithm (GA), and Artificial Bee Colony (ABC) are most commonly used to solve TSP. A brief review of past papers is presented below.

A hybrid method of Particle Swarm Optimization (PSO), Ant Colony Optimization (ACO), and 3-Opt was proposed by Mahi et al. [1]. PSO was used to find the optimum value of α and β, the weight parameters for ACO to select the next city in a path, while 3-Opt prevented the algorithm from falling into a local optimum.

A Discrete Symbiotic Organism Search (DSOS) was proposed by Ezugwu and Adewumi [2]. DSOS was inspired by symbiotic relationship between living organisms. DSOS added three mutation local searches to SOS—swap, insert, and reverse—to rebuild population, then applied the three original SOS phases to optimize the result.

A combination of PSO and Artificial Bee Colony (ABC) algorithms was presented by Sedighizadeh and Mazaheripour [3]. The hybrid algorithm focused on solving multi-objectives problems with precedence constraints and penalty method. PSO and ABC were used in parallel to construct good routes.

A hybrid of ACO and FA algorithms for solving VRP, HAFA, was presented by Goel and Maini [4]. The hybrid used ACO as the backbone, while FA was used to explored new solutions in the global search space. Then, there was pheromone shaking to adjust pheromone level to avoid getting trapped at a local optimum.

Chen et al. [5] introduced a hybrid two-stage Sweep Algorithm for solving CVRP. The algorithm used SA to cluster all customers for each vehicle based on the sequential order of the polar angle of the location of each customer. Then, the hybrid re-clusters the customers by adding customers from an adjacent cluster within the capacity of vehicle. Finally, a greedy search was used to determine the best sequence of cities for each vehicle.

A hybrid algorithm of ant colony optimization and 3-Opt algorithm was presented by Gülcü et al. [6]. The hybrid method used several ant colonies, each colony used ACO to generate solutions in a parallel system. 3-Opt was run on each colony to avoid getting trapped at a local optimum and shared the best solution with the other colonies.

The rest of this paper was organized into 4 sections: Sect. 2 briefly discusses background concepts—those of TSP, AS, and SA; Sect. 3 describes the proposed method; Sect. 4 reports the experiments and results; and finally, the last section concludes the paper.

2 Background Concepts

This section introduces background concepts of the TSP, AS, and SA.

2.1 Travelling Salesman Problem

The objective of Travelling Salesman Problem (TSP) is to find the shortest path for a salesman (or an ant in this sense) to visit all customers at various cities.

The objective function of TSP, Z, depends on the total distance traveled by a salesman. Z is defined as (1) below,

$$Z = \sum_{i=1}^{n} \sum_{j=1, j \neq i}^{n} d_{ij} \tag{1}$$

where n is the number of cities, and d_{ij} is the distance from city i to city j.

2.2 Ant System

Ant System (AS) [7] is a common algorithm for finding the shortest path from one location to another. This algorithm was inspired by the shortest path ants take in their foraging for food. Along their foraging path, ants leave pheromone trails for other ants to follow. Therefore, pheromone density increases with the number of ants that walk along that path. With passing time, if no other ants follow along that path, the pheromone density will gradually decrease.

Ants are assigned a different starting city. The next city to visit is selected based on the selection probability that depends on the distance between the starting city and the next city and the pheromone density along that edge, as defined in (2) below,

$$P_{ij}^k = \frac{(\tau_{ij})^\alpha (\eta_{ij})^\beta}{\sum_{l \in p} (\tau_{il})^\alpha (\eta_{il})^\beta} \tag{2}$$

where P_{ij}^k is the selection probability of the edge from city i to city j of ant k; τ_{ij} is the pheromone density on the edge from city i to city j; η_{ij} is the distance of the edge from city i to city j; α is the weight for the pheromone density, and β is the weight for the distance.

The city selection continues until the path covers all cities. Then, the total path distance is calculated. After all ants construct their own path, they are compared among each other to find the shortest path and recorded or updated as the globally shortest path. Next, the pheromone densities on all edges are updated based on (3) and (4) below,

$$\tau_{ij}' = (1 - \rho)\tau_{ij} + \sum_{k=1}^{K} \Delta \tau_{ij}^k \tag{3}$$

$$\Delta \tau_{ij}^k = \frac{1}{Z_k} \tag{4}$$

where ρ is the pheromone evaporation rate and Z_k is the length of the path that the ant k traveled. Ants that do not travel from city i to city j are not included in the updating process.

2.3 Sweep Algorithm

Sweep Algorithm (SA) was developed by Gillett and Miller [8]. It is typically used for clustering cities according to the relative polar angle of their locations. First, the sweep algorithm determines the polar angle for each city, then arranges cities based on the ascending degree of polar angle.

Next, to create the path, the salesman selects the first city to visit; then the next city is selected one by one in the sequential order of the ascending degree of the said polar angle.

3 Proposed Method

This section describes the proposed SA-MAntT algorithm for solving TSP. SA-MAntT consists of 3 modules: the first module, a Sweep Algorithm (SA) is used for generating a proper initial population; the second module, a Modified Ant System (MAS) iteratively finds the shortest path for visiting all cities; the third module consists of three local search algorithms, each of the 3 local search algorithms was executed once every iteration to prevent candidate paths from getting trapped at a local optimum or for getting them away from the trap at a local optimum. The steps of SA-MAntT are shown in Fig. 1.

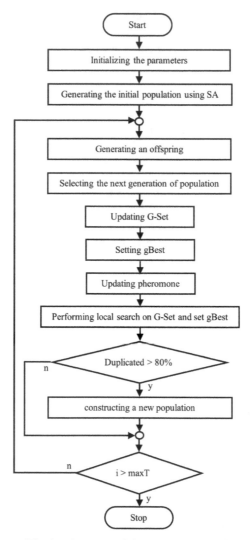

Fig. 1. Flowchart of the proposed method

3.1 Modified Ant System

Modified Ant System (MAS) is the main algorithm of SA-MAntT. MAS performs the following steps sequentially:

Step 1. Initialize the MAS parameters and the termination criterion, which in this case is the maximum number of iterations.

Step 2. Generate the initial population by using the Sweep algorithm, described in Sect. 3.2.

Step 3. Create offspring from each parent path in the current population until the termination criterion is met.

– Create an offspring:

- Randomly select a location in the parent path. The subpath from the selected location to the end of the parent path is used as the offspring's starting subpath.
- The cities next to the starting subpath are chosen in order until the offspring is complete. To choose the next city, first, the selection probability from the last city in the subpath to all unassigned cities, P_{ij}^k, are calculated. Then the city J with the highest selection probability, P_{ij}^k, is determined. Next, P_{ij}^k is compared to *Threshold*, a generated random number that helps make the new path more diverse. If $P_{ij}^k \leq$ *Threshold*, the city J is chosen as the next city. Otherwise, the city with the second-highest selection probability is chosen as the next city.

– Select the best n paths from both the parent population and the offspring population to be the population for the next MAS iteration. The set of the best n paths from these two populations are ranked in ascending order, and the top 70% from this set are selected. The other 30% are arbitrary chosen from the remaining set.
– If any selected paths are better than the paths in *G-Set*, then the *G-Set* is updated with these newly selected paths (*G-Set* is the set of m best global paths).
– Set the path with the shortest distance as *gBest*.
– Update the pheromone density [9] on all edges by using (5) and (6),

$$\tau'_{ij} = (1 - \rho)\tau_{ij} + \Delta\tau_{ij}^{best} \tag{5}$$

$$\Delta\tau_{ij}^{best} = \frac{1}{f(s^{best})} \tag{6}$$

where s^{best} consists of 3 paths: the *gBest*, the best path in this current iteration, and second-best path in this current iteration.

– Perform a local search on the paths in *G-Set* as described in Sect. 3.3.
– When 80% of the paths in the current iteration are identical, a new population is constructed in the following manner (Fig. 2).

- Randomly select a path from *G-Set*, and assign it to be the first parent, P_1. Then, randomly select another two paths from the paths in the new population for the next iteration and assign them to be the second and third parents, P_2 and P_3.
- The first city of P_1 is set to be the first city for two offspring, O_1 and O_2.
- Then, the second city is either a city in the path P_1, P_2, or P_3. The second city for O_1 will be the city with the shortest distance from the first city of P_1, and the second city for O_2 will be the city with the second shortest distance from the first city of P_1.
- For the rest cities of O_1 and O_2, the next city is either a city in the path P_1, P_2, or P_3, which is determined by the distance from the previous city.

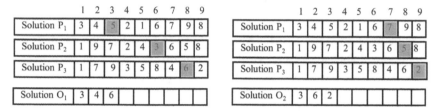

Fig. 2. Generation of the offspring when 80% of the paths are identical

Step 4. Return *gBest*.

3.2 Sweep Algorithm

Sweep algorithm creates the initial population by using the following steps.

Step 1. Calculate the polar angle of each city relative to the origin, by using (7).

$$
\theta_i = \begin{cases} \tan^{-1} \frac{y_i}{x_i}, & x_i > 0, y_i > 0 \\ \pi + \tan^{-1} \frac{y_i}{x_i}, & x_i < 0 \\ 2\pi + \tan^{-1} \frac{y_i}{x_i}, & x_i > 0, y_i < 0 \end{cases} \tag{7}
$$

where (x_i, y_i) is the coordinate of the location of the city i. After that, sort the cities according to the ascending order of polar angles. The example of the city location in polar coordinates is shown in Fig. 3.

Step 2. Generate the initial paths, each of which starts from different cities. For each path, the next city in the sorted list is selected until the path is completed.

3.3 Local Search

Local Search is used to prevent paths from getting trapped at a local optimum. In each iteration, one of three local search algorithms is used, starting from the first one to the third one, and then, for the next cycle of iterations, repeating them cyclically. The three algorithms are Swap, Insert, and Reverse.

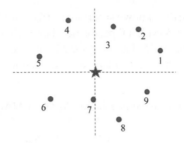

Fig. 3. Example of the city locations in polar coordinates

4 Experimental Results

The experiments were conducted on 15 standard TSP datasets, obtained from TSPLIB (Library of Traveling Salesman Problems) [10]. The datasets were 15 small- and medium-sized symmetric datasets, named eil51, berlin52, st70, eil76, pr76, rat99, kroA100, rd100, eil101, lin105, pr124, ch130, ch150, kroA150 and kroA200. The instance is in the format nameXX, name is the name of the dataset, XX is the number of cities.

Each data set was run 10 times with different initial populations each time. The assigned initial parameter values are shown in Table 1. τ is the initial pheromone density; ρ is the evaporation rate; α and β are the weights for pheromone density and distance, respectively; MaxIter is the maximum number of iterations, which depends on the size of the datasets.

Table 1. Initial parameter values

Parameter	Value
τ	3
ρ	0.05–0.5
α	1
β	2
MaxIter	5000, 10000

The experimental results of SA-MAntT are shown in Table 2. BKS is the best known solution. Best is the best result of SA-MAntT. $\Delta\%$ is the percent deviation of Best from BKS.

Table 3 shows comparative results of SA-MAntT and other three algorithms: GA-PSO-ACO proposed by Deng et al. [11], Hybrid VNS proposed by Hore et al. [12], and HAACO proposed by Tuani et al. [13].

SA-MAntT obtained the best-known solution in 10 out of 15 instances. In comparison to the above 3 algorithms, SA-MAntT performed better than or as good as GA-PSO-ACO on all 13 datasets. On all 15 datasets, SA-MAntT performed better than Hybrid VNS. On 8 datasets, SA-MAntT performed better or as good as HAACO.

Table 2. Experimental results of SA-MAntT

Instance	BKS	Best	Worst	Average	S.D	Δ%
eil51	426	**426**	437	431.60	3.84	**0.00**
berlin52	7542	**7542**	7775	7617.20	98.62	**0.00**
st70	675	**675**	687	682.30	3.65	**0.00**
eil76	538	**538**	556	548.90	6.05	**0.00**
pr76	108159	**108159**	112413	110266.50	1526.81	**0.00**
rat99	1211	**1211**	1288	1253.20	25.06	**0.00**
kroA100	21282	**21282**	21768	21473.20	176.21	**0.00**
rd100	7910	**7906**	8239	8069.60	104.35	**−0.05**
eil101	629	631	665	650.40	11.43	0.32
lin105	14379	**14379**	15017	14547.00	187.77	**0.00**
pr124	59030	**59030**	59887	59314.80	290.82	**0.00**
ch130	6110	6113	6408	6274.50	102.98	0.05
ch150	6528	6549	6938	6728.20	137.86	0.32
kroA150	26524	26601	27854	27211.80	339.88	0.29
kroA200	29368	29586	31483	30488.70	575.29	0.74

Bold indicates the results that are better than or equal to the current best known solutions

5   Conclusion

In this paper, we present SA-MAntT algorithm for solving the travelling salesman problem. SA-MAntT uses SA to generate initial population. Then, in every iteration, it uses AS, with updated pheromone density and a threshold to select next city, to create candidate paths. In addition, it uses three types of local search for preventing candidate paths from getting trapped at a local optimum. Experiments were conducted on 15 small- and medium-sized standard TSP datasets. The results were compared to other 3 algorithms: GA-PSO-ACO, Hybrid VNS, and HAACO. On 13 datasets, SA-MAntT outperformed or performed as good as those algorithms. However, it has not been tested with large-sized datasets. Our next research will try to improve the efficiency of the algorithm and test the algorithm with more challenging problems.

Table 3. Comparative results of GA-PSO-ACO, Hybrid VNS, HAACO and SA-MantT

Instance	BKS	GA-PSO-ACO	Hybrid VNS	HAACO	SA-MAntT
eil51	426	**426**	428.98	**426**	**426**
berlin52	7542	7544.37	7544.36	**7542**	**7542**
st70	675	679.6	677.11	**675**	**675**
eil76	538	545.39	545.39	**538**	**538**
pr76	108159	109206	**108159**	–	**108159**
rat99	1211	1218	1240.38	**1211**	**1211**
kroA100	21282	–	21618.2	**21282**	**21282**
rd100	7910	7936	7910.4	–	**7906**
eil101	629	633.07	642.31	**630**	631
lin105	14379	14397	14383	**14379**	**14379**
pr124	59030	59051	59030.74	–	**59030**
ch130	6110	6121.15	6140.66	–	**6113**
ch150	6528	–	6639.52	6566	**6549**
kroA150	26524	26676	26943.31	–	**26601**
kroA200	29368	29731	30300.56	**29483**	29586

Bold indicates the best results among the compared algorithms

References

1. Mahi, M., Baykan, O.K., Kodaz, H.: A new hybrid method based on particle swarm optimization, ant colony optimization and 3-Opt algorithms for traveling salesman problem. Appl. Soft Comput. **30**, 484–490 (2015)
2. Ezugwu, A.E.-S., Adewumi, A.O.: Discrete symbiotic organisms search algorithm for travelling salesman problem. Expert Syst. Appl. **87**, 70–78 (2017)
3. Sedighizadeh, D., Mazaheripour, H.: Optimization of multi objective vehicle routing problem using a new hybrid algorithm base on particle swarm optimization and artificial bee colony algorithm considering precedence constraints. Alexamdria Eng. J. **57**, 2235–2239 (2017)
4. Goel, R., Maini, R.: A hybrid of ant colony and firefly algorithm (HAFA) for solving vehicle routing problems. J. Comput. Sci. 28–37 (2018)
5. Chen, M.H., Chang, P.C., Chiu, C.Y., Annadurai, S.P.: A hybrid two-stage sweep algorithm for capacitated vehicle routing problem. In: Proceedings of the 2015 International Conference on Control, Automation and Robotics, pp. 195–199 (2015)
6. Gülcü, Ş, Mahi, M., Baykan, Ö.K., Kodaz, H.: A parallel cooperative hybrid method based on ant colony optimization and 3-Opt algorithm for solving traveling salesman problem. Soft. Comput. **22**(5), 1669–1685 (2016). https://doi.org/10.1007/s00500-016-2432-3
7. Dorigo, M., Maniezzo, V., Colorni, A.: Ant system: optimization by a colon of cooperating agents. IEEE Trans. Syst. Man Cybern. **26**, 29–41 (1996)
8. Gillett, B.E., Miller, L.R.: A heuristic algorithm for the vehicle-dispatch problem. Oper. Res. **22**, 340–349 (1974)

9. Zhao, G., Luo, W., Sun, R., Yin, C.: A modified max-min ant system for vehicle routing problem. In: Proceedings of the 4th International Conference on Wireless Communications, Networking and Mobile Computing, pp. 1–4 (2008)
10. TSPLIB. http://comopt.ifi.uni-heidelberg.de/software/TSPLIB95/tsp/, Accessed 30 Nov 2021
11. Deng, W., Chen, R., He, B., Liu, Y., Yin, L., Guo, J.: A novel two-stage hybrid swarm intelligence optimization algorithm and application. Appl. Soft Comput. **16**, 1707–1722 (2012)
12. Hore, S., Chatterjee, A., Dewanji, A.: Improving variable neighborhood search to solve the travelling salesman problem. Appl. Soft Comput. **68**, 83–91 (2018)
13. Tuani, A.F., Keedwell, E., Collett, M.: Heterogenous adaptive ant colony optimization with 3-Opt local search for the travelling salesman problem. Appl. Soft Comput. **97** (2020)

Determining All Pareto-Optimal Paths for Multi-category Multi-objective Path Optimization Problems

Yiming Ma[1], Xiaobing Hu[2(✉)], and Hang Zhou[1]

[1] Sino-European Institute of Aviation Engineering, Civil Aviation University of China, No. 2898, Jinbei Road, Dongli District, Tianjin, China
`2020122027@cauc.edu.cn` `hzhou@cauc.edu.cn`
[2] College of Safety and Engineering, Civil Aviation University of China, No. 2898, Jinbei Road, Dongli District, Tianjin, China
`xbhu@cauc.edu.cn,huxbtg@163.com`

Abstract. Increasing criteria have been involved in the path evaluation, and path optimization with multiplicative objectives becomes essential in the real world, such as the reliability of paths. Firstly, we define the multi-category multi-objective path optimization problem (MCMOPOP), in which each path simultaneously has multiple additive and multiplicative weights. Secondly, this paper proposes an agent-based and nature-inspired algorithm, the ripple-spreading algorithm (RSA), to solve the MCMOPOP. To the best of our knowledge, the newly proposed RSA is the first algorithm for the MCMOPOP that can find all Pareto-optimal paths. An illustrative example is provided to make the processes of the RSA more comprehensible. Comparative experiments demonstrate that the RSA outperforms other compared methods in computational efficiency and solution quality. Furthermore, the RSA maintains applicability when the number of objectives is large. The RSA can be expected to efficiently provide complete solutions for the practical applications modeled as the MCMOPOP.

Keywords: Multi-objective optimization · Path optimization · Ripple-spreading algorithm · Pareto-optimal path · Multiplicative weight

1 Introduction

The path optimization problem (POP) is classic in transportation research and computer science [1,2]. With the rapid development of network modeling, there are more criteria to evaluate paths. So the multi-objective path optimization problem (MOPOP) plays an essential role. Most MOPOP algorithms only minimize additive objectives, where all objectives are the summation of arcs'

N. Xiong et al. (Eds.): ICNC-FSKD 2022, LNDECT 153, pp. 327–335, 2023.
https://doi.org/10.1007/978-3-031-20738-9_37

weights [3]. So far, however, previous studies about the MOPOP have not dealt with the maximization of multiplicative objectives, which also have a wide range of real-world applications. For example, we need to maximize the reliability of paths derived by multiplying arcs' weights in evacuation [11]. Generally, the number of Pareto-optimal paths is more than one, and finding all of them is challenging as the MOPOP is NP-hard [1].

The MOPOP belongs to the multi-objective optimization problem (MOOP), whose algorithms are distinguished into the aggregate objective function (AOF), constrained objective function (COF), and multi-objective evolutionary algorithm (MOEA). AOF methods aggregate objectives into several single-objective optimization problems (SOOPs) [4]. The COF method optimizes one objective while treating the others as constraints [5]. To date, the NSGA-II and MOEA/D are two representative algorithms of the MOEA [6,7]. The major drawback of AOF and COF methods is subjectivity, on this account, they may miss Pareto-optimal solutions. MOEAs eliminate this drawback by treating all optimal solutions equally. However, due to their random features, they cannot guarantee optimality theoretically. In addition, there exist some specific algorithms for the MOPOP, such as the labeling method [8] and ranking method [9]. Most of these methods only focus on the bi-objective scenario. Recently, a ripple-spreading algorithm (RSA) has been proposed that can determine all Pareto-optimal paths [10].

In this paper, we attempt to define a new problem: the multi-category multi-objective path optimization problem (MCMOPOP), in which each path has multiple additive (e.g., the toll, length, and time) and multiplicative weights (e.g., reliability) simultaneously. Furthermore, we propose a novel RSA that can find all Pareto-optimal paths for the MCMOPOP. Theoretical analyses and comparative experiments demonstrate its efficiency and applicability. To the best of our knowledge, the RSA is the first optimal algorithm for the MCMOPOP.

The rest of this paper is organized as follows. Section 2 provides the mathematical description of the MCMOPOP. In Sect. 3, we adapt the RSA to the MCMOPOP and conduct theoretical analyses. Section 4 presents a toy example and comparative experiments. This paper ends with conclusions in Sect. 5.

2   Problem Description

Consider a network $G(V, E)$ where node set V contains N_N nodes, and arc set E contains N_A arcs. We denote the source node as s and the destination node as d. The adjacent matrix $A \in \mathbb{R}^{N_N \times N_N}$ is defined as: if there is an arc between node i and node j, $A(i, j) = 1$; otherwise, $A(i, j) = 0$. We use an integer vector P to denote a path with N_P nodes. If P is feasible, $A(P(i), P(i+1)) = 1, \forall i = 1, \cdots, N_P - 1$. There are N_{Obj} weights associated with each arc, which contain $N_{Obj,A}$ additive weights and $N_{Obj,M}$ multiplicative weights. For the arc between node i and node j, $C_{A,l}(i, j)$ denotes its lth additive weight, and $C_{M,m}(i, j)$ denotes its mth multiplicative weight. All weights are positive, and multiplicative weights range from 0 to 1. The lth additive objective of P is calculated as:

$$f_{A,l}(P) = \sum_{i=1}^{N_P-1} C_{A,l}(P(i), P(i+1)).$$ (1)

And the mth multiplicative objective of P is

$$f_{M,m}(P) = \prod_{i=1}^{N_P-1} C_{M,m}(P(i), P(i+1)).$$ (2)

In this study, we aim to minimize additive objectives and maximize multiplicative objectives. For simplicity, we take the opposite number of multiplicative objectives so that all objectives of the MCMOPOP are transformed into minimization problems. We define a mapping f from the path set to objective space:

$$f(P) = [f_{A,1}(P), \cdots, f_{A,N_{Obj,A}}(P), -f_{M,1}(P), \cdots, -f_{M,N_{Obj,M}}(P)].$$ (3)

We define the path set Ω_P containing all feasible paths from s to d. The mathematical description of the MCMOPOP is:

$$\min_{p \in \Omega_P} f(p).$$ (4)

The Pareto concept originates from economic research and becomes the core concept of the MOOP [12]. If P_1 Pareto dominates P_2, i.e., $P_1 \prec P_2$, then

$$\begin{cases} f(P_1)(i) \leq f(P_2)(i), & \forall i = 1, \cdots, N_{Obj}, \\ f(P_1)(j) < f(P_2)(j), & \exists j = 1, \cdots, N_{Obj}, \end{cases}$$ (5)

where $f(P)(i)$ denotes the ith component of $f(P)$. If a path P^* is Pareto-optimal, there is no path Pareto dominates P^*, i.e., $!\exists P \in \Omega_P, P \prec P^*$. As each Pareto-optimal path can be candidate to decision-makers, the primary goal of the MCMOPOP is finding all of them.

3 The RSA for the MCMOPOP

3.1 The Optimization Principle of the RSA

The following nature phenomenon inspires the RSA: if a point on a water surface is perturbed, a ripple will spread out in all directions at the same speed. After it reaches an obstacle (e.g., a stone), a new ripple is triggered and spreads at the same speed. This phenomenon contains an important optimization principle: a node is reached by ripples in the same order as their traveling distances.

The RSA was first proposed to tackle the shortest path problem (SPP) [13]. Although the RSA is deterministic, it can also be implemented as an agent-based algorithm, whose most outstanding merit is flexibility. Each agent has its behavior pattern, and all agents' combined behaviors can help solve the specific problem. We may modify agents' behaviors to adapt an agent-based algorithm to a new problem. In the RSA, all ripples and nodes can be deemed as agents. We can adapt their behaviors to solve various POPs [13,14].

3.2   Adapt the RSA to the MCMOPOP

This subsection adapts the ripple triggering condition and the termination judgment of the RSA for the MCMOPOP. We suppose the ripple R traveling along path P reaches a node n at time t. Before the end of t, other m ripples $R_1, ..., R_m$ reach node n traveling along paths $P_1, \cdots, P_m$, respectively. We define ripple R as a Pareto non-dominated ripple (PNDR) if and only if path P is not Pareto-dominated by any path in $[P_1, \cdots, P_m]$. The triggering condition of a new ripple at node n is modified as follows: if a ripple reaches node n and is assessed as a PNDR, it triggers a new ripple. The termination judgment of the RSA is changed as: if there is no active ripple or all active ripples are dominated by at least one PNDR at d, the algorithm terminates. We choose the network where the arcs' lengths are the first additive weights to spread ripples. To guarantee optimality, the ripple spreading speed v is set up as [13]:

$$v = \min(C_{A,1}(i,j)), \quad \forall i,j = 1, \cdots, N_N, \ A(i,j) = 1. \tag{6}$$

Table 1. The notation list.

Notations	Meanings
v	Ripple spreading speed subject to (6)
R_R	Radius set, the radius of ripple i is $R_R(i)$
E_R	Epicenter set, the epicenter of ripple i is $E_R(i)$
S_R	State set, $S_R(i) = 0/1$ denotes ripple i is inactive/active
P_R	Path set, $P_R(i)$ denotes the traveling path of ripple i from s to $E_R(i)$
O_R	Objective value set, $O_R(i) = f(P_R(i))$
Ω_{PNDR}	$\Omega_{PNDR}(n)$ denotes all PNDRs at node n

Figure 1 presents the flowchart of the RSA for the MCMOPOP, and Table 1 lists its notations. It is noteworthy that in **Step 3.3**, $C_{A,1}(E_R(i), n) \leq R_R(i) < C_{A,1}(E_R(i), n) + v$ denotes that ripple i reaches node n in this period t.

3.3   Optimality

This subsection conducts theoretical analyses to prove the optimality of the RSA.

Lemma 1. *The traveling paths of the PNDRs at d are Pareto-optimal.*

Proof. We assume that a ripple R reaches d at time t_1 along path P, which is assessed as a PNDR. According to the definition of PNDR in Sect. 3.2, P is not Pareto dominated by the traveling path of any ripple that reaches d before the end of t_1. Suppose another ripple R_1 reaches n at time t_2 along path P_1, and $t_1 < t_2$. According to the optimization principle stated in Sect. 3.1, the traveling distance of R from s to d is smaller than R_1, i.e., $f(P)(1) < f(P_1)(1)$. According to (5), it is impossible that $P_1 \prec P$. Therefore, the traveling path of ripple R is Pareto-optimal, and Theorem 1 holds.   □

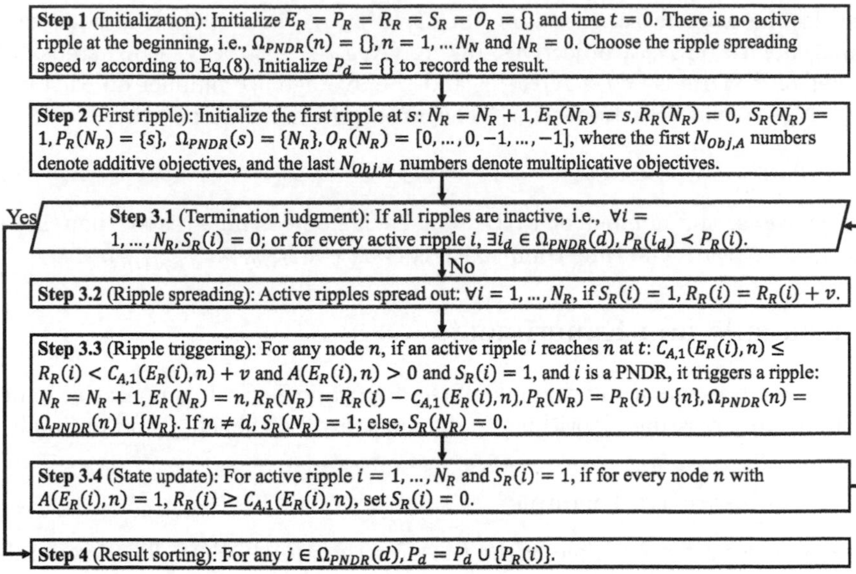

Step 1 (Initialization): Initialize $E_R = P_R = R_R = S_R = O_R = \{\}$ and time $t = 0$. There is no active ripple at the beginning, i.e., $\Omega_{PNDR}(n) = \{\}, n = 1, \dots N_N$ and $N_R = 0$. Choose the ripple spreading speed v according to Eq.(8). Initialize $P_d = \{\}$ to record the result.

Step 2 (First ripple): Initialize the first ripple at s: $N_R = N_R + 1, E_R(N_R) = s, R_R(N_R) = 0, S_R(N_R) = 1, P_R(N_R) = \{s\}, \Omega_{PNDR}(s) = \{N_R\}, O_R(N_R) = [0, \dots, 0, -1, \dots, -1]$, where the first $N_{Obj,A}$ numbers denote additive objectives, and the last $N_{Obj,M}$ numbers denote multiplicative objectives.

Step 3.1 (Termination judgment): If all ripples are inactive, i.e., $\forall i = 1, \dots, N_R, S_R(i) = 0$; or for every active ripple $i, \exists i_d \in \Omega_{PNDR}(d), P_R(i_d) \prec P_R(i)$.

Yes ← | No ↓

Step 3.2 (Ripple spreading): Active ripples spread out: $\forall i = 1, \dots, N_R,$ if $S_R(i) = 1, R_R(i) = R_R(i) + v$.

Step 3.3 (Ripple triggering): For any node n, if an active ripple i reaches n at t: $C_{A,1}(E_R(i), n) \leq R_R(i) < C_{A,1}(E_R(i), n) + v$ and $A(E_R(i), n) > 0$ and $S_R(i) = 1$, and i is a PNDR, it triggers a ripple: $N_R = N_R + 1, E_R(N_R) = n, R_R(N_R) = R_R(i) - C_{A,1}(E_R(i), n), P_R(N_R) = P_R(i) \cup \{n\}, \Omega_{PNDR}(n) = \Omega_{PNDR}(n) \cup \{N_R\}$. If $n \neq d, S_R(N_R) = 1$; else, $S_R(N_R) = 0$.

Step 3.4 (State update): For active ripple $i = 1, \dots, N_R$ and $S_R(i) = 1$, if for every node n with $A(E_R(i), n) = 1, R_R(i) \geq C_{A,1}(E_R(i), n)$, set $S_R(i) = 0$.

Step 4 (Result sorting): For any $i \in \Omega_{PNDR}(d), P_d = P_d \cup \{P_R(i)\}$.

Fig. 1. The flowchart of the RSA for the MCMOPOP.

Lemma 2. *For a Pareto-optimal path P^* containing N_P nodes, the sub-path $[P^*(1), \cdots, P^*(i)]$ is a Pareto-optimal path from s to node $P^*(i)$, $i = 1, \cdots, N_P$.*

Lemma 2 directly comes from the theorems in [10].

Theorem 1. *The RSA determines all Pareto-optimal paths for the MCMOPOP.*

Proof. Lemma 1 guarantees that all traveling paths of PNDRs at d are Pareto-optimal. From Lemma 2, we know that any sub-path of a Pareto-optimal path is also Pareto-optimal. According to the definition of PNDR, the ripple traveling along the Pareto-optimal sub-path from s to node n is a PNDR at n. The ripple triggering condition in **Step 3.3** is that any PNDR at a node triggers a new ripple, which makes the RSA never miss any Pareto-optimal sub-paths. Therefore, Theorem 1 holds, which states the optimality of the RSA. □

3.4 Time Complexity

This subsection calculates the time complexity of the RSA. From the flowchart of the RSA in Fig. 1, we know that there are three kinds of computational steps for each PNDR: (1) The active ripple spreads out. (2) Compare the arc's length and the ripple's radius. (3) The calculation of the objective value and the assessment of PNDR. We suppose each ripple takes average N_{ATU} time units to go through an arc. The first two computational steps execute each time, taking $(2 \times N_{ATU})$ steps. The third step executes only when a ripple reaches another node. In the worst case, there are already N_{PNDR} PNDRs at the node to be compared with on N_{Obj} objectives, taking $(N_{Obj} + N_{Obj} \times N_{PNDR})$ steps. The number of PNDRs is $N_N \times N_{PNDR}$, and the average arc number on each node is N_A/N_N, so the RSA takes N_{BCS} basic computational steps, where

$$N_{BCS} = (2 \times N_{ATU} + N_{Obj} + N_{Obj} \times N_{PNDR}) \times N_{PNDR} \times N_A. \quad (7)$$

In most cases, the number of PNDRs at each node is far greater than 1, i.e., $N_{PNDR} \gg N_{ATU}$. The time complexity is $O(N_A \times N_{Obj} \times N_{PNDR}^2)$.

4   Example and Experiments

This section presents a toy example of the RSA and several comparative experiments against existing algorithms to demonstrate its efficiency and optimality.

4.1   An Illustrative Example

Fig. 2 provides a toy example of the RSA for the MCMOPOP. The lower table in Fig. 2 lists all PNDRs' information, in which the first and second number denotes the additive and multiplicative objective, respectively. PNDRs at d are highlighted in the orange background. It is worth noticing that although R_2 has reached node 2 and node 4 (see Fig. 2(d)), it does not trigger new ripples. Take node 2 as an example. When R_2 reaches node 2, the objective value is $(77, 0.56)$, and it is Pareto-dominated by the objective value of R_3 at node 2. So R_2 is not a PNDR. At last, there are three PNDRs at d: R_4, R_7, and R_8. Their traveling paths are 1-3-5, 1-2-5, and 1-4-5, which are exactly all Pareto-optimal paths.

4.2   Comparative Experiments

This subsection conducts comparative experiments to evaluate the performance of the RSA against the AOF [4], COF [5], NSGA-II [6], and MOEA/D [7]. The test networks are generated as follows: randomly distribute N_N nodes in a square area defined by $[0, 100; 0, 100]$, and each node connects to its four closest nodes. The source s locates in the lower left corner, and the destination d locates in the upper right corner. Additive weights are uniformly distributed between 1 and 20, and multiplicative weights are uniformly distributed between 0.9 and 1.

We need to generate evenly distributed weight vectors for the AOF, COF, and MOEA/D to decompose the MCMOPOP. It is assumed that we select h evenly distributed numbers on each objective, the population size N_{POP} in MOEA/D and the number of SOOPs N_{SOOP} in the AOF and COF method are [5]:

$$N_{POP} = N_{SOOP} = \binom{N_{Obj} + h - 1}{N_{Obj} - 1}. \quad (8)$$

After decomposing the MCMOPOP into SOOPs, the objective function is a mixed additive and multiplicative function. We implement the genetic algorithm (GA) [15], and the parameters are set up as follows: the population

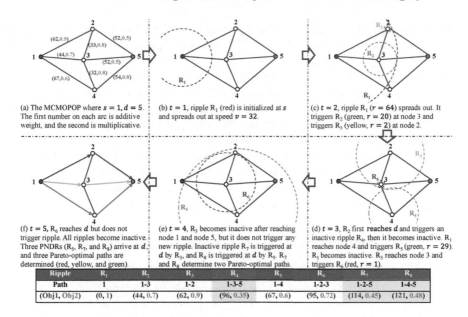

(a) The MCMOPOP where $s = 1, d = 5$. The first number on each arc is additive weight, and the second is multiplicative.

(b) $t = 1$, ripple R_1 (red) is initialized at s and spreads out at speed $v = 32$.

(c) $t = 2$, ripple R_1 ($r = 64$) spreads out. It triggers R_2 (green, $r = 20$) at node 3 and triggers R_3 (yellow, $r = 2$) at node 2.

(f) $t = 5$, R_6 reaches d but does not trigger ripple. All ripples become inactive. Three PNDRs (R_5, R_7, and R_8) arrive at d, and three Pareto-optimal paths are determined (red, yellow, and green).

(e) $t = 4$, R_3 becomes inactive after reaching node 1 and node 5, but it does not trigger any new ripple. Inactive ripple R_7 is triggered at d by R_3, and R_8 is triggered at d by R_5. R_7 and R_8 determine two Pareto-optimal paths.

(d) $t = 3$, R_2 first reaches d and triggers an inactive ripple R_4, then it becomes inactive. R_1 reaches node 4 and triggers R_5 (green, $r = 29$). R_1 becomes inactive. R_3 reaches node 3 and triggers R_6 (red, $r = 1$).

Ripple	R_1	R_2	R_3	R_4	R_5	R_6	R_7	R_8
Path	1	1-3	1-2	1-3-5	1-4	1-2-3	1-2-5	1-4-5
(Obj1, Obj2)	(0, 1)	(44, 0.7)	(62, 0.9)	(96, 0.35)	(67, 0.6)	(95, 0.72)	(114, 0.45)	(121, 0.48)

Fig. 2. A toy example of the RSA for the MCMOPOP.

size is 30, the generation number is 150, the mutation probability is 0.15, and the crossover probability is 1. Moreover, the population and generations of the NSGA-II and MOEA/D are set up as N_{POP} and 200, and the neighborhood size of the MOEA/D is 20. Table 2 lists the parameters of test cases.

Table 2. The parameters of comparative experiments.

N_{Obj}	$N_{Obj,A}$	$N_{Obj,M}$	h	N_{POP}
2	1	1	100	100
3	2	1	40	861
5	4	1	10	1001
10	5	5	4	715
20	10	10	3	1540
40	20	20	2	820

All algorithms were coded in Python 3.8, and all experiments were conducted on an ordinary computer with an Intel i7-10850H CPU and 32 GB RAM. Furthermore, all experimental data are obtained from 50 random experiments. After running all algorithms for the same experiment, all results are gathered. Using the non-domination-sort [6], all non-dominated solutions are true Pareto-optimal paths. We recorded three evaluation indicators: the computation time

(CT) in seconds, the number of Pareto-optimal paths (N_{PP}), and the number of true Pareto-optimal paths (N_{TPP}). The data are presented in Table 3, from which one may get the following observations: (1) In all experiments, the N_{PP} of the RSA equals N_{TPP}, which experimentally proves that the RSA can find all Pareto-optimal paths for the MCMOPOP. (2) The best data in Table 3 are shown in bold. What is striking about the RSA is its computational efficiency. When $N_{Obj} \leq 20$, the RSA is the most efficient algorithm. Even if N_{Obj} is huge, the RSA outputs all Pareto-optimal paths within reasonable computation time. (3) Each individual in the GA represents a feasible solution. However, it is difficult to estimate the number of optimal solutions and readjust the parameters of GA according to the problem scale. The RSA is a deterministic algorithm that can help the decision-makers avoid setting up the parameters.

Table 3. The results of comparative experiments.

N_{Obj}		AOF		COF		NSGA-II		MOEA/D		RSA
		Mean	Std	Mead	Std	Mean	Std	Mean	Std	
2	CT	2.63	0.57	2.96	2.95	3.72	0.11	2.77	0.12	**0.01**
	N_{PP}	6.88	1.33	4.98	1.18	6.32	2.27	4.69	1.57	**11.60**
	N_{TPP}	4.50	1.30	2.79	1.00	1.72	1.95	0.56	1.00	**11.60**
3	CT	20.11	0.25	27.97	7.51	750.34	1.63	0.84	0.03	**0.02**
	N_{PP}	5.60	1.85	3.24	1.17	13.23	1.33	3.64	0.49	**14.00**
	N_{TPP}	2.62	2.33	1.21	0.75	11.79	1.47	0.83	1.60	**14.00**
5	CT	144.81	0.58	326.29	70.41	1283.00	20.21	7.16	0.38	**0.19**
	N_{PP}	25.41	2.24	7.79	3.12	60.24	1.91	19.37	10.46	**82.02**
	N_{TPP}	16.78	2.23	2.83	3.76	56.58	1.67	2.84	3.43	**82.02**
10	CT	293.10	1.49	736.37	145.29	545.97	3.13	25.38	5.40	**15.40**
	N_{PP}	246.04	6.42	108.16	22.27	707.04	1.01	152.43	94.94	**1963.21**
	N_{TPP}	227.21	5.42	50.63	16.03	680.42	9.22	103.17	43.14	**1963.21**
20	CT	981.75	8.38	6593.72	130.64	3393.36	23.45	132.20	23.74	**121.60**
	N_{PP}	486.82	5.19	299.83	53.54	1521.44	0.49	305.57	120.54	**5518.04**
	N_{TPP}	444.59	7.12	37.57	5.82	1491.78	7.93	281.10	134.46	**5518.04**
40	CT	789.53	5.08	4277.40	69.01	1102.71	21.80	**96.22**	7.27	3420.29
	N_{PP}	621.96	19.72	624.83	35.44	789.57	1.85	341.62	62.65	**32848.79**
	N_{TPP}	614.62	20.19	295.97	21.00	732.83	5.49	334.05	63.05	**32848.79**

5   Conclusions and Future Work

The multiplicative weights of paths are of interest in the real world, such as the reliability of paths. However, the studies of MOPOPs considering multiplicative weights are insufficient. This paper defines this new kind of problem as the MCMOPOP. Each path can simultaneously have multiple additive weights to be minimized and multiplicative weights to be maximized. This paper proposes a novel nature-inspired and deterministic algorithm, the RSA, to solve the

MCMOPOP. We theoretically prove the optimality of the newly proposed RSA and calculate its time complexity. An illustrative example of the RSA is provided to make its processes more comprehensible. Comprehensive experiments are conducted against the other four multi-objective optimization algorithms. The results demonstrate that the RSA can find all Pareto-optimal paths of the MCMOPOP while maintaining high computational efficiency. We may extend the RSA to solve the MCMOPOP on dynamic networks by combining it with the co-evolutionary path optimization methodology for future work [14].

References

1. Zajac, S., Huber, S.: Objectives and methods in multi-objective routing problems: a survey and classification scheme. Euro. J. Oper. Res. **290**, 1–25 (2021)
2. Cormen, T.H., Leiserson, C.E., Rivest, R.L., Stein, C.: Intro. Algor. MIT Press, Cambridge (2022)
3. Cintrano, C., Chicano, F., Alba, E.: Facing robustness as a multi-objective problem: A bi-objective shortest path problem in smart regions. Inform. Sci. **503**, 255–273 (2019)
4. Gunantara, N.: A review of multi-objective optimization: methods and its applications. Cogent Eng. **5**(1), 1502242 (2018)
5. Messac, A., Ismail-Yahaya, A., Mattson, C.A.: The normalized normal constraint method for generating the Pareto frontier. Struct. Multidiscip. Optim. **25**(2), 86–98 (2003)
6. Deb, K., Pratap, A., Agarwal, S., Meyarivan, T.: A fast and elitist multiobjective genetic algorithm: NSGA-II. IEEE Trans. Evol. Comput. **6**(2), 182–197 (2002)
7. Zhang, Q., Li, H.: MOEA/D: a multiobjective evolutionary algorithm based on decomposition. IEEE Trans. Evol. Comput. **11**(6), 712–731 (2007)
8. Guerriero, F., Musmanno, R.: Label correcting methods to solve multicriteria shortest path problems. J. Optim. Theor. Appl. **111**(3), 589–613 (2001)
9. Hu, X.-B., Wang, M., Di, Paolo E.: Calculating complete and exact Pareto front for multiobjective optimization: a new deterministic approach for discrete problems. IEEE Trans. Cybernet. **43**(3), 1088–1101 (2013)
10. Hu, X.-B., Gu S.-H., Zhang C., Zhang G.-P., Zhang M.-K., Leeson, M.S.: Finding all Pareto optimal paths by simulating ripple relay race in multi-objective networks. Swarm Evol. Comput., 100908 (2021)
11. Stepanov, A., Smith, J.M.: Multi-objective evacuation routing in transportation networks. Euro. J. Oper. Res. **198**(2), 435–446 (2009)
12. Barr, N.: Economics Welfare State. Oxford University Press, USA (2020)
13. Hu, X.-B., Wang, M., Leeson, M.S., Di Paolo, E.A., Liu, H.: Deterministic agent-based path optimization by mimicking the spreading of ripples. Evol. Comput. **24**(2), 319–346 (2016)
14. Hu, X.-B., Zhang, M.-K., Zhang, Q., Liao, J.-Q.: Co-evolutionary path optimization by ripple-spreading algorithm. Transport. Res. Part B Methodol. **106**, 411–432 (2017)
15. Ahn, C.W., Ramakrishna, R.S.: A genetic algorithm for shortest path routing problem and the sizing of populations. IEEE Trans. Evol. Comput. **6**(6), 566–579 (2002)

Improved Whale Optimization Algorithm Based on Halton Sequence

Wenyu Zhang[1,2], Bingchen Zhang[1(✉)], Yongbin Yuan[1], Changyou Zhang[1], and Xining Jia[3]

[1] Xi'an University of Posts and Telecommunications, Xi'an 710061, China
zwy888459@xupt.edu.cn, zbc416@stu.xupt.edu.cn,13891881950@163.com
yyb123@stu.xupt.edu.cn, Changyou_zhang@stu.xupt.edu.cn
[2] China Research Institute of Aerospace Systems Science and Engineering, Beijing 100048, China
[3] Xi'an University of Science and Technology, Xi'an 710109, China
548732896@qq.com

Abstract. Since standard whale optimization algorithms doesn't have global and conflict-free search capabilities, an improved WOA based on the Halton Sequence (HGS-WOA) is proposed. The Halton sequence is introduced in the population initialization stage to obtain a more uniformly distributed search population; meanwhile, an adaptive weighting determinant is included in the whale position update strategy to improve the search algorithm as a whole by dynamically altering the effects of ideal and undesirable conditions; the variable cosine control coefficient is used to prevent the algorithm from adapting to local optimal solutions and to avoid low accuracy of the search of the algorithm in the future. With the help of simulation experiments on six optimization problems of complex high-dimensional functions, it is shown that HGS-WOA has excellent computational results.

Keywords: Whale optimization algorithm · Halton sequence · Adaptive strategy · Cosine control factor

1 Introduction

With the advent of modern swarm intelligent decision-making algorithms, meta-heuristic optimization methods [1] are gaining popularity in a wide range of applications. Seyedali Mirjalili introduced the Whale Optimization Algorithm [2]. And the WOA proved successful in resolving a variety of issues like feature optimization [3], the optimal image segmentation threshold [4] and efficient workshop scheduling [5]. The WOA is a local optimum and is easy to

Shaanxi Education and Scientific Research Project(08JK431).

discover with low coordinate precision as a novel intelligent optimization approach [6]. In response to the above problems, we introduce an enhanced whale
optimization algorithm (HGS-WOA) based on the variable cosine control factor
and the adaptive weights of the Halton Sequence. In this paper, Six complex
high-altitude experimental operations are chosen to simulation experiments and
comparative analysis, and the results verify that the HGS-WOA is significantly
improved in terms of optimal performance.

2 Basic WOA

2.1 Surrounding Prey

Almost whales hunt by encircling their prey, and this behavior is represented by
the next generation update position equation as follows:

$$\vec{D} = \left| \vec{C} \cdot \overrightarrow{X^*}(t) - \vec{X}(t) \right| \tag{1}$$

$$\vec{X}(t+1) = \overrightarrow{X^*}(t) - \vec{A} \cdot \vec{D} \tag{2}$$

where t denotes the round in question, $\vec{A}$ and $\vec{C}$ stand for coefficient-vectors,
X stands for the state vector and X^* is the state vector of the best currently
obtained solution. The expressions of $\vec{A}, \vec{C}$ are displayed:

$$\vec{A} = 2\vec{a} \cdot \vec{r} - \vec{a} \tag{3}$$

$$\vec{C} = 2 \cdot \vec{r} \tag{4}$$

$$a = \left(2 - \frac{2t}{T_{\max}} \right) \tag{5}$$

where $\vec{a}$ is the constriction, $\vec{r}$ is a random vector, with a value between 0 and
1, t is the current number of iters of search population, and $T_{\max}$ is the max
number of iters of WOA.

2.2 Updating Position by Spiral

A mathematical expression was created by the whale's spiral feeding movement [7] as follows:

$$\vec{X}(t+1) = \vec{D}' \cdot \exp(bl) \cdot \cos(2\pi l) + \overrightarrow{X^*}(t) \tag{6}$$

where b is a constant used to define the spiral motion, l is a random number
and its value in $[-1, 1]$, $\overrightarrow{D'} = \left| \overrightarrow{X^*}(t) - \vec{X}(t) \right|$ represents the prey's distance from
the i-th. The resulting equation for the location of the humpback whale spiral
update is is determined by the following equation:

$$\vec{X}(t+1) = \begin{cases} \overrightarrow{X^*}(t) - \vec{A} \cdot \vec{D} & \text{if } p < 0.5 \\ \vec{D}' \cdot \exp(bl) \cdot \cos(2\pi l) + \overrightarrow{X^*}(t) & \text{if } p \geq 0.5 \end{cases} \tag{7}$$

2.3   Searching for Quarry

The mathematical model for the stochastic searching step is:

$$\vec{D} = \left| \vec{C} \cdot \overrightarrow{X_{\text{rand}}} - \vec{X} \right| \tag{8}$$

$$\vec{X}(t+1) = \overrightarrow{X_{\text{rand}}} - \vec{A} \cdot \vec{D} \tag{9}$$

where $\overrightarrow{X_{\text{rand}}}$ is an arbitrary state vector, and it is selected from current location of the predation process.

3   Halton Global Searched WOA (HGS-WOA)

3.1   Halton Sequence to Initialize Populations

Halton sequence [8] is chosen to initialize the population in this paper. The definition of the Halton sequence is based on the root inverse function, and its defining function is shown below:

$$\phi_p(n) \equiv \frac{b_0}{p} + \frac{b_1}{p^2} + \cdots + \frac{b_m}{p^{m+1}} \tag{10}$$

where p is a prime number, $n = b_0 + b_1 p + \cdots + b_m p^m$ $(0 \leq b_j < p)$. The Halton sequence in S-dimension is shown in following equation:

$$X_n = (\phi_{p_1}(n), \phi_{p_2}(n), \cdots, \phi_{p_s}(n)) \tag{11}$$

where the multidimensional base variables $p_1, p_2, \ldots, p_s$ are mutually exclusive.

3.2   Adaptive Weighting Decision Variables

In most meta-heuristic optimization algorithms, the adaptive inertia weighting factor [9] is an important conditioning parameter. According to the principle of the WOA, the value of a will directly influence the universal search capabilities and the local optimization efficiency. On this page, the adaptive-weight decision variable ω is proposed, where expression of the adaptive weight decision variable ω is:

$$\omega = 0.02 + \frac{1}{0.8 + \exp\left(-\text{ fobj (Position }(t))/fit - 1\right)^{\frac{t}{2}}} \tag{12}$$

where fobj indicates the fitness function of the algorithm, Position(t) represents the current whale population position and fit_1 indicates the optimal whale position after the first algorithm iteration.

3.3 The Variable Cosine Control Factor

Due to the unique algorithmic mechanism of WOA, it is influenced by the value of $\vec{A}$ [10]. Therefore, this paper introduces a variational cosine control factor v and its expression is:

$$v = 2\cos\left(\left(1 + \frac{t - \text{Max\_iter}}{\text{Max\_iter}}\right) \cdot \frac{\pi}{2}\right) \tag{13}$$

Max_iter denotes the maximum number of repetitions, whereas t denotes the current number of repetitions. An improved surface optimization expression is shown below:

$$\vec{X}(t+1) = \omega \cdot \overrightarrow{X_{\text{rand}}} - v \cdot \vec{A} \cdot \vec{D} \tag{14}$$

$$\dot{X}(t+1) = \omega \cdot \overrightarrow{X^*}(t) - v \cdot \vec{A} \cdot \dot{D} \tag{15}$$

3.4 Algorithm Steps

According to the above improvement strategy, the pseudo-code for the execution step of the HGS-WOA is as follows:

Algorithm 1 Improved whale optimization algorithm(HGS-WOA)

Input: number of population N, variable dimension D, number of iterations *Max_iter*, Adaptation function $f(x)$

1: Based on Halton sequence initialize the position x_i ($i = 1, 2, ..., N$)
2: Let *iter*=1
3: **while** *iter* <= *Max_iter* **do**
4:     Calculate if the location of population S is beyond the feasible boundary
5:     Calculate the fitness function value of each individual and find X_{best}
6:     Sorting the populations S in ascending order by fitness size
7:     **for** $i = 0$ to N **do**
8:         Calculate the values of vectors$\vec{A}$ and$\vec{C}$ by Eq. (3) and Eq. (5)
9:         **if** $p < 0.5$ **then**
10:             **if** $|A| \geq 1$ **then**
11:                 Select any whale location as the initial solution and update the location randomly by Eq. (8) and Eq. (14)
12:             **else**
13:                 Select the current optimal solution as the initial solution and shrink the enclosing prey to update the position by Eq. (1) and Eq. (15)
14:             **end if**
15:         **else**
16:             Select the pre-optimal solution as the base solution for spiral prey capture by Eq. (7)
17:         **end if**
18:     **end for**
19:     Calculate the fitness function values of individuals in the population S and rank them according to their fitness values to obtain a new population N.
20: **end while**

Fig. 1. Pseudo-code of the HGS-WOA.

4   Algorithmic Performance Testing

4.1   Test Function and Parameter Setting

In the experiment, this paper introduces six complex high-order functional test groups presented in Table 1 for simulation tests to test the HGS-WOA, the standard WOA and four other metaheuristic optimization algorithms (PSO, GWO, ALO and MVO) are selected for comparison [11].

Table 1. Description of test functions.

Function	Range	$f_{\min}$
$F_1 = \max_i \{|X_i, 1 \leq i \leq D|\}$	$[-100, 100]$	0
$F_2 = \sum_{i=1}^{D} \left(-x_i \sin\left(\sqrt{|x_i|}\right)\right)$	$[-500, 500]$	-418.9829×5
$F_3 = \sum_{i=1}^{D} \left[x_i^2 - 10\cos\left(2\pi x_i\right) + 10\right]$	$[-5.12, 5.12]$	0
$F_4 \quad = \quad -20\exp\left(-0.2\sqrt{\frac{1}{n}\sum_{i=1}^{D} x_i^2}\right)$ $- \exp\left(\frac{1}{n}\sum_{i=1}^{D} \cos\left(2\pi x_i\right)\right) + 20 + e$	$[-32, 32]$	0
$F_5 = -\sum_{i=1}^{5} \left[(X - a_i)(X - a_i)^T + c_i\right]^{-1}$	$[0, 10]$	-10.1532
$F_6 = -\sum_{i=1}^{7} \left[(X - a_i)(X - a_i)^T + c_i\right]^{-1}$	$[0, 10]$	-10.1532

4.2   Simulation Test Results and Analysis

The test results of the six high-dimensional complex functions of the six optimization algorithms are shown in Tables 2 and 3. Through the analysis of Tables 2 and Table 3, WOA and ALO have strong search ability and high accuracy for solving single-peak function problems. In the case of multi-peak function optimization, their limitations are reflected, and their solution accuracy is not satisfactory and does not reach the optimal value. The HGS-WOA presented in this paper is much more accurate than the other five optimization algorithms for solving function problems for one or more peaks of size 100 or 300. For f_1, f_3, f_4, f_5 and f_6, at Dimensions-100 , the theoretical optimal values of 0 and -10.15 can be searched. The basic HGS-WOA outperforms the other five algorithms, but HGS-WOA can deviate from the local optimum. The accuracy of the HGS-WOA is the highest. Meanwhile, in terms of the search time, HGS-WOA also has an absolute advantage of less computation time than the other five algorithms.

Table 2. Experimental results of the six algorithms.

Function		Best	Dimension 100			Dimension 300		
			Avg.	Std.	Time/s	Avg.	Std.	Time/s
F_1	HGS-WOA	0	1.04E−81	2.99E−81	0.42	1.04E−81	2.99E−81	0.47
	WOA	52.32	57.46	17.46	0.56	57.46	17.46	0.76
	PSO	1.11	57.46	0.36	0.53	57.46	17.46	0.89
	GWO	7.04E−07	4.08E+04	1.22E+04	0.87	57.46	17.46	0.74
	ALO	17.73	8.63E-52	2.67E−51	1.32	0.43	1.46	0.92
	MVO	2.14	0.09	0.08	1.02	6.32	11.84	0.96
F_2	HGS-WOA	−12124.78	−1.19E+04	1.50E−01	0.62	−1.10E+04	805.15	0.42
	WOA	−10485.57	−1.05E+04	1.61E+03	0.86	−1.20E+04	1.61E+03	0.64
	PSO	−3310.53	−1.05E+04	1.61E+03	0.79	−1.05E+04	1.61E+03	0.77
	GWO	−6245.04	−1.05E+04	1.79E+03	1.22	−2.15E+04	0.81E+03	0.48
	ALO	−5.42E+03	−1.05E+04	1.03E+03	1.08	−1.17E+04	2.45E+03	0.88
	MVO	−7677.78	−1.05E+04	2.23E+03	0.97	−1.05e+04	1.61e+03	0.92
F_3	HGS-WOA	0	0	0	0.67	0	5.22E−05	0.44
	WOA	8.54E−16	2.14E−14	2.54E−14	1.23	5.32E−04	1.58E−04	0.62
	PSO	5.41	3.14E−14	4.54E−14	0.96	5.11E−04	1.42E−04	0.81
	GWO	1.13E−13	1.14E−14	0.54E−14	1.34	−2.84	0.11	0.58
	ALO	96.51	1.14E−14	2.54E−14	0.89	−1.8473	0.01	0.69
	MVO	1.16	1.14E−14	2.54E−14	1.52	−0.84	0.21	0.56
F_4	HGS-WOA	0	1.59E−15	1.49E−15	0.77	0.5E−24	0.4E−25	0.57
	WOA	4.44E−15	4.76E−15	1.83E−−15	0.89	3.2E−15	2.8E−15	0.92
	PSO	0.01	5.86E−15	1.36E−15	0.93	1.4E−15	1.9E−15	1.22
	GWO	1.00E−13	3.86E−15	1.03E−15	0.79	3.3E−15	1.1E−15	0.98
	ALO	1.93	9.86E−15	11.83E−15	0.98	3.8E−15	1.2E−15	1.34
	MVO	0.19	2.66E−15	1.83E−15	1.24	2.8E−15	1.6E−15	1.65
F_5	HGS-WOA	−10.15	−9.87	0.33	0.82	−10.17	2.91	0.92
	WOA	−10.08	−8.35	2.91	0.93	−9.88	0.91	1.44
	PSO	−10.13	−8.35	0.91	0.89	−8.98	0.88	1.63
	GWO	−5.10	−7.21	1.41	0.98	−5.35	0.12	1.67
	ALO	−10.15	−9.12	0.77	0.87	−9.86	0.33	0.98
	MVO	−5.05	−6.24	0.31	1.21	−6.08	0.22	1.87
F_6	HGS-WOA	−10.15	−10.14	0.39	0.53	−10.15	2.85	0.69
	WOA	−3.72	−4.69	2.85	0.82	−5.69	0.35	0.92
	PSO	−8.40	−7.69	2.33	0.78	−7.69	1.75	0.88
	GWO	−6.29	−7.34	2.12	0.69	−6.12	1.02	0.79
	ALO	−1.40	−1.84	0.01	0.88	−2.88	1.72	0.93
	MVO	−2.75	−3.00	1.39	0.92	−4.12	0.92	1.11

4.3 Convergence Analysis

For a more intuitive analysis of the impact optimization algorithm and the average accumulation curve, it was improved that f_2, f_3, f_5, and f_6 under the condition of D = 300 is chosen in Fig. 4. Take F2 and F3 as examples, it has been proved that HGS-WOA has improved accuracy and convergence speed than other optimization algorithms. Take F5 and F6 as examples, it indicates that HGS-WOA can find the best optimal value, while the remaining five algorithms all belong to the local optimal value. NO one can deny the fact that only the improved HGS-WOA algorithm has the advantage of searching more, which can not only start with the best locals solution to approach the global optimal solution but also the speed is far better than other optimization algorithms.

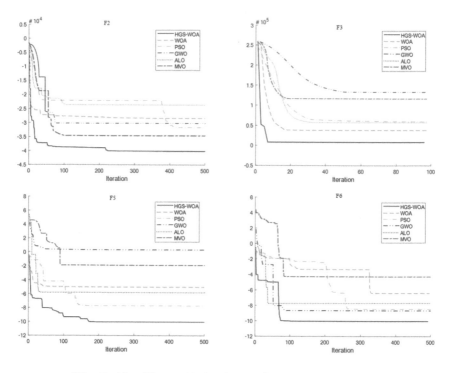

Fig. 2. Algorithm optimization performance comparison

5   Conclusion

This article suggests using HGS-WOA to resolve complex function optimization issues, which can boost overall search effectiveness and boost integration accuracy in the following stage. HGS-WOA provides an inertial weighted solution adaptive operator to control the Hatton sequence and population location update strategy, enabling improved global search capabilities. Secondly, the feature of different intervals of the cosine function with different function value increase and decrease is used to decide factor for the later position update to expand the scope of search target and integration efficiency. At last, the function simulation test results show that the HGS-WOA is more competitive. It can not only achieve higher accuracy, but also has the fastest convergence speed.

References

1. Dokeroglu, T., Sevinc, E., Kucukyilmaz, T.: A survey on new generation meta-heuristic algorithms. Comp. Indust. Eng. **137**, 106040 (2019). https://doi.org/10.1016/j.cie.2019.106040
2. Mirjalili, S., Lewis, A.: The whale optimization algorithm. Adv. Eng. Softw. **95**, 51–67 (2016). https://doi.org/10.1016/j.advengsoft.2016.01.008
3. Got, A., Moussaoui, A., Zouache, D.: A guided population archive whale optimization algorithm for solving multiobjective optimization problems. Exp. Syst. Appl. **141**, 112972 (2020). https://doi.org/10.1016/j.eswa.2019.112972
4. Pandey, A.C., Tikkiwal, V.A.: Stance detection using improved whale optimization algorithm. Compl. Intell. Syst. **7**(3), 1649–1672 (2021). https://doi.org/10.1007/s40747-021-00294-0
5. Jiang, T., Zhang, C., Zhu, H.: Energy-efficient scheduling for a job shop using an improved whale optimization algorithm. Mathematics **6**(11), 220 (2018). https://doi.org/10.3390/math6110220
6. Yan, Z., Zhang, J., Zeng, J.: Nature-inspired approach: An enhanced whale optimization algorithm for global optimization. Math. Comp. Simul. **185**, 17–46 (2021). https://doi.org/10.1016/j.matcom.2020.12.008
7. Chakraborty, S., Saha, A.K., Chakraborty, R., et al.: An enhanced whale optimization algorithm for large scale optimization problems. Knowl. Based Syst. **233**, 107543 (2021). https://doi.org/10.1016/j.knosys.2021.107543
8. Wang, Y., Chen, X.: Hybrid quantum particle swarm optimization algorithm and its application. Sci. China Inform. Sci. **63**(5), 1–3 (2020). https://doi.org/10.1007/s11432-018-9618-2
9. Zhou, J., Yao, X.: Multi-population parallel self-adaptive differential artificial bee colony algorithm with application in large-scale service composition for cloud manufacturing. Appl. Soft Comput. **56**, 379–397 (2017). https://doi.org/10.1016/j.asoc.2017.03.017
10. Gan, C., Cao, W., Wu, M., et al.: A new bat algorithm based on iterative local search and stochastic inertia weight. Exp. Syst. Appl. **104**, 202–212 (2018). https://doi.org/10.1016/j.eswa.2018.03.015
11. Becker, W.: Meta-functions for benchmarking in sensitivity analysis. Rel. Eng. Syst. Safe. **204**, 107189 (2020). https://doi.org/10.1016/j.ress.2020.107189

A Security-Oriented Assignment Optimization Model of Main Equipment and Facilities in Prefabricated Building

Chunguang Chang[1(✉)], Zhuo Zuo[1], and Hongbo Hou[2]

[1] School of Management, Shenyang Jianzhu University, Shenyang 110168, China
{ccg,Zuoz}@sjzu.edu.cn
[2] School of Business Administration, Shenyang Polytechnic College, Shenyang 110045, China
hhb@spc.edu.cn

Abstract. To improve the security guarantee level of prefabricated building (PB) construction, the safety-oriented assignment optimization model and the solving problem of the main equipment and facilities are studied. The objective function of maximizing the comprehensive safety utility of all assigned tower cranes, construction elevators and overall jack-up climbing frame is established. In the model constraints, the safety distance among tower cranes, the maximum working range, the requirements of the rated starting weight, the assigned amount requirement of different equipment and facilities in each building, the requirements of the transportation efficiency, the rent fee budget limit and so on are fully considered. The artificial immunity algorithm for solving above optimization model is designed, and the effectiveness of the optimization model and algorithm is verified by application.

Keywords: Safety · Prefabricated building · Equipment and facilities · Model

1 Introduction

The safety of the main equipment and facilities in prefabricated building (PB) is an important prerequisite to ensure the safety of PB construction. For main prefabricated construction equipment and facilities, under the premise of budget limit, how to realize the optimization of selection and assignment of them, improve the overall safety of the main PB construction safety degree is of great significance.

Some scholars have studied on equipment selection, security control, optimization model and algorithm. Niroomand designed a hybrid artificial electric field algorithm for equipment type selection [1]. By fuzzy theory, Alpay and Iphar proposed an equipment selection method [2]. Smith et al. proved the correlation between safety investment and accident rate [3]. Knyziak studied the relation between construction quality and security of PB construction [4]. Chang Chun-guang studied an optimization model of raw materials selective purchasing and designed an artificial immunity algorithm to solve it [5]. Above research work has enriched the relevant theories. However, there is a relative lack of research on the assignment optimization model of the main equipment and facilities in PB construction.

N. Xiong et al. (Eds.): ICNC-FSKD 2022, LNDECT 153, pp. 344–352, 2023.
https://doi.org/10.1007/978-3-031-20738-9_39

2 Model Establishing and Algorithm Designing

2.1 Variables and Parameters Setting

To describe the assignment optimization problem of tower crane, elevator and overall jack-up climbing frame, variables and parameters are set as follows:

$S_j^{(r)}$, $S_k^{(r)}$, $S_u^{(r)}$-The safety utility of j-type tower crane, k-type construction elevator, u-type overall jack-up climbing frame in building r.

R, n, K, U-Total amount of buildings to be built, optional tower crane types, construction elevator types, overall jack-up climbing frame types available.

$x_j^{(r)}$, $y_k^{(r)}$, $z_u^{(r)}$-Boolean variable of whether j-type tower crane, k-type construction elevator, u-type overall jack-up climbing frame is assigned in building r.

$q^{(r)}$-The total meter quantity of the overall jack-up climbing frame in building r.

L_j-Maximum working range of j-type tower crane.

$L_1^{(r)}$-The spacing between a building adjacent north and south to building r.

$L_0^{(r)}$-Maximum working range covered by tower crane required in building r.

d-The distance between the tower crane and the outside of the building.

d_1-Width of the tower body of the tower crane.

m_r-The number of segments of the working range in building r.

$w_i^{(r)}$-Maximum PB component weight in the i th range segment in building r.

$W_{ij}^{(r)}$-For j-type tower crane, the rated weight at the working range corresponding to the i th type PB components in building r.

ξ_j-The safety factor of the lifting weight of the tower crane type j.

$C_1^{(1)}$, $C_0^{(1)}$-The upper and the lower limit of the total cost budget that PB construction project can be used for tower crane rent fee.

$c_j^{(1)}$, $c_k^{(2)}$, $c_u^{(3)}$-Unit rent fee for j-type tower crane, k-type construction elevator, u-type overall jack-up climbing frame.

$C_1^{(2)}$, $C_0^{(2)}$-The upper and lower of the total cost budget that PB construction project can be used for construction elevator rent fee.

$C_1^{(3)}$, $C_0^{(3)}$-The upper and lower limit of the total cost budget that PB construction project can be used for overall jack-up climbing frame rent fee.

$V_0^{(r)}$-Minimum total efficiency of material transport required in building r.

v_j-The material lifting efficiency of j-type tower crane.

v_k-The material lifting efficiency of k-type construction elevator.

$Q_0^{(r)}$, $Q_1^{(r)}$-Minimum and maximum total meter quantity of the overall jack-up climbing frame in building r.

2.2 Optimization Model Establishing

The security-oriented assignment optimization model of main equipment and facilities for PB construction is established as follows.

$$\max f = \sum_{r=1}^{R}\left[\sum_{j=1}^{n} x_j^{(r)} S_j^{(r)} + \sum_{k=1}^{K} y_k^{(r)} S_k^{(r)} + \sum_{u=1}^{U} z_u^{(r)} S_u^{(r)} \frac{q^{(r)}}{Q_1^{(r)}}\right] / \sum_{r=1}^{R}\left(\sum_{j=1}^{n} x_j^{(r)} + \sum_{k=1}^{K} y_k^{(r)} + \sum_{u=1}^{U} z_u^{(r)}\right) \quad (1)$$

$$\text{s.t.} \quad L_0^{(r)} \leq \sum_{j=1}^{n} L_j x_j^{(r)} \leq L_1^{(r)} - 2d - d_1 \quad r = 1, 2, \cdots, R \tag{2}$$

$$\sum_{j=1}^{n} w_i^{(r)} x_j^{(r)} \xi_j \leq \sum_{j=1}^{n} W_{ij}^{(r)} x_j^{(r)} \quad i = 1, 2, \cdots, m_r; r = 1, 2, \cdots, R \tag{3}$$

$$\sum_{j=1}^{n} x_j^{(r)} = 1 \quad r = 1, 2, \cdots, R \tag{4}$$

$$C_0^{(1)} \leq \sum_{r=1}^{R} \sum_{j=1}^{n} c_j^{(1)} x_j^{(r)} \leq C_1^{(1)} \tag{5}$$

$$\sum_{j=1}^{n} x_j^{(r)} v_j + \sum_{k=1}^{K} y_k^{(r)} v_k \geq V_0^{(r)} \quad r = 1, 2, \cdots, R \tag{6}$$

$$\sum_{k=1}^{K} y_k^{(r)} = 1 \quad r = 1, 2, \cdots, R \tag{7}$$

$$C_0^{(2)} \leq \sum_{r=1}^{R} \sum_{k=1}^{K} c_k^{(2)} y_k^{(r)} \leq C_1^{(2)} \tag{8}$$

$$Q_0^{(r)} \leq q^{(r)} \leq Q_1^{(r)} \quad r = 1, 2, \cdots, R \tag{9}$$

$$\sum_{u=1}^{U} z_u^{(r)} = 1 \quad r = 1, 2, \cdots, R \tag{10}$$

$$C_0^{(3)} \leq \sum_{r=1}^{R} \sum_{u=1}^{U} c_u^{(3)} q^{(r)} z_u^{(r)} \leq C_1^{(3)} \tag{11}$$

$$\sum_{r=1}^{R} \sum_{j=1}^{n} c_j^{(1)} x_j^{(r)} + \sum_{r=1}^{R} \sum_{k=1}^{K} c_k^{(2)} y_k^{(r)} + \sum_{r=1}^{R} \sum_{u=1}^{U} c_u^{(3)} q^{(r)} z_u^{(r)} \leq C \tag{12}$$

$$x_j^{(r)} = 1 \text{ or } 0 \quad j = 1, 2, \cdots, n; \quad r = 1, 2, \cdots, R \tag{13}$$

$$y_k^{(r)} = 1 \text{ or } 0 \quad k = 1, 2, \cdots, K; \quad r = 1, 2, \cdots, R \tag{14}$$

$$z_u^{(r)} = 1 \text{ or } 0 \quad u = 1, 2, \cdots, U; \quad r = 1, 2, \cdots, R \tag{15}$$

$$q^{(r)} \geq 0 \quad r = 1, 2, \cdots, R \tag{16}$$

In above optimization model, formula (1) is the objective function which indicates maximizing the comprehensive safety utility of all tower cranes, construction elevators and overall jack-up climbing frame assignment for the project. Formula (2) limits the relationship and arm length of the tower cranes, with a safe distance between the lifting end of the tower crane and the tower body of another adjacent tower crane; the maximum working range of the j-type tower crane assigned in building r should reach the maximum working range length required by it. Formula (3) indicates that the lifting weight of the tower crane at different ranges should meet the rated lifting weight of the tower crane type at this range, and a certain degree of safety guarantee should also be considered. Formula (4) indicates that the assigned tower crane type is one of n types. Formula (5) indicate that the total rent fee of assigned tower crane should meet the minimum rent cost required by the project, and also meet the maximum value of the cost available for the project. Formula (6) indicates that the assignment of the elevator and the coordinated use of the tower crane transportation should meet the requirements of the transportation efficiency of personnel and small materials. Formula (7) indicates that the assigned elevator type is one of K types. Formula (8) indicate that the rent input of the elevator assigned should meet the minimum cost limit required by the project, and should meet the maximum cost limit of the cost available for the project. Formula (9) indicates that for building r, the installed meter length of the overall jack-up climb should meet its requirements. Formula (10) indicates that for building r, the overall jack-up climbing frame type assigned is one of the U types. Formula (11) indicates that the rent fee of the overall jack-up climbing frame assigned should meet the minimum cost limit and the maximum cost limit of the cost available for the project. Formula (12) indicates that the total rent fee input of tower cranes, elevators and overall jack-up climbing frame should not be higher than the total budget available for the three items. Formula (13) to formula (16) represent the value range of the variables.

2.3 Artificial Immunity Algorithm Designing

For above optimization model, $A = [a_{ij}]_{(n+K+U+1)\times R}$ is adopted to represent a solution as an antibody. The detail steps of artificial immune algorithm are designed as follows.

(1) The initial values of learning error threshold $\varepsilon_{\min}$, maximum iteration $N_{\max}$, antibody population size G, antigen density ρ_0, minimum threshold $\rho_{\min}$ of antibody density, threshold ρ_b of the b highest antibodies density ρ_b', activating ratio τ_1, controlling ratio τ_2, death ratio τ_3 of clone are evaluated.
(2) According to constraint formulas (2)–(16), generate G antibodies which forms antibody population in random. The initial value to density ρ_g of antibody g ($g = 1, 2 \cdots, G$) is evaluated.
(3) Calculate affinity degree ψ_g between antigen and antibody g by $\psi_g = 1/f$, , (f is the objective function value of the model). Calculate affinity degree ψ_{gh} between antibody g and antibody h by formula $\psi_{gh} = 1 - \left(\sum_{i=1}^{n+K+U+1} \sum_{j=1}^{R} |a_{ij}^{(g)} - a_{ij}^{(h)}| \right) \bigg/ \left(\sum_{i=1}^{n+K+U+1} \sum_{j=1}^{R} |a_{ij}^{(g)} + a_{ij}^{(h)}| \right)$.

(4) According to $\frac{d\rho_g}{dt} = \tau_1 \psi_g \rho_g \rho_0 - \frac{\tau_2}{G} \sum_{h=1}^{G} \psi_{gh} \rho_g \rho_h - \tau_3 \rho_g$, perform clone selection calculation for each antibody [6].

(5) If $\rho_g < \rho_{\min}$ $(g = 1, 2 \cdots, G)$, then delete antibody $A^{(g)}$, Simultaneously, the deleted antibodies amount is record to G_0.

(6) If iteration $N' < N_{\max}$, or $\rho_b' < \rho_b$, or learning error $\varepsilon > \varepsilon_{\min}$, then go to step (2), generate G_0 antibodies in random again, repeat steps from (3) to (6); else, the algorithm is end.

3 Model and Algorithm Application

The PB project (WD project) is taken as the application object, and above security-oriented assignment optimization model of the main equipment and facilities for PB construction is applied. WD project consists of 8 buildings in 2 rows, each with 4 buildings with 31 floors. Building 1, 2, 3 and 4 are located in the west area; building 5, 6, 7 and 8 are located in the east area. The west area and the east area are arranged axisymmetrically. The drawings for building1, 2, 7 and 8 are the same. The drawings for building 3, 4, 5 and 6 are the same.

Due to the limited construction site, tower cranes in building1, 3, 5 and 7 are arranged in the south side; tower cranes in building 2, 4, 6 and 8 are arranged in the north side. There are 3 types of tower crane; 4 types of construction elevators; and 5 overall jack-up climbing frame types can be selected. Above equipment and facilities are used for 10 months. The distance between the two adjacent building from north to south is 47 m. The distance between the tower crane and the outside of the building is 4m. The tower body of the tower crane is 2 m wide. Safety factor of tower crane is 1.25. The related data of different types of equipment and facilities are shown in Table 1. The total rent budget of tower crane, construction elevator and overall jack-up climbing frame is respectively [¥2.2 million, ¥2.3 million], [¥0.8 million, ¥0.9 million] and [¥3.5 million, ¥4.5 million]. The sum of them should not exceed ¥ 7.3 million.

The rated weight of varied type tower crane at different ranges is shown in Table 2.

The minimum rated weight (rated weight at the maximum range length) of the tower crane of QTZ125 (35m) is 3.53 tons. After considering the safety guarantee factor (1.25), the minimum safety rated weight is $3.53/1.25 = 2.824$ tons. In the same way, the minimum rated weight of the tower crane of QTZ130 (30m) is 5.63 tons. After considering the safety guarantee factor (1.25), the minimum safety rated weight is $5.63/1.25 = 4.504$ tons. The minimum rated weight of the tower crane of QTZ130 (35m) is 4.13 tons. After considering the safety guarantee factor (1.25), the minimum safety rated weight is $4.13/1.25 = 3.304$ tons. Therefore, for the above three types, when the actual lifting weight is under above minimum safety rated weight, it is not necessary to consider the rated weight constraint limit at a given range of each tower crane of each type. The relevant data of various PB components in varied buildings of the WD project can be seen respectively in Tables 3, 4, 5 and 6.

The relevant data is input into above optimization model, By solving, the optimal objective function value is 0.9336, and the comprehensive security degree is 93.36%. The optimal solution is shown in Table 7.

Table 1. The related data of different types of equipment and facilities.

Name	Type	Safety utility of building 1, 2, 7, 8	Safety utility of building 3, 4, 5, 6	Rent fee (¥10000/month)	Materials transport efficiency
Tower crane	QTZ125(35m)	0.89	0.87	2.7	0.15
	QTZ130(30m)	0.92	0.90	2.9	0.18
	QTZ130(35m)	0.95	0.93	3	0.20
Construction elevator	SC200/200	0.89	0.87	1	0.83
	SC200/200 FC	0.91	0.89	1.1	0.90
	SCD200/200	0.93	0.91	1.2	0.83
	SCD200/200 FC	0.95	0.93	1.3	0.90
Overall jack-up climbing frame	I	0.95	0.93	0.050	
	II	0.93	0.91	0.047	
	III	0.91	0.89	0.045	
	IV	0.89	0.87	0.042	
	V	0.87	0.85	0.040	

Table 2. The rated weight of varied type tower crane at different ranges.

Type	Item	Range and rated weight												
QTZ125 (35m)	Range	2.5–14.77	15	16	17	18	20	22	24	26	27.1	28	30	32	35
	Rated weight	10	9.82	9.11	8.49	7.94	7.02	6.27	5.64	5.12	4.87	4.68	4.29	3.96	3.53
QTZ130 (30m)	Range	2.5–18.36	19	20	22	25	27	30						
	Rated weight	10	9.62	9.06	8.12	6.99	6.38	5.63						
QTZ130 (35m)	Range	2.5–16.69	17	18	19	20	22	25	27	30	32	35		
	Rated weight	10	9.79	9.16	8.61	8.11	7.25	6.23	5.68	5.00	4.62	4.13		

4 Conclusion

In the process of PB construction, construction equipment and facilities are the important guarantee conditions for the implementation of PB construction. Under the construction budget limit, the selection and assignment optimization of construction equipment and facilities is an important work to ensure the construction security level of PB construction. The optimization model of main equipment and facilities of PB construction has important theoretical research and engineering practical significance. Tower crane, construction elevator and overall jack-up climbing frame are the most important types of construction equipment and facilities in the construction process of PB. Its input cost

Table 3. The relevant data of various PB components in building 1 and 7.

PB components name	No	Weight (t)	Horizontal distance (m)	QTZ125(35m) Rated weight(t)	QTZ130(30m) Rated weight(t)	QTZ130(35m) Rated weight(t)
PB external wall	WQ1	3.54	21	6.27	8.12	7.25
	WQ2	4.50	24	5.64	6.99	6.23
	WQ3	3.85	25	5.12	6.99	6.23
External fill wall with window	WTQ1	2.99	26	5.12	6.38	5.68
	WTQ3	3.90	24	5.64	6.99	8.61
PB bay window	WTC1	5.51	4	10	10	10
	WTC2	4.52	18	7.94	10	9.16
PB inner wall	NQ1	2.89	16	9.11	10	10

Table 4. The relevant data of various PB components in building 2 and 8.

PB components name	NO	Weight (t)	Horizontal distance (m)	QTZ125(35m) Rated weight(t)	QTZ130(30m) Rated weight(t)	QTZ130(35m) Rated weight(t)
PB external wall	WQ1	3.54	26	5.12	6.38	5.68
	WQ2	4.50	22	6.27	8.12	7.25
	WQ4	4.60	14	10	10	10
External fill wall with window	WTQ3	3.90	16	9.11	10	10
	WTQ4	2.99	23	5.64	6.99	6.23
PB bay window	WTC1	5.51	19	7.02	9.62	8.61
	WTC2	4.52	26	5.12	6.38	5.68

occupies the largest proportion in the equipment and facilities input of the whole PB construction process, and it also affects the safety level of the whole PB construction process very much. The author takes the selection and assignment of the above construction equipment and facilities as the research object, and the security-oriented assignment optimization model and algorithm of the main PB construction equipment and facilities established provides some reference for related practices. Future research on the optimal assignment problem that considers more equipment and facilities will be one of the future directions of research in this field.

Table 5. The relevant data of various PB components in building 3 and 5.

PB components name	NO	Weight (t)	Horizontal distance (m)	QTZ125(35m) Rated weight(t)	QTZ130(30m) Rated weight(t)	QTZ130(35m) Rated weight(t)
PB external wall	WQ1	3.54	23	5.64	6.99	6.23
	WQ2	4.50	26	5.12	6.38	5.68
External fill wall with window	WTQ1	3.29	28	4.68	5.63	5.00
	WTQ3	4.29	25	5.12	6.99	6.23
PB bay window	WTC1	6.06	4	10	10	10
	WTC2	4.54	18	7.94	10	9.16
PB inner wall	NQ1	3.18	17	8.49	10	9.79

Table 6. The relevant data of various PB components in building 4 and 6.

PB components name	NO	Weight (t)	Horizontal distance (m)	QTZ125(35m) Rated weight(t)	QTZ130(30m) Rated weight(t)	QTZ130(35m) Rated weight(t)
PB external wall	WQ1	3.54	27	4.87	6.38	5.68
	WQ2	4.50	24	5.64	6.99	6.23
	WQ3	3.85	13	10	10	10
	WQ4	4.60	15	9.82	10	10
External fill wall with window	WTQ3	4.29	16	9.11	10	10
	WTQ4	3.29	24	5.64	6.99	6.23
	WTQ5	2.99	25	5.12	6.99	6.23
PB bay window	WTC1	6.06	19	7.02	9.62	8.61
	WTC2	4.54	26	5.12	6.38	5.68
PB inner wall	NQ1	2.89	16	9.11	10	10

Table 7. The optimal solution.

Building No		1	2	3	4	5	6	7	8
Tower crane type	QTZ125(35m)	1	0	0	0	0	0	1	0
	QTZ130(30m)	0	1	1	1	0	0	0	1
	QTZ130(35m)	0	0	0	0	1	1	0	0
construction elevator type	SC200/200	0	0	0	0	0	0	0	1
	SC200/200 FC	1	0	1	1	0	1	1	0
	SCD200/200	0	1	0	0	1	0	0	0
	SCD200/200 FC	0	0	0	0	0	0	0	0
overall jack-up climbing frame type	I	0	0	0	0	0	0	0	0
	II	1	1	0	0	1	0	1	1
	III	0	0	1	1	0	1	0	0
	IV	0	0	0	0	0	0	0	0
	V	0	0	0	0	0	0	0	0
	Length (meter)	120	120	128	128	128	128	120	120

Acknowledgments. This work is supported by National Natural Science Foundation of China (51678375), Liaoning Provincial Colleges and Universities' Innovative Talents Support Plan (LR2020005), Liaoning Provincial Natural Science Foundation's Guiding Plan (2019-ZD-0683).

References

1. Niroomand, S.: Hybrid artificial electric field algorithm for assembly line balancing problem with equipment model selection possibility. Knowledge-Based Syst. **219**(11), 106905 (2021)
2. Alpay, S., Iphar, M.: Equipment selection based on two different fuzzy multi criteria decision making methods: Fuzzy TOPSIS and fuzzy VIKOR. Open Geosci. **10**(1), 661–677 (2018)
3. Smith, N.M., Ali, S., Bofinger, C., et al.: Human health and safety in artisanal and small-scale mining: an integrated approach to risk mitigation. J. Clean. Prod. **129**(5), 43–52 (2016)
4. Knyziak, P.: The impact of construction quality on the safety of prefabricated multi-family dwellings. Eng. Fail. Anal. **100**, 37–48 (2019)
5. Chun-guang, C.: Optimization of raw materials selective purchasing for refined copper strip producing by AIA. Control Eng. China **24**(12), 2496–2501 (2017)
6. Cayzer, S., Aickelin, U.: A recommender system based on the immune network, In: Proceedings of the 2002 Congress on Evolutionary Computation, pp. 807–813. IEEE, Honolulu, HI, USA (2002)

Mineral Identification in Sandstone SEM Images Based on Multi-scale Deep Kernel Learning

Mei Wang[1,2], Simeng Fan[1], Fei Han[2(✉)], Zhigang Liu[1], and Kejia Zhang[1]

[1] School of Computer and Information Technology, Northeast Petroleum University,
Daqing 163318, China
{wangmei,fansm,liuzhigang,zhangkejia}@nepu.edu.cn
[2] Artificial Intelligence Energy Research Institute, Northeast Petroleum University,
Daqing 163318, China
hanfei@nepu.edu.cn

Abstract. Identifying sandstone images and judging the types of minerals play an important role in oil and gas reservoir exploration and evaluation. Multiple kernel learning (MKL) method has shown high performance in solving some practical applications. While this method belongs to a shallow structure and cannot handle relatively complex problems well. With the development of deep learning in recent years, many researchers have proposed a deep multiple layer multiple kernel learning (DMLMKL) method based on deep structure. While the existing DMLMKL method only considers the deep representation of the data but ignores the shallow representation between the data. Therefore, this paper propose a multiple scale multiple layer multiple kernel learning (MS-DKL) method that "richer" feature data by fusing deep and shallow representations of mineral image features. Mineral recognition results show that MS-DKL algorithm is higher accuracy in mineral recognition than the MKL and DMLMKL methods.

Keywords: Mineral recognition · Deep kernel learning · Multiple scale · SLIC

1 Introduction

Mineral identification plays an important role in petrological research, and researchers can obtain important information about minerals through this method [1]. The identification of sandstone images can effectively obtain the structural characteristics of minerals and determine the types of minerals, which has important application value in oil and gas reservoir exploration.

Some studies [2–5] have adopted traditional image segmentation algorithms to segment sandstone SEM images. The traditional image segmentation method is simple and efficient, and the boundary position generated by image segmentation is relatively accurate. However, above method is sensitive to image noise and it is easy to misjudge the cleavage, twin crystals and other microstructures in minerals as grain boundaries.

In the field of sandstone image mineral recognition, some conventional machine learning algorithms have been widely used, including support vector machines (SVM)

© The Author(s), under exclusive license to Springer Nature Switzerland AG 2023
N. Xiong et al. (Eds.): ICNC-FSKD 2022, LNDECT 153, pp. 353–360, 2023.
https://doi.org/10.1007/978-3-031-20738-9_40

[6], etc. In order to solve the choice problem of a single kernel function, multiple kernel learning is proposed, which is a mainstream solution. It improves the recognition ability of kernels in a given classification task by learning the (sparse or convex) linear combination of elementary kernels [6, 7]. However, the standard MKL only considers the shallow combination of kernels, which will not be able to capture the "correct" similarity between highly semantic and variable content [8]. More and more experimental results show that the deep combination of learning basic kernels, that is, deep multilayer multiple kernel learning (DMLMKL) [9–11] can better improve classification capability.

For better identification of mineral image, we propose a multiple scale deep kernel learning algorithm. This study makes the following work:

- The research propose a deep kernel method and apply the method to sandstone image mineral recognition.
- We propose a vector concatenation method of fusion kernel matrix.
- We propose a multiple scale fusion method, which merges the deep representation and shallow representation of the data to obtain "richer" data representation.

The research plan is as follows. Section 2 describes the specific proposed MS-DKL. Section 3 evaluates and discusses the performance of the MS-DKL algorithm under the mineral data set by using various indicators. The paper is concluded in Sect. 4.

2 Multiple Scale Deep Multiple Learning Method

2.1 Algorithm Principle

Although MKL [12] is already very extensive, it is a simple combination of multiple kernel, and its structure is "shallow" [13]. In fact, the optimal kernel obtained is shallow, and it is impossible to get a deeper connection between the data. In a Deep Multiple layer Multiple Kernel Learning (DMLMKL) [14] architecture, each layer contains a set of kernels, where the output weighted linear combination of the kernel map of the previous layer is used as the input of the next layer.

The kernel K formed by the multiple layer deep kernel is defined as:

$$K^{(l)} = \left\{ K^{(l)}\left(K^{(l-1)}; \mu\right) = \sum_{t=1}^{M} \mu_t^{(l)} k_t^{(l)}\left(K^{(l-1)}\right) \middle| \mu_t^{(l)} \geq 0, l = 1, \ldots, L \right\} \quad (1)$$

$$K^{(1)} = K^{(1)}\left(x, x; \mu^{(1)}\right) = \sum_{t=1}^{M} \mu_t k_t(x, x_i) \quad (2)$$

where $k_t^{(l)}$ represents t^{th} base kernel at layer l and w represents the corresponding weight of the base kernel. The final decision function is defined as:

$$f(x) = \sum_{i=1}^{n} \alpha_i y_i K^{(L)}\left(K^{(L-1)}; \mu^{(L)}\right) + b \quad (3)$$

The nonlinear expressive ability of the algorithm is crucial to the performance of the model. The deep kernel maps the kernel function to the deep structure, and obtains the deep content between the data. However, the deep kernel only pays attention to the expression of deep-level mapping within the data, ignoring the expression of the previous layers between the data, making the expression between the data not "rich" enough. This paper proposes a multiple scale deep kernel (DS-DKL) algorithm to solve this problem, The DS-DKL architecture is shown in Fig. 1, where k_{Lp} represents the pth kernel at layer L and K_L represents the kernel function obtained by the weighted fusion of the L layer.

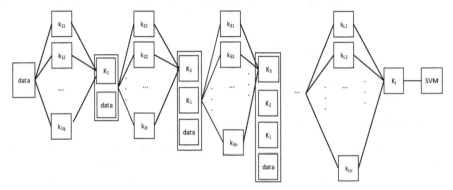

Fig. 1. Depiction of a MLMKL architecture.

where k_{Lp} represents the pth kernel at layer L and K_L represents the kernel function obtained by the weighted fusion of the L layer.

Set a set of datasets:

$$T = \{(x_i, y_i)|i = 1, \ldots n\} \tag{4}$$

where $x_i \in X \in R^d$ represents feature space, d represents feature dimension, and y_i represents label. The goal of training is to learn a multiple scale deep kernel classification.

As can be seen from Fig. 1, in the MS-DKL structure, the training set of the first layer kernel function is the input data, and the kernel function of the network layer except the first layer is the fusion of the output and input data of all layers before this layer as the input of this layer. The data fusion method in the model adopts Concatenate, that is, the vector splicing method, which splices two feature vectors of the same or different dimensions. Assuming that there are feature vectors $v_1 \in R^c$ and $v_2 \in R^t$, the two feature vectors v_1 and v_2 are fused, that is, the vectors are spliced, that is, $v = [v_1, v_2] \in R^{c+t}$.

In the MS-DKL structure, the kernel function input in the $L + 1(L \geq 2)$ network layer is:

$$x^{(L+1)} = \left[T, K^{(1)}, K^{(2)}, \ldots, K^{(L)}\right] \in R^{L \times d + n} \tag{5}$$

where $K^{(L)}$ is output at layer L, n is the quantity of samples in the training set, and d is the feature dimension.

According to the above introduction, the kernel K formed by the multiple scale deep kernel (the number of layers is greater than 2) is defined as:

$$K^{(1)} = \left\{ K^{(l)}\left(x^{(l-1)}; \mu\right) = \sum_{t=1}^{M} \mu_t^{(l)} k_t^{(l)}\left(x^{(l-1)}\right) \middle| \mu_t^{(l)} \geq 0, l = 1, \ldots, L \right\} \quad (6)$$

The first layer of synthetic kernel can be defined as:

$$K^{(1)} = K^{(1)}\left(x, x; \mu^{(1)}\right) = \sum_{t=1}^{M} \mu_t k_t(x, x_i) \quad (7)$$

where $k_t^{(l)}$ represents t^{th} at layer l, and $\mu_t^{(l)}$ is the corresponding weight of the base kernel. In the MS-DKL structure, model adopt SampleMKL [15] to optimize the kernel function weights of all network layers.

2.2 Algorithm Implementation Steps

Based on the above problems, the training process of the multi-scale deep kernel learning model proposed in this research can be composed of the following five steps:

Step 1:Initialize the model, including the base kernel function type, the base kernel method parameters, and the number of the model layers.

Step2: Use the input data as input to train the kernel function of the first layer and use SimpleMKL to optimize the weight of the kernel function, and linearly combine the kernel functions according to the weight to obtain the synthetic kernel of this layer.

Step 3: When the number of model layers than 2, the output and input data of all layers before the layer are fused by vector splicing, and as the input to this layer. The weight of the kernel function in the network layer is optimized by SimpleMKL, and the kernel function is linearly combined according to the weight to obtain the synthetic kernel of this layer.

Step 4: Iterate step 3 until the quantity of network layers reaches the quantity of model layers initialized in step 1.

Step 5: Input the finally obtained kernel function mapping data into SVM for classification, and obtain the optimal model hyperparameter C by grid search method.

3   Application of MS-DKL Method in Sandstone SEM Image Mineral Recognition

3.1   Data Set

Figure 2 gives the sandstone SEM image.

In the process of constructing the dataset, Pre-segmentation of sandstone images with SLIC superpixel algorithm, and the superpixels number is 400. Then, each small superpixel in the image is cut to form a superpixel sample. The method of cutting a single superpixel is to obtain the bounding rectangle of the superpixel according to the coordinates of the upper left corner and the lower right corner of the superpixel. Figures 3 and 4 gives the experimental results.

Fig. 2. Depiction of a MLMKL architecture.

Fig. 3. SLIC segmentation image

Fig. 4. Superpixel cutout image

Next, the samples formed by cutting are labeled according to the type of minerals, and a mineral image set is obtained. A total of 800 samples are obtained, including 284 samples of sandstone, 240 samples of pores, and 276 samples of feldspar. Finally, according to the differences between different minerals, select five feature algorithms to characterize the mineral images, extracted superpixel gray level histogram(Hist), Gray-level co-occurrence matrix(GLCM), Local Binary Patterns(LBP), Canny,and HU feature, form a 119 dimensional feature vector. Table 1 gives feature information.

Table 1. Feature information.

Feature	Number of dimensions
hist	32
GLCM	4
LBP	72
Canny	4
HU	7

3.2   Comparisons at Data with Different Algorithm

In this study, we select the MKL, MS-DKL, DNN, and DMLMKL image recognition algorithms that show the advantages of the proposed algorithm. In the experiment, the dataset 50% is the training set, and the remaining 50% is the test set.

In the following experiment, we applied four base kernels: (1) Linear kernel, (2) RBF kernel with $\gamma = 1$, (3) Two Poly kernels with two different degrees $d = 2, d = 3$ and $\beta = 1$, experiment with four base kernels and corresponding parameters.

In the DNN experiments, the hidden layer value is set to 3; the hidden layer node value is set to 256, 128, and 64, respectively; and activation function is set to relu function. The regularization parameter C of the SVM used in this paper is optimized using the grid search method, which range is set to [1, 1000]. In the experiment, 2-fold cross-validation is repeated 10 times, that is, p is set to 10 and k is set to 2 in RepeatedKFold. The quantity of layers L for DMLMKL and MS-DKL is set to 3.

In the experiment, we adopt the method of statistical confusion matrix to evaluate the performance of the classification algorithm. Every algorithm iterates 10 times, and the confusion matrix with the highest classification accuracy was obtained to evaluate the algorithm. Table 2 gives the calculation results.

Table 2. MS-DKL classification result confusion matrix.

	Sandstone	Pores	Feldspar
Sandstone	131	3	8
Pores	2	106	0
Feldspar	10	0	140
t	0.92	0.97	0.95

The row labels in Table 2 represent the true value labels, while the column labels represent the predicted value labels. The last row in the table shows the precision with which each mineral type was predicted.

In the experiment, the confusion matrix of the four classification algorithms was used to calculate the prediction precision of each mineral under different classification algorithms. Table 3 shows the classification results, where bold indicates the highest accuracy in each row.

Table 3. Comparison of recognition methods at SEM mineral image data sets.

	DNN	MKL	DMLMKL	MS-DKL
Sandstone	**0.93**	0.65	0.68	0.92
Pores	0.95	0.93	0.93	**0.97**
Feldspar	0.85	0.86	0.87	**0.95**

Table 3 shows that compared with other classification algorithms, the MS-DKL algorithm has the highest prediction accuracy for pores and feldspar, and the DNN with the highest accuracy rate for quartz is 0.93, while our algorithm is 0.92.

The algorithm proposed in this paper is applied to sandstone SEM image segmentation, which lays the foundation for region merging in the segmentation process. In the MS-DKL model classification process, Green represent pores, blue represent feldspar and red represent quartz. Sandstone SEM image mineral classified by MS-DKL model are shown in Fig. 5.

Fig. 5. MS-DKL model recognition result

3.3 Dissuasion

In the previous subsection, we compared the MKL, DNN, DMLMKL and MS-DKL on sandstone SEM image mineral recognition problem. In the mineral image recognition experiment, the algorithm with the highest recognition accuracy is our proposed algorithm in this research. The results of mineral image recognition give that the MS-DKL model constructed in this research is an effective tool for mineral image recognition.

4 Conclusion

The effective identification of minerals has a significant impact on the exploration and evaluation of oil and gas reservoirs Aiming at the problem of mineral image recognition, a deep kernel learning recognition method based on multiple scale is proposed in this study. According to the difference of different minerals in the mineral image, the appropriate feature extraction algorithm is used for feature extraction and using MS-DKL algorithm to establish recognition model. This model not only considers the use of a multi-layer model to bring more abstract expression of data, but also considers the combination of deep expression and shallow expression to obtain a "richer" expression of data. And improve the overall recognition model performance. The research results give that the MS-DKL model has a stronger ability to express data compared with the MKL model and the DMLMKL model, it can recognition minerals more accurately, and has stronger robustness and anti-interference ability.

References

1. El Haddad, J., de Lima Filho, E.S., Vanier, F.: Multiphase mineral identification and quantification by laser-induced breakdown spectroscopy. Miner. Eng. **134**(1), 281–290 (2019)
2. Heilbronner, R.: Automatic grain boundary detection and grain size analysis using polarization micrographs or orientation images. J. Struct. Geol. **22**(7), 969–981 (2000)
3. Ross, B.J., Fueten, F., Yashkir, D.Y.: Automatic mineral identification using genetic programming. Mach. Vision Appl. **13**(2), pp. 61–69 (2001)
4. Choudhury, KR., Meere.: Automated grain boundary detection by CASRG. J. Struct. Geol. **28**(3), pp. 363-375 (2006)
5. Adams, R.D., Bischof, L.: Seeded region growing. IEEE Trans. Pattern Anal. Mach. Intell. **16**(6), 641–647 (2002)
6. Jie, Z., Qizhi, T., Zhengyong, W.: Rock slice image segmentation based on support vector machine. J. Chengdu Univ. Inf. Technol. **22**, 186–189 (2007)
7. Kloft, M., Brefeld, U., Laskov, P.: Non-sparse multiple kernel learning. In: Proceedings of the workshop on kernel learning: automatic selection of optimal kernels. Whistler, Canada: the MIT Press, pp. 1–4 (2008)
8. Strobl, E.V., Visweswaran, S.: Deep multiple kernel learning. STATISTI 1 (2013)
9. Xiuyuan, C., Xiyuan, P., Ran, D.: Deep kernel learning method for SAR image target recognition. Rev. Sci. Instrum. **88**(10), 1–5 (2017)
10. Wilson, A.G., Hu, Z., Salakhutdinov, R., Xing, E.P.: Deep Kernel learning. Comput. Sci. (2015)
11. Bastian, B., Christian, R., Michael, G.: Representer theorem for deep kernel learning. J. Mach. Learn. Res. **20**(57–84), 1–32 (2019)
12. Shengbing, R., Wang-bo, S., Naeem, S.C.: Self-adaptive deep multiple kernel learning based on Rademacher complexity. Symmetry **11**(3), 325 (2019)
13. Sonnenburg, S., Rätsch, G., Schäfer, C.: A general and efficient multiple kernel learning algorithm. Adv. Neural Inf. Process. Syst. (2006)
14. Rebai, I., BenAyed, Y., Mahdi, W.: Deep multilayer multiple kernel learning. Neural Comput. Appl. **27**(8), 2305–2314 (2015). https://doi.org/10.1007/s00521-015-2066-x
15. Rakotomamonjy, A., Bach, F.R., Canu, S., et al.: SimpleMKL. J. Mach. Learn. Res. **9**(3), 2491–2521 (2008)

Optimal Selection of Left and Right Hand Multi-channel Pulse Features Based on Neighbourhood Component Analysis

Lin Fan[1,2,3], Yan Li[1(✉)], Jinsong Wang[1,2,3], Rong Zhang[1,2,3], and Ruiling Yao[1]

[1] School of Computer Science and Technology, Xi'an University of Posts and Telecommunications, Xi'an 710121, China
fanlin@xupt.edu.cn, liyan10@stu.xupt.edu.cn,linaldo_7@163.com,
wjs68@xupt.edu.cn, zhangr@xupt.edu.cn, 1158179725@stu.xupt.edu.cn
[2] Shannxi Key Laboratory of Network Data Analysis and Intelligent Processing, Xi'an 710121, China
[3] Xi'an Key Laboratory of Big Data and Intelligent Computing, Xi'an 710121, China

Abstract. The use of computer technology to analyze the pulse for disease diagnosis is an important research direction in the standardization of traditional Chinese medicine (TCM) diagnosis. TCM believes that the pulses at different positions of the left and right hands correspond to different organs. However, few studies have conducted pulse signal analysis based on this theory of TCM. Taking this theory into consideration, this research proposes a multi-channel feature selection method based on Neighbourhood Component Analysis(NCA), which uses multi-channel pulse features more comprehensively. First, select the most important features of each location through NCA, and construct multiple feature combinations. Then through the classification algorithm, the optimal feature combination under the corresponding algorithm is selected, and the feature subsets at different channels are constructed. The experimental results show that the left-hand three-channel can obtain better classification results by using pulse features of fewer channels, and has good performance, with a precision of 88.6% and a recall rate of 94.2%.

Keywords: Wrist pulse · TCM · NCA · Multi-channel · Hypertensive

1 Introduction

The positions of palpation in TCM are called Cun, Guan and Chi.Different pulse positions are related to different human organs. As shown in Fig. 1, the Cun of left hand is related to the heart and pericardium, the Guan of left hand is related to the liver and gallbladder, the Chi of left hand is related to the kidney and bladder, the Cun of right hand is related to the lung and Intestine, the Guan of right hand is related to the Spleen and Stomach, the Chi of right hand is related to the Mingmen and Middle Chest. TCM practitioners usually use the "Sanbu

© The Author(s), under exclusive license to Springer Nature Switzerland AG 2023
N. Xiong et al. (Eds.): ICNC-FSKD 2022, LNDECT 153, pp. 361–368, 2023.
https://doi.org/10.1007/978-3-031-20738-9_41

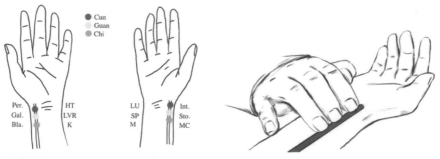

(a) The corresponding organs of Cun, Guan, and Chi of left and right hands.

(b) Sanbu Jiuhou.

Fig. 1. Schematic diagram of TCM theory.

Jiuhou" method to diagnose the state and disease information of different organs of the human body.

Although, TCM is widely used at present. However, TCM is limited by the personal subjective experience of practicing physicians and does not possess sufficient objectivity [1]. This relatively subjective diagnostic method makes it difficult to have a feasible scientific standard for pulse diagnosis [2,3]. With the improvement of the performance of signal sensors and machine learning algorithms [4–8], the application of computer and signal processing technology to explore pulse diagnosis standards has also been developed. Xu Lisheng et al. applied Approximate Entropy (ApEn) in pulse diagnosis% [9]. Wang Dimin et al. proposed a signal preprocessing framework for pulse analysis. Using this preprocessing framework in the diagnosis of diabetes can achieve a classification rate of up to 91% [10].

Most of the above pulse acquisition systems and pulse information analysis methods use the perspective of modern Western medicine to extract different pulse characteristics to study the diagnosis of diseases. In this study, combined with TCM theory, the performance differences of feature subsets of different channel combinations in disease diagnosis were investigated. First, a dataset of 46 hypertensive/healthy individuals was constructed through a self-developed acquisition device. Then, through the multi-channel feature selection method based on NCA, combined with disease characteristics, feature subsets of single-hand three-channel and six-channel were constructed. Finally, a number of classification models are trained according to the corresponding algorithms, which are classified based on the data sets of hypertensive/healthy people, and the models are evaluated and analyzed.

2   Method

This chapter mainly describes the acquisition device, the data acquisition and preprocessing plan, the content and meaning of feature extraction, the method of feature selection. Fig. 2 shows the overall working framework.

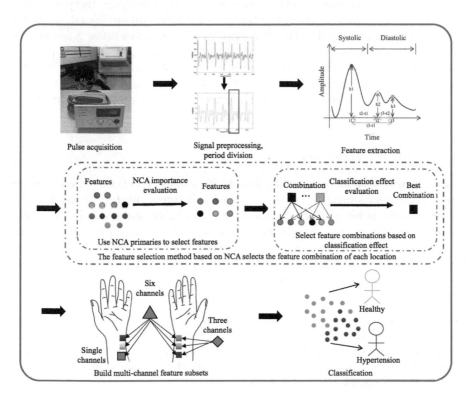

Fig. 2. Multi-channel pulse diagnosis framework based on TCM theory.

2.1   Materials

Data Acquisition This study recruiter 46 volunteers(35 healthy people and 11 hypertensive patients) to participate in our experiments with the approval of the research ethics boaed (REB) of the host institution. The experiment environment is comfortable and quiet.The pulse is collected using the device independently developed by the CM-01B sensor, and the sampling frequency is 100Hz. Before the acquisition, each subject was asked to sit and rest for 5 minutes to keep calm and normal. After that, use the OMRON HEM-712C sphygmomanometer to measure the blood pressure of the subjects, and help the subjects fill in the relevant information questionnaire and record the subjects' basic information.

Data Preprocessing The pulse signal is a very informative biological signal, and each pulse cycle is not exactly the same. At the same time, such signals are also susceptible to interference, resulting in inaccurate information analysis. Therefore, preprocessing is very important to obtain pulse information more accurately before further analysis.

In this study, according to the results of spectrum analysis, the parameters of the upper and lower limits of the Butterworth filter are optimized, so that it can retain the pulse signal to the maximum extent, suppress other noise signals, and keep the signal in the range as flat as possible. Then, the starting position of each pulse period is found based on the period division algorithm, and the single period is identified and segmented according to this position.

The original signal is preprocessed to extract pulse features more effectively and accurately.

2.2   Features Extraction

Time Domin Features and Statistical Features The time domain feature describes the relationship between the signal and time. Affected by cardiovascular diseases, the pulse signal changes significantly in time and space. In this study, the following features are selected for analysis:$h_1,h_2,h_3,H_1,t_1,t_2,t_3,T_1,T_2,T_3$. Mathematical statistics is a method of studying the regularity of random phenomena. In this study,the *Mean* and *Variance* were selected to analyze the fluctuation of the pulse waveform.

Energy Features The wavelet packet transform can decompose the low-frequency part of the signal as well as the high-frequency part, and this decomposition has neither redundancy nor omission, so the signal can be better time-frequency localized analysis.This study extracts the wavelet packet energy features in the frequency band of 0–8 Hz.

2.3   Features Selection

NCA feature importance evaluation Neighbourhood Component randomly selects the nearest neighbor components, and obtains the transformation matrix in the Mahalanobis distance by optimizing the result of the leave-one-out cross-check. Therefore, it can reduce the influence of the raw data distribution. In this calculation process, the importance index of the features is calculated, and finally the features are sorted.

The data set X=$x_1, x_2, \cdots, x_n$ has class labels in the space a, and defines the Mahalanobis distance transformation matrix $Q = A^{\mathrm{T}}A$. The Mahalanobis distance between two sample points is defined as follows:

$$d(x_i, x_j) = \sqrt{(x_i - x_j)^{\mathrm{T}}Q(x_i - x_j)}$$
$$= \sqrt{A(x_i) - A(x_j)^{\mathrm{T}}A(x_i - Ax_j)} \tag{1}$$
$$i, j = 1, 2, \cdots, n$$

The sample point x_i randomly selects a data as the nearest neighbor component x_j and inherits the probability P_{ij} of the class label c_j of the nearest neighbor component. The probability P_{ij} is defined as follows in the changing space using Euclidean distance:

$$
\begin{cases}
P_{ij} = \dfrac{exp(-\|Ax_i - Ax_j\|)}{\sum_{k \neq i} exp(-\|Ax_i - Ax_j\|)} & j \neq i \\[4mm]
P_{ij} = 0 & j = i
\end{cases}
\tag{2}
$$

Because each data point can be selected as a neighboring component, the selected data can inherit all class labels, and the probability that the sample point can be correctly classified is:

$$
P_i = \sum_{j \in C_i} P_{ij}
\tag{3}
$$

NCA search transformation matrix A, the objective function can be understood as maximizing the expectation of the number of correctly classified points, which is equivalent to minimizing the distance between classes:

$$
f(A) = \sum_i P_i = \sum_i \sum_{j \in C_i} P_{ij}
\tag{4}
$$

This unconstrained optimization problem uses the conjugate gradient method or the stochastic gradient method to find A, and uses the differential transformation matrix to define $x_{ij} = x_i - x_j$:

$$
\begin{aligned}
\frac{\partial f(A)}{\partial A} &= -2A \sum_i \sum_{j \in C_i} P_{ij} \left(x_{ij} x_{ij}^{\mathrm{T}} - \sum_k P_{ik} x_{ik} x_{ik}^{\mathrm{T}} \right) \\
&= 2A \sum_i \left(P_i \sum_k P_{ik} x_{ik} x_{ik}^{\mathrm{T}} - \sum_{j \in C_i} P_{ij} x_{ij} x_{ij}^{\mathrm{T}} \right)
\end{aligned}
\tag{5}
$$

After obtaining the gradient of the objective function to A, you can use the gradient descent method and the quasi-Newton method to continuously optimize the upper limit of the objective function, and obtain the importance evaluation of the feature when the objective function is optimal. Based on the importance evaluation, the feature selection can be done.

Performance evaluation of feature combination In order to reduce the effect of different feature parameters on different machine learning algorithms having different sensitivities, this study is based on the feature importance analysis results obtained by NCA, and obtains the top 6 items with high importance index for each part, and then constructs 20 different feature combinations for each part. Finally, based on multiple feature combinations, machine learning algorithms are used to learn part of the data set, and the feature combination with the best classification performance is selected.

According to the feature selection results, combined with the theory of TCM pulse diagnosis, feature subsets of different channels are constructed and used to train the model.

3   Result

3.1   Model Comparison of Different Algorithms

According to the feature selection method based on NCA, the features with better performance from the three different classifiers are selected to form three six-channel feature subsets, namely S1, S2, and S3, each with 18 features. S1 uses DT, S2 uses SVM, and S3 uses KNN. The performance of the model is tested through five-fold cross-validation. As shown in the Table 1, it shows the distinguishing performance of different machine learning algorithms for hypertension and health conditions. Further, three different dimensionality reduction methods are used to select the same number of features, and the model is trained based on the KNN machine learning algorithm. S3 uses the method proposed in this study, S4 uses PCA, and S5 uses FA. As shown in Table 2, the method proposed in this study performs relatively well, with the F1-Score being 3.9% and 3% higher than the other two methods, respectively.

Table 1. Performance indicators of six-channel classification models of different algorithms.

	Accuracy	Precision	Recall	Specificity	F1-score
DT	95.1	**90.8**	90.3	**96.8**	90.5
SVM	94.9	82.1	**98.1**	94.1	89.4
KNN	**96.2**	89.2	95.6	96.3	**92.3**

Table 2. Model performance indicators trained on feature subsets extracted by different feature dimensionality reduction methods.

	Accuracy	Precision	Recall	Specificity	F1-score
Our method	**96.2**	**89.2**	**95.6**	**96.3**	**92.3**
PCA	94.5	82.1	95.9	94.1	88.4
FA	94.8	85.1	94	95	89.3

3.2   Model Comparison of Three-Channel and Six-Channel Features

According to the method proposed in this research, the features that perform better under KNN are selected to construct feature subsets with different features at different locations. S4 is the feature set of all positions of the left and right hands, a total of 96 features. S5 is a feature subset composed of a combination

Table 3. Performance indicators of three-channel and six-channel classification models.

	Accuracy	Precision	Recall	Specificity	F1-score
S3	**96.2**	**89.2**	**95.6**	**96.3**	**92.3**
S4	94.4	88.3	89.8	94.5	89
S5	**95.7**	**88.6**	**94.2**	**96.2**	**91.4**
S6	89.5	77.5	80.7	92.3	79.1

of features of the three positions of the left hand with a better effect, with a total of 9 features.

In TCM theory, the common causes of hypertension are liver-yang hyperactivity type (Caused by liver problems) and qi stagnation and blood stasis type (Caused by obstruction of blood vessels). After suffering from high blood pressure, it may also damage the heart, kidneys and cerebral blood vessels. Combining the theory that different pulse diagnosis positions are related to different human organs, this study constructed the feature subset of left-hand three-channel (representing the heart, liver, and kidney), the feature subset of right-hand three-channel (representing the lung, spleen, and mingmen), and feature subset of the two-hand six-channel.Afterwards, the classification model is trained. The experimental results are shown in Table 3, the classification model constructed according to the left-hand three-channel is more suitable for the actual pulse objective diagnosis. At the same time, the theory that the position of pulse diagnosis in traditional Chinese medicine is related to human organs has also been verified in the field of hypertension.

3.3   Model Comparison of Single-Channel Features

This study also analyzes the classification performance of the single-position model.

According to Table 4, the model trained with features of the Cun of left hand performs better, and the Precision, Specificity and other indicators are the best. F1-Score also reaches 79%, which is also the highest score among the models trained by the above feature subsets.

Table 4. Performance indicators of single-channel classification models.

	Accuracy	Precision	Recall	Specificity	F1-score
LCun	**86.4**	**69.1**	92.3	**89.6**	**79**
LGuan	85.2	53.5	96.1	85.6	68.7
LChi	85.1	49.3	**97.4**	84.8	65.5
RCun	84.1	64.2	90.9	88	75.3
RGuan	80.9	56.1	89.4	85.5	68.9
RChi	81.2	40.4	95.3	82.2	56.7

The results show that Cun of left hand has a better effect in single-channel classification, which is consistent with the theory of Cun of left hand corresponding to the heart in TCM.

4 Conclusion

The standardization of pulse signal acquisition location and feature selection is an important research direction for the objectification of TCM. Some previous research did not consider the concept of TCM.

In this study, the NCA-based feature selection method is used to select the six-channel pulse feature, and the feature subsets of single-channel, single-handed three-channel, and six-channel are constructed, and then the classification model is trained and the performance of the model is analyzed. Extensive experiments show that this feature selection method is effective in multi-channel pulse selection. In the hypertension/health classification, the F1-Score of the six-channel and left-hand three-channel features-trained models reaches 92.3% and 91.4%, respectively. The number of channels in the left-hand three-channel selection is less, and the trained model also has a good classification effect, which is suitable for actual diagnosis scenarios. Future work will continue to in-depth study of the combination of pulse diagnosis and computer technology.

References

1. Xu, L., Wang, K., et al.: Objectifying researches on traditional Chinese pulse diagnosis. Inform. Med. Sloven. **8**(1), 56–63 (2003)
2. Cheng, Fafeng, Wang, Xueqian, et al.: Biologic basis of TCM syndromes and the standardization of syndrome classification. J. Trad. Chinese Med. Sci. **02**, 92–97 (2014)
3. Tyan, C.C., Liang, W.M., Shy, H.Y., et al.: How to standardize 3 finger positions of examiner for palpating radial pulses at wrist in traditional Chinese medicine. Acup. Electro-Therap. Res. **32**(1–2), 87 (2007)
4. Lee, J.Y.: The study on the intellectual analysis algorithm for oriental pulse parameters. J. Med. Syst. **31**(5), 345–349 (2007)
5. Xu, L.S., et al.: Baseline wander correction in pulse waveforms using wavelet-based cascaded adaptive filter. Comput. Biol. Med. **37**(5), 716–731 (2007)
6. Gayathri Devi, K., Balasubramanian, K., Anh Ngoc, L.: Machine Learning and Deep Learning Techniques for Medical Science. CRC Press (2022)
7. James, Z., Londa, S.: Ensuring that biomedical AI benefits diverse populations. EBioMedicine. 103358 (2021)
8. Rangaprakash, D., Narayana Dutt, D.: Study of wrist pulse signals using time domain spatial features. Comp. Electr. Eng. **45**, 100-107 (2015)
9. Lisheng, X., et al.: Morphology variability analysis of wrist pulse waveform for assessment of arteriosclerosis status. J. Med. Syst. **34**(3), 331-9 (2010)
10. Wang, D., Zhang, D., Lu, G.: A robust signal preprocessing framework for wrist pulse analysis. Biomed. Sign. Process. Contr. **23** (2016)

A Dissolving P System for Multi-objective Gene Combination Selection from Micro-array Data

Fan Liu[1,2], Shouheng Tuo[1,2(✉)], and Chao Li[1,2]

[1] School of Computer Science and Technology, Xi'an University of Posts and
Telecommunications, Xi'an, Shaanxi 710121, China
{l_nlxx,lichao}@stu.xupt.edu.cn, tuo_sh@xupt.edu.cn
[2] Shaanxi Key Laboratory of Network Data Analysis and Intelligent Processing, Xi'an,
Shaanxi 710121, China

Abstract. Currently, cancer is increasingly becoming one of the major threats to human health. It is important to precisely identify gene combinations associated with cancers. Therefore, a dissolving P system based on membrane computing theory is proposed in this study to detect effective gene combinations related to cancer from high-dimensional micro-array data. We employ three classification functions with penalty terms to form a multi-objective optimization problem to further guide the gene combination selection process. The proposed P system is capable of helping to balance the global exploration and local exploitation abilities, and the penalty term is used to screen out redundant genes by reducing the number of genes automatically in each selected gene combination. Taking two micro-array datasets of prostate cancer as examples, the experimental results show that our method can identify lower-order gene combinations with higher classification accuracy than existing algorithms. We further visualize the classification effect on two datasets comprehensively and validate the selected representative genes biologically. The proposed method successfully achieves the gene combination selection task. Moreover, we comprehensively investigate the performance of our method on three other benchmark datasets and obtain satisfactory results.

Keywords: Gene combinations · Dissolving P system · Penalty term · Micro-array data

1 Introduction

Cancer is a major public health problem worldwide [1], and many scholars have recently studied cancer. Existing studies have shown that cancer is caused by genetic changes that intricately influence cellular fitness [2], and only a small number of genes can be efficient biomarkers for a specific disease, such as cancer [3, 4]. Actually, it is a combination of gene biomarkers that can be useful for cancer diagnosis and treatment [4]. We can detect informative gene combinations with the help of gene expression micro-array data, which are formed by small samples available but with tens of thousands of genes consisting of redundant or irrelevant genes in cancer [5]. Feature selection technology is widely accepted to select the gene combination subset in micro-array data for cancer diagnosis and treatment [6]. The gene locus is analogized as a feature, and the process of gene combination selection is handled with an optimal problem to obtain high classification accuracy while irrelevant or redundant genes are abandoned as much as possible [3, 5]. The application of feature selection technology to gene expression data is accompanied by two main challenges. First, detection on micro-array data can be seen as a "large p, small n" problem, where p and n represent the number of gene loci and samples, respectively. It is easy to be prone to overfitting [7], which inevitably influences the classification accuracy precisely of the selected gene combinations consisting of many non-representative genes. Second, the high-dimensional micro-array data make it difficult to select the most informative gene combinations for cancer diagnosis and treatment. Ideally, the proposed optimal algorithm should explore the promising regions as extensively as possible while exploiting a satisfactory solution in a precise area [6, 7]. In short, the proposed algorithm should balance the capability of exploration and exploitation.

2 Related Work

The gene selection problem is always regarded as a single objective optimization problem with a constraint with a restricted number of genes to detect informative genes with high classification accuracy or a multi-objective optimization problem by identifying biomarker genes with high classification accuracy but with a few genes [8, 9]. The swarm intelligence (SI) algorithm is widely applied for identifying functional gene combination selection from micro-array data, such as particle swarm optimization (PSO) and the whale optimization algorithm (WOA) [11]. Almost all methods employ a single classifier only during the whole optimization process, similar to SVM and KNN. The SI algorithm has made huge progress in gene combination selection tasks from micro-array data.

However, the gene selection approach from high-dimensional micro-array data has much room for improvement. On the one hand, most of the existing algorithms are devoted to identifying the representative genes only [9–11], but an efficient combination of biomarkers can make a real difference in cancer diagnosis and treatment [4]. It is necessary to detect effective gene combinations related to cancer from high-dimensional micro-array data. On the other hand, the interpretation of final obtained gene combinations is intrinsically significant for cancer diagnosis and treatment [11]. The gene combination selection methods should perform so well that optimal gene combinations can clearly distinguish normal and diseased samples.

Membrane computing, as well as the P system, is a natural computing method that was first introduced by Gheorghe Păun in 1998 [12] and named the P system. Since the introduction of membrane computing, it has received much attention because of the traits of distributed and parallel computing [13] and has been employed successfully in many problems, especially some complex optimization problems [14]. Applying membrane computing reasonably is able to identify the gene combinations that can clearly distinguish the normal and diseased samples of corresponding cancer. By designing a unique P system, we can comprehensively address the gene combination problem in high-dimensional micro-array data.

3 Proposed Method

3.1 Multi-objective Function

In our study, the gene combination selection task in high-dimensional micro-array data is regarded as a multi-objective optimal problem. The objective function is defined as in (1).

$$\underset{\mathbf{x}}{\text{maximum }} \mathbf{F}(\mathbf{x}) = \left[f_1(\mathbf{x}), f_2(\mathbf{x}), f_3(\mathbf{x}) \right] \tag{1}$$

$$\text{s.t.} \begin{cases} f_1(\mathbf{x}) = f_{\text{SVM}}(\mathbf{x}) - \lambda \bullet |\mathbf{x}_l| / |\mathbf{x}| \\ f_2(\mathbf{x}) = f_{\text{KNN}}(\mathbf{x}) - \lambda \bullet |\mathbf{x}_l| / |\mathbf{x}| \\ f_3(\mathbf{x}) = f_{\text{RF}}(\mathbf{x}) - \lambda \bullet |\mathbf{x}_l| / |\mathbf{x}| \end{cases}$$

$\mathbf{x}$ represents a selected gene combination and can be expressed as $x = \{x_1, x_2, \cdots, x_D\}$, $x_i \in \{1, 2, \cdots, N\}$, where D is the maximum number of gene probes contained in a gene combination and N represents the total number of gene loci of the filtered data. $\{f_{\text{SVM}}(\mathbf{x}), f_{\text{KNN}}(\mathbf{x}), f_{\text{RF}}(\mathbf{x})\}$ represents the classification accuracy calculated by three common classifiers: support vector machine (SVM) [15], K-nearest neighbor (KNN) [16] and random forest (RF) [17]. $-\lambda \bullet |\mathbf{x}_l| / |\mathbf{x}|$ is a penalty term added to automatically reduce redundant genes. λ is a penalty factor set in [0,1]. $|\mathbf{x}_l|$ and $|\mathbf{x}|$ are the length of a gene combination after deduplication and before deduplication, respectively. During the optimization process, we comprehensively identified gene combinations with a maximum value of $\{f_1(\mathbf{x}), f_2(\mathbf{x}), f_3(\mathbf{x})\}$. Correspondingly, the fitness value of $\mathbf{F}(\mathbf{x})$ is a vector matrix containing three elements.

3.2 A Dissolving P System

Membrane computing is formed by three leading materials: the structure, objects and evolution rules [18]. The proposed P system contains a dissolvable membrane structure, which means that our membrane structure is changed because of membrane dissolution. Before membrane dissolution, the membrane structure is shown on the left of Fig. 1, and the structure on the right depicts the membrane structure after membrane dissolution. The numbers marked in Fig. 1 are the corresponding film numbers. There were six membranes in total before membrane dissolution. As shown in Fig. 1, the membrane structure before membrane dissolution is formed of three elementary membranes, membrane "1", membrane "2" and membrane "3", which lie in the innermost layer; two non-elementary membranes, membrane "4" and membrane "5", which are placed in the second and third layers, respectively; and a skin membrane in the outermost layer named membrane "6". The membranes are hierarchically arranged from the outside to the inside, as well as from the upper order layer to the lower order layer.

The membranes are independent and can perform in parallel in the proposed P system. Based on the aforementioned membrane structure, a unique membrane operation mechanism, such as objects and evolution rules, was designed to accomplish the multi-objective optimization task in high-dimensional micro-array data. All membranes separately contain their own harmony memories, which are regarded as objects. The evolutionary operation rules are specifically shown in Fig. 2.

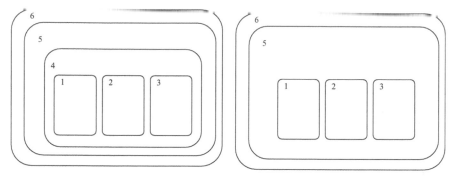

Fig. 1. The membrane structure of the proposed P system.

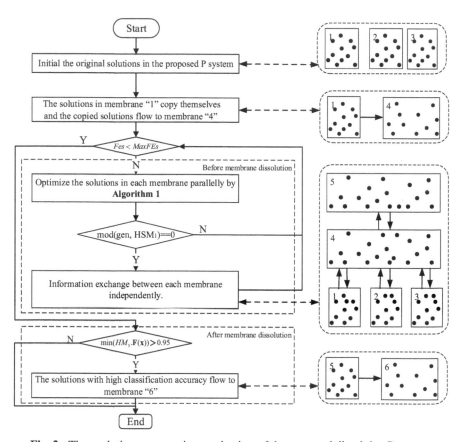

Fig. 2. The evolutionary operation mechanism of the proposed dissolving P system.

As shown in Fig. 2, information exchange between membranes occurs in different layers. Information exchange obeys a manner in which the membrane placed in the lower-order layer submits its best solution to the membrane contained in the upper-order layer, while the membrane in the upper-order layer delivers its worst solution to the lower-order layer. The best solution in a membrane is the one who dominates other memories in the corresponding harmony memory, and the worst solution is the one who is dominated by other solutions contrarily. The timing of the optimization process is prepared on the left of the dotted arrows, and the involved membranes are given on the right of the dotted arrows. *gen* and HSM_1 represent the evolutionary generation of the proposed P system and the capacity of harmony memories in three elementary membranes before membrane dissolution. When iteration *Fes* reaches the maximum number of iterations *MaxFEs*, the membrane evolution before membrane dissolution stops. After membrane dissolution, the membrane structure is changed as shown on the right side of Fig. 1. As shown in Fig. 2, according to the fitness value in membrane "5", $HM_5.\mathbf{F}(\mathbf{x})$, we defined that the solutions with high fitness value flow to membrane "6" and the flowed solutions are the optimal solutions we want.

In addition, the algorithm embedded in each membrane is a harmony search algorithm, which is a heuristic algorithm created by simulating the process of which the musicians constantly adjust the pitch and the orchestra to achieve a harmonic state during the performance [19]. **Algorithm 1** expresses the process of the harmony search algorithm in detail.

Algorithm 1: Harmony search algorithm

Input: M, HMS, $HMCR$, PAR, BW, F

Output: The updated HM

1:   **For** i ←1 to D
2:    **If** $rand(0, 1) < HMCR$
3:     a ←$ceil(rand(0, 1) \times D)$;
4:     $\mathbf{xnew}(i)$←$\mathbf{x}(ceil(rand(0, 1) \times HMS), a)$;
5:     **If** $rand(0, 1) < PAR$
6:      b←$BW \times (rand(0, 1) - 0.5)$;
7:      $\mathbf{xnew}(i)$←$round(\mathbf{xnew}(i) + b)$;
8:     **EndIf**
9:    **Else**
10:     Randomly produce a new solution as $\mathbf{xnew}$;
11:    **EndIf**
12:   **EndFor**
13: **If** $f_i(\mathbf{xnew}) > f_i(\mathbf{x})$ for $i= 1, 2, 3$.
14:   $\mathbf{x} = \mathbf{xnew}$;
15:   $\mathbf{F}(\mathbf{x}) = \mathbf{F}(\mathbf{xnew})$;
16: **EndIf**

The input of **Algorithm 2** includes the corresponding harmony memory (M) and the basic parameters of the harmony search algorithm, such as the size of the harmony memory (HMS), the harmony memory considering rate ($HMCR$), the pitch adjusting rate (PAR), the bandwidth (BW) and the objective function (F). Similarly, we replace the original harmony with the new harmony ($\mathbf{xnew}$) unless all values of the existing subobjective functions of $\mathbf{xnew}$ are correspondingly greater than the values of the original harmony, which is shown in line 13 of **Algorithm 1**. The complexity of our method is $O(D \times MaxFEs)$.

4 Experiments

4.1 The Efficiency on Prostate Cancer

In this work, we set D and λ as 5 and 0.1, respectively, and the parameters of the harmony search algorithm, such as *HMCR,PAR*, and *BW* are given as 0.98, 0.35, and 10. Three employed classifiers calculate the corresponding classification accuracy with 10-fold cross validation to obtain a believable gene combination subset. To ensure the consistency of gene combination selection in high-dimensional data, we provide the results after 25 runs for gene expression and DNA methylation data of prostate cancer. Taking the gene expression and DNA methylation data as examples, we compare the performance of our algorithm with four state-of-the-art algorithms, QSFS [20], RMA [21], WAO-CM [22] and ME-BPSO [23]. As shown in Table 1, our method has the best performance. Not only did we obtain optimal gene combinations containing a minimum number of genes but also with high classification accuracy.

Table 1. The classification accuracy and number of selected genes in the two datasets.

Methods	Gene expression data (8442 loci)		DNA methylation data (191789 loci)	
	Accuracy (%)	Number of genes	Accuracy (%)	Number of genes
RMA	**100±0.00**	7.40±0.76	99.93±0.00	7.60±0.50
WAO-CM	99.55±0.82	59.12±108.37	98.33±1.41	145.28±126.98
ME-BPSO	98.08±0.37	480.32±39.58	96.66±0.95	477.16±41.98
QSFS	**100±0.00**	4.24±1.39	**100±0.00**	4.70±1.42
MCHS	**100±0.00**	**3.66±0.47**	**100±0.00**	**3.67±0.53**

Bold indicates proposed method and the best results obtained for each column of data

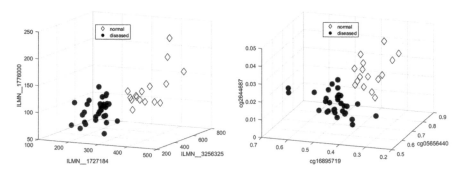

Fig. 3. The classification effect of selected gene combinations.

In Table 1, the most satisfactory effects on classification accuracy and number of genes are bolded. To further verify the correctness of our proposed algorithm, we visualize the classification effects of gene expression and DNA methylation data on the left and right sides of Fig. 3, respectively. It is obvious that the proposed method performs successfully in distinguishing the normal and diseased samples in both datasets.

In the experiment, the representative genes we found are summarized in Table 2, and all of them can be evaluated by Gene Ontologies (GO) terms and Kyoto Encyclopedia of Genes and Genomes (KEGG) pathways.

Table 2. Representative gene biomarkers.

Corresponding data	Probe ID	Gene name
Gene expression data	ILMN_1727184	WDR36 [24]
DNA methylation data	Cg16895719	EBF2 [25]
Both datasets	ILMN_1760153 cg08015883	GATA5 [26]

Bold indicates proposed method and the best results obtained for each column of data

4.2 The Application on Other Datasets

In this work, we also successfully apply our method to three other benchmark datasets about childhood tumors, ovarian cancer and leukemia. The details of the original datasets and the results after the gene combination selection process are given in Table 3.

Table 3. The details of the other three micro-array datasets.

Dataset	Features	No. selected features	Accuracy (%)
Childhood tumors	9946	4	100
Ovarian	15155	4	100
Leukemia	5148	4	100

Bold indicates proposed method and the best results obtained for each column of data

5 Conclusion

The dissolvable P system can be well integrated with intelligent search algorithms to further balance the global exploration and local exploitation ability. Our method can identify representative biomarker genes with small numbers but high classification accuracy from two datasets of prostate cancer. The selected gene combination can distinguish normal and diseased samples successfully and can be validated from a biological perspective. Moreover, our algorithm obtains satisfactory performance on the other three benchmark datasets. In short, the proposed P system is an efficient method to solve gene combination selection tasks from micro-array data.

References

1. Bray, F., Ferlay, J., Soerjomataram, I.: Global cancer statistics 2018: GLOBOCAN estimates of incidence and mortality worldwide for 36 cancers in 185 countries. CA Cancer J. Clin. **68**(6), 394–424 (2018)

2. Tollis, M., Schneider-Utaka, A.K., Maley, C.: The evolution of human cancer gene duplications across mammals. Mol. Biol. Evol. **37**(10), 2875–2886 (2020)
3. Gakii, C., Rimiru, R.: Identification of cancer related genes using feature selection and association rule mining. Inf. Med. Unlocked **24**, 100595 (2021)
4. Wang, L., Lipo, Y., Qing, C.: Feature selection methods for big data bioinformatics: a survey from the search perspective. Methods **111**, 21–31 (2016)
5. Lee, J., Choi, I.Y.: An efficient multivariate feature ranking method for gene selection in high-dimensional micro-array data. Expert Syst. Appl. **166**, 113971 (2021)
6. Wang, L., Chu, F., Xie, W.: Accurate cancer classification using expressions of very few genes. IEEE-ACM Trans. Bioinform. Comput. Biol. **4**(1), 40–53 (2007)
7. Liu, S., Wang, H., Peng, W.: A surrogate-assisted evolutionary feature selection algorithm with parallel random grouping for high-dimensional classification. In: IEEE Transactions on Evolutionary Computation (2022)
8. Zhang, G., Peng, Z., Li, X.: TABBA: a novel feature selection method based on binary bat algorithm and t test. In: 2021 IEEE 6th International Conference on Cloud Computing and Big Data Analytics (ICCCBDA), pp. 157–160. IEEE, Chengdu, China (2021)
9. Sayed, S., Nassef, M., Badr, A.: A nested genetic algorithm for feature selection in high-dimensional cancer micro-array datasets. Expert Syst. Appl. **121**, 233–243 (2019)
10. Bermingham, M.L., Pong-Wong, R., Spiliopoulou, A.: Application of high-dimensional feature selection: evaluation for genomic prediction in man. Sci. Rep. **5**(1), 1–12 (2015)
11. Păun, G.: Computing with membranes. J. Comput. Syst. Sci. **61**(1), 108–143 (2000)
12. Zhang, G., Shang, Z., Verlan, S.: An overview of hardware implementation of membrane computing models. ACM Comput. Surv. (CSUR) **53**(4), 1–38 (2020)
13. Elkhani, N., Muniyandi, R.C.: Membrane computing to model feature selection of micro-array cancer data. In: Proceedings of the ASE BigData & SocialInformatics, pp. 1–9. ACM Kaohsiung, Taiwan (2015)
14. Irizarry, R.A., Bolstad, B.M., Collin, F.: Summaries of affymetrix genechip probe level data. Nucleic Acids Res. **31**(4), e15–e15 (2003)
15. El-Naqa, I., Yang, Y., Wernick, M.N.: A support vector machine approach for detection of microcalcifications. IEEE Trans. Med. Imaging **21**(12), 1552–1563 (2002)
16. Cover, T.M., Hart, P.E.: Nearest neighbor pattern classification. IEEE Trans. Inf. Theory **13**, 21–27 (1967)
17. Breiman, L.: Bagging predictors. Mach. Learn. **24**(2), 123–140 (1996)
18. Paun, G.: Membrane computing. Scholarpedia **5**(1), 9259 (2010)
19. Tseng, V.S., Kao, C.-P.: Efficiently mining gene expression data via a novel parameter less clustering method. IEEE-ACM Trans. Comput. Biol. Bioinf. **2**, 355–365 (2005)
20. Ghosh, M., Sen, S., Sarkar, R., et al.: Quantum squirrel inspired algorithm for gene selection in methylation and expression data of prostate cancer. Appl. Soft Comput. **105**, 107221 (2021)
21. Ghosh, M., Begum, S., Sarkar, R.: Recursive memetic algorithm for gene selection in micro-array data. Expert Syst. Appl. **116**, 172–185 (2019)
22. Mafarja, M., Mirjalili, S.: Whale optimization approaches for wrapper feature selection. Appl. Soft Comput. **62**, 441–453 (2018)
23. Wei, J., Zhang, R., Yu, Z.: A BPSO-SVM algorithm based on memory renewal and enhanced mutation mechanisms for feature selection. Appl. Soft Comput. **58**, 176–192 (2017)
24. Huang, Y.H., Zhang, C.Z., Huang, Q.S.: Clinicopathologic features, tumor immune microenvironment and genomic landscape of epstein-barr virus-associated intrahepatic cholangiocarcinoma. J. Hepatol. **74**(4), 838–849 (2021)
25. Nikitina, A.S., Sharova, E.I., Danilenko, S.A., et al.: Novel RNA biomarkers of prostate cancer revealed by RNA-seq analysis of formalin-fixed samples obtained from Russian patients. Oncotarget **8**(20), 32990 (2017)
26. Zhang, J., Gao, K., Xie, H.: SPOP mutation induces DNA methylation via stabilizing GLP/G9a. Nat. Commun. **12**(1), 1–17 (2021)

Design and Implementation of Scalable Power Load Forecasting System Based on Neural Networks

Shu Huang[1](✉), Ze-san Liu[1], Hong-min Meng[1], Zhe-nan Xu[1], Ai-jun Wen[1], Shan Li[1], Di Liu[1], and Ge Ding[2]

[1] State Grid Information and Communication Industry Group Co., Beijing 100085, China
{huangshu,liuzesan,menghongmen,xuzhenan,wenaijun,lishan, liudi}@sgitg.sgcc.com.cn
[2] Beijing Sunwise Information Technology Co., Ltd, Beijing 100085, China
dingge@sunwiseinfo.com

Abstract. With the continuous advancement of science and technology, artificial intelligence has gradually been applied in various fields, and forecasting power and load with artificial intelligence technology is also widely used in the field of electric grids. At present, the mainstream load forecasting system mainly includes data collection, data processing, and forecasting using pre-trained models. Although in this way can complete the tasks of load forecasting, because the neural network model is pre-trained, it requires a lot of historical data, which cannot be satisfied in some scenarios. At the same time, the model structure cannot be dynamically modified and optimized after the training process and deployment is completed, and for the geographical expansion scenario, a model cannot adapt to the different climate, cultural and other factors in all regions, and must ask the system provider to retrain a new model. To solve the above problems, this paper proposes an adjustable load forecasting system based on artificial intelligence. With various ways of accessing user's data, users can customize a series of dimensions of the training process such as the type, scale, input, and output of the neural network model with their own data, so that they can dynamically adjusting the load forecasting model.

Keywords: Load forecasting · AI · *LSTM*

1 Introduction

With the continuous development of science and technology, the gradual improvement of hardware performance, especially the substantial improvement of GPU performance [1], provides a strong support for the application of artificial intelligence technology, allowing artificial intelligence to be used in image recognition [2], classification, natural language processing [3], and meteorological data prediction, autonomous driving [4] and other fields have been widely used. Thanks to the natural advantages of LSTM (long short-term memory) proposed by X Shi, Z Chen [5] in processing time series data,

N. Xiong et al. (Eds.): ICNC-FSKD 2022, LNDECT 153, pp. 377–385, 2023.
https://doi.org/10.1007/978-3-031-20738-9_43

the use of neural network models for power and load forecasting in the electric field has gradually become a Mainstream [6–8]. Compared the old way of forecasting future loads with previous of historical data, artificial intelligence-based load forecasting has the advantages of high accuracy, strong fault tolerance, and outstanding adaptability.

At present, there are many excellent systems that integrate neural network models for load forecasting, but most of them just integrate the trained models into the system, and only need to provide the input data required by the models to obtain the forecasting results. In this way, the training process is invisible, and the structure and parameters of the model are immutable, which greatly reduces the flexibility of the system. Due to the solidification of the structure and parameters of the model, it cannot be dynamically modified and optimized, so that the model not being able to self-adapt during use. When the historical data is insufficient and gradually accumulated, the model structure needs to be iterated frequently, or when the usage scenarios changes, such as area changes, climate changes, or even humanities changes, the prediction results of the model will become unreliable. Users can only turn to the system developers to retrain the network model according to the changes. It needs to be reintegrated into the system after training, which is a complex and tedious process. Load and power forecasting is the basis of a new power system, if the forecast results cannot be obtained normally and accurately, it will have a huge impact on the availability, reliability and timeliness of the entire electric system. In order to solve the above problems, this paper proposes a dynamic and adjustable power and load forecasting system. Through full-life-cycle data stream processing from bottom-level terminal data collection, data analysis and processing to neural network model training and deployment, users can independently train, iterate and re-apply models in the system. And achieve the purpose that the neural network model can be freely changed, optimized and deployed, and solve the problem of model solidification and inflexibility of the system. The test of the pilot project proves that the system proposed in this paper has high availability, can dynamically optimize the model, improve the prediction accuracy, and reduce the later maintenance cost of the system.

2 Background

Power and load forecasting is one of the important tasks of power system dispatching, power consumption, planning and other management departments. Improving the level of load forecasting technology is conducive to plan the management of power consumption, the rational arrangement of the power grid operation mode and the unit maintenance plan, the saving of coal, oil and the reduction of power generation costs, the formulation of reasonable power supply construction plans, and the improvement of electric power. Improving the level of load forecasting technology is conducive to power management, and is conducive to rationally arranging power grid operation modes and unit maintenance plans, and is conducive to saving coal, fuel and power generation costs, and is conducive to formulating reasonable power supply construction plans. At present, power and load forecasting technology can be divided into traditional technology and artificial intelligence-based forecasting technology:

1. Traditional techniques mainly use mathematical principles, such as nonlinear programming, genetic algorithm [9], least squares, regression analysis, etc. These mathematical methods can simulate and predict future loads according to historical load variation characteristics to a certain extent, but there are certain limitations.
2. The forecast technology based on artificial intelligence is an emerging technology in recent years. It simulates future load changes through the powerful computing capability of the hardware platform and historical data, mainly including algorithms such as *SVM*, *LSTM*, CNN [10], and ANN [11]. Practice has proved that load forecasting based on artificial intelligence can achieve better results than traditional way.

Most of the current load forecasting systems tend to use artificial intelligence-based forecasting techniques to achieve better forecasting effects, such as SVM used by Aqeel S. Jaber [12] in Short Term Load Forecasting, and power forecasting under the big data platform designed by B Xia [13], etc. The models used in these systems have all been trained in advance, and the main process is shown in Fig. 1.

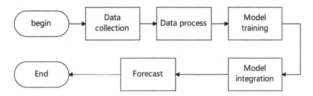

Fig. 1. The main process of model training in traditional load forecasting system.

3   System Design

In order to solve the above problems, this paper designs and implements an adjustable load forecasting system based on artificial intelligence. The system accesses the user's data, isolates the user's information, and allows the user to customize the training of their own models, so as to achieve the purpose of a flexible, scalable, and adaptable load forecasting system, and solve the problem that the user cannot quickly and easily expand their models. It mainly includes functional modules such as data import and processing, model training, model configuration, timing forecasting, and model evaluation.

3.1   Technology Architecture

The technical architecture of the system is shown in the Fig. 2.

Data process layer mainly uses Flink, ZigBee, Hive, Spark, Kettle and other technologies to process the data collected by the terminal device.

Data Store Layer uses a variety of technologies such as Mysql, Hbase, Opentsdb, Redis, Minio, etc. to store structured data, big data, and unstructured data respectively.

Technology layer is divided into two parts: java and python. Java mainly supports data reception and preprocessing of modules such as data import, data analysis, data processing, model selection and training, timing forecasting and accuracy evaluation. And realize timing forecasting through XXL-Job as the timing task scheduling center. Python provides API services such as data analysis, outlier repair, resampling, data analysis, load forecasting, filtering and noise reduction, model training and saving for the Java part in the form of web service through the FastAPI [14] framework. The data processing related parts mainly use the pandas toolkit, and the model training and prediction mainly use the tensorflow [15] neural network framework.

Micro Services Layer is mainly implemented based on Spring Boot and Spring Cloud, and is driven by the lower-level background API to provide a service foundation for upper-level applications.

Micro Application Layer provides a visual interface for the system, and micro-applications provide services to connect the systems to provide users with application services with an interactive interface. Users can view the system interface through browsers such as IE, Chrome, and FireFox.

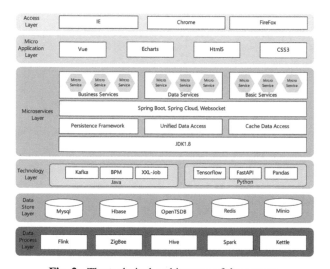

Fig. 2. The technical architecture of the system.

3.2 Functional Architecture

The system functional architecture is shown in the Fig. 3, which mainly includes several important modules such as data collection, data storage, data processing, model training, parameter configuration, real-time and timing forecasting, result display, accuracy evaluation, and remote interface.

Data Collection module is mainly divided into self-collected data, third-party platform access data and other access data. Self-collected data is the system that collects the information of internal terminal equipment in the system through IOT technology, while third-party platform access data is reserved through reservations. The third-party platform access data in system is from third-party platforms such as weather platforms,

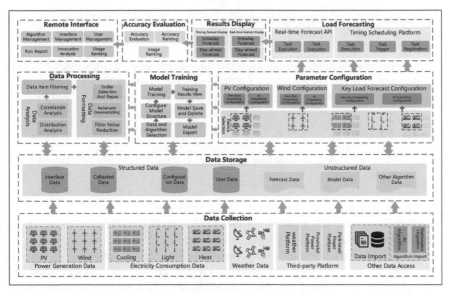

Fig. 3. The functional architecture of the system.

provincial power platforms or park-level power platforms. Other access data includes data imported through files and other algorithm models. The system can integrate a variety of algorithms to further expand the flexibility and availability.

Data Storage module provides the persistence function of various types of data, mainly including structured data (such as interface data, collection data, configuration data, user data, etc.) and unstructured data (such as forecast data, neural network model data, other algorithm data, etc.).

Data Processing module analyzes and processes the collected and imported data to provide a data basis for the training process. It mainly includes functions such as correlation analysis, distribution analysis, outlier screening and repair, resampling, downsampling, filtering and noise reduction.

Model Training is one of the core modules of this system. It provides a training function that can customize the model structure and parameters, allowing users to modify the model structure and parameters according to the actual situation. It mainly includes functions such as data and algorithm selection, model structure and training parameter configuration, model training, model result viewing, model saving and deletion.

Parameter Configuration module can configure different network models for different scenarios (wind power generation power, photovoltaic power generation power, and electricity load). Users can dynamically switch models for multiple scenarios to realize scene timing or real-time forecasting, and improve the scalability and flexibility of the system.

Power and Load Forecasting module mainly obtains relevant parameters through parameter configuration, calls the specified neural network model, and predicts power or load, mainly including real-time timing prediction.

Result Display and Accuracy Evaluation module mainly provides users with the functions of visual result display and comparison of prediction effects of different models.

Remote Interface module provides remote forecasting services for users who cannot deploy the system. By exposing relevant API interfaces, users can obtain prediction data in real time only by configuring parameters and transmitting prediction input data.

3.3 System Flow

The overall flow of the system is shown in the Fig. 4, which is mainly divided into seven steps:

1. As a multi-tenant system, the user first needs to log in.
2. Synchronize data to this system: upload in file form or synchronize import through system integration.
3. Perform data analysis on user data and view the relevant attributes of the data, including the data curve, mean, median, mode, variance, etc. At the same time, data cleaning can be performed on unsatisfactory data, mainly including outlier screening and repair (to remove erroneous or inappropriate data parts), resampling (to change data density and increase data density by interpolation), downsampling (reducing data density by averaging multiple points) and filtering noise reduction (for smoothing the original data, mainly using the sliding window smoothing method).
4. Model training: select the type of neural network, set the structure of the model (the number of layers, the type of network in each layer, the number of nodes in each layer, etc.), and set the relevant parameters during training (training times, step size, activation function, batch data, etc.) amount, learning rate, training set ratio, validation set ratio, test set ratio, etc.), train the model.
5. Model effect viewing and model saving: After the model is trained, the results of the test set and validation set during training will be output, which is used by the user to judge the effect of the model, and can choose whether to save the model.
6. Forecasting configuration and timing forecasting: Users can select the saved model for timing prediction configuration, bind the prediction input data required by the model with the existing data in the user system, and then realize timing prediction through the distributed timing task scheduling platform.
7. The system will evaluate the accuracy of all timing forecasting results configured by the user and present them on the user side, so that the user can select the most suitable model or use the model training module to freely optimize and update the model structure.

4 System Implement

According to the above system design in this paper, we have developed and implemented it. The B/S architecture is adopted, the front-end uses the VUE2.0 development framework, and ElementUI is used as the component library. The back-end development language is Java, the JDK version is JDK1.8, and microservices are built based

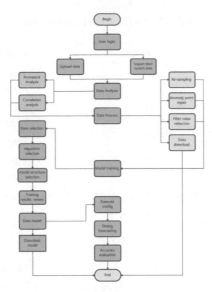

Fig. 4. System flow.

on SpringCloud scaffolding. Data processing and neural network-related logic use the TensorFlow1.13 framework and are developed using Python. For model training, we integrate the relatively mature LSTM algorithm, build a training framework with Tensorflow, and expose the model's layers, parameters, activation function, learning rate and other attributes during the training process to users, so that users can customize their operations. Table 1 is the detailed information of the LSTM models trained by the system. Figure 5 shows the effect verification of the models. The effect of model 1 is poor, so we obtained model 2 with better effect by modifying the parameters of model training. It can be seen that the model trained by the system custom parameters can be freely scaled to achieve the expected prediction effect.

5   Conclusion

In order to solve the problem that the network model of the traditional load forecasting system cannot be dynamically adjusted, resulting in poor system flexibility and low scalability, which affects power system scheduling, power consumption and planning, this paper designs and implements a load forecasting system that integrates neural network training, full life cycle data flow, and can arbitrarily change the network model. By integrating user data, the system provides training and deployment functions with custom network structures and parameters. Users can dynamically update and adjust network algorithms according to the on-site environment to improve system scalability and usability.

Table 1. LSTM model information

Parameter	Model1	Model1	Explanation
Layer	3	4	The number of layers in the model
Hidden	100	200	The number of hidden nodes in each layer
Steps	10	5	Step size during training
Batch size	200	200	The size of each batch of data during training
Input list	[Temp, Wind]	[Temp, Wind, humidity]	The input items of model
Output list	[Power]	[Power]	The output items of model
Train data ratio	0.8	0.8	The ratio of the training set to the data set
Test data ratio	0.1	0.1	The ratio of the testing set to the data set
Validate data ratio	0.1	0.1	The ratio of the validating set to the data set
Learning rate	0.02	0.02	The learning rate during training
Activation	Sigmoid	Sigmoid	The activation function of the model
Training iters	15000	30000	The number of training times

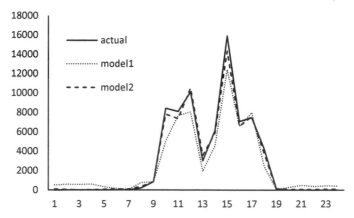

Fig. 5. Comparison of the forecasting results of photovoltaic power generation from the models trained by our system with the actual results.

References

1. Reddy Kuncham, G.K., Vaidya, R., Barve, M.: Performance study of GPU applications using SYCL and CUDA on Tesla V100 GPU. In: 2021 IEEE High Performance Extreme Computing Conference (HPEC), pp. 1–7 (2021). https://doi.org/10.1109/HPEC49654.2021.9622813.
2. Simonyan, K., Zisserman, A.: Very deep convolutional networks for large-scale image recognition. Computer Science (2014)
3. Ren, B.: The use of machine translation algorithm based on residual and LSTM neural network in translation teaching[. PLOS ONE **15** (2020)
4. Sohn, S.M., Yang, S.H., Park, H.M., et al.: SoC design of an auto-focus driving image signal processor for mobile camera applications. IEEE Trans. Consumer Elect. (2015)
5. Shi, X., Chen, Z., Wang, H., et al.: Convolutional LSTM network: a machine learning approach for precipitation nowcasting. MIT Press. (2015)
6. Hu, S., Zhao, Y., Tan, X., et al.: Saturated Load Forecast of Hebei Province Based on LSTM. IOP Conference Series Earth and Environmental Science **598**, 012014 (2020)
7. Guimaraes, T., Costa, L.M., Leite, H., et al.: A hybrid approach to load forecast at a micro grid level through machine learning algorithms. In: 2020 International Conference on Smart Energy Systems and Technologies (SEST) (2020)
8. Alavi, S.A., Mehran, K., Vahidinasab, V., et al.: Forecast based consensus control for DC microgrids using distributed long short-term memory deep learning models. IEEE Trans. Smart Grid **99**, 1 (2021)
9. Du, X., Feng, T., Tan, S.Q.: Study of power system short-term load forecast based on artificial neural network and genetic algorithm. In: International Conference on Computational Aspects of Social Networks. IEEE (2010)
10. Imani, M.: Electrical load-temperature CNN for residential load forecasting. Energy (2021)
11. Hua, X., Zhang, G., Yang, J., et al.: Theory study and application of the BP-ANN method for power grid short-term load forecasting. ZTE Comm. **3** (2022)
12. Jaber, A.S., Satar, K.A., Shalash, N.A.: Short term load forecasting for electrical dispatcher of Baghdad City based on SVM-PSO method. In: 2018 2nd International Conference on Electrical Engineering and Informatics (ICon EEI), pp. 140–143 (2018). https://doi.org/10.1109/ICon-EEI.2018.8784316
13. Xia, B., Yang, C., Zheng, K.: Design and implementation of power grid load forecasting system based on Big Data platform. Process Auto. Instrument. (2018)
14. FastAPI Homepage. https://fastapi.tiangolo.com/. Last accessed 2022/06/10
15. Abadi, M., Barham, P., Chen, J., et al.: TensorFlow: a system for large-scale machine learning. USENIX Association (2016)

Heave Compensation Sliding Mode Predictive Control Based on an Elman Neural Network for the Deep-Sea Crane

Zhimei Chen[✉], Yingbin Lu, Zhenyan Wang, Xuejuan Shao, and Jinggang Zhang

School of Electronic Information Engineering, Taiyuan University of Science and Technology, Taiyuan 030024, China
{1994003,2002047,1999017,1984032}@tyust.edu.cn,
S20190436@stu.tyust.edu.cn

Abstract. The heave movement may be generated with the interference of ocean waves, wind and other factors for a deep-sea crane, and it seriously affects the crane normal operation. A sliding mode predictive control method based on Elman neural network is proposed to improve the control accuracy of load displacement in the heave compensation system of a deep-sea crane. In order to improve the prediction accuracy, the sparrow search algorithm is used to optimize the network weights and thresholds and the beetle antennae search algorithm is adopted to optimize the sliding mode control law parameters. Simulation results prove that the proposed method has good control accuracy and robust performance.

Keywords: Heave compensation · Sparrow search algorithm · Elman neural network · Sliding mode predictive control · Beetle antennae search algorithm

1 Introduction

On the ocean, the hull of deep-sea crane is affected by ocean waves and ocean currents, which will produce complex free movement in six directions: heave, yaw, roll, horizontal movement, longitudinal movement and pitching, which seriously affect the normal operation of the deep-sea crane [1]. In order to eliminate the adverse effects of heave movement on the normal operation of deep-sea cranes, it is necessary to compensate the heave of deep-sea cranes.

Different control methods had been developed for the heave compensation of a deep-sea crane [2–4]. However, the control accuracy usually cannot satisfy the production demand with the PID and the traditional control methods. In order to improve the control accuracy of the heave compensation system of the deep-sea crane, a sliding mode predictive control method based on Elman neural network is proposed. The Elman neural network is adopted to construct the prediction model, and the sparrow search algorithm [5] was used to optimize the weights of the neural network. At the same time, the beetle antennae search algorithm [6] was utilized to optimize the control law to reduce the amount of calculation of rolling optimization.

© The Author(s), under exclusive license to Springer Nature Switzerland AG 2023
N. Xiong et al. (Eds.): ICNC-FSKD 2022, LNDECT 153, pp. 386–394, 2023.
https://doi.org/10.1007/978-3-031-20738-9_44

2   System Model

2.1   Deep Sea Crane Dynamics Model

The deep-sea crane heave compensation system is mainly composed of winch, composite hydraulic cylinder, electro-hydraulic proportional reversing valve and other components. Figure 1 is the heave compensation system schematic diagram [7].

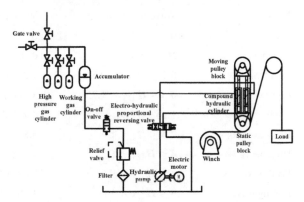

Fig. 1. The heave compensation system schematic diagram of the deep-sea crane

The dynamic model of the deep-sea crane heave compensation system is:

$$
y(s) = \frac{\frac{16k_s A_2 \beta_e \sqrt{P_s}}{V_t}}{s\left[(M_b + 16M_t)s^2 + \left(B_k + h_c A_1^2\right)s + \frac{4A_2^2 \beta_e}{V_t} + \frac{P_0 A_1^2}{V_0}\right]} \cdot I(s)
$$
$$
+ \left(\frac{(-4M_b - 16M_t)s^2}{(M_b + 16M_t)s^2 + \left(B_k + h_c A_1^2\right)s + \frac{4A_2^2 \beta_e}{V_t} + \frac{P_0 A_1^2}{V_0}} + 1\right) \cdot y_s(s) \qquad (1)
$$

where k_s is the proportional coefficient, A_1 and A_2 mean the area of the passive hydraulic cylinder and the active hydraulic cylinder separately, and β_e is the effective bulk elastic modulus. P_s is the oil pressure for oil supply, P_0 is the initial pressure inside the accumulator, M_t is the load mass, and M_b is the mass of the piston rod of the composite hydraulic cylinder. B_k is the hydraulic oil viscous damping coefficient, V_t is the internal volume of the active cylinder, and V_0 is the initial gas volume in the accumulator. $h_c = 128a\mu l/(\pi d^4)$. a is the number of the throttle valves, d is the diameter of the throttle valve orifice, and l is the length of the valve hole, μ is the oil viscosity coefficient.

In formula (1), $y(s)$ indicates the load displacement of the deep-sea crane, $y_s(s)$ represents the influence of the hull heave motion on the load displacement, which is the disturbance of the system. In order to compensate the influence of the hull heave motion on the load displacement, the heave compensation can be realized by regulating the input current $I(s)$ of the electro-hydraulic proportional reversing valve to control the expansion and contraction of the composite hydraulic cylinder.

2.2 Hull Heave Movement

The heave motion of the hull is mainly caused by ocean waves, and the numerical simulation of ocean waves is necessary. Table 1 shows the wave characteristics under different sea conditions.

Table 1. Characteristics of ocean waves

Sea state	Name	Wave height/m	Wave period/s
0	No wave	0	0
1	Micro wave	$H_{1/3} < 0.1$	0–1.2
2	Small wave	$0.1 \leq H_{1/3} < 0.5$	1.2–3.6
3	Light wave	$0.5 \leq H_{1/3} < 1.25$	3.6–4.7
4	Medium wave	$1.25 \leq H_{1/3} < 2.5$	4.7–5.8
5	Big wave	$2.5 \leq H_{1/3} < 4$	5.8–7.2
6	Huge wave	$4 \leq H_{1/3} < 6$	7.2–8.8
7	Wild wave	$6 \leq H_{1/3} < 9$	8.8–11.8
8	Raging wave	$9 \leq H_{1/3} < 14$	11.8–14.5
9	Angry wave	$14 \leq H_{1/3}$	>14.5

If ocean waves are regarded as the superposition of cosine waves with different initial phases and frequencies, the formula (2) is obtained.

$$\xi(x, t) = \sum_{i=1}^{N} a_i \cos(\omega_i t - \varepsilon_i) \tag{2}$$

where ξ is the wave height of a fixed position x of the wave relative to the stationary water surface, a_i is the amplitude of each cosine wave, ω_i is the angular velocity of each cosine wave, ε_i is the initial phase of each cosine wave.

In this paper, the PM spectrum is used as the ocean wave spectrum, and its spectral function is:

$$S(\omega) = \left(0.78/\omega^5\right) \exp\left[-3.12/\left(\omega^4 H_{1/3}^2\right)\right] \tag{3}$$

where $H_{1/3}$ stands for meaningful wave height, and the values of $H_{1/3}$ under different sea conditions can be seen in Table 1.

The heave motion of a ship can be regarded as a wave multiplied by a certain proportional coefficient. i.e.

$$y_s(t) = \mu y_h(t) \tag{4}$$

where $y_h(t)$ is the wave height relative to the still water surface, $y_s(t)$ indicate the heave displacement of the ship, and μ indicate the proportional coefficient between the heave motion of the ship and the waves.

3   Design of Sliding Mode Predictive Control System Based on SSA-Elman

3.1   Elman Neural Network

As a typical dynamic recurrent neural network, Elman neural network is generally divided into four layers: input layer, hidden layer, receiving layer and output layer [8]. Compared with the BP neural network, the added receiving layer is a one-step delay operator to achieve memory function. Elman neural network structure is shown in the Fig. 2.

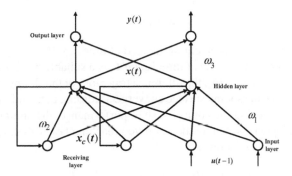

Fig. 2. Elman neural network structure diagram

3.2   Optimization of Elman Neural Network by Sparrow Search Algorithm

The sparrow search algorithm iteratively optimizes the sparrow through individual food search and anti-predation. Its advantages include less adjusting parameters, fast convergence, simple calculation and so on. In order to improve the prediction accuracy of the crane load displacement, here, the sparrow algorithm is used to optimize the initial weights and thresholds of the Elman neural network.

So, the SSA-Elman neural network prediction model is used to predict the load displacement of the deep-sea crane. From the deep-sea crane dynamics model, it can be known that the control input of the heave compensation system is the input current I (represented by u in this section) of the electro-hydraulic proportional reversing valve; the output is the load displacement y. In order to better control the system, it is necessary to carry out multi-step prediction of crane load displacement. In this paper, in order to perform P-step prediction, the trained SSA-Elman network needs to be used for recursive prediction of the crane load displacement. The prediction recursive formula can be expressed as:

$$\left.\begin{aligned}
y_n(k+1) &= g[y(k), \cdots, y(k-n+1), u(k), \cdots, u(k-m+1)] \\
y_n(k+2) &= g[y(k+1), \cdots, y(k-n+2), u(k+1), \cdots, u(k-m+2)] \\
y_n(k+P) &= g[y(k+P-1), \cdots, y(k+P-n), u(k+P-1), \cdots, u(k+P-m)]
\end{aligned}\right\}$$
$$(5)$$

where P is the network prediction step size, n is the output order of the prediction model, m is the input order of the prediction model, $y_n(k+1)$ is the predicted value of the crane load displacement at time $k+1$ by the network, $u(k)$, $y(k)$ are the input control and load displacement outputs of the crane at time k, respectively.

3.3 Sliding Mode Surface and Feedback Correction

The sliding mode surface are defined as:

$$\begin{cases} e(k) = y_a(k) - y(k) \\ s(k) = Ce(k) \end{cases} \tag{6}$$

where $y(k)$ is the actual displacement of the load at time k, $y_a(k) = y_r(k) - y_s(k)$ is the reference displacement of the load at time k, and C is a suitable dimensional matrix.

With the sliding surface in (6), the reference trajectory of the sliding surface needs to be established to make the sliding surface converge according to the reference trajectory. At the aiming of improving the control system performance, the sliding mode surface is designed and adopted combining the improved reaching law with the power function [9]. It can be expressed as follows:

$$\left. \begin{aligned} s_r(k+d) &= (1 - qT_s)s_r(k+d-1)\Phi(s_r(k+d-1)) \\ &\quad - \zeta T_s \Psi(s_r(k+d-1))f(s_r(k+d-1)) \\ s_r(k) &= s(k) \end{aligned} \right\} \tag{7}$$

$$\Phi(s_r(k)) = \begin{cases} 1, & |s_r(k)| > \eta \\ 0, & |s_r(k)| \le \eta \end{cases} \tag{8}$$

$$\Psi(s_r(k)) = \begin{cases} 1, & |s_r(k)| > \eta \\ \frac{|s_r(k)|^2}{\eta}, & |s_r(k)| \le \eta \end{cases} \tag{9}$$

where $s_r(k+d)$ is the expected value of the sliding surface at time $k+d$, $0 < \zeta T_s < 1$, $o < 1 - qT_s < 1$, $T_s > 0$, $\eta = \frac{\zeta T_s}{1-qT_s}$, $f(s_r(k))$ is a power function, which can be expressed as:

$$f(s, \alpha, \delta) = \begin{cases} |s|^\alpha \text{sign}(s) & |s| \ge \delta \\ \frac{s}{\delta^{1-\alpha}} & |s| < \delta \end{cases} \tag{10}$$

where $0 < \alpha < 1, 0 < \delta < 1$.

Considering the displacement prediction error will affect the system control accuracy, the correction at the time of prediction $k+p$ is made by introducing the error between the actual displacement and the predicted displacement of the crane load at time k:

$$y_p(k+i) = y_n(k+i) + h_i e(k) \tag{11}$$

where y_p is the predicted displacement of the corrected network, and h_i is the error correction coefficient.

3.4 Performance Index

Definition $S_p(k + 1) = [s_p(k+1), s_p(k+2), \ldots, s_p(k+P)]^T$, $S_r(k + 1) = [s_r(k+1), s_r(k+2), \ldots, s_r(k+P)]^T$ $Y_r(k + 1) = [y_r(k+1), y_r(k+2), \ldots, y_r(k+P)]^T$, $Y_n(k + 1) = [y_n(k+1), y_n(k+2), \ldots, y_n(k+P)]^T$, $U(k) = [u(k), u(k + 1), \ldots, u(k + P - 1)]^T$, $Q = diag[Q_1, Q_2, \ldots, Q_P]$.

where P is the prediction step, $S_p(k + 1)$ is the predicted value vector of sliding surface, $S_r(k + 1)$ is the reference value vector of sliding surface, $Y_r(k + 1)$ is the expected displacement vector of load, $Y_n(k + 1)$ is the predicted value vector of load displacement, $U(k)$ is the control law of the system at time k, and Q is the weighting coefficient matrix of tracking error.

In order to ensure that the system control system has good control performance, this paper designs the following performance indicators:

$$J = [S_p(k + 1) - S_r(k + 1)]^T Q [S_p(k + 1) - S_r(k + 1)] \tag{12}$$

3.5 Rolling Optimization of Beetle Antennae Search Algorithm

As a new bionic search algorithm, the beetle antennae search algorithm can effectively reduce the calculation amount of rolling optimization [10]. Let the position of the longicorn beetle m be x^m, and the fitness $f(x)$ of food smell at x is expressed by the performance index J.

In this paper, the performance index J is taken as the fitness function. Control the position of $U(k)$ as longicorn beetles. The control law is optimized by longicorn beetles algorithm, and the optimal control law of the system is obtained. Then the control output of the controller at time k is:

$$u(k) = [1, 0, \cdots, 0]U(k) \tag{13}$$

4 Simulation Results

To verify this control method at different levels under the waves heave compensation effect, take a load of 4 tons on the system at level 4, 5, and 6 sea condition of hull heave displacement with the crane load displacement after heave compensation simulation, get the system at different levels under the waves heave compensation effect, and with the literature [11] predictive control methods are compared, the simulation results are shown in Figs. 3, 4, 5, 6, 7 and 8.

As can be seen from Figs. 3, 4, 5, 6, 7 and 8, the heave compensation accuracy of the sliding mode predictive control method based on SSA-Elman neural network used in this paper is better than that of the literature method under the different level sea condition. In order to verify the effectiveness of the method, the heave compensation precisions are shown in Table 2.

It shows that the compensation accuracy with the SSA-Elman-SMPC compensation precision is higher than the literature method in different sea conditions, and the accuracy can reach more than 98%.

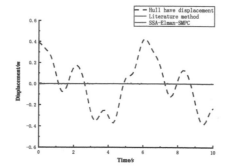

Fig. 3. Hull displacement and actual load displacement under grade 4 waves

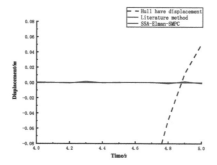

Fig. 4. Local ship displacement and actual load displacement under grade 4 waves

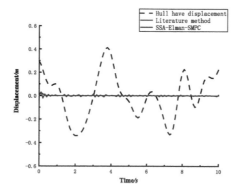

Fig. 5. Hull displacement and actual load displacement under grade 5 waves

4.1 Conclusions

In the proposed sliding mode predictive control method, the SSA-Elman neural network is used as the predictive model, and the longicorn beard algorithm is adopted to optimize the performance index. Simulation result show that the SSA-Elman neural network has higher prediction accuracy than the RBF neural network, and the sliding mode predictive

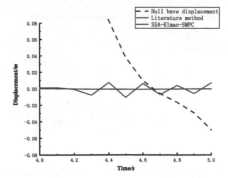

Fig. 6. Local ship displacement and actual load displacement under grade 5 waves

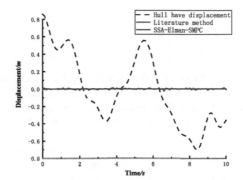

Fig. 7. Hull displacement and actual load displacement under grade 6 waves

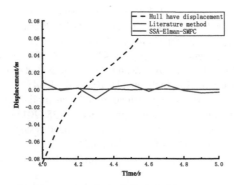

Fig. 8. Local ship displacement and actual load displacement under grade 6 waves

control method developed has higher compensation accuracy in the same sea condition and stronger robustness in different sea conditions.

Table 2. Heave compensation precisions

Wave level	MPC compensation precision/%	SSA-Elman-SMPC compensation precision/%
4	96.47	99.78
5	96.15	99.68
6	93.89	98.62

Acknowledgements. Project supported by the Shanxi Key Scientific Research Foundation (No. 202102020101013), the Shanxi Basic Research Foundation (No. 201901D111263), and the Doctor Foundation of Taiyuan University of Science and Technology (No. 20202070).

References

1. Wang, H.B., Wang, Q.F.: Design of underwater towing heave compensation system and its internal model robust control. J. Mech. Eng. **46**(08), 128–132 (2010)
2. Yang, J.M., Liu, L., Lv, H.N., et al.: Present situation and prospect of research and development of deep-sea mineral resources development equipment in China. Chinese Eng. Sci. **22**(6), 1–9 (2020)
3. Huang, L.M., Zhang, Y.T., Zhang, L., et al.: Design and simulation of offshore drilling active heave compensation draw works. Comp. Simul. **30**(11), 307–311 (2013)
4. Woodacre, J.K., Bauer, R.J., Irani, R.: Hydraulic valve-based active-heave compensation using a model-predictive controller with non-linear valve compensations. Ocean Eng. **152**, 47–56 (2018)
5. Li, Y.L., Wang, S.Q., Chen, Q.R., et al.: Comparative study of some new swarm intelligence optimization algorithms. Comp. Eng. Appl. **56**(22), 1–12 (2020)
6. Xu, D.X.: Improved longicorn beetle herd search algorithm and its application in ship pitching motion prediction. J. Guangdong Ocean Univ. **41**(03), 113–122 (2021)
7. Deng, Z.Y., Gao, J., Xie, J.H., et al.: Active and passive integrated heave compensation system and its control method. Ship Sci. Tech. **36**(11), 102–107 (2014)
8. Hou, Z.N., Wu, J.: Research on safety evaluation of lifting appliances based on Elman neural network. Safety technology of special equipment **04**, 33–35 (2016)
9. Wang, K.J., Zhang, X.X., Bai, L.N.: Networked control system based on sliding mode prediction. Elect. Measure. Tech. **44**(8), 76–81 (2021)
10. Jiang, X., Li, S.: BAS: Beetle antennae search algorithm for optimization problems. Inter. J. Robot. Cont. **1**(1), 1–5 (2018)
11. Yang, D.C.: Research on heave compensation control method of ship deck hoisting equipment. Huazhong University of Science and Technology, Wuhan (2018)

Multi-feature Fusion Flame Detection Algorithm Based on BP Neural Network

Jin Wu$^{(\boxtimes)}$, Ling Yang, Yaqiong Gao, and Zhaoqi Zhang

School of Electronic Engineering, Xi'an University of Posts and Telecommunications, Xi'an 710121, China

lifewujin@xupt.edu.cn, wujin1026@126.com, {yangling,gaoyaqiong, zhangzhaoqi}@stu.xupt.edu.cn

Abstract. In recent years, in order to ensure the safety of industrial boilers in production and improve the utilization rate of coal resources, a series of technical regulations on the detection of industrial boilers and related industrial emission regulations have been issued. In this paper, the traditional flame detection method has the problems of low accuracy, high failure rate and high maintenance cost caused by complicated detection equipment. A multi-feature fusion flame detection algorithm based on BP Neural Network is designed. For flame images with flickering characteristics, during the preprocessing of the data set, the principle of retaining more flame features is to use the sample matrix of four types of flame features, are used for training, and the proposed flame detection algorithm is applied to the actual flame sample test matrix to verify the timeliness of the algorithm proposed.

Keywords: Flame detection · Multi-feature fusion · BP neural network

1 Introduction

At present, large boilers are widely used in industrial environments. Therefore, furnace flame detection technology for boilers has always been a hot issue in the field of industrial automation and intelligence [1]. Boiler furnace flame detection methods can be generally classified into four types: sensor measurement methods [2], digital signal processing methods [3], image processing methods [4], and machine learning methods [5]. Now, the flame detection model based on deep learning can process more parameters, complete the fire detection work independently according to the demand, and can also ensure the timeliness and accuracy of the network while processing huge parameters [6]. At present, several types of deep learning network models that are widely used mainly include: Radial Basis Function (RBF) Neural Network [7], Convolutional Neural Network (CNN) [8–13], Recurrent Neural Network (RNN) [14], Back Propagation (BP) Neural Network. BP is short for Back Propagation Neural Network, and its network structure layer is composed of nonlinear transformation units, feedforward networks and some very simple, highly connected processing units. It is currently widely used in supervised learning algorithms for various decision classifications [15].

© The Author(s), under exclusive license to Springer Nature Switzerland AG 2023
N. Xiong et al. (Eds.): ICNC-FSKD 2022, LNDECT 153, pp. 395–401, 2023.
https://doi.org/10.1007/978-3-031-20738-9_45

2 BP Flame Detection Network Design

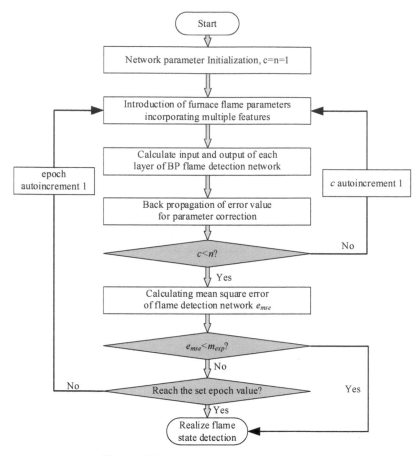

Fig. 1. BP flame detection network flow chart

The design flow of the BP flame detection algorithm is shown in Fig. 1. First, the network parameters need to be preset, c represents the number of flame samples that have been used, n is the total number of input flame samples, m_{exp} is the expected value of mean square error of the network, e_{mse} is the actual mean square error of the network, epoch means the number of rounds that the BP flame detection algorithm needs to iterate.

3 Network Training and Testing

As shown in Fig. 2 (a), when the sampling frequency is 9 Hz, that is, the sampling interval of the image is 1/9 s, the obtained 6 continuous flame images are constantly changing "on" and "off". The flame image data of 100% preserves the characteristics of flame

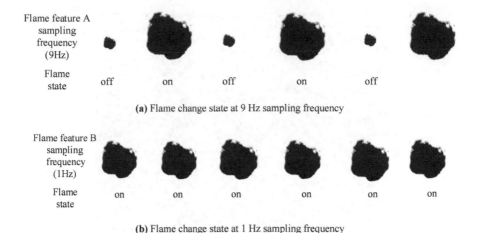

Flame feature A sampling frequency (9Hz)

Flame state   off   on   off   on   off

(a) Flame change state at 9 Hz sampling frequency

Flame feature B sampling frequency (1Hz)

Flame state   on   on   on   on   on   on

(b) Flame change state at 1 Hz sampling frequency

Fig. 2. Variation of flame samples at different sampling intervals

flickering in continuous video signals. Figure 2 (b) shows a group of flame changing states with a sampling frequency of 1 Hz. It can be clearly seen that the flame is always in a "on" state, which ignores the flame flickering factor in the furnace.

A total of 2000 sets of flame sample data are classified, of which 90% are used for the BP Neural Network training and 10% for the network testing, which becomes 1800 sets of training samples and 200 sets of test samples. The flame data distribution is shown in Fig. 3. It can be seen from the sample data diagram that after processing and transforming the original flame image data, the obtained flame sample is composed of a 4-dimensional array matrix with a size of 500×4. The flame data of the 4 dimensions respectively represent the 4 combustion states of the flame in the furnace.

Figure 4 (a), (c) and (e) are the training results of BP flame detection network designed in this paper on the flame dataset. It can be seen that when MSE size is different, the output value of the BP flame detection network is also significantly different from the target curve of the training sample. When MSE = 0.0001, the network has the highest training accuracy, and the network error at this time is close to 0.0001. From the three sub-graphs of Fig. 4 (b), (d), and (f), it can be found that the test accuracy of the BP flame detection network is also significantly improved.

In the test results shown in Fig. 4 (b), there are 105 samples with large data errors. Compared with the results in Fig. 4 (b), the test samples in Fig. 4 (d) have higher fitting ability. There are 20 samples with large errors, while the training error in Fig. 4 (f) is the smallest, and 11 samples have large test errors, and the rest of the samples can fit the sample objective function well.

4   Experimental Results and Analysis

Comprehensive analysis and comparison, when MSE = 0.0001, the BP flame detection algorithm has the strongest fitting ability. However, at the same time, the timeliness of

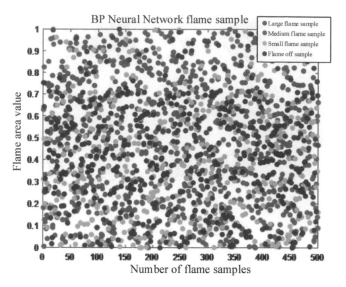

Fig. 3. Flame sample data processing results

the algorithm and the complexity of the model need to be considered. The specific result parameters are shown in Table 1.

During the training process, due to the randomness of the test results, the performance of BP Neural Network flame detection algorithm is tested by using the method of control variables. In the BP Neural Network, the number of neurons is 10, and the optimal results of network after multiple training and testing are taken for comparison. Among them, with the continuous decrease of MSE, the error value of network judgment result also decreases, but the number of algorithm iterations also increases. When MSE = 0.01, the network iterates 335 times to complete the convergence. When MSE = 0.0001, the algorithm needs to iterate 582 times to reduce the error to the target value. The running time of the algorithm is also increased from 35 to 95s, and the running rate is reduced by nearly 1.7 times. However, when MSE = 0.001, the network training times jump out of the loop in 426 times, reaching the target accuracy, which is only 91 more iterations than when MSE = 0.01, and the running time of the algorithm is improved 17s. Therefore, considering the performance of BP flame detection network under the three errors, when MSE = 0.001 and the number of hidden layer neurons is 10, the error is small and the running rate is relatively good.

The variation curve of MSE with the number of training rounds is shown in Fig. 5. After 426 rounds of training, the final MSE reached 0.0009, which is less than 0.001 required for training. Therefore, the proposed flame detection algorithm based on BP Neural Network has achieved high recognition accuracy.

The convergence speed of the BP flame detection network is not very fast, and it takes at least 426 iterations to reach the minimum error value, which is mainly affected by the network structure and the distribution of flame sample data. But in general, the BP Neural Network flame detection algorithm has good training accuracy, small error,

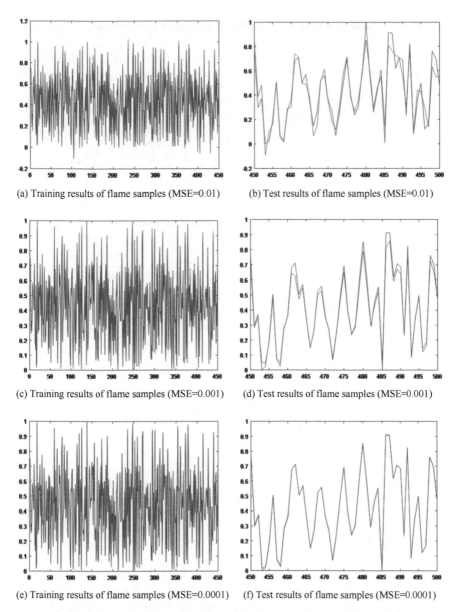

(a) Training results of flame samples (MSE=0.01)

(b) Test results of flame samples (MSE=0.01)

(c) Training results of flame samples (MSE=0.001)

(d) Test results of flame samples (MSE=0.001)

(e) Training results of flame samples (MSE=0.0001)

(f) Test results of flame samples (MSE=0.0001)

Fig. 4. Judgment results of BP flame detection network with different MSE

and also has a good performance for judging unfamiliar samples in the data set, and has good generalization ability.

Table 1. BP Neural Network performance result parameters

Mean square error value	Number of hidden layer neurons	Actual error value	Number of training rounds (epoch)	Algorithm running time (s)	Algorithm running speed (fps)
MSE = 0.01	10	0.00872	335	35	107
MSE = 0.001	10	0.00089	426	52	68
MSE = 0.0001	10	0.00009	582	95	40

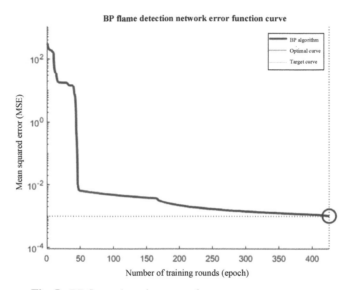

Fig. 5. BP flame detection network mean square error curve

5 Conclusion

By analyzing the research results of current detection methods, a multi-feature fusion flame detection algorithm based on BP Neural Network for the purpose of improving the detection accuracy of boiler flame is proposed. On the premise, the sample data of four types of flame characteristics are used, namely large flame samples, medium flame samples, small flame samples and flame off samples. Train and test the BP detection network, and the flame detection judgment is finally realized through the processing of the hidden layer. The experimental results show that the BP flame detection network proposed achieves a good detection accuracy of MSE < 0.001.

Acknowledgments. Supported by Shaanxi Province Key Research and Development Project (2021GY-280, 2021GY-029); Shaanxi Province Natural Science Basic Research Program Project (2021JM-459).

References

1. Wang, W., Peng, C., Mi, H., et al.: Furnace flame recognition based on improved particle swarm optimization algorithm. Proceedings of the Institution of Mechanical Engineers, Part I: J. Syst. Cont. Eng. **234**(8), 888–899 (2020)
2. Nefedev, A.I., Konovalenko, A.A.: Flame ionization detector for boiler control system. In: 2020 International Conference on Industrial Engineering. Applications and Manufacturing (ICIEAM), pp. 1–6 (2020)
3. Cui, F., Ji, S., Xu, Q.: Design of flame end points detection system for refuse incineration based on ARM and DSP, pp. 1243–1253. Wireless Communications, Networking and Applications (2016)
4. Qiu, X., Xi, T., Sun, D., et al.: Fire detection algorithm combined with image processing and flame emission spectroscopy. Fire Technol. **54**(5), 1249–1263 (2018)
5. Fan, R.S., Wang, Y., Wei, D.X., et al.: Research on visual monitoring method of boiler furnace flame based on BP Neural Network. Comp. Appl. Soft. **8**(2), 101–104 (2015)
6. Chung, Y.L., Chung, H.Y., Chou, C.W.: Efficient flame recognition method based on a deep Convolutional Neural Network and image processing. In: 2019 IEEE 8th Global Conference on Consumer Electronics (GCCE), pp. 573–574 (2019)
7. Wen, Z., Xie, L., Feng, H., et al.: Robust fusion algorithm based on RBF neural network with TS fuzzy model and its application to infrared flame detection problem. Appl. Soft Comput. **76**, 251–264 (2019)
8. Qi, R.Y., Liu, Z.Q., et al.: Extraction and classification of image features for fire recognition based on Convolutional Neural Network. Traitement du Signal **38**(3), 895–902 (2021)
9. Badža, M.M., Barjaktarović, M.Č: Classification of brain tumors from MRI images using a convolutional neural network. Appl. Sci. **10**(6), 1999 (2020)
10. Ker, J., Wang, L., Rao, J., et al.: Deep learning applications in medical image analysis. IEEE Access **6**, 9375–9389 (2018)
11. Yadav, S.S., Jadhav, S.M.: Deep convolutional neural network based medical image classification for disease diagnosis. J. Big Data **6**(1), 1–18 (2019). https://doi.org/10.1186/s40537-019-0276-2
12. Singh, S.P., Wang, L., Gupta, S., et al.: 3D deep learning on medical images: a review. Sensors **20**(18), 5097 (2020)
13. Singh, S.P., Wang, L., Gupta, S., et al.: Shallow 3D CNN for detecting acute brain hemorrhage from medical imaging sensors. IEEE Sens. J. **21**(13), 14290–14299 (2021)
14. Dua, M., Kumar, M., Charan, G.S., et al.: An improved approach for fire detection using deep learning models. In: 2020 International Conference on Industry 4.0 Technology (I4Tech), pp. 171–175 (2021)
15. Liu, T., Cai, Z., Wang, N., et al.: Prediction method of coal dust explosion flame propagation characteristics based on principal component analysis and BP Neural Network. Mathematical Problems in Engineering (2022)

Single Infrared Image Non-uniformity Correction Based on Genetic Algorithm

Gaojin Wen[1], Changhai Liu[2], Hongmin Wang[1], Pu Huang[1], Can Zhong[1], Zhiming Shang[1], and Yun Xu[3(✉)]

[1] Beijing Institute of Space Mechanics and Electricity, Beijing, China
766300930qq.com, wanghongmin317@qq.com, huangpu508@163.com,
lunch_1234@163.com, shang_zhi_ming@163.com
[2] The Troop of 63768, Xi'an, Shaanxi, China
lch2000719@163.com
[3] Institute of Applied Physics and Computational Mathematics, Beijing, China
xu_yun@iapcm.ac.cn

Abstract. Non-uniformity correction for single infrared image is typical multi-extremum problem, which can be solved by genetic algorithms suitably and efficiently. Based on the idea of genetic algorithms (GAs), in this paper a new single image stripe non-uniformity correction method has been established successfully. Non-uniformity correction parameters (gain and offset coefficients) are turned into individual of GAs. Special constraint are designed to form the fitness function, which is efficient for optimization process of GAs. To validate our method, comparisons with other traditional methods are presented through real infrared images. Experimental results demonstrate the proposed method can eliminate stripe non-uniformity of infrared images as well as high consistency of original image.

Keywords: Non-uniformity correction · Genetic algorithm

1 Introduction

Due to the limitation of material quality and manufacturing process, impedance, capacities reactance, photosensitive area and resistance temperature of each pixel in the infrared detector will vary. These subtle differences will lead to different response transfer function for each pixel. This non-uniform response results in fixed pattern stripe noise in the captured infrared images [1]. This kind of noise seriously degrades the image quality and greatly hinders the application of infrared imaging in medical, military, agricultural and forestry monitoring and other fields [2,3]. Therefore, when carrying out infrared remote sensing applications, it is necessary to perform non-uniformity correction to improve the image quality of infrared images.

After decades of research in the field of infrared remote sensing applications, two major types of infrared image non-uniformity correction (NUC) methods have been developed: calibration-based methods and scene-based methods.

© The Author(s), under exclusive license to Springer Nature Switzerland AG 2023
N. Xiong et al. (Eds.): ICNC-FSKD 2022, LNDECT 153, pp. 402–409, 2023.
https://doi.org/10.1007/978-3-031-20738-9_46

Calibration-based methods mainly include two-point method and multi-point method [4,5], using standard radiation source black body as a reference, the detector response is calibrated by measuring radiation at two or more temperatures. The advantage of this method is that the algorithm is simple and easy to implement. However, a standard radiation source is required, and the calibration environment and the application environment need to be consistent, and the limitations are relatively large in practical applications. Scene-based NUC methods mainly include multi-frame NUC method [6–9] and single-image NUC method [1,10–17].

The multi-frame NUC method use various iterative filters to improve the estimation of NUC parameters, such as Kalman filter [6], edge-preserving nonlinear spatial filter [7], quadratic filters [8] and temporal-spatial nonlinear filter [9]. The multi-frame NUC methods generally require multiple frames of images, which result in high computational complexity and large storage requirements. Experiments and comparison are presented to demostrate the efficiency of the proposed method. As to single-image NUC methods, it can be divided into three categories in terms of implementation approaches, filter methods [10–13], machine learning methods [14,15] and numerical optimization methods [1,16,17].

In this paper, we propose a single image stripe non-uniformity correction method based on genetic algorithms(GAs). The key point of our method is to parameterize the individual of GAs by non-uniformity correction parameters, and to construct a specific designed fitness function. For the rest part, it is presented as follows: The background and problem definition are given in Sect. 2. Main framework of the proposed method is provided in Sect. 3. We discuss the experiments and analysis in Sect. 4, and summary the conclusions briefly in Sect. 5.

2   Problem Definition

For a line scanning sensor with n pixels, a raw image $\mathbf{X}$ with size $m \times n$ is formed after m scanning. The non-uniformity correction for each pixel $x_{i,j}$ can be mathematically described as follows:

$$\mathbf{y}_{i,j} = a_j \mathbf{x}_{i,j} + b_j, \quad 1 \leq j \leq n, 1 \leq i \leq m, a_j \neq 0 \tag{1}$$

where gain coefficients a_j and offset coefficients b_j can be obtained by non-uniformity correction method.

The problem of single image nonuniformity correction can be formulated as minimization of a squared error function as following,

$$\min E = \sum_{j=2}^{n} \sum_{i=1}^{m} (y_{i,j} - y_{i,j-1})^2 \tag{2}$$

Inserting Eq.(1) into Eq.(2) follows

$$\min_{\mathbf{A},\mathbf{B}} E(\mathbf{A}, \mathbf{B}) = \sum_{j=2}^{n} \sum_{i=1}^{m} (a_j x_{i,j} + b_j - a_{j-1} x_{i,j-1} - b_{j-1})^2 \tag{3}$$

where $\mathbf{A} = [a_1, a_2, \ldots, a_n]$, $\mathbf{B} = [b_1, b_2, \ldots, b_n]$, $a_j \neq 0$.

Due to lacking of constraint to X, Eq.(3) may generate result image that deviates from the input image X. So different constraints were proposed to modify Eq.(3), e.g. Qian et al. [16] add an average change constraints between processed image and original raw image within threshold T as follows:

$$\frac{1}{n*m} \sum_{j=1}^{n} \sum_{i=1}^{m} \|a_j x_{i,j} + b_j - x_{i,j}\|_2 \leq T \tag{4}$$

Constrained optimization problem like Eq.(3) with constraints Eq.(4) can be solved by constrained numerical optimization methods, which usually results in a local extremum.

3   Single Infrared Image Non-uniformity Correction Based on Genetic Algorithm

3.1   Individual Representation

The gene for GA is non-uniformity correction parameters $\mathbf{A}$ and $\mathbf{B}$. They can be parameterized using a real vector.

$$\mathbf{P} = \{a_1, a_2, \ldots, a_n, b_1, b_2, \ldots, b_n\} \tag{5}$$

In population initialization step, $\{a_i\}_{i=1}^{n}$ is generated with a uniform distribution in the range [1-s,1+s] and $\{b_i\}_{i=1}^{n}$ is random generated with a uniform distribution in the range [-t,t].

3.2   Fitness Function

We propose an efficient constraint to constrain Eq.(3), and construct the fitness function as the following equation.

$$f(\mathbf{P}) = E(\mathbf{P}(1:n), \mathbf{P}(n+1:2*n)) \tag{6}$$
$$= \sum_{j=2}^{n} \sum_{i=1}^{m} (a_j x_{i,j} + b_j - a_{j-1} x_{i,j-1} - b_{j-1})^2 + \lambda \sum_{j-2}^{n} \sum_{i-1}^{m} (a_j x_{i,j} - x_{i,j})^2 \tag{7}$$

where λ is a weighting parameter which can be empirically determined.

3.3   GA Operators

Note that the code of gene be a real vector, it is obvious that the classical mutation operators such as Gaussian mutation, and the classical crossover operators including single point crossover and two point crossover, can be adopted directly.

4   Experiment and Discussion

The single-image non-uniformity correction method proposed in this paper is implemented by C++, and compared with midway infrared equalization (MIRE) [10], guided filter (GF) [11], constrained numerical optimization (CNO) [16], weighted least-squares (WLS) [12], fitting nearest pixels of columns (FNPC) [17] methods. Ten real images with non-uniformity noise are used to compare these methods.

The roughness index ρ is a classical metric to measure the non-uniformity of infrared image, which is defined as follows:

$$\rho = \frac{\|h_1 * X\|_1 + \|h_2 * X\|_1}{\|X\|_1} \tag{8}$$

where $h_1 = [1, -1]$ and $h_2 = [1, -1]^T$ are horizontal and vertical differencing filters, respectively. $\| \cdot \|_1$ is L_1 norm of matrix. The smaller ρ means lower non-uniformity, so smaller values of ρ are expected in NUC images.

As listed in Table 1, for genetic algorithms the parameters set is used in the simulation The comparison of NUC results for one real image are shown in Fig. 1. There is still some residual stripe noise in the processed images of MIRE, GF and CNO methods, while nearly no stripe noise can be found in the image processed by WLS and our method. There are minor differences between the result images by WLS and our method visually. In fact, our method can achieve the smallest value of ρ, which means that our method outperforms all other five methods.

Table 1. The parameters for genetic algorithms

Parameter	Value
Population size	100
Crossover frequency	0.8
Mutation frequency	0.06
Generation number	100
s	0.2
t	300
λ	0.5

The convergence of our method for the real image is displayed in Fig. 2. Our method has shown a robust convergence.

The statistical NUC results for the real image dataset has been shown in Fig. 3. Our method can achieve the smallest value of ρ robustly.

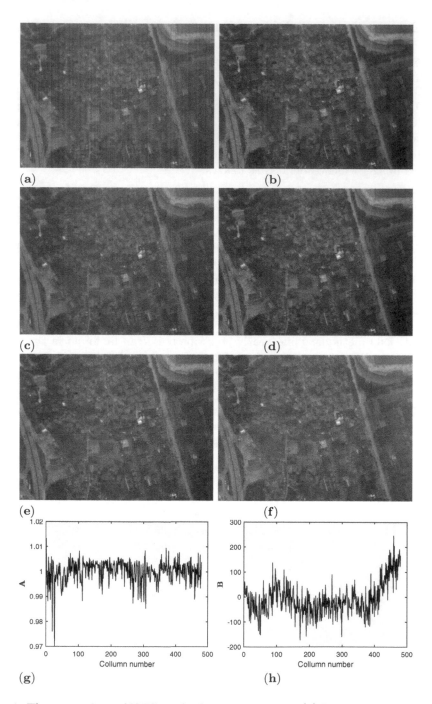

Fig. 1. The comparison of NUC results for one real image. (**a**) Raw image $\rho = 0.01458$ (**b**) MIRE : $\rho = 0.01777$ (**c**) GF: $\rho = 0.01348$. (**d**) CNO: $\rho = 0.01605$. (**e**) WLS: $\rho = 0.01330$ (**f**) Ours: $\rho = 0.01324$. (**g**) The result gain coefficients **A** obtained by our method. (**h**) The result offset coefficients **B** generated by our method.

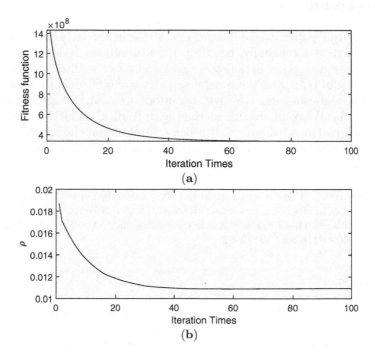

Fig. 2. The convergence of our method for real image. (**a**) The convergence of the fitness function. (**b**) The convergence of the roughness index.

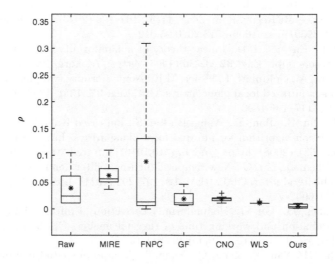

Fig. 3. The comparison of statistical results for real image dataset.

5 Conclusion

To summary this paper, we propose a new single image stripe non-uniformity correction method successfully, based on the idea of genetic algorithms(GAs). Non-uniformity correction parameters (gain and offset coefficients) are turned into individual of GAs, and the roughness index is modified as the fitness function. Special mutation and crossover operators are designed for optimization process of GAs. We compare this method with traditional methods for validation, by using real infrared images. It is demonstrated that the proposed method can eliminate stripe non-uniformity of infrared images as well as high consistency of original image.

Acknowledgment. The authors gratefully acknowledge the support to this work from all our colleagues in Beijing Engineering Research Center of Aerial Intelligent Remote Sensing Equipment. This work is supported by National Natural Science Foundation of China No. 60903116 and 12172062.

References

1. Wen, G., Wang, H., Zhong, C., Shang, Z.: An optimization method for infrared nonuniformity parametric correction based on image entropy. Spacecraft Recov. Rem. Sens. **42**(4), 91–98 (2021). (In Chinese) https://doi.org/10.3969/j.issn.1009-8518.2021.04.011

2. Huang, Y., Xu, J., Bai, S., Gao, J., Li, A., Gao, H.: Satellite infrared remote sensing technology and its application in disaster prevention and relief. Spacecraft Recov. Rem. Sens. **41**(5), 118–126 (2020). (In Chinese) https://doi.org/10.3969/j.issn.1009-8518.2020.04.014

3. Li, Y., Wu, Y., He, H.: The ship wake characterization study based on GF-5 infrared images. Spacecraft Recov. Rem. Sens. **41**(5), 102–109 (2020). (In Chinese) https://doi.org/10.3969/j.issn.1009-8518.2020.05.012

4. David, L.P., Eustace, L.D.: Linear theory of nonuniformity correction in infrared staring sensors. Opt. Eng. **32**, 32-36 (1993). https://doi.org/10.1117/12.145601

5. Friedenberg, A., Goldblatt, I., Kruer, M.R.: Nonuniformity two-point linear correction errors in infrared focal plane arrays. Opt. Eng. **37**, 1251–1253 (1998). https://doi.org/10.1117/1.601890

6. Zhou, H., Qin, H., Jian, Y., Wang, B., Liu, S.: Improved Kalman-filter nonuniformity correction algorithm for infrared focal plane arrays. Infrared Phys. Technol. **51**(6), 528–531 (2008). https://doi.org/10.1016/j.infrared.2008.04.002

7. Zuo, C., Chen, Q., Gu, G.: New temporal high-pass filter nonuniformity correction based on bilateral filter. OPT. REV. **18**, 197–202 (2011). https://doi.org/10.1007/s10043-011-0042-y

8. Sui, X., Chen, Q., Gu, G.: Nonuniformity correction of infrared images based on infrared radiation and working time of thermal imager. Optik. **124**(4), 352–356 (2013). https://doi.org/10.1016/j.ijleo.2011.12.055

9. Li, J., Qin, H., Yan, X., Zeng, Q., Yang, T.: Temporal-spatial nonlinear filtering for infrared focal plane array stripe nonuniformity correction. Symmetry **11**, 673 (2019). https://doi.org/10.3390/sym11050673

10. Tendero, Y., Landeau, S., Gilles, J.: Non-uniformity correction of infrared images by midway equalization. Image Process Line **2**, 134–146 (2012). https://doi.org/10.5201/ipol.2012.glmt-mire
11. Cao, Y., Yang, M.Y., Tisse, C.L.: Effffective strip noise removal for low-textured infrared images based on 1-D guided filtering. IEEE Trans. Circuits Syst. Video Technol. **26**, 2176–2188 (2016). https://doi.org/10.1109/TCSVT.2015.2493443
12. Li, F., Zhao, Y., Xiang, W.: Single-frame-based column fixed-pattern noise correction in an uncooled infrared imaging system based on weighted least squares. Appl. Opt. **58**, 9141–9153 (2019). https://doi.org/10.1364/AO.58.009141
13. Wang, E., Jiang, P., Li, X.: Infrared stripe correction algorithm based on wavelet decomposition and total variation-guided filtering. J. Eur. Opt. Soc.-Rapid Publ. **16**, 1 (2020). https://doi.org/10.1186/s41476-019-0123-2
14. Kuang, X., Sui, X., Chen, Q., Gu, G.: Single infrared image stripe noise removal using deep convolutional networks. IEEE Photonics J. **9**, 1–13 (2017). https://doi.org/10.1109/JPHOT.2017.2717948
15. He, Z., Cao, Y., Dong, Y., Yang, J., Cao, Y., Tisse, C.L.: Single-image-based nonuniformity correction of uncooled long-wave infrared detectors: a deep-learning approach. Appl. Opt. **57**, D155–D164 (2018). https://doi.org/10.1364/AO.57.00D155
16. Qian, W., Chen, Q., Gu, G., Guan, Z.: Correction method for stripe nonuniformity. Appl. Opt. **49**, 1764–1773 (2010). https://doi.org/10.1364/AO.57.00D155
17. Ren, J., Chen, Q., Qian, W.: Efficient single image stripe nonuniformity correction method for infrared focal plane arrays. OPT. REV. **19**, 355–357 (2012). https://doi.org/10.1007/s10043-012-0056-0
18. Cao, Y., Li, Y.S.: Strip non-uniformity correction in uncooled long-wave infrared focal plane array based on noise source characterization. Opt. Commun. **339**, 236–242 (2015). https://doi.org/10.1016/j.optcom.2014.10.041

Load Forecasting of Electric Vehicle Charging Station Based on Power Big Data and Improved BP Neural Network

Hao Sun[1(✉)], Shan Wang[2], and Chunlei Liu[1]

[1] Baoding Power Supply Company of State Grid Hebei Electric Power Co., Ltd, Baoding, Hebei Province 071000, China
2485212226@qq.com, 15931838627@163.com
[2] Limited Skills Training Center, State Grid Jibei Electric Power Co., Ltd, Baoding 071000, China
6766387@qq.com

Abstract. In order to accurately predict the load of electric vehicle charging stations, this paper extracts power data from the enterprise data center through the power big data analysis method, cleans and processes the data, and then analyzes the main factors affecting the charging load. The load prediction model is established for the station and all charging stations in the area, and finally the reliability of the model is verified according to the field measured data.

Keywords: Electric car · Big data · Load forecasting · Neural networks

1 Introduction

With the advancement of industrialization, environmental pollution and energy shortages have intensified, and electric vehicles as a low-carbon means of transportation have received extensive attention [1]. The charging and discharging behavior of electric vehicles is highly random, which affects both the power quality and the reliability of power grid operation. Reasonable guidance of its charging and discharging process can not only adjust the peak-to-valley difference of the power grid and optimize resource allocation; it can also be used as a distributed energy storage resource to assist power grid frequency regulation. To achieve this goal, it is necessary to ensure accurate prediction of charging load.

Reference [2] analyzes the driving laws in some European countries, and uses a Monte Carlo model to fit the charging laws. Reference [3] analyzes the load data information of the motor when the motor is not fully loaded according to the driving law and charging data, and based on the charging state and charging time of the initial operation. In the era of big data, in order to fully explore the value of data, extract processing historical charging data, and introduce multi-domain data at the same time, apply machine learning [4, 5] and deep learning [6] to result optimization. Reference [7] proposed the use of a recurrent neural network (NEE) model to solve the problems of gradient disappearance and gradient explosion that occur in the conventional RNN training process.

In this paper, a load prediction model of electric vehicle charging station is established based on the data processing in the data middle platform and the optimized BP neural network algorithm. First, the load data of the middle station is preprocessed, and then an improved neural network is used to dynamically model the standardized time series to complete the load prediction of the electric vehicle charging station.

2 Basic Theory

2.1 Power Grid Data Middle Platform

The proposal and development of concepts such as smart grid, energy internet, power internet of things, and new power systems have promoted the in-depth digital transformation of State Grid enterprises. In the fields of equipment, personnel, and business, enterprises have accumulated rich data resources. By establishing a data middle platform (see Fig. 1), mining data value, improving data utilization and sharing adaptability. The data center is an information sharing and publishing platform that integrates data from various systems.

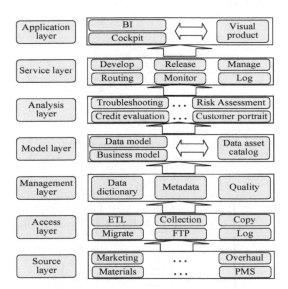

Fig. 1. Power grid data middle platform

1. Data access layer: Extract the source layer data and synchronize it to the previous layer. 2. Data governance layer: Responsible for metadata, data dictionary management and data quality control. 3. Data model layer: Establish various data models including personnel data, marketing data, production data, etc., and manage them through the data asset catalog. 4. General analysis layer: Build prediction, diagnosis, early warning and other analysis models according to business scenarios. 5. Data service layer: Provide data visualization scenarios and service tools.

2.2 BP Neural Network

The artificial network establishes the topological relationship by simulating the biological neural network, and forms an information storage and mapping system. Includes connections, summation units, and activation functions.

Express the above process mathematically:

$$
\begin{cases}
u_k = \sum_{j=1}^{n} w_{kj} x_j \\[2mm]
v_k = u_k - \theta_k \\[2mm]
y_k = \varphi(v_k)
\end{cases}
\tag{1}
$$

where $x_1, x_2 \ldots x_n$ are the input signals; $w_{k1}, w_{k2} \ldots w_{kn}$ are the weights of neuron k; u_k is the linear combination result; θ_k is the threshold; φ is the activation function; y_k is the neuron The output of element k. The activation function φ can be a threshold function, a piecewise linear function and a sigmoid function.

Perception includes an input layer and an output layer to deal with linearly separable problems. For nonlinear problems, a hidden layer can be added between the two layers of neurons, as shown in Fig. 2.

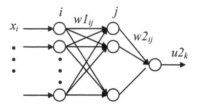

Fig. 2. Multilayer neural network

$$
\begin{cases}
a1_j = f1\left(\sum_{j=1}^{n} w1_{ij} p_i + b1_j \right) \\[4mm]
u2_k = f2\left(\sum_{j=1}^{m} w2_{kj} a1_j + b2_k \right)
\end{cases}
\tag{2}
$$

In the above formula, $a1_j$ is the hidden layer output, $u2_k$ is the output layer output, $w1$ is the hidden layer weight, $w2$ is the output layer weight.

The error backpropagation (BP) algorithm solves the weights through the gradient algorithm, which is suitable for the training optimization of the multi-layer network. BP learning includes forward propagation and back propagation of errors. In the process of forward propagation, the actual output of the output layer does not match the expectation. The output error is back propagated to the input layer through the hidden layer, and the error signal of each layer unit is used as the correction weight.

The weight threshold vector adjustment formula is:

$$x(k + 1) = x(k) + \alpha\Delta(k) \tag{3}$$

In the formula, $x(k + 1)$ and $x(k)$ are the weights or threshold vectors at the k + 1 and k iterations, respectively, α is the learning rate, and $\Delta(k)$ is the negative gradient.

In order to overcome the slow convergence speed of the BP algorithm, a momentum factor η $(0 < \eta < 1)$ is introduced based on the gradient descent algorithm.

$$\Delta x(k + 1) = \eta\Delta x(k) + \alpha(1 - \eta)\frac{\partial E}{\partial x(k)} \tag{4}$$

E is the mean square error. The algorithm's previous correction results affect the current correction results.

Due to the unique advantages of neural network in dealing with complex problems, it can be applied to the fields of load forecasting, safety analysis, and fault diagnosis of power systems. Use neural networks for load forecasting, learning from training samples, and automatically build accurate forecasting models.

3 Load Prediction Model

3.1 Load Influencing Factors

There are many factors (see Fig. 3) that affect the load of electric vehicle charging station. This paper divides it into classification factors and direct factors according to the actual situation. The classification factors include weather, temperature, date type, and epidemic situation. The direct factors are the load data of the past three days and the number of electric vehicles. The non-numerical factors need to be quantified (Table 1).

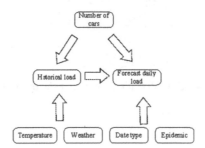

Fig. 3. Factors affecting load

In this paper, the number of electric vehicles is selected as the numerical input of the neural network, and the load prediction model of all charging stations in the area is modeled. The input can select the number of all electric vehicles in the range. For the load prediction of a single charging station, the calculation method of electric vehicle ownership is shown in Fig. 4: find the charging station B and charging station C closest to the target charging station A, and divide the circular area between the three equidistantly. According to the type statistics, this paper focuses on bus charging stations, ignoring the impact of electric buses.

Table 1. Quantitative value of influencing factors

Influencing factors	Classification	Value
Date type	Working day	0
Date type	Off day	1
Date type	Holidays	1.5
Weather	Sunny, Cloudy	0
Weather	Mild fog, Rain, Snow, etc	0.5
Weather	Bad weather	1
Weather	Comfortable	0
Temperature	Hot	0.5
Temperature	Cold	0.8
Temperature	Too cold	1
Epidemic	None	0
Epidemic	Mild to moderate	0.5
Epidemic	Serious	1

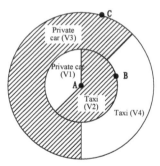

Fig. 4. The calculation method of electric vehicle

3.2 Data Processing

Due to the abnormal communication between the terminal or the master station, the source data needs to be processed, including the whole process of data initialization, selection and preprocessing. During the initialization process, conduct data quality screening, determine data structure, and find data problems; then select and evaluate valid data; preprocessing includes deletion of duplicate and abnormal data (including data on the day of sudden changes such as power outages, maintenance, and major events), Missing data imputation and data normalization. There are three main methods for dealing with missing values [8]: direct deletion, mean or median imputation, and machine learning-based imputation. Because the load curve is smooth, the acquisition success rate is high, and the amount of missing data is relatively small, the mean interpolation method can be used to meet the engineering requirements.

$$L_i = \frac{L_{i+1} + L_{i-1}}{2} \tag{5}$$

In the formula: L_i is the missing data; L_{i-1} is the previous load value, and L_{i+1} is the next load value.

Normalizing the data can speed up the neural network convergence and improve the prediction accuracy, and normalize the load data and each feature in the input layer. In this paper, the terminal data is used to normalize the data.

$$x' = \frac{(x - x_{min})}{(x_{max} - x_{min})} \tag{6}$$

x' is the normalized value, x is the value to be normalized; x_{max} is the maximum value in the data set; x_{min} is the minimum value in the data set.

In order to objectively evaluate the effect of the strategy from the level of data indicators, and to reduce the influence of a certain load data on the results, this paper selects the root mean square error as the evaluation indicator.

$$\eta_{RMSE} = \sqrt{\frac{1}{\Lambda} \sum_{\lambda=1}^{\Lambda} \left(x'(\lambda) - x'_p(\lambda) \right) \left(x'(\lambda) - x'_p(\lambda) \right)^T} \tag{7}$$

3.3 Load Forecast of All Charging Stations in an Area

Select a circular area with a radius of 5 km in a certain area, obtain the geographical location of all charging stations within the range and data information in the past 1 year, and complete the normalization of the original data. The input X is a two-dimensional matrix with n rows and 9 columns, and the neural network parameters are set as follows: 1. The training data accounts for 70%, and the test data accounts for 30%. 2. Single hidden layer, unit range 1–50. 3. Hyperbolic tangent activation function. 4. Optimized conjugate gradient algorithm (Table 2, Figs. 5 and 6).

According to the Table 3, by comparing the importance of the model input, the weather conditions in the non-numerical factors have a greater impact on the prediction results, and the numerical factors have a greater impact on the previous day's load and vehicle ownership. According to the prediction results, it shows that the model has a high fitting ability, and the error is within 10%, which meets the engineering requirements.

3.4 Load Forecasting of a Single Charging Station

Select a charging station in the region, obtain the geographical location of the charging station and various data in the past 1 year, and standardize the data. The input X is a two-dimensional matrix with n rows and 12 columns, and the neural network parameters are set as above (Figs. 7 and 8).

According to the experimental results, the model has a high fitting ability, and the error also meets the engineering requirements.

Table 2. Neural network parameter settings

Parameter	Value
Number of hidden layers	1
Number of units in hidden layer	48
Activation function	Hyperbolic tangent
Dependent variable	T
Number of units	1
Activation function	Identity
Error function	Sum of square

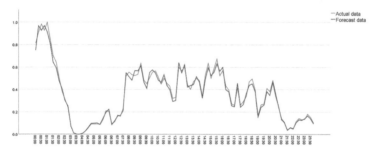

Fig. 5. Load forecast curve of all charging stations in the area

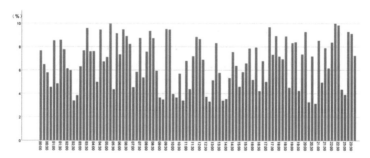

Fig. 6. Forecast error distribution

4 Conclusion

Compared with traditional methods, the load forecasting model based on power big data and improved BP neural network has strong advantages in big data analysis and learning, model training and engineering applications. This paper uses the power big data, extracts historical load data, analyzes the main factors affecting the charging load, and completes the construction of the model. Finally, the model training and actual data verification are carried out through the measured data, which proves the feasibility of the proposed method in the load prediction of electric vehicle charging stations.

Table 3. Importance calculation

Input volume	Importance
Weather	0.122
Temperature	0.111
datetype	0.105
epidemic	0.102
T-1	0.192
T-2	0.105
T-3	0.097
V_{all}	0.166

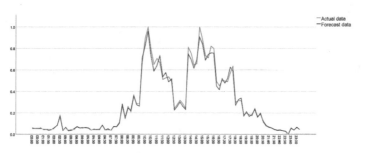

Fig. 7. Load forecast curve of a single charging station

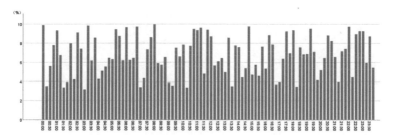

Fig. 8. Forecast error distribution

References

1. Cazzola, P., Gorner, M., Tattini, J., et al.: Global EV outlook 2019 - scaling up the transition to electric mobility (2019)
2. Jordan, M.I.: Serial order: a parallel distributed processing approach. Adv. Psychol. **121**(97), 471–495 (1997)
3. Ashtari, A., Bibeau, E., Shahidinejad, S., et al.: PEV charging profile prediction and analysis based on vehicle usage data. IEEE Trans. Smart Grid **3**(1), 341–350 (2012)

4. Majidpour, M., Qiu, C., Chu, P., et al.: Modified pattern sequence-based forecasting for electric vehicle charging stations. In: IEEE International Conference on Smart Grid Communications, pp. 710–715. IEEE (2014)
5. Xiaobo, X., Liu, W., Xi, Z., et al.: Short-term load forecasting for the electric bus station based on GRA-DE-SVR. In: 2014 IEEE Innovative Smart Grid Technologies - Asia (ISGT ASIA). IEEE (2014)
6. Zhu, J., Yang, Z., Chang, Y., et al.: A novel LSTM based deep learning approach for multi-time scale electric vehicles charging load prediction. In: 2019 IEEE Innovative Smart Grid Technologies - Asia (ISGT Asia). IEEE (2019)
7. Hadley, S.W., Tsvetkova, A.A.: Potential impacts of plug-in hybrid electric vehicles on regional power generation. Electricity J. **22**(10), 56–68 (2009)
8. Kaiser, J.: Dealing with missing values in data. J. Syst. Integr. **5**(1), 42–51 (2014)

A Probabilistic Fuzzy Language Multi-attribute Decision Making Method Based on Heronian Operator and Regret Theory

Wenyu Zhang[1,2], Xue Gao[1(✉)], Yongbin Yuan[1], Weina Luo[1], and Jiahao Zeng[3]

[1] Xi'an University of Posts and Telecommunications, Xi'an 710061, China
zwy888459@xupt.edu.cn, gx876645@stu.xupt.edu.cn, yyb123@stu.xupt.edu.cn,
17392189719@stu.xupt.edu.cn
[2] China Research Institute of Aerospace Systems Science and Engineering, Beijing 100048, China
[3] Northwest University, Xi'an 710127, China
202132501@stumail.nwu.edu.cn

Abstract. Aiming at the multi-attribute decision-making problem with unknown attribute and expert weight, this paper proposes a multi-attribute decision-making method based on Heronian operator and regret theory. Firstly, the probability language term set (PLTS) is used to describe the evaluation information of decision-makers to form an initial matrix. Then the distance between experts is calculated, and the expert weight is obtained by using the node tightness. Secondly, integrate the decision information of all experts and construct the initial comprehensive matrix. Then define an impact factor parameter to reduce the impact of unreasonable schemes in the polymerization process and form the final comprehensive evaluation matrix. Thirdly, construct perceived utility matrix using Regret Theory. Finally, the information entropy theory is used to determine the attribute weight, and the three parameter weighted Heronian average operator is used to aggregate the decision values as weighted Heronian average operator is used to aggregate the decision values as verified by an example of evaluating the innovation ability of scientific research verified by an example of evaluating the innovation ability of scientific research institutions, and the results obtained by this method are reasonable and reliable.

Keywords: Probabilistic linguistic term sets · Regret theory · Heronian operator · Information entropy · Node tightness

Shaanxi Education and Scientific Research Project(08JK431).

N. Xiong et al. (Eds.): ICNC-FSKD 2022, LNDECT 153, pp. 419–427, 2023.
https://doi.org/10.1007/978-3-031-20738-9_48

1   Introduction

Decision making is a process in which people make the best decision or choice of things. Objective things have complexity and strong uncertainty, so experts often like to use fuzzy language terms [1] such as "good", "medium" and "poor" to express their evaluation of alternatives. Zadeh [2] put forward the concept of language variable, However, people may hesitate between several possible language terms. In order to solve this problem, Rodriguez [3] et al. put forward the concept of hesitant fuzzy language term set, but the importance of each language information is evenly distributed, which can not express that decision makers prefer a language term. Considering this situation, Pang [4] et al. proposed the concept of probabilistic language term set (PLTS), which can clearly express the weight of each possible language term. Considering the lack of generality of single fuzzy language term and the hesitant fuzzy language term set can not express the weight, this paper adopts the probabilistic language term set to evaluate various attributes of alternative information.

The emotion of decision-makers is an important factor affecting the decision making results, which forms a series of methods related to psychological factors, such as cardinal utility theory [5], prospect theory [6], expected utility theory [7], Regret Theory [8], etc. Among them, the regret theory proposed by Bell,Loomes and Sugden [9, 10] is a decision-making method considering the regret and rejoice of decision makers. For the integration of decision information, the traditional linear weighting method is not suitable for information integration. The three-parameter weighted Heronian mean operator (TPWHM) defined by Liu Jinjun et al. [11] can reflect the interaction between attributes. To sum up, this paper proposes a multi-attribute decision making method based on Heronian operator and regret theory for probabilistic language multi-attribute decision making.

2   Related Concepts

Definition 1. If $S = \{s_t \mid t \in [0, 1, \cdots, \tau]\}$ is a language term set, the probabilistic language term set (PLTs) can be defined as:

$$PL(p) = \left\{ L_k\,(p_k) \mid L_k \in S, p_k \geq 0, k = 1, 2, \cdots, \#L(p), \sum_{k=1}^{\#L(p)} p_k \leq 1 \right\} \quad (1)$$

Where, $L_k\,(p_k) = s_t^k\,(p_k)$ is a probability language element, using probability p_k and s_t^k to form the kth probabilistic language element, and the set of probabilistic language elements generates the probabilistic language set. $\#L(p)$ represents the number of different language terms in the probabilistic language term set.

Definition 2. Let the decision maker set be $D = \{d_l \mid l = 1, 2, \cdots, h\}$, expert d_l evaluation of scheme O_i on attribute C_j can be expressed as:

$$PL_{ij}^l(p) = \left\{ L_k^l\left(p_k^l\right) \mid L_k^l \in S, p_k^l \geq 0, k = 1, 2, \cdots, \#L_{ij}^k(p), \sum_{k=1}^{\#L_{jj}^k(p)} p_k^l \leq 1 \right\} \tag{2}$$

Definition 3. Let the weight vector of h experts be $W = \{\omega_1, \omega_2, \cdots \omega_h\}$, then the group evaluation can be represented by a probabilistic language term set:

$$PL_{ij}^*(p) = \left\{ L_k\left(\widehat{p_k}\right) \mid L_k \in S, \widehat{p_k} \geq 0, k = 1, 2, \cdots, \#L(p), \widehat{p_k} = \sum_{l=1}^{m} p_k^l \omega^l \right\} \tag{3}$$

Definition 4. Let the probability language term set be $PL(p) = \{L_k(p_k) \mid L_k \in S, p_k \geq 0, k = 1, 2, \cdots, \#L(p)\}$, then its complement is:

$$PL(p)^c = \{L_{-k}(1 - p_k) \mid L_k \in S, p_k \geq 0, k = 1, 2, \cdots, \#L(p)\} \tag{4}$$

Definition 5. Let the probabilistic language term sets
$PL_i(p)$ $=$
$\left\{ L_k(p_{ik}) \mid L_k \in S, p_{ik} \geq 0, k = 1, 2, \cdots, \#L_i(p), \sum_{k=1}^{\#L_i(p)} p_{ik} \leq 1 \right\}$
and $PL_j(p) = \left\{ L_k(p_{jk}) \mid p_{jk} \geq 0, k = 1, 2, \cdots, \#L_j(p), \sum_{k=1}^{\#L_j(p)} p_{jk} \leq 1 \right\}$
and $\#L_i(p) = \#L_j(p)$, define the distance between $PL_i(p)$ and $PL_j(p)$ as:

$$d\left(PL_i(p), PL_j(p)\right) = \sqrt{\sum_{k=1}^{\#L_i(p)} \frac{1}{\#L_i(p)} \left(\varphi_k\left(p_{ik} - p_{jk}\right)\right)^2} \tag{5}$$

Where $\#L_i(p)$ is the number of different language terms contained in $PL_i(p)$ and $PL_j(p)$, and φ_k is the subscript of L_k.

Definition 6. According to the language information and probability characteristics of PLTS, the score function $F(PL(p))$ of PLTS is defined as:

$$F(PL(p)) = \bar{\alpha} - \frac{\sum_{k=1}^{\#L(p)} p_k(\varphi_k - \bar{\alpha})^2}{\text{var}(S)\sum_{k=1}^{\#L(p)} p_k} \quad \text{Where } \bar{\alpha} = \frac{\sum_{k=1}^{\#L(p)} \varphi_k p_k}{\sum_{k=1}^{\#L(p)} p_k}, \text{var}(S) = \frac{\tau(\tau + 1)}{3} \tag{6}$$

Definition 7. Regarding each expert as a node, $V = \{v_1, v_2, \cdots v_h\}$ is the set of nodes, then the node tightness can be defined as:

$$c(v_i) = \frac{m - 1}{\sum_{i=1, i \neq i}^{m} d\left(PL_i(p), PL_j(p)\right)} \tag{7}$$

Definition 8. Formula for calculating rejoice value:

$$G_{ij} = 1 - \exp\left(-\delta d\left(PL_{ij}(p), PL_j^-(p)\right)\right) \tag{8}$$

Where, $PL_j^-(p) = \min\left(F\left(PL_{ij}(p)\right), i = 1, 2, \cdots, m\right)$ is the negative ideal point under attribute j.

Definition 9. Formula for calculating regret value:

$$R_{ij} = 1 - \exp\left(\delta d\left(PL_{ij}(p), PL_j^+(p)\right)\right) \tag{9}$$

Where, $PL_j^+(p) = \max\left(F\left(PL_{ij}(p)\right), i = 1, 2, \cdots, m\right)$ is the positive ideal point under attribute j.

Definition 10. Perceived utility function

$$S_{ij}(a_{ij}) = G_{ij} + R_{ij} \tag{10}$$

Definition 11. Let $v_i(i = 1, 2, \cdots, n)$ is a set of non-negative numbers, $\mu = \{\mu_1, \mu_2, \cdots \mu_n\}$ is the weight vector of $v_i(i = 1, 2, \cdots, n)$, and $p, q, r \geq 0, \mu_i \geq 0, \sum_{i=1}^n \omega_i = 1$, and TPWHM are called three-parameter language weighted Heronian average operators. The formula is as follows:

$$TPWHM_\omega^{p,q,r}(a_1, a_2, \cdots a_n) = \left(\frac{1}{\lambda} \sum_{i=1}^n \sum_{j=i}^n \sum_{l=j}^n (\omega_i a_i)^p (\omega_j a_j)^q (\omega_l a_l)^r\right)^{\frac{1}{p+q+r}} \tag{11}$$

Where

$$\lambda = \sum_{i=1}^n \sum_{j=i}^n \sum_{l=j}^n \omega_i^p \omega_j^q \omega_l^r$$

3   Construction of a Multi-attribute Decision-making Method Based on Heronian Operator and Regret Theory

In this paper, there are m alternatives, n evaluation criteria and h evaluation experts to participate in the decision-making of probabilistic language terms with multiple attributes and multiple experts. The optional scheme set is $O = \{O_1, O_2, \cdots O_m\}$, the evaluation criteria set is $C = \{C_1, C_2, \cdots C_n\}$, the set of review experts is $D = \{d_1, d_2, \cdots d_h\}$. Expert weight vector $\omega = \{\omega_1, \omega_2, \cdots \omega_h\}$ and attribute weight vector $\mu = \{\mu_1, \mu_2, \cdots \mu_n\}$ are unknown, satisfy $\sum_{l=1}^h \omega_l = 1, \sum_{j=1}^n \omega_j = 1$.

Step 1: Collect the initial evaluation information of each expert and construct a standardized initial decision matrix.

The expert d_l involved in decision-making, provides the evaluation value of alternative O_i relative to attribute c_j by using PLTS, and the evaluation matrix of each expert can be obtained by using Formula (2) as follows:

$$D_l = \left[PL_{ij}^l(p)\right]_{m \times n} = \begin{bmatrix} L_{11}^l(p) & L_{12}^l(p) & \cdots & L_{1n}^l(p) \\ L_{21}^l(p) & L_{22}^l(p) & \cdots & L_{2n}^l(p) \\ \vdots & \vdots & \ddots & \vdots \\ L_m^l(p) & L_{m2}^l(p) & \cdots & L_{mn}^l(p) \end{bmatrix}$$

In order to avoid the non-unique calculation result of each element of the evaluation information due to different positions, this paper sorts the elements of the initial probabilistic language term set. According to the subscript of the language term, the probabilistic language elements are sorted in ascending order to standardize the PLTS. At the same time, for the cost attribute j_c, the complement of the estimated PLTS value is selected; for the benefit attribute j_b, the estimated PLTS value remains unchanged. which is:

$$D'_l = \left[PL^l_{ij}(p)\right]'_{m \times n} = \begin{cases} PL^l_{ij}(p)^c, j \in j_c \\ PL^l_{ij}(p), j \in j_b \end{cases} \tag{12}$$

Where, $PL^l_{ij}(p)^c$ is the complement of $PL^l_{ij}(p)$, which can be calculated by formula (4). Based on the normalized PLTS value, the initial evaluation matrix of each expert is constructed.

Step 2: Calculate the weight of the expert through the tightness of the node.

In the decision-making problems, each expert can be regarded as a node. Formula (5) is used to calculate the distance of PLTS at the corresponding position of each two experts, and then formula (13) is used to calculate the distance of each two experts.

$$d_{12} = \frac{1}{m \times n} \sum_{i=1}^{m} \sum_{j=1}^{n} d\left(PL^{l_1}_{ij}(p), PL^{l_2}_{ij}(p)\right) \tag{13}$$

Then formula (7) is used to calculate the tightness of each expert node, and formula (14) is used to calculate the weight of each expert.

$$\omega_i = \frac{c(v_i)}{\sum_{i=1}^{m} c(v_i)} \tag{14}$$

Step 3: Construct the initial comprehensive decision matrix. According to the influence factors of each scheme, the PLTS value is modified to form the final comprehensive decision matrix.

The evaluation opinions of experts are gathered by formula (3) to construct the initial comprehensive decision matrix $S = [PL_{ij}(p)]_{m \times n}$. In decision-making, the impact of those who may not be the most preferred alternative should be reduced. The impact coefficient I_i defines [8] as follows:

$$I_i = \exp\left(\sum_{j=1}^{n} \varphi_k \frac{p^k_{ij} - p^k_{j\max}}{p^k_{j\max}}\right) \tag{15}$$

Where φ_k is the subscript corresponding to L_k, p^k_{ij} represents the kth element of the PLTS in the attribute j of the scheme i, and $p_{j^k_{\max}}$ is the maximum probability value except p_{ij}.

Use the influencing factors of the alternative scheme to adjust its influence on the result and adjust the corresponding probability value. The specific formula is as follows:

$$p^{k'}_{ij} = I_i p^k_{ij} \tag{16}$$

In this way, the final comprehensive decision matrix $S' = [PL_{ij}(p)']_{m \times n}$ is formed.

Step 4: The perceived utility matrix is constructed by using regret theory.

The dominance of each scheme is calculated through the score function of formula (6). Use formula (5) to calculate the distance from each attribute of each scheme to positive and negative ideal schemes. Then use formula (8) to calculate the rejoice value, formula (9) to calculate the regret value, and finally construct the perceived utility matrix $S^* = (v_{ij})_{m \times n}$ according to formula (10).

Step 5: use information entropy to calculate attribute weight.

Step 5.1: The deviation value of each attribute is defined by the following formula

$$D_{ij} = |v_{ij} - \overline{v_j}| \tag{17}$$

Where, $\overline{v_j}$ is the average value of the jth attribute.

Step 5.2: Information entropy of each attribute

$$En_j = -\frac{1}{\ln(m)} \sum_{i=1}^{m} \left(\frac{D_{ij}}{D_j^{ot}} \ln \left(\frac{D_{ij}}{D_j^{ot}} \right) \right) \tag{18}$$

Where, D_j^{tot} is the total deviation value of the jth attribute.

Step 5.3: Normalize entropy to obtain attribute weight $\mu = \{\mu_1, \mu_2, \cdots \mu_n\}$

$$\mu_j = \frac{En_j}{\sum_{j=1}^{m} En_j} \tag{19}$$

Step 6: Based on the three-parameter weighted Heronian average operator, the perceived utility values are aggregated and sorted.

4   Case Analysis

4.1   Example Analysis

A city needs to make a decision on the innovation capabilities of the four scientific research institutions $O = \{O_1, O_2, O_3, O_4\}$, and invite four experts to set up an expert group, starting from innovation basic capabilities (c_1), innovation investment The four attributes of ability (c_2) and innovation output ability (c_3) are evaluated.

Step 1: According to the PLTS evaluation value given by the experts, sort each PLTS and construct the evaluation matrix table of each expert. As shown in Table 1.

Step 2: Calculate the distance between each expert according to formula (5). Calculate the node tightness of each expert node by formula (7), and then calculate the weight of each expert node according to formula (14). The expert weight is: $\omega = \{0.241, 0.239, 0.254, 0.266\}$.

Step 3: Construct a comprehensive decision matrix according to Definition 4. The initial comprehensive decision matrix is shown in Table 2 below. The calculation result of the influence coefficient is: $I_1 = 2.092E - 3, I_2 = 1.110E - 2, I_1 =$

Table 1. Expert 1 initial decision value

	c_1	c_2	c_3
O_1	$\{s_1(0.5), s_2(0.3), s_3(0.2)\}$	$\{s_0(0.3), s_1(0.7)\}$	$\{s_2(0.1), s_3(0.4), s_4(0.5)\}$
O_2	$\{s_2(0.5), s_3(0.2), s_4(0.3)\}$	$\{s_2(0.4), s_3(0.6)\}$	$\{s_1(0.3), s_3(0.4), s_4(0.3)\}$
O_3	$\{s_2(0.2), s_3(0.8)\}$	$\{s_1(0.5), s_3(0.5)\}$	$\{s_1(0.2), s_3(0.6), s_4(0.2)\}$
O_4	$\{s_1(0.2), s_2(0.2), s_3(0.6)\}$	$\{s_0(0.1), s_1(0.4), s_2(0.5)\}$	$\{s_2(0.5), s_3(0.3), s_4(0.2)\}$

Table 2. Initial comprehensive decision matrix

	c_1	c_2	c_3
O_1	$\{s_1(0.27), s_2(0.22), s_3(0.43), s_4(0.08)\}$	$\{s_0(0.27), s_1(0.57), s_2(0.13), s_3(0.02)\}$	$\{s_1(0.05), s_2(0.18), s_3(0.45), s_4(0.33)\}$
O_2	$\{s_1(0.16), s_2(0.22), s_3(0.48), s_4(0.14)\}$	$\{s_0(0.05), s_1(0.33), s_2(0.47), s_3(0.14)\}$	$\{s_1(0.24), s_2(0.18), s_3(0.38), s_4(0.20)\}$
O_3	$\{s_1(0.08), s_2(0.53), s_3(0.31), s_4(0.08)\}$	$\{s_0(0.17), s_1(0.56), s_2(0.05), s_3(0.22)\}$	$\{s_1(0.28), s_2(0.20), s_3(0.47), s_4(0.05)\}$
O_4	$\{s_1(0.12), s_2(0.33), s_3(0.47), s_4(0.08)\}$	$\{s_0(0.02), s_1(0.30), s_2(0.50), s_3(0.17)\}$	$\{s_1(0.07), s_2(0.41), s_3(0.47), s_4(0.05)\}$

$6.609E - 4, I_4 = 2.416E - 3$. The decision value is modified by formula (16) to obtain the final comprehensive decision matrix.

Step 4: According to Definition 7, calculate the regret value from each plan to the optimal plan and the joy value from each plan to the worst plan, where δ is 0.3. Use Equation (10) to construct the perceived utility matrix.

Step 5: Using the method of information entropy, formula (17) calculates the deviation value, The attribute information entropy calculated is: $En_1 = 1.251, En_2 = 1.185, En_3 = 0.961$ Attribute weight calculated according to formula (19) is: $\mu = \{0.368, 0.349, 0.283\}$.

Step 6: Use $\widehat{v_{ij}} = v_{ij} + \varphi$ to correct the perceived utility value, $\varphi = 0.1, p = q = r = 1$. The calculation results are as follows: $\lambda = 0.3737, \text{TPEHM}_\omega^{p,q,r}(O_1) = 0.0597, \text{TPEHM}_\omega^{p,q,r}(O_2) = 0.0546, \text{TPEHM}_\omega^{p,q,r}(O_3) = 0.0622, \text{TPEHM}_\omega^{p,q,r}(O_4) = 0.0508$. Therefore, the ranking results of the innovation capabilities of the four scientific research institutions can be obtained as follows: $O_3 > O_1 > O_2 > O_4$. the scientific research institution with the best innovation capability is Institution 3 .

4.2   Sorting Comparison of the Three Methods

In this paper, the generalized Regret Theory [8] and prospect theory [6] are compared with the methods proposed in this paper. The ranking results are shown in Table 3 : As can be seen from the above table, the ranking results of the three decision-making methods are consistent with the optimal results, indicating the feasibility of the decision-making method in this paper. The generalized regret theory in the existing literature has made some corrections to the perceived utility value, but in the process of assembling each scheme, the commonly used linear weighting method is used, which does not take into account that there may be certain correlations between various attributes. Prospect theory involves many parameters and complex calculations, so it is not suitable for rapid decision-making. The language and decision-making ideas used in this paper can better integrate the real ideas of experts into the data, and the result is reasonable.

Table 3. Comparison of results of different methods

Method	Sort	Optimal
Generalized regret theory	$O_3 > O_1 > O_2 > O_4$	O_3
Prospect theory	$O_3 > O_1 > O_2 > O_4$	O_3
Three-parameter weighted Heronian average operator and improved regret theory	$O_3 > O_1 > O_2 > O_4$	O_3

5   Conclusion

In order to solve the influence of the decision maker's psychological characteristics on the positive and negative ideal solutions and the interaction between different attributes, this paper integrates regret theory and Heronian operator into multiattribute decision-making. The initial assessment information given by the decision-maker may not be completely accurate, and it should be revised to reduce the influence of the obvious unreasonable plan on the final result. In terms of weight, it uses node density and information entropy calculation to overcome the defect of subjective weighting. This paper fully considers the regret avoidance psychology of decision makers, determines the positive and negative ideal schemes, defines the perceived utility function that conforms to human thinking, and uses the Heronian operator to gather information for sorting. In the case of unknown attribute weights and unknown expert weights, the algorithm can also make decisions, and can also be applied to rapid decision-making, enriching and developing multi-attribute decision-making.

References

1. Zhou, H.J., Mao, X.Y.: A review of research on hesitant fuzzy sets in decision theory. Adv. Math. Res. **24**(04), 1–5 (2021)
2. Zadeh, L.A.: The concept of a linguistic variable and its application to approximate reasoning-I. Information Sciences **8**(3), 199–249 (1975)
3. Rodriguez, R.M., Martinez, L., Herrera, F.: Hesitant fuzzy linguistic term sets for decision making. IEEE Trans. Fuzzy Syst. **20**(1), 109–119 (2012)
4. Pang, Q., Wang, H., Xu, Z.: Probabilistic linguistic term sets in multi-attribute group decision making. Inf. Sci., 128–143 (2016)
5. Meng, L.: The application of cardinal utility theory in the operation of B2C e-commerce platform. Mod. Mark. (late issue), 130 (2017)
6. Wu, H., Ang, S.: Two stage DEA cross efficiency evaluation model based on prospect theory. Oper. Res. Manage. **30**(11), 53–59 (2021)

7. Yang, J.P., Shi, C.X., Chiew, D., Qiu, J., Treepongkaruna, S.: Fund rating method based on expected utility-entropy model and its application in Chinese fund rating. China Manage. Sci. **27**(12), 1–10 (2019)
8. Zhang, F.F., Wang, W.M.: Linguistic multi-attribute decision-making method based on regret theory and DEMATEL. China Manage. Sci. **28**(06), 201–210 (2020)
9. De Bell.: Regret in decision making under uncertainty. Oper. Res. **30** (1982)
10. Graham, L., Robert, S.: An alternative theory of rational choice under uncertainty. Econ. J. **368**, 805–824 (1982)
11. Li, J.J., Li, T.T., Bao, Y.E., Chen, M.H.: Interval Pythagorean fuzzy power geometric heronian mean operator and its application in multi-attribute group decision making. Oper. Res. Manage. **30**(12), 78–83 (2021)

TODIM Multi-attribute Decision-Making Method Based on Spherical Fuzzy Sets

Wenyu Zhang[1,2], Weina Luo[1(✉)], Xue Gao[1], Changyou Zhang[1], and Keya Wang[3]

[1] Xi'an University of Posts and Telecommunications, Xi'an 710016, China
zwy888459@xupt.edu.cn,
17392189719@stu.xupt.edu.cn,gongmliang@gmail.com,17392189719@163.com,
gx876645@stu.xupt.edu.cn, changyou_zhang@stu.xupt.edu.cn
[2] China Research Institute of Aerospace Systems Science and Engineering, Beijing 100048, China
[3] Lianhu District Urban Management and Comprehensive Law Enforcement Bureau, Xi'an 710082, China

Abstract. To solve the problem of wider fields of preference, taking into account the influence of risk attitudes of decision makers on decision outcomes and their membership degree, non-membership degree and hesitation degree squared less than or equal to 1, this article proposes to deploy the TODIM method extends to the spherical fuzzy set (SFS) to solve the problem of multi-attribute decision making. First, construct the initial decision matrix of alternatives on attributes and the weight matrix of decision makers on attributes; Then select the reference attributes to get the relative weight of each attribute relative to the reference attribute; The size of the score function value is used as the judgment basis to obtain the dominance among the alternatives; finally, the overall relative dominance of each alternative is calculated to obtain the ranking result. By selecting an example of a green supplier, the feasibility and effectiveness of the method are checked, comparison with the SFS-TOPSIS, SFS-VIKOR and SFS-DEMATEL methods. The results show that the method proposed in this paper does not ignore the decision makers. The impact of risk attitudes on decision making outcomes is in line with the actual situation of decision makers and is effective and achievable.

Keywords: Multi-attribute decision-making · Spherical fuzzy set · TODIM

1 Introduction

Multi-attribute decision making (MADM) is a decision-making problem in modern decision science where multiple decision makers choose the best alternative under multiple attributes. It has been deeply studied by many researchers and

Shaanxi Education and Scientific Research Project (08JK431).

widely used in practical problems such as education, military, engineering, society, etc. [1]. In the MADM problem, the risk attitude of the decision maker is an indispensable consideration factor, and TODIM considers the psychological behavior of the decision maker under the risk factor, which is a very valuable method to deal with this factor. To fully describe uncertain information, Zadeh (1965) proposed fuzzy sets characterized by the description of fuzzy phenomena and the resolution of fuzzy problems [2]. Atanassov (1986) proposed intuitionistic fuzzy (IFSs), which is characterized by the fact that the degree of membership and non-membership is taken into account and that the sum is limited to a maximum of 1 [3]. Yager (2017) proposed Pythagorean Fuzzy Sets (PyFSs), satisfying that the sum of the squares of the membership degree and the nonmembership degree is less than 1 [4]. However, to expand the scope of the definition of decision makers, Kutlu Gündoğdu and Fatma (2019) lead into the concept of spherical fuzzy set (SFS) [5] in the display of life, which resolved that the decision maker's hesitation degree can be independent of the degree of membership and nonmembership. For most spherical fuzzy decision problems, the current multi-attribute decision-making methods for ranking alternatives include SFTOPSIS [5], SF-VIKOR [6], SF-AHP [7], SF-DEMATEL [8], etc. method. Shen Lingling (2019) uses the score function to propose a TODIM method based on probabilistic language terminology [9]. Liu Haidong (2019) proposed the distance measure of standard deviation preference on the basis of hesitating intuitionistic fuzzy language set and language scale function [10].

In this paper, the TODIM method is extended to spherical fuzzy sets (SFS) to solve the multi-attribute decision problem when attribute weights are uncertain. This method fully takes into account the risky psychological behavior of the decision maker and effectively improves the accuracy of decision-making.

2   Spherical Fuzzy Set

Definition 1. Spherical fuzzy set, let the universe set be U, is said to be SFS on U.

$$\tilde{A}_S = \left\{ \langle u, \left(\mu_{\tilde{A}_s}(u), v_{\tilde{A}_s}(u), \pi_{\tilde{A}_s}(u) \right) \middle| u \in U \right\} \tag{1}$$

where $\mu_{\tilde{A}_s} : U \to [0,1], v_{\tilde{A}_s} : U \to [0,1], \pi_{\tilde{A}_s} : U \to [0,1]$, are respectively represent the membership degree, non-membership degree, and hesitation degree of element u belonging to $\tilde{A}_S$. The following condition: $(\forall u \in U)$, and $(0 \leq \mu_{\tilde{A}_s}^2(u) + v_{\tilde{A}_s}^2(u) + \pi_{\tilde{A}_s}^2(u)1)$.

Definition 2. Assuming that $\tilde{A}_S = \left(\mu_{\tilde{A}_s}, v_{\tilde{A}_s}, \pi_{\tilde{A}_s} \right)$ and $\tilde{B}_S = \left(\mu_{\tilde{B}_s}, v_{\tilde{B}_s}, \pi_{\tilde{B}_s} \right)$ be any two spherical fuzzy numbers (SFNs), in which $\omega = (\omega_1, \omega_2, \cdots\cdots, \omega_n)$ be the weight vector, with $\omega_i \in [0,1]$ and $\sum_{i=1}^{N} \omega_i = 1$, then Spherical weighted arithmetic mean (SWAM) describes as:

$$\text{SWAM}_\omega \left(\tilde{A}_{S_1}, \cdots \cdots, \tilde{A}_{S_n} \right) = \omega_1 \tilde{A}_{S_1} + \omega_2 \tilde{A}_{S_2} + \cdots \cdots + \omega_n \tilde{A}_{S_n}$$

$$= \left\{ \left[1 - \prod_{i=1}^{n} \left(1 - \mu_{\tilde{A}_{S_i}}^2 \right)^{\omega_i} \right]^{\frac{1}{2}}, \prod_{i=1}^{n} v_{\tilde{A}_{s_i}}^{\omega_i}, \right.$$

$$\left. \left[\prod_{i=1}^{n} \left(1 - \mu_{\tilde{A}_{s_i}}^2 \right)^{\omega_i} - \prod_{i=1}^{n} \left(1 - \mu_{\tilde{A}_{s_i}}^2 - \pi_{\tilde{A}_{s_i}}^2 \right)^{\omega_i} \right]^{\frac{1}{2}} \right\} \tag{2}$$

Definition 3. The score functions and the precision function of the SFS sort are defined by

$$\text{Score}\left(\tilde{A}_S \right) = \left(2\mu_{\tilde{A}_s} - \pi_{\tilde{A}_s} \right)^2 - \left(v_{\tilde{A}_s} - \pi_{\tilde{A}_s} \right)^2 \tag{3}$$

Note that: $\tilde{A}_S < \tilde{B}_S$ if and only if $\text{Score}\left(\tilde{A}_S \right) < \text{Score}\left(\tilde{B}_S \right)$ or $\text{Score}\left(\tilde{A}_S \right) <$ $\text{Score}\left(\tilde{B}_S \right)$ and $\text{Accuracy}\left(\tilde{A}_S \right) < \text{Accuracy}\left(\tilde{B}_S \right)$.

Definition 4. [5]. The equation of Euclidean distance between $\tilde{A}_S$ and $\tilde{B}_S$ on the surface of the sphere:

$$\text{dis}\left(\tilde{A}_S, \tilde{B}_S \right) = \sqrt{\left(\left(\mu_{\tilde{A}_s} - \mu_{\tilde{B}_s} \right)^2 + \left(v_{\tilde{A}_s} - v_{\tilde{B}_s} \right)^2 + \left(\pi_{\tilde{A}_s} - \pi_{\tilde{B}_s} \right)^2 \right)} \tag{4}$$

We have that $0 \, dis\left(\tilde{A}_S, \tilde{B}_S \right) 1$ and $\mu_{\tilde{A}}^2 + v_{\tilde{A}}^2 + \pi_{\tilde{A}}^2 = 1, \mu_{\tilde{R}}^2 + v_{\tilde{R}}^2 + \pi_{\tilde{R}}^2 = 1$.

3   TODIM Multi-Attribute Decision-Making Method Based on Spherical Fuzzy Sets

This section presents a multi-attribute TODIM decision-making method based on spherical fuzzy sets (SFS).

Let $X_i(i = 1, 2, \cdots, m)$ is a limited set of alternatives in decision-making, $C_j(j = 1, 2, \cdots, n)$ is a limited set of properties. For spherical (SFS)of the MADM problem can be solved with a spherical fuzzy decision matrix $D = (X_{ij})_{m \times n}$ express, $X_{ij} = (\mu_{ij}, v_{ij}, \pi_{ij})$ indicating alternatives X_i about the criteria C_j evaluation value. Use $\omega = (\omega_1, \omega_2, \cdots, \omega_n)^T$ weight vector representing the attribute, where ω_j is an attribute C_j weight, satisfying $0 \leq \omega_j \leq 1$ and $\sum_{j=1}^{n} \omega_j = 1$.

Step 1: According to Table 1, construct the initial decision matrix about the attributes of the alternatives $D = (X_{ij})_{m \times n}$.

Table 1. Language term set and matching spherical fuzzy numbers.

Language term	(μ, v, π)	Language term	(μ, v, π)
Definitely more important (AMI)	$(0.9, 0.1, 0.1)$	Slightly less important (SLI)	$(0.4, 0.6, 0.4)$
Very high importance (VHI)	$(0.8, 0.2, 0.2)$	Low importance (LI)	$(0.3, 0.7, 0.3)$
High importance (HI)	$(0.7, 0.3, 0.3)$	Very low importance (VLI)	$(0.2, 0.8, 0.2)$
Slightly more important (SMI)	$(0.6, 0.4, 0.4)$	Absolutely low importance (ALI)	$(0.1, 0.9, 0.1)$
Equal importance (EI)	$(0.5, 0.5, 0.5)$		

Step 2: The initial decision matrix criterion is normalized as $H = (h_{ij})$, the cost attribute Ω_c is transformed into a benefit attribute Ω_b.

$$h_{ij} = (\mu_{ij}, v_{ij}, \pi_{ij}) = \begin{cases} (\mu_{ij}, v_{ij}, \pi_{ij}), & C_j \in \Omega_b \\ (\pi_{ij}, \mu_{ij}, v_{ij}), & C_j \in \Omega_c \end{cases} \tag{5}$$

Step 3: Use Definition 2 Spherical Weighted Average Operator (SWAM) Aggregate decision matrix.

Step 4: According to Table 1, the decision maker's weight on attributes is constructed, and the spherical weighted average operator is defined according to 2(SWAM) aggregate attribute weights.

Step 5: Aggregated attribute weights are defuzzified and normalized using a score function.

$$\omega_j^S = (2\mu_{ij} - \pi_{ij})^2 - (v_{ij} - \pi_{ij})^2 \tag{6}$$

$$\bar{\omega}_j^S = \frac{\omega_j^S}{\sum_{j=1}^n \omega_j^S} \tag{7}$$

Step 6: Select reference properties C_r (Usually select the attribute with the largest weight as the reference attribute) calculate each attribute C_j relative to the reference property's C_r relative weight.

$$\omega_{jr}^* = \frac{\bar{\omega}_j^S}{\bar{\omega}_r^S} \tag{8}$$

Step 7: Calculate each property C_j below, alternative X_i the dominance between them. In, θ the lower the value, the greater the loss aversion of decision makers. $d(h_{ij}, h_{kj})$ express h_{ij} and h_{kj} the distance between.

$$\varphi_j(X_i, X_k) = \begin{cases} \sqrt{\frac{\omega_{jr}^*}{\sum_{i=1}^n \omega_{jr}^*} d(h_{ij}, h_{kj})}, & S(h_{ij}) - S(h_{kj}) > 0 \\ 0, & S(h_{ij}) - S(h_{kj}) = 0 \\ -\frac{1}{\theta} \sqrt{\frac{(\sum_{i=1}^n \omega_{jr}^*) d(h_{ij}, h_{kj})}{\omega_{jr}^*}}, & S(h_{ij}) - S(h_{kj}) < 0 \end{cases} \tag{9}$$

Step 8: Calculate alternatives' X_i overall dominance between them.

$$\eta(X_i, X_k) = \sum_{j=1}^{n} \varphi_j(X_i, X_k) \tag{10}$$

Step 9: Calculate the overall relative dominance for each alternative.

$$\xi(X_i) = \frac{\sum_{i=1}^{m} \eta(X_i, X_k) - \min_i \{\sum_{i=1}^{m} \eta(X_i, X_k)\}}{\max_i \{\sum_{i=1}^{m} \eta(X_i, X_k)\} - \min_i \{\sum_{i=1}^{m} \eta(X_i, X_k)\}}, i = 1, 2, \cdots, m \tag{11}$$

Step 10: Rank the options according to their overall relative advantage and determine the best option. $\xi(X_i)$ The larger the value, the optimal solution.

4 Examples

4.1 Example Analysis

An existing enterprise has four alternative green suppliers $X = (X_1, X_2, X_3, X_4)$ Selection decisions, by three experienced experts $DM = (DM_1, DM_2, DM_3)$ To form a green supplier judge, consider the following four attributes $C = (C_1, C_2, C_3, C_4)$ assessment, C_3 is a cost attribute. the three experts with different experience levels were weighted at $0.3, 0.4$, and 0.3, respectively. All evaluation results are shown in Table 2.

Step 1: Build an initial decision matrix of experts on alternatives $D = (X_{ij})_{4 \times 4}$.
   (1) Construct expert judgments of alternatives based on the linguistic terms in Table 1 and construct the initial decision matrix are shown in Table 2.

Table 2. DM_1, DM_2, DM_3 initial decision matrix.

Decision maker	Options	C_1	C_2	C_3	C_4
DM_1	X_1	$(0.7, 0.3, 0.3)$	$(0.2, 0.8, 0.2)$	$(0.4, 0.6, 0.4)$	$(0.9, 0.1, 0.1)$
	X_2	$(0.3, 0.7, 0.3)$	$(0.8, 0.2, 0.2)$	$(0.7, 0.3, 0.3)$	$(0.9, 0.1, 0.1)$
	X_3	$(0.5, 0.5, 0.5)$	$(0.6, 0.4, 0.4)$	$(0.5, 0.5, 0.5)$	$(0.7, 0.3, 0.3)$
	X_4	$(0.3, 0.7, 0.3)$	$(0.7, 0.3, 0.3)$	$(0.9, 0.1, 0.1)$	$(0.7, 0.3, 0.3)$
DM_2	X_1	$(0.3, 0.7, 0.3)$	$(0.6, 0.4, 0.4)$	$(0.5, 0.5, 0.5)$	$(0.6, 0.4, 0.4)$
	X_2	$(0.8, 0.2, 0.2)$	$(0.7, 0.3, 0.3)$	$(0.7, 0.3, 0.3)$	$(0.8, 0.2, 0.2)$
	X_3	$(0.4, 0.6, 0.4)$	$(0.6, 0.4, 0.4)$	$(0.9, 0.1, 0.1)$	$(0.5, 0.5, 0.5)$
	X_4	$(0.2, 0.8, 0.2)$	$(0.4, 0.6, 0.4)$	$(0.4, 0.6, 0.4)$	$(0.8, 0.2, 0.2)$
DM_3	X_1	$(0.6, 0.4, 0.4)$	$(0.7, 0.3, 0.3)$	$(0.5, 0.5, 0.5)$	$(0.6, 0.4, 0.4)$
	X_2	$(0.8, 0.2, 0.2)$	$(0.6, 0.4, 0.4)$	$(0.8, 0.2, 0.2)$	$(0.5, 0.5, 0.5)$
	X_3	$(0.8, 0.2, 0.2)$	$(0.2, 0.8, 0.2)$	$(0.8, 0.2, 0.2)$	$(0.7, 0.3, 0.3)$
	X_4	$(0.7, 0.3, 0.3)$	$(0.3, 0.7, 0.3)$	$(0.4, 0.6, 0.4)$	$(0.9, 0.1, 0.1)$

Step 2: Standardize the initial decision matrix to construct a normalized decision matrix $H = (h_{ij})_{m \times n}$.

Table 3. Aggregation decision matrix.

	C_1	C_2	C_3	C_4
X_1	$(0.558, 0.459, 0.342)$	$(0.570, 0.452, 0.339)$	$(0.473, 0.528, 0.477)$	$(0.745, 0.264, 0.295)$
X_2	$(0.724, 0.291, 0.226)$	$(0.713, 0.290, 0.298)$	$(0.274, 0.266, 0.733)$	$(0.793, 0.214, 0.250)$
X_3	$(0.609, 0.409, 0.368)$	$(0.527, 0.492, 0.373)$	$(0.312, 0.200, 0.782)$	$(0.636, 0.368, 0.380)$
X_4	$(0.468, 0.573, 0.281)$	$(0.509, 0.510, 0.344)$	$(0.343, 0.351, 0.653)$	$(0.819, 0.183, 0.239)$

Step 3: According to Eq. (2) SWAM. The operator aggregates the normalized decision matrix are shown in Table 3.

Step 4: Calculate attribute weights. Construct attribute evaluation values according to Table 1 and use formula (2) SWAM. The operator aggregates attribute weights are shown in Table 4.

Table 4. Attribute weight matrix and aggregate attribute weight.

	Attribute weight matrix			Aggregate attribute
	DM_1	DM_2	DM_3	weights
C_1	EI	VHI	HI	$(0.708, 0.297, 0.318)$
C_2	HI	LI	SLI	$(0.503, 0.518, 0.335)$
C_3	VHI	EI	HI	$(0.681, 0.326, 0.348)$
C_4	EI	AMI	SLI	$(0.743, 0.277, 0.289)$

Step 5: The attribute weights aggregated in Table 4 are de-fuzzified and normalized according to Eqs. (6) and (7) are shown in Table 5.

Table 5. Defuzzification and normalization of attribute weights.

	C_1	C_2	C_3	C_4
Defuzzification and normalization of attribute weights	0.3136	0.1085	0.2051	0.3728

Step 6: Pick the maximum attribute weight C_4 as the reference attribute, the relative weight of each attribute relative to the reference attribute is calculated according to Eq. (8) are shown in Table 6.

Table 6. Relative weights.

	C_1	C_2	C_3	C_4
Relative weight relative to C_4	0.8412	0.2909	0.5501	1

Step 7: Calculate the degree of dominance among alternatives under each attribute, taking $\theta = 1$.

(1) According to the definition 4 score function, calculate the value of the score function of the normalized decision matrix Table 4, as the judgment condition for calculating the dominance degree between the alternatives are shown in Table 7.

(2) According to formula (9), calculate C_j, the degree of dominance between the alternatives. E.g: Table 8 is about the calculate C_1.

Table 7. Aggregate decision matrix score function values.

	C_1	C_2	C_3	C_4
X_1	0.5865	0.6292	0.2179	1.4287
X_2	1.4901	1.2716	−0.1846	1.7859
X_3	0.7200	0.4479	−0.3146	0.7959
X_4	0.3427	0.4269	−0.0902	1.9527

Table 8. About calculate C_1, the degree of dominance between the alternatives.

	C_1	C_2	C_3	C_4
X_1	0	−0.9155	−0.4910	0.2224
x_2	0.2871	0	0.2610	0.3474
X_3	0.1540	−0.8322	0	0.2703
x_4	−0.7093	−1.1078	−0.8620	0

Table 9. Scheme X_i overall dominance.

	X_1	X_2	X_3	X_4
X_1	0	−2.5179	0.1622	−0.0145
X_2	−0.8092	0	0.8821	−0.6158
X_3	−2.8489	−3.9478	0	−1.5460
X_4	−1.1387	−2.4915	−0.9228	0

Step 8: Calculate the alternative according to Eq. (10). X_i the overall dominance between them is shown in Table 9.

Step 9: From the data in Table 9, calculate the overall relative dominance of each alternative according to Eq. (11) $\xi(X_i)$. Which is $\xi(X_1) = 0.7657, \xi(X_2) = 1, \xi(X_3) = 0, \xi(X_4) = 0.4859$.

Step 10: Due to this $\xi(X_2) > \xi(X_1) > \xi(X_4) > \xi(X_3)$, therefore $X_2 > X_1 > X_4 > X_3$, is the result of the ranking of the four alternatives, that the second green supplier is the best choice.

4.2   Analysis of Results

Illustrate the effectiveness and feasibility of the decision-making method in this article, the existing literature proposed SF-TOPSIS [5], SF-VIKOR [6], SFDE-MATEL [8]. The methods were compared and analyzed, and the results are shown in Table 10.

The comparison of the three results above shows that the ranking results obtained with different decision methods are not very different and the optimal solutions are all the same, demonstrating the effectiveness and feasibility of the decision. The SF-TODIM method used in this paper obtains $X_4 > X_3$, while

Table 10. Comparison of methods.

Method	Sort	Optimal solution
SF-TOPSIS	$X_2 > X_1 > X_3 > X_4$	X_2
SF-VIKOR	$X_2 > X_1 > X_3 > X_4$	X_2
SF-TODIM	$X_2 > X_1 > X_4 > X_3$	X_2
SF-DEMATEL	$X_2 > X_1 > X_3 > X_4$	X_2

the other methods get $X_3 > X_4$. The reason for the different results is that this paper considers the psychological and behavioral factors of decision makers under risk.

5 Conclusion

Aiming at the decision maker's psychological behavior and its membership degree, non-membership degree, and hesitant degree, the sum of squares is less than or equal to 1. This paper will TODIM method extended to spherical fuzzy sets (SFS), proposed a method to solve the multi-attribute decision-making problem. Fully considering the psychological risk behavior of decision makers. At the same time, a spherical fuzzy weighted average operator is introduced to aggregate the normalized decision matrix and attribute matrix respectively. A score function is introduced to defuzzify the aggregated attribute weights and calculate the dominance among alternatives under each attribute. In dealing with multi-attribute decisionmaking problems, the method proposed in this paper fully considers the decision maker's psychological behavior under risk and reduces the impact on the decisionmaking result caused by F ignoring the decision maker's risk attitude. This method is closer to the real situation of the decision maker, so the decision results are more reliable and efficient.

References

1. Meng, F.Y., Chen, X.H., Zhang, Q.: Multi-attribute decision analysis under a linguistic hesitant fuzzy environment. Inf. Sci.: Int. J. **267**, 287–305 (2014)
2. Liu P., You X.: Bidirectional projection measure of linguistic neutrosophic numbers and their application to multi-criteria group decision making. Comput. Ind. Eng. **128**(2), 447–457(2018)
3. Atanassov, K.T.: Intuitionistic fuzzy sets. Fuzzy Sets Syst. **20**(1), 87–96 (1999)
4. Yager, R.R.: Pythagorean membership grades in multicriteria decision making. IEEE Trans. Fuzzy Syst. **22**(4), 958–965 (2014)
5. Kutlu Gündoğdu, F., Kahraman, C.: Spherical fuzzy sets and spherical fuzzy TOPSIS method. J. Intell. Fuzzy Syst. 1–16 (2018)
6. Gündodu, F.K., Kahraman, C.: A novel VIKOR method using spherical fuzzy sets and its application to warehouse site selection. J. Intell. Fuzzy Syst. **37**(1), 1–15 (2019)

7. Gündodu, F.K., Kahraman, C.: A novel spherical fuzzy analytic hierarchy process and its renewable energy application. Soft Comput. **24**(5) (2020)

8. Gül, S.: Spherical fuzzy extension of DEMATEL (SF-DEMATEL). Int. J. Intell. Syst. 35 (2020)

9. Shen, L.L., Pang, X.D., Zhang, Q., Qian, G.: TODIM method based on probabilistic language term set and its application. Stat. Decis. **35**(18), 80–83 (2019)

10. Liu, D.H., Liu, Y.Y., Chen, X.H.: Research on multiple attribute decision making based on hesitant intuitive fuzzy language set mean-standard deviation preference distance. China Manage. Sci. **27**(01), 174–183 (2019)

Modelling of Fuzzy Discrete Event Systems Based on a Generalized Linguistic Variable and Their Generalized Possibilistic Kriple Structure Representation

Shengli Zhang[1(✉)] and Jing Chen[2]

[1] School of Information Technology, Minzu Normal University of Xingyi, Xingyi, Guizhou 562400, China
{zhangshengli,chenjing}@xynun.edu.cn,zsl8203@163.com
[2] College of Economics and Management, Minzu Normal University of Xingyi, Xingyi, Guizhou 562400, China

Abstract. In order to provide a model checking method and convenient and efficient modeling mechanism for the fuzzy discrete event system (FDES), we apply the generalized possibilistic Kriple structure and generalized linguistic variable with three different forms of negations to FDES, respectively. First, we study how to transfer a FDES to the generalized possibilistic Kriple structure and propose the transformation method. Second, the linguistic variable is extended to the generalized linguistic variable by introducing three different forms of negations, so that FDES is modeled conveniently and efficiently based on the generalized linguistic variable. Finally, some cases are given to demonstrate the methods that are presented in this paper.

Keywords: Fuzzy discrete event system · Generalized possibilistic kriple structure · Negative information · Opposite negation · Medium negation · Generalized linguistic variable

1 Introduction

An important theorem of discrete event systems (DESs, in short) has been used and gives one of the formal methods for a lot of engineering systems [1]. There are, however, a number of situations where the state transfers of a system are, in some sense, fuzzy, imprecise, and uncertain. To deal with the indeterminacy occurring in states and state transfers of DESs, Lin and Ying [2] combined a fuzzy finite automaton model with DES and proposed fuzzy DES (FDES).

As we know, the classical model checking [3] is a formal verification technique which has been applied in many social and natural fields. However, a lot of uncertainty and inconsistency information inevitably appears in most large and complex systems. In order to handle the systematic verification on uncertain systems

© The Author(s), under exclusive license to Springer Nature Switzerland AG 2023
N. Xiong et al. (Eds.): ICNC-FSKD 2022, LNDECT 153, pp. 437–446, 2023.
https://doi.org/10.1007/978-3-031-20738-9_50

with non-deterministic information in possibility theory, Li etc. [4,5] introduced quantitative computation tree logic and linear-time logic model checking in which the uncertainty is modeled by the generalized possibilistic Kriple structure. Subsequently, Zhang and Li [6] further studied the expressiveness of linear-temporal logic on the basis of generalized possibility measures.

Although the study of the FDES theory has been broaden to address various issues (see [7] for more detail and references therein), there are many important issues to solve. The first important problem is how to transfer a FDES to the generalized possibilistic Kriple structure (see below for more detail) proposed by Li etc. in order to provide a model checking method for the FDES. The second problem is how to model a fuzzy DES efficiently and conveniently. As we know, it is nontrivial to study whether there are effective mathematical model to solve FDES modelling problems. The surprising result of this paper is that the FDES is modeled conveniently and efficiently based on the generalized linguistic variable as stated below. In this paper the complete study of the above two problems will be given.

2   Preliminaries

2.1   Fuzzy Information and Its Different Negation

Negative information plays a very important role in knowledge representation and reasoning. In the past decades, some scholars suggested that uncertain information processing requires different forms of negations in various fields [8–14]. Given arbitrary concept, denoted as w, the opposite negation of w could be written as $\lrcorner w$, in which the symbol $\lrcorner$ is referred to as an opposite negative operator. Consequently, a pair of the opposite negative concepts are represented by using w and $\lrcorner w$. As widely known, the formal symbol $\neg$ is illustrated as "not" indicating the classical negation. So, the symbol $\neg w$ is used to depict the contradictory negation of a concept w. Therefore, a pair of contradictory negative concepts are depicted by w and $\neg w$.

On the other hand, in our surrounding there are a large number of medium concepts between the two opposite concepts, e.g. "dusk" is a medium state from day to night, zero may be viewed as a intermediate state between the positive and negative and so on. Therefore, Pan et al. [12–14] proposed that the medium fuzzy notion(concept) should be viewed as a new negation of a pair of opposite fuzzy notions, referred to as "medium negation" of the opposite fuzzy concepts. We denote the medium negation of any concept w by $\sim w$, in which the symbol $\sim$ is called a medium negative operator and pronounced "to have partially". Consequently, $\sim w$ is used to express the medium negation of a concept w(or $\lrcorner w$). Furthermore, the medium negative relationship is represented by the relationship between $\sim w$ and a pair of opposite negations $w, \lrcorner w$.

2.2 Fuzzy Discrete Event Systems and a Generalized Possibilistic Kriple Structure

The FDES is modeled by max-product fuzzy automata in the literature [2]. However, in this paper, a fuzzy DES is modeled by a max-min fuzzy automaton in order to represent a fuzzy DES by using the generalized possibilistic Kriple structure proposed below, that is,

$$\tilde{G} = (\tilde{Q}, \tilde{\Sigma}, \tilde{\delta}, \tilde{q}_0)$$

where is $\tilde{Q}$ a fuzzy state space, $\tilde{\Sigma}$ is the set of fuzzy events, each $\tilde{q} \in \tilde{Q}$ can be denoted by a vector $(x_0, x_1, \ldots, x_{n-1})$, where x_i indicates the membership grade of the system being in state q_i for $i = 0, 1, \ldots, n-1$. A fuzzy event $\tilde{\sigma}$ is a fuzzy matrix $[a_{ij}]_{n \times n}$ in which $[a_{ij}] \in [0, 1]$ shows the possibility of the system transferring from the state q_i to q_j when $\tilde{\sigma}$ happens. The partial state transition function $\tilde{\delta} : \tilde{Q} \times \tilde{\Sigma} \to \tilde{Q}$ is defined as $\tilde{\delta}(\tilde{q}, \tilde{\sigma}) = \tilde{q} \odot \tilde{\sigma}$, where $\odot$ is the max-min operator for an $n \times m$ matrix $A = [a_{ij}]_{n \times m}$ and an $m \times k$ matrix $B = [B_{ij}]_{m \times k}$; then, an element in the $n \times k$ matrix $C = A \odot B$ is $c_{ij} = \max_{l=1}^{m} \min\{a_{il}, b_{lj}\}$.

Definition 1. [4,6] A generalized possibilistic Kripke structure, written as GPKS briefly, is defined by a tuple $M = (S, P, I, AP, L)$, where

(1) S represents a nonempty, countable set of states;
(2) a map $P : S \times S \longrightarrow [0, 1]$, referred to as possibilistic transition distribution;
(3) a possibilistic initial distribution is the following function $I : S \longrightarrow [0, 1]$;
(4) AP shows a set of atomic propositions;
(5) a map $L : S \to [0, 1]^{AP}$ is called a generalized labeling function, where $[0, 1]^{AP}$ indicates all fuzzy sets within AP.

In addition, we call $M = (S, P, I, AP, L)$ a finite generalized possibilistic Kripke structure if the sets S and AP are finite.

3 Generalized Possibilistic Kripke Structure Representation of Fuzzy DESs

Let $\tilde{G} = (\tilde{Q}, \tilde{\Sigma}, \tilde{\delta}, \tilde{q}_0)$ be a fuzzy DES, the corresponding generalized possibilistic Kriple structure $M = (S, P, I, AP, L)$ is constructed in the following ways.

Step 1. Use the state set S, atomic proposition set AP and generalized labeling function L to depict the fuzzy state space $\tilde{Q}$. More precisely, let $|S| = n = $ the dimension number of fuzzy state space $\tilde{Q} = |Q|$, $L = \bigvee_{\tilde{\sigma} \in \tilde{\Sigma}} \{ \bigwedge_{k=1}^{\max - \tilde{\sigma}} L_{\tilde{\sigma}}^{(k)} \}$, where $|\Delta|$ represents the number of elements in set $\Delta (\Delta \in \{S, Q\})$, Q is a crisp state space corresponding to $\tilde{Q}$, and $L_{\tilde{\sigma}}^{(k)}$ is a generalized labeling function induced by using the fuzzy event $\tilde{\sigma}$ for the kth time, called sub-generalized labeling function (see subsequent explanation), max-$\tilde{\sigma}$ is the maximum time of employing $\tilde{\sigma}$ in a fuzzy DES. Given a fuzzy event $\tilde{\sigma} \in \tilde{\Sigma}$, for each $q_i \in Q$, i.e., $\tilde{q} \in \tilde{Q}$, we take the state $s_i \in S$, $A_{\tilde{q}_i} \in [0,1]^{AP}$ and $L_{\tilde{\sigma}}^{(k)}$ such that $s_i = q_i$ and $A_{\tilde{q}_i} = L_{\tilde{\sigma}}^{(k)}(s_i) = \tilde{q}_i = (x_0, x_1, \ldots, x_{n-1})$ for $i = 0, 1, \ldots, n-1$. Additionally, let the initial distribution $I(s_0)$ be equal to one otherwise zero, where s_0 corresponds to the initial state $\tilde{q}_0 \in \tilde{Q}$.

Step 2. For each fuzzy event $\tilde{\sigma} \in \tilde{\Sigma}$, construct a possibilistic transition relation $P_{\tilde{\sigma}}$(called sub-generalized transition possibility distribution, sub-transition distribution for short) satisfying $P_{\tilde{\sigma}} = \tilde{\sigma}$. Assume $P = \bigcup_{\tilde{\sigma} \in \tilde{\Sigma}} P_{\tilde{\sigma}}$. If we assume $\bigcup$ takes $\vee$, thus, P equips with a concrete meaning, that is, P can be viewed as a the least upper bound (or supremum) a collection of the possibilities of transitions driven by fuzzy events.

Since the membership grades of a fuzzy event simultaneously belonging to the observable, the unobservable and failure event set, are different, so does a sub-transition distribution. We write the unobservable sub-transition distribution fuzzy subset, the observable sub-transition distribution fuzzy subset and the failure sub-transition distribution fuzzy subset as $P_{uo} : P \to [0, 1]$, $P_o : P \to [0, 1]$ and $P_f : P \to [0, 1]$, respectively. Similarly, $P_{uo}(P_{\tilde{\sigma}}) + P_o(P_{\tilde{\sigma}}) = 1$ and $P_f(P_{\tilde{\sigma}})$ represents the possibility of failure occurring on $P_{\tilde{\sigma}} \subseteq P$.

Step 3. The partial state transition function $\tilde{\delta}$ can be redefined as $\tilde{\delta}(\tilde{q}, \tilde{\sigma}) = L_{\tilde{\sigma}}^{(k)}(q) \odot P_{\tilde{\sigma}} = A_{\tilde{q}} \odot P_{\tilde{\sigma}}$ with $\odot$ being the max-min operator, for any $\tilde{q} \in \tilde{Q}$, $\tilde{\sigma} \in \tilde{\Sigma}$, $k = 1, 2, \ldots, \max - \tilde{\sigma}$.

From the above transformation procedure, one can readily get the result as follows.

Theorem 1. *In terms of modeling systems with fuzzy uncertainty, the generalized possibilistic Kriple structure is more powerful than a fuzzy DES.*

4   Modeling of Fuzzy DESs Based on a Generalized Linguistic Variable

4.1   Generalized Linguistic Variables

Definition 2. Given any universal discourse U and finite numerical district D, namely, it has the following forms:$[a, b]$, $(a, b]$, $[a, b)$, (a, b), or $\{a = x_1 < x_2 < - < x_n = b\}$, where $a, b \in \mathbb{R}$, referred to as the left and right end of D, respectively. Then the function $f : U \to D$ is called a finite quantized district function.

Definition 3. Given any universal discourse U, the function f is identical to Definition 2. A generalized linguistic variable is depicted as a tuple $(X, T(x), f(U), G, M)$ where X indicates the name of variables; $T(x)$(or breafly T)indicates the term set of X; $f(U)$ is a finite numerical district and is range of X taking on values, where a, b is, respectively, its left and right end; G expresses a syntax rule for generating the name of linguistic values of X; and M is a semantic rule for associating with each linguistic atomic word $t \in T(x)$. In addition, $\neg t$, $\exists t$ and $\sim t$ denote the contradictory, opposite and medium negation w.r.t. t, respectively, and their semantic meanings are defined by Eqs. (1, 2 and 3)

$$M(\neg t)(u) = n(M(t)(u)); \tag{1}$$
$$M(\exists t)(u) = M(a + b - u) \text{ and } M(t)(u) + M(\exists t)(u) \leq 1; \tag{2}$$
$$M(\sim t)(u) = M(\neg t)(u) * M(\neg \exists t)(u) = n(M(t)(u)) * n(n(M(t)(u))), \tag{3}$$

where $u \in f(u)$, $*$ is a t-norm, and n is a complement.

4.2 Modeling Process

In the sequel, we only consider the case of a single generalized linguistic variable, that is, each atomic fuzzy proposition A in AP belongs to the terms set of the identical generalized linguistic variable. Given a fuzzy event $\tilde{\sigma}$, the computational core of the membership degrees of all atomic propositions over all states can be described as a three-steps process as follows:

Step 1. Select an appropriate finite quantized district mapping f, which maps the state space S to a finite numerical district D;
Step 2. Determine membership function(s) of some (several) appropriate atomic proposition(s) according to practical experience, in general, the membership function is related to the states;
Step 3. Compute membership functions of the other propositions using the semantic rule of this generalized linguistic variable and an appropriate linguistic hedge.

Clearly, it is easy to see that the above computational process is suitable for multiple generalized linguistic variables and multiple fuzzy events. Indeed, one can repeat the above computational process simply.

5 Illustrative Cases

Case 1. Assume that an animal is getting sick because of a new disease. For the novel disease, the doctor is not enough familiar with it, however, by experience, he (or she) trusts that the following drugs, written as $\tilde{\alpha}$, $\tilde{\beta}$, $\tilde{\gamma}$ and $\tilde{\theta}$, respectively, have a possible effect on the disease.

For convenience, suppose that the doctor divides roughly the animal's condition to be three states, namely, "good", "fair", and "poor". As mentioned above, it is fuzzy when the animal's state is said to be "good", "fair", or "poor", because the animal's state can simultaneously belong to "good", "fair" or "poor" with respective membership degrees in the real-life situation [2]. Hence, when the generalized possibilistic Kriple structure representation of a fuzzy DES is used to model the treatment process of the animal, a sub-generalized labeling function is naturally written as a 3-D vector

$$L_{\tilde{\sigma}}(s_{\tilde{q}}) = A_{\tilde{q}} = [\overset{\text{high}}{a_1}, \overset{\text{medium}}{a_2}, \overset{\text{low}}{a_3}],$$

which is represented as the respective possibility degrees of some drug's symptom property index "high", "medium" and "low" (surely, other symptom property indexes are also possible for some given drug, such as "negative", "zero" and "positive") over state $s_{\tilde{q}}$.

Analogously, after a drug treatment (i.e., fuzzy event), it is inaccurate to say at what point precisely the animal has transformed from one state to another, because each drug event, when it occurs, may change from a state to multistates with different corresponding membership degrees. Therefore, in the treatment process modeled by the generalized possibilistic Kriple structure representation of a fuzzy DES, a sub-transition distribution (corresponding to a fuzzy event) is represented as a 3×3 matrix

$$P_{\tilde{\sigma}} = \begin{matrix} poor \\ fair \\ good \end{matrix} \begin{bmatrix} \overset{poor}{a_{11}} & \overset{fair}{a_{12}} & \overset{good}{a_{13}} \\ a_{21} & a_{22} & a_{23} \\ a_{31} & a_{32} & a_{33} \end{bmatrix}.$$

In the sequel, we only give the modeling procedure of the sub-transition distributions $P_{\tilde{\alpha}}$, $P_{\tilde{\beta}}$ (corresponding to drug events theophylline, ipratropium bromide, respectively), similar are others. Consider the drug event theophylline $\tilde{\alpha}$. For simplicity, we use the symbols h, m and l to represent the atomic propositions "high", "medium" and "low", respectively. According to the above-stated generalized linguistic variable, one can see that the atomic propositions l and m are opposite and medium negations of h, respectively, i.e., $l = \lrcorner h$, $m =\sim h$. Suppose the possibility distribution of the atomic proposition h over states "poor", "fair" and "good" for the fuzzy event $\tilde{\alpha}$ is the following vector:

$$L_{\tilde{\sigma}}(s_{\tilde{q}}) = A_{\tilde{q}} = [\overset{poor}{0.8}, \overset{fair}{0.3}, \overset{good}{0.1}].$$

According to the modeling process, we can take the finite quantized district function f as being $f(\text{poor}) = 1$, $f(\text{fair}) = 2$ and $f(\text{good}) = 3$. So it is not difficult to calculate the other atomic propositions as $l = \lrcorner h = (0.1, 0.3, 0.8)$, $f =\sim h = (0.2, 0.7, 0.2)$. Suppose the initial sub-generalized labeling function just is $h = (0.8, 0.3, 0.1)$, the drug event $\tilde{\alpha}$ is evaluated by use of medical theory and the doctor's experience as follows:

$$h \odot P_{\tilde{\alpha}} = f$$

where $\odot$ is the max-min operator. By the solution of fuzzy relation equation [15], we can obtain infinite many solutions by solving the above equation, where the greatest solution is the following:

$$P_{\tilde{\alpha}} = h \rightarrow_G f = \begin{bmatrix} 0.2 & 0.7 & 0.2 \\ 0.2 & 1 & 0.2 \\ 1 & 1 & 1 \end{bmatrix},$$

where $\rightarrow_G$ is Gödel implication, that is, $a \rightarrow_G b = \{\begin{array}{l} 1, \; a \leq b \\ b, \; \text{otherwise} \end{array}, \forall a, b \in [0, 1].$
By the medical theory and the doctor's experience, we can take $P_{\tilde{\alpha}}$ as

$$P_{\tilde{\alpha}} = \begin{bmatrix} 0.2 & 0.6 & 0.2 \\ 0 & 0.5 & 0.2 \\ 0 & 0 & 0.2 \end{bmatrix}.$$

For the drug event $\tilde{\beta}$, suppose the possibility distribution of the atomic proposition h (maybe corresponding to another symptom property index) over states "poor", "fair" and "good" is the following vector: $h = (0.2, 0.3, 0.6)$. Analogously, we can calculate the other atomic propositions as $l = \lnot h = (0.6, 0.3, 0.2)$, $f = \sim h = (0.4, 0.7, 0.4)$, and evaluate the drug $\tilde{\beta}$ according to doctor's experience and medical theory as follows: $f \odot P_{\tilde{\beta}} = l$. Similarly to the just above computational process, one can obtain infinite many solutions in which the greatest solution is as follows:

$$P_{\tilde{\beta}} = \begin{bmatrix} 1 & 0.3 & 0.2 \\ 0.6 & 0.3 & 0.2 \\ 1 & 0.3 & 0.2 \end{bmatrix}.$$

By the medical theory and the doctor's experience, we can take $P_{\tilde{\beta}}$ as

$$P_{\tilde{\beta}} = \begin{bmatrix} 0.3 & 0 & 0 \\ 0.6 & 0.3 & 0 \\ 0.3 & 0.3 & 0.2 \end{bmatrix}.$$

In the sequel, one can calculate the other sub-transition distributions by use of the above method as follows:

$$P_{\tilde{\gamma}} = \begin{bmatrix} 0.9 & 0.9 & 0.4 \\ 0 & 0.4 & 0.4 \\ 0 & 0 & 0.4 \end{bmatrix}, P_{\tilde{\theta}} = \begin{bmatrix} 0.5 & 0 & 0 \\ 0.1 & 0.1 & 0 \\ 0.1 & 0.1 & 0.1 \end{bmatrix}.$$

Case 2. It is supposed that one can establish the treatment process model of the animal by the following a fuzzy DES, shown as Fig. 1, in which the initial state vector is $\tilde{q}_0 = [0.9, 0.1, 0]$, fuzzy events just are the above-stated $\tilde{\alpha}$, $\tilde{\beta}$, $\tilde{\gamma}$ and $\tilde{\theta}$ in Case 1. By the max-min operation and just calculated fuzzy events, we can compute the other states as

$\tilde{q}_1 = [0.2, 0.6, 0.2], \tilde{q}_2 = [0.6, 0.3, 0.2], \tilde{q}_3 = [0.6, 0.6, 0.4], \tilde{q}_4 = [0.5, 0.1, 0],$
$\tilde{q}_5 = [0.2, 0.5, 0.2], \tilde{q}_6 = [0.5, 0.3, 0.2], \tilde{q}_7 = [0.5, 0.5, 0.4].$

As the above-stated transformation process, we can construct the generalized possibilistic Kriple structure $M = (S, P, I, AP, L)$ corresponding to the fuzzy DES in this case as $S = \{$poor, fair, good$\}$; $I = [1, 0, 0]$; $AP = \{h, m, l\}$, where h, m and l represent the atomic propositions "high", "medium" and "low", respectively, and $[0,1]^{AP} = \{A_0 = \tilde{q}_0, A_1 = \tilde{q}_1, A_2 = \tilde{q}_2, A_3 = \tilde{q}_3, A_4 = \tilde{q}_4, A_5 = \tilde{q}_5, A_6 = \tilde{q}_6, A_7 = \tilde{q}_7\}$;

$$P = P_{\tilde{\alpha}} \vee P_{\tilde{\beta}} \vee P_{\tilde{\gamma}} \vee P_{\tilde{\theta}} = \begin{bmatrix} 0.9 \; 0.9 \; 0.4 \\ 0.6 \; 0.5 \; 0.4 \\ 0.3 \; 0.3 \; 0.4 \end{bmatrix};$$

$L = \bigvee_{\tilde{\sigma} \in \tilde{\Sigma}}\{L_{\tilde{\sigma}}\}$, where $L_{\tilde{\sigma}} = \bigwedge_{k=1}^{\max-\tilde{\sigma}} L_{\tilde{\sigma}}^{(k)}$, $\tilde{\Sigma} = \{\tilde{\alpha}, \tilde{\beta}, \tilde{\gamma}, \tilde{\theta}\}$, max-$\tilde{\sigma}$ is the maximum time of using $\tilde{\sigma}$. From Fig. 1 and the above computational outcomes, one can readily obtain the following results: $L_{\tilde{\alpha}}^{(1)}(poor_{ini}) = A_0$, $L_{\tilde{\alpha}}^{(1)}(good) = A_1$, $L_{\tilde{\alpha}}^{(2)}(fair) = A_3$, $L_{\tilde{\alpha}}^{(2)}(good) = A_1$, $L_{\tilde{\alpha}}^{(3)}(poor) = A_4$, $L_{\tilde{\alpha}}^{(3)}(good) = A_5$, $L_{\tilde{\alpha}}^{(4)}(fair) = A_7$, $L_{\tilde{\alpha}}^{(4)}(good) = A_5$, $L_{\tilde{\beta}}^{(1)}(good) = A_1$, $L_{\tilde{\beta}}^{(1)}(poor) = A_2$, $L_{\tilde{\beta}}^{(2)}(good) = A_5$, $L_{\tilde{\beta}}^{(2)}(poor) = A_6$, $L_{\tilde{\gamma}}^{(1)}(poor) = A_2$, $L_{\tilde{\gamma}}^{(1)}(fair) = A_3$, $L_{\tilde{\gamma}}^{(2)}(poor) = A_6$, $L_{\tilde{\gamma}}^{(2)}(fair) = A_7$, $L_{\tilde{\theta}}^{(1)}(poor_{ini}) = A_0$, $L_{\tilde{\theta}}^{(1)}(poor) = A_4$, where $poor_{ini}$ denotes the initial state. Furthermore,

$$L_{\tilde{\alpha}}(poor) = L_{\tilde{\alpha}}^{(1)}(poor_{ini}) \wedge L_{\tilde{\alpha}}^{(3)}(poor) = A_0 \wedge A_4 = [0.5, 0.1, 0],$$
$$L_{\tilde{\beta}}(poor) = L_{\tilde{\beta}}^{(1)}(poor) \wedge L_{\tilde{\beta}}^{(2)}(poor) = A_2 \wedge A_6 = [0.5, 0.3, 0.2],$$
$$L_{\tilde{\gamma}}(poor) = L_{\tilde{\gamma}}^{(1)}(poor) \wedge L_{\tilde{\gamma}}^{(2)}(poor) = A_2 \wedge A_6 = [0.5, 0.3, 0.2],$$
$$L_{\tilde{\theta}}(poor) = L_{\tilde{\theta}}^{(1)}(poor_{ini}) \wedge L_{\tilde{\theta}}^{(1)}(poor) = A_0 \wedge A_4 = [0.5, 0.1, 0],$$

and hence

$$L(poor) = L_{\tilde{\alpha}}(poor) \vee L_{\tilde{\beta}}(poor) \vee L_{\tilde{\gamma}}(poor) \vee L_{\tilde{\theta}}(poor) = [0.5, 0.3, 0.2].$$

Analogously, we can calculate the other generalized labeling functions as

$$L(fair) = [0.5, 0.5, 0.4], L(good) = [0.2, 0.5, 0.2].$$

Accordingly, one can easily draw the figure of the just above-established generalized possibilistic Kriple structure $M = (S, P, I, AP, L)$ as shown in Fig. 2.

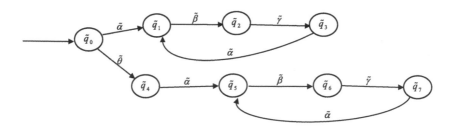

Fig. 1. Treatment process model of the animal modeled by a fuzzy DES.

At last, we can perform model checking of the above fuzzy DES by means of the approach stated in [4, 5].

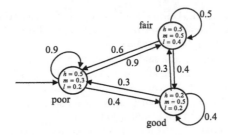

Fig. 2. Generalized possibilistic Kriple structure representation of Fig. 1.

6   Conclusion

The main work of this paper is the following. We have studied how to transfer a FDES to the generalized possibilistic Kriple structure and given the transformation method. Therefore, we think the generalized possibilistic Kriple structure is more powerful than the FDES with respect to their expressiveness of systems with fuzzy uncertainty. In order to provide a simple and efficient modeling mechanism for the fuzzy discrete event system (FDES), we proposed modeling method of FDES using the generalized linguistic variable with three different forms of negations. The illustrative cases demonstrated that the modelling method is effective and simple.

Acknowledgement. This research is supported by the Guizhou Provincial Science and Technology Foundation, Grant No. 1458 [2019] Contract Foundation of the Science and Technology department of Guizhou Province.

References

1. Cassandras, C.G., Lafortune, S.: Introduction to Discrete Event Systems. Springer, US (2010)
2. Lin, F., Ying, H.: Modeling and control of fuzzy discrete event systems. IEEE Trans. Syst., Man, Cybern., Cybern., **32**(4), 408–415 (2002)
3. Baier, C., Katoen, J.P.: Principles of Model Checking. The MIT Press, Cambridge (2008)
4. Li, Y.M., Ma, Z.Y.: Quantitative computation tree logic model checking based on generalized possibility measures. IEEE Trans. Fuzzy Syst. **23**(6), 2034–2047 (2015)
5. Li, Y.M.: Quantitative model checking of linear-time properties based on generalized possibility measures. Fuzzy Sets Syst. **320**, 17–39 (2017)
6. Zhang, S.L., Li, Y.M.: Expressive power of linear-temporal logic based on generalized possibility measures. In: Proeedings of 2016 IEEE International Conference on Fuzzy Systems, pp. 431–436, IEEE (2016)
7. Luo, M.N., Li, Y.M., Sun, F.C., Liu, H.P.: A new algorithm for testing diagnosability of fuzzy discrete event systems. Inf. Sci. **185**(1), 100–113 (2012)
8. Wagner, G.: Web rules need two kinds of negation. In: Bry, F., Henze, N., Maluszynski, J. (eds.) Proceedings of the 1st International Workshop on Principles and Practice of Semantic Web Reasoning, Heidelberg, pp. 33–50. Springer LNCS 2901 (2003)

9. Analyti, A., Antoniou, G., Damasio, C., Wagner, G.: Negation and negative information in the W3C resource description framework. Ann. Math., Comput. Teleinformatics (AMCT) **1**(2), 25–34 (2004)

10. Ferré, S.: Negation, opposition, and possibility in logical concept analysis. In: Ganter, B., Kwuida, L. (eds.) Proceeding of the fourth International Conference on Formal Concept Analysis, LNAI, vol. 3874, pp. 130–145. Heidelberg, Springer (2006)

11. Kaneiwa, K.: Description logics with contraries, contradictories, and subcontraries. New Gener. Comput. **25**(4), 443–468 (2007)

12. Pan, Z.H.: One logical description of different negative relations in knowledge. Prog. Nat. Sci. **18**(12), 1491–1499 (2008) (in Chinese)

13. Pan, Z.H.: Three kinds of fuzzy knowledge and their base of set. Chin. J. Comput. **35**(7), 1421–1428 (2012) (in Chinese)

14. Zhang, S.L., Li, Y.M.: Algebraic representation of negative knowledge and its application to design of fuzzy systems. Chin. J. Comput. **39**(12), 2527–2546 (2016) (in Chinese)

15. Li, Y.M.: Finite automata theory with membership values in lattices. Inf. Sci. **181**(5), 1003–1017 (2011)

Design of AUV Controllers Based on Generalized S-Plane Function and AFSA Optimization

Chunmeng Jiang, Jiaying Niu[✉], Shupeng Li, Fu Zhu, and Lanqing Xu

Wuhan Institute of Shipbuilding Technology, Wuhan 430050, China
{JiangCM,NiuJY,LiSP,ZhuF,XuLQ}@mail.wspc.edu.cn

Abstract. To enhance the control quality of intelligent underwater vehicles, a capacitor-plate model is introduced and generalized S-plane control is put forward in this paper. In accordance with the theory of fuzzy control, the foresaid controller draws on the simple structure of PD control and adopts artificial fish swarm algorithm (AFSA) for the optimization of control parameters. Contrastive studies are carried out between the classical S-plane controller and capacitor-plate controller. The simulation experiments and sea tests showed that the generalized S-plane controller, with nonlinear sigmoid function of better convergence and control parameters optimized by AFSA, promised more admirable control results. The feasibility and effectiveness of the proposed controller are verified.

Keywords: Generalized S-plane controller · Classical S-plane controller · AFSA optimization · Intelligent underwater vehicle · Motion control

1 Introduction

With a promising prospect in ocean exploitation and military application, autonomous underwater vehicles (AUV) have become an important means that complete a variety of underwater operation tasks [1]. For complex nonlinear systems, their controller is expected to make corresponding adjustments on control methods according to the state of the system. Linear controllers complete the foresaid adjustments in a global range based on fixed structure and specifications, which fails to meet the control requirements and needs of the system. Although the intelligent control methods can achieve flexible adjustments on the control strategy, the construction of the control decision is too complicated.

A control method known as the classical S-plane controller is put forward [2]. It has been widely used in AUV motion control and has achieved satisfactory results in sea trials. Based on the variation similarity between the control surface of the phase plane and the field strength in-between the capacitor plates, a novel controller called capacitor-plate model is proposed [3]. With simple configuration of PD control and few inputs, both these two control strategies can provide flexible gain tactics in accordance with the system state, highlight the theory of fuzzy logic control, and meet the requirements of nonlinear system control.

N. Xiong et al. (Eds.): ICNC-FSKD 2022, LNDECT 153, pp. 447–454, 2023.
https://doi.org/10.1007/978-3-031-20738-9_51

With comparative analysis of the two control models, a more generalized S-plane controller is presented in this paper. The substitute of the sigmoid function in classical S-plane method with other nonlinear sigmoid functions will lead to novel S-plane controllers of various kinds.

2 AUV Profile

2.1 Equation of Motion

With the presupposed balance between buoyancy and gravity, the 5-DOF equation of AUV motion [4] is expressed as follows with the external disturbance neglected.

$$M\ddot{X} = F_{\text{vis}}(\dot{X}) + F_{\text{T}} \tag{1}$$

M is the mass matrix. $\ddot{X}$, $\dot{X}$ and X are respectively the acceleration matrix, velocity matrix and displacement matrix in the five degrees of freedom. F_{vis} is the viscous hydrodynamic matrix and F_T is the matrix of thrusting force.

2.2 Buffeting Decrease Strategy

The research object in this paper is composed of basic motion control system, inertial navigation system, intelligent planning system, vision system, emergency handling system and surface monitoring system. All the systems are embedded in the AUV body except for the surface monitoring system [1]. The structure of the AUV is shown in Fig. 1. It is used for two major purposes -testing of facilities to provide a platform for motion control, navigation and planning, and inspection of submarine cables, oil pipelines and dams.

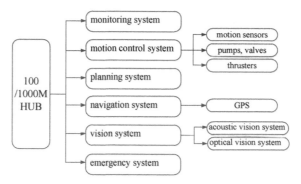

Fig. 1. AUV structure.

3 Two Controllers

3.1 Classical S-Plane Controller

A smooth curve can be formed with numerous broken lines of infinitesimal length. Numeric values on the leading diagonal in the rule list of common fuzzy logic controller can be fitted by a sigmoid function [2].

$$y = 2.0 / (1.0 + \exp(-kx)) - 1.0 \tag{2}$$

The classical S-plane controller is designed as follows.

$$u = 2.0 / (1.0 + \exp(-k_1 e_r - k_2 \dot{e}_r)) - 1.0 \tag{3}$$

e_r and $\dot{e}_r$ are normalized deviation and deviation variation rate respectively. u is normalized control output. k_1 and k_2 are the control parameters corresponding to e_r and $\dot{e}_r$, the adjustment of which determines the convergence speed of the controller. k_1 and k_2 are the only two parameters that need to be adjusted.

3.2 Capacitor-Plate Controller

In a capacitor-plate model as shown in Fig. 2, $-e_{x0}$ and e_{x0} are the coordinate boundaries. $-U_0$ and $+U_0$ are the voltage values carried by the negative and positive plates. e_x is the coordinate of a certain point P in-between the two poles. l_1 and l_2, the distances from the two plates to P, are used to represent the state of the system. The control effect u is calculated by way of quantified l_1 and l_2. Therefore, the capacitor-plate model is a negative feedback control system.

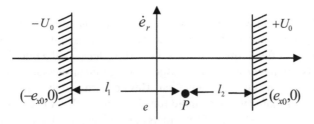

Fig. 2. Capacitor-plate model.

With $U_0 = 1$ and u the effect upon the controlled object in view of the error at P, hence

$$u = f(l_2, l_1) - f(l_1, l_2) \tag{4}$$

where $f(*, *)$ is the constraint function that reflects the relation between the field strength and the error's distance to the plate. $f(*, *)$ is expressed as

$$f(x_1, x_2) = \frac{x_1^{-k}}{x_1^{-k} + x_2^{-k}} \tag{5}$$

Based on Eqs. (4) and (5), the control output is calculated as follows, the plate model function.

$$u = \frac{(e_{x0} + e_x)^k - (e_{x0} - e_x)^k}{(e_{x0} + e_x)^k + (e_{x0} - e_x)^k} \tag{6}$$

With respect to the plate model function, the plate controller of AUV is designed.

$$\begin{cases} u_{e_x} = \dfrac{2}{1 + (\frac{e_{x0} - e_x}{e_{x0} + e_x})^{k_1}} - 1 \\[4mm] u_{\dot{e}_x} = \dfrac{2}{1 + (\frac{e_{x0} - \dot{e}_x}{e_{x0} + \dot{e}_x})^{k_2}} - 1 \\[4mm] F = K_{e_x} u_{e_x} + K_{\dot{e}_x} u_{\dot{e}_x} \end{cases} \tag{7}$$

In practice, $K_{ei} = K_{\dot{e}i} = K_i$. K_i is the maximum thrusting force provided in the ith degree of freedom. At this point, Eq. (7) is accordingly changed into

$$\begin{cases} u_{e_{xi}} = \dfrac{2}{1 + \left(\frac{1.1 - e_{xi}}{1.1 + e_{xi}}\right)^{ki1}} + \dfrac{2}{1 + \left(\frac{1.1 - \dot{e}_{xi}}{1.1 + \dot{e}_{xi}}\right)^{ki2}} - 2 \\[4mm] F_i = K_i^* u_{e_{xi}} \end{cases} \tag{8}$$

where $i = 1,2,3,5,6$. The movement of rolling is insignificant in practice, so it is usually ignored, thus $i = 4$.

4 Contrastive Analysis

It can be seen from the above analysis that the classical S-plane controller is similar to PD controller in equation, except that the former is nonlinear and the latter is linear. In the design process, it is often loose at the two sides while dense in the middle, which means the controller is relatively loose with respect to significant deviation while relatively strict with respect to insignificant deviation. This conforms to the variation tendency of sigmoid function. In this regard, sigmoid function reflects the concept of fuzzy control.

$e_{x0} = 1.1$ and $e_x \in [-1, +1]$. The outputs of the plate model with different values of k are shown in Fig. 3. The plate model varies in conformity with the sigmoid function. In Eq. (7), when is the standard PD control.

For contrast between the plate model and sigmoid function, $e_{x0} = 1.1, k = 6$ in Eqs. (2) and (6). The results are displayed in Fig. 4. It can be seen that the plate model shows faster convergence than the sigmoid function together with quicker response to small deviations.

The following conclusions can be drawn from the above contrastive analysis.

I. The list of fuzzy rules can be embodied by sigmoid functions, which can refine the rules and lower the complexity of parameter adjustment as well.

II.   Both the classical S-plane controller and plate controller are based on the concept of fuzzy control. Different types of nonlinear sigmoid functions can produce different S-plane controllers. In this regard, the plate-model controller is actually an S-plane controller, called a generalized S-plane controller.

III. With the same parameters, nonlinear sigmoid functions with attribute of faster convergence can enable the controller to respond in a faster manner. For example, Eq. (3) can be extracted of square root or cube root (called sub-S-plane) [5] to obtain Eq. (9). The sub-S-plane control models are more sensitive and responsive to small deviations.

$$
\begin{cases}
u_0 = 2.0 / \left(1.0 + \exp(-k_1 e_x - k_2 \dot{e}_x)\right) - 1.0 \\
u = sign(u_0)\sqrt{|u_0|}
\end{cases}
\tag{9}
$$

IV. Generalized S-plane controllers are a kind of nonlinear PD controller. Therefore, it follows the empirical criteria as below [6]. With significant error and insignificant variation rate of error, the controller is expected to provide large outputs to drive the system toward the desired value. With insignificant error and significant variation rate of error, the controller is expected to provide small or reverse outputs to restrain the system from overshoot.

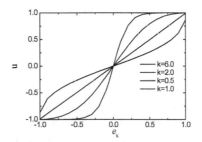

Fig. 3. Capacitor plate model with different k

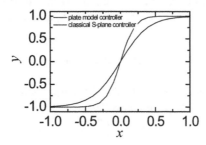

Fig. 4. Contrast of plate model curve and sigmoid function curve

5 AFSA Optimization

Appropriate control parameters determine the quality of AUV control. The existing methods of parameter optimization are highly dependent on human experience and are very likely to be stuck with local extremum. Therefore, it is proposed to optimize the proportional parameter k_1 and differential parameter k_2 for the generalized S-plane controllers by virtue of the admirable global searching ability of Artificial Fish Swarm Algorithm (AFSA).

In the solution space, the food concentration $\mathbf{X}$ at the position where an artificial fish stays is a potential solution. The individual fish perceives the state of the other fishes $\mathbf{X}_V$ within its vision *Visual*. If $\mathbf{X}_V$ is superior to $\mathbf{X}$, the foresaid artificial fish will move toward $\mathbf{X}_V$ with distance *Step* based on the judgment on behaviors of prey

$$\mathbf{X}_i^{t+1} = \mathbf{X}_i^t + \frac{\mathbf{X}_j - \mathbf{X}_i^t}{\|\mathbf{X}_j - \mathbf{X}_i^t\|} \cdot Step \cdot rand(), \text{ following } \mathbf{X}_i^{t+1} = \mathbf{X}_i^t + \frac{\mathbf{X}_j - \mathbf{X}_i^t}{\|\mathbf{X}_j - \mathbf{X}_i^t\|} \cdot Step \cdot rand(), \text{ or}$$

clustering $\mathbf{X}_i^{t+1} = \mathbf{X}_i^t + \frac{\mathbf{X}_c - \mathbf{X}_i^t}{\|\mathbf{X}_c - \mathbf{X}_i^t\|} \cdot Step \cdot rand()$. If $\mathbf{X}_V$ is not superior to $\mathbf{X}$, the foresaid artificial fish continues search within its vision for a higher food concentration.

The AFSA optimization of generalized S-plane control parameters complies with the model constructed below in Fig. 5.

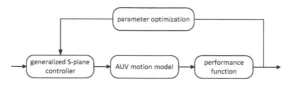

Fig. 5. Model for generalized S-plane parameter optimization with AFSA

The range of control parameters (namely the solution space of AFSA) are first determined based on practical field trials of the classical S-plane controller. The simulations of AUV motion control based on a generalized S-plane controller are then carried out with the performance function $\Phi_p = \int_0^t t|e(t)|dt$ for evaluation of the simulation results. AFSA continues the search in accordance with the evaluation feedback till the performance function converges or reaches the required times of iteration. At this point, the records on the AFSA bulletin board are the optimal parameters.

6 Simulation Experiments and Sea Tests

Simulation experiments and sea tests of AUV motion control based on the classical S-plane controller and capacitor plate controller were achieved to verify the validity and feasibility of the generalized S-plane controller.

6.1 Simulation Experiments

The controllers are respectively designed based on Eqs. (4) and (9) for the control over velocity in the surge direction, depth, heading angle and trimming angle. The control parameters k_1 and k_2 for each degree of freedom are determined based on practical experience and actual condition of the trials.

The simulations proceeded on the self-developed test platform. The SGI workstation realized visual simulation, hydrodynamic calculation and motion control simulation. The test platform provides 10 outputs per second and the motion controller works with a frequency of 2 Hz.

The control results of depth and velocity in the surge direction are shown in Fig. 6, with the same k_1 and k_2 for both controllers.

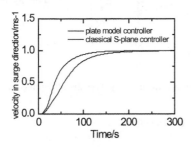
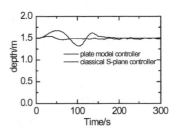

(a) results of velocity control in surge direction   (b) results of depth control

Fig. 6. Results of AUV simulation experiments

With the same parameters, the generalized S-plane controller provided better control effect than that of the classical S-plane controller, with faster response and smaller overshoot. This is important to AUV systems that have real-time requirements.

6.2 Sea Tests

TO verify the performance of the presented controller mentioned above, sea tests were carried out in open waters of rapid and complex currents. The environment is shown in Fig. 7. The trials included control over velocity, position and fault diagnosis of long-range navigation [7].

Fig. 7. Environment of sea trials

Figure 8 displays the control results of AUV velocity control based on plate model controller and AFSA-optimized parameters, with desired depth of 1.5 m and desired velocity of 2.5 m/s. It can be seen from the results that the classical S-plane controller with empirical parameters showed a satisfying control effect with depth deviation of ± 1.0 m and heading deviation of ± 3°. The generalized S-plane controller with AFSA-optimized parameters outstood with depth deviation of ± 0.5 m and heading deviation of ± 2°.

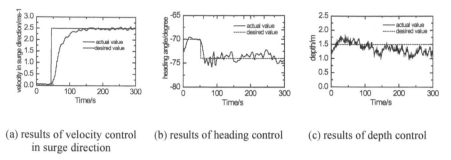

(a) results of velocity control in surge direction   (b) results of heading control   (c) results of depth control

Fig. 8. Results of AUV sea tests

7   Conclusions

A novel control method–generalized S-plane controller–is brought forward in this paper based on the contrastive analysis of the classical S-plane controller and plate model controller. With the attribute of faster convergence, the generalized S-plane controller based on the plate function ensures more admirable control performance than the classical S-plane controller with the same parameters. In the meanwhile, Artificial Fish Swarm Algorithm is adopted for the optimization of k_1 and k_2 in the generalized S-plane controller. Both the simulation experiments and sea trials have verified the effectiveness and superiority of the generalized S-plane controller based on AFSA-optimized parameters.

References

1. Wang, M.H., Zeng, B., Wang, Q.J.: Study of motion control and a virtual reality system for autonomous underwater vehicles. Algorithms **14**(3), 93 (2021)
2. Jiang, C.-M., Wan, L., Sun, Y.-S.: Design of motion control system of pipeline detection AUV. J. Cent. S. Univ. **24**(3), 637–646 (2017). https://doi.org/10.1007/s11771-017-3464-2
3. Ziaeefard, S., Page, B.R., Pinar, A.J., Mahmoudian, N.: Effective turning motion control of internally actuated autonomous underwater vehicles. J. Intell. Rob. Syst. **89**(1–2), 175–189 (2017). https://doi.org/10.1007/s10846-017-0544-3
4. Zhang, S., Fan, L.T., Gao, J.W.: Fault diagnosis of underwater vehicle and design of intelligent self-rescue system. J. Coast. Res. **83**(sp1), 872–875 (2018)
5. Wang, Y., Li, Y.R.: Fault diagnosis of underwater vehicle propulsion system based on deep learning. J. Coast. Res. **107**(sp1), 65–68 (2020)
6. Khodayari, M.H., Balochian, S.: Modeling and control of autonomous underwater vehicle (AUV) in heading and depth attitude via self-adaptive fuzzy PID controller. J. Mar. Sci. Technol. **20**(3), 559–578 (2015). https://doi.org/10.1007/s00773-015-0312-7
7. Sun, Y.S., Luo, X.K., Ran, X.R.: A 2D optimal path planning algorithm for autonomous underwater vehicle driving in unknown underwater canyons. J. Mar. Sci. Eng. **9**(3), 252 (2021)

Online Tuning of PID Controllers Based on Membrane Neural Computing

Nemanja Antonic, Abdul Hanan Khalid, Mohamed Elyes Hamila, and Ning Xiong[(✉)]

School of Innovation, Design and Engineering, Mälardalen University, 72123 Västerås, Sweden
nemanja.antonic@mail.polimi.it, {akd21001,
mha21002}@student.mdu.se, ning.xiong@mdu.se

Abstract. PID controllers are still popular in a wide range of engineering practices due to their simplicity and robustness. Traditional design of a PID controller needs manual setting of its parameters in advance. This paper proposes a new method for online tuning of PID controllers based on hybridized neural membrane computing. A neural network is employed to adaptively determine the proper values of the PID parameters in terms of evolving situations/stages in the control process. Further the learning of the neural network is performed based on a membrane algorithm, which is used to locate the weights of the network to optimize the control performance. The effectiveness of the proposed method has been demonstrated by the preliminary results from simulation tests.

Keywords: PID controller · Online gain tuning · Neural network · Membrane algorithm

1 Introduction

Nowadays in the era of Industry 4.0 there is an increasing demand to improve process control to reach high performance. The well known proportional-integral-derivative (PID) controllers are still popular to be used in a wide range of engineering practices. The attractiveness of PID controllers is attributed to their simplicity in structure and robustness in performance against variations of operational conditions. The three parameters that need to be specified for a PID controller are: proportional gain (K_p), integral time constant (K_i), and derivative time constant (K_d). Finding an appropriate setting of these three parameters has been an important issue for realization of process control systems of high quality.

Generally the methods of parameter setting of PID controllers can be divided into two categories. In the first category one determines the PID parameters in advance by employing some heuristic schemes such as Ziegler-Nichelos tuning rules and Chien-Hrons-Reswick method [1–3]. The predetermined parameters are then used, and they remain constant in runtime. But, when there is a change in the underlying process, the parameters need to be tuned again to adapt to new properties of the process. The methods in the second category aim for self-tuning of the PID parameters along with the control process. The rationale is given by the fact that the proportional, integral and derivative components of the PID controller play different roles to achieve the control objective, and consequently their intensity can be advantageously adjusted according to varying situations and various stages. A representative of the works in this category lies in fuzzy PID controllers [4], which operate by online deciding the values of the PID parameters via fuzzy rule-based reasoning. However, constructing the knowledge base for fuzzy PID is not a trivial task. It may require extensive efforts to collect adequate knowledge to guide dynamic setting of the PID parameters based on input situations.

This paper proposes a new method for online tuning of PID controllers based on hybridized neural membrane computing. The basic idea is to utilize a neural network as the universal approximator to adaptively determine the proper values of the PID parameters in terms of evolving situations/stages in the control process. Further the learning of the neural network is performed based on the membrane algorithm, which is used to locate the weights of the network to maximize the satisfaction of the control objective. The merit of this proposed method has been verified through preliminary results from a set of simulation tests. The main contribution of our work is highlighted as follows:

- We made the first attempt of learning neural networks for self–tuning PID by employing a bio-inspired membrane algorithm, which is considered as more powerful in global optimization than the traditional gradient descent method and standard genetic algorithms.
- We realized the design of a self-tuning PID controller without prior knowledge and experience. In our system the decision maker (neural network) is optimized in terms of an objective function such that no previous experience in the form of labelled training data is required.

2   Related Work

Various metaheuristics were used to find the parameter values of PID controllers to optimize the control performance, including Ant Colony Optimization [5], Fruit Fly Optimization [6], Genetic Algorithm [7], and Elephant Herding Optimization [8], to mention a few. Notably augmented Lagrangian Particle Swarm Optimization [9] was proposed for robust tuning of the gains of PID controllers. It has shown to work well to acquire PID controller parameters meeting the H∞ criteria. A common characteristic of all these works is that the PID parameter settings optimized will be used and remain constantly in a real-time control process.

Online tuning of the gains for PID controllers provides an opportunity for further improving control performance, given that the parameters have varying importance depending on different situations. Zhao et al. [10] proposed a framework that utilized an extra fuzzy controller to decide the appropriate values of the PID gains during a running process. This fuzzy controller's rules and fuzzy set membership functions were learned by a genetic algorithm in [11] to acquire optimal policy of online gain tuning. However, the total amount of fuzzy rules may exponentially grow with the number of input variables of the fuzzy controller.

The Membrane Algorithm was first introduced by Nishida [12], whose work has given promising results in dealing with parameter optimization. It provides a framework in which a set of sub-algorithms (such as genetic algorithm and particle swarm optimization) can run simultaneously and communicate with each other. A membrane algorithm has been used for locating centroids of data clusters in smart case mining [13]. In this paper we aim to exploit the strength of membrane algorithms in global optimization to foster the learning of neural networks responsible for PID gain tuning.

3   Membrane Neural System for PID Tuning

The parameters of the PID controller is yielded by a feed-forward Neural Network, which takes error and change in error of the closed-loop control system as its inputs. The Membrane Algorithm is used to find the best weights of the Neural Network. As the PID parameters are calculated based on the error and change of error, they can be adapted online according to situations of the control process.

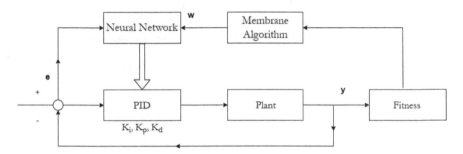

Fig. 1. The proposed system structure

3.1   The Proposed System Structure

As can be seen in Fig. 1, the proposed membrane-neural system for PID tuning consists of a neural network and membrane algorithm, in combination with the closed-loop PID control system. The neural network will decide the values of the PID parameters K_i, K_p and K_d in real time, in terms of the error and change in error of the control system. The membrane algorithm is used for optimization to find the weights of the neural network. The fitness of the neural network is assessed by analyzing the performance of the PID controller in manipulating the plant in simulation. Then, the fitness score is passed to the membrane algorithm, which progressively locates better sets of weights of the neural network until a given number of iterations is reached.

3.2   Neural Network Yielding PID Gains

After the weights being found by MA (membrane algorithm), the neural network (NN) utilizes them in a forward direction for calculating PID parameters. The NN in our system consists of three layers: input layer, hidden layer, and output layer, as depicted in Fig. 2. The input layer has two input variables: error and change of error of the PID controller. The hidden layer has five units, and the number of units of the output layer is three (corresponding to the three parameters of the PID controller which are to be calculated). Each neuron in the hidden layer receives the input data, processes them using the activation function, and then passes the result to the successive output layer.

The calculation in each node of the hidden and output layer has two steps: pre-activation (a) and activation (h). In pre-activation the weighted sum of the inputs is calculated. In activation, the sigmoid function is used to produce the output for a unit based on the weighted sum of its inputs.

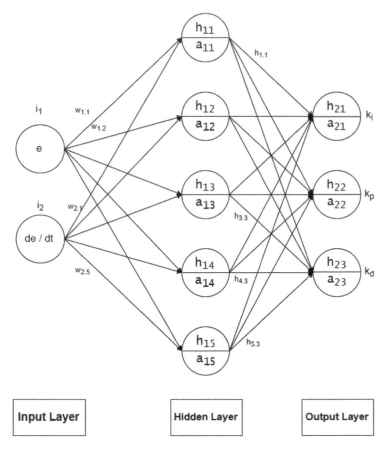

Fig. 2. Feed Forward Neural Network

3.3 Membrane Algorithm to Optimize Neural Network

The membrane algorithm (MA) is an optimization algorithm designed by putting sub-algorithms into layers and executing each of the latter twice. In this paper MA is used to find the optimal values of the connection weights of the NN. Each layer of MA contains a sub-algorithm for optimization, in our case, it is the genetic algorithm (GA). Each instance of the GA is executed twice per layer, yielding 2 solutions per layer. The number of layers is arbitrary. The execution of the algorithm is iterative, and at each iteration every layer will generate two solutions, which are the best solutions of the two instances of GA.

Let b denote the best solution and w the worst one. At each intermediate layer i, b is sent to the layer above $i - 1$ and w to the layer underneath $i + 1$. Layer 0 only sends w to layer 1 and layer $n - 1$ discards w.

When a layer receives all its solutions, it tests them and keeps the best one among the two, using it as an initial solution for the next iteration of its sub-algorithm. The exchange of solutions between layers is depicted in Fig. 3.

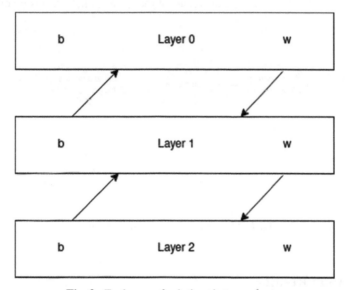

Fig. 3. Exchange of solutions between layers

The aim of MA is to find the weights of the NN which minimizes the cost function (opposite to fitness) as defined as:

$$Cost(W) = \sum_{t=0}^{20} e^2(t) + \sum_{t=0}^{20} u^2(t) \tag{1}$$

where e is the error defined as $e = r - y$ (with r as the reference input and y as the actual output of the plant), and u denotes the action of the PID controller.

Each individual in GA is a set of real numbers representing the weights of the NN. The starting population is 2 and it is increased if the minimum cost in the population doesn't decrease by a given threshold ratio (0.1) every generation. If the best individual doesn't decrease its cost, the whole population will be mutated, producing mutants by adding to each real number a small random disturbance. The parents are chosen by roulette selection and the probability of each individual to be selected as parent is given by the inverse of its cost divided by the sum of the inverse of the costs of all individuals:

$$P_i = \frac{\frac{1}{Cost(i)}}{\sum_j \frac{1}{Cost(j)}} \tag{2}$$

where $Cost(i)$ denotes the cost of individual i in the population. The crossover is done by applying BLX-α, which combines two parents and creates a new individual by taking, for each element of the genome, a random number in the interval

$$[C_{min} - I \cdot \alpha, C_{max} - I \cdot \alpha]$$

where $\alpha = 0.1$, C_{min} and C_{max} are the minimum and maximum of the corresponding weight, and $I = C_{max} - C_{min}$. The flowchart of the genetic algorithm is given in Fig. 4.

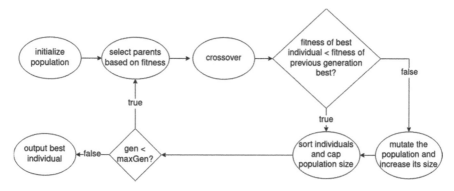

Fig. 4. Flowchart of the genetic algorithm

4   Simulation Results

Simulation tests were conducted to examine the efficacy of the proposed method of applying the mixture of membrane and neural computing for online tuning of the parameters of the PID controller. The whole system was implemented by the combination of Python code with MATLAB and Simulink.

Several different plant models have been used in simulation to compare the performance of our proposed method with other traditional methods. The results are shown via the figures below.

4.1   Plant 1

$$G(s) = \frac{s+1}{s^3 + 2s^2 + 2s + 1}$$

The graph for the membrane-neural model (Fig. 5) shows a larger overshoot but shorter settling time than those obtained from the Ziegle-Nichols method, as shown in Fig. 6.

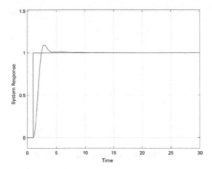

Fig. 5. System Response of the MA-NN model on Plant 1

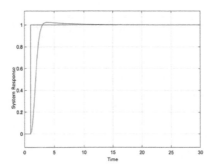

Fig. 6. Ziegle-Nichols system response [2] on Plant 1

4.2 Plant 2

$$G(s) = \frac{400}{s^2 + 50s}$$

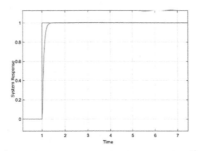

Fig. 7. System Response of the MA-NN model on Plant 2

Comparing to the system response of using genetic algorithm for PID parameter optimization as shown in Fig. 8, our membrane-neural computing model obtained very similar PID control performance as depicted in Fig. 7.

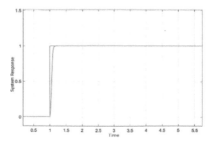

Fig. 8. System Response of using GA on Plant 2

4.3 Plant 3

$$G(s) = \frac{4.228}{(s + 0.5)(s^2 + 1.648s + 8.456)}$$

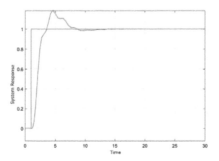

Fig. 9. System Response of the MA-NN model on Plant 3

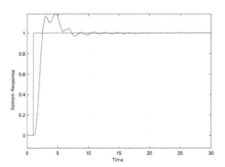

Fig. 10. Ziegle-Nichols system response on Plant 3

As shown in Fig. 9, the online PID gain tuning enabled by our membrane-neural model resulted in shorter settling time as compared to the Ziegle-Nichols method for PID parameter tuning (Fig. 10).

4.4 Plant 4

$$G(s) = \frac{27}{(s+1)(s+3)^3}.$$

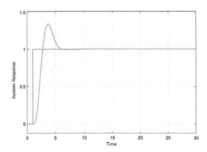

Fig. 11. System Response of the MA-NN model on Plant 4

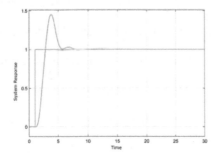

Fig. 12. Ziegle-Nichols system response on Plant 4

Figure 11 shows that the system response created by our membrane-neural model has both lower overshoot and shorter settling time than the response of the Ziegle-Nichols method as shown in Fig. 12.

Further, we compare the costs of the online tuning PID control performance based on our membrane-neural computing model with other traditional methods in Table 1. We can see that only on Plant 1, the cost of our system is slightly higher. On the other three plants, we achieved lower cost for our system, which indicated and verified the effectiveness of the proposed method for online PID gain tuning.

5 Conclusion

The main contribution of this paper is the proposal of a new and simple method for online tuning of the PID controller parameters when manipulating a certain plant. This new method combines the Membrane Algorithm and Neural Network to reach an optimal policy of PID gain tuning. The simulation test results show that the proposed membrane neural computing method can obtain desirable results compared to Ziegle-Nichols method and some other methods for several plants in tests.

Table 1. Comparison of the cost values

	Plant 1		Plant 2	
	MA-NN	Ziegle-Nichols	MA-NN	GA
$\sum u^2$	10555	10465	10698	11072
Cost	10574	10486	10719	11087
	Plant 3		Plant 4	
	MA-NN	Ziegle-Nichols	MA-NN	Ziegle-Nichols
$\sum u^2$	10107	10161	10190	10309
Cost	10129	10182	10217	10337

References

1. Ziegler, J.G., Nichols, N.B.: Optimum settings for automatic controllers. J. Dyn. Sys. Meas. Control **115**(2B), 220–222 (1993)
2. Cohen, G.H., Coon, G.A.: Theoretical consideration of retarded control. Trans. ASME **75**, 827–834 (1953)
3. Chien, K.L., Hrons, J.A., Reswick, J.B.: On the automatic control of generalized passive systems. Trans. Am. Soc. Mech. Eng. **74**, 175–185 (1972)
4. Kim, J.H., Oh, S.J.: A fuzzy PID controller for nonlinear and uncertain systems. Soft. Comput. **4**(2), 123–129 (2000)
5. Chiha, I., Liouane, N., Borne, P.: Tuning PID controller using multiobjective ant colony optimization. Appl. Comput. Intell. Soft Comput. (2012)
6. Han, J., Wang, P., Yang, X.: Tuning of PID controller based on fruit fly optimization algorithm. In: 2012 IEEE International Conference on Mechatronics and Automation, pp. 409–413. IEEE (2012)
7. Jaen-Cuellar, A.Y., et al.: PID-controller tuning optimization with genetic algorithms in servo systems. Int. J. Adv. Rob. Syst. **10**(9), 324 (2013)
8. Gupta, S., Singh, V.P., Singh, S.P., et al.: Elephant herding optimization based PID controller tuning. Int. J. Adv. Technol. Eng. Explor. **3**(24), 194 (2016)
9. Kim, T.H., Maruta, I., Sugie, T.: Robust PID controller tuning based on the constrained particle swarm optimization. Automatica **44**(4), 104–1110 (2008)
10. Zhao, Z.Y., Tomizuka, M., Isaka, S.: Fuzzy gain scheduling of PID controllers. IEEE Trans. Syst. Man Cybern. **23**(5), 1392–1398 (1993)
11. Xiong, N.: Fuzzy tuning of PID controller based on genetic algorithm. Syst. Sci. **25**(3), 65–75 (1999)
12. Nishida, T.Y.: Membrane algorithm: an approximate algorithm for NP-complete optimization problems exploiting P-systems. In: 6th International Workshop on Membrane Computing, pp. 26–43 (2005)
13. Holmberg, J., Xiong, N.: Smart case mining based on membrane clustering. In: 2019 IEEE Symposium Series on Computational Intelligence (SSCI), pp. 2527–2533. IEEE (2019)

Fuzzy Logic Based Energy Management Strategy for Series Hybrid Bulldozer

Cong feng Tian[1,3], Jia jun Yang[2(✉)], Ru wei Zhang[1,3], Jin dong Xu[1,3], and Yong Zhao[2]

[1] Shantui Construction Machinery Co., Ltd, Jining 272073, China
{tiancf,zhangrw,xujd}@shantui.com

[2] Key Laboratory of Highway Construction Technology and Equipment of Ministry of Education, Chang'an University, Xi'an 710064, Shaanxi, China
{2020225128,zhaoyongn1107}@chd.edu.cn

[3] Shandong Community Construction Machinery Co., Ltd, Jining 272073, China

Abstract. In order to improve the fuel economy of series hybrid bulldozer, the fuzzy energy management strategy is proposed. Firstly, the fuzzy logic controller is designed considering the state of charge of the super-capacitor in the current state. Then the inputs of the fuzzy controller are the required power and the state of charge of the super-capacitor. The outputs of the fuzzy controller are the target output power of the engine, the output power of the generator and the demand power of the super-capacitor. The results show that the strategy can not only distribute the output power of engine and discharge power of super-capacitor reasonably according to the demand power, but also maintain the SOC fluctuation in a reasonable range. It saves 13.3% fuel and improves the fuel economy of the whole machine. Therefore, this method is a feasible and effective energy management strategy.

Keywords: Series hybrid · Bulldozer · Fuzzy · Energy management

1 Introduction

Bulldozer has complex working conditions and large load fluctuations in the driving process and has disadvantages such as high energy consumption, high emissions and high noise in the driving process, which seriously affects the environment [1, 2]. The research on reducing the fuel consumption and emission of bulldozer has become the focus of the construction machinery industry. From the current research technology, hybrid technology is one of the effective methods to improve the emission and fuel economy of construction machinery [3–5]. In 2010, Caterpillar introduced the world's first hybrid bulldozer D7E, which improves the fuel economy by 25%. In 2012, Shantui Construction Machinery Co., Ltd., Tongji University, Jilin University, Weichai Group, Shandong University and Beijing Institute of Technology jointly developed the first series hybrid bulldozer in China. At present, the prototype has been successfully trial-produced. It is well known that the energy management strategy (EMS) is the key technology for the hybrid bulldozer. Energy management strategies mainly include rule-based EMS

N. Xiong et al. (Eds.): ICNC-FSKD 2022, LNDECT 153, pp. 465–474, 2023.
https://doi.org/10.1007/978-3-031-20738-9_53

[6–8] and optimized based EMS [9]. Rule-based energy management strategies mainly include deterministic rules and fuzzy rules. The energy management strategy based on deterministic rules is to control the working point of the engine by setting the threshold value of the battery power according to the universal characteristics of the engine and the driver's experience. However, the engine load changes frequently and the fuel efficiency of the whole machine cannot be improved effectively. Energy management strategy based on fuzzy rules is expressed by natural language based on expert experience. It has good robustness and adaptability, and can realize reasonable energy distribution and improve fuel economy. Therefore, this paper proposes a fuzzy-based energy management strategy for series hybrid bulldozer, which can reasonably allocate energy according to the working conditions.

2 System Structure

The structure of the series hybrid studied is shown in Fig. 1. It includes the front power output chain and the rear power output chain. The front power output chain includes the engine, the whole machine controller, the generator, the generator controller, the super-capacitor, the super capacitor controller. The rear power output chain includes the motor controller, the drive motor and the side drive.

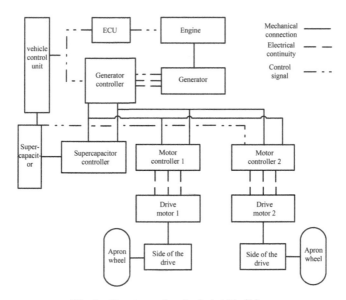

Fig. 1. Structure of series hybrid bulldozer

3 Fuzzy Energy Management Strategy

Fuzzy energy management strategy is an energy management strategy expressed by natural language by expert experience. Unlike the traditional method, it does not rely on the mathematical model of the accused object, has good robustness and adaptability and improve the fuel economy performance of hybrid bulldozer.

The implementation of fuzzy energy management strategy mainly depends on a fuzzy controller. The design of fuzzy controller mainly includes the structure of fuzzy controller, the selection of language variables and the design steps of fuzzy rules.

3.1 The Structure of Fuzzy Controller

InputS of fuzzy control are the value of the vehicle's demand power and super-capacitor. The control of the fuzzy rule library obtains the target output power of the engine, the output power of the generator, and the demand power of the super-capacitor. The structure of the fuzzy controller is shown in Fig. 2.

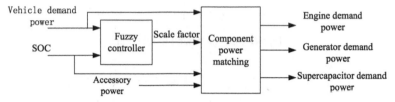

Fig. 2. Structure of fuzzy controller

3.2 Selection of Language Variables

In order to simplify fuzzy logic and improve operational efficiency, we selected very small (low), small (low), medium, large (high), very large (high) 5 fuzzy language variables to describe the demand power, super-capacitor charge, engine output power. Through the quantitative treatment of the fuzziness of each variable, the domain of the demand power of the whole vehicle can be obtained as [0,2], 0 indicates that the demand power of the vehicle is zero, 2 indicates the maximum demand power. The vehicle demand power input function is selected as a trapezoidal membership function notation. The super-capacitor domain is [0,1]and its input membership function is represented by a trapezoidal function expression. The field of fuzzy output T is[0,2], which means that the vehicle demands power of 0, the engine output power is also 0, and 2 indicates that the vehicle demands the maximum power, At this point, the engine and super-capacity also match the corresponding maximum output power, the output membership function selected Gaussian type. The membership function of the variables are analyzed as shown in the Figs. 3, 4 and 5:

3.3 Fuzzy Rule Design

According to the driving characteristics of the series hybrid bulldozer, the optimal fuel consumption of the engine is the control target, considering the working characteristics of the engine and super-capacitors, formulating the optimal control rules, and reasonably distributing the multi-power source of the hybrid bulldozer. The establishment of the rule library refers to the following requirements:

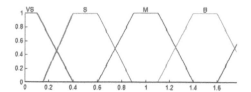

Fig. 3. Membership function of input variable R

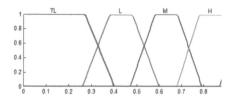

Fig. 4. SOC membership function

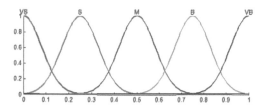

Fig. 5. Membership function of output variable

(1) When the value of the super-capacitor is low and the vehicle demand power is less than the optimal output power of the engine, adjust the engine output power to the optimal fuel consumption area. The output energy of the engine is used for the travel of the tractor and the charging of super-capacitors respectively.

(2) When the value of the super-capacitor is low, the demand power of the vehicle and the engine's optimal operating power are similar, at this time adjust the output power of the engine to the best working area, the engine output power drive tractor travel, excess power to charge the super-capacitor.

(3) When the value of the super-capacitor is low, the demand power of the tractor is higher than the optimal output power of the engine, at this time, if the maximum output power of the engine can meet the demand power of the tractor, the engine alone provides power for the tractor.

(4) When the value of the super-capacitor is moderate, according to the vehicle demand power and the size of the engine's optimal output power to determine whether the super-capacitor is charged and discharged.

(5) When the super-capacitor value is large, the vehicle demand power is less than or equal to the optimal power of the engine, the engine provides power for the bulldozer driving demand power, at this time the engine according to the power size of the tractor demand for power matching.

(6) When the super-capacitor value is large, the demand power of the tractor is greater than the optimal output power or maximum output power of the engine, adjust the output power of the engine to the optimal output power or maximum output power, and the super-capacitor discharge compensates for the lack of engine power.

(7) In the formulation of fuzzy control rules, on the basis of ensuring the performance of tractor driving operation, the maximum control of the engine and super-capacitor operation in the optimal state, reduce the fuel consumption of the whole machine, improve the efficiency of each tractor components. Table 1 is the fuzzy logical rules.

Table 1. Fuzzy logical rules

SOC	VS	S	R M	B	VB
TL	M	M	B	VB	VB
L	M	M	B	VB	VB
M	S	MS	M	B	VB
H	S	S	M	B	VB
TH	S	S	S	B	VB

4 Simulation Analysis

In order to verify the fuzzy energy management strategy, the vehicle simulation platform shown is built based on Matlab/simulink. It is analyzed from multi-power source distribution, super-capacitor, engine torque working point and so on. Its main simulation parameters are as follows: engine maximum power Pe = 172 kW, generator maximum power Pg = 180 kW, super-capacitor capacity C = 3F, drive motor maximum power Pm = 105 kW, super-capacitor SOC optimal range [0.5,0.8], SOC initial value 0.65. According to the conditions of the bulldozer driving operation, the comprehensive simulation conditions shown in Fig. 6 are established:

4.1 Torque Analysis

From Fig. 7, the output torque of the two-sided motor can meet the torque demand of the vehicle under different operating conditions. In 0 ~ 30 s, the bulldozer changed from uniform acceleration to uniform deceleration, two-sided motor output torque is equal. In 30 ~ 57 s, the two-sided motor output torque reached a maximum of 670 N.m. In 68 ~ 92 s, the left motor output torque is − 338 N.m, the right motor output torque is 500 N·m. The B/2 steering was achieved. In 92 ~ 113 s, due to ramp resistance, the actual vehicle speed is slightly lower than the target speed of 0.6 km/h, the two-sided motor output torque is the same as 477 N·m. In 113 ~ 130 s, the situ center steering was achieved, the two-sided motor is ± 550 N·m.

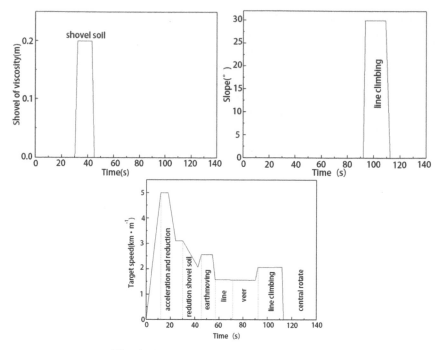

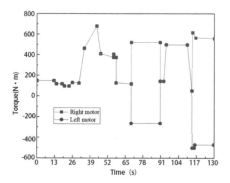

Fig. 6. Comprehensive simulation conditions

Fig. 7. Torque of double side motor

4.2 Multi-power Source Distribution Simulation Analysis

As can be seen from Figs. 8 and 9, fuzzy energy management strategies can reasonably distribute engine power and super-capacitor power according to demand power. In 0–8, 18–30, 45–57, 57–68, 90–92 s, engine power is greater than the demand power of the machine, super capacitor charging. The SOC value was increased from 0.52 to 0.68. When the bulldozer is driving evenly, within 8–18 s, the engine power is slightly less than the demand power of the whole machine, the supercapacitor discharges, the SOC value is reduced from 0.68 to 0.61.When the bulldozer is in deceleration shoveling

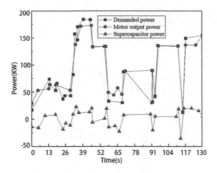

Fig. 8. Power distribution curve

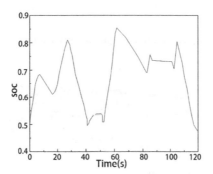

Fig. 9. Super-capacitor SOC

conditions, it is 30–45 s, the demand power of the whole machine exceeds the engine's optimal output power ceiling of 160 kW, the SOC value increases from 0.61 to 0.8, the super-capacitor discharge and the engine together power the bulldozers. When the bulldozer turns, the SOC value increases from 0.58 to 0.82 in the 68–90 s, the adjustment of power operation in the optimal fuel economic zone by the super-capacitor appropriate discharge to supplement the demand power of the whole machine. When the bulldozer straight climb, in 92–113 the SOC value is basically unchanged, the demand power of the whole machine and the output power of the engine is basically the same, the super-capacitor is neither charged nor discharged, When the bulldozer is in situ center rotation, in 113–130 s the SOC value increased from 0.7 to 0.8, the demand power of the whole machine was supplemented by the appropriate discharge of the super-capacitor. In the 30, 68 and 113 s, when the vehicle demand power suddenly increased, the engine output power can not quickly provide energy for the vehicle, at this time the super-capacitor rapid discharge and engine together provide power for the vehicle. Under the whole comprehensive operating conditions, the engine output power is basically controlled between 50 and 160 kW, the engine operates at a maximum output power of 172 kW, only under deceleration shoveling conditions, and the super-capacitor "cuts peaks" the output power of the engine, controlling the engine operation in the optimal output power area.

4.3   Economic Simulation Analysis

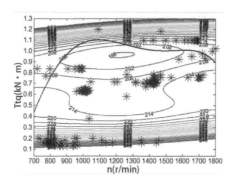

Fig. 10. Engine torque the original model

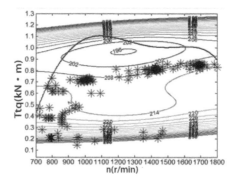

Fig. 11. Engine torque of hybrid

From Figs.10 and 11 show that under the same conditions, the engine output torque of hybrid tractor is smaller than that of the engine of the original model, and the engine torque working point tends to be lower fuel consumption rate and more fuel-efficient.

As can be seen from Fig. 12, the fuel consumption of the original model in the 130 s cycle conditions is 1520 g, the fuel consumption of the hybrid fuel is 1318 g, the fuzzy control energy management strategy can save up to 13.3%. Compared to the original model, the fuel economy of the whole machine can be significantly improved.

5   Results

(1) Based on the fuzzy control theory, the fuzzy controller is designed. The vehicle required power and the SOC of the super-capacitor are taken as the input of the fuzzy controller and the target output power of the engine.The output power of the generator and the required power of the super-capacitor are taken as the output and 25 control rules are designed.

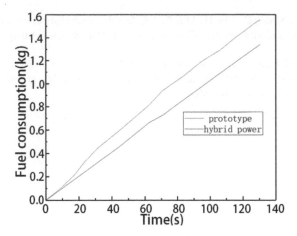

Fig. 12. Fuel consumption comparison of engine

(2) A vehicle simulation platform is established based on Matlab/Simulink, and 130 s typical working conditions are used to verify the strategy. The strategy can effectively distribute the power engine. The hybrid bulldozer can work in a lower fuel consumption region than that of the original vehicle. The fuel consumption of the original model is 1520 g, and that of the hybrid bulldozer is 1318 g. compared with the original model, the fuzzy logic based energy management strategy can save 13.3% fuel and the fuel economy is improved obviously.

References

1. Li, T., Huang, L., Liu, H.: Energy management and economic analysis for a fuel cell supercapacitor excavator. Energy **172**(4), 840–851 (2019)
2. Yu, Y., Cuong-do, T., Park, Y., Kwan-Ahn, K.: Energy saving of hybrid hydraulic excavator with innovative powertrain. Energy Convers. Manag. **244**(7), 114447 (2021)
3. Gan, M., Hou, H., Wu, X., Liu, B., Yang, Y., Xie, C.: Machine learning algorithm selection for real time energy management of hybrid energy ship. Energy Rep. **8**(14), 1096–1102 (2022)
4. Liu, H., Yao, Y., Wang, J., Li, T.: A control architecture to coordinate energy management with trajectory tracking control for fuel cell/battery hybrid unmanned aerial vehicles. Int. J. Hydrogen Energy **47**(34), 15236–15253 (2022)
5. Thien, T., Axelsen, H., Merten, M., Sauer, D.U.: Energy management of stationary hybrid battery energy storage systems using the example of a real-world 5 MW hybrid battery storage project in Germany. J. Energy Storage **51**(3), 104257 (2020)
6. Aguila-Leon, J., Vargas-Salgado, C., Chiñas-Palacios, C., Díaz-Bello, D.: Energy management model for a standalone hybrid microgrid through a particle swarm optimization and artificial neural networks approach. Energy Convers. Manage. **267**(7), 115920 (2022)
7. Jain, D.K., Neelakandan, S., Veeramani, T., Bhatia, S., Memon, F.H.: Design of fuzzy logic based energy management and traffic predictive model for cyber physical systems. Comput. Electr. Eng. **102**(7), 108135 (2022)

8. Bo, L., Han, L., Xiang, C., Liu, H., Ma, T.: A Q-learning fuzzy inference system based online energy management strategy for off-road hybrid electric vehicles. Energy **252**(4), 123976 (2022)
9. Rezaei, H., Abdollahi, S.E., Abdollahi, S., Filizadeh, S.: Energy management strategies of battery-ultracapacitor hybrid storage systems for electric vehicles: review, challenges, and future trends. J. Energy Storage **53**(7), 105045 (2022)

A Hybrid Improved Multi-objective Particle Swarm Optimization Feature Selection Algorithm for High-Dimensional Small Sample Data

Xiaoying Pan[1,2], Jun Sun[1(✉)], and Yufeng Xue[1]

[1] Xi'an University of Posts and Telecommunications, Xi'an 710121, China
panxiaoying@xupt.edu.cn, 15389096817@163.com, 18829210417@163.com
[2] Key Laboratory of Network Data Analysis and Intelligent Processing, Xi'an 710121, China

Abstract. Traditional feature selection methods have great limitations for processing high-dimensional small sample data, and it is difficult to accurately and efficiently propose the optimal feature subset. To improve the feature selection ability to deal with high-dimensional data, this paper proposes a hybrid improved multi-objective particle swarm optimization feature selection algorithm based on the classification accuracy and feature selection number, called HIMOPSO. The algorithm uses the inherent attributes of the data to compute and filter the features quickly and uses multi-objective particle swarm optimization to perform a fine search. The parameters of particle velocity are modified nonlinearly according to the number of iterations. The explosion of new particles creates a pool of candidate particles to select the new particles as the next generation of initial particles. This increases population diversity, allows particles to explore more potential areas, and improves the overall population quality. The final feature subset is classified on the gene expression profile data. Experimental results show that this method is superior to the comparison algorithm in classification accuracy and feature number and can balance these two objectives.

Keywords: Feature selection · Particle swarm optimization · High-dimensional small sample · Classification

1 Introduction

In recent years, a large number of gene expression profile data have been published. However, the data have the characteristics of high dimensionality and small samples, and there are many redundant and irrelevant data, which will reduce the performance and effectiveness of the feature selection algorithm. Therefore, feature selection on these data can reduce the impact of irrelevant redundant data on learning performance, improve model performance, and help researchers in professional fields gain deeper knowledge [1].

Feature selection [2] selects some features from the original features to improve the data classification ability. At present, researchers have performed much research

© The Author(s), under exclusive license to Springer Nature Switzerland AG 2023
N. Xiong et al. (Eds.): ICNC-FSKD 2022, LNDECT 153, pp. 475–482, 2023.
https://doi.org/10.1007/978-3-031-20738-9_54

on the feature selection algorithm [3, 4], among which the random heuristic algorithm applied to the feature selection problem has achieved good results [5, 6]. Tang et al. [7] proposed a five-way joint mutual information (FJMI) feature selection method, which addresses the interaction of high-dimensional data through joint mutual information, and uses the 'maximum of the minimum' method to avoid overestimating features. Zhang et al. [8] proposed a return-cost-based binary firefly algorithm (Rc-BBFA), which designs a return-cost-based index and an adaptive jumping binary movement operator, and utilizes the Pareto dominance strategy to improve the algorithm's performance. Ma et al. [9] proposed a two-stage hybrid ACO for high-dimensional feature selection (TSHFS-ACO) method. The method calculates the number of selected features in the first stage, and performs feature reduction in the second stage. The method guides the search for ACO through feature correlation, symmetric uncertainty, and classification performance so that the algorithm can select the final feature subset. Rostami et al. [10] proposed a multi-objective PSO with the node centrality (MPSONC) feature selection method, which improved the initial population and mutation operator and used three methods of feature similarity, separability index, and feature subset size to perform particle selection.

The contributions of this method are summarized as follows:

(1) A hybrid new feature selection method based on a two-stage framework is proposed. The proposed method uses the union of two filter algorithms and an improved multi-objective particle swarm optimization(MOPSO) method to improve the performance of feature selection.
(2) The algorithm improves the traditional parameters for calculating particle velocity, and adjusts the parameters nonlinearly according to the number of iterations.
(3) The algorithm uses the generation strategy of particle explosion to generate potential particles to make up for the shortage of traditional particle updates and improve the convergence accuracy.

2 Methodology

A two-stage hybrid improved multi-objective particle swarm optimization feature selection algorithm (HIMOPSO) is proposed. At first, Fisher Score [11] and MIC [12] were used to filter the original feature set respectively, and then combines the two filtered subsets to obtain the feature subset after preliminary screening. Secondly, the MOPSO is used to perform a fine search for the initial screening feature subset, the particle swarm algorithm is improved by combining the parameter adjustment of the nonlinear mechanism and the particle generation strategy based on explosion, and the algorithm search ability is improved.

2.1 Particle Parameter Adjustment

The core idea of particle swarm optimization [13] is to find the solution set by calculating the velocity and position of particles in the iterative process. The update formulas of these two attributes are shown in formulas (1) and (2).

$$x_{id}^{t+1} = x_{id}^t + v_{id}^{t+1} \tag{1}$$

$$v_{id}^{t+1} = w * v_{id}^t + c_1 * r_1 * \left(p_{id} - x_{id}^t\right) + c_2 * r_2 * \left(p_{gd} - x_{id}^t\right) \tag{2}$$

where t represents the number of iterations, w represents the inertia weight. c_1 and c_2 represent learning factors. r_1 and r_2 are two random numbers. P_{id} and p_{gd} represent pbest and gbest. There are two main parameters in particle updating: the inertia weight coefficient and the learning factor. Therefore, the nonlinear optimization of these parameters based on the number of iterations is carried out.

Nonlinear optimization is designed for w, making it large in the early stage and small in the later stage, as shown in formulas (3) and (4).

$$w_e = w_s - \left(1 - \sqrt[n]{\ln\left(e - t_{iter}(e - 1)/T_{max}\right)}\right)\left(w_s - w_f\right) \tag{3}$$

$$w_l = w_s - \sqrt[n]{\ln\left(1 + t_{iter}(e - 1)/T_{max}\right)}\left(w_s - w_f\right) \tag{4}$$

where w_e represent the early w, and w_l represents the late w. w_s represents the initial value of w, w_f represents the final value of w, t_{iter} is the current number of iterations, and T_{max} is the total number of iterations.

Different nonlinear calculation methods of c_1 and c_2 are used for particles with different iterations, as shown in formulas (5) and (6).

$$\begin{cases} c_{1e} = c_{1\,max} - \left(1 - \sqrt[n]{\ln\left(e - t_{iter}(e - 1)/T_{max}\right)}\right)(c_{1\,max} - c_{1\,min}) \\ c_{2e} = c_{2\,min} + \left(1 - \sqrt[n]{\ln\left(e - t_{iter}(e - 1)/T_{max}\right)}\right)(c_{2\,max} - c_{2\,min}) \end{cases} \tag{5}$$

$$\begin{cases} c_{1l} = c_{1\,max} - \sqrt[n]{\ln\left(1 + t_{iter}(e - 1)/T_{max}\right)}(c_{1\,max} - c_{1\,min}) \\ c_{2l} = c_{2\,min} + \sqrt[n]{\ln\left(1 + t_{iter}(e - 1)/T_{max}\right)}(c_{2\,max} - c_{2\,min}) \end{cases} \tag{6}$$

The early learning factors were represented by c_{1e} and c_{2e} respectively, and the later learning factors were represented by c_{1l} and c_{2l} respectively. c_{1max}, c_{1min}, c_{2max}, and c_{2min} represent the maximum and minimum values of c_1 and c_2, respectively.

2.2 Candidate Particle Generation Strategy Based on Explosive Particles

In the strategy of generating sub-generations through particle explosion, the most important thing is to determine the number and range of the generated particles. The number of generated particles is given a fixed number to prevent a particle from generating a large number of sub-generations and increase the complexity of the algorithm.

First, the range of the generated sub-generations is determined, as shown in formula (7). The displacement of the particles can only be generated within this range.

$$A_i = \hat{A}\left(\left(f(x_i) - y_{min} + \varepsilon\right)\middle/ \left(\sum_{i=1}^{N}\left(f(x_i) - y_{min}\right) + \varepsilon\right)\right) \tag{7}$$

where $\hat{A}$ represents the maximum vibration amplitude, $f(x_i)$ is the current evaluation of the individual, y_{min} represents the best particle performance in the current population, and ε is a constant to avoid errors.

A displacement operation is performed on each position of the particle, as shown in Formula (8).

$$\triangle x_i^k = x_i^k + rand(0, A_i) \tag{8}$$

where $rand(0, A_i)$ represents that a random number within the range of displacement.

A mutation operation is also performed on the particles to improve their diversity, as shown in formula (9).

$$x_i^k = x_i^k g \tag{9}$$

where $g \sim N(1, 1)$, that is, g follows a random number with a mean and variance of 1.

2.3 Particle Selection Strategy

For the updated original particles, the sub-generations generation strategy is used to obtain the sub-generations particle swarm. At this time, the candidate particle is obtained by mixing the original particle and its sub-generations particles. The candidate particles are selected to screen out the particles used in the next iteration.

During the screening of candidate particles, the elite retention strategy is adopted to retain the better particles. To reduce the cost of computation, this operation is carried out over a certain interval of iterations.

2.4 External Archive Update

The essence of the feature selection problem is to solve a multi-objective problem, selecting the classification accuracy and the number of features as the two objective fitness functions in the algorithm. After each iterative calculation of the two fitness functions, the relationship between the generated solutions is judged. If the two fitness values of the solutions are inferior to other solutions, that is, they are dominated by other solutions, delete the dominated solutions, add the non-dominated solutions to the archive, and update the external archive. When the archive limit is reached, the crowding distance of the archive is calculated and sorted, and the extra solutions are removed so that the archive can add new non-dominated solutions.

2.5 Algorithm Flow

Table 1 describes the pseudo-code of the HIMOPSO feature selection algorithm.

3 Experiments and Analysis

3.1 Datasets

The experiments used 4 typical public gene expression datasets. These datasets are all typical high-dimensional small sample data, as shown in Table 2 for details.

Table 1. Pseudo-code of HIMOPSO algorithm.

Algorithm1 HIMOPSO

Input: Original dataset
Output: Optimal feature subset
Process:
1. Sort and filter features according to the Fisher value to generate feature subset D_1
2. Sort and filter the features according to the MIC value to generate a feature subset D_2
3. Generate feature subsets $D \leftarrow D_1 \cup D_2$
4. Initialize particles positions and velocities for feature subset D
5. Initialize external archive to store non-dominated solutions
6. Initialize the number of iterations $t = 0$
7. **WHILE** maximum number of iterations is not reached **DO**
8.   **FOR** each particle **DO**
9.   Update the velocity and position of particles with a nonlinear parameter adjustment strategy
10.   **IF** $t\%10 == 0$ **DO**
11.   **FOR** each position of the particle **DO**
12.   Calculate the position of the new particle according to the proposed particle explosion strategy, and generate sub-generations particle
13.   **END**
14.   **END**
15. **END**
16. Generate candidate particle swarms
17. Select particles with good performance to form new population
18. Calculation of particle fitness value
19. Maintain and update external archive
20. **END**
21. Generation of optimal feature subset

Table 2. Datasets.

Dataset	#Features	#Instances	#Classes	% Sample distribution
Colon	2000	62	2	[22,40]
SRBCT	2308	83	4	[29,11,18,25]
Leukemia	7129	72	2	[47,25]
Lung	12,600	203	5	[139,17,6,21,20]

3.2 Experimental Results and Analysis

In this section, we first conduct experiments on the variation of the classification accuracy and the number of features of the HIMOPSO feature selection algorithm with the number of iterations under four datasets.

Figure 1 shows the change in the accuracy and feature number of the HIMOPSO feature selection algorithm with iteration times under four datasets. Among them, the

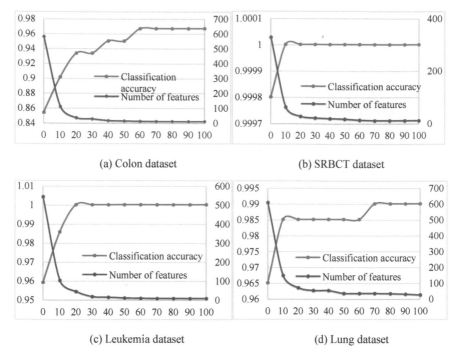

(a) Colon dataset

(b) SRBCT dataset

(c) Leukemia dataset

(d) Lung dataset

Fig. 1. Iterative results of the HIMOPSO algorithm under 4 datasets.

classification accuracy of the Colon dataset has obvious step-like changes every 10 iterations. The remaining datasets can achieve the best classification accuracy in approximately 20 generations. The number of features of all datasets before 20–40 generations has decreased significantly. Figure 1 show that the algorithm can balance the classification accuracy and the number of features, and can get better solutions for these two objectives.

To verify the superiority of the algorithm, HIMOPSO is compared with FJMI [7], Rc-BBFA [8], MPSONC [10], MOSPO [14], HSGA2 [15], and AltWOA [16].

Table 3 lists the classification accuracy of the HIMOPSO method and six comparison algorithms under the SVM classifier. Compared with the six feature selection algorithms, this algorithm obtains the optimal evaluation indexes under all four datasets and achieves 100% classification accuracy on two datasets (SRBCT and Leukemia). In all datasets, HIMOPSO is significantly better than FJMI, Rc-BBFA, MPSONC, and AltWOA. In the Colon dataset, the accuracy of the HIMOPSO algorithm is 4.04% higher than that of MOPSO and 2.5% higher than that of HSGA2. On the leukemia dataset, the classification accuracy of the HIMOPSO algorithm is on average 1.38% higher than that of the two comparison algorithms, MOPSO and NSGA2. On the Lung dataset, the classification accuracy of HIMOPSO is about 1.5% higher than that of the two comparison algorithms, MOPSO and NSGA2.

Table 4 lists the number of features selected by the six algorithms. The number of features selected by the proposed algorithm is the lowest in all datasets because the

Table 3. Classification accuracy of 6 feature selection methods using the SVM classifier.

Dataset	Colon (%)	SRBCT (%)	Leukemia (%)	Lung (%)
FJMI	84.51	79.6	84.34	87.42
Rc-BBFA	90.7	96.18	84.15	55.36
MPSONC	87.89	81.79	90.16	91.81
AltWOA	80.64	76.47	92.19	85.36
MOPSO	91.79	**100**	98.57	97.5
HSGA2	93.33	**100**	98.67	97.49
HIMOPSO	**95.83**	**100**	**100**	**98.99**

algorithm pre-screens the data. The FJMI is a filter algorithm that selects the number of features with high classification accuracy. Rc-BBFA selects the most features, and the single-stage improved evolutionary algorithm has certain limitations on feature selection. Although the number of features selected by several other algorithms is not significantly different, the proposed algorithm can achieve the highest classification accuracy.

Table 4. The size of the final feature subset.

Dataset	Colon	SRBCT	Leukemia	Lung
FJMI	20	20	20	30
Rc-BBFA	1002	1130.8	3547.5	6395.8
MPSONC	18.42	17.58	19.31	28.19
AltWOA	34.5	40.2	25.5	33
MOPSO	16.5	13.2	9	40
HSGA2	17.9	14	8.5	44
HIMOPSO	**12.5**	**7.5**	**7.5**	**23.5**

4   Conclusion

The HIMOPSO algorithm proposed in this paper is a two-stage feature selection algorithm. In the first stage, the Fisher Score and MIC are used for the initial screening of features. In the second stage, the improved multi-objective particle swarm optimization algorithm is used for further searching of the obtained feature subset, to obtain the final feature subset. In the second phase of the search, the particle update strategy is changed to maintain the diversity of the population and improve the quality of the population, which provides an effective search for particles. Experimental results show that the proposed algorithm can achieve higher accuracy with a small number of features in high-dimensional small sample datasets. Compared with the six feature algorithms, optimal results can be obtained.

Acknowledgements. This work was supported by the National Key R&D Program of China (Item NO: 2019YFC0121502) and the Special Fund of Shaanxi Key Laboratory of Network Data Analysis and Intelligent Processing.

References

1. ÖZCAN ŞİMŞEK, N.Ö., ÖzgÜr, A., GÜrgen, F.: A novel gene selection method for gene expression data for the task of cancer type classification. Biol. Direct **16**(1), 1–5 (2021)
2. Anuar, N.K., Bakar, A.A., Ahmad, A.R., et al.: Privacy preserving features selection for data mining using machine learning algorithms. In: Proceedings of the International Conference on Information Technology and Multimedia, pp. 108–113. IEEE (2020)
3. Xiong, N.: A hybrid approach to input selection for complex processes. IEEE Trans. Syst. Man Cybern. A **32**(4), 532–536 (2002)
4. Kumar, V., Minz, S.: Feature selection: a literature review. SmartCR **4**(3), 211–229 (2014)
5. Rostami, M., Berahmand, K., et al.: Review of swarm intelligence-based feature selection methods. Eng. Appl. Artif. Intell. **100**, 104210 (2020)
6. Sekhar, P.R., Sujatha, B.: A literature review on feature selection using evolutionary algorithms. In: Proceedings of the International Conference on Smart Structures and Systems, pp. 1–8. IEEE (2020)
7. Tang, X., Dai, Y., Xiang, Y.: Feature selection based on feature interactions with application to text categorization. Expert Syst. Appl. **120**, 207–216 (2019)
8. Zhang, Y., Song, X., Gong, D.: A return-cost-based binary firefly algorithm for feature selection. Inf. Sci. **418**, 561–574 (2017)
9. Ma, W., Zhou, X., Zhu, H., et al.: A two-stage hybrid ant colony optimization for high-dimensional feature selection. Pattern Recogn. **116**, 107933 (2021)
10. Rostami, M., Forouzandeh, S., et al.: Integration of multi-objective PSO based feature selection and node centrality for medical datasets. Genomics **112**(6), 4370–4384 (2020)
11. Gan, M., Zhang, L.: Iteratively local fisher score for feature selection. Appl. Intell. **51**(8), 6167–6181 (2021). https://doi.org/10.1007/s10489-020-02141-0
12. Kinney, J.B., Atwal, G.S.: Equitability, mutual information, and the maximal information coefficient. Proc. Natl. Acad. Sci. **111**(9), 3354–3359 (2014)
13. Marini, F., Walczak, B.: Particle swarm optimization (PSO). A tutorial. Chemom. Intell. Lab. Syst. **149**, 153–165 (2015)
14. Annavarapu, C., Dara, S., Banka, H.: Cancer microarray data feature selection using multi-objective binary particle swarm optimization algorithm. EXCLI J. **15**, 460–473 (2016)
15. Deb, K., Pratap, A., Agarwal, S., Meyarivan, T.: A fast and elitist multiobjective genetic algorithm: NSGA-II. IEEE Trans. Evolutionary Computation **6**(2), 182–197 (2002)
16. Rohit, K., et al.: AltWOA: Altruistic whale optimization algorithm for feature selection on microarray datasets. Comput. Biol. Med. **144**, 105349 (2022)

Exploiting Inhomogeneities of Subthreshold Transistors as Populations of Spiking Neurons

Etienne Mueller[1(✉)], Daniel Auge[1,2], and Alois Knoll[1]

[1] Department of Informatics, Technical University of Munich, Munich, Germany
etienne.mueller@tum.de, 3534244889@qq.com, daniel.auge@tum.de,
alois.knoll@tum.de
[2] Infineon Technologies AG, Munich, Germany

Abstract. As machine learning applications are becoming increasingly more powerful and are deployed to an increasing number of different appliances, the need for energy-efficient implementations is rising. To meet this demand, a promising field of research is the adoption of spiking neural networks jointly used with neuromorphic hardware, as energy is solely consumed when information is processed. The approach that maximizes energy efficiency, an analog layout with transistors operating in subthreshold mode, suffers from inhomogeneities such as device mismatch which makes it challenging to create a uniform threshold necessary for spiking neurons. Furthermore, previous work mainly focused on spiking feedforward or convolutional networks, as neural networks based on rectified linear units translate well to rate coded spiking neurons. Consequently, the processing of continuous sequential data remains challenging, as neural networks, based on long short-term memory or gated recurrent units as recurrent cells, utilize sigmoid and tanh as activation functions. We show how these two disadvantages can compensate for each other, as a population of spiking neurons with a normally distributed threshold can reliably represent the sigmoid and tanh activation functions. With this finding we present a novel method how to convert a long short-term memory recurrent network to a spiking neural network. Although computationally expensive in a simulation environment, this approach offers a significant opportunity for energy reduction and hardware feasibility as it leverages the often unwanted process variance as a design feature.

Keywords: Spiking neural networks · Conversion · Long short-term memory · Subthreshold analog neuromorphic hardware

N. Xiong et al. (Eds.): ICNC-FSKD 2022, LNDECT 153, pp. 483–492, 2023.
https://doi.org/10.1007/978-3-031-20738-9_55

1   Introduction

In recent time, the use of neural networks in power-restricted environments is continuing to increase. Decision processes in autonomous vehicles or locally processed speech recognition tasks in modern-day smartphones, for example, have to evaluate a large quantity of data with the smallest amount energy as possible. Reducing the energy required for these tasks directly results in a longer range of electric vehicles or battery life in edge devices.

To heavily reduce the energy consumption in neural networks the current focus is on developing new architectures and optimizing hardware. In the long term, another direction of research is the utilization of biologically inspired neurons that communicate with short pulses instead of continuous-valued activation functions. These spiking neural networks (SNNs) provide superior computational power compared to today's common analog neural networks (ANNs) [14] and promise ultra-low powered hardware feasibility due to their simple integrate-and-fire computations. In spite of these advantages, SNNs are not widely used in today's machine or deep learning tasks. Due to temporal dynamics and non-differentiable activation functions, training SNNs still present a challenge and lags behind the performance of backpropagation-trained ANNs. The prinipal approaches for training SNNs are as follows: (1) unsupervised learning approaches (e.g. spike-timing-dependent plasticity [15]), (2) supervised learning approaches, that aim to adapt backpropagation based on gradient descent in different ways [21], and (3) conversion approaches, which this work focuses on.

Instead of direct learning of SNNs, the ideal performing approach is to train analog networks and convert the weights to a network of spiking neurons [24]. As the firing rate of spiking neurons increase linearly with its input, the conversion of sigmoid-activated networks in early work resulted in a large approximation error [20]. However, with the predominant use of rectified linear units (ReLUs) [19], in which activation also increases linearly with its input, this approach results in a practically lossless conversion [5]. Ever since, conversion of feedforward and convolutional neural networks has been well-theorized [23]. Although generating a reliable output signal needs a certain amount of time to enable even the lowest activated neurons to propagate their signal through the network, different optimization methods have been presented which reduce the inference time to a reasonable amount [18].

The conversion of recurrent neural networks (RNNs), however, remains a challenge. Reports show performance losses of more than 10% points during the conversion of Elman RNNs [7] to SNNs [6]. As this basic form of RNNs has shown to be vulnerable to vanishing gradients over time, different memory cells have been developed, most notably long short-term memory (LSTM) cells [9] and gated recurrent units (GRU) [4]. However, a viable conversion method for these cells is currently lacking, as both cells are based on the S-shaped sigmoid and tanh activation functions, which translate poorly to rate coded spiking neurons. To leverage the advantages of short-term memory in SNNs, non-conversion-based implementations have been presented [1,25], although they still lag behind the performance of ANNs.

Neuromorphic hardware is the silicon realization of SNNs, which promises to drastically reduce energy consumption by only consuming energy when information in the form of spikes is being transmitted. The most optimistic estimates range between $10^1 - 10^5$ times lower energy consumption compared with classic computing approaches [22]. This broad range can be attributed to the different approaches used to realize this type of hardware. The most common approach is digital implementation, where the dynamics of spiking neurons are implemented using digital CMOS logic, as has been done with SpiNNaker [8] and by IBM with TrueNorth [17]. An even more promising approach can be found in analog implementations, as their characteristics are well suited for biologically inspired networks. In itself, analog neuromorphic hardware can again be divided into two groups: (1) superthreshold operating mode, which allows for higher rates as demonstrated by BrainScaleS [16] and (2) subthreshold operating mode, which offers the highest possible energy efficiency as presented with Neurogrid [2]. Strictly speaking, both chips fall within the mixed analog/digital family, as they rely on a surrounding digital communication framework.

An important property of the subthreshold analog neuromorphic hardware involves inhomogeneities in the silicon transistors, including device mismatch as a result of process variance as well as shot and thermal noise, which can lead to an inaccurate spiking threshold of the neurons [11]. The sources of mismatch need either be minimized at the device level [12], or else they can be exploited for computational purposes [3]. In this paper, we show that this property can be used to solve the previously mentioned problems of converting LSTM-based RNNs. By using a population of spiking neurons with a normally distributed threshold, we can accurately replicate S-shaped activation functions such as sigmoid and tanh. Additionally, as the population works in parallel, it reduces the inference settling time after each input change to a minimum.

2   Methodology

LSTM units utilize a cell state variable to preserve information for later time steps. Multiple input-dependent non-linear functions, therefore, influence the input such that the cell state is updated whenever the desired information is present. In total, the standard LSTM cell consists of three sigmoid activation functions for its input, output, and forget gate, as well as two tanh activation functions for the hidden and carry state. A direct conversion of these two types of activation functions into a spiking representation has been shown to be challenging, as they exhibit S-shaped curves, while the spiking frequency of pulsing neurons increases linearly with the input [20]. Previous research has focused on adapting the spike frequency of single neurons, so they can exhibit a similar activation [26] or developing new SNN architectures with different approaches for short-term memory [1].

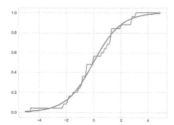
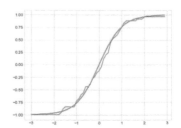

(a) Sigmoid function (dark) and the representation through the accumulation of spikes over the input of an excitatory spiking population of 25 neurons with a normally distributed, randomly initialized threshold with mean $\mu = 0$ and standard deviation $\sigma_{\text{sigmoid}} = 1.7$ (light).

(b) Tanh function (dark) and the representation through the accumulation of spikes over the input of an excitatory and an inhibitory spiking population of 25 neurons each with normally distributed, randomly initialized thresholds with mean $\mu = 0$ and standard deviation $\sigma_{\text{tanh}} = 0.88$ (light).

Fig. 1. Sigmoid and tanh function represented by the accumulation of spikes.

2.1   Conversion of Sigmoid and Tanh

In our approach, we make use of inhomogeneous thresholds in subthreshold operating analog neuromorphic hardware. As the cumulative distribution function of a normal distribution also produces an S-shaped curve, it is possible to use this property to closely approximate the sigmoid and tanh activation functions. Consequently, when feeding input to a sufficiently large population of spiking neurons whose thresholds are randomly initialized with a normal distribution, the number of resulting spikes can also be predicted with the formula for the cumulative distribution function depending on the input current I:

$$F(I) = P(V_{\text{thresh}} \leq I) \tag{1}$$

where P represents the probability that the spiking threshold V_{thresh} is less or equal to input I.

To represent the sigmoid function as accurately as possible, the sum of the spikes per population needs to be normalized to return a value in the same range between $[0, 1]$. Moreover, the value of the standard deviation has to be set so the cumulative distribution function closely maps the sigmoid function. This value can be calculated by minimizing the error between both functions and results in $\sigma_{\text{sigmoid}} = 1.75$. With as few as 25 neurons in the population this value can already produce a reasonable outcome (see Fig. 1a).

As the output of the tanh function ranges between $[-1, 1]$, the conversion to a spiking population requires further steps. Additionally, for a population that emits excitatory (positive) spikes, a second population with inhibitory (negative) spikes as output is needed. This population is also initialized with the same random threshold as before, but with a reversed sign. The optimal standard deviation can be derived as before and is calculated as $\sigma_{\text{tanh}} = 0.88$. With this value, a small population of 50 neurons (25 for each excitatory and inhibitory population) can visibly approximate the tanh function (see Fig. 1b).

2.2 Compensating External Factors

In the event the normal distribution is influenced by external conditions and cannot be adjusted, the input current to the population has to be modified. If the actual mean is $\mu_{\text{real}} \neq 0$, then μ_{real} has to be added to the input. Additionally, if the standard deviation of the sigmoid population is $\sigma_{\text{real}} \neq \sigma_{\text{sigmoid}}$ or the standard deviation of the tanh population is $\sigma_{\text{real}} \neq \sigma_{\text{tanh}}$, then the input current for each population has to be multiplied with the ratio between the actual value and the target value of the respective standard deviation. The formula would then read:

$$I_{\text{sigmoid}} = I \left(\frac{\sigma_{\text{pop}}}{0.88} \right) + \mu_{\text{real}} \tag{2}$$

for the input current to the sigmoid population and

$$I_{\text{tanh}} = I \left(\frac{\sigma_{\text{pop}}}{1.75} \right) + \mu_{\text{real}} \tag{3}$$

for the tanh population, accordingly. As our experiments are performed in a simulation environment, there is no need to use the compensation.

3 Experiments

To evaluate our proposed method, we trained an LSTM-based neural network on a sentiment classification dataset. In the first experiment, we assess performance and accuracy loss after conversion of the single activation functions of the cell. With an estimate of the necessary population sizes, we then convert and evaluate the entire network in the second experiment.

3.1 Network Architecture and Dataset

For our experiments, we use the IMDB movie review sentiment classification dataset [13], which is composed of 25,000 movie reviews encoded as a list of word indices and labeled with positive or negative sentiment. The dataset is split in half into a training set and a test set. Only the 2500 most frequent words are used and every review is limited to 500 words, with shorter reviews being filled with zeros. To represent the input as spikes, the data is encoded as one-hot vectors, with one word being fed into the network at every time step.

The baseline network to be converted comprises of an input layer, a single hidden layer with 25 LSTM cells, and a densely connected single output neuron for the desired binary classification. The network is trained with the Adam optimizer [10] for four epochs and reaches an accuracy of 92.6% on the training set and 89.8% on the testing set. Because of the large populations after conversion, the test set used to evaluate the resulting performance of the converted spiking network is reduced to 1000 reviews to reduce the computational need.

3.2   Experiment 1

In the first experiment, each individual activation function of the different gates and states is replaced with populations of spiking neurons. The population sizes range from a single neuron up to 10,000 neurons. Since the system's classification accuracy is directly influenced by the random initialization of the spike thresholds, each simulation is conducted 50 times to average the performance over different random seeds.

Table 1. Results from the first experiment: the mean accuracy μ_{acc} and the standard deviation σ_{acc} in percentage points after conversion of the individual activation function in the LSTM cell to populations of different sizes. Averaged over 50 passes with random initialization of the spiking thresholds.

Population	Input gate		Output gate		Forget gate		Hidden state		Carry state	
	$\mu_{acc}(\%)$	σ_{acc}	$\mu_{acc}(\%)$	σ_{acc}	$\mu_{acc}(\%)$	σ_{acc}	$\mu_{acc}(\%)$	σ_{acc}	$\mu_{acc}(\%)$	σ_{acc}
1	75.7	9.8	70.6	11.2	52.6	3.3	54.2	7.8	53.5	7.1
10	84.2	3.8	84.0	3.9	65.4	12.9	61.4	10.9	55.7	7.9
100	88.9	1.0	88.9	1.0	87.0	2.4	75.8	9.2	68.5	12.7
1000	89.6	0.2	89.5	0.3	86.2	1.2	82.3	2.4	77.8	6.9
10,000	89.7	0.3	89.7	0.2	85.8	0.5	83.4	0.7	83.2	1.0

As can be seen in Table 1, exchanging a single activation function in the LSTM cell with a population of spiking neurons, the mean accuracy μ_{acc} increases with the growing number of neurons while the standard deviation σ_{acc} decreases.

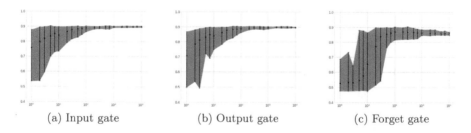

      (a) Input gate   (b) Output gate   (c) Forget gate

Fig. 2. Overview of the accuracy range after converting each of the three sigmoid-activated gates in relation to the population sizes. The exemplary population sizes are run 50 times with randomly initialized thresholds.

The conversion of the sigmoid activated gates (see Fig. 2) works successfully, as a population of 200 neurons for the input and output gate already perform close to the original classification accuracy of 89.8% with only a small deviation over the random seeds. The populations that activate the forget gate already reach the converted network's peak mean accuracy of 87.0% with a size of 100 neurons, but from 1,000 neurons and up it the scattering decreases. It should be noted that the accuracy decreases slightly, but it is assumed that this could be resolved by averaging over more passes.

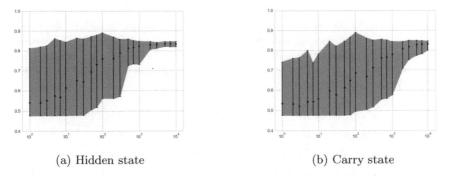

(a) Hidden state   (b) Carry state

Fig. 3. Overview of the accuracy range after converting each of the two tanh-activated states in relation to the number of neurons in the inhibitory and excitatory populations. The exemplary population sizes are run 50 times with randomly initialized thresholds.

The conversion of the tanh activated cell states (see Fig. 3) performs worse than for sigmoid. The conversion of the hidden state evens out at about 83% with 2,000 neurons for each inhibitory and excitatory population. As the deviation also stays roughly consistent, it is assumed that a further increase of the population will not benefit the conversion. The populations representing the activation function of the carry state however need 20,000 neurons in total to reach over 83% accuracy. Although the accuracy seems to even out from this point on, a further increase of the population might still decrease the divergence.

By replacing all three sigmoid functions with a population of 1,000 neurons each, the network performs with a mean accuracy of 85.8% and a standard deviation of 1.4% points. This is only slightly below the single conversion of the forget gate and shows that the sigmoid conversion performs considerably. However, replacing the two tanh functions with a population of two times 10,000 neurons (inhibitory and excitatory) each results in a mean accuracy of 80.9% with a standard deviation of 0.9% points. This exhibits a larger loss compared to the single tanh conversions, which means that the tanh function is not represented as good by the spiking population.

3.3   Experiment 2

In the second experiment, we simultaneously evaluate the network with all converted activation functions. The sizes of the populations are determined based on the best sufficient individual performance in the first experiment. For the sigmoid activated gates, the population sizes are 200 neurons for the input and output gates and 1000 neurons for the forget gate. The tanh activation functions are replaced with 2×2000 for the hidden state and $2\times10,000$ for the carry state.

Again, the experiment is run 50 times with randomly initialized thresholds each time. To determine the impact of different population sizes, the experiment is repeated with half and twice the number of neurons.

Table 2. Results of the second experiment: Overview of the mean accuracies, standard deviation in percentage points, as well as maximum/minimum values for the conversion of an LSTM cell to population coded spiking networks. The baseline conversion consists of 200 neurons for input and output gate, 1,000 for the forget gate, and $2 \times 2,000$ and $2 \times 10,000$ for the hidden and carry state, respectively.

	$\mu_{acc}(\%)$	σ_{acc}	Min (%)	Max (%)
Complete conversion	78.0	1.35	72.8	80.6
Halved populations	77.7	2.99	67.2	84.4
Doubled populations	78.1	0.73	76.2	80.4

The resulting conversion (as shown in Table 2) has a mean accuracy of 78.0% with a standard deviation of 1.35%. That results in an average conversion loss of 13.2% compared to the 89.8% accuracy of the original ANN. Both, halving and doubling the population sizes keep the mean accuracies roughly the same, with the halving performing slightly worse and showing more extreme outliers. The divergence can be strongly reduced by using larger populations. In this example, doubling the populations leads to an approximate halving of the standard deviation.

4   Discussion

In this work, we introduced a novel conversion method for sigmoid and tanh activated neural networks to population coded SNNs with normally distributed, randomly initialized spiking thresholds. In our experiment, we trained an LSTM-based recurrent network for sentiment classification and evaluated the conversion of the single activation functions and subsequently the entire converted network. To achieve reasonable performance, each LSTM-cell was replaced by 25,400 neurons, totaling 635,000 neurons in our example. The accuracy of the converted network showed a conversion loss of 13.2%; doubling the size of the population reduced the scattering of the accuracy in different trials of the same experiment without increasing the overall performance.

As the conversion of the tanh activation functions resulted in the largest accuracy loss, improving this specific conversion method might lead to the most improvement. In addition, the use of other cells such as GRUs, which only contain a single tanh function, might lead to increased performance. Alternatively, training the network with the cumulative normal distribution function for the gates can significantly reduce the conversion loss and thus improve performance.

The sheer number of necessary neurons makes this approach impractical in a simulation environment, as the comparatively small example network increases the computational needs many times over. The main benefit of our approach is that a potential drawback of neuromorphic hardware can be used as an advantage. Furthermore, each input only has to be processed once to achieve results, in contrast to rate coded SNNs, where inputs have to be presented for multiple time steps. For a comprehensive evaluation, the conversion approach must be assessed on compatible hardware, which is not available at this scale. Nonetheless, our approach shows successful conversion and can yield a simple conversion method for future subthreshold-operating analog neuromorphic hardware.

Acknowledgement. The authors would like to thank Infineon Technologies AG for supporting this research.

References

1. Bellec, G., Salaj, D., Subramoney, A., Legenstein, R., Maass, W.: Long short-term memory and learning-to-learn in networks of spiking neurons. In: Advances in Neural Information Processing Systems, vol. 31. Curran Associates, Inc. (2018)
2. Benjamin, B.V., Gao, P., McQuinn, E., Choudhary, S., Chandrasekaran, A.R., Bussat, J.M., Alvarez-Icaza, R., Arthur, J.V., Merolla, P.A., Boahen, K.: Neurogrid: a mixed-analog-digital multichip system for large-scale neural simulations. Proc. IEEE **102**(5), 699–716 (2014)
3. Chicca, E., Badoni, D., Dante, V., D'Andreagiovanni, M., Salina, G., Carota, L., Fusi, S., Del Giudice, P.: A VLSI recurrent network of integrate-and-fire neurons connected by plastic synapses with long-term memory. IEEE Trans. Neural Networks **14**(5), 1297–1307 (2003)
4. Cho, K., van Merrienboer, B., Gulcehre, C., Bahdanau, D., Bougares, F., Schwenk, H., Bengio, Y.: Learning Phrase Representations using RNN Encoder-Decoder for Statistical Machine Translation. arXiv:1406.1078 [cs, stat] (2014)
5. Diehl, P.U., Neil, D., Binas, J., Cook, M., Liu, S., Pfeiffer, M.: Fast-classifying, high-accuracy spiking deep networks through weight and threshold balancing. In: 2015 International Joint Conference on Neural Networks (IJCNN), pp. 1–8 (2015)
6. Diehl, P.U., Zarrella, G., Cassidy, A., Pedroni, B.U., Neftci, E.: Conversion of artificial recurrent neural networks to spiking neural networks for low-power neuromorphic hardware. In: 2016 IEEE International Conference on Rebooting Computing (ICRC), pp. 1–8 (2016)
7. Elman, J.L.: Finding structure in time. Cogn. Sci. **14**(2), 179–211 (1990)
8. Furber, S.B., Lester, D.R., Plana, L.A., Garside, J.D., Painkras, E., Temple, S., Brown, A.D.: Overview of the SpiNNaker system architecture. IEEE Trans. Comput. **62**(12), 2454–2467 (2013)
9. Hochreiter, S., Schmidhuber, J.: Long short-term memory. Neural Comput. **9**(8), 1735–1780 (1997)
10. Kingma, D.P., Ba, J.: Adam: A Method for Stochastic Optimization. arXiv:1412.6980 [cs] (2017)
11. Liu, S.C., Delbruck, T., Indiveri, G., Whatley, A., Douglas, R.: Event-Based Neuromorphic Systems. Wiley (2014)
12. Liu, S.C., Kramer, J., Indiveri, G., Delbrück, T., Douglas, R.: Analog VLSI: Circuits and Principles. MIT Press (2002)

13. Maas, A.L., Daly, R.E., Pham, P.T., Huang, D., Ng, A.Y., Potts, C.: Learning word vectors for sentiment analysis. In: Proceedings of the 49th Annual Meeting of the Association for Computational Linguistics, p. 9 (2011)
14. Maass, W.: Networks of spiking neurons: the third generation of neural network models. Neural Networks **10**(9), 1659–1671 (1997)
15. Markram, H., Lübke, J., Frotscher, M., Sakmann, B.: Regulation of synaptic efficacy by coincidence of postsynaptic APs and EPSPs. Science **275**(5297), 213–215 (1997)
16. Meier, K.: A mixed-signal universal neuromorphic computing system. In: 2015 IEEE International Electron Devices Meeting (IEDM), pp. 4.6.1–4.6.4. IEEE, Washington, DC (2015)
17. Merolla, P.A., Arthur, J.V., Alvarez-Icaza, R., Cassidy, A.S., Sawada, J., Akopyan, F., Jackson, B.L., Imam, N., Guo, C., Nakamura, Y., Brezzo, B., Vo, a., Esser, S.K., Appuswamy, R., Taba, B., Amir, A., Flickner, M.D., Risk, W.P., Manohar, R., Modha, D.S.: A million spiking-neuron integrated circuit with a scalable communication network and interface. Science **345**(6197) (2014)
18. Mueller, E., Hansjakob, J., Auge, D., Knoll, A.: Minimizing inference time: optimization methods for converted deep spiking neural networks. In: International Joint Conference on Neural Network, p. 8 (2021)
19. Nair, V., Hinton, G.E.: Rectified linear units improve restricted Boltzmann machines. In: Proceedings of the 27th International Conference on Machine Learning, p. 8 (2010)
20. Perez-Carrasco, J.A., Zhao, B., Serrano, C., Acha, B., Serrano-Gotarredona, T., Chen, S., Linares-Barranco, B.: Mapping from frame-driven to frame-free event-driven vision systems by low-rate rate coding and coincidence processing-application to feedforward ConvNets. IEEE Trans. Pattern Anal. Mach. Intell. **35**(11), 2706–2719 (2013)
21. Pfister, J.P., Toyoizumi, T., Barber, D., Gerstner, W.: Optimal spike-timing-dependent plasticity for precise action potential firing in supervised learning. Neural Comput. **18**(6), 1318–1348 (2006)
22. Poon, C.S., Zhou, K.: Neuromorphic silicon neurons and large-scale neural networks: challenges and opportunities. Front. Neurosci. **5** (2011)
23. Rueckauer, B., Lungu, I.A., Hu, Y., Pfeiffer, M., Liu, S.C.: Conversion of continuous-valued deep networks to efficient event-driven networks for image classification. Front. Neurosci. **11** (2017)
24. Sengupta, A., Ye, Y., Wang, R., Liu, C., Roy, K.: Going deeper in spiking neural networks: VGG and residual architectures. Front. Neurosci. **13** (2019)
25. Shrestha, A., Ahmed, K., Wang, Y., Widemann, D.P., Moody, A.T., Van Essen, B.C., Qiu, Q.: A spike-based long short-term memory on a neurosynaptic processor. In: 2017 IEEE/ACM International Conference on Computer-Aided Design (ICCAD), pp. 631–637 (2017)
26. Zambrano, D., Bohte, S.M.: Fast and Efficient Asynchronous Neural Computation with Adapting Spiking Neural Networks. arXiv:1609.02053 [cs] (2016)

Backpropagation Neural Network with Adaptive Learning Rate for Classification

Rujira Jullapak[✉] and Arit Thammano

Computational Intelligence Laboratory, Faculty of Information Technology, King Mongkut's
Institute of Technology Ladkrabang, Bangkok, Thailand
63606002@kmitl.ac.th, arit@it.kmitl.ac.th

Abstract. This research aims to improve the classification accuracy by modifying an original backpropagation neural network. In the proposed BPNN-ZMP,
the learning rates were automatic tuned to improve the classification accuracy.
Breast Cancer Coimbra dataset and Banknote Authentication dataset were used
for testing the model performances. The results demonstrate that BPNN-ZMP
improved over the original backpropagation neural network by 12.12 and 11.46%
for Breast Cancer Coimbra dataset and Banknote Authentication dataset respectively. Although BPNN-ZMP could improve the model accuracy, the high accuracy
in neural network backpropagation has been challenged in future work.

Keywords: Classification · Backpropagation neural network · Adaptive learning
rate

1 Introduction

Data classification is important role in many applications such as industrial [1, 2], agricultural [3], business [4], and medical [5]. Classification techniques are supervised learning
that defines by their labeled to train algorithms, classify data, or predict outcomes accurately. The classification process begins with feeding all training data into the model;
the model is then updated its weights until the error is within an acceptable range.
Finally, the trained model is used to predict unseen data. There are several classification
techniques such as decision tree, Bayesian, Rule-based, association ruled classifications, and neural network. In recent years, neural network technique is mostly used in
classification. However, the accuracy improvement has been challenging. Jullapak and
Thammano [6] introduced BPNN-ALR which had an ability to automatically adjust the
learning rate (ALR). They concluded that BPNN-ALR can solve the imbalanced dataset
with high accuracy when compared with original BPNN. Inspire by their success, this
study proposes the BPNN-ZMP method by modifying original BPNN to improve model
accuracy.

2 Related Works

In the last decade, many studies proposed techniques to improve the performance of
the neural network. For instance, Singh et al. [7] introduced 3 algorithms including

N. Xiong et al. (Eds.): ICNC-FSKD 2022, LNDECT 153, pp. 493–499, 2023.
https://doi.org/10.1007/978-3-031-20738-9_56

Adaptive gradient descent (AGD), Gradient descent (GD), and Gradient descent with GDM to identify 40 different mammary tumors. The results showed that the Adaptive gradient descent (AGD) outperformed other algorithms; it had the highest classification accuracy of 84.6%.

Poynton and McDaniel [8] examined the performance of the backpropagation neural network in distinguishing the current from the former smokers in the 2000 National Health Interview Survey (NHIS) sample adult file. The results showed that the BPNN classifier outperformed the random chance with the confident level of 95%.

Anwani and Rajendran [9] developed a learning method for the spiking neural network (SNN). Normalized Approximate Descent (NormAD) based spatio-temporal error backpropagation was employed to derive the iterative synaptic weight update rule.

3 Proposed Method

First, the learning process of the original backpropagation neural network is briefly discribed as follows:

Step 1 Initialize the learning rate, z, to a small number between 0 and 1.
Step 2 Determine the initial values of all parameters and define the termination criteria, e.g., the maximum number of iterations.
Step 3 Randomly initialize the weights W_{ji} and W_{kj}.
Step 4 For each of the training data, perform Steps 4 to 7. Compute the output values of all hidden nodes,H_j, by using Eqs. (1) and (2).

$$net_j = \sum_{i=1}^{n} W_{ji} X_i \tag{1}$$

$$H_j = \frac{1}{1 + e^{-net_j}} \tag{2}$$

where X_i is the value of the node i in the input layer. W_{ji} is the weight that connects the node i in the input layer and the node j in the hidden layer.
Step 5 Compute the output values of the nodes in the output layer with Eqs. (3) and (4).

$$net_k = \sum_{j=1}^{m} W_{kj} H_j \tag{3}$$

$$O_k = \frac{1}{1 + e^{-net_k}} \tag{4}$$

where W_{kj} is the weight that connects the node j in the hidden layer and the node k in the output layer.
Step 6 Compare O_k with Y_k and calculate the error by using Eq. (5).

$$E = \frac{1}{2} \sum_{k=1}^{p} (y_k - O_k)^2 \tag{5}$$

Step 7 The result from Eq. (5) is used to update the weight W_{kj} by using Eqs. (6) and (7). Similarly, the weight W_{ji} is updated by using Eqs. (8) and (9).

$$\frac{\partial E}{\partial W_{kj}} = -(y_k - O_k)O_k(1 - O_k)H_j \tag{6}$$

$$W_{kj}^{new} = W_{kj}^{old} - z\frac{\partial E}{\partial W_{kj}} \tag{7}$$

$$\frac{\partial E}{\partial W_{ji}} = -\sum_{k=1}^{p}(y_k - O_k)O_k(1 - O_k)W_{kj}H_j(1 - H_j)X_i \tag{8}$$

$$W_{ji}^{new} = W_{ji}^{old} - z\frac{\partial E}{\partial W_{ji}} \tag{9}$$

Step 8 Go back to Step 4 until the maximum number of iterations is reached.

In this paper, we developed the BPNN-ZMP algorithm. Most part of the BPNN-ZMP algorithm is similar to that from the original BPNN algorithm and the BPNN-ALR algorithm. The learning process of the BPNN-ZMP is as follows:

Step 1 Initialize the learning rate for each class c, z_c, to a small number between 0 and 1.

Step 2 Determine the initial values of all parameters and define the termination criteria, e.g., the maximum number of iterations.

Step 3 Randomly initialize the weights W_{ji} and W_{kj}.

Step 4 For each of the training data, perform Steps 4 to 7. Compute the output values of all hidden nodes,H_j, by using Eqs. (1) and (2).

Step 5 Compute the output values of the nodes in the output layer with Eqs. (3) and (4).

Step 6 Compare O_k with Y_k and calculate the error by using Eq. (5).

Step 7 Update the weight W_{kj} by using Eqs. (6) and (10). Similarly, the weight W_{ji} is updated by using Eqs. (8) and (11).

$$W_{kj}^{new} = W_{kj}^{old} - z_c\frac{\partial E}{\partial W_{kj}} \tag{10}$$

$$W_{ji}^{new} = W_{ji}^{old} - z_c\frac{\partial E}{\partial W_{ji}} \tag{11}$$

Step 8 Compute the learning rate for each class by using Eq. (12).

$$z_c = \begin{cases} z_c + 0.05; \ if \ \frac{A_c}{N_c} < 0.5 \\ z_c - 0.05; \ if \ \frac{A_c}{N_c} \geq 0.5 \wedge \frac{P_c}{N_c} > 1 \end{cases} \tag{12}$$

where A_c is the number of correctly predicted class c data. P_c is the number of data predicted as class c. N_c is the number of data in class c.

Step 9 If the maximum number of iterations is not reached, go back to Step 4.

4   Results and Discussions

Breast Cancer Coimbra dataset and Banknote Authentication dataset published on the UCI Machine Learning Repository [10] were used to test the proposed algorithm.

The Breast Cancer Coimbra dataset was comprised of two groups including 52 healthy samples and 64 breast cancer samples collected from the University Hospital Centre of Coimbra in 2018. Table 1 shows the sample of Breast Cancer Coimbra dataset. Table 2 shows the Breast Cancer Coimbra dataset details.

The banknote authentication images collected from counterfeit banknote samples in 2013. These datasets consist of 610 actual banknotes and 762 fake banknotes. Table 3 shows the sample of Banknote Authentication dataset. Table 4 explains the Banknote Authentication image details.

Table 1. Sample of Breast Cancer Coimbra dataset.

Age	BMI	Glucose	Insulin	...	MCP.1	Class
48	23.5	70	2.707	...	417.114	1
83	20.69049	92	3.115	...	468.786	1
82	23.12467	91	4.498	...	554.697	1
68	21.36752	77	3.226	...	928.22	1
86	21.11111	92	3.549	...	773.92	1
45	26.85	92	3.33	...	268.23	2
62	26.84	100	4.53	...	330.16	2
65	32.05	97	5.73	...	314.05	2
72	25.59	82	2.82	...	392.46	2
86	27.18	138	19.91	...	90.09	2

For testing the proposed classification model, this research used the 10-fold cross validation method to test the performance of the model. First, 10-fold cross-validation procedure was divided a dataset into 10 non-overlapping folds. Then, each of the 10 folds was given an opportunity to be used as a test set, while all other folds collectively were used as a training set.

Table 5 show classification accuracy of Banknotes Authentication dataset. Table 6 show classification accuracy of Breast Cancer Coimbra Dataset. For Banknotes Authentication dataset, BPNN-ZMP shows higher accuracy than the original BPNN. In addition, BPNN-ZMP also shows higher accuracy when compared with BPNN-ALR. BPNN-ZMP outperforms the original BPNN and BPNN-ALR by 11.46 and 3.87% respectively. For Breast Cancer Coimbra Dataset, BPNN-ZMP outperforms the original BPNN by 12.12%. The superior performance of BPNN-ZMP is due to the ability of the adaptive learning rate in reducing the bias toward the majority class.

Table 2. Attributes of the Breast Cancer dataset.

Attribute	Value
Age	24–89 years
BMI	18.37–38.58 kg/m^2
Glucose	60–201 mg/dL
Insulin	2.43–58.46 µU/mL
HOMA	0.47–25.05
Leptin	4.31–90.28 ng/mL
Adiponectin	1.65–38.04 µg/mL
Resistin	3.21–82.10 ng/mL
MCP.1	45.84–1698.44 pg/dL
Class	1, 2

Table 3. Sample of Banknote Authentication dataset.

Variance	Skewness	...	Entropy	Class
3.6216	8.6661	...	− 0.44699	0
4.5459	8.1674	...	− 1.4621	0
3.866	− 2.6383	...	0.10645	0
− 2.2987	− 5.227	...	0.91722	1
− 1.239	− 6.541	...	− 0.033204	1
0.75896	0.29176	...	0.83834	1

Table 4. Attributes of the Banknote Authentication dataset.

Attribute	Value
Variance	− 7.0421–6.8248
Skewness	− 13.7731–12.9516
Curtosis	− 5.2861–17.9274
Entropy	− 8.5482–2.4495
Class	0, 1

5   Conclusions

This research aimed to optimize the accuracy of data classification by modifying a backpropagation neural network (BPNN) so that it can automatically adjust, increases or decreases, the learning rates by itself and becomes a more accurate classifier. This

Table 5. Comparative results of BPNN, BPNN-ALR and BPNN-ZMP algorithms on Banknote Authenfication dataset.

	Accuracy (%)		
	BPNN	BPNN-ALR	BPNN-ZMP
Fold 1	44.53	52.55	58.39
Fold 2	44.53	51.09	58.39
Fold 3	44.53	49.64	59.12
Fold 4	44.53	53.28	54.74
Fold 5	44.53	52.55	56.93
Fold 6	44.53	51.09	54.01
Fold 7	44.53	49.64	56.20
Fold 8	44.53	48.18	57.66
Fold 9	44.53	58.39	51.09
Fold 10	44.53	54.74	53.28
Mean	44.53	52.12	55.99
S.D	0.00	2.94	2.62

Table 6. Comparative results of BPNN, BPNN-ALR and BPNN-ZMP algorithms on Breast Cancer Coimbra Dataset.

	Accuracy (%)		
	BPNN	BPNN-ALR	BPNN-ZMP
Fold 1	33.33	45.45	45.45
Fold 2	33.33	45.45	45.45
Fold 3	33.33	45.45	45.45
Fold 4	33.33	45.45	45.45
Fold 5	33.33	45.45	45.45
Fold 6	33.33	45.45	45.45
Fold 7	33.33	45.45	45.45
Fold 8	33.33	45.45	45.45
Fold 9	33.33	45.45	45.45
Fold 10	33.33	45.45	45.45
Mean	33.33	45.45	45.45
S.D	0.00	0.00	0.00

new modified classifier is called BPNN-ZMP. We tested its classification accuracy on a medical dataset, Breast Cancer Coimbra Dataset. The original backpropagation neural network obtained the accuracy of 33.33% while BPNN-ZMP method was much more accurate, 45.45%. On another dataset of counterfeit banknote samples, BPNN-ZMP was also more accurate, 55.99%, than the original backpropagation neural network method, 44.53%. In the future work, we will test the proposed BPNN-ZMP with more diverse datasets and further improve its performance.

References

1. Yu, W., Patros, P., Young, B., Klinac, E., Walmsley, T.G.: Energy digital twin technology for industrial energy management: classification, challenges and future. Renew. Sustain. Energy Rev. **161**, 112407 (2022)
2. Borg, D., Sestito, G., Silva, M.: Machine-learning classification of environmental conditions inside a tank by analyzing radar curves in industrial level measurements. Flow Meas. Instrum. **79**, 101940 (2021)
3. Peng, Y., Wang, Y.: An industrial-grade solution for agricultural image classification tasks. Comput. Electron. Agric. **187**, 106253 (2021)
4. Krajsic, P., Franczyk, B.: Semi-supervised anomaly detection in business process event data using self-attention based classification. Procedia Comput. Sci. **192**, 39–48 (2021)
5. Li, X., Wang, J., Hao, W., Wang, M., Zhang, M.: Multi-layer perceptron classification method of medical data based on biogeography-based optimization algorithm with probability distributions. Appl. Soft Comput. **121**, 108766 (2022)
6. Jullapak, R., Thammano, A.: Backpropagation neural network with adaptive learning rate for classification of imbalanced data. Suthiparithat J. **35**(2), 130–146 (2021)
7. Singh, B.K., Verma, K., Thoke, A.S.: Adaptive gradient descent backpropagation for classification of breast tumors in ultrasound imaging. Procedia Comput. Sci. **46**, 1601–1609 (2015)
8. Poynton, M.R., McDaniel, A.M.: Classification of smoking cessation status with a backpropagation neural network. J. Biomed. Inform. **39**(6), 680–686 (2006)
9. Anwani, N., Rajendran, B.: Training multi-layer spiking neural networks using NormAD based spatio-temporal error backpropagation. Neurocomputing **380**, 67–77 (2020)
10. Dua, D., Graff, C.: UCI machine learning repository. http://archive.ics.uci.edu/ml. University of California, School of Information and Computer Science (2019)

Bidirectional Controlled Quantum Teleportation of Two-Qubit State via Eight-Qubit Entangled State

Jinwei Wang[(⊠)]

School of Mathematics and Statistics, Guizhou University of Finance and Economics,
Guiyang 550025, China
201601009@mail.gufe.edu.cn,{jinweiwang888,alfredkpa}@gmail.com

Abstract. In this paper, a bidirectional controlled quantum teleportation via an eight-qubit state is given. In the scheme, Alice teleports a
two-qubit entangled state to Bob and Bob teleport an arbitrary two-
qubit state to Alice at the same time. This quantum teleportation task
is supervised by a controller. Furthermore, only Bell-state measurements
and single-qubit measurements are needed in our scheme.

Keywords: Bidirectional teleportation · Quantum information ·
Bell-base state

1 Introduction

Quantum teleportation is an important technique for communicating quantum
information. In 1993, Bennett [1] proposed the first scheme via the EPR state,
in which a agent wishes to communicate an unknown quantum qubit to the
other agnet. From then on, many teleportations via entangled states were investigated [2–6]. Bidirectional quantum teleportation(BQT) is a quantum scheme,
in which two agents can teleport quantum information simultaneously. Zha [7]
proposed a BQT method to teleport two single-qubit states via a five-qubit state
in 2013. Later, Binayak [8] and Zadeh [9] proposed a BQT method to teleport
two-qubit states via ten-qubit and eight-qubit states. Recently, Mohammad [10]
gave another quantum protocol for communicating arbitrary numbers of qubits.
Then again, Hassanpour [11] proposed a scheme to teleport EPR state. In the
quantum scheme, GHZ states are utilized as channel. This type of teleportation can transmit more quantum information with less consumption of quantum
resources. Later, Yang [12] proposed a protocol of transmitting two three-qubit
entangled states between two users, and Zhou proposed two methods to teleport
a two-qubit entangled state [13] or a two-qubit entangled state and a three-qubit
entangled state [14]. Furthermore, Zadeh [15] proposed a method to transmit two
n-qubit entangled states.

In 2015, Sang [16] proposed an asymmetric bidirectional quantum telepor-
tation(ABQT) protocol via a five-qubit state. In the protocol, a agent, Alice,

N. Xiong et al. (Eds.): ICNC-FSKD 2022, LNDECT 153, pp. 500–508, 2023.
https://doi.org/10.1007/978-3-031-20738-9_57

wishes to teleport a two-qubit entangled state to the other agnet, Bob, in the meantime Bob also wishes to teleport single-qubit state back to Alice. Different from the BQT scheme, the number of qubit state transmitted in ABQT scheme are not in accord. Later, Li [17] proposed an controlled ABQT scheme via a six-qubit state for communicating a two-qubit entangled and single-qubit state. Zhou [18] and Vikram [19] also proposed two ABQTschemes via two sets of GHZ states. Binayak [20] proposed a bidirectional teleportation of three- and two-qubit entangled states via a nine-qubit state. All of the above asymmetric schemes transmit a multi-qubit entangled state and a single-qubit state. Furthermore, Hong [21] and Zhang [22] proposed an controlledABQT scheme via a seven-qubit state and eight-qubit state, respectively. Later, Long [23] proposed a scheme to teleport an arbitrary single-qubit state and three-qubit state via a nine-qubit genuine entangled state, and Huo [24] proposed teleportation of two- and three-qubit states by using eleven-qubit state as channel. As stated above, Binayak [8] proposed a scheme to transmit two two-qubit states via ten-qubit state. A year later, Zadeh [9] also provided a scheme to transmit two two-qubit states via an eight-qubit state. It is easy to see that the latter is more efficient and economic. These two articles inspired us to determine a method to transmit quantum information efficiently. Based on the study above, we propose a bidirectional controlled quantum teleportation to teleport two-qubit and two-qubit entangled states via an eight-qubit genuine entangled state.

The organization of this paper is outlined as follows. In Sect. 2, an eight-qubit entangled state is produced in theory and an bidirectional controlled quantum teleportation scheme is proposed. Next, we give a comparison and discussions in Sect. 3. Finally, conclusions are given in Sect. 4.

2   Bidirectional Quantum Teleportation of Two-Qubit State

To teleport a two-qubit quantum state, we need to produce an eight-qubit entangled state, which is shared among three users as a quantum channel in advance. The eight-qubit entangled state is given by

$$
|\phi\rangle = \frac{1}{2\sqrt{2}}(|00000000\rangle + |00011110\rangle + |00001011\rangle + |00010101\rangle
$$
$$
+ |11100001\rangle + |11101010\rangle + |11110100\rangle + |11111111\rangle)_{12345678}. \tag{1}
$$

As shown in Fig. 1, the state $|\phi\rangle$ can be produced from the state $|\phi_0\rangle = |00000000\rangle$. Here, we use the genuine-multipartite-entanglement (GME) concurrence [25] as a measure for genuine entanglement, which is defined as

$$
C_{GME}(|V\rangle) := \min_{r_i \in r} \sqrt{2[1 - Tr(\rho_{A_{r_i}}^2)]}, \tag{2}
$$

where $r = \{r_i\}$ represents the set of all possible bipartitions $\{A_i|B_i\}$ of $|V\rangle$ and $\rho_{A_{r_i}}$ is the reduced density matrix of A_{r_i}. Then, the GME concurrence of the

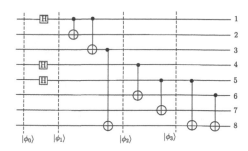

Fig. 1. A quantum circuit for producing an eight-qubit cluster state.

state $|\phi\rangle$ is $C_{GME}(|\phi\rangle) = \sqrt{2[1 - (\frac{1}{2})]} = 1$. Therefore, the eight-qubit state $|\phi\rangle$ is a genuine entangled state.

Let introduce our bidirectional controlled quantum teleportation scheme. Suppose there are three agents, Alice, Bob and Charlie. Alice wishes to teleport an arbitrary two-qubit entangled state $|\psi_1\rangle$ to Bob, at the same time, Bob have a two-qubit quantum information $|\psi_2\rangle$ to transmit to Alice. The third agent, Charlie, is a supervisor who can terminate the quantum teleportation if she thinks the communication process is not secure. The two quantum states are given by

$$|\psi_1\rangle_{A_1 A_2} = \sum_{i=0}^{1} a_i|ii\rangle), \ |\psi_2\rangle_{B_1 B_2} = b_0|00\rangle + b_1|01\rangle + b_2|10\rangle + b_3|11\rangle, \quad (3)$$

where a_i and b_j are complex coefficients, which should satisfy the following conditions: $\sum_{i=0}^{1} |a_i|^2 = 1, \sum_{j=0}^{3} |b_j|^2 = 1$. A_1, A_2, B_1, B_2 means the qubits held by Alice and Bob, respectively. To achieve BQT scheme, they should share the eight-qubit entangled state $|\phi\rangle$ as a quantum channel in advance. Now, the total quantum information system can be described by

$$|\Omega\rangle = |\psi_1\rangle_{A_1 A_2} \otimes |\phi\rangle_{12345678} \otimes |\psi_2\rangle_{B_1 B_2} \quad (4)$$

where particles A_1, A_2, 1, 4, and 5 held by Alice, Bob posses particles B_1, B_2, 2, 3 6, and 7, and Charlie have particle 8.

The BQT scheme consists of the following steps:

Step 1. Firstly, Alice performs a C-NOT operation on particle A_1 and A_2 with particle A_1 as target qubit and A_2 as control qubit. Then the state $|\psi_1\rangle$ is transformed to the state $|\psi_1\rangle_{A_1 A_2} = a_0|00\rangle + a_1|01\rangle$ and we can drop the particle A_1 because it does not contain anything about the state $|\psi_1\rangle$. Next, Alice has to perform a Bell-state measurement on her particles A_2 and 1. Bob also performs two Bell-state measurements on his particles B_2, 7 and B_1, 6. they both inform their outcomes to others via a classical channel. According to all the results of their quantum measurements, we rewrite the system state Equation (4) into the following equation:

$$|\Omega\rangle = \frac{1}{8}[|\Phi^i\rangle_{A_21}|\Phi^j\rangle_{B_27}|\Phi^k\rangle_{B_16}(a_0b_0|00000\rangle + (-1)^i a_1b_0|11001\rangle + (-1)^j a_0b_1|00011\rangle$$

$$+ (-1)^{i+j}a_1b_1|11010\rangle + (-1)^k a_0b_2|00101\rangle + (-1)^{i+k}a_1b_2|11100\rangle$$

$$+ (-1)^{j+k}a_0b_3|00110\rangle + (-1)^{i+j+k}a_1b_3|11111\rangle)_{23458}$$

$$+ |\Phi^i\rangle_{A_21}|\Phi^j\rangle_{B_27}|\Psi^k\rangle_{B_16}(a_0b_0|00101\rangle + (-1)^i a_1b_0|11100\rangle + (-1)^j a_0b_1|00110\rangle$$

$$+ (-1)^{i+j}a_1b_1|11111\rangle + (-1)^k a_0b_2|00000\rangle + (-1)^{i+k}a_1b_2|11001\rangle$$

$$+ (-1)^{j+k}a_0b_3|00011\rangle + (-1)^{i+j+k}a_1b_3|11010\rangle)_{23458}$$

$$+ |\Phi^i\rangle_{A_21}|\Psi^j\rangle_{B_27}|\Phi^k\rangle_{B_16}(a_0b_0|00011\rangle + (-1)^i a_1b_0|11010\rangle + (-1)^j a_0b_1|00000\rangle$$

$$+ (-1)^{i+j}a_1b_1|11001\rangle + (-1)^k a_0b_2|00110\rangle + (-1)^{i+k}a_1b_2|11111\rangle$$

$$+ (-1)^{j+k}a_0b_3|00101\rangle + (-1)^{i+j+k}a_1b_3|11100\rangle)_{23458}$$

$$+ |\Phi^i\rangle_{A_21}|\Psi^j\rangle_{B_27}|\Psi^k\rangle_{B_16}(a_0b_0|00110\rangle + (-1)^i a_1b_0|11111\rangle + (-1)^j a_0b_1|00101\rangle$$

$$+ (-1)^{i+j}a_1b_1|11100\rangle + (-1)^k a_0b_2|00011\rangle + (-1)^{i+k}a_1b_2|11010\rangle$$

$$+ (-1)^{j+k}a_0b_3|00000\rangle + (-1)^{i+j+k}a_1b_3|11001\rangle)_{23458}$$

$$+ |\Psi^i\rangle_{A_21}|\Phi^j\rangle_{B_27}|\Phi^k\rangle_{B_16}(a_0b_0|11001\rangle + (-1)^i a_1b_0|00000\rangle + (-1)^j a_0b_1|11010\rangle$$

$$+ (-1)^{i+j}a_1b_1|00011\rangle + (-1)^k a_0b_2|11100\rangle + (-1)^{i+k}a_1b_2|00101\rangle$$

$$+ (-1)^{j+k}a_0b_3|11111\rangle + (-1)^{i+j+k}a_1b_3|00110\rangle)_{23458}$$

$$+ |\Psi^i\rangle_{A_21}|\Phi^j\rangle_{B_27}|\Psi^k\rangle_{B_16}(a_0b_0|11100\rangle + (-1)^i a_1b_0|00101\rangle + (-1)^j a_0b_1|11111\rangle$$

$$+ (-1)^{i+j}a_1b_1|00110\rangle + (-1)^k a_0b_2|11001\rangle + (-1)^{i+k}a_1b_2|00000\rangle$$

$$+ (-1)^{j+k}a_0b_3|11010\rangle + (-1)^{i+j+k}a_1b_3|00011\rangle)_{23458}$$

$$+ |\Psi^i\rangle_{A_21}|\Psi^j\rangle_{B_27}|\Phi^k\rangle_{B_16}(a_0b_0|11010\rangle + (-1)^i a_1b_0|00011\rangle + (-1)^j a_0b_1|11001\rangle$$

$$+ (-1)^{i+j}a_1b_1|00000\rangle + (-1)^k a_0b_2|11111\rangle + (-1)^{i+k}a_1b_2|00110\rangle$$

$$+ (-1)^{j+k}a_0b_3|11100\rangle + (-1)^{i+j+k}a_1b_3|00101\rangle)_{23458}$$

$$+ |\Psi^i\rangle_{A_21}|\Psi^j\rangle_{B_27}|\Psi^k\rangle_{B_16}(a_0b_0|11111\rangle + (-1)^i a_1b_0|00110\rangle + (-1)^j a_0b_1|11100\rangle$$

$$+ (-1)^{i+j}a_1b_1|00101\rangle + (-1)^k a_0b_2|11010\rangle + (-1)^{i+k}a_1b_2|00011\rangle$$

$$+ (-1)^{j+k}a_0b_3|11001\rangle + (-1)^{i+j+k}a_1b_3|00000\rangle)_{23458}]$$

$$(5)$$

where i, j, k = 0, 1, and $|\Phi^{0,1}\rangle = \frac{1}{\sqrt{2}}(|00\rangle \pm |11\rangle)$, $|\Psi^{0,1}\rangle = \frac{1}{\sqrt{2}}(|01\rangle \pm |10\rangle)$. Equation (5) shows the whole state of the quantum information system will collapse to one of sixty-four possible states with equal probability when Alice and Bob have finished their measurements. For example, assume Alice and Bob's measurement result is $|\Psi^0\rangle_{A_21}|\Phi^1\rangle_{B_27}$ $|\Phi^0\rangle_{B_16}$. The state of quantum system (5) collapses into the following state:

$$|\Omega\rangle = (a_0b_0|11001\rangle + a_1b_0|00000\rangle - a_0b_1|11010\rangle - a_1b_1|00011\rangle$$
$$+ a_0b_2|11100\rangle + a_1b_2|00101\rangle - a_0b_3|11111\rangle - a_1b_3|00110\rangle)_{23458} \quad (6)$$

Step 2. The controller, Charlie, can do nothing to her own particle 8 to interrupt the teleportation if she does not allow them to exchange quantum information in any secure causes. Otherwise, she can perform a Hadamard operation and single-qubit measurement with classical bases $\{|0\rangle, |1\rangle\}$

on the particle. Of course, she also needs to tell the measurement result to the others. Let

$$\alpha_{0,1} = a_0|00\rangle \pm (-1)^i a_1|11\rangle; \quad \alpha_{2,3} = a_0|11\rangle \pm (-1)^i a_1|00\rangle$$

$$\beta_{0,1} = b_0|00\rangle \pm (-1)^j b_1|01\rangle \pm (-1)^k b_2|10\rangle + (-1)^{j+k} b_3|11\rangle$$

$$\beta_{2,3} = b_0|10\rangle \pm (-1)^j b_1|11\rangle \pm (-1)^k b_2|00\rangle + (-1)^{j+k} b_3|01\rangle \quad (7)$$

$$\beta_{4,5} = b_0|01\rangle \pm (-1)^j b_1|00\rangle \pm (-1)^k b_2|11\rangle + (-1)^{j+k} b_3|10\rangle$$

$$\beta_{6,7} = b_0|11\rangle \pm (-1)^j b_1|10\rangle \pm (-1)^k b_2|01\rangle + (-1)^{j+k} b_3|00\rangle.$$

Then, we rewrite the Equation (5)

$$
\begin{aligned}
|\Omega\rangle = \frac{1}{8\sqrt{2}} [& |\Phi^i\rangle_{A_21} |\Phi^j\rangle_{B_27} |\Phi^k\rangle_{B_16} (|0\rangle_8 \alpha_0 \otimes \beta_0 \\
& + |1\rangle_8 \alpha_1 \otimes \beta_1) + |\Phi^i\rangle_{A_21} |\Phi^j\rangle_{B_27} |\Psi^k\rangle_{B_16} (|0\rangle_8 \alpha_0 \otimes \beta_2 \\
& - |1\rangle_8 \alpha_1 \otimes \beta_3) + |\Phi^i\rangle_{A_21} |\Psi^j\rangle_{B_27} |\Phi^k\rangle_{B_16} (|0\rangle_8 \alpha_0 \otimes \beta_4 \\
& - |1\rangle_8 \alpha_1 \otimes \beta_5) + |\Phi^i\rangle_{A_21} |\Psi^j\rangle_{B_27} |\Psi^k\rangle_{B_16} (|0\rangle_8 \alpha_0 \otimes \beta_6 \\
& + |1\rangle_8 \alpha_1 \otimes \beta_7) + |\Psi^i\rangle_{A_21} |\Phi^j\rangle_{B_27} |\Phi^k\rangle_{B_16} (|0\rangle_8 \alpha_2 \otimes \beta_0 \\
& - |1\rangle_8 \alpha_3 \otimes \beta_1) + |\Psi^i\rangle_{A_21} |\Phi^j\rangle_{B_27} |\Psi^k\rangle_{B_16} (|0\rangle_8 \alpha_2 \otimes \beta_2 \\
& + |1\rangle_8 \alpha_3 \otimes \beta_3) + |\Psi^i\rangle_{A_21} |\Psi^j\rangle_{B_27} |\Phi^k\rangle_{B_16} (|0\rangle_8 \alpha_2 \otimes \beta_4 \\
& + |1\rangle_8 \alpha_3 \otimes \beta_5) + |\Psi^i\rangle_{A_21} |\Psi^j\rangle_{B_27} |\Psi^k\rangle_{B_16} (|0\rangle_8 \alpha_2 \otimes \beta_6 \\
& - |1\rangle_8 \alpha_3 \otimes \beta_7)],
\end{aligned}
\tag{8}
$$

The eq. (8) shows that the finial state held by Alice and Bob is a tensor product of α_i and β_j, which means they can recover the quantum state by performing some unitary operations on their qubit respectively.

Suppose Charlie allows the other two agents to teleport their quantum information and they have finished their operations in step 2. Equation (6) can be rewritten into eq. (9), and the final quantum state collapses to one of four states when all of agents have finished their single-qubit quantum words in the former step.

$$
\begin{aligned}
|\Omega\rangle = & \frac{1}{\sqrt{2}} |0\rangle_8 (a_0|11\rangle + a_1|00\rangle)_{23} \otimes (b_0|00\rangle - b_1|01\rangle + b_2|10\rangle - b_3|11\rangle)_{45} \\
& - \frac{1}{\sqrt{2}} |1\rangle_8 (a_0|11\rangle - a_1|00\rangle)_{23} \otimes (b_0|00\rangle + b_1|01\rangle - b_2|10\rangle - b_3|11\rangle)_{45}
\end{aligned}
\tag{9}
$$

Step 3.   When Alice and Bob have received the measurement outcomes informed from the others, they can recover the quantum information by performing appropriate unitary operations on their own particles according to the measurement results. The unitary operations are Pauli operations. Now, our quantum scheme is finished successfully.

Consider the example above. Suppose the quantum measurement results of Alice and Bob are $|\Psi^0\rangle_{A_21}$, $|\Phi^1\rangle_{B_27}$, $|\Phi^0\rangle_{B_16}$, $|0\rangle_8$. The final state of the quantum information system can be represented by

$$|\Omega\rangle = (a_0|11\rangle + a_1|00\rangle)_{23} \otimes (b_0|00\rangle - b_1|01\rangle + b_2|10\rangle - b_3|11\rangle)_{45}. \tag{10}$$

Then, Alice should perform a unitary operation $I_4 \otimes Z_5$ on particles 4 and 5, and Bob perform a operation $X_2 \otimes X_3$ on particles 2 and 3 to recover the quantum information from the other one. On the other hand, if Charlie does not allow agents to communicate their quantum information, she can do nothing with her own particle 8. The whole quantum system state held by the users is $|\Omega\rangle = \frac{1}{\sqrt{2}}|0\rangle_8\alpha_2 \otimes \beta_0 - \frac{1}{\sqrt{2}}|1\rangle_8\alpha_3 \otimes \beta_1$. It is easy to see that, when the users have finished their operations on their own particles, particle 8 is still entangled with the other four qubits 2, 3, 4 and 5. In other words, Alice and Bob cannot obtain the quantum information transmitted from the other user without Charlie's help.

Notably, we can always choose appropriate operations to recover quantum information with the controller's help if the result is different from that in the example above.

3   Comparison

Table 1. Comparison of the six schemes.

SC	QC	NO	CRC	TQIT	η
B	2EPR, 2GHZ	4BM, 2SM	10	2Q, 2Q	1/5
D	Eight-Q ES	1BM, 1FM, 2SM	8	2Q, 1Q	3/16
Z	4EPR	4CN, 8SM	8	2Q, 2Q	1/4
R	2GHZ	1GM, 1BM, 1CN, 1CCN	5	3QE, 1Q	4/11
Y	Six-Q ES	2BM, 2SM, 1CCN	6	2QE, 1Q	1/4
Our	Eight-Q ES	3BM, 1SM, 2CN	7	2QE, 2Q	4/15

In the table, the B, D, Z, R, and Y schemes are from Refs. [8,9,18,22] and [17], respectively. SC denotes the scheme, QC denotes the quantum channel, ES denotes the entangled state, NO denotes the necessary operations, BM and GM denotes the Bell and GHZ measurements, SM and FM denotes the single-qubit and four-qubit measurements, CN denotes the C-NOT operation, CCN denotes the co-C-NOT operation, CRC denotes the classical resource consumption, TQIT denotes the type of quantum information transmitted, and QE denotes qubit entangled state. The intrinsic efficiency [26] is defined as $\eta = \frac{q_s}{q_u+q_t}$, where q_s is the number of qubits that consist of quantum information to be shared, q_u is the number of the qubit that is used as a channel and q_t is the classical bits transmitted.

Here, six schemes are compared in the Table 1. Schemes B, D and Z teleport two two-qubit states or two-qubit and single-qubit states. Scheme Z, which consumes 8 c-bits and 8 e-bits, possesses higher intrinsic efficiency. Schemes B, D but Z are controlled quantum schemes, and the controllers in the two schemes both consume two e-bits and two c-bits to supervise their quantum teleportation, which causes their lower intrinsic efficiency. Schemes R, Y and our scheme teleport an entangled multiqubit state and a two- or single-qubit state. Among these three schemes, Scheme R is the only teleportation scheme without a controller, and it has the highest intrinsic efficiency among the six schemes. Schemes Z, Y and our scheme have equal intrinsic efficiency. For their operation complexities, Scheme Z, which needs to perform 4 C-NOT operations and 8 single-qubit measurements, is simpler than the others. In addition, Schemes R and Y both need to perform cooperation during the communicating process. The cooperation is an unlocal operation and difficult to implement. Therefore, the two schemes are more difficult and complex than the others.

Finally, we compare our scheme with the other five schemes. Although only Bell-state and classical base measurements are needed in our scheme, Scheme Z, which, similar to our scheme, uses only an eight-qubit state as a quantum channel, is simpler than our scheme. However, as a controlled scheme (the controller possesses one qubit of the quantum channel), our scheme is efficient and economic. Furthermore, the intrinsic efficiency of our scheme is higher than that of Scheme Y and is less than that of Scheme R. However, the two schemes both need to perform unlocal operation. Overall, our scheme, as a bidirectional controlled quantum scheme, is better than the other schemes above.

4   Conclusion

In this work, we introduce a quantum teleportation via eight-qubit state. As mentioned above, our scheme is efficient and simple for bidirectional controlled quantum teleportation. Although,Bell-state measurements, C-NOT operations and single-qubit unitary operations have been reported in experiments [27,28], the eight-qubit state $|\phi\rangle$, has not been reported in experiments. With advances in quantum information technology, we hope that our scheme will be implemented experimentally in the future.

Acknowledgement. The author thanks the anonymous reviewer for their constructive comments and suggestions. This work is supported by the Science and Technology Foundation of Guizhou Province (No. [2018]1019).

References

1. Bennett, C.H., Brassar, G., et al.: Teleporting an unknown quantum state via dual classical and Einstein-Podolsky-Rosen channels. Phys. Rev. Lett. **70**, 1895 (1993)
2. Karlsson, A., Bournnane, M.: Quantum teleportation using three-particle entanglement. Phys. Rev. A **58**, 4394–4400 (1998)
3. Rigolin, G.: Quantum teleportation of an arbitrary two-qubit state and its relation to multipartite entanglement. Phys. Rev. A **71**, 032303 (2005)

4. Cao, Z.L., Song, W.: Teleportation of of a two-particle entangled state via W class states. Phys. A **347**, 177–183 (2005)

5. Wang, J.W., Shu, L., Mo, Z.W., et al.: Controlled teleportation of a qudit state by partially entangled GHZ states. Int. Theor. Phys. **53**, 2867–2873 (2014)

6. Zou, Z.Z., Yu, X.T., et al.: Multihop teleportation of two-qubit state via the composite GHZ-bell channel. Phys. Lett. A **381**, 76–81 (2017)

7. Zha, Z.W., Zou, Z.C., et al.: Bidirectional quantum controlled teleportation via five-qubit cluster state. Int. J. Theor. Phys. **55**(6), 3008–3016 (2013)

8. Binayak, S.C., Arpan, D.: A bidirectional teleportation protocol for arbitrary two-qubit state under the supervision of a third party. Int. J. Theor. Phys. **55**, 2275–2285(2016)

9. Zadeh, M.S.S., Houshmand, M., et al.: Bidirectional teleportation of a two-qubit state by using eight-qubit entangled state as a quantum channel. Int. J. Theor. Phys. 56 (2017)

10. Mohammad, S.S., Monireh, H., et al.: Bidirectional quantum teleportation of an arbitrary number of qubits over noisy channel. Quantum Inf. Process. **18**, 353 (2019)

11. Hassanpour, S., Houshmand, M.: Bidirectional teleportation of a pure EPR state by using GHZ states. Quantum Inf. Process. **15**, 905–912 (2016)

12. Yang, G., Lian, B.W., Nie, M., Jin, J.: Bidirectional multi-qubit quantum teleportation in noisy channel aided with weak measurement. Chin. Phys. B. **26**(4), 04305 (2017)

13. Zhou, R.G., Qian, C., et al.: Cyclic and bidirectional quantum teleportation via pseudo multi-qubit states. IEEE Access **7**, 42445–42449 (2019)

14. Zhou, R.G., Li, X., et al.: Quantum bidirectional teleportation 2–2 or 2–3 qubit teleportation protocol via 6-qubit entangled state. Int. J. Theor. Phys. **59**, 166–172 (2020)

15. Zadeh, M.S.S., Houshmand, M., et al.: Bidirectional quantum teleportation of a class of n-qubit states by using (2n+2)-qubit entangled states as quantum channel. Int. J. Theor. Phys. 57 (2018)

16. Bidirectional quantum teleportation by using five-qubit cluster state: MH. Sang. Int. J. Theor. Phys. **55**, 1333–1335 (2015)

17. Li, Y., Nie, L., et al.: Asymmetric bidirectional controlled teleportation by using six-qubit cluster state. Int. J. Theor. Phys. **55**, 3008–3016 (2016)

18. Zhou, R.G., Xu, R., Lan, H.: Bidirectional quantum teleportation by using six-qubit cluster state. IEEE Access **7**, 44269–44275 (2019)

19. Vikram, V.: Bidirectional quantum teleportation by using two GHZ-states as the quantum channel. IEEE. Commun. Lett. **25**(3) (2021)

20. Binayak, S.C., Soumen, S.: Asymmetric bidirectional 3–2 qubit teleportation protocol between Alice and Bob via 9-qubit cluster state. Int. J. Theor. Phys. 56 (2017)

21. Hong, W.: Asymmetric bidirectional controlled teleportation by using a seven-qubit entangled state. Int. J. Theor. Phys. **55**, 384–387 (2016)

22. Zhang, D., Zha, X.W., et al.: Bidirectional and asymmetric quantum controlled teleportation via maximally eight-qubit entangled state. Quantum Inf. Process. 14 (2015)

23. Long, Y.X., Shao, Z.L.: Bidirectional controlled quantum teleportation by a genuine entangled 9-qubit state. Sci. Sin-Phys. Mech. Astrom. 49 (2019)

24. Huo, G., Zha, X., et al.: Controlled asymmetric bidirectional quantum teleportation of two-and three-qubit states. Quantum Inf. Process. **20**(24) (2021)

25. Ma, Z.H., Chen, Z.H., Chen, J.L.: Measure of genuine multipartite entanglement with computable lower bounds. Phys. Rev. A **83**, 062325 (2011)
26. Yuan, H., Liu, Y.M., et al.: Optimizing resource consumption,operation complexity and efficiency in quantum-state sharing. J. Phy. B **41**, 145506 (2008)
27. Boschi, D., Branca, S., et al.: Experimental realization of teleportating an unknown pure quantum state via dual classical and Einstein-Podolsky-Rosen channels. Phys. Rev. Lett. **80**, 1121 (1998)
28. Gao, W.B., Xu, P., et al.: Experimental realization of a controlled-Not gate with four-photon six-qubit cluster states. Phys. Rev. Lett. **104**, 020501 (2010)

Incremental Bayesian Classifier for Streaming Data with Concept Drift

Peng Wu[1,2], Ning Xiong[2(✉)], Gang Li[3], and Jinrui lv[4]

[1] Computer Engineering Department, Taiyuan Institute of Technology, Taiyuan 030008, China
14112078@bjtu.edu.cn

[2] School of Innovation, Design and Engineering, Malardalen University, 72123 Vasteras, Sweden
ning.xiong@mdu.se

[3] College of Software, Taiyuan University of Technology, Jinzhong 030600, China
ligang@tyut.edu.cn

[4] Department of Information Engineering, Taiyuan City Vocational College, Taiyuan 030027, China
ljr19840802@163.com

Abstract. Classification is an important task in the field of machine learning. Most classifiers based on offline learning are invalid for open data streams. In contrast, incremental learning is feasible for continuous data. This paper presents the Incremental Bayesian Classifier "*Incremental_BC*", which continuously updates the probabilistic information according to each new training sample via recursive calculation. Further, the *Incremental_BC* is improved to deal with the flowing data whose distribution and property evolve with time, i.e., the concept drift. The effectiveness of the proposed methods has been verified by the results of simulation tests on benchmark data sets.

Keywords: Incremental learning · Online learning · Bayesian classifier · Concept drift

1 Introduction

Classification is a popular topic in the field of machine learning, and many engineering problems can be attributed to it, such as intelligent diagnosis [1], decision-making [2] and product identification [3]. Traditional supervised learning needs to read all the training samples at once, and then construct the classifier after multiple acsesses of the training samples [4]. However, some data belong to infinite data streams, such as traffic data flow, product in assembly line [5]. It is impossible to read all the training samples from infinite data flow at once. Furthermore, the distribution and property of samples in the stream may fluctuate with time [6], which brings many challenges to traditional offline learning methods.

Naive Bayesian classifier is a highly efficient and fast tool to solve many classification problems [7]. This paper studies incremental and online learning of Bayesian

N. Xiong et al. (Eds.): ICNC-FSKD 2022, LNDECT 153, pp. 509–518, 2023.
https://doi.org/10.1007/978-3-031-20738-9_58

classifier based on open data streams. We present the Incremental Bayesian Classifier, *Incremental_BC*, which is able to continuously learn from new data in a one-pass manner. Further, for handling concept drift [8–10] with fluctuating data streams, we propsed *Improved_IBC*, which is an improvement of the *Incremental_BC* by weighing more on recent data in the recursive derivation of probabilistic information. Simulation tests have been conducted on benchmark data sets to show the feasibility of the proposed methods.

2 Related Work

Classification is a popular branch in machine learning. Classifiers are specific classification algorithms or tools. Technically, an efficient classifier is often constructed by learning or training from existing data and it maps the data in the dataset to a given category. Hence, accuracy of classification is the significant indicator of evaluating classifiers.

The traditional offline learning methods for classification have been well established. As the result, various classifiers have been proposed by scholars, such as Support Vector Machine [11], Naive Bayes, Decision Tree [12], K-Nearest Neighbor [13], Deep Learning [14], as well as their improvements [15]. However, they require all training data to be stored in advance, they are infeasible for being applied to infinite data streams.

According to literatures [16, 17], the features of incremental learning suitable for data streams are as follows: (1) only one training sample is learned at a time; (2) while learning new knowledge, most of the previously learned knowledge can be preserved; (3) samples already processed are discarded.

Currently, some online incremental learning methods for classifiers were proposed, such as online learning of fuzzy classifiers [18, 19]. However they only consider time-invariant streams. So far there are few reports on online Bayesian classifiers targeting time-varying data streams in dynamic environments.

3 Proposed Method

We first introduce the principle of Naive Bayes classifier in Sect. 3.1. Section 3.2 concentrates on the incremental Bayesian classifier, and Sect. 3.3 gives the improved method of the incremental Bayesian classifier for handling concept drift.

3.1 Bayesian Classifier

Naive Bayesian classifier is quite suitable for the classification problems with large amount of samples and still has many applications at present [18, 19]. Bayesian formula is shown as formula (1), where x represents a sample and C_i represents a class. Assuming that there are n possible classes, the different class probabilities $P(C_i|x)(i=1...n)$ can be obtained according to (1) and the required probabistic quantities can be acquired from the set of training examples.

$$P(C_i|x) = \frac{P(x|C_i) \cdot P(C_i)}{\sum_{i=1}^{n} P(C_i|x) \cdot P(x)} \tag{1}$$

We want to choose the class that gets the highest probablility in (1). Since the denominator on the right side of formula (1) is same for all possible classes, it does not to be calculated in comparison. Hence, we decide on the class according to formula (2).

$$C = \text{argmax}\{P(x|C_i) \cdot P(C_i)\} \tag{2}$$

However, for online classification, the training data are usually infinite streams, such as data from production lines in manufacturing, which leads to the invalidation to count all the training samples at once. For the data in streams, the prior probabilities need to be continuously updated according to the online samples. The following section presents the incremental Bayesian classifier suitable for infinite data streams.

3.2 Incremental Learning of Bayesian Classifier

Incremental learning needs to continuously absorb knowledge from new samples and preserves most of the knowledge already learned. It is very similar to human cognitive processes. With increasing age, people receive new data every day, while not forgetting the knowledge they have learned previously. In manufacturing, the production line can be regarded as the source of infinite sample flow [20]. In many cases, products need to be identified and categorized in real time. Since it is impossible to obtain all samples at one time, the traditional offline methods are infeasible for learning classifiers from data streams of production lines.

In the incremental Bayesian classifier, the probability information needed in formula (2) will be updated one by one given arrival of each new sample. In this paper, it is assumed that all the attributes of the samples follow the Gaussian distribution. Hence only the mean and variance of the samples are required to be updated with recursive computing. Suppose the mean and variance of the first n samples are μ_n and σ_n^2, respectively. Now a new sample x_{n+1} has arrived. The relationship between μ_n and μ_{n+1} is given by the following formula (3).

$$\mu_{n+1}(\mu_n, x_{n+1}) = \frac{x_1 + x_2 + \dots + x_{n+1}}{n+1} = \frac{n \cdot \mu_n + x_{n+1}}{n+1} = \frac{n}{n+1}\mu_n + \frac{x_{n+1}}{n+1} \tag{3}$$

The relationship between σ_n^2 and σ_{n+1}^2 is given by the following formula (4).

$$\sigma_{n+1}^2\left(\sigma_n^2, \mu_n, x_{n+1}\right) = \frac{n\sigma_n^2}{n+1} + \frac{n(\mu_n - x_{n+1})^2}{(n+1)^2} \tag{4}$$

The derivation of formula (4) can be seen in the the following.

$$\sigma_{n+1}^2\left(\sigma_n^2, \mu_n, x_{n+1}\right) = \frac{x_1^2 + x_2^2 + \dots + x_{n+1}^2}{n+1} - \mu_{n+1}^2$$

$$= \frac{x_1^2 + x_2^2 + \dots + x_{n+1}^2}{n+1} - \left(\frac{n}{n+1}\mu_n + \frac{x_{n+1}}{n+1}\right)^2$$

$$\sigma_n^2 = \frac{x_1^2 + x_2^2 + \dots + x_n^2}{n} - \mu_n^2 \Leftrightarrow \frac{n\sigma_n^2 + n\mu_n^2 + x_{n+1}^2}{n+1} - \left(\frac{n}{n+1}\mu_n + \frac{x_{n+1}}{n+1}\right)^2$$

$$= \frac{n\sigma_n^2}{n+1} + \frac{n(n+1)\mu_n^2 + (n+1)x_{n+1}^2 - (n\mu_n + x_{n+1})^2}{(n+1)^2}$$

$$= \frac{n\sigma_n^2}{n+1} + \frac{n\mu_n^2 + nx_{n+1}^2 - 2n\mu_n x_{n+1}}{(n+1)^2}$$

$$= \frac{n\sigma_n^2}{n+1} + \frac{n(\mu_n - x_{n+1})^2}{(n+1)^2} \qquad \square$$

From formulas (3) and (4), the new probability information can be acquired by updating the previous mean and variance with the new sample. The algorithm for updating the mean and variance with new samples is given below.

In "*Updating_mean_variance()*", ε is a small floating-point real number (10^{-9} in this article). Actually, when calculating the initial probability density, the variance of the samples may be 0, which results in infinity for values of the Gaussian density. Because the Bayesian classifier takes the class with the largest "probability" as the predicted result, adding ε to each attribute at the same time will not change the result of the function *argmax* (given in Sect. 3.1). Besides, ε also ensures that the program will not throw an exception due to infinity in calculation.

Updating_mean_variance(): the algorithm for updating μ and σ^2 by a new sample

---

1: **begin**

2: $n = 1, \boldsymbol{\varepsilon} = \left[\varepsilon_1 = 10^{-9}, \varepsilon_2 = 10^{-9}, ..., \varepsilon_m = 10^{-9} \right]$;

3: $\mu = first\_sample()$;

4: $\sigma^2 = 0 + \boldsymbol{\varepsilon}$;

5: $x = get\_new\_sample()$;

6: **while** (x)

| **begin while**

7:     $\mu = \frac{n}{n+1}\mu + \frac{x}{n+1}$;

8:     $\sigma^2 = \frac{n\sigma^2}{n+1} + \frac{n(\mu-x)^2}{(n+1)^2}$;

9:     $n++$;

10:     $x = get\_new\_sample()$;

| **end while**

11: **end**

According to "*Updating_mean_variance()*", class probability and conditional probability can be calculated from stream. As the result, we can get "*Incremental Bayesian classifier()*" which is to be used for classifying new instances.

Incremental Bayesian classifier(): the function of classifying new instances

---

1: thread.start_new_thread(*Updated_mean_variance()*, &, &μ, &σ^2);

2: **while** $(x = get_next_one_TestSample())! = $ NULL

3:   **begin while**

3:     Max $= 0.0$;

(*continued*)

(continued)

7:	**for** i in range(class_number)
8:	**begin for**
9:	**if** P(x|C$_i$) · P(C$_i$) >Max
10:	**begin if**
11:	Max =P($x|C_i$) · $P(C_i)$;
12:	prediction_class = i;
13:	**end if**
14:	**end for**
15:	**end while**

3.3 Handling Concept Drift

In the incremental Bayesian classifier as introduced in Sect. 3.1, all historical samples are treated equally. This is good in the aspect of presereving the knowledge learned before. But, in a dynamic environment, data streams may change their distribution over time. Consequently the mean and variance of future samples will deviate from those of past samples. How to keep the previously learned knowledge while continuously learning new knowledge to adapt to concept drift is a challenging issue that will be addressed in this section.

In formula (3) and (4), n represents the total number of samples that have been received and processed. In this way, all samples (including those that have been generated early and those that have been generated recently) are treated equally. However, when the samples change their distribution, the new samples may have significantly different variance and mean values from the old samples. As a result, the importance of past samples declines. This is similar to the human learning process. People always remember the recent events more than the old ones. We propose to adjust "$1/(n+1)$" to a fixed value α (for example, $\alpha = 0.01$) to increase the influence of new samples that arrive later. In this way, formula (3) and formula (4) can be modified into the following formula (5) and formula (6).

$$\mu_{n+1}(\mu_n, x_{n+1}) = (1-\alpha)\mu_n + \alpha x_{n+1} \tag{5}$$

$$\sigma_{n+1}^2\left(\sigma_n^2, \mu_n, x_{n+1}\right) = (1-\alpha)\sigma_n^2 + (1-\alpha)\alpha(\mu_n - x_{n+1})^2 \tag{6}$$

The conversion into formula (5) is obvious. The explanation of how formula (6) derived is given below:

$$
\begin{aligned}
\sigma_{n+1}^2\left(\sigma_n^2, \mu_n, x_{n+1}\right) &= \frac{n\sigma_n^2}{n+1} + \frac{n(\mu_n - x_{n+1})^2}{(n+1)^2} \\
&= \left(1 - \frac{1}{n+1}\right) \cdot \sigma_n^2 + \left(1 - \frac{1}{n+1}\right) \cdot \frac{1}{n+1} \cdot (\mu_n - x_{n+1})^2
\end{aligned} \tag{7}
$$

Replacing $1/(n+1)$ in formula (7) with α, formula (6) can be obtained. Hence, the algorithm "*Updating_mean_variance()*" in Sect. 3.1 is modified to "Improved *Updating ()* " shown as below.

"*Improved Updating _()*": improved algorithm of updating μ and σ^2
1:    **begin**
2:    $n = 1, \boldsymbol{\varepsilon} = \left[\varepsilon_1 = 10^{-9}, \varepsilon_2 = 10^{-9}, ..., \varepsilon_m = 10^{-9} \right], \alpha = 0.01;$
3:    $\mu = first\_sample();$
4:    $\sigma^2 = 0 + \boldsymbol{\varepsilon};$ 5: $x = get\_new\_sample();$ 6:   **while** (x) | **begin while**
7:            $\mu = (1 - \alpha)\mu + \alpha x;$
8:            $\sigma^2 = (1 - \alpha)\sigma^2 + (1 - \alpha)\alpha(\mu - \mathbf{x})^2;$
9:            $n + +;$ 10: $x = get\_new\_sample();$ | **end while**
11:    **end**

In "*Improved Updating ()*", α is set as 0.01, which means that a new sample always has the weight of 1% for influencing the updated results, regardless of the amount of samples analyzed before. With "Improved *Updating ()*", we get an improved version of *Incremental Bayesian classifier()*, referred to as *Improved_IBC*.

For the situations in which samples are not fluctuating, "*Improved_IBC*" will still perform well. The reason is that the new samples will have the same variance and mean as the past samples. Therefore increasing the influence of new samples will not lead to changes in the updated probability information.

4   Empirical Evaluation

4.1   Experimental Setup

The experimental tool used is Python. The experiments involved 8 real-world datasets on UCI [21]. Table 1 summarizes te information of these datasets.

Some values in datasets need preprocessing before experiment. For example, the label of "Iris-setosa" class in dataset Iris is replaced by the number "0" so as to be more suitable for the algorithms proposed in this paper. Furthermore, all datasets above are shuffled with the same random seed (seed = 3 in this paper) and the K-fold cross-validation method [22] is used in order to compare the results between different methods under the same standard. Detailed evaluation is introduced in Sect. 4.2.

4.2   Results and Evaluation

During the learning process, the performance of the "*Icremental_BC*" changes with the proportion of cumulatively processed training samples. Figure 1 shows that the

Table 1. Detail information about the benchmark datasets

Datasets name	Attributes number	Classes number	Samples number
Zoo	16	7	101
Iris	4	3	150
Monk-2	6	2	432
Skin-Nonskin	3	2	245,057
Vertebral	6	3	310
Tic-Tac-Toe	9	2	958
Breast cancer	9	2	683
Banknote	4	2	1372

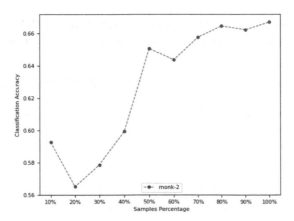

Fig. 1. Evolution of test accuracy on dataset Monk-2 by "*Inrementl_BC*"

classification accuracy on test data increases while dealing with the training samples of Monk-2 one by one.

Similar to Fig. 1, Fig. 2 presents the classification accuracy on test data when learning from the training data stream of Vertebral.

For time-invariant data streams, the classification accuracy of "*Incremental _BC*" usually converges with processing enough samples. Whereas, for fluctuating data streams, *Improved_IBC* is more suitable than *Incremental BC*. Table 2 gives the test accuracy of *Improved_IBC* and *Incremental _BC* after the learning has been completed with the whole data streams, as well as the result of the *Naive Bayesian classifier*, on 8 real-world datasets.

As discussed before, the fixed α in (5) and (6) grants the new, late samples larger weight to handle the concept drift in time-varying streams. Notably this practice will have no influence on the accuracy on time-invariant streams, where the mean and variance of samples hardly change over time. As shown in Table 2, the classification accuracy of

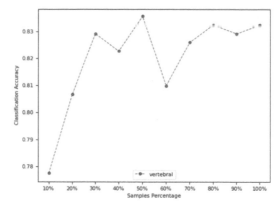

Fig. 2. Evolution of test accuracy on dataset Vertebral by "*Incremental_BC*"

Incremental BC and *Improved IBC* is very close, which implies that *Improved IBC* can be applied not only to stable samples but also to time-varying streams.

Table 2. Test accuracy of "*Incremental BC*" and "*Improved_IBC*" on some benchmark datasets

Datasets name	Naive Bayesian classifier		Cumulative_BC		Improvement_CBC	
	Mean	Var	Mean	Var	Mean	Var
Zoo	0.95	0.0065	0.95	0.0065	0.96	0.0044
IRIS	0.96	0.0028	0.96	0.0028	0.96	0.00196
Monk-2	0.6668	0.0056	0.6668	0.0056	0.6529	0.0065
Skin-Nonskin	0.9239	4.0165	0.9239	4.0165	0.9258	2.5481
Vertebral	0.8323	0.0035	0.8323	0.0035	0.8226	0.0028
Tic-Tac-Toe	0.6941	0.0023	0.6941	0.0023	0.7149	0.0033
Breast cancer	0.9634	0.0003	0.9634	0.0003	0.9605	0.00039
Banknote	0.8396	0.00026	0.8396	0.00026	0.8425	0.00047

5   Conclusion

The data in continuous flow makes the traditional classifier failed. Our main contribution is the proposal of the incremental and online learning for Bayesian classifiers, giving rise to *Incremental BC* and *Improved IBC* that are able to learn from open data streams. Simulation tests on some real-world datasets show that *Improved IBC* is not only suitable for handling time-varying data streams with concept drift but also performs well on other streaming situations that are time invariant.

References

1. Liu, N., Qi, E.S., Xu, M., Gao, B., Liu, G.Q.: A novel intelligent classification model for breast cancer diagnosis. Inf. Process. Manage. **56**(3), 609–623 (2019)
2. Zare, M., et al.: Multi-criteria decision making approach in E-learning: a systematic review and classification. Appl. Soft Comput. **45**, 108–128 (2016)
3. Fuchs, K., Grundmann, T., Fleisch, E.: Towards identification of packaged products via computer vision: convolutional neural networks for object detection and image classification in retail environments. In: Proceedings of the 9th International Conference on the Internet of Things, pp. 1–8. ACM, New York (2019). https://doi.org/10.1145/3365871.3365899
4. Wu, X., et al.: Top 10 algorithms in data mining. Knowl. Inf. Syst. **14**(1), 1–37 (2008)
5. Rossit, D.A., Tohmé, F., Frutos, M.: An Industry 4.0 approach to assembly line resequencing. Int. J. Adv. Manuf. Technol. **105**(9), 3619–3630 (2019). https://doi.org/10.1007/s00170-019-03804-0
6. Hoi, S.C., Sahoo, D., Lu, J., Zhao, P.: Online learning: a comprehensive survey. Neurocomputing **459**, 249–289 (2021)
7. Yang, F.J.: An implementation of Naive Bayes classifier. In: 2018 International Conference on Computational Science and Computational Intelligence, pp. 301–306 (2018). https://doi.org/10.1109/CSCI46756.2018.00065
8. Gama, J., Žliobaitė, I., Bifet, A., Pechenizkiy, M., Bouchachia, A.: A survey on concept drift adaptation. Comput. Surv. **46**(4), 1–37 (2014)
9. Lu, J., Liu, A., Dong, F., Gu, F., Gama, J., Zhang, G.: Learning under concept drift: a review. IEEE Trans. Knowl. Data Eng. **31**(12), 2346–2363 (2018)
10. Žliobaitė, I., Pechenizkiy, M., Gama, J.: An overview of concept drift applications. In: Big Data Analysis: New Algorithms for a New Society, pp. 91–114 (2016)
11. Suthaharan, S.: Support vector machine. In: Machine Learning Models and Algorithms for Big Data Classification, pp. 207–235. Springer, Boston, MA (2016)
12. Myles, A.J., Feudale, R.N., Liu, Y., Woody, N.A., Brown, S.D.: An introduction to decision tree modeling. J. Chemom. **18**(6), 275–285 (2004)
13. Kramer, O.: K-nearest neighbors. In: Dimensionality Reduction with Unsupervised Nearest Neighbors, pp. 13–23. Springer, Berlin, Heidelberg (2013)
14. LeCun, Y., Bengio, Y., Hinton, G.: Deep learning. Nat. **521**, 436–444 (2015)
15. Sarker, I.H.: Machine learning: algorithms, real-world applications and research directions. SN Comput. Sci. **2**(3), 1–21 (2021). https://doi.org/10.1007/s42979-021-00592-x
16. Geng, X., Smith-Miles, K.: Incremental learning. In: Encyclopedia of Biometrics, pp. 731–735. Springer, Boston, MA (2009). https://doi.org/10.1007/978-0-387-73003-5_304
17. Gepperth, A., Hammer, B.: Incremental learning algorithms and applications. In: European Symposium on Artificial Neural Networks (2016)
18. Visa, S., Ralescu, A.: Towards online learning of a fuzzy classifier. In: Meeting of the North American Fuzzy Information Processing Society. Cincinnati, USA (2005)
19. Yavtukhovskyi, V., Abukhader, R., Tillaeus, N., Xiong, N.: An incremental fuzzy learning approach for online classification of data streams. In: International Conference on Soft Computing and Pattern Recognition (SoCPaR 2020). AISC, vol. 1383, pp. 583–592. Springer, Switzerland (2021)
20. Abhilash, P.M., Chakradhar, D.: Sustainability improvement of WEDM process by analysing and classifying wire rupture using kernel-based naive Bayes classifier. J. Braz. Soc. Mech. Sci. Eng. **43**(2), 1–9 (2021). https://doi.org/10.1007/s40430-021-02805-z
21. Jayachitra, S., Prasanth, A.: Multi-feature analysis for automated brain stroke classification using weighted Gaussian naïve Bayes classifier. J. Circuits Syst. Comput. **30**(10), 2150178 (2021)

22. Chen, W.: Intelligent manufacturing production line data monitoring system for industrial internet of things. Comput. Commun. **151**, 31–41 (2020)
23. UCI Machine Learning Repository. http://archive.ics.uci.edu/ml. Accessed 10 Apr 2022
24. Refaeilzadeh, P., Tang, L., Liu, H.: Cross-validation. Encycl. Database Syst. **5**, 532–538 (2009). https://doi.org/10.1007/978-1-4899-7993-3_565-2

DDoS Detection Method Based on Improved Generalized Entropy

Jiaqi Li[1]([✉]), Xu Yang[2], Hui Chen[1], Haoqiang Lin[2], Xinqing Chen[2], and Yanhua Liu[1]([✉])

[1] College of Computer and Data Science, Fuzhou University, Fuzhou 350108, China
200320072@fzu.edu.cn,1062999098@qq.com, 864655674@qq.com, lyhwa@fzu.edu.cn

[2] State Grid Info-Telecom Great Power Science and Technology CO., LTD., Fuzhou 350000, China
13541563@qq.com, 18650741647@163.com, 344260545@qq.com

Abstract. With the rapid development of network technology, network security is facing serious problems. Distributed Denial of Service (DDoS) attack is one of the most difficult security threats to guard against. In this paper, we propose a DDoS detection method based on improved generalized entropy. The model includes a preliminary detection module based on improved generalized entropy and a DDoS detector based on deep neural networks (DNN). The preliminary detection module filters as much normal traffic as possible while ensuring the accuracy of the model by calculating the generalized entropy threshold of the traffic. The DNN-based DDoS detector takes the filtered data as input and detects DDoS attacks more accurately. The experimental results show that the method achieves more than 99% accuracy, precision, and recall on the dataset of this paper.

Keywords: Distributed denial of service · Attack detection · Improved generalized entropy · Deep neural networks

1 Introduction

With the rapid development of network technologies such as big data and cloud computing, applications based on network information technology are developing rapidly. However, events that endanger the security of cyberspace are also occurring continuously at the same time [1]. Especially distributed denial of Service attacks (DDOS) is still one of the main threats to the Internet [2]. Flood DDoS attacks and low-rate distributed denial of service (LDDoS) [3] are the two most common types of DDoS attacks. LDDoS attacks have the characteristics of high concealed attack dispersion degree and great periodicity. There are many studies on the detection of DDoS attacks [4], and the detection of flood DDoS attacks has achieved some success. However, LDDoS attack is relatively hidden and difficult to be detected, it poses a greater threat to the Internet [5]. The

N. Xiong et al. (Eds.): ICNC-FSKD 2022, LNDECT 153, pp. 519–526, 2023.
https://doi.org/10.1007/978-3-031-20738-9_59

detection of DDoS attacks has gradually become an important research topic for researchers around the world [6].

There are three existing DDoS detection technologies: DDoS detection based on statistics [7,8], DDoS detection based on machine learning [9,10], and DDoS detection based on deep learning [11]. The detection method based on statistics can achieve simple and fast detection, but the detection accuracy is low because the model is relatively simple and has fewer features [12]. The detection method based on machine learning is more accurate, but the implementation is relatively complex and the detection at a slower rate [13]. Deep learning, unlike machine learning, can learn complex features from data autonomously. Deep learning has been successfully studied and applied in the field of network security, and it has emerged as one of the new DDOS detection methods [14]. However, existing DDoS attack detection struggles to effectively assess the features of DDoS attack dispersion and periodicity, as well as meet the demands of high real-time and high detection rates.

To solve the above problems, we propose a DDoS detection method based on improved generalized entropy. The main work and contributions of this paper are summarized below:

(1) The one-hot-encoding technology is used to reconstruct network traffic string fields such as the source and destination IP address.
(2) We propose a DDoS detection method based on improved generalized entropy that enables normal traffic filtering and reduces the data scale.
(3) Detecting the filtered traffic data based on a deep neural network to improve the performance of the DDoS detection model. Experimental results show that the accuracy, recall rate, and F1_Score of the model are above 99%.

2   Our Method

We propose a DDoS detection method based on improved generalized entropy due to the varying degrees of dispersion and periodicity of DDoS attacks. Firstly, feature extraction is applied to the traffic data, followed by the one-hot encoding of the source and destination IP addresses. Then, The preliminary detection module learns the DDoS feature's distribution rule by optimizing settings automatically and filters some normal traffic. Finally, the filtered data is input into The detection module based on deep neural networks (DNN) to detect DDoS attacks more accurately.

2.1   Data Preprocessing

In this paper, the one-hot-encoding method is utilized to reconstruct string fields in network traffic data, such as source IP address and destination IP address. N columns are used to encode N separate IP addresses in the dataset.

Algorithm 1: DDoS preliminary detection algorithm based on improved generalized entropy.

Input: DDoS training set and DDoS test set.
Output: Suspected DDoS data set.

1 Begin;
2 **for** $g \leftarrow 1$ **to** n **do**
3   **for** $\alpha \leftarrow 1$ **to** m **do**
4   Formula (1) is used to calculate the generalized entropy for training sets;
5   Determine the maximum entropy value as threshold T;
6   Formula (1) is used to calculate the generalized entropy for test set;
7   Filter DDoS test sets;
8   **if** $accuracy == 1$ **then**
9              $Ans = (\alpha, g)$;
10   **end**
11   **end**
12 **end**

Due to a large amount of data in DDoS attacks and a large number of different IP addresses, using one-hot-encoding directly will cause too many new data columns. To avoid this situation, we add a threshold value which is the number of one IP address to the one-hot-encoding technology. One-hot-encoding is performed for specific IP addresses whose number is greater than or equal to the threshold. If the number of IP addresses is smaller than the threshold, they are classified as a category called other IP addresses. Through several experiments, 100 is chosen as the threshold in this paper.

2.2 The Preliminary Detection Module Based on Improved Generalized Entropy

Information entropy is a measure of uncertainty associated with random variables [15]. Generalized entropy is a generalization of information entropy and one of the function families that quantify the uncertainty or randomness of the diversity of systems.The generalized entropy formula of IP address $x = (x_1, x_2, \ldots, x_n)$ is formula (1).

$$H_\alpha (x) = \frac{1}{1 - \alpha} \log_2 \left(\sum_{i=1}^{n} p_i^\alpha \right).$$ (1)

where, p_i is the probability of x_i, $p_i \geq 0$ and p_i satisfies $\sum_{i=1}^{n} p_i = 1$, $\alpha \geq 0, \alpha \neq 1$.

Compared with information entropy, the value of generalized entropy depends on α, which can increase the deviation between different probability distributions and effectively reflect the dispersion and periodicity of DDoS attacks. However, the parameter values greatly affect the detection results of DDoS attacks.

To solve the above problems, we improve the generalized entropy and propose a parameter self-training process. The DDoS preliminary detection algorithm based on improved generalized entropy is shown as Algorithm 1.

The process of parameter self-training is to divide the training set and test set into g groups on average, and calculate the generalized entropy of each group by using formula (1). The maximum generalized entropy value in the data group of the training set is taken as the threshold value, and the generalized entropy value of each group in the test set is compared with the given threshold value. Set the parameters and repeatedly, calculate the accuracy of the data group greater than the threshold value.

2.3   The DDoS Detector Based on DNN

The detection module based on DNN takes the filtered data as input and classifies the data to achieve more accurate DDoS attack detection. The DNN is a fully connected neural network that consists of an input layer, multiple hidden layers, and an output layer. The input layer is used to receive a large number of data features, the hidden layer is a layer consisting of a large number of neurons and connections between the layers, and the output layer outputs a one-dimensional vector, which is the result of the detection of the traffic data.

The loss function used in this article is binary cross-entropy, and the formula is shown below.

$$L\left(\widehat{y}, y\right) = -\sum_{i=1}^{m}\left(y_i \log\left(\widehat{y_i}\right) + (1 - y_i) \log\left(1 - \widehat{y_i}\right)\right). \tag{2}$$

where $\widehat{y}$ is the model classification result and y is the real result. The loss function is used to measure the difference between the model classification results and the real results.

3   Experimental Evaluation

3.1   Dataset

To verify the efficiency of this model for DDoS attack detection, this experiment constructs the LDDoS dataset for the experiment. The LDDoS dataset comes from the CSE-CIC-IDS2018-AWS dataset [16] and CICIDS2017 [17]. It contains 332841 network traffic data, of which 17 percent are LDDoS attack samples and 83 percent are benign samples.

3.2   Metrics

The accuracy rate (ACC), the precision (P), recall (R), and F1_Score (F_1) are used to evaluate the results. ACC is the percentage of predicted correct detection results in total data samples. P is the proportion of correctly predicted normal flow to all predicted normal flow. R is the proportion of correctly predicted normal traffic to actual normal traffic. F_1 is the harmonic mean of accuracy and recall. They can be defined as shown in formula (2).

$$\begin{cases} ACC = \frac{TP+TN}{TP+TN+FP+FN} \\ P = \frac{TP}{TP+FP} \\ R = \frac{TP}{TP+FN} \\ F_1 = 2\frac{P*R}{P+R} \end{cases} \qquad (3)$$

In the above formula, the case that normal traffic recognized by the model is defined as True Positive (TP). The case that normal traffic is judged by the model as DDoS is defined as False Negative (FN). The case that DDoS traffic is judged as normal by the model is defined as False Positive (FP). The case that DDoS attack traffic identified by the model is defined as True Negative (TN).

3.3   Experimental Results

The experiment in this part is used to verify the advantages of our method and the effectiveness of the preliminary detection module. Considering the correspondence between source and destination IP addresses, this experiment considers each pair of source and destination IP addresses in the dataset as a tuple and calculates the generalized entropy value.

The DDoS Attack Threshold Different parameters g and α were selected in the experiment to calculate the DDoS attack threshold based on generalized entropy. Where, g is the number of data sets grouped, and α is the generalized entropy parameter. The threshold results are shown in Fig. 1.

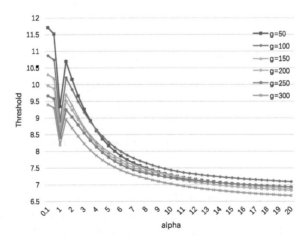

Fig. 1. The DDoS attack threshold varies with g and alpha (α)

The Preliminary Detection Performance In this experiment, the generalized entropy value of each group was firstly calculated and then compared with the threshold value of the corresponding parameter calculated in Fig. 1. The experimental results are shown in Table 1.

Table 1. The preliminary detection result of LDDoS dataset

g	α	P	R	Filter data
50	1	1	0.0009	49
50	6	1	0.2727	11958
100	1	0	0	0
100	20	1	0.1143	6313
150	1	0	0	0
150	20	1	0.0755	4169

When $\alpha \to 1$, the generalized entropy converges to the information entropy. Comparing the experimental results of information entropy and generalized entropy, the DDoS detection method based on the improved generalized entropy can filter more normal samples.

The DDoS Detector Results To verify the effectiveness of the proposed method, the proposed method was compared with the DNN-based detection model and detection mode based on information entropy and DNN. In this experiment, we choose the parameters $g = 50$ and $\alpha = 6$, which are obtained in the Preliminary Detection Performance. The experimental results are shown in Tabel 2.

Table 2. The preliminary detection result of LDDoS dataset

Model	Time/s	$ACC/(\%)$	$P/(\%)$	$R/(\%)$	$F_1/(\%)$
DNN	48.4	99.90	99.92	99.50	99.71
Information Entropy and DNN	49.7	99.91	99.90	99.58	99.74
Our method	**40.1**	**99.98**	**99.93**	**99.98**	**99.96**

Experimental results show that the detection model based on improved generalized entropy can filter more normal traffic, and its accuracy, precision, and recall rate are better than the comparative model we used. Experiments verify the effectiveness of the proposed method and the superiority of the preliminary detection module based on improved generalized entropy.

4 Conclusion

This paper proposes a DDoS detection model based on improved generalized. The preliminary detection module based on improved generalized entropy can analyze the dispersion and periodicity of DDoS attacks and filter more normal traffic. The model based on DNN can achieve high-precision DDoS detection. Finally, the results of the experiments suggest that the improved generalized entropy approach can filter more normal traffic and improve the detection rate. DDoS attacks can be detected more quickly and accurately using the method described in this study. As a future research effort, we aim to further design the DNN model to improve the detection results.

Acknowledgement. This work is supported by the National Natural Science Foundation of China (Grant No.62072109, and No.U1804263), the Natural Science Foundation of Fujian Province(Grant No.2021J01625, No.2021J01616), Major Science and Technology project of Fujian Province(Grant No.2021HZ0115).

References

1. Liu, Jian, Purui, Su., Yang, Min, He, Liang, Zhang, Yuan, Zhu, Xueyang, Lin, Huimin: Software and cyber security—a survey. J. Softw. **29**(1), 42–68 (2018)
2. Wang, A., Chang, W., Chen, S., Mohaisen, A.: Delving into internet DDoS attacks by botnets. IEEE/ACM Trans. Networking (TON) **26**(6), 2843–2855 (2018)
3. Liu, X., Ren, J., He, H., et al.: Low-rate DDoS attacks detection method using data compression and behavior divergence measurement. Comput. Secur. **100**, 102107 (2021)
4. Luo, W., Cheng, J.: Hybrid DDoS attack distributed detection system based on hadoop architecture. Netinfo Secur. **21**(2), 61–69 (2021)
5. Chen, M., Chen, J., Wei, X., et al.: Is low-rate distributed denial of service a great threat to the internet. IET Inf. Secur. **15**(5), 351–363 (2021)
6. Tang, D., Feng, Y., Zhang, S., et al.: Fr-red: fractal residual based real-time detection of the LDoS attack. IEEE Trans. Reliab. **70**(3), 1143–1157 (2021)
7. Yuhua, X., Sun, Z.: Research development of abnormal traffic detection in software defined networking. J. Softw. **31**(01), 183–207 (2020)
8. Ding, D., Savi, M., Siracusa, D.: Tracking normalized network traffic entropy to detect DDoS attacks in P4. IEEE Trans. Dependable Secure Comput 1–1 (2021)
9. Giray, G.: A software engineering perspective on engineering machine learning systems: state of the art and challenges. J. Syst. Softw. **180**, 111031 (2021)
10. Tang, D., Tang, L., Dai, R., et al.: MF-adaboost: LDoS attack detection based on multi-features and improved adaboost. Future Gener. Comput. Syst. **106**, 347–359 (2020)
11. Nazih. W., Hifny, Y., Elkilani, W.S., et al.: Countering DDoS attacks in SIP based VoIP networks using recurrent neural networks. Sensors **20**(20), 5875 (2020)
12. Alekseev, I.V.: Detection of distributed denial of service attacks in large-scale networks based on methods of mathematical statistics and artificial intelligence. Autom. Control Comput. Sci. **54**(8), 952–957 (2020)
13. Zhao, S., Chen, S.: Review: traffic identification based on machine learning. Comput. Eng. Sci. **40**(10), 1746–1756 (2018)

14. Haghighat, M.H., Jun, L.: Intrusion detection system using voting-based neural network. Tsinghua Sci. Technol. **26**(4), 484–495 (2021)
15. Xiang, Y., Li, K., Zhou, W.: Low-rate DDoS attacks detection and traceback by using new information metrics. IEEE Trans. Inf. Forensics Secur. **6**(2), 426–437 (2011)
16. CSE-CIC-IDS2018 on AWS. https://www.unb.ca/cic/datasets/ids-2018.html. Accessed 10 June 2022
17. Intrusion Detection Evaluation Dataset (CIC-IDS2017). https://www.unb.ca/cic/datasets/ids-2017.html. Accessed 10 June 2022

Subway Train Time-Energy Alternative Set Generation Considering Passenger Flow Data Based on Adaptive NSGA-II Algorithm

Benzun Huang[1,3](✉), Dewang Chen[2,3](✉), Yunhu Huang[1,3], and Wenzhu Lai[1,3]

[1] College of Computer and Data Science, Fuzhou University, Fuzhou 350108, China
`200327163@fzu.edu.cn`, `N190310001@fzu.edu.cn`, `200327042@fzu.edu.cn`
[2] School of Transportation, Fujian University of Technology, Fuzhou 350118, China
`dwchen@fjut.edu.cn`
[3] Key laboratory of Intelligent Metro of Universities in Fujian Province, Fuzhou University, Fuzhou 350108, China

Abstract. Aiming at the problem that the energy consumption and time indexes are affected by the dynamic mass of train in the operation of automatic train operation system (ATO), a method of generating time-energy consumption alternative sets for subway trains based on the spatio-temporal data of passenger flow is proposed. Firstly, this paper integrates time periods and stations to cluster passenger flows using K-means algorithm to get the classification of passenger mass between different stations under different time periods, and calculates the approximate mass of train operation based on the classification, replacing the commonly used fixed mass of AW0-AW3 with a train mass that is closer to reality. Then, an improved adaptive NSGA-II algorithm is proposed to optimize the operating curves of trains with dynamic mass to obtain the Pareto frontier time-energy consumption alternative set of time and energy consumption, which can be further used for the optimization of multi-objective train schedules.

Keywords: ATO · Metro · Time · Energy · Multi-objective optimization · NSGA- II

1 Introduction

The subway has become an important aspect of urban transportation as a comfortable and convenient means of public transportation [1]. Although metro is more energy efficient than other means of transportation, its energy consumption still needs to be reduced [2].

ⓒ The Author(s), under exclusive license to Springer Nature Switzerland AG 2023
N. Xiong et al. (Eds.): ICNC-FSKD 2022, LNDECT 153, pp. 527–535, 2023.
https://doi.org/10.1007/978-3-031-20738-9_60

To achieve the lowest energy consumption, the first aspect is to optimize the sequence of train operating conditions between two stations to produce an energy-saving train speed recommendation curve that meets the constraint. The second issue is to optimize the overall schedule of the line network and redistribute the inter-station operation time. The recommended curve acquired after optimizing the first component of the schedule can be utilized to optimize the second aspect of the schedule for energy savings.

2   Related Work

Howlett [3] using the maximum value concept demonstrated that the best driving technique on feeder roads is a combination of four modes, including traction, cruise, inactivity, and braking. Under a time constraint, Anh [4] computed the energy-efficient operating curve. An ant colony technique was utilized by Fernández [5] to achieve a multi-objective optimization of the railway speed curve. Despite the fact that academics have undertaken extensive research on the optimization of recommended curves, there are still certain flaws. On the one hand, existing studies typically calculate train mass as a fixed constant, ignoring changes in train mass caused by dynamic passenger flow during inter-station operation, some researchers who considered dynamic train mass used random mass [6], and some researchers who used passenger flow data did not optimize the operation curve separately [7]; on the other hand,some scholars used the weighting method to transform the multi-objective problem into a single-objective problem [8,9].

Based on the current research, this paper classifies passenger flow quantity by analyzing passenger flow data to generate a more realistic dynamic mass of metro trains, which it then uses to improve the train energy saving optimization model. The solution is based on an improved adaptive NSGA-II algorithm, and its efficiency is confirmed using a Matlab simulation environment. The findings suggest that the alternative set of train running time energy consumption obtained can serve as an adequate foundation for timetable creation.

3   Problem Description and Model Building

3.1   Problem Description

The alternative set generation problem for train time energy consumption explored in this study is a two-stage problem. The initial stage of the train dynamic mass solution can be reduced to a clustering issue, in which the number of passengers between two stations at different times is classified using AFC data clustering, and then the train mass between two stations at that moment is calculated. Train speed profile optimization can be reduced to a multi-objective optimization problem with several nonlinear constraints in the second stage.

Multi-objective optimization problems are common, these problems have multiple objectives, but these objectives are in conflict with one another, it is impossible to optimize multiple objectives at the same time, so the decision maker can only choose a better solution to achieve the best possible result for multiple objectives. The train operation curve optimization problem is a multi-objective optimization problem with many typical objective functions, such as operating energy consumption, operating time, passenger comfort, parking accuracy, and so on. Because these functions are in conflict, no single curve can achieve the minimum value.

In this paper, we choose energy consumption and running time as the main objectives to build a multi-objective optimization model. The ideal state of train operation is to minimize the train running time and minimize the energy consumption, but there is a negative correlation between the running time and energy consumption, and the two indicators conflict with each other, so it is impossible to reach the optimal solution at the same time. Therefore, the objective of this paper is to find a set of relative optimal solutions, which is the time-energy alternative set of train operation.

3.2   Assumptions

The following assumptions are made in this research to simplify the model.

Assumption 1. The train is considered as a single mass point model.

Assumption 2. The line slope is taken as a reasonable random value.

Assumption 3. No additional resistance to tunnels are considered.

Assumption 4. Regenerative braking energy is not considered, only traction energy consumption is considered.

Assumption 5. Adopt traction-idle alternate strategy instead of cruising strategy.

3.3   Train Kinematic Modeling

The operation process of the train is very complex, and there are many factors to be considered, so generally the force analysis is carried out by Newton's second law after appropriate simplification. As mentioned in the previous assumptions, the train is considered as a single mass point model in this paper, and the transverse force is not considered, and the kinematic equations of the train can be obtained after force analysis.

$$
\begin{cases}
\frac{dv}{dt} = \frac{a \cdot F(v) - b \cdot B(v) - F_f(v)}{M_{i,j}^t} \\
\frac{ds}{dt} = v \\
F_f(v) = F_0 + F_1 + F_2 \\
F_0 = A + Bv + Cv^2 \\
F_1 = M_{i,j}^t \cdot G \cdot \sin\theta \\
F_2 = \frac{H}{R}
\end{cases}
\tag{1}
$$

where, v is the train running speed; t is the train running time; a is the tractive force use factor; $F(v)$ is the tractive force; b is the braking force use factor; $B(v)$ is the braking force; s is the position of the train; $F_f(v)$ is the train running resistance; F_0 is the basic unit resistance of the train running, usually calculated by the *Davis* formula, in which A, B, C are constant related to the appearance and mechanical structure of the train; F_1 is the unit ramp additional resistance, θ is the slope; F_2 is the unit curve additional resistance, R is the curve radius of the curve, H is the curve additional resistance constant; $M_{i,j}^t$ is the total mass of the train between two stations during the time period, calculated by the following formula.

$$M_{i,j}^t = M_{avg} \cdot N_{i,j}^t + M_{train} \tag{2}$$

where M_{avg} is the average mass of passengers, $N_{i,j}^t$ is the number of passengers between stations i, j during the time period, and M_{train} is the unloaded mass of the train.

3.4   Multi-objective Optimization Modeling

The train running curve optimization problem is a multi-objective optimization problem with multiple nonlinear constraints: to search for the optimal condition switching point to obtain the optimal speed-distance curve while the train is running, under the constraints of running distance, speed limit, and working condition switching. The Multi-objective optimization model is as follows.

Decision Variables

$$X = (point, x_1, x_2, \ldots x_{point}, u_1, u_2, \ldots, u_{point+1}) \tag{3}$$

where *point* is the number of switching points; $x_1, x_2, \ldots x_{point}$ are the positions of the switching points; $u_1, u_2, \ldots, u_{point+1}$ are the control sequence of the switching points.

Objective Function

$$\min\{E, T\} \tag{4}$$

$$\begin{cases} E = \int_0^T F \cdot v dt = \sum_{t=0}^{T} F \cdot v \cdot \Delta t \\ T = \sum_{i=1}^{k} T_i \end{cases} \tag{5}$$

where, E is the total traction energy consumption; F is the traction force; v is the train speed; Δt is the discrete unit time of the system; T is the total running time; T_i is the running time of the train on each interval; k is the number of intervals after discretization.

Constraints

$$\begin{cases} S = (0, x_1, x_2, \ldots, x_{point}, S_{rail}) \\ |u_{i+1} - u_i| = 1, u_1 = 1, u_{point+1} = 3; i = 1, 2, \ldots, point \\ 0 \le v(s) \le v_{\lim}(s) \\ a_{\min} \le a \le a_{\max} \\ v_0 = v_T = 0 \end{cases} \tag{6}$$

where, S is the array of working condition switching point positions; u_i is the working condition, $u_i = 1$ is the traction state, $u_i = 2$ is the idling state, $u_i = 3$ is the braking state; $v_{\lim}(s)$ is the train interval speed limit; $a_{\max}$ is the maximum acceleration of the train; v_0 is the starting speed of the train; v_T is the speed of the train when it stops running.

4 Method

4.1 K-means Clustering Algorithm

The classical k-means clustering algorithm [10] is selected for the solution of train mass, which categorizes samples by calculating the distance between each sample in the dataset and the center of each category, and continuously updates the center of each category through multiple iterations of calculation until the termination condition set in advance by the algorithm is reached to obtain the final clustering results. In this paper, Euclidean distance is selected as the formula for calculating the distance between samples.

In order to determine the initial number of clusters k, this paper uses the commonly used elbow method and the contour coefficient two methods to select the number of clusters k. The calculation principle of the elbow method is the cost function, the cost function is the sum of the degree of category distortion, the value increases in the process, the position corresponding to the greatest decline in the improvement effect of the degree of distortion is the elbow. The contour coefficient is a way to evaluate the goodness of the clustering effect, which combines two factors: the degree of cohesion and the degree of separation, and the larger the contour coefficient is the more reasonable the corresponding k value.

4.2 Adaptive NSGA-II Algorithm

In this paper, an improved adaptive NSGA-II algorithm is proposed to solve the train time-energy alternative set. NSGA-II is a non-dominated ranking genetic algorithm based on elite strategy proposed by Deb et al. [11]. In the field of multi-objective optimization, NSGA-II is a mature and widely used algorithm. However, NSGA-II still has some drawbacks, such as poor distribution of Pareto solution sets due to its insufficient global search capability and poor fast convergence capability. To address the disadvantages of NSGA-II and the specific situation of this study, an improved adaptive NSGA-II algorithm (A-NSGA-II) is proposed in this paper to accelerate the convergence to the optimal Pareto frontier by improving the crossover operator to enhance the diversity of the final Pareto solution set.

Population Individual Design. The decision variables selected in this paper are the number of condition changeover points, the location of the condition changeover points and the condition after switching, and the number of condition changeover points and the condition after switching are coded as integers, while the location of the condition changeover points is coded as real numbers.

Crossover Operator Improvement. The crossover operator used in traditional NSGA-II is a simulated binary crossover operator, and its crossover ability depends on the empirical parameter, the larger the parent gene is, the more similar to the offspring gene. Therefore, a smaller one should be chosen at the beginning of the iteration to explore new possibilities, and a larger one should be chosen at the later stage to increase the convergence speed, but the determination is very subjective. In this paper, we use an adaptive crossover operator (ASBX), which can change the value with the number of iterations, so as to open up the search space as much as possible in the early stage of the algorithm and converge faster in the later stage of the algorithm.

$$
\begin{cases}
rand = R(0,1) \\
\gamma = \dfrac{2}{gen_{\max} - gen + 1} \\
\alpha = \begin{cases} (2rand)^{\frac{1}{\gamma+1}}, rand \le 0.5 \\ [2(1 - rand)]^{\frac{1}{\gamma+1}}, rand > 0.5 \end{cases} \\
x_{1,j} = 0.5\left[(1 + \alpha)\, p_{1,j} + (1 - \alpha) p_{2,j}\right] \\
x_{2,j} = 0.5\left[(1 - \alpha)\, p_{1,j} + (1 + \alpha) p_{2,j}\right]
\end{cases}
\tag{7}
$$

Algorithm Process. In summary, the flow of the proposed A-NSGA-II algorithm is as follows.

- *Step 1*: Algorithm parameters are set: population size N , mutation operator ϕ , total number of iterations $gen_{\max}$.
- *Step 2*: Real numbers encode individuals and generate primitive populations.
- *Step 3*: Calculating the fitness values of the n generation population. The $n + 1$ generation population is obtained after selection, crossover, mutation and merging and fast non-dominated sorting of the n generation population.
- *Step 4*: The iteration is terminated when the maximum number of iterations is reached, otherwise the iteration is cycled through step 4.

The Pareto frontier of the final population obtained after terminating the iterations is the number of work transition points, the coordinates of the work transition points, and the set of control sequence solutions obtained.

5   Experiment

5.1   Train Dynamic Mass Clustering

This phase of the experiment is based on the weekday AFC data of Fuzhou Metro Line 2 in May 2019 to cluster the passenger flow, and since the maximum train departure interval is 8 min, the number of passengers in each train can be roughly obtained by segmenting the daily 6–24 h with 8 min as the unit time period. After splitting, 135 working periods can be obtained for each working day, and the number of passengers between each two stations at different times of the day is counted, and then the average of the number of passengers between each two stations at different times of all working days is taken for clustering. The effect of the Elbow Method (as in Fig. 1(a)) and the Silhouette Coefficient (as in Fig. 1(b)) showed that the initial k value of 3 was more appropriately chosen for the k-means algorithm. The data were then clustered using the k-means algorithm, and the three classes of the numbers of people in the train obtained N_0, N_1, N_2 were 52, 230, and 548, respectively.

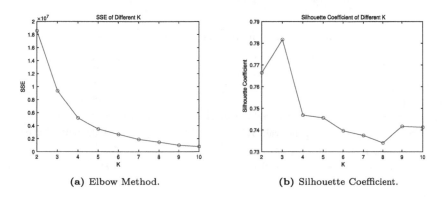

(a) Elbow Method.   (b) Silhouette Coefficient.

Fig. 1. Determine the value of K

5.2   Train Time Energy Consumption Alternative Set Generation

In this phase of experiment, Fuzhou Metro Line 2 is used as the research object, and Ninghua-Xiyang station on Line 2 is selected for the experiment, on which the effectiveness of the model and algorithm is verified. The parameters of the line are shown in Table 1, the train parameters are shown in Table 2, and the remaining algorithm preset parameters are shown in Table 3, where the train mass is obtained using the results obtained from the previous clustering step.In this paper, the curved time energy consumption alternative set between two stations for three time periods is solved to reflect the effect of different masses on the train time energy consumption.

As shown in Fig. 2, the time and energy consumption of a train running between two stations is affected to some extent by the number of passengers, and the counting of the number of passengers helps to select a more reasonable train running time and train running curve. And the distribution of the improved algorithm is better than the original algorithm.

Table 1. Section parameters from Ninghua to Xiyang station.

Start mileage (m)	End mileage (m)	Limit speed (Km/h)	Slope (°)	Curve radius (m)
0	132.37	50	5	/
132.37	161.2	50	−2.2	/
161.2	887.21	50	−2.2	330
887.21	1032.37	50	1.5	/
1032.37	1141.51	80	−3	/
1141.51	1332.82	62	2.2	300
1332.82	1385.13	62	2.1	/

Table 2. Fuzhou metro line 2 train parameters

Train parameters	Value
Train unloaded mass (t)	198.6
Train length (m)	118.66
Number of passengers carried by the train at different times $N_0 N_1 N_2$	52 230 548
Basic resistance parameters of the train ABC	4.407 0.05715 0.000657
Average mass of passengers M_{avg}(kg)	60
Maximum train operating speed (Km/h)	80
Curve additional resistance constant	600
Maximum acceleration of train (m/s^2)	1

Table 3. Algorithm preset parameters.

Parameters	Value
Population size N	100
Probabilistic selection experience λ	1
Mutation operator ϕ	0.2
Number of evolutionary iterations gen_{max}	200

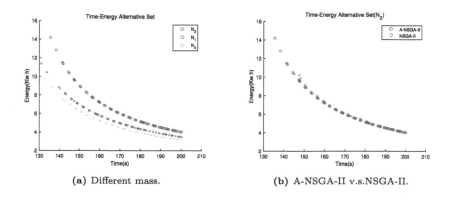

(a) Different mass.   (b) A-NSGA-II v.s.NSGA-II.

Fig. 2. Time-energy alternative set.

6   Conclusion

The optimization of energy-efficient train operation is a classical research problem, but there are still problems such as fixed train mass and only limited operation curves can be obtained. In this paper, two improvements are made to address the above problems. The first point is to obtain the classification of the number of passengers between different stations in different time periods by clustering analysis of real passenger flow data, and then to be able to use the mass during inter-station operation for a more accurate calculation of train energy consumption; the second point is to propose an improved adaptive NSGA-II algorithm and use it to solve the time-energy consumption alternative set. It provides a reliable basis for the overall design of train schedules.

Acknowledgements. This work was supported by the National Natural Science Foundation of China (NSFC) under grants 61976055 and 61906043.

References

1. Yang, X., Ning, B., Li, X., Tang, T.: A two-objective timetable optimization model in subway systems. IEEE Trans. Intel. Transp. Syst. **15**(5), 1913–1921 (2014)
2. Feng, X., Zhang, H., Ding, Y., Liu, Z., Peng, H., Xu, B.: A review study on traction energy saving of rail transport. Discrete Dyn. Nature Soc. (2013)
3. Howlett, P.G., Milroy, I., Pudney, P.: Energy-efficient train control. Control Eng. Pract. **2**(2), 193–200 (1994)
4. Anh, A.T.H.T., Van Quyen, N.: Energy-efficient speed profile: an optimal approach with fixed running time. TELKOMNIKA (Telecommun. Comput. Electron. Control) **20**(3), 663–671 (2022)
5. Fernández, P.M., Font Torres, J.B., Sanchís, I.V., Franco, R.I.: Multi objective ant colony optimisation to obtain efficient metro speed profiles. In: Proceedings of the Institution of Mechanical Engineers, Part F: Journal of Rail and Rapid Transit, p. 09544097221103351 (2022)
6. Yang, X., Chen, A., Ning, B., Tang, T.: A stochastic model for the integrated optimization on metro timetable and speed profile with uncertain train mass. Transp. Res. Part B Methodol. **91**, 424–445 (2016)
7. Wang, H., Wu, P., Yao, Y., Zhuo, X.: Bi-objective subway timetable optimization considering changing train quality based on passenger flow data. In: Green Connected Automated Transportation and Safety, pp. 149–161. Springer (2022)
8. ShangGuan, W., Yan, X.H., Cai, B.G., Wang, J.: Multiobjective optimization for train speed trajectory in ctcs high-speed railway with hybrid evolutionary algorithm. IEEE Trans. Intel. Transp. Syst. **16**(4), 2215–2225 (2015)
9. Lu, S., Weston, P., Zhao, N.: Maximise the regenerative braking energy using linear programming. In: 17th International IEEE Conference on Intelligent Transportation Systems (ITSC), pp. 2499–2504. IEEE (2014)
10. Hartigan, J.A., Wong, M.A.: Algorithm as 136: a k-means clustering algorithm. J. R. Stat. Soc. Ser. C (Applied Statistics) **28**(1), 100–108 (1979)
11. Deb, K., Pratap, A., Agarwal, S., Meyarivan, T.: A fast and elitist multiobjective genetic algorithm: Nsga-ii. IEEE Trans. Evolut. Comput. **6**(2), 182–197 (2002)

Research on FPGA Accelerator Optimization Based on Graph Neural Network

Jin Wu[✉], Xiangyang Shi, Wenting Pang, and Yu Wang

School of Electronic Engineering, Xi'an University of Posts and Telecommunications, Xi'an 710121, China
lifewujin@xupt.edu.cn, wujin1026@126.com, {SXY,ardentpang, wyf9722}@stu.xupt.edu.cn

Abstract. In this paper, Field Programmable Gate Array (FPGA) hardware structure is designed to accelerate the graph neural network model, and suitable hardware structure is designed for different computing modules to accelerate it, so that the function of the whole system can be improved. The performance of computing intensive applications can be improved by using heterogeneous systems, which are composed of various processor architectures, such as FPGA. Graph Neural Network (GNN) is a framework that uses deep learning to learn graph structure data directly in recent years. Its excellent performance has attracted scholars' high attention and in-depth exploration. By formulating certain strategies on the nodes and edges of the graph, the graph neural network converts the graph structure data into a standard representation, and inputs it into a variety of different neural networks for training. Good results have been achieved in the task of node classification, edge information dissemination and graph clustering.

Keywords: FPGA · Accelerator · Graph Neural Network

1 Introduction

In recent years, the research enthusiasm on GNN in the field of deep learning is growing day by day. Graph neural network has become the research hotspot of each deep learning summit. GNN has made new breakthroughs in its outstanding ability to process unstructured data in network data analysis, recommendation system, physical modeling [1], natural language processing and graph combination perfect. In the real world, not all things can be represented as a sequence or a grid, such as social networks, knowledge maps, etc., that is to say, many things are unstructured. Compared with simple text and image, the unstructured data of this network type is very complex, which promotes the emergence and development of graph neural network. Hardware acceleration [2] refers to the use of hardware modules instead of software algorithms to make full use of the inherent fast characteristics of hardware. The hardware is much faster in performing various operations, for example, when calculating complex mathematical functions, data is transferred from one place to another through the same operation many times. If FPGA is used in the system design, customized hardware can be added at any time in the design cycle. So FPGA is suitable for accelerator.

N. Xiong et al. (Eds.): ICNC-FSKD 2022, LNDECT 153, pp. 536–542, 2023.
https://doi.org/10.1007/978-3-031-20738-9_61

2   Architecture of FPGA Accelerator

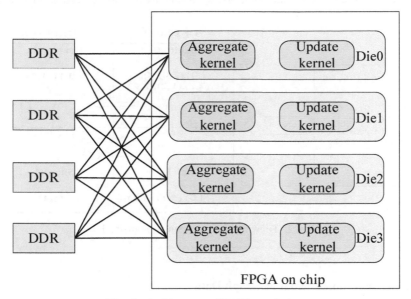

Fig. 1. Architecture of FPGA accelerator

Many modern FPGAs are composed of multiple chips, and the number of inter-connection lines across chips is limited. Therefore, multiple copies of the kernel are implemented and distributed among multiple chips, as shown in Fig. 1 Multiple chips and multiple DDR channels are connected through the full interconnection generated by the tool, and the input characteristic matrix is evenly divided into DDR channels. In order to use multiple computing cores for a single small batch, task division is performed for small batch training.

3   Figure Neural Network Hardware Accelerator Optimization

3.1   Graph Segmentation

Many real-world graphics are too large to be placed in a given level of memory hierarchy [3], which seriously reduces the performance of graphics processing algorithms. There-fore, it is extremely difficult to develop any reuse, especially when high-dimensional features are associated with nodes or edges. In order to meet this challenge, similar to the operation in CNN target recognition, the graph is processed hierarchically [4, 5] or the picture is divided into small blocks. The graph is usually divided into smaller parts. In this way, each sub graph can adapt to the memory hierarchy of a given level. This process is called graph slicing [6]. Similar to [6], this paper adopts the two-dimensional segmentation paradigm, as shown in Fig. 2. In this example, the edge list of a graph is divided into multiple slices. Each slice contains at most edges. Among them, is an adjustable parameter to limit the number of target nodes of each slice. The graph slicing algorithm divides the edge list into multiple slices (subgraphs), and then processes them in the way that the source is still (blue arrow) or the target is still (green oblique arrow).

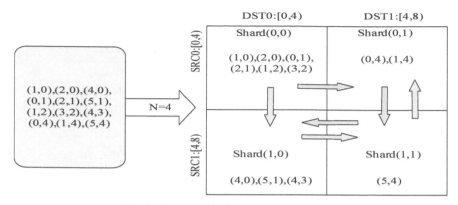

Fig. 2. Drawing block segmentation method

3.2   Representation Format of Graph Data

Three common storage methods of sparse matrix are introduced below:

COO: this storage method is the simplest one. Each element in the matrix needs to be represented as row number [7, 8], column number and value by a ternary array, which is represented by row, col and data. COO format is very convenient for graph data storage. It is usually the original graph format in real-time applications, but its performance is poor for calculation.

CSC: the compressed sparse column format is to store a sparse matrix by column, where the data in indptr represents the index of the data stored in each column in the matrix corresponding to the beginning and end of the data in the data. The data in indices represents the number of rows of the corresponding data data in its corresponding column.

CSR: in the compressed sparse row format, the sparse matrix is stored by row [9], similar to CSC. The data in Indptr represents the index corresponding to the beginning and end of each row of the matrix in data, while the data in indices represents the number of columns of the rows in the matrix in the corresponding data. Under the same conditions of indptr and indices, the sparse matrices obtained by CSC and CSR storage methods are transposed matrices.

The original input graph is stored in COO format and converted to CSR or CSC format as required. When the graph flows to FPGA, the data storage format is updated for all GNN layers.

3.3   Data Flow Optimization

The sparsity and intensive operation of graph data will lead to very complex data flow and cause great difficulties to the accelerator. The data flow can be optimized with the help of parallel architecture and fine-grained instructions.

Memory computing deployment: in the aggregation phase, the accumulation operation is used to aggregate the features from neighboring nodes. When the accumulation command is received, the PE controller will penetrate the accumulation mode. If the data delay is greater than or equal to the delay of the adder, the data cache can be used to make the memory computing deployment very simple. The memory delay can completely cover the calculation delay. In this scenario, there will be continuous execution. The GNN model used in this paper adopts batch processing, which requires aggregation of neighbor features on a batch of nodes. These neighbor nodes may overlap and lead to redundant memory access. Use the same DRAM address for continuous addition but different cache addresses. Then, the control module will notify the calculation module to reuse the input data from the DRAM, so as to eliminate redundant memory access.

Data blocking: the update phase also takes up a majority of GNN execution time. This stage mainly performs rule vector operations. Except for matrix operations, all these operations are memory limited. Therefore, it is necessary to speed up vector operations in memory, but this does not mean giving up the matrix operations of MGNN accelerator. The number of on-chip storage and computing units of FPGA [10] is enough. In order to provide higher performance and efficiency than CPU, the method of correctly using these resources for matrix operation can be adopted. Here, the data can be expanded and divided into a single computing unit to realize matrix multiplication. As shown in Fig. 3, it shows how to avoid the weight matrix to solve the problem of insufficient data cache. Therefore, matrix multiplication can calculate the memory occupation through formulas (1):

$$Memory_{accessed} = L \times M + L \times N \times N_M + (2N_L - 1) \times N \times M \qquad (1)$$

Under the limitation of on-chip cache capacity ($M' \times L' \leq BRAM\_size$), the value of the sum that minimizes Eq. (1) can be easily determined.

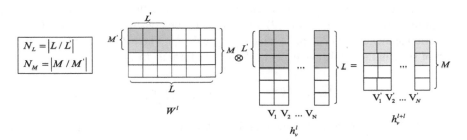

Fig. 3. Data expansion method when the on-chip storage is smaller than the weight matrix

4   Accelerator Performance Evaluation Criteria

For the sake of simplifying the symbols, it can be assumed that $f^{(l)} = f, \forall 0 \leq l \leq L$, therefore, the weight matrix can be expressed as $W_o^{(l)} \in R^{f \times \frac{1}{2}f}$ and $W_*^{(l)} \in R^{f \times \frac{1}{2}f}$, in which the factor is obtained due to the cascade in the forward transfer. For the classifier, it is assumed that a single-layer MLP has $W_{MLP} \in R^{f \times f}$. Regarding the resource utilization of FPGA, it can be assumed that both the accumulator and multiplier are implemented by DSP and have the same hardware cost. Here, it is assumed that the target FPGA can implement a number of R_{DSP} accumulators or multipliers, store R_{BRAM} words, and exchange words with external memory R_{BM} in each cycle. One word here represents an element of the eigenvector or weight matrix. The following performance determines the training parameters:

(1) Small batch and GCN: $|V_s|, \overline{d}, f$;
(2) Reduce redundancy: $\gamma_{add}, \gamma_{read}, \sum_{M \in M_a} |M|$;
(3) FPGA Architecture: P_{agg}, P_{sys}. Use $d \ll F \ll |V_s|$ to simplify the analysis.

Calculation and analysis: suppose each graph volume layer performs 3 feature aggregations and 6 weighted products, in which two feature aggregations operate on a vector with a length of and the other operates on a vector with a length of. Since feature aggregation is only parallel along the feature dimension, features with aggregate length of need to be accurately read for $\gamma_{read} \cdot |V_s| \cdot \overline{d} \cdot f / P_{agg}$ cycles. The weights of all six weighted products have the same complexity, and each operation requires a $\frac{1}{2} \cdot |V_s| \cdot f^2 / P_{sys}^2$ cycle. In order to reduce the time of feature aggregation, operations can be defined as Eqs. (2) and (3).

$$\gamma_{read} \cdot |V_s| \cdot \overline{d} \cdot f \cdot \frac{1}{P_{agg}} = (1 - \frac{2P_{sys}}{f}) \cdot \frac{1}{2} \cdot |V_s| \cdot f^2 \cdot \frac{1}{P_{sys}^2} \tag{2}$$

$$P_{agg} + 2P_{sys}^2 = R_{DSP} \tag{3}$$

$$P_{agg} \leq f \tag{4}$$

Among them, the $1 - \frac{2P_{sys}}{f}$ coefficient is very small, and the impact on performance is negligible. By solving Formula (2) and (3) under the constraint of Formula (4), the structural parameters and can be obtained P_{agg}^* and P_{sys}^*. The FPGA completes a small batch of steps in the whole cycle can be defined as formula (5):

$$T_{batch} = (3L + 3) \cdot |V_s| \cdot f^2 \cdot \frac{1}{(P_{sys}^*)^2} \tag{5}$$

The GNN accelerator is designed, and the designed hardware architecture is optimized to remove redundancy and reduce delay, so as to improve the performance of the accelerator. The designed hardware accelerator is implemented in hardware, and the experiment is carried out under the board to obtain various performance indicators of the accelerator. Figure 4 shows the hardware experiment platform.

Table 1 indicates the various performances of the accelerator before optimization and after optimization such as cycle expansion. From the table, it can be seen that 2531658 clock cycles are required to deploy GNN directly on the hardware. After some instruction optimization, such as pipelining and cycle expansion, the experiment found that the optimized delay is reduced to 281300 clock cycles, which significantly reduces the delay and accelerates GNN.

Fig. 4. Ultra96-V2 development board

Table 1. Comparison of accelerator delay before and after optimization

Optimization strategy	No. optimization	Loop expansion	Pipelined
Delay(cycle)	2531658	843886	281300

5 Conclusion

FPGA mainly deals with the task of intensive computing. The division of labor and cooperation between the two improves the work efficiency of the accelerator. For the main computing module CMCore, it is improved, which is divided into two parts: aggregation module and update module. It is better suitable for the mixed execution mode of GNN, and then the parallel kernel is used to further improve the performance of the accelerator. It is optimized to divide the input graph into small blocks, optimize the data flow, reduce the delay of the accelerator, reduce the redundancy of data reuse, and make the accelerator more perfect. Finally, some commonly used performance evaluation index calculation methods are proposed, which lays a solid foundation for the next hardware implementation.

Acknowledgments. Supported by Shaanxi Province Key Research and Development Project (2021GY-280, 2021GY-029); National Natural Science Foundation of China (No. 61834005,61772417,61802304).

References

1. Battaglia, P., Pascanu, R., Lai, M., et al.: Interaction networks for learning about objects, relations and physics. Adv. Neural. Inf. Process. Syst. **29**(4), 346–358 (2016)
2. Ma, Y., Cao, Y., Vrudhula, S., et al.: An automatic RTL compiler for high-throughput FPGA implementation of diverse deep convolutional neural networks. In: 2017 27th International Conference on Field Programmable Logic and Applications (FPL). IEEE, pp. 1–8 (2017)
3. Bordes, A., Usunier, N., Garcia-Duran, A., et al.: Translating embeddings for modeling multi-relational data. Adv. Neural. Inf. Process. Syst. **26**(4), 258–272 (2013)
4. Li, S., Hanson, E., Qian, X., et al.: Boosting the efficiency of sparse CNN accelerator with kernel decomposition. In: 54th Annual IEEE/ACM International Symposium on Microarchitecture, pp. 992–1004 (2021)
5. Lin, Y.C., Zhang, B., Prasanna, V.: GCN inference acceleration using high-level synthesis. In: 2021 IEEE High Performance Extreme Computing Conference (HPEC). IEEE, pp. 1–6 (2021)
6. Serafini, M., Guan, H.: Scalable graph neural network training: the case for sampling. ACM SIGOPS Oper. Syst. Rev. **55**(1), 68–76 (2021)
7. Parravicini, A., Sgherzi, F., Santambrogio, M.D.: A reduced-precision streaming SpMV architecture for Personalized PageRank on FPGA. In: 2021 26th Asia and South Pacific Design Automation Conference (ASP-DAC). IEEE, pp. 378–383 (2021)
8. Wang, D., Shen, J., Wen, M., et al.: Efficient implementation of 2d and 3d sparse deconvolutional neural networks with a uniform architecture on FPGAS. Electronics **8**(7), 803 (2019)
9. Zhang, Y., Yang, X., Lei, W., et al.: Hierarchical synthesis of approximate multiplier design for field-programmable gate arrays (FPGA)-CSRmesh system. Int. J. Comp. Appl. **180**(17), 1–7 (2018)
10. Brusati, M., Camplani, A., Cannon, M., et al.: Mitigated FPGA design of multi-gigabit transceivers for application in high radiation environments of high energy physics experiments. Measurement **108**(2), 171–192 (2017)

A Weighted Naive Bayes for Image Classification Based on Adaptive Genetic Algorithm

Zhurong Wang[✉], Qi Yan, Zhanmin Wang, and Xinhong Hei

School of Computer Science and Engineering, Xi'an University of Technology, Xi'an, China
{wangzhurong,Zhanmin,heixinhong}@xaut.edu.cn, yanqi14451@163.com

Abstract. Naive Bayes (NB) is a simple and widely used classification model, but due to the conditional independence assumption, the accuracy of NB is not very competitive in the field of image recognition. Therefore, this paper proposes a Weighted Naive Bayes classification algorithm with an Adaptive Genetic Algorithm (AGA_WNB), which is used to reduce the impact of this assumption. First, reduce the dimensionality of the image and binarize the image. Then, the initial weights of the features are used as the initial population, and the classification accuracy of the Weighted Naive Bayes (WNB) model is used as the fitness function. Adjust the crossover probability and mutation probability according to the fitness function, and select the better chromosome to enter the next generation. Finally, the optimal weights are selected by iteration. The experimental results on the public dataset MNIST show that under the same environment, the average accuracy of AGA_WNB is 3.25% higher than Weighted Naive Bayes based on Genetic Algorithm (GA_WNB) and 9.7% higher than NB. The single digit accuracy of AGA_WNB is 19% higher than NB. Compared with the comparison methods, the accuracy of AGA_WNB is also improved, and it has a good application prospect.

Keywords: Adaptive genetic algorithm (AGA) · Weighted Naive Bayes (WNB) · Image classification · Data dimensionality reduction

1 Introduction

There are many mature algorithms in the field of image classification, such as NB, Logistic Regression (LR), support vector machines (SVM), and convolutional neural networks (CNN), etc. Among them, NB is relatively mature in theory and has been widely used in various classification problems. However, NB is based on the assumption of conditional independence, and the features of high-dimensional image data are rarely independent, which makes NB uncompetitive in the accuracy of image recognition problems. But NB has some unique advantages that it is simpler. Therefore, removing this assumption or weakening the effect of this assumption can further improve the classification accuracy of NB and increase the competitiveness of NB in image classification.

In recent years, domestic and foreign researchers have done a lot of research work on improving the accuracy of NB. V. Jain et al. compared the performance of NB,

© The Author(s), under exclusive license to Springer Nature Switzerland AG 2023
N. Xiong et al. (Eds.): ICNC-FSKD 2022, LNDECT 153, pp. 543–550, 2023.
https://doi.org/10.1007/978-3-031-20738-9_62

SVM, Decision Tree, and LR on MNIST, and the accuracy of NB reached 73.25% [1]. K. Wang et al. verified that the conditional independence assumption required by NB may not be well suited for MNIST-like image recognition problems. But for those samples that were scored incorrectly, the sub-optimal answer was often the correct answer [2]. M. M. A. Taha et al. proposed a Naive Bayesian Nearest Neighbor Algorithm (NBNN), and verified that applying dynamic variable priors in NB is effective [3]. A. Kaushik et al. applied feature engineering to NB, and tested with a new dataset processed by feature selection, and the accuracy increased from 46.09 to 72.65% [4]. Pedro et al. proposed a WNB classifier to improve NB by adopting a weighted strategy. The experimental results show that WNB has a good application prospect [5]. Gregory F. Cooper et al. proposed an adaptive attribute weighting method based on the artificial immune system for NB [6]. Zhang H et al. proposed five WNB algorithms to evaluate the degree of influence on classification according to different weighting directions. These algorithms weaken the conditional independence of features, and the performance of NB is optimized to a certain extent [7]. J. Yang et al. applied the cuckoo search algorithm to WNB classification and searched for a more reasonable weighting coefficient according to the influence of different weight values on the classification results. The experimental results show that this algorithm has good performance and can improve classification accuracy [8]. To overcome the shortcomings of Long Short Term Memory (LSTM)'s vanishing gradient and high training cost, K. Chen et al. adopted the AGA to optimize the network parameters of LSTM neural network. The experimental results show that AGA is an effective parameter optimization method [9]. X. Lei et al. proposed a dilated CNN model which is built through replacing the convolution kernels of traditional CNN by the dilated convolution kernels. The experimental results show that the accuracy of the model on the training set is as high as 98%, and it is very competitive in the field of image classification [10].

In image classification problems, images of the same type have the same or similar features. Therefore, the features of such problems are correlated, which will affect the classification accuracy of NB. To weaken the influence of the conditional independence assumption on NB, this paper combines AGA with NB, and uses the excellent optimization ability of AGA to find the optimal weight that can represent the correlation of features. Experiments are carried out on the public handwritten digit dataset, and the results show that the method can reduce the influence of the conditional independence assumption and improve the classification accuracy, which is an effective method.

The rest of this paper is shown as follows. Section 2 introduces the implementation principles of AGA and AGA_WNB. Section 3 introduces the experimental data, model building, and parameter settings. Section 4 introduces the experimental results and comparative analysis. Section 5 presents the research conclusions and future work.

2   Algorithm Principle

2.1   Adaptive Genetic Algorithm

The choice of crossover probability and selection probability in GA will directly affect the convergence speed and performance of the algorithm. If the crossover probability is too high, the inheritance pattern will be disrupted. If the mutation probability is too high, the genetic algorithm becomes a random search algorithm. If the crossover probability and mutation probability are too small, the convergence speed of the algorithm will slow down or even stagnate. Therefore, the choice of crossover probability and mutation probability is very critical.

The crossover probability and mutation probability of AGA can be changed automatically with fitness. When the fitness of an individual is lower than the average fitness, increase the mutation probability and crossover probability to increase the probability of generating new individuals. When the individual fitness is higher than the average fitness, the mutation probability and crossover probability are reduced to protect the individual from entering the next generation.

2.2 Weighted Naive Bayes Classification Algorithm Based on Adaptive Genetic Algorithm

WNB is a classification method based on Bayes theorem and Bayesian estimation. For the training set, it learns the joint probability distribution model of the input and output, and then based on this model, for the given input x, the Bayes theorem is used to obtain the category y with the largest posterior probability as the output. In this paper, the prior probability and conditional probability are calculated by using the Bayesian estimation method with Laplace, and 1 is added to the frequency of each value of the random variable to avoid the probability value being estimated as 0.

AGA_WNB uses the excellent optimization ability of AGA to find the optimal weight of WNB. The algorithm description of AGA_WNB is shown in Algorithm 1. The flow chart of the AGA_WNB algorithm is shown in Fig. 1.

Algorithm 1 AGA_WNB.

(1) Chromosome coding. Integer coding method is adopted in this paper, and the number of genes is equal to the number of features in practical problems.
(2) Initial population, crossover probability and mutation probability.
(3) Add one gene for each feature, and as the weights of the features, n genes for n features make up a chromosome. The classification result of each picture is calculated as follows:

$$f(x) = \arg \max_{c_k} \prod_{j=1}^{n} P(c_k) P(x^j | c_k)^{\omega^j} \tag{1}$$

Among them, define: x is the input, which is described by n features. c_k is the output category, and k is the number of categories. ω is the weight of features.
(4) The classification accuracy of WNB on the dataset was taken as fitness function, and the higher the accuracy, the greater the chromosome fitness. The accuracy is calculated as follows:

$$A = \frac{N_{TP} + N_{TN}}{N_{TP} + N_{TN} + N_{FP} + N_{FN}} \tag{2}$$

Among them, define: N_{TP} is true positive prediction sample; N_{FP} is false positive prediction sample; N_{TN} is a true negative prediction sample; and N_{FN} is a false negative prediction sample.

(5) The crossover probability and mutation probability were adjusted according to the adaptive value. The crossover probability and mutation probability are calculated as follows:

$$P = \begin{cases} 0, & f_i = f_{\max} \\ e^{-(k \times f_i)}, & f_i \neq f_{\max} \end{cases} \tag{3}$$

Among them, define: f_i is the adaptive value of chromosome i, $f_{\max}$ is the maximum adaptive value in the population, and k is a constant.

(6) If the termination condition is reached, the test set is given to the trained model and the classification result is obtained. Otherwise repeat (3).

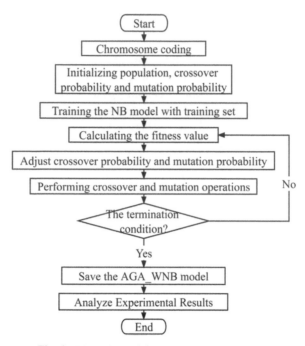

Fig. 1. Flow chart of the AGA_WNB algorithm.

3 Data and Model Building

3.1 Data Sample Introduction

In order to verify the effectiveness and accuracy of the proposed AGA_WNB algorithm, the test data set is selected from the public MNIST.The MNIST is an open-source database from the National Institute of Standards and Technology (NIST), download link: http://www.cs.nyu.edu/~roweis/data.html. It contains 60,000 training set data and 10,000 test set data in total. The 60,000 training set data consists of 0–9 numbers handwritten by 250 different people, 50% of whom are high school students and 50% from Census Bureau staff. The source and proportion of handwritten digit data in the test set is related to the training set the same. Each handwritten number picture is a 28×28 grayscale image, and the numbers in all pictures are similar in size and basically centered.

3.2 Data Preprocessing

The steps of data preprocessing are as follows:

Step 1: Divide the dataset into training set and test set, of which 85% is training set and 15% is test set.
Step 2: Reduce 784-dimensional grayscale images to 49 dimensions.
Step 3: Binarize 49-dimensional grayscale images into black and white images with a global average threshold.

3.3 Parameter Setting and AGA_WNB Construction

The experimental operating environment is Windows 10 Intel(R) Core(TM) i5-8250U CPU @ 1.60GHz 1.80 GHz. The implementation of the algorithm is programmed with Python 3.8, and PyCharm2021.2.1 is used as the integrated development environment.

First, import the downloaded MNIST dataset and divide the dataset into two parts: training set and test set. Then perform dimensionality reduction and binarization operations on the dataset. The prior probability, conditional probability with initial weights, and posterior probability are then calculated. The population is then selected, crossed and mutated using the MNIST classification accuracy as a fitness function. The better population is selected to enter the next generation until the termination condition is reached, and the final AGA_WNB model is obtained.

Regarding the AGA_WNB model parameters, the values are: the 784 features were reduced to 49, the global average threshold was 50, the number of genetic iterations was 1000, the population size was 50, and the training set and test set was 1/10 of the original data.

4 Results and Analysis

In the experimental tests, we use the NB model, the GA_WNB model and the model obtained by the proposed AGA_WNB method to compare data sets.

Table 1 shows the accuracy of the three classification algorithms on MNIST data sets. In order to make the results fairer, MNIST was tested 10 times with three algorithms, and finally the maximum (Max), minimum (Min), mean (Mean) and standard deviation (Std.) of the 10 results were obtained respectively. Among them, the maximum, minimum and mean accuracy are calculated by Eq. 2. As can be seen from Table 1, under the same experimental environment, compared with NB, GA_WNB improves the mean accuracy by 6.45%. Compared with GA_WNB and NB, AGA_WNB improves the mean accuracy by 3.25% and 9.7% respectively.

Figure 2 shows the convergence learning curve of AGA_WNB on 10 single numbers. It can be seen that the accuracy of 10 numbers has improved, the first 400 iterations improved rapidly, and the 400th to 800th iterations improved relatively slowly. There will be no better solutions after the 800th generation, and it has reached a state of convergence.

Figure 3 shows the mean accuracy of 10 single numbers using the NB algorithm, GA_WNB algorithm, and AGA_WNB algorithm respectively under the same environment. It can be seen that GA_WNB slightly improves the performance of the NB algorithm, while AGA_WNB proposed in this paper significantly improves the performance of the NB algorithm. GA_WNB compared with NB, the three most significant numbers are 5, 2, and 8, which increased by 18%, 16%, and 10% respectively. AGA_WNB compared with NB, the three most significant numbers are 2, 5, and 8, which are increased by 19%, 18%, and 16.9% respectively.

In order to further study the convergence of the AGA_WNB algorithm, convergence learning curves have been drawn to show the variation of the classification accuracy and the number of iterations of the two algorithms on the dataset. Figure 4 shows the convergence learning curves of GA_WNB and AGA_WNB. It can be seen that the convergence speed of the GA_WNB algorithm is slow, and sometimes there is a phenomenon of regression. The improved AGA_WNB algorithm has better performance in convergence speed and jumping out of the local search, and there is no regression phenomenon.

Table 1. The accuracy of the three classification algorithms on MNIST data sets

Accuracy	Mean (%)	Max (%)	Min (%)	Std
NB	72.65	72.65	72.65	0
GA_WNB	79.10	80.05	77.63	0.0104
AGA_WNB	82.35*	84.11	81.30	0.0117

5 Conclusion

In order to reduce the influence of conditional independence assumption on NB and improve the accuracy of image recognition, this paper proposes a Weighted Naive Bayes classification algorithm with an Adaptive Genetic Algorithm (AGA_WNB). The experimental results show that the algorithm improves the average accuracy by 9.7% on the MNIST dataset, and reduces the impact of the conditional independence assumption on the classification accuracy.

At the same time, we find that in the wrong classification samples, the sub-optimal category output by the AGA_WNB classifier is usually the correct answer fact, and it is not unreasonable for AGA_WNB to classify them wrong if consider the structural similarity between the right category and the wrong category in shape.

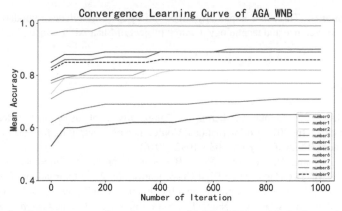

Fig. 2. The convergence learning curve of AGA_WNB on 10 single numbers.

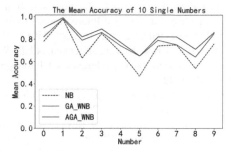

Fig. 3. The mean accuracy of 10 single numbers using the NB, GA_WNB, and AGA_WNB respectively.

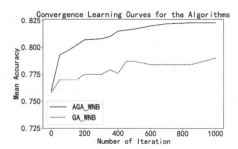

Fig. 4. The convergence learning curves of GA_WNB and AGA_WNB.5 conclusion.

In future work, we will consider more influencing factors and try to use more methods to reduce the impact of the conditional independence assumption on NB, thereby improving the classification accuracy of NB in the field of image recognition and expanding the application scope of NB.

Acknowledgments. This work is partially supported by Natural Science Foundation of China (No. 61773313), Science and technology research project of Shaanxi Province (2021JM-346).

References

1. Jain, V., Dubey, A., Gupta, A., Sharma, S.: Comparative analysis of machine learning algorithms in OCR. In: 2016 3rd International Conference on Computing for Sustainable Global Development (INDIACom), pp. 1089–1092 (2016)
2. Wang, K., Zhang, H.: A Novel Naive Bayesian Approach to Inference with Applications to the MNIST Handwritten Digit Classification. In: 2020 International Conference on Computational Science and Computational Intelligence (CSCI), pp. 1354–1358 (2020)
3. Taha, M.M.A., Teuscher, C.: Naive Bayesian inference of handwritten digits using a memristive associative memory. In: IEEE/ACM International Symposium on Nanoscale Architectures (NANOARCH), 139–140 (2017)
4. Kaushik, A., Gupta, H., Latwal, D.S.: Impact of feature selection and engineering in the classification of handwritten text. In: 2016 3rd International Conference on Computing for Sustainable Global Development (INDIACom), pp. 2598–2601 (2016)
5. Pedro, S.D.D.S., Hruschka, E.R., Hruschka, E.R., Ebecken, N.F.F.: WNB: A Weighted Naïve Bayesian Classifier. Seventh International Conference on Intelligent Systems Design and Applications (ISDA), pp. 138–142 (2007)
6. Cooper, G.F.: The computational complexity of probabilistic inference using Bayesian belief networks. Amsterdam: Elsevier Sci. Publ. Ltd. **42**(2–3), 393–405 (1990)
7. Zhang, H., Sheng, S.: Learning weighted naive Bayes with accurate ranking. In: IEEE International Conference on Data Mining. IEEE, pp. 567–570 (2005)
8. Yang, J., Ye, Z., Zhang, X., Liu, W., Jin, H.: Attribute weighted Naive Bayes for remote sensing image classification based on cuckoo search algorithm. In: 2017 International Conference on Security, Pattern Analysis, and Cybernetics (SPAC), pp. 169–174 (2017)
9. Chen, K.: An online retail prediction model based on AGA-LSTM neural network. In: 2020 2nd International Conference on Machine Learning, Big Data and Business Intelligence (MLBDBI), pp. 145–149 (2020)
10. Lei, X., Pan, H., Huang, X.: A dilated CNN model for image classification. IEEE Access **7**, 124087–124095 (2019)
11. Zhao, W., et al.: Superpixel-based multiple local CNN for panchromatic and multispectral image classification. IEEE Trans. Geosci. Remote Sens. **55**(7), 4141–4156 (2017)
12. Zhang, H., Jiang, L.X., Yu, L.J.: Attribute and instance weighted Naive Bayes. Pattern Recogn. **111**, 107674–107684 (2021)

Quantum Voting Protocol Based on Blind Signature

Qiang Yuwei$^{(\boxtimes)}$, Chen Sihao, Li Na, and Bai Qian

Xi'an University of Posts and Telecommunications, Xi'an 710000, China
445499@stu.xupt.edu.cn, 19960713@xpu.edu.cn, {210311029,
41706130119}@stu.xpu.edu.cn

Abstract. Aiming at the problem of high technical difficulty in the implementation of quantum voting protocol, a quantum voting protocol based on blind signature is proposed, which avoids the use of complex quantum fingerprint function and realizes the voting process of quantum blind signature by using the five particle entangled state in quantum cryptography. Among them, voter Alice transmits voting information by measuring particle state, and Bob, Charlie and Trent complete signature, verification and audit by measuring single particle state through Z-basis. Since the signed message is invisible to the signer, it is a quantum voting protocol to protect the anonymity of signed messages. Therefore, all keys in this protocol are issued by BB84 protocol, and one-time encryption algorithm is adopted, which has unconditional security, Compared with the existing quantum voting protocols, the proposed voting protocol selects a simpler measurement base, has simple measurement operation, easy implementation and high execution efficiency. It is a safe and efficient quantum voting protocol.

Keywords: Quantum voting protocol · Blind signature · Quantum cryptography · Five particle entangled state

1 Introduction

The traditional electronic voting is mostly realized by the technology based on blind signature and group signature [1, 2]. The classical blind signature and group signature schemes are mostly based on the computational difficulties such as factorization, discrete logarithm and quadratic residue [3, 4]. However, with the increasing computing power of attackers, especially the proposal of quantum computing, these algorithms and protocols become increasingly insecure. With the continuous development of technology, quantum cryptography began to be concerned, which can effectively make up for this defect. Because the security of quantum cryptography is based on the physical characteristics of quantum states rather than computational complexity, such as Heisenberg uncertainty theorem and quantum non cloning, many quantum cryptography protocols have been proved to be unconditionally secure [5, 6].

Based on quantum cryptography, quantum voting uses the physical characteristics of quantum itself to overcome the security problems existing in classical cryptography,

© The Author(s), under exclusive license to Springer Nature Switzerland AG 2023
N. Xiong et al. (Eds.): ICNC-FSKD 2022, LNDECT 153, pp. 551–558, 2023.
https://doi.org/10.1007/978-3-031-20738-9_63

ensure the security of ballot information and the legitimacy of participant identity, and improve the reliability of voting activities. Vaccaro et al. defined the standard of quantum voting protocol in 2007 [7]. The first two quantum voting protocols were proposed by Hillery, including mobile voting scheme and distributive voting scheme [8]. Since then, various quantum voting protocols have been proposed successively. Tian proposed a quantum voting scheme based on entangled states [9]. In 2008, Li proposed a secure anonymous voting scheme based on the principle of quantum entanglement [10]. In 2011, Horoshko et al. Designed a voting scheme for anonymous verification based on Bell state [11]. Based on the traditional vote counting scheme proposed by Bell [12]. In 2015, a vote counting scheme based on the particle entanglement model was completed. In 2016, Naseri et al. Proposed an anonymous voting method based on GHz entangled states [13]. In this scheme, there are only two voters. When the number of voters increases, the number of votes increases accordingly. Meanwhile, Wang et al. Proposed a quantum anonymous voting protocol assisted by two entangled quantum states [14]. In 2017, Liu et al. Proposed a voting scheme based on GHZ state [15], which set up a double supervision mode to enhance security. However, if there are a large number of voters, it is necessary to repeatedly prepare four particle GHZ state, and the consumption of quantum prepared state is large. Zhang et al. Proposed a quantum voting scheme based on quantum proxy blind signature [16], which needs to prepare both EPR pairs and GHZ States, resulting in low quantum bit efficiency. At present, the quantum voting scheme based on strong blind signature has attracted the attention of academia. Using the quantum coherence of GHZ three particle entangled states, document [12] proposed a strong blind signature protocol to construct a quantum voting scheme. In order to prevent the ticket checker from cheating, the scheme designs the audit program through the quantum fingerprint function to supervise the post. However, the technical implementation of quantum fingerprint function is difficult, and the audit procedure is carried out afterwards, so cheating can not be found immediately. In order to make up for this defect, a blind signature quantum voting protocol is proposed, which prevents cheating by introducing a vote monitor. The quantum fingerprint function and audit procedure are abandoned, and the real-time supervision is realized.

Aiming at the problems existing in the existing voting protocols, this paper uses the five particle entangled state as the quantum channel to realize the transmission of information, and designs a simple and efficient quantum voting protocol that meets the anonymity, non repeatability, fairness, integrity and unconditional security. The proposed voting protocol mainly discards the complex quantum fingerprint function and quotes Charlie, the scrutineer, to supervise Trent, the scrutineer, in order to improve the security of the protocol. It makes the protocol simple and efficient in the design process. In addition, it also uses $Z \otimes Z$ and Z measurement bases to realize measurement. The selection of measurement bases is simple and convenient for measurement.

2 Basic Principles

The voting protocol in this paper is based on the five particle entangled state [17], which is expressed as follows:

$$|\xi\rangle_{12345} = \frac{1}{4}(|00000\rangle + |00110\rangle - |01010\rangle + |01100\rangle$$

$$+ |10010\rangle + |10100\rangle + |11000\rangle - |11100\rangle$$
$$+ |10001\rangle + |10111\rangle - |11011\rangle + |11101\rangle$$
$$+ |00011\rangle + |00101\rangle + |01001\rangle - |01111\rangle)_{12345} \qquad (1)$$

By executing the protocol, Alice owns particle 1 and particle 2, Bob owns particle 3, Trent owns particle 4 and Charlie owns particle 5. If Alice performs $Z \otimes Z$ measurement $\{ |00\rangle, |01\rangle, |10\rangle, |11\rangle \}$ on particle 1 and particle 2, the rest of them perform Z-based measurement on their own particles, and get the corresponding results by measuring each other's particles. For example, Alice completes her quantum voting after base $Z \otimes Z$ measurement, Charlie and Bob accept Alice's quantum voting, Bob and Charlie encode their own particles after Z-based measurement, encrypt and send the coding results to Trent. According to the measurement results sent to him by Bob and Charlie, combined with the results of Z-based measurement coding of his particles, Trent can open Alice's vote and know that their measurement results are highly correlated.

3 Voting Agreement

The four roles in the scheme are defined as follows:

(1) Voter Alice is the owner of voters and voting information;
(2) Bob, the voting management center, will check the qualification of voters, distribute ballot papers and be the signatory.
(3) Trent is a teller who will verify the message and signature.
(4) Charlie is a ticket supervisor. He supervises Trent's behavior.

3.1 Initial Stage

The initialization scheme of the voting scheme is as follows:

(1) Trent shares keys K_{TB} and K_{TC} with Bob and Charlie respectively. The shared key process uses BB84 quantum key distribution protocol.
(2) Bob of the voting management center prepares $Q(Q >> n)$ five particle entangled states as shown in formula (1) as quantum channels, and sends particles (1, 2) to Alice, particle 4 to Trent and particle 5 to Charlie. Particle 3 remains in its own hands.
(3) Quantum channel security detection. Bob did a wiretap. He randomly selected Q-n group particle 3, recorded its position, measured it on Z-basis, and then announced its position. According to Bob's published position, Alice measured her corresponding particle (1, 2), Trent measured his corresponding particle 4, Charlie measured his corresponding particle 5, and then published their measurement results. Here, the order of declarations is random. If their measurement results are correlated or the uncorrelated error rate is less than the threshold, it is proved that the quantum channel is safe. Otherwise, they terminate the agreement and start again [18].

3.2 Registration Phase

Voter Alice sends her identity information to Bob of the voting management center, and Bob verifies whether Alice's identity is legal and whether it is her first vote. If so, the voting management center sends a random sequence bit string f_{ID} to Alice as Alice's registration information [19].

3.3 Voting Stage

(1) Alice performs an $Z \otimes Z$ measurement on the particles (1, 2) in her hand, and then she gets the message as follows:

$$m' = \{m'(i), i = 1, 2, \cdots, n\} \quad m'(i) \in \{|00\rangle, |01\rangle, |10\rangle, |11\rangle\} \tag{2}$$

(2) According to the message, Alice can get a 2N bit voting message sequence (including her voting ID, voting content, etc.). The coding rules are as follows: If $m'(i)=|00\rangle$, $m'(i)=|01\rangle$, $m'(i)=|10\rangle$, $m'(i)=|11\rangle$, Corresponding to $m(i)=00$, $m(i)=01$, $m(i)=10$, $m(i)=11$. Then Alice gets the message: $m = \{m(i), i = 1, 2, \cdots, n\}$, and $m(i) \in \{00, 01, 10, 11\}$.

Voter Alice has completed her own voting operation. Bob and Charlie follow the following steps to receive and calculate Alice's quantum voting.

3.4 Signature Phase

The voting stage is as follows:

(1) Bob performs Z-based measurement on his particle 3 and then encodes it. The coding rules are as follows: the $|0\rangle$ state corresponds to the classical bit 0; The $|1\rangle$ state corresponds to classical bit 1. After this coding, we get:

$$\beta_B = \{\beta(i)_3, i = 1, 2, ...n\} \tag{3}$$

Among them $\beta_B(i) \in \{0, 1\}$.

(2) Bob encrypts β_B with key K_{TB} to get $Sig(m) = E_{K_{TB}}(\beta_B)$, because Bob doesn't know the content of message m, and signature $Sig(m)$ is blind.

(3) Bob sends $Sig(m)$ to Trent.

3.5 Supervision and Counting Stage

The signing and supervision stages are as follows:

(1) Charlie performs Z-basis measurement on particle 5 and then encodes it. The coding rules are the same as Bob's. Get $\beta_C = \{\beta(i)_5, i = 1, 2, \cdots, n\}$ and $\beta(i)_5 \in \{0, 1\}$.

(2) Charlie encrypts K_{TC} with key β_C, gets $S_C = E_{K_{TC}}(\beta_C)$ and sends it to Trent.

(3) Trent performs Z-basis measurement on particle 4, and then encodes it. The coding rules are the same as Bob's to get $\beta_T = \{\beta(i)_4, i = 1, 2, \cdots, n\}$ and $\beta(i)_4 \in \{0, 1\}$.

(4) Trent decrypts K_{TB} and K_{TC} with keys $Sig(m)$ and S_C to get β_B and β_C. Combined with their own measurement results, Alice's message m is inferred. Open Alice's vote according to the coding rules.

(5) All voting results of the agreement will be published on the bulletin board. If the voter finds that her ballot has been tampered with or lost, she can apply for re voting [20, 21].

4 Safety Analysis

Our quantum voting protocol satisfies all the following characteristics:

(1) Anonymity

In the whole voting process, we must ensure that voters' privacy and voting content are not disclosed, that is, only voters know what votes they vote for, and no one can connect voters with votes. After Alice finished voting by measuring her particles (1, 2), she automatically withdrew from the voting process. In this scheme, the ballot is blinded to m' by Alice, and then encoded into a classical bit string m. The registration information f_{ID} is issued to Alice by Bob, the vote management center. It is a random sequence. Except for the voters themselves and Bob, no one knows the corresponding relationship between f_{ID} and voters. When the final vote results are published, only voters can verify their votes according to f_{ID}. Therefore, no one can track voters, that is, the anonymity of the agreement is maintained [22, 23].

(2) Non repeatability

Each voter has only one chance to vote. In our scheme, only qualified voters can vote. The process of any voter being issued with ballot f_{ID} is unique. Repeated voting will be found by Bob of the ballot management center. When Alice re applies for registration information to Bob of the ballot management center, Bob will verify whether Alice's identity information matches the corresponding f_{ID}. if it exists, Bob will think Alice is dishonest and refuse to issue a second ballot, so no voter can vote again. f_{ID} is randomly assigned by the ballot management center, and voters cannot forge f_{ID}, so it meets the non repeatability of the agreement [24, 25].

(3) Fairness

No one can affect the election process, especially the intermediate results of the election can not be disclosed, otherwise it will affect the voting tendency of other voters. If Alice divulges the content of her vote to other voters, it will affect the voting tendency of other voters and affect the agreement. There is no such situation in this agreement, because each voter Alice only contacted bob of the ballot management center and withdrew from the agreement process after completing the voting process, so there is no contact with other voters, as well as Trent, the ticket inspector and Charlie, the supervisor.

For Trent and Charlie, Alice's vote is blind, while Bob's signature on the vote and Charlie's measurement of particles for exercising supervision are blind. Therefore, it can be ensured that the intermediate result of the election is secret, that is, the election is fair and fair. Charlie, the supervisor, restricted the power of the teller Trent and maintained the fairness of the voting agreement.

(4) Integrity

All legitimate ballots can be counted and proved correctly. If the ballots are tampered with or omitted, they are easy to be found. First, if Trent wants to tamper with the ballot, because the measurement results of all participants in the protocol are highly correlated, his malicious tampering is easy to be found. Second, because the vote results are public, all voters Alice will check whether their votes are omitted or tampered with on the bulletin board. If voters find that their vote results are omitted or tampered with, she will apply to Bob, the vote management center, and the vote will be invalidated and the voting and counting process will be restarted. Make the protocol meet the integrity.

(5) Unconditional security

In the process of protocol implementation, all keys are issued by BB84 protocol. In the initial stage, Trent, Bob and Charlie share keys K_{TB} and K_{TC}, and adopt quantum key distribution protocol or measurement device independent QKD protocol. In the process of quantum channel security detection, the four parties publish the participation results and evaluate the correlation through threshold, so as to detect the security of quantum channel and meet the perfection of security. The scheme is based on quantum key distribution and one-time encryption algorithm, which meets the unconditional security.

5 Conclusion

In this paper, the five particle quantum entangled state is used to realize the quantum blind signature voting scheme. The scheme realizes voting, counting and auditing by measuring the particle state. According to the security analysis, the scheme meets the requirements of anonymity, non repeatability, fairness, integrity and unconditional security. Compared with the traditional quantum voting scheme, the proposed scheme has certain advantages. Firstly, the eavesdropping inspection process is arranged to provide a secure quantum channel. Secondly, only $Z \otimes Z$ and Z-basis are used in the measurement of particle states, which is easy to realize under the existing experimental and technical conditions. At the same time, the use of one-time encryption algorithm and quantum key distribution protocol meets the security of the protocol. Compared with other quantum voting protocols, it avoids the use of complex quantum fingerprint functions, which reduces the difficulty of design and implementation. Therefore, the implementation of quantum voting protocol technology in this paper is less difficult and practical.

Acknowledgments. Fund projects: Shaanxi Natural Science Basic Research Program (NO. 2021JM-462).

References

1. Ahn, B.: Implementation and early adoption of an ethereum-based electronic voting system for the prevention of fraudulent voting. Sustainability **14**(5), 212–219 (2022)
2. Adewale, O.S., Boyinbode, O.K., Salako, E.: An innovative approach in electronic voting system based on fingerprint and visual semagram. Int. J. Inf. Eng. Electron. Business (IJIEEB) **13**(5), 178–183 (2021)
3. Gang, X., Yibo, C., Xu, S, Y., et al.: A novel post-quantum blind signature for log system in blockchain. Comp. Syst. Sci. Eng. **41**(3), 321–327 (2022)
4. Wang, J.H., Li, Y.X., Meng, W., et al.: Protocol of quantum blind signature based on two—qubit and three-qubit maximally entangled states. Laser Optoelectron. Progr. **58**(7), 133–143 (2021)
5. Hu, X.M., Chen, F.S., Ma, C., et al.: Comment on security and improvement of partial blind signature scheme and revocable certificateless signature scheme. J. Phys. Conf. Ser. **1827**(1), 56–63 (2021)
6. Nedal, M.T., Ashraf, T., Ramzi, A., et al.: Design of identity-based blind signature scheme upon chaotic maps. Int. J. Online Biomed. Eng. (iJOE) **16**(5), 262–273 (2020)
7. Vaccaro, J.A., Spring, J., Chefles, A.: Quantum protocols for anonymous voting and surveying. Phys. Rev. A **75**(1), 123–131 (2007)
8. Hillery, M.: Quantum voting and privacy protection. Int. Soc. Opt. Eng. **63**(12), 16–23 (2006)
9. Tian, J.H., Zhang, J.Z., Li, Y.P.: A voting protocol based on the controlled quantum operation teleportation. Int. J. Theor. Phys. **55**(5), 2303–2310 (2016)
10. Li, Y., Zeng, G.: Quantum anonymous voting systems based on entangled state. Opt. Rev. **15**(5), 219–223 (2008)
11. Horoshko, D., Kilin, S.: Quantum anonymous voting with anonymity check. Phys. Lett. A **375**(8), 1172–1175 (2011)
12. Cao, H.J., Ding, L.Y., Yu, Y.F., et al.: A electronic voting scheme achieved by using quantum proxy signature. Int. J. Theor. Phys. **55**(9), 4081–4088 (2016)
13. Naseri, M., Gong, L.H., Houshmand, M., et al.: An anonymous surveying protocol via greenberger-horne-zeilinger states. Int. J. Theor. Phys. **55**(10), 4436–4444 (2016)
14. Wang, Q., Yu, C., Gao, F., et al.: Self-tallying quantum anonymous voting. Phys. Rev. A **94**(2), 223–231 (2016)
15. Liu, X.C.: Universal three-qubit entanglement generation based on linear optical elements and quantum non-demolition detectors. Int. J. Theor. Phys. **56**(2), 427–436 (2017)
16. Zhang, J.L., Xie, S.C., Zhang, J.Z.: An elaborate secure quantum voting scheme. Int. J. Theor. Phys. **56**(10), 3019–3028 (2017)
17. He, Y.F., Ma, W.P.: Two-party quantum key agreement with five-particle entangled states. Int. J. Quant. Inf. **15**(3), 175–186 (2017)
18. Niu, X.F., Zhang, J.Z., Xie, S.C., et al.: An improved quantum voting scheme. Int. J. Theor. Phys. **57**(10), 3200–3206 (2018)
19. Zhang, Q., Li, C., Li, Y., et al.: Quantum secure direct communication based on four-qubit cluster states. Int. J. Theor. Phys. **52**(1), 22–27 (2013)
20. Wang, J., Xu, G.B., Jiang, D.H.: Quantum voting scheme with Greenberger-Horne-Zeilinger states. Int. J. Theor. Phys. **59**(8), 2599–2605 (2020)
21. Cao, H.J., Yu, L., Feng, R.C.: A four-particle cluster state is used to implement quantum electron voting. J. Liaoning Normal University (Natural Science Edition) **42**(14), 179–186 (2019)
22. Zhou, B.M., Zhang, K.J., Zhang, X., et al.: The cryptanalysis and improvement of a particular quantum voting model. Int. J. Theor. Phys. **59**(4), 1109–1120 (2020)

23. Zhang, X., Zhang, J.Z., Xie, S.C.: A secure quantum voting scheme based on quantum group blind signature. Int. J. Theor. Phys. **59**(3), 719–729 (2020)
24. He, Y.F., Li, C.Y., Guo, J.R., et al.: Passive measurement-device-independent quantum key distribution based on heralded pair coherent states. Chinese J. Lasers **47**(9), 26–32 (2020)
25. He, Y.F., Guo, J.R., Li, C.Y., et al.: Fluctuation analysis of key distribution protocol based on heralded single photon source and orbital angular momentum. Chinese J. Lasers **47**(4), 168–177 (2020)

Computer Vision (17)

Face Detection and Tracking Algorithm Based on Fatigue Driving

Zhe Li and Jing Ren[✉]

School of Electronic Engineering, Xi'an University of Posts and Telecommunications, Xi'an, China
xytx03@xupt.edu.cn, 5464@stu.xupt.edu.cn, 3534244889@qq.com

Abstract. Aiming at the problem that drivers may have traffic accidents due to fatigue driving for a long time, a method of detecting and analyzing the facial features of drivers is proposed to study fatigue driving. The preprocessing operation of the video image is completed before detection, and the driver's face in the video image is detected by RestNet10-SSD target detection algorithm. The experimental results show that this method can better detect the face when the front, side and part of the driver's face are blocked. However, the detection rate of the algorithm is relatively slow and can not meet the real-time requirements, On this basis, Dlib tracking algorithm is used to optimize face detection. It turned out that the driver's face detection and tracking through the fusion algorithm not only has a high detection as well as recognition rate, but also meets the real-time requirements, and plays a key role in the subsequent fatigue driving detection.

Keywords: Fatigue driving detection · ResNet10-SSD · Face detection · Dlib

1 Introduction

Drowsy driving is one of the main causes of road accidents. According to statistics, traffic accidents give rise to fatigue driving account for 21% of the totality traffic accidents in China, and the death rate of traffic crash leads to fatigue driving is also very high [1]. By identifying the driver's fatigue condition and giving an alarm, it can effectively remind the driver to pay attention to rest, so as to avoid traffic accidents and guarantee the safety of people's lives and property as far as possible. From this point of view, the research on fatigue driving detection is provided with great significance. The traditional fatigue driving detection methods mainly check the driver's physiological characteristics (such as EEG [2], ECG [3], EEG [4]) and vehicle behavior (such as steering wheel angle, vehicle speed [5], steering wheel grip [6], etc.), but these two methods need to be connected with wearing equipment, which has high cost and is not practical, so they are not conducive to popularization. The fatigue driving detection method based on visual feature analysis obtains and processes the real-time video image of the driver, and acquires the feature parameters of the driver's eyes, mouth and head, so as to judge whether the driver is tired [7, 8].

N. Xiong et al. (Eds.): ICNC-FSKD 2022, LNDECT 153, pp. 561–567, 2023.
https://doi.org/10.1007/978-3-031-20738-9_64

According to the characteristics of driver fatigue, combined with the deep learning model, Opencv and image processing, the fatigue driving algorithm is designed. In the face detection part, firstly, the video image is preprocessed, the target is detected through RseNet10-SSD, and the face area is tracked by Dlib tracking algorithm, so that the detection speed is improved. While ensuring the detection accuracy, the method improves the real-time detection of the system, so as to meet the real-time needs of fatigue driving.

2   Image Preprocessing

The video image collected by the camera may interfere with the face detection due to the background mutation or the influence of external noise. Therefore, the image processing method is used to complete the video image preprocessing operation, thereby improving the robustness of information detection, and enhancing feature extraction and image recognition. Accuracy.

The contrast is improved by the histogram equalization method. The method is to map the gray level of the original image evenly in the entire gray level range, so that an image with uniform gray level distribution can be obtained. The method of histogram equalization is used to adjust the local contrast of the video image to achieve image enhancement without affecting the overall contrast.

Due to the interference of the image acquisition equipment and the driving environment, the clarity of the image is reduced and the picture becomes blurred. The median filtering method is used to complete the denoising of the video image. The median value of this point is used to replace the value of the point itself, so that the surrounding pixel value is close to the real value of the pixel, so that the isolated noise points can be eliminated and the details of the image can be preserved to a greater extent.

3   ResNet-SSD Face Detection Algorithm

3.1   ResNet10-SSD Model Structure

The SSD target extraction and detection network framework [9] and the trained ResNet10 face detection model of OpenCV and its training weight are used as the basic network for face detection. The SSD model structure is shown in Fig. 1 below.

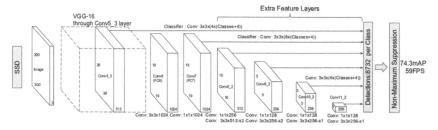

Fig. 1. SSD model structure.

SSD, also known as single shot multi box detector, has the advantages of quickly detection speed and high detection precision. Its backbone network feature network is vgg16 [10] to collect the details of the input image, convert FC6 and FC7 into convolution layers, detach all dropout layers and FC8 layers, and add four convolution layers: conv8, conv9, conv10 and conv11. Using conv4_3, FC7, Conv8_2, Conv9_2, Conv10_ 2 and conv11_2. These six improved feature layer models extract the features of the input picture and generate multiple feature maps. Finally, the results are given by using the non maximum suppression NMS (non maximum suppression) algorithm combined with the information of multi-layer feature map. For the acquired features, a priori frame matching mechanism is adopted to match the real frame. The scale of its priori frame follows the linear increasing rule, as shown in formula (1).

$$S_k = S_{min} + \frac{S_{max} - S_{min}}{m - 1}(k - 1), k \in [1, m] \tag{1}$$

where m represents the number of feature maps, and the values of S_{min} and S_{max} are 0.2 and 0.9 respectively, indicating the proportion of the bottom feature map and the top feature map in the original image. After obtaining the prior frame size, determine the width and height of the prior frame according to Eqs. (2) and (3).

$$W_k^\alpha = S_k \sqrt{\alpha_\gamma} \tag{2}$$

$$h_k^\alpha = S_k \sqrt{\alpha_\gamma} \tag{3}$$

W_k^α and h_k^α respectively represent the height of the a priori frame, and s_k are the dimensions of the a priori frame, $\alpha_k \in \{1, 2, 3, \frac{1}{2}, \frac{1}{3}\}$.

3.2 Analysis of Model Test Results

Through the algorithm results, it is found that SSD network can better detect the face under the conditions of front, side and occlusion. The front detection result of the driver are shown in Fig. 2.

Fig. 2. Driver front detection result diagram.

The detection results of the driver's side face are shown in Fig. 3.

Fig. 3. Driver's side face detection result diagram.

Driver's detection result diagram under face occlusion in Fig. 4.

Fig. 4. Face occlusion detection results.

4 Dlib Target Tracking

Experiments show that the algorithm based on ResNet10-ssd has good detection accuracy, but the detection speed of the model is low and unable to achieve the real-time effect. With a view to improving the efficiency of face detection, Dlib tracking algorithm is introduced.

Using the Dlib tracking method studied by Martin et al. [11], when tracking the target, first create a tracking class and use Dlib. Correlation_ Tracker() function; At start_ track Set the face frame to be tracked in the video in track; Track the next frame through the update() function; Finally, get_ position() function obtains the position of the tracked face in the new frame.

Dlib tracker is based on Mosse tracker. The size of Mosse tracker may change during target tracking, resulting in poor tracking effect. Based on this, Dlib adopts "scale pyramid" method to estimate the proportion of target position more accurately.

In Dlib tracker, the target position is located in the video image through the correlation filter. Firstly, one-dimensional filter is used to estimate the size of the target. Secondly, two-dimensional filter is used to translate the target. Finally, three-dimensional filter is used to complete the specific position of the target in space. Taking the feature picture of dimension as the signal, the face rectangular box extracted from the feature picture is, and the rectangular boxes of different dimensions are used to represent, in which. Each feature dimension contains a correlation filter. An optimal correlation filter is found by minimizing the square error, as shown in Eq. (4).

$$\varepsilon = \left\| \sum_{l=1}^{d} h^l \times f^l - g \right\|^2 + \lambda \sum_{l=1}^{d} \left\| h^l \right\|^2 \tag{4}$$

where g and f is the expected output associated with, and the parameter $\lambda (\lambda > 0)$ is used to control the influence of regularization. Equation (4) is applicable to one training sample. Solve the above equation to obtain Eq. (5).

$$H^l = \frac{\overline{G} F^l}{\sum_{k=1}^{d} \overline{F^k} F^k + \lambda} \tag{5}$$

By minimizing the output errors of all training blocks, the optimal correlation filter can be obtained. However, in this case, the linear equations of each pixel need to be solved, which undoubtedly increases the amount of calculation. In order to obtain a robust approximation, the molecules and denominators in formula (5) are optimized respectively, expressed by A_t^l and B_t respectively.

$$A_t^l = (1 - \eta) A_{t-1}^l + \eta \overline{G_t} F_t^l \tag{6}$$

$$B_t = (1 - \eta) B_{t-1} + \eta \sum_{k=1}^{d} \overline{F_t^k} F_t^k \tag{7}$$

where η represents the learning rate parameter. Use Eq. (8) to calculate the correlation score y at the rectangular box of the feature map, and then find the new motion state of the target through maximizing the correlation score y.

$$y = \mathcal{F}^{-1} \left\{ \frac{\sum_{l=1}^{d} \overline{A^l} Z^l}{B + \lambda} \right\} \tag{8}$$

Mosse tracker is limited to estimating the translation of the target, which means that the tracking effect of the tracker is not very good in the case of obvious scale transformation. In the target tracking scene, Dlib tracker locates the current frame by means of the position filter and estimates the target size of the current frame through the scale filter, so as to complete the target tracking and scale transformation. In addition, because the position filter and scale filter are relatively independent, different feature types and calculation methods can be selected during training and testing, so as that the robustness of tracking effect can be improved.

5 Algorithm Fusion

The fusion of face detection algorithm based on ResNet10-ssd and Dlib tracking algorithm makes face region location more accurate and efficient. The fusion algorithm flow is shown in Fig. 5.

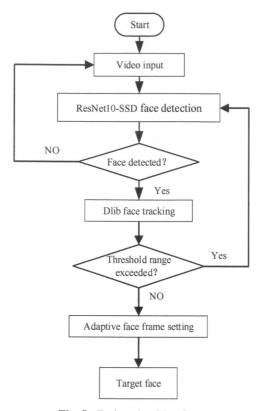

Fig. 5. Fusion algorithm flow.

When the algorithm is used to detect the face, the system will judge whether the face is detected. If the face is detected, the system will track the face through Dlib tracking algorithm. During tracking, it will judge whether the face area rises above the face confidence threshold value setting. If it exceeds, it will go back to the face detection part to detect the face once more. If not, the tracked facial frame will be transmitted to the next frame of image as the target face for subsequent detection. If no face is detected, that is to say that valid face information may not exist in the video, so it is necessary to return the video to detect it again.

Table 1 lists the comparison results of detecting ratio by algorithm. Based on the comparison results, can see taht the fusion algorithm can ensure a better test results, but also greatly reduce the time of detection and improve the real-time performance of system detection.

Compared with KCF, Boosting, CSRT, TLD, MIL and Mosse, it is found that the Dlib method can track the face well in the case of occlusion and side face, and the tracking success rate is close to 1; When using the methods of Boosting, CSRT and TLD, the face can be tracked, but the detection speed is far lower than that of Dlib method; When using MTL, KCF and Mosse methods for tracking, when the driver is pitching, head

Table 1. Algorithm detection speed comparison

Method	FPS/(f s^{-1})
ResNet10-SSD	23.0
Fusion algorithm	40.0

tilting or simulating the driver's side head turning out of the window and returning to normal driving, these three methods will lose tracking.

6 Conclusion

This paper uses ResNet10-SSD and Dlib tracking method to detect as well as track the face. From the experimental results, we can see that the face detection algorithm could detect the face well in the case of occlusion and side face, and the tracking algorithm can improve the detection speed while ensuring the detection accuracy. For further verifying the performance of the algorithm, compared with the six tracking methods of KCF, Boosting, CSRT, TLD, MIL and Mosse, this algorithm has better tracking effect than the first six methods., which provides great convenience for the next step of fatigue driving detection.

References

1. Tefft, B.C.: Acute Sleep Deprivation and Risk of Motor Vehicle Crash Involvement. AAA Foundation for Traffic Safety, Washington DC (2016)
2. Jia, H., Liu, Z., Wang, P., Zhou, G.: Research on fatigue driving identification method based on human vehicle characteristics. Chinese Sci. Tech. Paper **11**(07, 751–753+759 (2016)
3. Czermiński, R., Yasri, A., Hartsough, D.: Use of support vector machine in pattern classification: application to QSAR studies. QSAR Comb. Sci. **20**(3), 227–240 (2015)
4. Huo, X.: Research on Fatigue Driving Detection Based on EEG and Forehead Eye Fusion. Shanghai Jiaotong University (2017)
5. Baek, J.W., Han, B.G., Kim, K.J., et al.: Real-time drowsiness detection algorithm for driver state monitoring systems. In: Proceedings of the 2018 Tenth International Conference on Ubiquitous and Future Networks (ICUFN 2018), July 3–6, 2018, Prague, Czech Republic, IEEE, pp. 73–75 (2018)
6. Sha, C., Li, R., Zhang, M.: Research on fatigue driving detection based on steering wheel grip strength. Sci. Technol. Eng. **16**(30), 299–304 (2016)
7. Wu, K., Yu, S., Xu, W., Fan, H.: Research on fatigue driving simulation based on BFR algorithm and head posture. Comp. Simul. **38**(08), 152–156+268 (2021)
8. Feng, X.: Fatigue driving detection method based on fusion of eye, mouth and head features. J. Safe. Environ. **22**(01), 263–270 (2022). https://doi.org/10.13637/j.issn.1009-6094.20201232
9. Liu, W., Anguelov, D., Erhan, D., et al.: SSD: single shot multibox detector. In: Proceeding of the 2016 European Conference on Computer Vision. Springer, Amsterdam, pp. 21–37 (2016)
10. Simonyan, K., Zisserman, A.: Very deep convolutional networks for large-scale image recognition (2014). arXiv preprint arXiv:1409.1556
11. Danelljan, M., Hager, G., Khan, F.S., et al.: Accurate scale estimation for robust visual tracking. In: British Machine Vision Conference (2014)

Algorithm Application Based on YOLOX Model in Health Monitoring of Elderly Living Alone

Zhe Li🆔 and Jing Dang$^{(\boxtimes)}$ 🆔

School of Electronic Engineering, Xi'an University of Posts and Telecommunications,
Xi'an 710061, China
xytx03@xupt.edu.cn, djnb@stu.xupt.edu.cn, 759636570@qq.com

Abstract. Due to the outbreak of the COVID-19 novel coronavirus, the restrictions on population entry and exit have resulted in most elderly people staying at home alone, causing them a lot of inconvenience. Aiming at the problem that the elderly living alone at home may cause various diseases due to negative emotions but cannot be detected and solved in time, a method of detecting the facial expressions of the elderly is proposed to determine whether the elderly need timely care. YOLOX is the latest generation of YOLO series target detectors released by Megvii Technology in July 2021. It adopts the latest technology in the industry in recent years and surpasses existing similar products in performance and accuracy. If the YOLOX detector can be applied to the health monitoring of the elderly living alone under the current epidemic situation, it will be of great significance to improve detection rate and accuracy of detection and reduce labor costs.

Keywords: Elderly living alone · YOLOX · Expression recognition · Deep learning

1 Introduction

With the development of social economy and the improvement of medical level, the phenomenon of population aging and "empty nesting" of the elderly population has followed, and real-time care for the emotional state of empty nest elderly people and timely detection and treatment of physical abnormalities of empty nest elderly people have become hot spots and difficulties in social research. The outbreak of the new coronavirus in 2020, so that most of the elderly living alone are trapped in their homes, bringing them a lot of inconvenience, according to medical and psychological research, if people are in a negative attitude for a long time, it is easy to cause various diseases of the body and affect their health. In particular, for the empty nest elderly who have low immunity and cannot get the careful care of their partners and children, they will not only be in negative emotions for a long time, but also cause various elderly diseases due to weak resistance, and even face the occurrence of disease pain, and fail to be found and treated by their families in time, resulting in death and many other huge risks.

Based on the above background, it is of great significance to monitor the facial expressions of the elderly living alone in the lens in real-time monitoring through the

© The Author(s), under exclusive license to Springer Nature Switzerland AG 2023
N. Xiong et al. (Eds.): ICNC-FSKD 2022, LNDECT 153, pp. 568–574, 2023.
https://doi.org/10.1007/978-3-031-20738-9_65

camera. YOLO (You Only Look Once) series of algorithms is an algorithm that uses neural networks to provide real-time object detection, due to the speed and accuracy of the algorithm itself, once released, YOLOX is the latest generation of Megvii technology released in July 2021 YOLO series of object detectors, compared with other members of the series, with better performance and better balance of speed and accuracy, has a broader application prospect for face detection of the elderly living alone.

2   Research Background

2.1   Population Aging and the Current Situation of Empty Nesters

According to the data in the 2021 China Statistical Yearbook, in the seven national censuses between 1953 and 2020, the number of elderly people over the age of 65 in China is increasing year by year, and the change in the number of elderly people is shown in Table 1. It is predicted that the world's elderly population will reach 20.2 billion in 2050, of which China's elderly population will reach 480 million, accounting for about a quarter of the world's elderly population, when the elderly population will move to a new level.

Table 1. Shows the proportion of the elderly population in the seven national censuses

Index	1953	1964	1982	1990	2000	2010	2020
0–14 years old	36.28	40.69	33.59	27.69	22.89	16.60	17.95
15–64 years old	59.31	55.75	61.50	66.74	70.15	74.53	68.55
65 years old+	**4.41**	**3.56**	**4.91**	**5.57**	**6.96**	**8.87**	**13.50**

2.2   Emoticon Recognition

With the development of machine learning and deep neural networks and the popularization of smart devices, face recognition technology is undergoing unprecedented development. At present, the accuracy of face recognition has exceeded the human eye, and the basic conditions for large-scale popularization of software and hardware have also been met, the application market and field demand are very large, and the market development and specific applications based on this technology are showing a vigorous development trend. As an important part of face recognition technology, facial expression recognition (FER) has received widespread attention in the fields of human-computer interaction, robot manufacturing, medical treatment, driving, safety, automation and other fields, and has become a research hotspot in academia and industry. Facial expressions are the result of one or more movements or states of facial muscles that express an individual's emotional state toward the observer, are part of the body's language, and are a psychological and physiological response. Among them, abnormal facial expressions can intuitively reflect the current physical reaction and emotional state of the elderly, and the identification of the expression of the elderly can timely determine whether the elderly need timely care, and prevent the empty nest elderly from living alone due to physical problems and cannot be rescued in time.

3   YOLOX

3.1   Improvements to YOLOX

YOLO series algorithm detection of targets is very fast, YOLO v3 in the v2 made further optimizations, such as Darknet53, multi-scale prediction, cross-scale feature fusion, etc., but still has a low recall, positioning accuracy needs to be optimized, crowded target detection effect is not good waiting for improvement points, so the most widely used YOLO v3-SPP as the basis, and finally got a new generation of YOLOX detector. The main improvements are: using understanding coupling, Anchor-free, and brute force data enhancements.

Decoupled Head, Decoupling Head. Using the enhanced feature extraction network can obtain three effective feature layers, first of all, the input of the feature layer using convolutional standardization plus activation function for feature integration, and then the prediction is divided into classification and regression two parts, the classification part of the use of two convolutional standardization plus activation function to determine the type of object belongs, the regression part of the same operation to obtain the regression prediction results and whether the feature point has a corresponding object. Experiments have shown that decoupling the head is computationally inexpensive and effectively improves the accuracy and speed of the model.

Anchor-free, no anchor frame. The Anchor-based mechanism brings about the general problem of specific anchor frames facing different data sets, increases the complexity of detecting heads and the number of parameter generations, and improves the computational complexity and logic complexity. The Anchor-free mechanism can significantly reduce the number of parameters and related step techniques, making the detector's work in the training and decoding stages quite simple, and can obtain the performance of the anchor frame mechanism.

Strong data augmentation, strong data enhancement. Data enhancement methods widely used in other detectors such as YOLO v4 and v5 are used: Mosaic and the modified MixUp data enhancement strategy.

YOLOX achieves a better trade-off between speed and accuracy over all model sizes than other corresponding object detection algorithms.

3.2   Performance of YOLOX

YOLOX gets its best results on most models. It applies some of the industry's latest and most advanced technologies to detectors.

Training 300epoch on the COCO training set and performing a 5epoch warm-up, YOLOX-DarkNet53 increased the best performance of the COCO dataset from 44.3% AP of YOLOOv3 to 47.3% AP, and the optimal performance increased to 50.0% AP at a resolution of 640*640, which was 1.8% higher than that of YOLOV5-L. YOLOX-Tiny and YOLOX-Nano (Only 0.91M parameters and 1.08G FLOPs) are 10% APs and 1.8% more AP than their counterparts YOLoOv4-Tiny and NanoDets, respectively.

This shows that even in very small-scale models, YOLOX still has a very bright performance. Compared with the YOLOv3, YOLOv4, and YOLOv5 series, YOLOX improves optimal performance and optimizes inference speed (Fig. 1).

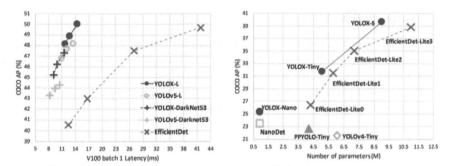

Fig. 1. YOLOX compares performance to the rest of the detectors

4 Experimental Results

4.1 Dataset Selection

The selected dataset is the open data set of the elderly face generated by style-GAN, and on the basis of the dataset, the image is enlarged by rotating, flipping, cropping and panning, and a total of 1430 images of the elderly face are obtained as positive samples, and the landscape photos taken during the trip are taken as negative samples, a total of 2200 images. Set the seven expressions as follows: angry, disgust, fear, happy, sad, surprised, and normal, and define happy and normal as natural expressions, and the rest as unnatural expressions. When an unnatural expression is recognized, an early warning begins.

4.2 YOLOX Training and Optimization

Using mobiletv2 lightweight convolutional neural network transfer learning, using deep-wise separable convolution to reduce the amount of arguments and the amount of operations, and using shortcut network, increasing the number of channels, getting more features, using Linear to prevent Relu breaking features. Usually, in the residual block, the feature is first compressed and then convoluted and finally expanded, while mobiletv2 expands first and then convolutes feature and finally compresses.

After the transfer learning ends, through the label sequence number corresponding to the class name, neutral = 0, disgust = 1, fear = 2, happy = 3, sad = 4, surprise = 5, angry = 6, define the prediction function to adjust the picture size, first store the picture in images, adjust the size of the picture to 224 * 224, the image is converted to tensor type, and adjust the input network image size, through GPU acceleration, call the trained mobiletv2 model, and pass it in, Returns the maximum probability value and subscript, followed by the class name, label, and display of the result.

In the predict_photo.py, add print ('The number of people currently detected is greater than 1, not detected'), so faces are not detected when two or more people appear. In predict_video.py, define capture = cv2. Video Capture (0) reads the path of the camera, reads one of the frames and performs a format shift, from BGR to RGB to meet the OpenCV format, performs detection, sets fill = (255, 0, 0), detects the red box of the face.

4.3 Experimental Results

The processed dataset of elderly faces performed very well on the YOLOX model, and the results were shown in the figure, and the test results could be obtained quickly. Under the camera that comes with the computer, the face can be detected in real time, and the accurate expression category can be obtained. The detection frame rate is about 25.0fps, and the confidence level is above 0.91 and above, and the result is shown in Figs. 2, 3 and 4.

Fig. 2. Seven classification expression recognition corresponding serial numbers

5   Conclusion

In this paper, an algorithm to improve the YOLOX model is proposed to detect and identify the facial expressions of the elderly, and the experimental results show that the facial features of the elderly can be detected by the face detection algorithm, and the recognized expressions can be classified. The improved model slightly increases the model size, but the accuracy and recognition time are better.

Face expression recognition has been widely used in many fields such as human-computer interaction, security monitoring, artificial intelligence, and social entertainment, and has huge development and application space. Although the current face expression recognition research has achieved more results, but the research on the elderly group is not enough, the accuracy rate obtained is not high, the effectiveness and robustness of the algorithm is not good enough, and there are still certain challenges from meeting the requirements of practical application.

Fig. 3. Static image expression recognition

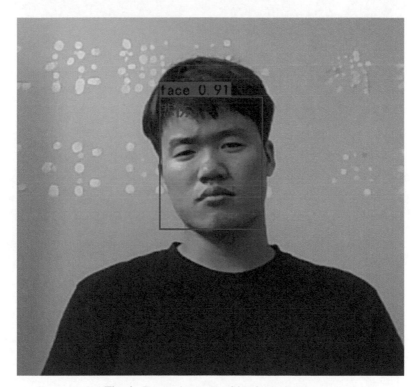

Fig. 4. Expression recognition in the camera

References

1. Ge, Z., Liu, S., Wang, F.: YOLOX: Exceeding YOLO Series in 2021 (2021)
2. Pengxiang, J., Dan, L.: Application of YOLOX-based model in face recognition under the background of epidemic. Yangtze River Inform. Commun. **35**(01), 1–2+6 (2022)
3. Min, C., Guanmao, W.: Detection and recognition of traffic signs based on improved YOLIX. Mod. Inform. Technol. **6**(02), 101–103+106 (2022)
4. Jianfei, Z., Sai, K.: Research on improving YOLOX fire scene detection method. Comp. Dig. Eng. **50**(02), 318–322+349 (2022)
5. Xinyue, Z.: The current situation characteristics and development trend of population aging in China and its countermeasures. China Manage. Informat. **23**(05), 195–199 (2020)
6. Shuo, L., Xuan, C., Rui, F.: YOLOv3 network based on improved loss function. Comp. Syst. Appl. **28**(02), 3–9 (2019)
7. Kaihan, L.: Research on Face Detection and Segmentation Method Based on Deep Learning. Guangdong Normal University of Technology (2020)
8. Xugan, S., Xiaoming, Z., Shiqing, Z.: New progress in face expression recognition. Labor. Res. Explor. **33**(10), 103–107 (2014)

Abdominal Multi-organ Localization
with Adaptive Random Forest in CT Images

Ruihao Wang, Jiaxin Tan, Laquan Li, and Shenhai Zheng$^{(\boxtimes)}$

Chongqing University of Posts and Telecommunications, Chongqing 400065, China
{S200231031,S200231216}@stu.cqupt.edu.cn, {lilq,
zhengsh}@cqupt.edu.cn

Abstract. Medical image localization plays an important role in digital medical research, therapy planning, and delivery. However, the presence of noise and low contrast renders automatic abdominal multi-organ localization an extremely challenging task. In this study, we focus on an adaptive weighted random forest method for abdominal organ localization. Different from the traditional random forest, the proposed method first trains the weighted random regression forest, then iterates multiple weighted random forests to form a stronger one with the Adaptive Boosting technique, and finally performs the final regression through the random forest regressor to get the final bounding box. Our adaptive random forest algorithm can efficiently realize the direct mapping from voxel to organ position and size. Through the quantitative verification of CT scanning data, our method has higher accuracy than the original random forest method and AdaBoost method.

Keywords: Multi-organ localization · Adaptive random forest · CT images

1 Introduction

Medical CT image detection and location is the premise of many medical image analysis tasks [1]. As the key data obtained by patients after examination, medical CT image [2] plays a decisive role in analyzing the health status of organs. Medical image processing is the precursor of accurate organ positioning and medical diagnosis. Medical CT images may be obtained after the whole scanning of the human body, or they may only be targeted at a certain part. However, in most cases, medical image slices include many tissues and organs. Figure 1 (green is the liver, blue is the pancreas, yellow is the kidney and red is the spleen) shows four abdominal CT images, in which tissues and organs are densely distributed. For medical analysis or processing of certain organs, one needs to find out the regional position of the organ in the image first.

In practice, clinicians still need to manually observe the location and lesions of various organs of patients, while the intelligent positioning and diagnosis services of CT medical images are slightly weak. So, medical image multi-organ localization automatically plays a very important role in medical image processing [3].

© The Author(s), under exclusive license to Springer Nature Switzerland AG 2023
N. Xiong et al. (Eds.): ICNC-FSKD 2022, LNDECT 153, pp. 575–583, 2023.
https://doi.org/10.1007/978-3-031-20738-9_66

However, there is a lot of noise in the 3D-dimensional medical CT image, the morphology of each organ is different, and the intensity between organs and tissues is similar, which leads to the misleading of different organs and the presence of lesions in organs, such as large tumors, liver cirrhosis or scars left after partial hepatectomy.

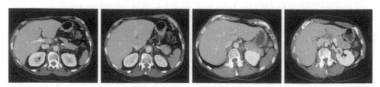

Fig. 1. Samples of healthy abdominal CT images arising from the AbdomenCT-1K dataset, provided with ground truth organ delineations.

Therefore, these render automatic abdominal multi-organ localization an extremely challenging task [4] in CT images. In this paper, we study an adaptive weighted random forest method for abdominal multi-organ localization with a 3D bounding box to palliate the drawbacks of the original random forest. Our adaptive random forest model realizes rough positioning first and then precise positioning. First, we train the weighted random regression forest, then iterate the weighted random forest, and finally realize the frame regression through the random forest regression. Our adaptive random forest can learn a nonlinear mapping from the input feature to the output 3D bounding box for organ positioning. Experiments show that our adaptive random forest multi-organ location method is more robust and accurate than the original random forest algorithm.

2 Related Work

In order to achieve more accurate organ localization and detection, many scholars have studied different methods according to the characteristics of medical CT images. The methods mainly include the registration method, marker point detection method, supervised classification method, regression method, and deep learning method.

By registering the template containing the target organ [5] into the test image, traversing and exhaustively searching all positions of the whole image for registration. Another method is supervised classification, which can better consider the differences between individuals. Most detection-based methods establish a classifier and input local features into the classifier to judge the organ to which the voxel belongs [6]. Luchao of Siemens also adopts a similar idea [7]. However, only considering the local features loses the global context information. To solve this problem, Criminisi and others use the generalized Haar-feature to model the context information [8]. Compared with classifying each voxel, some scholars believe that finding a vector parameter to describe the position of the organ can be used to describe the surface of the organ or a simpler bounding box. Then, Criminisi [9] used the random regression forest method in machine learning to model the multi-organ location and transformed the multi-organ location problem into a regression model for predicting the six values of the boundary bounding box, which has been improved in time and accuracy. Subsequently, Gauriau [10] improved the positioning effect by cascading the initial random regression forest model. In recent years, with the advent of deep learning [11], the performance of many tasks of image processing has been improved. For example, Humpire-Mamani [12, 13] used three convolutional neural networks to predict the position of organs. Compared with the method of de Vos [14], their convolutional neural network uses several adjacent slices to predict the organ distribution location of the central slice. However, deep learning needs big data and powerful hardware resources, high training time cost, and poor interpretability of the model.

Based on the above research, in order to avoid the shortcomings of traditional random forests in the multi-organ localization of medical images, an efficient adaptive random forest multi-organ localization model is proposed in this paper. The proposed method combines the enhanced random forest by optimizing the weight of all random forest decision trees and iterating multiple random regression forests with the Adaptive Boosting technique and then uses the random forest regression to carry out the final frame return to improve the effect of multi-organ localization. The experiment shows that the adaptive random regression forest model greatly enhances the regression performance and achieves a better generalization effect.

3   Materials and Methods

3.1   Dataset

In this work, we conducted an experiment on an abdominal clinical CT dataset (AbdomenCT-1K). In this dataset, the resolution of the CT scan is 512×512, any size of spacing, layer thickness between 1.25 and 5 mm, CT is obtained by GE multidetector spiral. The dataset contains annotations for all four organs in each dataset, including the liver, kidney (left and right), pancreas, and spleen. We have the manually labeled segmentation labels of these cases. The coordinate value of the 3D bounding box is obtained by image slice scanning of segmented cases, which is used for the label of multi-organ localization. In order to demonstrate the generalization ability of the model, 80 cases of organs of different sizes and shapes were selected for verification, including 60 cases for training and 20 cases for testing.

3.2   Preprocessing

All medical CT images are adjusted to the dimension of $100 \times 512 \times 512$, and the Hu value of voxels is truncated to -200 to 200. At the same time, we also adjust the window of these medical images to make the Hu value of voxels better reflect the characteristics of various organs. Noise in medical CT images often affects the detection and positioning effect of abdominal multiple organs. In this paper, the BM3D [15] denoising algorithm is used to smooth and denoise all medical CT images.

3.3   Feature Extraction

In order to extract efficient and robust features, this paper combines 3D Haar-like feature, HOG feature, context feature, image moment feature, and gray feature. Experiments show that this combined feature can bring a better multi-organ localization effect to our adaptive random forest model.

Haar-like feature. We extend 2D Haar-like features [16] to 3D organ detection. This feature can better reflect local features, edge features, and central features. In the calculation of 3D Haar-like features, given the origin of the coordinate system, the integral value at position (x, y, t) includes the sum of pixels whose position index is less than the current position. As shown in Eq. (1).

$$h(x, y, t) = \sum_{x' \leq x, y' \leq y, t' \leq t} i(x', y', t'), \tag{1}$$

where $h(x, y, t)$ is the integral body. By the integral volume, any triple sum can be calculated in eight array references. Since two cuboid cubic filters involve adjacent cuboids, they can be calculated in twelve array references. Because of the fast addition and subtraction operation speed, the feature extraction can be processed quickly.

HOG feature. It has good detection and recognition effect in pedestrian detection. Therefore, this feature can better describe the gradient information of local area of 3D object. In CT images, local gradient information is of great significance for organ detection. It can often better characterize the contour and morphological information of organs. In Eqs. (2)–(5), $I(x, y, z)$ Represents the pixel value of a point in the image, $G(x, y, z)$ Represents directional gradient information.

$$G_x(x, y, z) = I(x + 1, y, z) - I(x - 1, y, z), \tag{2}$$

$$G_y(x, y, z) = I(x, y + 1, z) - I(x, y - 1, z), \tag{3}$$

$$G_z(x, y, z) = I(x, y, z + 1) - I(x, y, z - 1), \tag{4}$$

$$G(x, y, z) = \sqrt{G_x(x, y, z)^2 + G_y(x, y, z)^2 + G_z(x, y, z)^2}, \tag{5}$$

Context feature. At the same time, we take the regular voxel pixel values on the rays from the central point of the 3D-dimensional medical image to the eight vertices as the spatial context feature, which can detect the approximate distribution trend of each organ to a certain extent.

Image moment feature. The first moment feature is the average intensity of each color, and the second moment reflects the variance of regional color, the first-order moment and second-order moment features of these sub regions are calculated by regularly sampling the sub regions in the 3D-dimensional medical image. Equation (6) is the calculation process of the first-order moment and the second-order moment. P_{ij} represents the pixel information of the image.

$$F_1 = \frac{1}{N} \sum_{j=1}^{N} P_{ij}, \quad F_2 = \sqrt{\frac{1}{N} \sum_{j=1}^{N} (P_{ij} - F_1)^2}, \tag{6}$$

Gray feature. Extract the gray matrix of the central slice, and the statistical values of the matrix include energy, entropy and contrast. Energy can reflect the uniformity of image gray distribution and a certain degree of texture. Entropy can represent the complexity of an image. Contrast can reflect the clarity of the image. In the following formula, *ASM* represents energy, *ENT* represents entropy, *CON* represents contrast, $P(i,j)$ represents image pixel value.

$$ASM = \sum_i \sum_i P(i,j)^2, \tag{7}$$

$$ENT = -\sum_i \sum_j P(i,j) \log(P(i,j)), \tag{8}$$

$$CON = \sum_i \sum_j (i-j)^2 P(i,j), \tag{9}$$

3.4 Methods

As we all know, the random forest model is integrated by multiple decision trees in the way of bagging. Different decision trees are constructed by randomly sampling samples and randomly selecting features. A small number of decision trees may perform poorly. At this time, other decision trees perform relatively well, which can reduce the impact of these "failed" decision trees when the final voting results.

Theoretically, it is possible to train a random forest classifier to judge which organ and category each voxel belongs to. In fact, a medical image has tens of millions of voxel points, which will take many costs to finish this task. Therefore, this paper does the multiple regression task of multi-organ location and applies the concept of "weight adaptation" to the random regression forest model, which greatly reduces the impact of some "failed" decision trees. Macroscopically, our adaptive random forest model realizes rough positioning first and then precise positioning. First, we train the weighted random regression forest, then iterate the weighted random forest, and finally realize the frame regression through the random forest regression. Our method can learn a nonlinear mapping from the input feature to the output 3D bounding box. Figure 2 shows the whole process of multi-organ localization of our adaptive random forest.

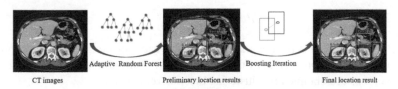

Fig. 2. The flow of abdominal organ detection

Weighted random forest. We first train the adaptive random forest. Figure 3 shows the algorithm flow of weighted random forest.

Specifically, the specific construction process of random forest is as follows: a) Using bootstrap method to select some samples from the training set. b) Some features

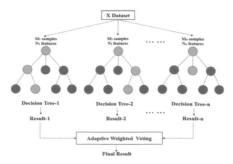

Fig. 3. The algorithm flow of weighted random forest

are randomly selected, and the minimum mean square error (MSE) is used to segment the nodes of the decision tree. c) Repeat the above steps repeatedly to get multiple decision trees. d) Gather multiple decision trees into a random forest, and use adaptive weighted voting to determine the final regression result.

In particular, this paper adopts the continuous form of information gain *Entropy(D)* and uses the form of differential entropy function $G(x, v)$ to replace the Shannon entropy of discrete probability distribution. The calculation method is as follows:

$$Entropy(D) = - \int_{y \in Y} p(y) \log(p(y)) dy, \tag{10}$$

$$G(x, v) = \frac{1}{N_s} \left(\sum_{y_i \in X_L} (y_i - y_L)^2 + \sum_{y_i \in X_R} (y_i - y_R)^2 \right), \tag{11}$$

Method explanation. Through the previously extracted features and obtained labels, we can train the basic random forest regressor. The model potentially finds that some voxel clusters can confidently predict the location of organs. These voxels are used as markers to locate these organs, and leaf nodes store the offset vectors from different voxel points to different organs, use the following formula to calculate the offset vector $d(v, c)$ between voxels and organs:

$$d(v, c) = v - b(c), \tag{12}$$

where c is organs, $b(c)$ is the bounding box vector, and v made from the voxel position (v_x, v_y, v_z), $d(v, c)$ represent the offset of voxels. At result, the location and size of the required organs are obtained by clustering these voxel clusters.

Each decision tree in the random forest obtains subspace samples through bootstrap. By randomly selecting features, the optimal splitting node is selected according to the mse of the features, and the left and right subtrees are divided. The split samples are stored in the left and right subtrees. Through continuous recursion, a decision tree will be generated finally and the leaf's nodes store different distributions of organs. At the test stage, when the new test image arrives, the model can predict the different distribution of different organs according to the extracted features and the learned model.

After the training stage, we trimmed the branches of the decision tree appropriately, Fig. 4 shows the distribution of all the features of importance and the difference in accuracy caused by the different number of decision trees in the training process of the random regression forest. In the experiments, 100 trees were trained in the proposed adaptive random regression forest detection.

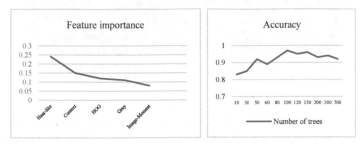

Fig. 4. The left image is a random forest built-in feature importance and the right image is different numbers of decision trees bring different accuracy.

Iterative optimization. The second step of the adaptive random forest model is to iterate the weighted random forest trained in the first step. In addition, this paper iterates the random forest regressor through the Adaptive Boosting technique, so as to learn a better regressor and achieve better generalization performance.

Bounding box regression. This process is also the last step of our adaptive random forest. We find the relationship between the prediction bounding box and the real box through the random regression forest. Due to the complexity of multiple organs, we combine the previously extracted robust features with the prediction bounding box as the input and the label of the ground-truth bounding box as the output, so as to improve the generalization performance of bounding box regression.

4 Experiment Results

To evaluate the performance of multi-organ localization, we calculate the widely reported metrics, i.e. the mean and standard deviation of the absolute wall distance (Wall dist) and centroid distance between the predicted 3D bounding box and the ground-truth 3D bounding box. Figure 5 shows some visualization results of the different organs. The red box represents the liver, the orange box represents the pancreas, the blue box represents the right kidney, the green box represents the left kidney, and the sky blue represents the spleen.

The quantitative experimental results are shown in Table 1. In this paper, five abdominal organs are located and detected, the average Wall dist value is optimized from 14.08 to 7.96 mm in the most primitive random forest. The experiment shows that this method has achieved certain results. Similarly, under the experiment of a small amount of data, our adaptive random forest multi-organ location method is more robust and accurate than the traditional random forest and boost method. The improvement in performance is due to our method will automatically learn the weight of the decision tree in the random

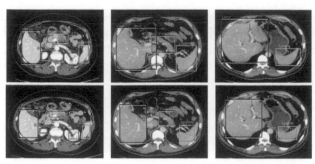

Fig. 5. Visualization results of the proposed detection method. The first row represents ground truth and the second row shows the detection results.

forest according to the diversity of CT organs to achieve a better generalization effect, Obviously, our method can greatly improve the positioning effect on five organs.

Table 1. Quantitative evaluation on four abdominal organs.

Methods	Liver	Spleen	Pancreas	L-kidney	R-kidney
Random forest	14.6	13.2	16.5	12.5	13.6
Adboost	12.3	13.6	17.3	11.7	12.8
Ours	8.6	7.5	9.8	7.2	6.7

5   Conclusion

In this paper, the adaptive random forest model is realized by improving the traditional random forest algorithm. Combined with the robust and efficient features, we can learn the nonlinear mapping between the features and the 3D bounding box of each organ, and we can predict the medical image we have never seen before. According to its characteristics, we can effectively locate the location of each organ. Experimental results show that this method has high detection accuracy and localization accuracy.

Acknowledgements. This work was supported in part by the National Natural Science Foundation of China (Grant Nos. 61902046 and 61901074) and the Science and Technology Research Program of Chongqing Municipal Education Commission (Grant Nos. KJQN201900631 and KJQN201900636) and China Postdoctoral Science Foundation (Grant No. 2021M693771) and Chongqing postgraduates innovation project (CYS21310).

References

1. Guan, H., Liu, M.: Domain adaptation for medical image analysis: a survey. In: IEEE Transactions on Biomedical Engineering, pp. 1173–1185 (2022)

2. Diwakar, M., Kumar, M.: A review on CT image noise and its denoising. In: Biomedical Signal Processing and Control, pp.73–88 (2018)
3. Tournier, J.D., Smith, R., Raffelt, D., et al.: MRtrix3: a fast, flexible and open software framework for medical image processing and visualisation. Neuroimage **202**, 116–137 (2019)
4. Fu, Y., Lei, Y., Wang, T., et al.: A review of deep learning based methods for medical image multi-organ segmentation. Phys. Med. **85**, 107–122 (2021)
5. Martin, S., Daanen, V., et al.: Atlas-based prostate segmentation using an hybrid registration. Int. J. Comp. Assis. Radiol. Surg. **3**(6), 485–492 (2008)
6. Montillo, A., Shotton, J., Winn, J., et al.: Entangled decision forests and their application for semantic segmentation of CT images. In: Biennial International Conference on Information Processing in Medical Imaging. Springer, Berlin, Heidelberg, pp. 184–196 (2011)
7. Lu, C., Zheng, Y., Birkbeck, N., Zhang, et al.: Precise segmentation of multiple organs in CT volumes using learning-based approach and information theory. In: International Conference on Medical Image Computing and Computer-Assisted Intervention. Springer, Berlin, Heidelberg, pp. 462–469 (2012)
8. Criminisi, A., Shotton, J., Bucciarelli, S.: Decision forests with long-range spatial context for organ localization in CT volumes. In: Medical Image Computing and Computer-Assisted Intervention (MICCAI), pp.69–80 (2009)
9. Criminisi, A., Robertson, D., Konukoglu, E., et al.: Regression forests for efficient anatomy detection and localization in computed tomography scans. Med. Image Anal. **17**(8), 1293–1303 (2013)
10. Gauriau, R., et al.: Multi-organ localization with clascaded global-to-local regression and shape prior. Med. Image Anal. **23**(1), 70–83 (2015)
11. Xu, X., Zhou, F., Liu, et al.: Efficient multiple organ localization in CT image using 3D region proposal network. IEEE Trans. Med. Imaging **38**(8), 1885–1898 (2019)
12. Mamani, G., Setio, A., Ginneken, B.V., et al.: Organ detection in thorax abdomen CT using multi-label convolutional neural networks. In: Medical Imaging 2017: Computer-Aided Diagnosis (2017)
13. Humpire-Mamani, G.E., Van Ginneken, B., et al.: Efficient organ localization using multi-label convolutional neural networks in thorax-abdomen CT scans. Phys. Med. Biol. **63**(8) (2018)
14. De Vos, B.D., Wolterink, J.M., De Jong, P.A., et al.: 2D image classification for 3D anatomy localization: employing deep convolutional neural networks. In: Medical Imaging 2016: Image Processing, pp. 517–523 (2016)
15. Zhao, T., Hoffman, J., et al.: Ultra-low-dose CT image denoising using modified BM3D scheme tailored to data statistics. Med. Phys. **46**(1), 190–198 (2019)
16. Adouani, A., Lachiri, Z., et al.: Comparison of Haar-like, HOG and LBP approaches for face detection in video sequences. In: 2019 16th International Multi-Conference on Systems, Signals and Devices (SSD), IEEE (2019)

An Indoor Floor Location Method Based on Minimum Received Signal Strength (RSS) Dynamic Compensation and Multi Label Classification

Mingzhi Han[✉] and Yongyi Mao

School of Electronic Engineering, Xi'an University of Posts and Telecommunications, Xi'an 710061, China
djnb@stu.xupt.edu.cn, 1610645735@qq.com, maoyongyi@263.net

Abstract. In high-rise buildings, floor identification is the premise and basis of indoor two-dimensional plane positioning. Accurate floor judgment can effectively reduce the search space in the matching stage. In order to eliminate the need of complex data pre/post-processing and less parameter adjustment in the traditional fingerprint algorithm, and also for the phenomenon of missing signal, this paper proposes an indoor floor positioning method based on the Received signal intensity (RSS) minimum dynamic compensation and multi-label classification. In the offline stage, complete the construction of fingerprint database; In the online stage, the RSS of the missing signal in the fingerprint is dynamically compensated, and then the feature space dimension is reduced by stacked auto-encoder (SAE) and the feed-forward classifier converts the multi label classification results into multi class classification results, so as to realize floor location. The suggested strategy outperforms previous deep neural network-based systems, according to experimental results using the UJIIndoorLoc data set, which estimate building and floor with 99.24 and 98.73% accuracy.

Keywords: Multi-floor indoor localization · dynamic compensation · Multi-label classification

1 Introduction

Due to the quick advancement and spread of mobile Internet technology, the positioning service industry is booming, and people's demand for location-based services (LBS) is increasing sharply [1]. For modern buildings, it is equally important to accurately locate the user's vertical position and horizontal position.

WLAN fingerprint identification [2, 3] is a promising and widely used indoor floor positioning method. Radar system [4] is the first fingerprint location system based on Wi-Fi. Bahl et al. Combined with RSS ranging and fingerprint location method, realized the location accuracy with a median error of 2~3 m. Then, an improved radar system [5] is proposed, and the positioning accuracy is improved by 33%. After the radar system,

© The Author(s), under exclusive license to Springer Nature Switzerland AG 2023
N. Xiong et al. (Eds.): ICNC-FSKD 2022, LNDECT 153, pp. 584–591, 2023.
https://doi.org/10.1007/978-3-031-20738-9_67

many Wi-Fi fingerprint positioning systems have emerged [6, 7]. The proposal of these positioning systems makes the WLAN Indoor Positioning Technology more and more perfect.

Fingerprint database is the key of fingerprint location. When establishing fingerprint database, too many wireless access points will reduce the efficiency of location. Traditional fingerprint algorithms [8] rely on labor-intensive manual parameter adjustments and complicated filtering, which makes them unsuitable for large-scale indoor environments and takes a lot of time. In recent years, researchers have adopted the method of deep learning for indoor location.

A deep neural network (DNN) model for classifying a building's floor was put forth in [9]. This effort has a 92% success rate. Multi-label classification was introduced in [10] as a solution to the scalability issue. The floor hit rate for this piece of work is 91.27%. In [11], a random forest filter and an SAE were used to minimize the number of dimensions in the data collection. Filtered data is categorized using four main classifiers, and the secondary classifier then predicts the class based on those four values, achieving a floor hit rate of 95.13%. A hierarchical recursive neural network is presented in [12], where the RNN predicts locations sequentially from a general to a specific one to realize floor location in multi-building and multi-floor situations.

At the same time, considering the difficulties of data collection in indoor environment and the impact of RSS missing phenomenon on fingerprint location, in [13], three different RSSI data enhancement methods based on Multi-Output Gaussian Process (MOGP) are studied for multi-building and multi-floor indoor location, and the correlation between RSSI observations of multiple access points (APS) is processed collectively. In [14], an indoor weighted k-nearest neighbor (WKNN) fingerprint location method based on RSS minimum dynamic compensation is proposed, which improves the location accuracy and stability to a certain extent.

In this paper, we propose an indoor floor location method based on minimum received signal strength (RSS) dynamic compensation and multi label classification, and use UJIIndoorLoc dataset [15] to evaluate its location performance.

2 Proposed Algorithm Architecture

2.1 A Subsection Sample

The main process of indoor floor location method based on minimum received signal strength dynamic compensation and multi label classification is as follows:

In the off-line stage, it is responsible for fingerprint data acquisition, data preprocessing, spatial location mapping and constructing fingerprint database; In the online stage, the RSS dynamic compensation of the missing signal in the fingerprint is carried out, and then the stacked automatic encoder is used to self-encode a certain amount of crowdsourcing data to obtain the data features hidden in the crowdsourcing data, so as to reduce the dimension of the feature space. Finally, the multi label classification results are transformed into multi class classification results through the feed forward classifier to realize the accurate positioning of the floor. The process is shown in Fig. 1.

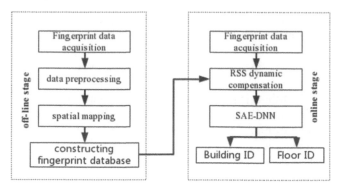

Fig. 1. Process of indoor floor location method

2.2 RSS Minimum Value Dynamic Compensation

In the offline stage, the fingerprint database is constructed as $D = (L_i, F_i)$, $i = 1, 2,..., n$, n is the number of reference points, L_i is the spatial position vector of the ith reference point, and F_i is the corresponding signal fingerprint matrix, which is called the reference fingerprint. As shown in Eq. (1), RSS_j represents the RSS mean value corresponding to MAC_j collected for many times, and m is the number of AP signals collected at the ith reference point.

$$F_i = \begin{bmatrix} MAC_1 & \overline{RSS_1} \\ MAC_2 & \overline{RSS_2} \\ \vdots & \vdots \\ MAC_m & \overline{RSS_m} \end{bmatrix} \tag{1}$$

The fingerprint data f collected in the online stage is called the test fingerprint. As shown in Eq. (2), mac and rss are used to represent the physical address and received signal strength respectively, and the number of AP signals collected is t.

$$f = \begin{bmatrix} mac_1 & rss_1 \\ mac_2 & rss_2 \\ \vdots & \vdots \\ mac_m & rss_m \end{bmatrix} \tag{2}$$

Let U_{mac} be the union of the physical addresses in the reference fingerprint and the test fingerprint, the number of physical addresses is u, and the value range is [max (m, t), m + t]; Rss_{min} is the minimum received signal strength of all APS in the reference fingerprint, and rss_{min} is the minimum received signal strength of all APS in the test fingerprint; Rss^{tp} is the received signal strength of the reference fingerprint participating in the matching and positioning calculation, and rss^{tp} is the received signal strength of the test fingerprint participating in the matching and positioning calculation. Traverse the physical address in U_{mac}. If the reference fingerprint contains the physical address, RSS^{tp} is used to represent the received signal strength corresponding to the physical address in the reference fingerprint, otherwise RSS^{tp} is assigned with rss_{min}; If the physical address is included in the test fingerprint, rss^{tp} is used to represent the received signal strength corresponding to the physical address in the test fingerprint, otherwise rss_{min} is used to assign rss^{tp}.

2.3 Network Architecture

Stacked automatic encoder is a kind of neural network. The self-coding network can be regarded as composed of two parts: encoder and decoder. The decoder of the network is only partially linked after the unsupervised learning of SAE weights is finished, as seen in Fig. 2, and the normal full connection layer of the deep network is connected to the encoder's output, which we refer to as a classifier.

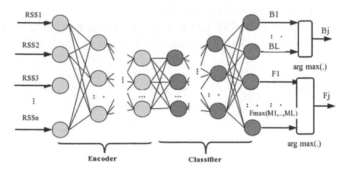

Fig. 2. Proposed network architecture based on DNN and SAE.

3   Experiment and Result Analysis

3.1   Experimental Data

The public fingerprint database given in reference [15] is used for simulation verification. Table 1 gives the relevant details of the three buildings, including the number of floors F of each building, the number of training samples Nf, the number of test samples Nt and the number of AP connected to each building Nap.

Table 1. Details of fingerprint data in three buildings

Building	F	Nf	Nt	Nap
T0	4	5249	536	200
T1	4	5196	307	207
T2	5	9492	268	203

The database includes 19937 training data and 1111 test data. The original verification set is used as the test set in the experiment. There are 520 different wireless access
points (WAPs), and the received signal strength of each wireless access point ranges
from negative integer −104 to 0 dBm.

In order to facilitate the subsequent normalization, using a fixed value to represent the
undetected WAP may lead to deviation in the distance similarity between fingerprints,
resulting in positioning error. Therefore, the RSS with missing signals in the fingerprint
is dynamically compensated.

3.2 Experimental Evaluation

To measure performance, we employ classification accuracy: We calculate hit rates for
buildings, floors, and buildings/floors, the latter of which only takes into account the
accurate identification of both building and floor IDs.

With the use of the gathered RSS dataset, Fig. 3 illustrates the training and validation
accuracy of the DNN-based indoor localization system's floor level position estimation.

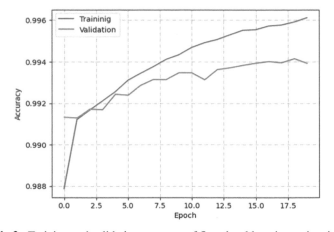

Fig.3. Training and validation accuracy of floor-level location estimation

As shown in the figure, after five epochs, the DNN architecture for building/floor
classification can achieve an accuracy of more than 99.2%. Through training and verification, we apply the system to the training data set of 520 fingerprints. The label is the
three IDs of the corresponding building, floor and reference point. We can see that this
is a multi-label classification problem.

For buildings, floors, reference point, we can use the pd.get_dummies function in
the pandas library to get the corresponding one-hot code, and then merge it into 118-bit
label, where the first three digits represent the building, the next five digits represent the
floor, and the back 110 digits are the reference point. So the label for all training data
is 19937 x 118. Then the arg max function returns the index of the maximum value of
each vector as a classification class.

Table 2 summarizes the hyper parametric values of the experiment. These parameter values are selected through the experiment and used in the whole experiment. We divide the training data into new training data and verification data in a ratio of 7:3, and take the verification data as test data to evaluate the performance of the model. The SAE neural network hidden layer is from 256 neurons to 64 neurons, and then from 64 neurons to 256 neurons. This design is to better learn the signal information of each layer of each building. At the same time, Adam optimizer is used for rapid learning. A total of 20 rounds of training and 10 training batches are trained. In order to prevent over fitting of the network during classification, dropout rate is 0.2.

Table 2. Hyper parameters

Parameter	Value
Ratio of training data to overall data	0.7
Number of epochs	20
Batch size	10
SAE hidden layers	256-128-64-128-256
SAE activation	Rectified Linear(Relu)
SAE optimizer	Adam
SAE loss	Mean squared error (Mse)
Classifier hidden layers	128–128
Classifier optimizer	Adam
Classifier loss	Binary_crossentropy
Classifier dropout rate	0.2

Table 3 compares our findings to those of other techniques using the same UJIIndoor-Loc data set. It is clear from Table 3 that our algorithm is more accurate in locating floors than the other four types of algorithms, with a 98.7% accuracy rate. The method suggested in this research exhibits excellent effectiveness and accuracy for floor positioning when compared to previous algorithms.

Table 3. The results are compared

Approach	Floor hit rate (%)
Proposed	98.73
DNN [10]	92.00
Scalable DNN [11]	91.27
RF + SAE + stacking [12]	95.13
RNN [13]	95.23

4 Conclusions

In this paper, we propose an indoor floor location method based on minimum received signal strength (RSS) dynamic compensation and multi-label classification. In the online positioning stage, the RSS of the missing signal in the fingerprint is dynamically compensated in order to reduce the impact of the missing signal on the positioning and improve the positioning accuracy and stability. Better scalability is obtained in the context of multi label classification employing the hierarchical nature of building/floor estimates and the label construction of the system.

The experimental results show that the proposed floor discrimination method can obtain high floor discrimination accuracy in multi floor indoor location scene.

Acknowledgements. This work is supported by Key R&D project of Shaanxi Province (2022GY-094), Research on indoor floor recognition and location technology based on sensor data fusion.

References

1. Dwiyasa, F., Lim, M.H.: A survey of problems and approaches in wireless based indoor positioning. In: International Conference on Indoor Positioning and Indoor Navigation, pp. 1–7 (2016)
2. Khalajmehrabadi, A., Gatsis, N., Akopian, D.: Modern WLAN fingerprinting indoor positioning methods and deployment challenges. IEEE Commun. Surv. Tutor. **19**(3), 1974–2002 (2017)
3. He, S., Chan, S.: Wi-Fi fingerprint-based indoor positioning: recent advances and comparisons. IEEE Commun. Surv. Tutor. **18**(1), 466–490 (2017)
4. Bahl, P., Padmanabhan, V.N.: RADAR: an in-building RF-based user location and tracking system. INFOCOM 2000. In: Nineteenth Annual Joint Conference of the IEEE Computer and Communications Societies, pp. 775–784. Tel Aviv, Israel (2000)
5. Bahl, P., Padmanabhan, V.N., Balachandran, A.: Enhancements to the RADAR user location and tracking system. Micros. Res. **2**(MSR-TR-2000–12), 775–784 (2000)
6. Youssef, M., Agrawala, A.: The Horus WLAN location determination system. Wireless Netw. **14**(3), 357–374 (2005)
7. Zhong, Y., Wu, F., Zhang, J., et al.: Wifi indoor localization based on k-means. In: 2016 International Conference on Audio, Language and Image Processing (ICALIP), pp. 663–667 (2016)
8. Ge, X., Qu, Z.: Optimization WIFI indoor positioning KNN algorithm location-based fingerprint. In: 2016 7th IEEE International Conference on Software Engineering and Service Science (ICSESS), pp. 135–137 (2016)
9. Nowicki, M., Wietrzykowski, J.: Low-effort place recognition with WiFi fingerprints using deep learning. In: International Conference Automation, Springer, Cham, pp. 575–584 (2017)
10. Kim, K.S., Lee, S., Huang, K.: A scalable deep neural network architecture for multi-building and multi-floor indoor localization based on Wi-Fi fingerprinting. Big Data Anal. **3**(1), 4 (2017)
11. JunLin, G., Xin, Z., HuaDeng, W., et al.: WiFi fingerprint positioning method based on fusion of autoencoder and stacking mode. In: 2020 International Conference on Culture-Oriented Science and Technology (ICCST), pp. 356–361 (2020)

12. Elesawi, A.E.A., Kim, K.S.: Hierarchical multi-building and multi-floor indoor localization based on recurrent neural networks. In: 2021 Ninth International Symposium on Computing and Networking Workshops (CANDARW), pp. 193–196 (2021)
13. Tang, Z., Li, S., Kim, K.S., et al.: Multi-output Gaussian Process-Based Data Augmentation for Multi-Building and Multi-Floor Indoor Localization. arXiv preprint arXiv:2202.01980 (2022)
14. Jingxue, B., Yunjia, W., Donghua, J., et al.: Indoor WKNN fingerprint localization method for RSS minimum dynamic compensation. Surv. Mapping Sci. **45**(12), 22–27 (2020)
15. Torres-Sospedra, J., Montoliu, R., Martinez-Uso, A., et al..: UJIIndoorLoc: a new multi-building and multi-floor database for WLAN fingerprint-based indoor localization problems. In: 2014 International Conference on Indoor Positioning and Indoor Navigation (IPIN), pp. 261–270 (2014)

An Improved Multi-dimensional Weighted K-nearest Neighbor Indoor Location Algorithm

Yongyi Mao and Rong Liu[✉]

Xi'an University of Posts and Telecommunications, Xi'an 710061, China
maoyongyi@263.net, djnb@stu.xupt.edu.cn, 1158989567@qq.com

Abstract. Aiming at the defects of low positioning accuracy and poor stability of the traditional WiFi location fingerprint indoor positioning method based on the K-nearest neighbor (KNN) algorithm, an improved multi-dimensional weighted K-nearest neighbor (MDW-KNN) indoor positioning algorithm is proposed in this paper. The algorithm improves the three basic elements of KNN algorithm: distance metric, the number of nearest neighbors K and decision-making method, introduces multi-dimensional weight coefficients based on similarity, fingerprint distance and physical distance to reduce the spatial ambiguity in localization. Moreover, the number of nearest neighbors K is dynamically changed by the method of range restriction, which improves the stability of the algorithm. Simulation using MATLAB software, experiments show that under the same test environment, the location accuracy of the MDW-KNN is nearly 46% higher than that of the traditional KNN algorithm, and the MDW-KNN's location estimation point trajectory is smooth, with good stability and fault tolerance.

Keywords: Indoor positioning · Location fingerprint · K-nearest neighbor (KNN) · Multi-dimensional weight coefficients

1 Introduction

With the continuous development of wireless communication technology and the continuous improvement of the performance of intelligent terminals, people's demand for location services is increasing rapidly. The increasing number of urban "canyons" also puts forward higher requirements for indoor positioning. The current mainstream indoor positioning technologies [1] include Ultra-Wide Bandwidth, Near Field Communication, WiFi, RFID, Bluetooth, etc. Due to the popularization of WiFi technology and the completeness of infrastructure, the indoor location method of fingerprint based on WiFi received signal strength indication (RSSI) [2] has become the most commonly used technical means at present.

The fingerprinting localization system is usually divided into two stages: offline stage and online stage. The main problem is the establishment of offline fingerprint database and the location matching and location of online fingerprint data. The traditional location fingerprint matching algorithms mainly include: K-nearest neighbor (KNN), artificial neural network (ANN) [3] and support vector machine (SVM) [4]. Based on the

N. Xiong et al. (Eds.): ICNC-FSKD 2022, LNDECT 153, pp. 592–599, 2023.
https://doi.org/10.1007/978-3-031-20738-9_68

consideration of algorithm complexity and accuracy, this paper uses KNN algorithm for location estimation. However, due to the uncertainty of online acquisition environment and spatial ambiguity, the traditional KNN positioning error fluctuates greatly, which can not meet the current requirements of positioning accuracy.

In recent years, researchers have proposed different improvement measures to improve the positioning accuracy of KNN algorithm. In order to measure the similarity difference between different reference points and the points to be located, the weighted K-nearest neighbor (WKNN) algorithm proposed by Brunato et al. [5] gives weight to the reference points according to the Euclidean distance of the fingerprint, which greatly reduces the positioning error of KNN, but only using the Euclidean distance as the weight increases the theoretical and empirical model conversion error. Therefore, He et al. [6] proposed to fuse the weight coefficients of different dimensions in Euclidean distance, but this method is not applicable to all indoor scenes. Considering the mobile ability of mobile users in indoor environment, Hoang et al. [7] proposed to fuse the previous positioning estimation point in Euclidean distance to reduce the spatial ambiguity of positioning. The positioning accuracy of this algorithm is greatly improved compared with WKNN, but the fault tolerance of the algorithm is poor. In the research on the K value of KNN algorithm, Li [8] proposed to dynamically change the number of nearest neighbors K, which has little effect and high requirements for the actual scene. In addition, in the decision algorithm of KNN, Guo [9] used the fingerprint mean to measure the contribution of the reference point to the location point. Tian [10] deduced a weight calculation method based on fingerprint timing features. Although the location accuracy is lower than that of Hoang's SRL-KNN algorithm, it provides more research ideas for the improvement of KNN.

All of the above methods provide acceptable accuracy, but the stability and fault tolerance of the algorithm are not well solved. In this paper, the matching algorithm and location estimation algorithm of fingerprint database are studied. Aiming at the defects of low positioning accuracy, poor stability and fault tolerance of the current KNN algorithm, a multi-dimensional weight coefficient is introduced to dynamically adjust the position estimation algorithm, so as to improve the positioning accuracy of WiFi location fingerprint indoor positioning.

2 Algorithm Description

2.1 The Classical KNN Algorithm

The KNN algorithm first matches the RSSI of the point to be located with the RSSI corresponding to each reference point in the fingerprint database, that is, calculate the fingerprint Euclidean distance between the point to be located and the reference point:

$$d_i = \sqrt{\sum_{j=1}^{n} \left(RSSI^j - rssi_i^j \right)^2} \tag{1}$$

where $RSSI^j$ is the signal strength of the j-th AP received at the point to be located, $rssi_i^j$ is the signal strength of the j-th AP received at the i-th reference point.

Then select the first K reference points with the smallest Euclidean distance, and calculate the average of their positions as the final positioning result. The positioning result is expressed as:

$$(x, y) = \frac{1}{k} \sum_{i=1}^{k} (x_i, y_i) \tag{2}$$

where (x_i, y_i) is the coordinate position of the i-th reference point.

2.2 Multi-dimensional Weighted KNN (MDW-KNN) Algorithm

KNN algorithm takes the average as the positioning result, ignoring the similarity difference between different nearest neighbors and the points to be located, resulting in large fluctuation of positioning error. In order to solve the problems of low positioning accuracy and poor stability of KNN algorithm, this paper studies and improves the three basic elements of KNN algorithm: distance metric, the number of nearest neighbors K and decision-making method, and proposes a multi-dimensional weighted K-nearest neighbor (MDW-KNN) algorithm to improve the positioning accuracy.

Improvement of distance metric. The traditional fingerprint location method measures the similarity difference between two points by fingerprint Euclidean distance, but the simple Euclidean distance ignores the possible measurement error in the actual acquisition environment. If there is a large error in the acquisition of one RSSI in the collected data of a certain point, the rest of RSSI are close to the nearest neighbor's reference point, and the calculated Euclidean distance will be large, so that the reference point is not selected as the nearest neighbor. In order to reduce the impact of acquisition error, the Euclidean distance of the point to be located is given a weight coefficient in this paper. Based on the similarity of fingerprint data, the probability that the difference of AP signal strength between the point to be located and the reference point is less than 1 dBm is calculated, and the reciprocal of this probability is used as the weight coefficient of Euclidean distance. The improved Euclidean distance calculation formula is:

$$D_i = \frac{1}{P_i} \sqrt{\sum_{j=1}^{n} \left(RSSI^j - rssi_i^j \right)^2} \tag{3}$$

$$P_i = \frac{n_{(RSSI-rssi)<1}}{n_{AP}} \tag{4}$$

where P_i represents the probability that the difference in signal strength is less than 1 dBm, $n_{(RSSI-rssi)<1}$ is the number of differences in signal strength less than 1 dBm, n_{AP} is the number of AP.

Improvement of K value. K is a super parameter of KNN algorithm, which means the number of reference points in the decision-making stage. When K value is small, the generalization ability is weak, and easy to cause the over fitting; When the value of K is large, it has strong generalization ability and is prone to under fitting. The traditional K value selection is generally fixed to a certain value, but due to spatial ambiguity, the reference points with long distance may also have similar fingerprint information with the points to be located, or the fingerprint Euclidean distance is small, so it is selected as the nearest neighbor to reduce the positioning accuracy. In order to select more valuable reference points near each point to be located, this paper uses the dynamic K value to adjust the number of adjacent points, that is, the number of adjacent points in a certain range is taken as the value of K. This range can be selected as a circle in which the previously estimated point of a certain position is the center of the circle and R is the radius, in which R is determined by the distance between two consecutive points to be located. Due to the limited moving track of the user, the maximum moving distance in the two sampling times is 2 m, so R = 2 m.

Improvement of decision-making method. The traditional KNN decision-making method takes the mean value of the nearest neighbor points as the positioning result, ignoring the contribution of each nearest neighbor point to the positioning estimation, that is, the closer the nearest neighbor point is to the positioning estimation, the greater the contribution of the nearest neighbor point to the positioning estimation. The decision-making method with fingerprint Euclidean distance as the weight is the most commonly used improvement measure at present, that is, the weighted average is carried out according to the fingerprint Euclidean distance between the first K nearest neighbors estimated in KNN and the point to be located. The weight coefficient based on fingerprint Euclidean distance is as follows:

$$\lambda_i = \frac{\frac{1}{D_i}}{\sum_{i=1}^{k} \frac{1}{D_i}} \tag{5}$$

where D_i is the improved Euclidean distance between the neighbor point (x_i, y_i) and the point to be located.

In order to overcome the spatial ambiguity of KNN algorithm, the physical distance between the nearest neighbor and the point to be located can be selected as an auxiliary calculation to measure the contribution of the nearest neighbor to the location estimation. Since the actual position of the point to be located is unknown and the user's moving distance is limited in the process of continuous time sampling, the physical distance between the user's previous estimated point and the nearest neighbor point can be used as the calculation basis. The weight coefficient based on the physical distance is as follows:

$$l_i = \frac{\frac{1}{\sqrt{(x_i - x_{pre})^2 + (y_i - y_{pre})^2}}}{\sum_{i=1}^{k} \frac{1}{\sqrt{(x_i - x_{pre})^2 + (y_i - y_{pre})^2}}} \tag{6}$$

where (x_{pre}, y_{pre}) is the coordinate of the previous estimated point.

In summary, the final positioning result is expressed as:

$$(x, y) = \sum_{i=1}^{k} l_i \lambda_i (x_i, y_i) \tag{7}$$

In the initial stage of positioning, (x_{pre}, y_{pre}) in the weight based on the physical distance is unknown, so the weight l_i is ignored in the calculation of the initial point, and the value of K is set to 5,the positioning result of the initial point is:

$$(x, y) = \sum_{i=1}^{5} \lambda_i(x_i, y_i) \tag{8}$$

3   Experiment and Analysis

In order to verify the performance of the improved KNN algorithm (MDW-KNN), the experiment uses MATLAB software to simulate the algorithm, and compares with traditional KNN and WKNN algorithms to further analyze the advantages of the improved algorithm.

3.1   Construction of Experimental Environment

The simulation experiment is carried out by simulating the real scene and positioning data. Using MATLAB to simulate a 21m by 11m room, five wireless APs are arranged in the room. The RSSI distance loss model is used to simulate the propagation of wireless signals and obtain the offline fingerprint database data. Simulate the user's moving track and randomly generate a track with 100 sampling points in the room to obtain online fingerprint data.

3.2   Analysis of Experimental Results

In order to more intuitively compare the positioning results of each algorithm, the positioning estimation point trajectory of each algorithm is simulated by MATLAB and compared with the real trajectory. Figure 1 illustrates the trajectory comparison, and Fig. 2 compares the cumulative distribution function (CDF) of localization errors for different algorithms.

As can be seen from Fig. 1, the deviation between the estimated location point and the real point of MDW-KNN algorithm is small, and the trajectory is smooth and closer to the real trajectory. Figure 2 illustrates that the CDF curve of MDW-KNN is better than other algorithms, and the error distance converges to about 1.2 m, while the convergence distance of KNN algorithm is about 1.8 m. In addition, Table 1 lists all the average localization errors, the average error of MDW-KNN is less than KNN and WKNN. The average positioning accuracy is 46% higher than KNN and 14% higher than WKNN.

3.3   Fault Tolerance Analysis

To verify the data fault tolerance of the new algorithm, it can be assumed that there is a large error in the data collected from a certain location, and the location estimation is completed by using KNN, WKNN and MDW-KNN, respectively. It is assumed here that there is an error in the acquisition of the second AP signal strength at the 50th position

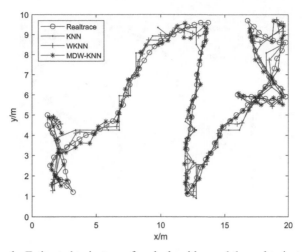

Fig. 1. Estimated trajectory of each algorithm and the real trajectory

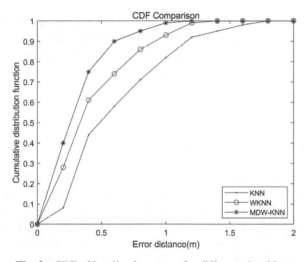

Fig. 2. CDF of localization errors for different algorithms

Table 1. Average localization error

Algorithm	KNN	WKNN	MDW-KNN
Average error (m)	0.613618	0.382342	0.329701

(change its RSSI from −54 to −64 dBm). In the case of acquisition errors, location estimation point tracks for each algorithm are shown in Fig. 3.

Figure 3 illustrates the locating errors of the KNN and WKNN algorithms are large at the 50th position, and the locating estimates are discrete, but the deviation between

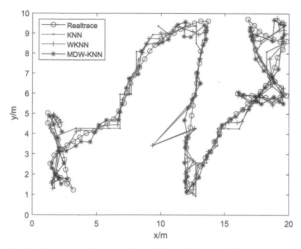

Fig. 3. Estimated trajectory of each algorithm and the real trajectory

the locating estimates of MDW-KNN and the real points is small, and it can output a smooth locating track. This proves that the MDW-KNN algorithm has a strong data fault tolerance, when data error occurs, it still has good positioning performance.

4 Conclusions

This paper describes the theory of location fingerprint positioning, and deeply studies the KNN algorithm in location fingerprint positioning. In view of the existing short-comings of the algorithm, improvement on three basic elements of KNN algorithm: distance metric, K value selection and decision method, and a KNN algorithm with multidimensional weights is presented. Through simulation experiments, the improved KNN algorithm has good fault tolerance and stability, and the positioning accuracy is improved by nearly 46% compared with the traditional KNN algorithm. However, the computational load of the algorithm is large and the computational time is long. How to reduce the computational load while improving the positioning accuracy is still a direction worth studying.

Acknowledgements. This work is supported by Key R&D project of Shaanxi Province (2022GY-094), Research on indoor floor recognition and location technology based on sensor data fusion.

References

1. Yan, D.Y., Song, W., Wang, X.D.: Review of development status of indoor location technology in China. J. Navig. Position. **7**(4), 5–12 (2019)
2. Xie, S.C., Yu, X.X., Zhao, J.X., et al.: An improved WIFI location fingerprint indoor positioning algorithm. J. Hefei Univ. Technol. (Nat. Sci.) **44**(6), 753–757 (2021)

3. Chen, L., Shi, Z.C., Zhang, X.: RFID vehicle indoor positioning algorithm based on PSO-ANN. Transducer Micros. Technol. **39**(12), 127–133 (2020)
4. Dang, X.C., Ru, C.R., Hao, Z.J.: An indoor positioning method based on CSI and SVM regression. Comp. Eng. Sci. **43**(5), 853–861 (2021)
5. Brunato, M., Battiti, R.: Statistical learning theory for location fingerprinting in wireless LANs. Comp. Netw. **47**(6), 825–845 (2005)
6. He, Z.S., Wan, Y.P., Zhao, Q., et al.: Research and improvement of indoor positioning technology based on WiFi. J. Univ. South China (Sci. Technol.) **32**(3), 60–64 (2018)
7. Li, Y.F., Zhang, P., Wang, Y.P.: An indoor and outdoor seamless localization algorithm with weight adjusted k nearest neighbor. Microprocessors **2**, 30–36 (2019)
8. Hoang, M.T., Zhu, Y., Yuen, B., et al.: A soft range limited k-nearest neighbors algorithm for indoor localization enhancement. IEEE Sens. J. **18**(24), 10208–10216 (2018)
9. Guo, X.G., Hu, L.: WiFi fingerprint location algorithm based on k-means and improved k-nearest neighbor. J. Changchun Univ. Technol. **39**(1), 73–79 (2018)
10. Tian, Z.Y., Yu, X., Huang, J.: Indoor positioning algorithm based on TS-KNN. J. Chongqing Univ. **43**(5), 93–103 (2020)

Image Caption Description Generation Method Based on Reflective Attention Mechanism

Qiao Pingan[1,2], Li Yuan[1,2(✉)], and Shen Ruixue[1,2]

[1] Xi'an University of Posts and Telecommunications, Xi'an 710121, China
744284130@qq.com
[2] The Key Laboratory of Network Data Analysis of Shaanxi Province, Xi'an 710121, China

Abstract. Aiming at the problem that the existing image caption model cannot model the dependence between the generated words, this paper introduces a reflective attention mechanism that uses the context vector generated at each time step to model the relationship between words, so that the model can fuse the previous words information when generating image description words, and finally generate more accurate and coherent image description sentences. The experimental results on the MSCOCO dataset show that the method is effective and has a great improvement in each evaluation metric.

Keywords: Image caption · Attention mechanism · Long short-term memory (LSTM)

1 Introduction

The purpose of image caption is to automatically generate an appropriate image description from an image. It is a cross-modal domain involving computer vision and natural language processing. It is attracting more and more people's attention.

Early image caption methods, such as literature [1], are retrieval-based and template-based methods. The retrieval-based method is to search the most similar images in the dataset, and the template-based method is to generate image descriptions with fixed rules. With the development of deep learning, inspired by the successful application of sequence to sequence [2] in machine translation, the encoder-decoder structure is widely used to solve this problem.

The image caption model of encoder-decoder structure firstly extracts meaningful image information from the image. Secondly, in the decoder, RNN or LSTM generates a predicted word at each time step based on the image feature information provided by the encoder and the generated words. However, the traditional LSTM model tends to focus on words closer to the current time step while ignoring words farther away. Today's mainstream decoder models cannot solve this long-term dependency problem, which will generate inaccurate image descriptions to a large extent.

This paper introduces a decoder network based on the reflective attention mechanism [3], which enhances the long-time sequence modeling ability of the decoder. Different from the previous methods to improve the model capability, including improving the visual attention mechanism [4, 5] or making the encoder provide more abundant representation [6, 7], the network in this paper combines the attention mechanism in the field of vision and text.

2 Related Work

2.1 Long Short-Term Memory Network

To reduce the long-term dependency problems, the recurrent neural network (RNN) adds hidden state to retain past information, but the initial information disappears with the increase of length. The long short-term memory network (LSTM) further retains the past information through memory cells to enhance the relationship dependence, which effectively solves the problem of gradient disappearance and has wide applications in visual-linguistic tasks.

2.2 Encoder-Decoder

The encoder-decoder structure has a successful application in sequence-sequence tasks. Inspired by this structure, Oriol et al. [8] first applied the encoder-decoder framework to image caption creatively. The encoder extracts image features from the picture, and the decoder decodes the image features provided by the encoder. The LSTM generates only one word at each time step until the sentence terminator is generated. The encoder-decoder structure has been widely used in the field of image caption. In this paper, Faster R-CNN [9] is used as the encoder and LSTM as the decoder.

2.3 Attentional Mechanism

Recently, attention mechanisms have been introduced into the encoder-decoder structure, which is a more in-depth modeling of image caption. The attention mechanism simulates that people will pay attention to different positions of images when they say different words, which better simulates the process of the human brain thinking to generate languages. When generating a word at each time step, the attention mechanism first calculates the weight for the image features generated by the encoder and calculates new visual features through this weight to generate the caption. This attention-based approach can effectively guide the generation of descriptive words.

3 Image Caption Description Generation Method Based on Reflective Attention Mechanism

3.1 Encoder

The purpose of the encoder is to extract meaningful expressions from the input image. In this paper, Faster R-CNN [9] in target detection is used to extract the expression of the regional level. The expression of the regional level extracted by the encoder is expressed as $\{a_i\}_{i=1}^k, a_i \in R^D$, k represents the number of extracted image regions, D represents the feature dimension of each image region, and a_i represents the average pooling feature of the extracted image regions. This is a bottom-up attention mechanism, compared with the traditional top-down attention. The object-level encoder pays more attention to the salient regions in an image.

3.2 Decoder

Given the image feature set generated by the encoder, the purpose of the decoder is to generate the text description $S = \{S_1, S_2, \cdots S_n\}$. As shown in Fig. 1, the LSTM model based on attention mechanism adopts bottom-up attention mechanism to selectively focus on the feature vector of the image regions. It is used as the basic image caption decoder, and the reflective attention mechanism assists the basic decoder to enhance the quality of the text description so that the generated text description is more coherent.

The decoder's input x_t^1 at the time step t contains three parts, average pooling image feature $\bar{a} = \prod_{i=1}^{k} a_i$, word embedding vector $\Pi_t W_e$, context vector at the last time step h_{t-1}^2, and the overall structure of the network is shown in Fig. 1. Among them, $\bar{a}$ represents the context information of the input image, $W_e \in \mathbb{R}^{D_o \times E}$ is the word embedding matrix of one-hot vector, Do is the size of vocabulary, E is the size of word embedding, and the update formula of LSTM layer 1 is as follows:

$$\left(h_t^1, m_t^1\right) = LSTM\left(\left[\Pi_t W_e, a + h_{t-1}^2\right], \left(h_{t-1}^1, m_{t-1}^1\right)\right) \tag{1}$$

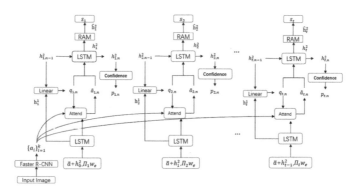

Fig. 1. Overall network structure.

3.3 Adaptive Attention Time Attention Mechanism (AAT)

In this paper, the AAT attention mechanism module [10] is used to weight and sum image features many times. To let the AAT module decide how many attention times to implement, an additional confidence network is added to the output of LSTM₂. The confidence network is a two-layer feed-forward network. The feed-forward network propagates the preceding item as shown in formula (2):

$$p_{t,n} = \begin{cases} \sigma\left(\max\left(0, h_t^1 W_1 + b_1\right) W_2 + b_2\right) & n = 0 \\ \sigma\left(\max\left(0, h_{t,n}^2 W_1 + b_1\right) W_2 + b_2\right) & n > 0 \end{cases} \tag{2}$$

The total number of attention time required is determined by the following formula:

$$N(t) = \min\left\{ n' : \prod_{n=0}^{n'} (1 - p_{t,n}) < \epsilon \right\} \tag{3}$$

ϵ is a threshold, in this paper $\epsilon = 1e - 4$.

The input $q_{t,n}$ of AAT is determined by the formula (4), where $h_{t,0}^2 = h_{t-1}^2$.

$$q_{t,n} = \left[h_t^1, h_{t,n-1}^2 \right] W_q + b_q \tag{4}$$

AAT outputs the weighted sum of image feature vectors and inputs $\hat{a}_{t,n}$ and $q_{t,n}$ into the second layer of LSTM. Finally, LSTM2's hidden state and memory cell are calculated as follows:

$$\begin{cases} h_t^2 = \beta_{t,0} h_t^1 + \sum_{n=1}^{N(t)} \beta_{t,n} h_{t,n}^2 \\ m_t^2 = \beta_{t,0} m_{t-1}^2 + \sum_{n=1}^{N(t)} \beta_{t,n} m_{t,n}^2 \end{cases} \tag{5}$$

$$\beta_{t,n} = \begin{cases} p_{t,0} & n = 0 \\ p_{t,n} \prod_{n-1=0}^{n-1} (1 - p_{t,n}) & n > 0 \end{cases} \tag{6}$$

Adding h_t^1 indicates that when the network decides not to pay attention to the image feature vectors, we can directly obtain the context vector from the input module, m_{t-1}^2 indicates that when the network does not pass the image features to the attention mechanism, the memory cells can maintain the previous states and remain un-updated.

3.4 Reflective Attention Mechanism (RAM)

The reflective attention module includes a reflective attention layer Att_{ref} and the second layer LSTM, the second layer LSTM is used to output language description.

The reflective attention module is based on the hidden state h_t^2 of the current time step and the hidden state set $\{h_1^2, h_2^2, \cdots, h_{t-1}^2\}$ of the previous time steps. The reflective attention layer calculates the standardized weights distribution α_t^{ref} based on the hidden state generated by all the time steps in the current second layer LSTM. The formula definition is as follows:

$$\alpha_{i,t}^{ref} = W_h^2 \tanh\left(W_{h2h}^2 h_i^2 + W_{hhh}^2 h_t^1\right) \tag{7}$$

$$\alpha_t^{ref} = \text{softmax}\left(\alpha_t^{ref}\right), \alpha_t^{ref} = \left\{\alpha_{i,t}^{ref}\right\}_{i=1}^{t} \tag{8}$$

where $W_h^2 \in \mathbb{R}^{1 \times D}$, $W_{h_2 h}^2 \in \mathbb{R}_f^{D} \times D_h$, $W_{h_1 h}^2 \in \mathbb{R}^{D_f \times D_h}$, α_t^{ref} represents the attention weight probability corresponding to the hidden state at time step t, $\alpha_{i,t}^{ref}$ represents the.

pairwise correlation between the words predicted by the past time steps and the word predicted by the present time step, so we can calculate the weighted hidden state $\hat{h}_t^2 = \sum_{i=1}^{t} \alpha_{i,t}^{ref} h_i^2$.

The output $\hat{h}_t^2$ of Att_{ref} is used as a context vector to predict word s_t:

$$p(s_t|s_{1:t-1}) = \text{softmax}\left(W_s \hat{h}_t^2 + b_s\right) \tag{9}$$

where $W_S \in \mathbb{R}^{D_o \times D_h}$ is trainable weights and $b_s \in \mathbb{R}^{D_o}$ is bias. Reflective attention module models the dependence correlation between two words at different time steps, which enhances the long-term dependence of the LSTM model. The hidden state of LSTM at all time steps is taken into account to make the generated image caption description more coherent. The mechanism of reflective attention is shown in Fig. 2:

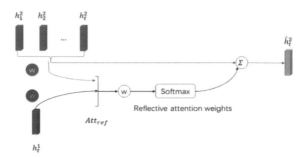

Fig. 2. Reflective attention mechanism.

4   Experiments and Analysis

4.1   Dataset, Settings, and Metrics

Dataset. The dataset used in all experiments in this paper is the most popular MSCOCO dataset. MSCOCO dataset contains 1,23,287 pictures, each containing five sentence descriptions, of which 82,783 pictures are used for training and 40,504 pictures are used for verification. This paper uses 'Karpathy' [11] data split to divide the dataset, including 1,13,287 pictures for training, 5K pictures for verification, 5K pictures for testing.

Settings. We convert all sentences into lowercase. If the word appears less than 5 times in all sentences, this word will be discarded, and the length of the sentence is limited to the maximum length of 16. Finally, we get a vocabulary containing 10,369 words.

We use Faster R-CNN as the encoder, and we use ImageNet and Visual Genome to pretrain the encoder. The dimension of the extracted image feature vectors is d = 2048, and the dimension of the feature vectors is transformed into d = 1024. The dimension of the hidden state of LSTM in the decoder is 1024 and the dimension of word embedding is 1024.

For the training process of the network, the first 20 epochs of the network are trained by the cross-entropy loss function. The batch size is set to 10. To avoid falling into local optimum, the experiments adopt ADAM optimization strategy. The learning rate is set to 1e-4 and anneals by 0.8 every two epochs. To reduce the exposure bias, scheduled sampling [12] is adopted. During training, the model used all words of ground truth sentences as input and gradually increased the probability of words predicted at the previous time step as input, increasing the probability of 0.05 for every 3 epochs. Then we use self-critical sequence training (SCST) [13] to train the other 20 epochs to optimize the CIDEr-D score, and the initial learning rate of reinforcement learning is 1e-5. When the CIDEr-D score does not improve in several epochs on the validation set, the learning rate will decay by 0.5.

Metrics. To evaluate the quality of the resulting sentences of our model objectivly, we use five widely accepted automatic evaluation indicators, including CIDEr-D, SPICE, BLEU, METEOR, and ROUGE-L. CIDEr-D is an improvement of CIDEr. If there is no hint, CIDEr in the following is CIDEr-D.

4.2 Experimental Results and Analysis

To study the effect of reflective attention mechanism, we set up the following experiments: (1) Baseline: Bottom-UP and Top-Down network [14] is selected as the baseline network; (2) AAT: the attention mechanism module in the baseline network is replaced by the AAT module; (3) based on adding AAT module, the reflective attention mechanism is added after the second layer of LSTM.

Table 1. Experimental results.

Cross-entropy loss						CIDEr score optimization					
	B@1	B@4	M	R	C	S	B@1	B@4	M	R	C	S
BaseLine	76.8	36.2	27.0	56.4	113.5	20.3	79.8	36.3	27.7	56.9	120.1	21.4
AAT	76.9	36.5	28.0	57.0	115.7	21.0	80.1	**38.5**	28.3	58.3	126.8	21.9
RAN	**77.6**	**37.2**	**28.0**	**57.2**	**117.1**	**21.2**	**80.3**	38.4	**28.3**	**58.4**	**127.3**	**22.1**

According to Table 1, we can see that the reflective attention mechanism improves the performance of the image caption model in almost all evaluation metrics. This result shows that the reflective attention mechanism enhances the long-term dependence between the words. By combining the AAT module and the reflective attention mechanism, the reflective attention network (RAN) exceeds the benchmark network in terms of evaluation metrics scores. Among them, the CIDEr and SPICE scores reach 117.1 and 21.2 without reinforcement learning optimization. In the case of reinforcement learning, CIDEr and SPICE scores reached 127.3 and 22.1, improved by 6.0 and 3.3%.

Table 2. Comparison of results between different models.

| | Cross-Entropy Loss | | | | | | CIDEr Score Optimization | | | | | |
	B@1	B@4	M	R	C	S	B@1	B@4	M	R	C	S
LSTM	–	29.6	25.2	52.6	94.0	–	–	31.9	25.5	54.3	106.3	–
SCST	–	30.0	25.9	53.4	99.4	–	–	34.2	26.7	55.7	114.0	–
LSTM-A	75.4	35.2	26.9	55.8	108.8	20.0	78.6	35.5	27.3	56.8	118.3	20.8
RFNet	76.4	35.8	27.4	56.5	112.5	20.5	79.1	36.5	27.7	57.3	121.9	21.2
GCN	77.3	36.8	27.9	57.0	116.3	20.9	80.5	38.2	**28.5**	58.3	127.1	22.0
NBT	75.5	34.7	27.1	–	108.9	20.1	–	–	–	–	–	–
CNM	77.3	36.5	27.6	57.0	116.4	20.7	80.3	38.3	28.2	**58.6**	126.4	21.5
SGAE	77.6	36.9	27.7	57.1	116.4	20.9	80.2	38.4	28.2	58.4	127.2	22.0
MN	76.9	36.1	–	56.4	112.3	20.3	–	–	–	–	–	–
RAN	**77.6**	**37.2**	**28.0**	**57.2**	**117.1**	**21.2**	**80.3**	**38.4**	28.3	58.4	**127.3**	**22.1**

The method in this paper and other methods, for example, shown in Table 2, where '-' indicates that the original paper did not provide data. Without reinforcement learning optimization, BLEU-1, BLEU-4, METEOR, ROUGE-L, CIDEr-D, and SPICE scores reached 77.6, 37.2, 28.0, 57.2, 117.1, and 21.2, respectively. In the case of reinforcement learning, BLEU-1, BLEU-4, METEOR, ROUGE-L, CIDEr-D, and SPICE scores reached 80.3, 38.4, 28.3, 58.4, 127.3, and 22.1, respectively. The experimental results show that the method proposed in this paper is partially improved compared with the comparison method. Especially under the CIDEr and SPICE evaluation metrics, CIDEr and SPICE are the evaluation metrics especially designed for image caption. SPICE is closer to the human evaluation standard. When the SPICE score of the multimodal image caption model is higher, the semantic correctness of the model is higher.

Table 3 shows the image captions generated by the benchmark network and reflective attention network on the test set and the corresponding ground truth image sentence descriptions. It is found that the sentence descriptions generated by the reflective attention network are closer to the real sentence descriptions generated by human beings, and the generated sentences are more accurate and coherent. For example, in the first graph, the benchmark network fails to identify the book on the bench, while the reflective attention network accurately identifies the book, and can accurately describe the position relationship between the book and the bench. In the second graph, the benchmark network fails to describe the state of the traffic light, otherwise RAN notices this tiny detail. Although in the third graph, the benchmark network describes the plants in the graph, the sentence is not natural and coherent, and the sentence generated by RAN is more coherent. In the fourth graph, the benchmark network fails to describe this as a

Table 3. Image caption generated by baseline and reflective attention network and corresponding ground truth.

Image	Captions
	Base model: a wooden bench sitting on top of a wooden bench RAN (ours): a book sitting on top of a wooden bench GT1: a book on a wooden bench on a street GT2: a book sitting on top of a wooden desk GT3: a big entitled california is on the bench
	Base model: a traffic light on the side of a street RAN (ours): a green traffic light on the side of a street GT1: a couple of traffic lights hanging over a city Street GT2: traffic lights shine over an empty intersection at twilight GT3: a street with traffic lights, wall and buildings
	Base model: a picture of a garden with a plant in a window RAN (ours): a collage of photos of plants and trees in a field GT1: colorful collection of pictures of a wooded area GT2: a collage of several photos show outdoor scenes GT3: VARIOUS types of pictures that are of different plants

(continued)

Table 3. (*continued*)

Image	Captions
	Base model: a mountain with a mountain in the background RAN(Ours): a black and white photograph of mountains GT1: a black and white picture of hills or mountains GT2: a black and white photo of mountains and the sky GT3: a black and white photograph of mountains with snow

black-and-white picture and the sentence does not conform to the usage habit. RAN correctly describes this as a black-and-white picture and generates a sentence description that conforms to the usage habit of human language.

5 Conclusion

To enhance the long-term dependence of the decoder network, the reflective attention mechanism is introduced in this paper. The reflective attention mechanism is used to model the relationship dependence between two generated words. The hidden states of all historical time steps are involved in the generation of predicted words, which is consistent with the human language system. The results show that the reflective attention mechanism makes the generated sentences more continuous and intelligent, and makes the network generate more accurate image caption descriptions.

Acknowledgments. This paper was supported by the National Natural Science Foundation of China (Item NO: 61105064), Scientific Research Project of Shaanxi Provincial Department of Education (Item NO:16JK1689), and the Key Laboratory of Network Data Analysis of Shaanxi Province.

References

1. Kulkarni, G., Premraj, V., Ordonez, V., et al.: Babytalk: understanding and generating simple image descriptions. IEEE Trans. Pattern Anal. Mach. Intell. **35**(12), 2891–2903 (2013)
2. Sutskever, I., Vinyals, O., Le, Q.V.: Sequence to sequence learning with neural net- works. Adv. Neural Inf. Process. Syst., 3104–3112 (2014)

3. Ke, L., Pei, W., Li, R., et al.: Reflective decoding network for image captioning. In: Proceedings of the IEEE/CVF International Conference on Computer Vision, 8888–8897 (2019)
4. Xu, K., Ba, J., Kiros, R., et al.: Show, attend and tell: neural image caption generation with visual attention. In: International Conference on Machine Learning. PMLR, pp. 2048–2057 (2015)
5. Chen, L., Zhang, H., Xiao, J., et al.: SCA-CNN: spatial and channel-wise attention in convolutional networks for image captioning. In: Proceedings of the IEEE Conference on Computer Vision and Pattern Recognition, pp. 5659–5667 (2017)
6. Yao, T., Pan, Y., Li, Y., et al.: Boosting image captioning with attributes. In: Proceedings of the IEEE International Conference on Computer Vision, pp. 4894–4902 (2017)
7. Gan, Z., Gan, C., He, X., et al.: Semantic compositional networks for visual captioning. In: Proceedings of the IEEE Conference on Computer Vision and Pattern Recognition, pp. 5630–5639 (2017)
8. Vinyals, O., Toshev, A., Bengio, S., et al.: Show and tell: a neural image caption generator. In: Proceedings of the IEEE Conference on Computer Vision and Pattern Recognition, pp. 3156–3164 (2015)
9. Ren, S., He, K., Girshick, R., et al.: Faster r-cnn: towards real-time object detection with re-region proposal networks. Adv. Neural. Inf. Process. Syst. **28**, 91–99 (2015)
10. Huang, L., Wang, W., Xia, Y., et al.: Adaptively aligned image captioning via adaptive attention time. Adv. Neural. Inf. Process. Syst. **32**, 8942–8951 (2019)
11. Karpathy, A., Fei-Fei, L.: Deep visual-semantic alignments for generating image descriptions. In: Proceedings of the IEEE Conference on Computer Vision and Pattern Recognition, 3128–3137 (2015)
12. Bengio, S., Vinyals, O., Jaitly, N., et al.: Scheduled sampling for sequence prediction with recurrent neural networks. arXiv preprint arXiv:1506.03099 (2015)
13. Rennie, S.J., Marcheret, E., Mroueh, Y., et al.: Self-critical sequence training for image captioning. In: Proceedings of the IEEE Conference on Computer Vision and Pattern Recognition, pp. 7008–7024 (2017)
14. Anderson, P., He, X., Buehler, C., et al.: Bottom-up and top-down attention for image captioning and visual question answering. In: Proceedings of the IEEE Conference on Computer Vision and Pattern Recognition, pp. 6077–6086 (2018)

Research on Image Description Generation Method Based on Visual Sentinel

Pingan Qiao[1,2], Ruixue Shen[1(✉)], and Yuan Li[1]

[1] School of Computer Science and Technology, Xi'an University of Posts and
Telecommunications, Xi'an, China
`paqiao@xupt.edu.cn, snow@stu.xupt.edu.cn, 903678724@qq.com,`
`744284130@qq.com`
[2] Shaanxi Key Laboratory of Network Data Analysis and Intelligent Processing, Xi'an, China

Abstract. Attention-based encoder-decoder framework has been currently utilize for image description. Many methods are to keep visual attention dynamic for each produced word. But the decoder does not need to obtain any visual information from the picture to forecast non-visual words at every moment. In addition, the development of non visual vocabulary may mislead and reduce the overall effectiveness of visual signals in generating guidelines.So we provides a new model AoANet++ to solve the above problems. At each time step, the model decides when to display the picture and when the language model generates the next word to produce a better sentence.The experimental results on MSCOCO dataset prove the effectiveness of this method, which improves the best CIDEr-D and SPICE scores of the initial report by 8.6% and 4.7% respectively.

Keywords: Image description · Visual sentinel · Encoder-decoder · Attention on attention

1 Introduction

Image description is a more complex cognitive task than the basic perception task [1–5]. Not only does it require identifying prominent objects in an image and understanding how they interact, it also needs to express them in natural language in order to make the computer understand the given image. And with the development of the Internet, multimedia data presents an explosive growth trend, and more and more image data can be obtained from various information sources. Due to the massive and unstructured characteristics of image data, how to organize, store and retrieve images quickly and effectively has become an important research topic, and complete image semantic understanding is the key problem. Therefore, the automatic generation of image description has become a prominent interdisciplinary research problem [1]. And image description also has very broad application prospects. Firstly, image description generation technology can be applied to automatic image indexing, which is of great significance to improve the effect and efficiency of image retrieval. Therefore, image description generation can be applied to many application fields of image retrieval, including medical treatment,

N. Xiong et al. (Eds.): ICNC-FSKD 2022, LNDECT 153, pp. 610–620, 2023.
https://doi.org/10.1007/978-3-031-20738-9_70

commerce, military, education, digital library and so on; Secondly, image description generation technology can help social media platforms (such as Facebook, twitter, etc.) generate natural language descriptions for images, including important information such as where we are, what we wear and what we do, which can directly help and guide our daily life; Finally, image description generation technology can also play a key role in the application fields of robot interaction, preschool education and visual impairment assistance.

In order to generate high-quality subtitles, the image description generation model involves a challenge—the model needs to contain fine-grained visual cues from the image. Inspired by neural machine translation, attention mechanism is widely used in the image description codec framework, which generates the weighted average of the coding vector at each time step to direct the description decoding course. In 2019, Lun Huang et al. [6] proposed AoA (Attention on Attention) module, which expands the traditional attention mechanism to decide the correlation between attention queries and results, and creats a better description than previous models. Like most attention models used for image description, AoA (Attention on Attention) module only focuses on the picture at each time step, regardless of which word is the next [7]. But, not all words in produced descriptions have relevant visual signals, and staying away from non-visual words can be misleading and reduce effectiveness of visual signals in generating sentences.

To solve the above problems and generate better sentences, this paper proposes AoANet++ model, which generates better descriptions by introducing AoA module, "visual sentinel" and adaptive attention model into AoANet++.

The work done in this paper has the following points:

- We provide a new structure called AoANet++, which automatically aim when to depend on the language model and when to look at the image to generate the next word.
- On this basis, we developed an AoANet++ model with "visual sentinel", which helps to extract meaningful information in order to continuously generate words.

2 Related Work

2.1 Image Description

The methods of generating image descriptions can be divided into three categories by the way of generating sentences [8].

Template-based [9] method mainly uses the formulated sentence template to generate sentences. This method is intuitive, but it lacks enough flexibility to generate meaningful sentences.

Retrieval-based method [10] is to retrieve similar images from the annotated image description dataset and then transfer the description of the retrieved image to the query image. But the generated sentences sometimes do not express the content of the image correctly.

The widely used method now is based on the encoder-decoder framework [11, 12]. Although the efficiency of this structure is very high, the decoding is inaccurate because the decoder cann't get enough input information. On this basis, Dmitry bandana et al. [13] advanced an attention-based encoder-decoder framework (shown in Fig. 1), which solves the above problems by introducing attention weight to make the encoder dynamically pay attention to the more important part of the input sequence. It can make better use of input information. Compared with the first two methods, the method based on the neural network can generate more accurate sentences. Based on this framework, this paper studies the proposed problems.

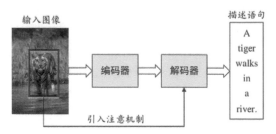

Fig. 1. Attention-based encoder-decoder framework.

2.2 Attention

The attention mechanism [14] is the result of research on human vision. In cognitive science, information processing bottlenecks lead people to follow only a part of all info, while eliding other selectively visible information. And its essence is to give positive weight to each part. Show the importance of these parts.

In this paper, we apply adaptive attention to image descriptions to generate higher-quality sentences.

3 Image Description Generation Method Based on Visual Sentinel

3.1 Attention on Attention (AoA)

This paper uses the AoA (shown in Fig. 2) to gauge the correlation between attention results and queries.

This module takes the attention query and the result through two linear transformations to generate an attention gate **g** and an information vector **i**, both conditioned on the present context (i.e. the **q**) and the attention result:

$$g = \sigma \left(W_q^g q + W_v^g \hat{v} + b^g \right) \tag{1}$$

$$i = W_q^i q + W_v^i \hat{v} + b^i \tag{2}$$

where σ represents the sigmoid activation function, b^g, $b^i \in R^D$, W_q^i, W_v^i, W_q^g, $W_v^g \in R^{D \times D}$, $\hat{v} = f_{att}(Q, K, V)$ is an attention result, D are the dimension of **v** and **q**, and f_{att} is an attention module.

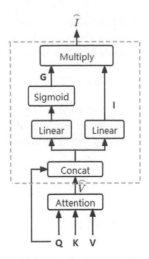

Fig. 2. Attention on attention.

Then, AOA uses element by element multiplication to apply the attention gate **g** to the information vector **i** to increase another attention and get the attention information $\hat{\mathbf{i}}$:

$$\hat{i} = g \odot i \tag{3}$$

where $\odot$ represents element-wise multiplication. The overall calculation formula of the AoA model is as follows:

$$AoA(f_{att}, \boldsymbol{Q}, \boldsymbol{K}, \boldsymbol{V}) = \sigma \left(W_q^g \boldsymbol{Q} + W_v^g f_{att}(\boldsymbol{Q}, \boldsymbol{K}, \boldsymbol{V}) + b^g \right)$$
$$\odot \left(W_q^i \boldsymbol{Q} + W_v^i f_{att}(\boldsymbol{Q}, \boldsymbol{K}, \boldsymbol{V}) + b^i \right) \tag{4}$$

The purpose of applying it in the encoder and decoder is to better model the relationship between various objects in the picture in the encoder, filter out irrelevant attention results in the decoder and keep useful results.

3.2 Visual Sential

Visual sential is a new component extracted from the decoder. The model can use this component to decide when to focus on the image. In this paper, the LSTM is extended to obtain the "visual sential" vector $\mathbf{s_t}$:

$$g_t = \sigma(W_x x_t + W_h h_{t-1}) \tag{5}$$

$$s_t = g_t \odot \tanh(m_t) \tag{6}$$

Among σ For the logistic sigmoid activation function, **xt** is the import of LSTM at the time step t, Wh and Wx are the weight parameters to be learned, and **gt** is the gate used to the storage unit **mt**.

On the basis of visual sential, a new adaptive context vector $\hat{c}_t$ is further proposed. The purpose is to weigh the new information considered by the network from the image and the information it already knows (i.e. visual sential). It is defined as follows:

$$\hat{c}_t = \beta_t s_t + (1 - \beta_t)c_t \tag{7}$$

Among β_t is the fresh sentry gate at time t. It generates a scalar quantity in the range [0, 1]. A value of 0 means that exclusively spatial image message is utilized when generating the next word, while a value of 1 means that exclusively visual sential message is utilized. It is defined as:

$$\beta_t = \alpha_t[k + 1] \tag{8}$$

$$\hat{\alpha}_t = \text{softmax}\left(\left[z_t; w_h^T \tanh\left(W_s s_t + \left(W_g h_t\right)\right)\right]\right) \tag{9}$$

$$z_t = w_h^T \tanh\left(W_v V + \left(W_g h_t\right)1^T\right) \tag{10}$$

where Ws is the weight parameter, [·; ·] represents the connection, and $1 \in R^k$ is a vector with all elements set to 1. wh $\in R^k$, Wg, Wv $\in R^{k \times d}$ is the parameter to be learned.

3.3   Encoder

In the encoder (as shown in Fig. 3), for the image, a Faster-RCNN-based network is first utilized to draw a group of feature vectors $A = \{a_1, a_2, ..., a_k\}$, where $a_i \in R^D$, D is the dimension of each vector and k is the number of vectors in A. We build a network containing AoA modules to optimize their representation without inputing these vectors directly to the decoder.

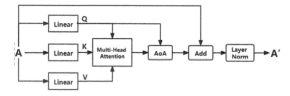

Fig. 3. Encoder structure.

We employ a multi-head attention function, calling the AoA module in the encoder AoA^E, where V, K and Q are three separate linear projections of the feature vector A. After this comes residual connections and layer normalization:

$$A' = LayerNorm(A + \\ AoA^E\left(f_{mh-att}, W^{Q_e}A, W^{K_e}A, W^{V_e}A\right)) \tag{11}$$

where $f_{mh\text{-}att}$ and $f_{dot\text{-}att}$ are the multihead attention and dot product attention function respectively, W^{Ve}, W^{Ke}, $W^{Qe} \in R^{D \times D}$ are the three linear transformation matrices. V,

K and **Q** are then divided into H = 8 slices by the channel dimension using $f_{mh\text{-}att}$, and attention is computed for each slice using $f_{dot\text{-}att}$, and then the results of each slice are concatenated together to form the final attention vector.

$$f_{mh-att}(\boldsymbol{Q}, \boldsymbol{K}, \boldsymbol{V}) = Contact(head_i, ..., head_H) \tag{12}$$

$$head_i = f_{dot-att}(\boldsymbol{Q}_i, \boldsymbol{K}_i, \boldsymbol{V}_i) \tag{13}$$

$$f_{dot-att}(\boldsymbol{Q}_i, \boldsymbol{K}_i, \boldsymbol{V}_i) = \text{softmax}\left(\frac{\boldsymbol{Q}_i \boldsymbol{K}_i^T}{\sqrt{d}}\right) \boldsymbol{V}_i \tag{14}$$

In this optimization module, the self-attention multi-head attention module looks for interactions between objects in the picture and uses AoA to judge the degree of correlation between them. After optimization, update the eigenvector $\mathbf{A} \rightarrow \mathbf{A}'$.

3.4 Decoder

In the decoder (shown in Fig. 4), the optimized feature vector $\mathbf{A}'$ is used to generate a description **y** sequence. The context vector $\mathbf{c}_t$ holds the newly acquired information and decoding state, which is created from the output $\mathbf{h}_t$ of the LSTM and the attention feature vector $\hat{\mathbf{a}}_t$ (where ât is an result of attention, which comes from the attention module).

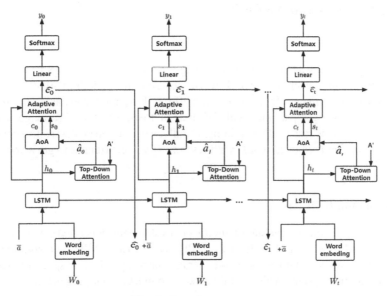

Fig. 4. Decoder structure.

The decoder uses its internal LSTM to model the description decoding process.Its input has a visual vector $(\mathbf{c}_{t-1} + \bar{\mathbf{a}})$ and the embedding of the import word for the present time step, where $\mathbf{c_{t-1}}$ represents the context vector of the former time step, $\bar{\mathbf{a}} = 1/k\sum_i \mathbf{a_i}$ represents the average pooling of A:

$$x_t = \left[W_e \Pi_t, \bar{a} + c_{t-1}\right] \tag{15}$$

$$h_t, m_t = LSTM\left(x_t, h_{t-1}, m_{t-1}\right) \tag{16}$$

where $\prod_t$ is the one-hot encoding used to input the word ω_t at time step t, and $W_e \in R^{E \times |\Sigma|}$ is the word embedding matrix.

As shown in Fig. 4, we obtain $\mathbf{c_t}$ from the AoA module, labeled AoA^D:

$$c_t = AoA^D\left(f_{mh-att}, W^{Q_d}[h_t], W^{K_d}A, W^{V_d}A\right) \tag{17}$$

where $W^{Ve}, W^{Ke}, W^{Qe} \in R^{D \times D}$; $\mathbf{m_t}, \mathbf{h_t} \in R^D$ are the hidden states of the $\mathbf{h_t}$ and LSTM used as attention query. Then we model the new context vector $\hat{\mathbf{c}}_t$ to count conditional probabilities on the word list:

$$p\left(y_t | y_{1:t-1}, I\right) = \text{softmax}\left(W_p \hat{c}_t\right) \tag{18}$$

where $|\sum|$ is the size of the word list and $W_P \in R^{D \times |\Sigma|}$ is the weight parameter to be learned.

4   Experiments and Analysis

4.1   Dataset

The dataset we used was the benchmark image description dataset MS COCO, which has 1,23,287 images, each with five different descriptions, including 40,504 validation and 82,783 training images. In this paper, the offline 'Karpathy' data partition is used for offline performance comparisons, with 5K images used for testing, 5K images used for validation, and all the rest used for training. Finally, we replaced all descriptive sentences with lower case and removed words with less than five occurrences to get a vocabulary of 10,369 words.

4.2   Evaluation Indicators

We use standard automatic evaluation metrics to evaluate the generated sentence quality, namely METEOR, CIDEr-D, ROUGE-L, SPICE and BLEU. SPICE is based on scene map matching, while CIDEr is based on n-gram matching, which is specially designed to evaluate subtitle system, which is more likely to meet human judgment.

4.3 Implementation Details

In this paper, We adopt the Faster-RCNN model to extract the feature vectors of images and use ImageNet and Visual Genome to pre-train the Faster-RCNN model. The original dimension of the vector is 2048 and we shadow it into a new space of dimension D = 1024. During training, we trained AoANet++ under XE loss for 25 epochs, with batch size set to 10, using the ADAM optimiser, setting the learning rate initially to 2e-4, decaying to the previous 0.8 every 3 epochs, and increasing the planned sampling probability by 0.05 every 5 epochs. SCST was then used for the CIDEr-D scores were then optimized for a new 15 epochs, setting the initial learning rate to 2e-5, and if the validation split did not improve, we decayed it to the previous 0.5.

4.4 Experimental Results and Analysis

To better study the effect of our proposed model, we have done the following comparative experiments: (1) **Baseline**: Bottom-up and top-Dow networks are selected as the benchmark network; (2) **AoA**: Introduce AoA module into the benchmark network; (3) **AoANet++**: On the basis of adding AoA module, add visual sentry after LSTM.

Table 1. Compared with the baseline network.

Cross-entropy loss							CIDEr score optimization					
	B@1	B@4	S	C	R	M	B@1	B@4	S	C	R	M
BaseLine	77.2	36.2	20.3	113.5	56.4	27.0	79.8	36.3	21.4	120.1	56.9	27.7
AoA	77.4	37.2	21.3	119.8	57.5	28.4	80.2	38.9	22.4	129.8	58.8	29.2
AoANet++	**78.0**	**37.3**	**21.4**	**120.1**	**58.0**	**28.7**	**80.5**	**39.4**	**22.4**	**130.4**	**58.9**	**29.6**

According to Table 1 it can be found that the addition of visual sentinel improves the performance of the image description model in almost all evaluation metrics. This result shows that the visual sentinel mechanism enhances the effectiveness of the image understanding model. By combining the AoA module and the visual sentinel mechanism, the AoANet++ model comprehensively surpasses the benchmark network in terms of evaluation index scores. Among them, Without reinforcement learning optimization, BLEU-1, BLEU-4, SPICE, CIDEr-D, ROUGE-L and METEOR scores reached 78.0, 37.3, 21.4, 120.1, 58.0 and 28.7, increasing by 1, 3, 5.4, 5.8, 2.8 and 5.9%. In the case of reinforcement learning, BLEU-1, BLEU-4, SPICE, CIDEr-D, ROUGE-L and METEOR scores reached 80.5, 39.4, 22.4, 130.4, 58.9 and 29.6, increasing by 0.9, 8.5, 4.7, 8.6, 3.5 and 6.9%.

With Table 2 we can see some examples of images and sentences generated by the AoANet++ model, as well as the corresponding sentences and real sentences of the benchmark network. From these examples, it can be found that the sentences generated by the benchmark network conform to the language logic, but the image content described is not accurate, while the sentences generated by AoANet++ model are not

Table 2. Experimental result

Image	Description
	Baseline: a young man holding a tennis ball on a court **AoANet++**: a young boy hitting a tennis ball with a racquet **GT**: a boy hitting a tennis ball on the tennis court
	Baseline: a bird sitting on top of a tree **AoANet++**: two birds sitting on top of a giraffe **GT**: two birds going up the back of a giraffe
	Baseline: a cat is looking out of a window **AoANet++**: a cat looking at itself in a mirror **GT**: a cat looking at itself in a mirror
	Baseline: a black and white cat laying on top of a bed **AoANet++**: a couple of cats laying on top of a bed **GT**: a couple of cats laying on top of a bed

only grammatically correct but also accurately described. For example, in the first example, we can know that it is "boy" rather than "man"; In the second example, we can know that two birds are not one bird, and the bird is on a giraffe rather than a tree; In the third example, we can know that the cat is looking in the mirror rather than outside the window; In the fourth example, we can identify two cats.

Table 3. Results comparison among different models.

	Cross-entropy loss						CIDEr score optimization					
	B@1	S	C	R	M	B@4	B@1	S	C	R	M	B@4
gLSTM	67.0	–	81.3	–	22.7	26.4	–	–	–	–	–	–
Stack-Cap	76.2	–	109.1	–	26.5	35.2	78.6	20.9	120.4	56.9	27.4	36.1
RFNet	76.4	20.5	112.5	56.5	27.4	35.8	79.1	21.2	121.9	57.3	27.7	36.5
SGAE	77.6	20.9	116.4	57.1	27.7	36.9	**81.0**	22.2	129.1	58.9	28.4	39.0
VASS	76.9	20.8	114.0	56.5	27.9	36.5	80.5	21.7	126.7	58.8	28.3	38.9
GCN	77.3	20.9	116.3	57.0	27.9	36.8	80.5	22.1	127.8	58.6	28.4	38.4
SRT	77.1	21.3	116.9	56.9	28.0	36.6	80.3	22.4	129.1	58.4	28.7	38.5
AoANet++	**78.0**	**21.4**	**120.1**	**58.0**	**28.7**	**37.3**	80.5	**22.4**	**130.4**	**58.9**	**29.6**	**39.4**

The comparison between the model in this paper and other models is shown in Table 3. The place of "–" indicates that the original paper does not provide data. Compared with other models, without reinforcement learning optimization, BLEU-1, SPICE, CIDEr-D, ROUGE-L, METEOR and BLEU-4 scores reached 78.0, 21.4, 120.1, 58.0, 28.7 and 37.3, increased by 1.2, 0.5, 2.7, 1.9, 2.5 and 1.91%. In the case of reinforcement learning, BLEU-4, CIDEr-D, ROUGE-L and METEOR scores reached 39.4, 130.4, 58.9 and 29.6, increased by 2.3, 1, 0.9 and 3.1%. The experimental results show that most indicators of our model are improved compared with other models, and the improvement of indicators also shows the advantages of our AoANet++ model.

5 Conclusion

In this paper, we present an AoANet++ encoder-decoder framework. It provides a backoff option for the decoder, by introducing a new LSTM extension, which allows it to generate an additional "visual sentinel" to help the model decide when to depend on the language model and when to look at images to generate the next a word, extract more meaningful information for sequence word generation and generate better sentences. Finally, the effectiveness of our AoANet++ model can be demonstrated through experimental results. It improves the best initial score reported on the MS COCO dataset and improves the best SPICE and CIDEr-D scores of the initial report by 4.7% and 8.6% respectively.

Acknowledgment. This paper was supported by the National Natural Science Foundation of China (project number: 61105064), the Shaanxi Provincial Department of Education Research Project (project number: 16JK1689), and the Shaanxi Provincial Key Laboratory of Network Data Analysis.

References

1. Fang, H., Gupta, S., Iandola, F., Srivastava, R.K., Deng, L., Dollár, P., Gao, J., He, X., Mitchell, M., Platt, J.C., et al.: From captions to visual concepts and back. In: CVPR (2015)
2. He, K., Zhang, X., Ren, S., Sun, J.: Deep residual learning for image recognition. In: Proceedings of the. IEEE International Conference on Computer Vision, pp. 770–778 (2016)
3. Tang, S., Li, Y., Deng, L., Zhang, Y.-D.: Object localization based on proposal fusion. IEEE Trans. Multimedia **19**(9), 2105–2116 (2017)
4. Redmon, J., Divvala, S., Girshick, R., Farhadi, A.: Y ou only look once: unified, real-time object detection. In: Proceedings of the IEEE Conference on Computer Vision Pattern Recogonition, pp. 779–788 (2016)
5. Radenovi´c, F., Tolias, G., Chum, O.: CNN image retrieval learns from bow: unsupervised fine-tuning with hard examples. In: Proceedings of the European Conference on Computer Vision, Springer, Berlin, Germany, pp. 3–20 (2016)
6. Huang, L., Wang, W., Chen, J., Wei, X.-Y.: Attention on attention for image Captioning. In: ICCV (2019)
7. Yang, Z., He, X., Gao, J., Deng, L., Smola, A.: Stacked attention networks for image question answering. In: CVPR (2016)
8. Xiong, C., Merity, S., Socher, R.: Dynamic memory networks for visual and textual question answering. In: ICML (2016)

9. Farhadi, A., et al.: Every picture tells a story: generating sentences from images. In: Proceedings of the European Conference on Computer Vision, Berlin, Germany, Springer, pp. 15–29 (2010)

10. P. Kuznetsova, V. Ordonez, A. C. Berg, T. L. Berg, and Y. Choi, "Collective generation of natural image descriptions," inProc. 50th Annu. Meet. Assoc. Comput. Linguistics, Long Papers-V olume 1, 2012, pp. 359–368

11. Szegedy, C., Vanhoucke, V., Ioffe, S., Shlens, J., Wojna, Z.: Rethinking the Inception Architecture for Computer Vision (2015). arXiv:1512.00567

12. Cho, K., et al.: Learning phrase representations using RNN encoder-decoder for statistical machine translation. Empirical Meth. Natural Lang. Process. 1724–1734 (2014)

13. Bahdanau, D., Cho, K., Bengio, Y.: Neural machine translation by jointly learning to align and translate. Proc. Int. Conf. Learn. Represent. (2015)

14. Rensink, R.A.: The dynamic representation of scenes. Visual Cogn. **7**:17–42 (2000)

Improved Weakly Supervised Image Semantic Segmentation Method Based on SEC

Xingya Yan, Zeyao Zheng[⊠], and Ying Gao

Xi'an University of Posts and Telecommunications, Xi'an 710121, China
{zhengze,gaoyin}@stu.xupt.edu.cn, 1770110491@qq.com

Abstract. The study of weakly supervised image semantic segmentation based on weak supervision is received increasing attention. The segmentation algorithm based on seed, expansion, and constraint (SEC) principles is relatively effective. However, this method also suffers from sparsity and discontinuity of initialized seeds. It causes over-growth or under-growth of seeds. This paper proposes an improved SEC weakly supervised image semantic segmentation method. This method enhances the initial seed leads generated by SEC and guides the seed growth through the saliency map generated by the self-attentive network. We also introduce a new loss function to enhance the model's ability to acquire edge information and optimize the final segmentation network. The effectiveness of the proposed method is verified by training the segmentation network on the PASCAL VOC 2012.

Keywords: Semantic segmentation · Weakly supervised · Regional growth introduction

1 Introduction

Semantic segmentation of images has been a crucial topic in computer vision in recent years. It enables simultaneous segmentation and recognition of targets and can provide fine-grained and high-level semantic information for subsequent vision tasks such as image analysis and understanding [1]. Current methods that achieve good semantic segmentation results are based on full supervision, but fully supervised image semantic segmentation methods rely on large-scale pixel-level label training data [2], and this time-consuming and costly large-scale pixel-level label [3] annotation work severely limits the further improvement of image semantic segmentation performance and the scalability of practical applications [4]. Image-level tags can be accurately and efficiently annotated compared to pixel-level tags, which significantly reduces the time and cost of data annotation.

Currently, most approaches are implemented with the help of class activation maps (CAM) localization networks or classification networks [5]. The steps of this class of methods can be divided into generating discriminative sparse seed regions based on class activation graph localization networks or other classification network models [6]; mining and expansion of discriminative seed regions to non-discriminative seed regions;

N. Xiong et al. (Eds.): ICNC-FSKD 2022, LNDECT 153, pp. 621–628, 2023.
https://doi.org/10.1007/978-3-031-20738-9_71

and fusing pseudo-labeled data from discriminative and non-discriminative regions to construct segmentation network models [7]. Kolesnikov et al. [3] proposed a new loss function under the principle of seed, expansion and constraint (SEC), and adopted the method of global weighted pooling to expand the seed region, and constructed a boundary constraint model based on fully connected conditional random fields; Tong et al. [8] using guidance information from the target and network domains to progressively mine fine-grained regional semantic information; Fan et al. [9] use intra-class discriminatory to differentiate between foreground and background info; Ahn et al. [10] proposed a deep neural network and an affinity network based on random fields to expand the target area; Oh et al. [11] used saliency as prior information to guide the expansion of seed regions; Huang et al. [12] used the region growth method to expand the seed regions and integrate them into a deep segmentation network; Jo et al. [13] proposed Puzzle-CAM, a method consisting of puzzle module and two regularization terms for discovering the most integrated regions in an object. However, activation-like graph localization networks or other pre-trained classification networks are only able to identify sparse, high-response discriminative regions in an image and lack an accurate description of the boundaries where the target regions are located.

This paper proposes an improved SEC weakly supervised image semantic segmentation method. This method enhances the initial seed leads generated by SEC. The self-attentive network is used to generate the saliency map, and then the saliency map is used to guide the seed region growth method to be extended to solve the problems of undergrowth and overgrowth in the SEC method.

2 SEC Method

The SEC (Seed, Expand, Constrain) segmentation network model is a classical, weakly supervised image semantic segmentation model based on image-level labeling proposed by Kolesnikov et al. [3]. The algorithm usually uses the CAM method to localize segmented objects to generate seeds. This method allows accurate localization and thus obtains the position information of the object. However, the sparse position information obtained by this method cannot be directly used as supervisory information, so subsequent expansion of the region based on the seed is required. Kolesnikov et al. [3] defined the seed loss function as Eq. 1

$$L_{\text{seed}} (f(X), T, S_c) = -\frac{1}{\sum\limits_{c \in T} |S_c|} \sum_{c \in T} \sum_{u \in S_c} \log f_{u,c}(X) \tag{1}$$

The above equation denotes the set of locations labeled as class c, while $f_{u,c}(X)$ denotes the probability that region u is predicted as class c in image X. T is a subset of the category labels and represents the object class labels contained in the currently processed image [2]. This loss function can constrain the training of the neural network to motivate the segmentation network to focus only on the highlighted focal regions that are more discriminative for the target object during the learning process and ignore the irrelevant regions to improve the classification prediction.

In the process of seed growth, SEC first obtains a coarse semantic mask through the segmentation network and then calculates the seed loss by combining the initial seed. At the same time, the global weight pooling operation is performed on the mask to calculate the growth loss [14], and finally, the segmentation mask is updated by combining the seed loss and the growth loss to realize the growth of the seeds. To expand the initial sparse segmentation mask, Kolesnikov et al. [3] designed to expand the initial sparse segmentation mask, Kolesnikov et al. [3] designed the corresponding growth loss function using $G_c(f(X), d_+)$, $G_c(f(X), d_-)$, $G_{c^{bg}}(f(X), d_{bg})$ to denote the decay parameter of the global weight pooling:

$$L_{expand}(f(X), T) = -\frac{1}{|T|} \sum_{c \in T} \log G_c(f(X); d_+)$$
$$-\frac{1}{|c' \backslash T|} \sum_{c \in C' \backslash T} \log(1 - G_c(f(X); d_-)) - \log G_{c^{bg}}(f(X); d_{bg}) \tag{2}$$

If there are only seeds and expansion, then as the network is trained, the area of the mask will keep increasing until it exceeds the size of the object itself, and the results obtained will not meet the requirements [15]. So, in general, there will also be constraints on the boundaries of the objects, limiting their expansion. I.e., limiting the boundaries of the segmented objects so that they do not cross the boundary [16]. Kolesnikov et al. [3] designed the constraint loss function to achieve the objective of the boundary constraint. Equation 3:

$$L_{constrain}(X, f(X)) = \frac{1}{n} \sum_{u=1}^{n} \sum_{c \in C} Q_{u,c}(X, f(X)) \log \frac{Q_{u,c}(X, f(X))}{f_{u,c}(X)} \tag{3}$$

This loss function enhances the quality of the labeled mask by minimizing the computation of the difference between the neural network and the CRF output so that the mask output from the segmentation network and the CRF output converge as much as possible.

3   Improve the SEC

3.1   Seed Generation

The initial seed cues are generated by further processing the saliency map generated by the CAM method or the Grad-CAM method on the VGG16 classification network model, which has been proved to be equivalent to the weight parameters trained by the CAM method using mathematical formulas [17]. The initial seed leads generated by this class of methods suffer from sparsity problems. In this paper, the Grad-CAM++ method [2] is introduced to generate the initial seed leads.

By further improving the way the weights have been created, the GRAD-CAM++ model provided in the literature [2] enables more precise localization of the generated saliency zones. The GRAD-CAM++ model models each pixel's contribution to the final feature map by applying the gradient of that class score to each pixel value in the feature map to the relu function [18] and then weighting the average to capture the importance of the weights to a specific activation map while removing the negative impact on the specific class scores. GRAD-CAM++ proposes a method for obtaining the gradient weights α_{ij}^{kc} for a particular category c and activation graph k. The score Y^c for a particular category c is obtained as follows.

$$Y^c = \sum_k \left\{ \sum_a \sum_b \alpha_{ab}^{kc} \cdot relu\left(\frac{\partial Y^c}{\partial A_{ab}^k} \right) \right\} \left[\sum_i \sum_j A_{ij}^k \right] \tag{4}$$

In the above equation (i, j) and (a, b) are iterators on the same activation graph A^k to avoid confusion. $\frac{\partial Y^c}{\partial A_{ab}^k}$ is feature map pixels that contribute to the presence of the object. Where relu is the Rectified Linear Unit activation function.

In this paper, after obtaining the saliency map of the target object by GRAD-CAM++ weakly supervised localization algorithm [19], the corresponding foreground seed cues are generated by calculating the pixel positions corresponding to the weights larger than a certain threshold.

3.2 Seed Region Growth

The generated seeds have high accuracy and confidence levels. However, there is an obvious problem that the initial seed segmentation mask is small and cannot form a semantic segmentation of the whole image [20]. Therefore, to get more information, we have to expand the seeds. The most direct way is to treat localization information as seeds growing in unmarked areas, which is also known as the seed area growth method.

A saliency map, which displays the saliency value of each pixel, often serves as a representation of saliency information, an intrinsic quality of images [19]. The significant zones that were segregated from the saliency map have clearly defined borders, making them useful for directing seed development. For a certain image and the corresponding saliency map.

$$sim_{i,j} = w_{i,j} \| I(x_i, y_i) - I(x_j, y_j) \| \tag{5}$$

The difference between the significance values of two pixels is computed using $w_{i,j}$. The seed development process can then be affected by $w_{i,j}$ to increase the likelihood that pixels with identical significance values would spread [21]. The information from the HSV color space is utilized to compute the similarity. These are the significance weights $w_{i,j}$:

$$w_{i,j} = \exp(|S(x_i, y_i) - S(x_j, y_j)|) \tag{6}$$

The saliency mapping pixel at position (x, y) has a value of $S(x, y)$. The growth similarity criteria P is thus provided by:

$$P_{i,c}(\theta) = \begin{cases} True & \text{if } sim_{i,c} < \theta \\ False & \text{otherwise.} \end{cases} \tag{7}$$

c stands for the seed pixel with a label. If pixel i at position (x_i, y_i) is next to pixel c and only the value of $P_{i,c}$ is true, pixel i can be given the same label as pixel c. To ensure that the developing region takes on the shape of the significant object, significance weights will make it simpler to broadcast pixels of equal significance.

3.3 Boundary Constraints

After the expansion process of the seed region, the target region has problems such as boundary ambiguity and boundary-blurring. For the problem of unclear target boundary segmentation, the fully connected conditional random field (Dense CRF) is used to optimize the segmentation model [22]. Previous applications of a conditional random field in computer vision have mainly been used to smooth the noise and couple the corresponding nodes in the image together so that the neighboring pixel points have the same label [24], and its main function is to remove the false prediction of weak classifiers, however, for deep convolutional neural networks, the ultimate goal is to obtain segmentation-accurate boundaries

$$Loss = -\sum_{c=1}^{C} w_c y_c \log(p_c) \tag{8}$$

Using a cross-entropy loss function with weights, the boundary constraint is achieved. This function is bottom of the cross-entropy loss with a weight parameter added for each category [23], as illustrated in Eq. 8. Where $W_c = (N - N_c)/N$ represents the weight, N the total number of pixels, and N_c the number of pixels belonging to category c in the pixel-level semantic annotation. The segmentation outcome is adjusted to more closely match the target's actual shape [25].

4 Experiments

To further illustrate the effectiveness of the algorithm, it was validated on the PASCAL VOC 2012 [20].

4.1 The Experimental Setup

Evaluation metrics. The intersection over union ratio IoU (intersection over union) is the most often used evaluation metric in the weakly supervised image semantic segmentation method based on picture-level labeling [15], which is the ratio of the sum of correctly segmented positive samples to incorrectly segmented positive samples, incorrectly segmented negative samples, and correctly segmented positive samples. To evaluate the overall segmentation accuracy of the algorithm for all target classes [26], the average intersection ratio MIoU of all target classes is usually used for representation. Where MIoU is defined as:

$$MIoU = \frac{1}{c+1} \sum_{i=0}^{c} \frac{p_{ii}}{\sum_{j=0, j\neq i}^{c} (p_{ij} + p_{ji}) + p_{ii}} \tag{9}$$

Dataset. The semantic segmentation dataset of PASCAL VOC 2012 is used in this experiment.

4.2 Results and Analysis

Our approach is contrasted with many weakly supervised linguistics segmentation methods that supervise using image-level labels. According to Table 1, the segmentation rates for the validation set and test set are 61.8% and 62.6% respectively. The MIoU performance of the suggested strategy improved 11.1% in the Pascal VOC 2012 validation set and 11.5% in the test set when compared to the SEC model. Despite STC [13] using additional training data, the findings of the approach in this paper outperform STC on the validation and test sets by 12.0 and 11.4 percent, respectively. DCSP [24], MCOF [25] and RDC [15] has good segmentation results. In comparison to this documented method, the methodology in this paper yields better outcomes. Although the results of the method are slightly lower than DSRG [11] on the test, this method is taller than DSRG [15] on the validation. In summary, the method in this paper outperforms some advanced weakly supervised image semantic segmentation algorithms, which validates the feasibility of the proposed method.

Table 1. Model cross-comparison results.

Methods	Val	Test
DCSP [24]	60.8	61.9
DSRG [11]	61.4	63.2
DCSM [7]	44.1	45.1
MCOF [25]	60.3	61.2
RDC [15]	60.4	60.8
SEC [3]	50.7	51.1
STC [13]	49.8	51.2
Ours	61.8	62.6

Fig. 1. Results of the suggested method's visual segmentation.

Some of the sample qualitative results are shown in Fig. 1. In which, the first row is the original image in the dataset. The second row is the manually true labeled segmented image. The third row is the segmentation result of the method of this paper. We analyze the different categories and segmentation results. From the segmentation results of the dog, bird, and bus, we can see that the method can get clear segmentation results with more accurate and detailed boundaries; from the segmentation results of the boat and human, we can see that small parts missing in the segmentation results, but the overall contour is more accurate; from the segmentation results of sheep, we can see that the small objects, in the distance, are segmented and the segmentation results of large objects in the near are more accurate. In summary, the above experimental results demonstrate the effectiveness of the weakly supervised semantic segmentation method improved in this paper.

5   Conclusion

During seed growth, salience values square measure accustomed weight the similarity between pixels to manage the expansion. Experiments were conducted with information in PASCAL VOC 2012 and also the results show that the planned methodology during this paper is effective. Currently, weak supervised linguistics segmentation supported image-level annotation has become a hot topic of current analysis. More refinement of this methodology are meted out on this basis within the future, and its application to medical image segmentation and different fields are tried.

References

1. Sun, F., Li, W.: Saliency Guided Deep Network for Weakly Supervised Image Segmentation (2018)
2. Chattopadhyay, A.: Grad-CAM++: Improved Visual Explanations for Deep Convolutional Networks: Computer Vision and Pattern Recognition (2017)
3. Kolesnikov, A., Lampert, C.H.: Seed, expand and constrain: three principles for weakly-supervised image segmentation. In: Leibe, B., Matas, J., Sebe, N., Welling, M. (eds.) ECCV 2016. LNCS, vol. 9908, pp. 695–711. Springer, Cham (2016). https://doi.org/10.1007/978-3-319-46493-0_42
4. Zheng, S., Jayasumana, S., Romera-Paredes, B.: Conditional Random Fields as Recurrent Neural Networks. IEEE (2015)
5. Shimoda, W., Yanai, K.: Distinct class-specific saliency maps for weakly supervised semantic segmentation. In: Leibe, B., Matas, J., Sebe, N., Welling, M. (eds.) ECCV 2016. LNCS, vol. 9908, pp. 218–234. Springer, Cham (2016). https://doi.org/10.1007/978-3-319-46493-0_14
6. Pathak, D., Krhenbühl, P., Darrell, T.: Constrained Convolutional Neural Networks for Weakly Supervised Segmentation. IEEE (2015)
7. Tong, S., Lin, G., Shen, C.: Bootstrapping the performance of Weakly supervised semantic segmentation. In: IEEE/CVF Conference on Computer Vision and Pattern Recognition. IEEE (2018)
8. Fan, J., Zhang, Z., Song, C.: Learning integral objects with intra-class discriminator for Weakly-supervised semantic segmentation. In: 2020 IEEE/CVF Conference on Computer Vision and Pattern Recognition (CVPR). IEEE (2020)
9. Khoreva, A., Benenson, R., Hosang, J.: Simple does it: weakly supervised instance and semantic segmentation. In: 2017 IEEE Conference on Computer Vision and Pattern Recognition (CVPR) (2016)

10. Oh, S.J., Benenson, R., Khoreva, A.: Exploiting Saliency for Object Segmentation from Image Level Labels. IEEE, 2017
11. Huang, Z., Wang, X., Wang, J.: Weakly-supervised semantic segmentation network with deep seeded region growing. In: 2018 IEEE/CVF Conference on Computer Vision and Pattern Recognition (CVPR), IEEE (2018)
12. Jo, S., Yu, I.J.: Puzzle-CAM: Improved Localization Via Matching Partial and Full Features (2021)
13. Wei, Y., Liang, X., Chen, Y.: STC: a simple to complex framework for Weakly-supervised semantic segmentation. IEEE Trans. Pattern Anal. Mach. Intell. **39**(11), 2314–2320 (2017)
14. Chattopadhyay, A., Sarkar, A., Howlader, P.: Grad-CAM++: Generalized Gradient-based Visual Explanations for Deep Convolutional Networks (2017)
15. Wei, Y., Xiao, H., Shi, H.: Revisiting Dilated Convolution: A Simple Approach for Weakly- and Semi-Supervised Semantic Segmentation. IEEE (2018)
16. Chang, Y.T., Wang, Q., Hung, W.C.: Weakly-Supervised Semantic Segmentation via Sub-category Exploration (2020)
17. Ahn, J., Kwak, S.: Learning Pixel-level Semantic Affinity with Image-level Supervision for Weakly Supervised Semantic Segmentation. IEEE (2018)
18. Jiang, P.T., Hou, Q., Cao, Y., Cheng, M., Wei, Y., Xiong, H.: Integral object mining via online attention accumulation. In: 2019 IEEE/CVF International Conference on Computer Vision (ICCV), pp. 2070–2079 (2019)
19. Pathak, D., Shelhamer, E., Long, J.: Fully Convolutional Multi-class Multiple Instance Learning Computer (2014)
20. Everingham, M., Eslami, S., Gool, L.V.: The pascal visual object classes challenge: a retrospective. Int. J. Comput. Vision **111**(1), 98–136 (2015)
21. Di, L., Dai, J., Jia, J.: Scribble-supervised: scribble-supervised convolutional networks for semantic segmentation. In: 2016 IEEE Conference on Computer Vision and Pattern Recognition (CVPR), IEEE (2016)
22. Wei, Y.: Object region mining with adversarial erasing: a simple classification to semantic segmentation approach. In: 2017 IEEE Conference on Computer Vision and Pattern Recognition (CVPR), IEEE (2017)
23. Zhang, X., Wei, Y., Feng, J.: Adversarial Complementary Learning for Weakly Supervised Object Localization. IEEE (2018)
24. Chaudhry, A., Dokania, P., Torr, P.: Discovering class-specific pixels for Weakly-supervised semantic segmentation. In: BMVC (2017)
25. Wang, X., You, S., Li, X.: Weakly-Supervised Semantic Segmentation by Iteratively Mining Common Object Features. IEEE (2018)
26. Hong, S., Yeo, D., Kwak, S.: Weakly Supervised Semantic Segmentation Using Web-Crawled Videos. IEEE Computer Society (2017)
27. Fan, J., Zhang, Z., Song, C., Tan, T.: Learning integral objects with intra-class discrimina-tor for weakly-supervised semantic segmentation. In: CVPR, pp. 4283–4292 (2020)

A Document Image Quality Assessment Algorithm Based on Information Entropy in Text Region

Zongrui Zhang[1,2]([✉]), Jian Qiu[1,2], and Hao He[1,2]([✉])

[1] State Key Lab of Advanced Optical Communication System and Network, School of Electronic Information and Electrical Engineering, Shanghai Jiao Tong University, Shanghai, China
zzr965586083@sjtu.edu.cn, qiujian1223@sjtu.edu.cn, hehao@sjtu.edu.cn
[2] China Institute for Smart Court, Shanghai Jiao Tong University, Shanghai, China

Abstract. The quality of the image is critical to Optical Character Recognition (OCR), poor quality images will lead OCR to generate unreliable results. There are relative high ratio of low quality images in practical OCR-based application scenarios, how to evaluate quality of image and filter out unqualified images by document image quality assessment (DIQA) algorithms effectively is a big challenge for these scenarios. Current DIQA algorithms mainly focus on the overall image features rather than the text region, while the quality of the text region is dominant factor for OCR. In this paper, we propose a document image quality assessment algorithm based on information entropy in text region of image. Our algorithmic framework mainly consists of three networks to detect, extract and evaluate text region in image respectively. We build a quality prediction network based on HyperNet, and use the information entropy of the text region as the score weight, so that the final score can reflect the quality of the text region better. Finally, testing results on benchmark dataset SmartDoc-QA and our constructed dataset DocImage1k demonstrate that the proposed algorithm achieves excellent performance.

Keywords: DIQA · OCR · HyperNet · Information entropy

1 Introduction

With the advancement of digitization, each industry needs to intelligently get information from the document images. Optical character recognition(OCR) has played an important role because it can extract text information from the paper document images. However, the quality of the image is crucial to the results of OCR recognition. Document images with noise make OCR misidentified. How to find low-quality images in a large number of document images quickly and

N. Xiong et al. (Eds.): ICNC-FSKD 2022, LNDECT 153, pp. 629–638, 2023.
https://doi.org/10.1007/978-3-031-20738-9_72

efficiently is a challenge for OCR. Therefore, we need document image quality assessment (DIQA) to give quality scores for document images. The image with high score will be input into OCR for text extraction and processing, and the image with low score will remind the user to re-upload.

The goal of the image quality assessment (IQA) task is different from that of DIQA. The downstream tasks of IQA are mostly target recognition or classification tasks, while the downstream tasks of DIQA are to ensure the effectiveness of OCR recognition. IQA and DIQA also focus on different features. IQA focuses on the overall image features, while DIQA focuses on features of text regions. In fact, the features of the text area determine the quality of the document image. Therefore, we use information entropy to evaluate the information content and weight of different text regions.

In this work, we propose a DIQA algorithm based on information entropy in text region of image and a new dataset named DocImage1k in this paper. Our algorithmic framework mainly consists of three networks to detect, extract and evaluate text region in image respectively. We locate text regions through the text region detection network and calculate the information entropy of each text region as weights. Then the quality feature vector of the text region is obtained through the quality feature extraction network. Finally, we build a quality score prediction model based on the HyperNet [1], and use the information entropy as the weight to calculate the quality score. We complete experiments on benchmark dataset SmartDoc-QA [2] and our proposed dataset DocImage1k. Our proposed algorithm evaluates the results closest to the real scores, achieving excellent performance.

2   Related Works

Early scholars mainly implemented the DIQA algorithm based on handcrafted features. Blando [3] proposed that the quality of the document image can be determined by the number of white connected components and white patches. Similarly, Souza [4] computes document image quality scores by collecting information on the number of connected components. Cannon [5] analyze background blobs and broken characters to evaluate the quality scores of document images. Peng [6] used stroke gradients and average aspect ratios to obtain quality. Kumar [7] developed a DIQA method based on median filtering and the difference between the original grayscale images. Bui [8] used a similar approach. Although DIQA based on handcrafted features is sensitive to a certain type of distortion features, it cannot evaluate complex document images and lacks efficiency.

In recent years, Convolutional Neural Networks (CNNs) have been successfully applied to various computer vision tasks, while many results have been achieved in DIQA. Li [9] proposed that a text line quality evaluation model can be constructed based on CNN. By detecting the OCR detection results of each text line in the document image, the quality score of each page of the document is comprehensively given by the OCR scores of all text lines. Peng and

Wang [10] proposed a Siamese CNN-based method to extract deeper document image quality features. Lu [11] proposed a deep convolutional neural network DIQA model based on OCR accuracy descriptors, in which the first stage of the network invokes the rich knowledge learned in natural scenes to help the model learn the quality features of document images. Rodin [12] proposed a two-stage DIQA model. The first stage extracts the blur and size features of the text, and the second stage evaluates the quality of OCR according to the extracted features.DIQA based on deep learning often pays attention to the features of the overall image and ignores the features of the text region. In fact, the features of the text region more affect the quality score of the document image.

3   Method

In this section, we will introduce the overall architecture of our model and then explain the details of individual networks.

3.1   Overall Architecture

The proposed algorithm framework for DIQA based on text region quality features is mainly composed of three parts, namely text detection network, text region quality feature extraction network, and quality prediction network based on HyperNet and information entropy.

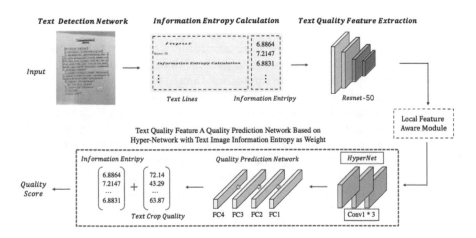

Fig. 1. The framework of proposed algorithm.

3.2   Text Detection Network

The front end of the DIQA network we built is a text detection network based on DBNET [13]. The structure diagram is shown in Fig. 2. We first use CNN to extract features from the input image and get feature maps. We use the lightweight network MobileNetv3 as the backbone network and use the FPN structure to obtain multi-scale features. Here we extract 4 feature maps of different scales for splicing, and then use the structure of one convolutional layer and two transposed convolutional layers to obtain the predicted probability map and threshold map. Finally, we use the DB method to get an approximate binary graph.

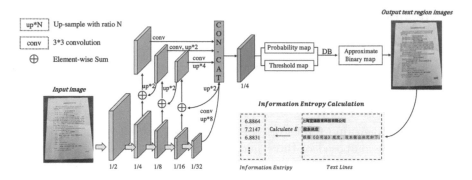

Fig. 2. The framework of text detection network and information entropy calculation process.

In a document image, some areas are significantly more important than others. For example, the title area is more important than a random sentence in the text and a dense sentence is more important than a sentence with less content. Therefore, we need to make a quantitative indicator for the importance of the text area as a weight. The quality scores of text regions with high weights are more important than the quality scores of text regions with low weights. The spatial feature quantity of the gray distribution we define is given by the neighborhood gray values of the image. In this process, the feature two-tuple formed by the gray value of the image pixel is denoted as (i, j), where i represents the gray value of the pixel, and j represents the mean gray value of the neighborhood. The comprehensive features of the gray value at a pixel position and its surrounding pixel gray distribution are shown in formula (1):

$$P_{ij} = f(i, j)/N^2 \tag{1}$$

Among them, $f(i, j)$ represents the frequency of (i, j) , N represents the scale of the image, and the discrete image information entropy calculation formula is shown in formula (2). We denote the computed image information entropy by E.

$$E = \sum_{i=0}^{255} p_{ij} \log p_{ij} \tag{2}$$

As shown in Fig. 2, after we detect the text line images named crops in the document image based on the text detection network, we use formula (1) to calculate the information entropy of crops. These information entropies will be used as the corresponding weights for their quality scores in the subsequent prediction networks.

3.3   The Text Region Quality Feature Extraction Network

The quality feature extraction network in this paper focuses on analyzing the quality of the input text line image and outputs multi-scale quality features to the quality prediction network. We design a text region quality feature extraction module to obtain multi-scale quality features. As shown in Fig. 3.

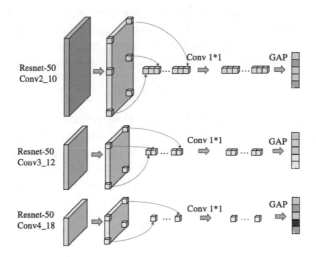

Fig. 3. The framework of text region distortion aware module.

Different from image quality assessment algorithms, our designed text region distortion perception module aims to effectively extract the quality features of text regions in images.We use the $conv2\_10$, $conv3\_12$, $conv4\_18$ layers of ResNet50 to extract document image multi-scale features, divide the document image multi-scale feature map into non-overlapping patches, stack the patches along the channel dimension, perform 1×1 convolution and global average pooling into vectors. Then we output the multi-scale quality feature vector v_x based on the average pooling layer and the fully connected layer.

3.4   The Quality Prediction Network Based on HyperNet and Information Entropy

Traditional deep learning document image quality prediction models take input images and map them directly to quality scores. This predictive model means extracting the same type of quality features to predict different images. Therefore, this traditional quality assessment method is flawed because different images cannot be assessed for their quality with the same set of rules.

Therefore, we can solve this problem based on HyperNet, which dynamically constructs different rules to evaluate the quality of the image based on its content. The learning of the HyperNet-based DIQA model we built is as follows:

$$\varphi\left(x, \zeta_x\right) = q \tag{3}$$

where ζ_x is the quality-aware rule of the super-network based quality score prediction network, x is the document image, q is the document image quality score, the ζ_x is not fixed and depends on the document image content itself.

And we express the dynamically changing quality-aware ruleζ_x as follows:

$$\zeta_x = H(\Gamma(\boldsymbol{x}), \boldsymbol{\Psi}) \tag{4}$$

where H is the HyperNet mapping function, $\boldsymbol{\Psi}$ represents HyperNet parameters, and $\Gamma\left(x\right)$ represents the quality features extracted from the input image x. Therefore, the function of HyperNet is to learn the mapping from document image content to the rule of how to judge document image quality.Then we use a HyperNet to learn a quality-aware rule mapping from document image content to document image quality, obtaining a document image quality score.

In order to reduce the training cost, we use the output of the text region distortion aware module mentioned in Sect. 3.3 as the input of the HyperNet, which is expressed as follows:

$$v_x = \Gamma_{ms}(\boldsymbol{x}) \tag{5}$$

Our final HyperNet-based IQA model announcement in this paper is shown below:

$$(v_x, H(\Gamma(x), \boldsymbol{\Psi})) = q \tag{6}$$

In this work, we construct the structure of the HyperNet through weight generating branches and 1×1 convolutional layers. We generate two types of network parameters, fully connected layer weights and biases. We determine the output channels of the convolutional and fully connected layers according to the size of the corresponding layers in the target network. The final weights we get are rules for document image quality perception, and these rules will help guide the target network to predict image quality.

The quality prediction network we designed mainly consists of two steps. In the step 1, we design to obtain a preliminary quality score by constructing a network consisting of four fully-connected layers that receive multi-scale quality feature vectors as input and propagate through weight-determined layers. In the

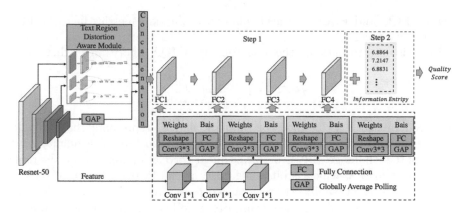

Fig. 4. The framework of quality prediction network.

step 2, we use the information entropy algorithm mentioned in Sect. 3.2 , take the information entropy of each text line image as its dynamic weight, and combine the quality score of the first step to give the final quality score by weighting and normalization.

4   Experiment

4.1   Dataset

In this work, we test the performance of our model on the publicly available document image dataset SmartDoc-QA as well as on our own constructed dataset DocImage1k, comparing with other DIQA work. We divide the dataset into training set and test set according to the ratio of 4:1.

SmartDoc-QA contains more than 2000 document images and their corresponding quality scores as labels, where the labeled quality scores are scored by the OCR engine.

DocImage1k contains 1276 real document images as well as document images with different noise and blur added and corresponding quality scores. First, we obtained document images and the text content in the document images based on real office scenes (after manual verification). Based on these clean document images, we construct document images with varying degrees of blur and noise using an image degradation algorithm. Finally, we identify the document image content based on the OCR engine, and the recognition result after manual re-examination is used as the quality score label of these document images.

4.2   Evaluation Metrics

We use Spearman's rank order correlation coefficient (SRCC) and Pearson's linear correlation coefficient (PLCC) to measure the data consistency of the quality scores generated by the model with the real quality labels of the document images. SRCC and PLCC are also common evaluation metrics in the field of image quality access.

SRCC is used to measure the monotonicity of IQA algorithm prediction.

$$SRCC = 1 - \frac{6 \sum_{i=1}^{N} d_i^2}{N(N^2 - 1)} \tag{7}$$

where N represents the number of samples, and d_i represents the difference between the subjective quality score ranking and the objective quality score ranking of the ith image. SRCC performs linear correlation analysis on the ranks of two target arrays, which is often regarded as the Pearson linear correlation coefficient after the two objects are arranged respectively, and has a wide range of applications. The value ranges from 0 to 1. When the performance value is equal to 1, it indicates that the two sets of data are completely consistent.

PLCC evaluates the correlation between the subjective score (MOS) and the objective score after nonlinear regression. Before calculating PLCC, a nonlinear regression operation needs to be performed on the objective score and the subjective score to establish the objective score. The logistic function for nonlinear regression is calculated as follows:

$$p(Q) = \beta_1 \left[\frac{1}{2} - \frac{1}{1 + e^{(\beta_2(Q-\beta_3))}} \right] + \beta_4 Q + \beta_5 \tag{8}$$

In Eq. 8, Q represents the original objective quality score, β_1, β_2, β_3, β_4, and β_5 are model parameters (parameters that need to be fitted), and p is the objective quality score after regression operation.

$$PLCC = \frac{\sum_{i=1}^{N} (s_i - \bar{s})(p_i - \bar{p})}{\sqrt{\sum_{i=1}^{N} (s_i - \bar{s})^2 \sum_{i=1}^{N} (p_i - \bar{p})^2}} \tag{9}$$

In Eq. 9, s_i and p_i represent the subjective quality score and objective quality score of the ith image, respectively, and $\bar{s}$ and $\bar{p}$ represent the average subjective quality score and the average objective quality score, respectively.

4.3   Result

We completed performance experiments on DIQA using datasets SmartDoc-QA and DocImage1k, comparing the performance with other DIQA models. The evaluation metrics are PLCC and SRCC. When the values of PLCC and SRCC are closer to 1, the scores predicted by the model are closer to the true scores. The results are shown in Table 1. Our algorithm has better performance on the metrics PLCC and SRCC. Compared with other algorithms, our algorithm can extract the quality features of the text area in the document image, so the calculated quality score is closer to the real quality of the image.

Table 1. Results on SmartDoc-QA and DocImage1k

DIQA model	SmartDoc-QA		DocImage1k	
	PLCC	SRCC	PLCC	SRCC
CORNIA [15]	0.9370	0.8620	0.7457	0.7913
MetricNR [16]	0.8867	0.8207	0.7120	0.6585
CG-DIQA [17]	0.9063	0.8565	0.7945	0.8316
Our algorithm	0.9410	0.8647	0.8432	0.8610

5 Conclusion

In this paper, we propose a new network architecture to deal with the difficult problem of traditional DIQA task, which is to extract quality features of text regions. Our network consists of a text detection network, a text region quality feature extraction network, and a quality prediction network based on HyperNet and information entropy. We also propose a new DIQA dataset DocImage1k. Our proposed algorithm achieves excellent performance on benchmark dataset SmartDoc-QA and our constructed dataset DocImage1k.

References

1. Kong, T., Yao, A., Chen, Y., Sun, F.: Hypernet: Towards accurate region proposal generation and joint object detection. In: 2016 IEEE Conference on Computer Vision and Pattern Recognition (CVPR), pp. 845–853 (2016)
2. Nayef, N., Luqman, M. M., Prum, S., Eskenazi, S., Chazalon, J., Ogier, J. M.: SmartDoc-QA: a dataset for quality assessment of smartphone captured document images-single and multiple distortions. In: 2015 International Conference on Document Analysis and Recognition (ICDAR), pp. 1231–1235 (2015)
3. Blando, L. R., Kanai, J., Nartker, T. A.: Prediction of OCR accuracy using simple image features. In: 1995 International Conference on Document Analysis and Recognition, pp. 319–322 (1995)
4. Souza, A., Cheriet, M., Naoi, S., Suen, C. Y.: Automatic filter selection using image quality assessment. In: 2003 International Conference on Document Analysis and Recognition, pp. 508–512 (2003)
5. Cannon, M., Hochberg, J., Kelly, P.: Quality assessment and restoration of typewritten document images. Int. J. Doc. Anal. Recogn. 80–89 (1999)
6. Peng, X., Cao, H., Subramanian, K., Prasad, R.,Natarajan, P.: Automated image quality assessment for camera-captured OCR. In: 2011 IEEE International Conference on Image Processing, pp. 2621–2624 (2011)
7. Kumar, J., Chen, F., Doermann, D.: Sharpness estimation for document and scene images. In: 2012 International Conference on Pattern Recognition (ICPR), pp. 3292–3295 (2012)
8. Bui, Q. A., Molard, D., Tabbone, S.: Predicting mobile-captured document images sharpness quality. In: 2018 International Workshop on Document Analysis Systems (DAS), pp. 275–280 (2018)

9. Li, H., Zhu, F., Qiu, J.: Towards document image quality assessment: a text line based framework and a synthetic text line image dataset. In: 2019 International Conference on Document Analysis and Recognition (ICDAR), pp. 551–558 (2019)

10. Peng, X., Wang, C.: Camera captured DIQA with linearity and monotonicity constraints. In: 2020 International Workshop on Document Analysis Systems, pp. 168–181 (2020)

11. Lu, T., Dooms, A.: A deep transfer learning approach to document image quality assessment. In: 2019 International Conference on Document Analysis and Recognition (ICDAR), pp. 1372–1377 (2019)

12. Rodin, D., Loginov, V., Zagaynov, I., Orlov, N.: Document image quality assessment via explicit blur and text size estimation. In: 2021 International Conference on Document Analysis and Recognition(ICDAR), pp. 281–292 (2021)

13. Liao, M., Wan, Z., Yao, C., Chen, K., Bai, X.: Real-time scene text detection with differentiable binarization. In: 2020 Conference on Artificial Intelligence, pp. 11474–11481 (2020)

14. Kim, J., Zeng, H., Ghadiyaram, D., Lee, S., Zhang, L., Bovik, A.C.: Deep convolutional neural models for picture-quality prediction: challenges and solutions to data-driven image quality assessment. IEEE Signal Process. Mag. **34**(6), 130–141 (2017)

15. Ye, P., Kumar, J., Kang, L., Doermann, D.: Unsupervised feature learning framework for no-reference image quality assessment. In: 2012 Conference on Computer Vision and Pattern Recognition(CVPR), pp. 1098–1105 (2012)

16. Nayef, N., Ogier, J. M.: Metric-based no-reference quality assessment of heterogeneous document images. In: Document Recognition and Retrieval XXII, vol. 9402, p. 94020L (2015)

17. Li, H., Zhu, F., Qiu, J.:CG-DIQA: no-reference document image quality assessment based on character gradient. In: 2018 International Conference on Pattern Recognition (ICPR), pp. 3622–3626 (2018)

RetinaHand: Towards Accurate Single-Stage Hand Pose Estimation

Zilong Xiao[1], Luojun Lin[1(✉)], Yuanxi Yang[2], and Yuanlong Yu[1]

[1] College of Computer and Data Science, Fuzhou University, Fuzhou, China
{211020024,ljlin}@fzu.edu.cn
[2] Minjiang University, Fuzhou, China
3192701115@stu.mju.edu.cn

Abstract. Due to the high joint flexibility and deformation degree of hands, hand pose estimation is more challenging in the detection task. In order to ensure the accuracy of prediction, two-stage algorithms are proposed recently, which requires a huge and redundant model structure and is difficult to implement end-to-end deployment. In this paper, we propose a novel dynamic single-stage CNN (RetinaHand) for end-to-end 2D handpose estimation of RGB images based on RetinaNet. RetinaHand firstly extracts image features through the backbone with dynamic convolutional layers. In the neck module, we propose Context Path Aggregation Network (CPANet) that fuse different scale features and expands context information to improve performance. In addition, we use the idea of multi-task learning to add a keypoints heatmap regression branch on the basis of the existing classification and bounding box regression branch, and use multi-task loss training model. Experimental results on the Eric.Lee and Panoptic datasets consistently show that our proposed RetinaHand has comparable performance to existing hand pose estimation methods at more efficient inference rates.

Keywords: Hand pose estimation · Multi-task learning

1 Introduction

In the process of interaction between humans and devices, hands are frequently used and highly flexible, and the posture of hands can convey a lot of information. Therefore, detection of hand information is an important topic for many applications such as Human-Computer Interaction (HCI), Virtual Reality (VR) and Augmented Reality (AR). Early hand pose estimation methods used wearable sensor devices to track hands and obtain position information.

Hand pose estimation is the location of key points of gesture recognition in the image. The traditional machine learning method is to make full use of the color, texture, contour and other features of the image, extract the hand feature information through Principal Component Analysis (PCA), Histogram of oriented gradient (HOG), scale-invariant feature and Local Binary Patterns

(LBP), and then classify and regression hand pose. Among them, many methods represented by CPM [1] adopt a two-stage method that predicts the hand mask first and then predicts the keypoint heatmap. However, such methods often require more complex models and greater time overhead. In order to reduce the model complexity of the hand pose estimation task and improve the efficiency of inference, we will not use the prior knowledge provided by the hand mask but perform multi-task learning based on a single-stage detection framework.

In this paper, we propose a single-stage RetinaHand for efficient end-to-end detection of 2D hand keypoints in RGB images. We designed our model based on the classical single-stage detection framework RetinaNet. In our method, we use deep neural network with dynamic convolutional layers to extract features, and solve the problem of instance differentiation caused by large degree of hand deformation in different images. Compared with the object detection task, hand pose estimation needs to regress the coordinates of 21 points, which will bring great difficulty to the localization performance of the model. Therefore, in the neck module, we propose CPANet to replace Feature Pyramid Networks (FPN) [2] in order to propagate semantic information of high-level features and fuse location information of low-level features.

Overall, our contributions can be summarized as follows:

(1) We propose a new single-stage hand pose estimation method based on RetinaNet. In this method, a multi-task learning strategy is adopted to predict the object category, hand bounding box and hand keypoints position.
(2) We use the dynamic convolutional layers in the feature extractor, so that the model with instance awareness capability can adapt to the deformation differences caused by the high flexibility of hands in different samples.
(3) In order to integrate features of different scales and increase the context awareness of the model, we proposed Context Path Aggregation Network (CPANet) in the neck module.
(4) Experimental results on Eric.Lee [3] and Panoptic datasets [4] demonstrate the effectiveness of our method.

2   Related Work

Because depth sensors are often limited in indoor environment, the application range of hand posture estimation is limited. In recent years, hand pose estimation based on RGB image has gradually become a research hotspot, and many methods based on deep learning framework have been proposed [5–7]. However, due to the dexterity of hand joints, similarity between fingers and self-occlusion, hand pose estimation still faces some challenges. Deeppose [8] first introduced convolutional neural network (CNNs) in hand pose estimation. It is well known that deep learning has high dependence on training data.

In this paper, 2D hand pose estimation based on RGB image is studied [9]. Proposed a 2D hand pose estimation method based on multi-scale heatmap

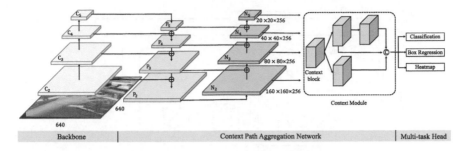

Fig. 1. Architecture of RetinaHand. The input image firstly extracts features through the backbone, then through context path aggregation network (CPANet) to strengthen feature fusion and expand the context information between pixels, and finally predict through the multi-task detection head. Context module consists of four context blocks.

regression, and used hand skeleton information to provide constraints for regression problems. In each estimation stage, this method trains each network separately, so that each stage achieves local optimal, but not global optimal. In order to overcome this defect, [10] adopt hierarchical network structure to obtain higher precision hand pose in a pose estimation cascade approach with multiple iterations. However, the method of obtaining the hand mask first and then predicting the key points often requires multi-stage model and large time cost. To solve this problem, Wang et al. [11] proposed a new method for real-time 2D hand pose estimation. It includes mask prediction stage and pose estimation stage, and the prediction results of these two stages are mutually beneficial.

3 Method

3.1 Overview

Our proposed overview of RetinaHand is shown in Fig. 1. We modify the single-stage detector RetinaNet to fit the task of hand pose estimation. In backbone, we use dynamic filter to replace original filter in feature extractor. In the neck module, we propose a new CPANet to fuse features of different scales and expand local context information. In the detection header, we added a heatmap regression branch for multi-task learning.

3.2 Backbone with Dynamic Convolutional Layers

Due to the high flexibility and self-occlusion of finger joints, different hand postures will lead to great differences in hand shape. The distance and angle between the keypoints vary greatly, which poses great challenges for detectors that are not suitable for hand masks. In order to cope with the variable shape of hand samples, backbone is designed as a network with dynamic convolutional layer, which adaptively challenges the parameters in the convolution kernel with the method of instance awareness.

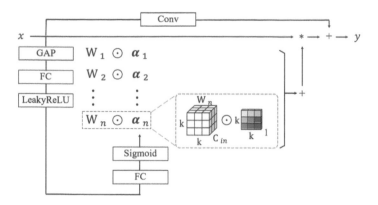

Fig. 2. Structure illustration of the basic block for backbone of RetinaHand. We use the attention mechanism (GAP + FC + LeakyReLU + FC + Sigmoid) to generate attention matrices with the same dimension as the filter space to weight a set of filters so that the model generates dynamic filter parameters based on the input instances.

As shown in Fig. 2, we use the attention mechanism to generate n attention matrices to dynamically weight n filters, so that the convolutional operation of linear combination filters is highly dependent on the input. Dynamic convolution operation is mathematically defined as

$$\triangle \boldsymbol{x} = (\boldsymbol{\alpha}_1 \boldsymbol{W}_1 + \cdots + \boldsymbol{\alpha}_n \boldsymbol{W}_n) * \boldsymbol{x}_0 \tag{1}$$

where $\boldsymbol{x}_0 \in \mathbb{R}^{h \times w \times c_{in}}$ and $\triangle \boldsymbol{x} \in \mathbb{R}^{h \times w \times c_{out}}$ denote the input features and the output features, c_{in} and c_{out} represent the input channel and output channel respectively; $\boldsymbol{W}_i \in \mathbb{R}^{k \times k \times c_{in}}$ denotes the ith convolutional filter; $\boldsymbol{\alpha}_i \in \mathbb{R}^{k \times k \times 1}$ is the attention matrix for weighting $\boldsymbol{W}_i$; $*$ denotes the convolution operation. For conciseness, here we omit the bias term.

In particular, in order to enhance the perception ability of feature extractor in position space, we specially designed $\boldsymbol{\alpha}_i$ as a $k \times k \times 1$ matrix, and weighted filter $\boldsymbol{W}_i$ in spatial dimension. Using the dynamic convolution layer with spatial attention to perceive the hand instances with obvious shape differences is of great help to the subsequent scale feature fusion.

For the stability of the model, the output feature is obtained by adding the feature map after dynamic convolution and the feature map after ordinary convolution, which is defined as

$$\boldsymbol{y} = \boldsymbol{W}_0 * \boldsymbol{x}_0 + \triangle \boldsymbol{x} \tag{2}$$

where $\boldsymbol{x}_0$ denotes the ordinary convolutional filter; $\boldsymbol{y}$ denotes the output features.

3.3   Context Path Aggregation Network

Since the hand attitude estimation task needs to predict the coordinates of 21 keypoints, it is a greater challenge than object detection task and requires the model to perceive the position information more accurately and specifically. In order to solve this problem, we propose Context Path Aggregation Network (CPANet) to propagate position information from bottom to top in the neck module, and uses local context information to assist positioning tasks, which is demonstrated in Fig. 1. We use $\{C_2, C_3, C_4, C_5\}$ to denote feature levels of backbone at different stages, and we use $\{P_2, P_3, P_4, P_5\}$ to represent the four layers of features generated by FPN. The top-down fusion path starts from C_5 and gradually approaches C_2. From C_5 to C_2, the space size is up-sampled with factor two inch by inch. Then, we represent new feature maps $\{N_2, N_3, N_4, N_5\}$ based on $\{P_2, P_3, P_4, P_5\}$ and set a bottom-up path. Similarly, from P_2 to P_5 the space size is down-sampled with factor two inch by inch.

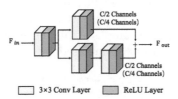

Fig. 3. Structure illustration of the context block. Context block consists of a parallel structure composed of 3×3 convolutional layers and ReLU layers. It does not change the height and width of the input feature map F_{in}, but reduce the channel dimension by half or a quarter.

Specifically, in order to reduce the spatial size of each feature map N_i we use 3×3 convolutional layer with stride 2. Then obtain the down-sampled feature map P_i through the horizontal channel. We use the same method to generate a subsequent series of subnetworks N_{i+1}. We use channel 256 of feature maps for these blocks, and all convolutional layers are followed by a ReLU.

Hand pose estimation needs to detect many keypoints in a smaller spatial size. After feature fusion, context module is used to further assist the model to perceive more accurate underlying position information, and context block is demonstrated in Fig. 3. $F_{in} \in \mathbb{R}^{H \times W \times 256}$ represents the feature map to be input for context block, which is composed of convolutional layers and ReLU layers. After two parallel path which including context blocks, we can obtain two fuature maps with 64 channels and one feature map with 128 channels respectively. We concatenate the three feature maps on the channel dimension to obtain the output $F_{out} \in \mathbb{R}^{H \times W \times 256}$ of the context module.

3.4 Multi-task Loss

In the detection head module we inherit the classification and bounding box regression subnets of RetinaNet, and we add the branch of heatmap prediction of hand keypoints. The loss of each part will be introduced in detail below.

Classification Loss. We use a binary classification problem to distinguish between hands and backgrounds. Due to the serious imbalance between negative samples (background samples) and positive samples (target samples) in target detection, class balance should be considered. In this project, we use hard example mining losses, so that the proportion of negative samples and positive samples is stable. The loss function can be formulated as:

$$L_{cls} = -\sum_i^m x_{pos}^* (i) \log x_{pos} (i) - \sum_j^n x_{neg}^* (j) \log x_{neg}' (j) , \tag{3}$$

where x_{pos} denotes positive samples; x_{neg}^* denotes first n samples hard-to-negative samples; x_{pos}^* and y_{neg}^* represent labels for hard-to-negative samples and positive samples respectively; m denotes the number of positive samples, and n denotes the number of hard-to-negative samples. n is three times as large as m. Specifically, the negative sample prediction probability values are sorted from small to large, which can be formulated as:

$$x_{neg}' = sort (x_{neg}) , \tag{4}$$

the first n negative samples are taken as hard-to-negative samples, and the remaining negative samples are discarded.

Box Regression Loss. We simply use L1 Loss for hand bounding box regression loss, which is formulated as:

$$L_{loc} = \sum_i^m |y^* (i) - y (i)| , \tag{5}$$

where y^* denotes ground truth box labels; y denotes predicted bounding box; m denotes the number of positive samples.

Heatmap Loss. We define the ground truth KCM of keypoint k is a 2D Gaussian, which center is standard deviation σ_{KCM}, p_k^* is a pixel point that the ground truth KCM of keypoint k is a 2D Gaussian centered around the keypoint. The loss function can be formulated as:

$$L_h = \sum_{k=1}^{21} \sum_{p \in I} \|C^* (p \mid k) - C (p \mid k)\|_2^2 , \tag{6}$$

where k is a hand keypoint, p is a hand pixel and I is the set of p, $C (p \mid k)$ is the predicted result of hand pose. $C^* (p \mid k)$ is the hand pose label, which is formulated as:

$$C^* (p \mid k) = exp \left\{ -\frac{\|p - p_k^*\|_2^2}{2\sigma_{KCM}^2} \right\} . \tag{7}$$

In general, for any training anchor i and hand keypoint k, we minimise the follwing multi-task loss:

$$L = \alpha L_{cls} (x_i, x_i^*) + \beta L_{loc} (y_i, y_i^*) + \gamma L_h (C (p \mid k), C^* (p \mid k)), \tag{8}$$

where α, β, γ are Balancing factors for these losses. In particular, for positive anchors, the proposed multi-task loss is calcuated, but for negative anchors, only classification loss is applied.

4 Experiments

4.1 Datasets

In this paper, we evaluate RetinaHand on two publicly datasets of hand poses: the CMU Panoptic Hand dataset (Panoptic) [4] and the Eric.Lee dataset [3].
Panoptic. The CMU Panoptic Dataset is a large-scale multi-view and multi-person 3D attitude Dataset. Currently, it contains 65 sequences and 1.5 million 3D skeletons. They built an impressive 360-degree motion capture dome that included 480 VGA cameras (25FPS), 31 HD cameras (30FPS), 10 Kinect2 sensors (30FPS), and 5 DLP projectors. In particular, it includes social scenes with multiple people. Multi-person 3D pose estimation methods usually extract partial data for evaluation.
Eric.Lee. Eric.Lee is a new multi-view hand pose dataset that provides color images of hands and different types of annotations for each sample, namely 2D and 3D positions on the boundary boxes and hand joints. It contains 49,062 images of hands in different and complex situations

4.2 Experimental Setting

Experimental Details. All our experiments are conducted on RTX 3090 GPU with PyTorch 1.9.0. We train the RetinaFace using SGD optimiser (momentum at 0.9, weight decay at 0.0005, batch size of 8 × 4). We use ResNet-50 [12] as the backbone. We replace the normal convolutional layers in the original network structure with dynamic convolutional layers. During training, anchors are matched to a ground-truth box when IoU is larger than 0.5, and to the background when IoU is less than 0.3. The learning rate starts from 10^{-4}, rising to 10^{-3} after 20 epochs, then divided by 10 at 100 and 155 epochs. The training process terminates at 200 epochs.
Evaluation Metric. To make a quantitative comparison, We use the standard Averaged Precision (AP) [13] and Percentage of Correct Keypoint (PCK) [14] as the evaluation metrics for bounding box regression and 2D hand pose, respectively. The PCK is a classical evaluation index of hand pose estimation algorithm, which represents the accuracy of the predicted key points position.

4.3 Performance Comparison

To prove the performance of our model, we compare RetinaHand with CPM [1], Real Mask [16] and NSRM [15] in terms of prediction accuracy and time efficiency. As shown in Tables 1 and 2, our method is compared with three different threshold of σ_{PCK} in datasets Panoptic and Eric.Lee. On the Panoptic dataset,

Table 1. Detailed numerical results of PCK (in percent) evaluated at different thresholds on the Panoptic testing data.

Method	Time (ms)	$\sigma_{PCK} = 0.04$	$\sigma_{PCK} = 0.08$	$\sigma_{PCK} = 0.12$	PCK_{avg}
CPM [1]	131	55.25	81.45	88.80	75.17
NSRM-LDM(G1) [15]	81	59.20	83.54	89.81	77.49
NSRM-LPM(G1) [15]	83	59.81	*84.16*	90.26	78.08
Real mask [16]	115	56.22	80.76	85.15	74.04
Ours	*56*	*60.12*	83.63	*90.74*	*78.16*

Table 2. Detailed numerical results of PCK (in percent) evaluated at different thresholds on the Eric.Lee testing data.

Method	Time (ms)	$\sigma_{PCK}=0.04$	$\sigma_{PCK}=0.08$	$\sigma_{PCK}=0.12$	PCK_{avg}
CPM [1]	131	53.58	80.46	89.21	74.42
NSRM-LDM(G1) [15]	81	56.20	82.45	90.81	76.49
NSRM-LPM(G1) [15]	83	57.36	83.47	*92.37*	77.73
Real mask [16]	115	51.22	77.76	84.75	71.24
Ours	*56*	*58.79*	*84.03*	91.26	*78.02*

the PCK of our model is significantly improved when the threshold of σ_{PCK} is 0.04 and 0.12, and the average PCK exceeds that of other methods. On the Eric.Lee dataset, our model achieves optimal results for thresholds of σ_{PCK} is 0.04 and 0.08, and also the average PCK exceeds that of other methods. It is worth noting that in terms of reasoning time, our method only needs 56ms for a single image, which is a considerable progress compared with other methods, which will bring faster feedback and better experience for users.

5   Conclusion

In this paper, we propose a novel dynamic single-stage CNN (RetinaHand) for end-to-end 2D handpose estimation of RGB images. We use the dynamic convolutional layers in the feature extractor, so that the model with instance awareness capability can adapt to the deformation differences caused by the high flexibility of hands in different samples. In order to integrate features of different scales and increase the context awareness of the model, we proposed Context Path Aggregation Network (CPANet) in the neck module to achieve the state-of-the-art performance on different benchmarks. The results prove that our method can promote the development of one-stage hand pose estimation.

Acknowledgments. This work was supported by the Fujian Provincial Youth Education and Scientific Research Project (No. 650722) and the University-Industry Cooperation Project of Fujian Provincial Department of Science and Technology (No. 2020H6101).

References

1. Wei, S.E., Ramakrishna, V., Kanade, T., Sheikh, Y.: Convolutional pose machines. In: Proceedings of the IEEE Conference on Computer Vision and Pattern Recognition, pp. 4724–4732 (2016)
2. Lin, T.Y., Dollár, P., Girshick, R., He, K., Hariharan, B., Belongie, S.: Feature pyramid networks for object detection. In: Proceedings of the IEEE Conference on Computer Vision and Pattern Recognition, pp. 2117–2125 (2017)
3. EricLee: https://codechina.csdn.net/ericlee/handpose x (2021)
4. Joo, H., Liu, H., Tan, L., Gui, L., Nabbe, B., Matthews, I., Kanade, T., Nobuhara, S., Sheikh, Y.: Panoptic studio: a massively multiview system for social motion capture. In: Proceedings of the IEEE International Conference on Computer Vision, pp. 3334–3342 (2015)
5. Sun, X., Wei, Y., Liang, S., Tang, X., Sun, J.: Cascaded hand pose regression. In: Proceedings of the IEEE Conference on Computer Vision and Pattern Recognition, pp. 824–832 (2015)
6. Tang, D., Jin Chang, H., Tejani, A., Kim, T.K.: Latent regression forest: Structured estimation of 3d articulated hand posture. In: Proceedings of the IEEE conference on computer vision and pattern recognition, pp. 3786–3793 (2014)
7. Tompson, J., Stein, M., Lecun, Y., Perlin, K.: Real-time continuous pose recovery of human hands using convolutional networks. ACM Trans. Graph. (ToG) 33(5), 1–10 (2014)
8. Toshev, A., Szegedy, C.: Deeppose: Human pose estimation via deep neural networks. In: Proceedings of the IEEE Conference on Computer Vision and Pattern Recognition, pp. 1653–1660 (2014)
9. Kourbane, I., Genc, Y.: Skeleton-aware multi-scale heatmap regression for 2d hand pose estimation (2021). arXiv:2105.10904
10. Zhang, M., Zhou, Z., Deng, M.: Cascaded hierarchical cnn for 2d hand pose estimation from a single color image. Multimed. Tools Appl. 1–19 (2022)
11. Wang, Y., Zhang, B., Peng, C.: Srhandnet: real-time 2d hand pose estimation with simultaneous region localization. IEEE Trans. Image Process. 29, 2977–2986 (2019)
12. He, K., Zhang, X., Ren, S., Sun, J.: Deep residual learning for image recognition. In: Proceedings of the IEEE Conference on Computer Vision and Pattern Recognition, pp. 770–778 (2016)
13. Shrivastava, A., Gupta, A., Girshick, R.: Training region-based object detectors with online hard example mining. In: Proceedings of the IEEE Conference on Computer Vision and Pattern Recognition, pp. 761–769 (2016)
14. Yang, Y., Ramanan, D.: Articulated human detection with flexible mixtures of parts. IEEE Trans. Pattern Anal. Mach. Intel. 35(12), 2878–2890 (2012)
15. Chen, Y., Ma, H., Kong, D., Yan, X., Wu, J., Fan, W., Xie, X.: Nonparametric structure regularization machine for 2d hand pose estimation. In: Proceedings of the IEEE/CVF Winter Conference on Applications of Computer Vision, pp. 381–390 (2020)
16. Wang, Y., Peng, C., Liu, Y.: Mask-pose cascaded cnn for 2d hand pose estimation from single color image. IEEE Trans. Circuits Syst. Video Technol. 29(11), 3258–3268 (2018)

Land Use/Cover Change Estimation with Satellite Remote Sensing Images and Its Changing Pattern of Chebei Creek in the Rapid Urbanization Process

Junxiang Liu[⊠], Hongbin Wang, Zhong Xu, Weinan Fan, Wenxiong Mo, and Lin Yu

Guangzhou Power Supply Bureau, Guangdong Power Grid Co., Ltd, Guangzhou 510275, China
{liujunxiang,Wanghongbin,Xuzhong,Fanweinan,Mowenxiong,
Yulin}@guangzhou.csg.cn, 331787138@qq.com

Abstract. In order to quantitatively assess the impact of urbanization on the land use/cover change of the watershed, the Chebei Creek in Guangzhou City was selected as the study case. Based on the remote sensing images of Landsat satellites from 1978 to 2020, the quantitative estimation of land use/cover change in Chebei Creek was carried out by using Gradient Boosting Regression Tree (GBRT) algorithm. The land use/cover spatial-temporal dynamic change in Chebei Creek was analyzed. The results show that the urban land increased from 0.35% in 1978 to 33.78% in 2020, the forestry land changed from 41.90% to 36.08%, the farm land changed from 44.42% to 12.28%, while the grassland, bare land and water remained in a small area.

Keywords: Urbanization · Land use/cover change · Remote sensing interpretation

1 Introduction

Urbanization is a global trend in the past century, and the world's urban population exceeded the rural population for the first time in 2009 [1]. Developed countries started the process of urbanization earlier, but urbanization has not yet ended [2]. In contrast, developing countries started to urbanize later, but urbanization is growing faster. With the deepening of China's reform and opening-up policy, urbanization has got into a period of rapid development in China. In 1980, the total urban population in China was less than 200 million, accounting for only 19.39%. By 2011, the urban population surpassed the rural population for the first time [3]. At present, China has formed several giant urban agglomerations, among which the urbanization speed of the Pearl River Delta urban agglomeration ranks first among the major urban agglomerations in China. There are many highly urbanized watersheds in the Pearl River Delta urban agglomeration, and the Chebei Creek watershed is a typical case, which has also experienced rapid urbanization in the past four decades.

Urbanization has changed the land use/cover (LUC) type of watershed, and in most cases, it converts the permeable surface (vegetation land) to the impervious surface (urban land). Landsat satellite remote sensing images are an important data source for quantitative estimation of large-scale and long-series LUC [4]. This makes it possible to map watershed LUC changes spaning a large range of urbanization at low cost, particularly for those in developing countries where significant urbanization starts after 1978. For a large number of satellite remote sensing data, the automatic classification algorithm of remote sensing images is mainly used for interpretation [5]. Automatic classification algorithms are divided into supervised classification methods, semi-supervised classification methods and unsupervised classification methods. In this paper, the automatic classification algorithm based on gradient boosting tree is used to quantitatively estimate the LUC in the Chebei Creek watershed.

2   Study Area and Data

2.1   Study Area

Chebei Creek is a highly urbanized watershed in Guangzhou, with a drainage area of 65 km², a main stream of 20.4 km in length and an average slope of about 1.5%. The upstream are dominated by mountains and hills, while the downstream are dominated by plains. The terrain is high in the north and low in the south, and the height difference varies greatly, ranging from 2.0 to 390.0 m. Since the Chebei Creek watershed directly discharges the floodwater into the city center, there is a huge threat to the flood control safety of Guangzhou. Figure 1. Shows the location of Chebei Creek.

Fig. 1. Location and sketch map of Chebei Creek.

The Chebei Creek watershed is in the subtropical oceanic monsoon area, with an annual average temperature of 22.3 °C and an annual average rainfall of 1720 mm. Due to the influence of heavy rainfall in summer, flood disasters occur frequently in this area. The rapid urbanization of Guangzhou in the past three decades has largely changed the LUC in Chebei Creek watershed. At present, the main land use type in the basin is urban land. Rapid urbanization has accelerated the flooding process in the basin, causing many flood disasters in recent years [6, 7].

2.2 Data

The data needed in this paper are mainly digital terrain elevation model (DEM) and remote sensing image data of the watershed.

1. DEM data

The DEM data of the Chebei Creek watershed is produced based on the 1:10,000 vector topographic map of Guangzhou City, using ArcGIS 10.2 software, with a spatial resolution of 30m as shown in Fig. 2. DEM data was used to generate watershed area, also in ArcGIS 10.2 software. The average, highest and lowest elevations of Chebei Creek are 66.13 m, 390.08 m and 2.10 m respectively.

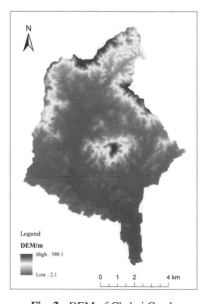

Fig. 2. DEM of Chebei Creek

2. Remote sensing image data

In this study, the archived Landsat satellite imageries have been downloaded from the US Geological Survey (USGS) website (http://glovis.usgs.gov/) to estimate the LUC changes of the studied watershed. Three imageries taken between 1978 and 2020 have been obtained at an average 20 year interval, basic information of these imageries is listed in Table 1.

Table 1. Information of Landsat imageries used

Platform	Sensor	Bands	Spatial resolution (m)	Acquisition year
Landsat 2	MSS	4, 5, 6, 7	60	1978
Landsat 5	TM	1, 2, 3, 4, 5, 7	30	2000
Landsat 8	OLI	2, 3, 4, 5, 6, 7	30	2020

3 Methodology

3.1 A Subsection Sample

Gradient Boosting Regression Tree (GBRT) constructs N weak classifiers and finally combines them into a strong classifier after multiple iterations [8, 9]. It iteratively improves the original model, so that the model generated next time has a smaller error than the previous model, and a new combined model is established in the gradient direction with the reduced residual error. The types of decision trees include regression and classification. GBRT can accumulate the output of all decision trees as the final output, but the characteristics of decision trees cannot be accumulated. Therefore, in GBRT, the classifiers basically use regression trees [10].

Assuming that $\{x_i, y_i\}_{i=1}^{N}$ is the training sample, $R = \{\beta_n, \alpha_n\}_1^N$ represents the parameter set, β is the weight of each classifier, and α is the parameter in the classifier, then the x function with R as the parameter is

$$F(X, R) = \sum_{n=1}^{N} \beta_n I(x, \alpha_n) \qquad (1)$$

In the formula, each $I(x, \alpha_n)$ represents a very small regression tree. By iterating N times, the parameter set of R is obtained, and the model is finally obtained by merging

$$F_n(x) = F_{n-1}(X) + \rho_n I_n(x, \alpha_n) \qquad (2)$$

4 Results

4.1 LUC Estimation Based on GBRT Algorithm

In this study, the Gradient Boosting Regression Tree (GBRT) classification algorithm is used to automatically classify the land use/cover of the above-mentioned three periods of remote sensing image data, and the land use/cover change data of the Chebei Creek watershed since 1978 are produced, including 1978, 2000 Year and 2020, a total of 3 years of data. The average accuracy of product classification reaches 93.1%, and the average Kappa coefficient reaches 0.908. The results are generally reasonable and accurate, and can be used for the analysis of urbanization process characteristics and other related studies. The land use/cover types are divided into 6 LUC types, including urban land (impervious surface), water body, forestry land, farm land, grass land and bare land. The results are shown in Fig. 3.

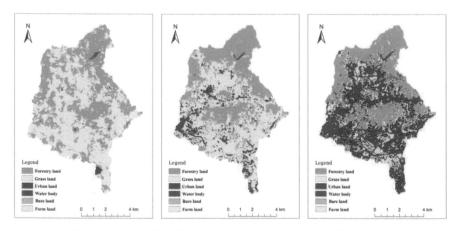

Fig. 3. Estimated LUC results of Chebei Creek from 1978 to 2020.

4.2   Analysis on the Temporal and Spatial Variation Characteristics of the LUC of Chebei Creek Watershed

In the interpretation results of land use/cover in the Chebei Creek watershed from 1978 to 2020, the land use mainly includes forestry land, grass land, farm land, urban land, water body and bare land. Over a period of about 40 years, land use/cover types have changed dramatically, as shown in Fig. 4.

During the urbanization process of the Chebei Creek watershed, the spatial and temporal pattern of forestry land changed greatly. The forestry land around the built-up areas of towns in the Chebei Creek watershed were first occupied by urban land during the process of urbanization expansion and transformed into urban land. At the same time, with the gradual expansion of urban land, some scattered small forestry land were developed, and the distribution of forestry land became more concentrated. The forestry land area has decreased significantly, from 29.78 km^2 in 1978 to 25.58 km^2 in 2020. In 1978, the Chebei Creek watershed had the largest forestry land area, accounting for 41.90% of the whole area.

Farm land has been the main land use type in the Chebei Creek watershed for a long time. In 1978, the area of farm land in the Chebei Creek watershed was 31.57 km^2, accounting for 44.42% of the area. In 2020, the area of farm land was 31.41 km^2,

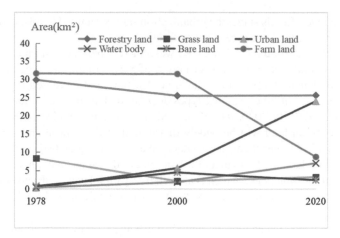

Fig. 4. LUC change line chart of Chebei Creek.

accounting for 44.30% of the area. By 2020, the area of farm land dropped to 8.71 km^2, accounting for 12.28%, a decrease of 22.86% compared to 1978 km^2.

From 1978 to 2020, the Chebei Creek watershed experienced a violent urbanization process. The urban land of the watershed increased from 0.25 km^2 in 1978 to 23.95 km^2 in 2020, an increase of 95.8 times. The area ratio has increased from 0.35% in 1978 to 33.78% in 2020.

In general, in the urbanization process of the Chebei Creek watershed, the area of urban land, forestry land and farm land changed the most. As one of the indicators of the urbanization process, the area of urban land has been showing a rapid growth trend during the study period, with a total increase of 23.70 km^2. The main sources of its growth are forestry land and farm land. Corresponding to the increase of urban land, the area of forestry land and farm land decreased all the way. During the study period, the forestry land decreased by 4.2 km^2, and the farm land decreased by 22.86 km^2, which contributed to the development of the urban land. Grass land fluctuates greatly in the process of urbanization, but the area has always been small. Bare land and water body have always been smaller and generally less fluctuating.

4.3 Urbanization Development Model of Chebei Creek Watershed

The Chebei Creek watershed has its own unique urban development model. In 1978, China was still in the early stage of urbanization. The overall urbanization rate of the Chebei Creek watershed was extremely low, and the watershed was mainly a rural area dominated by forestry land and farm land. In 1987, the Chebei Creek watershed began to urbanize, but until 2000, its urbanization process has been in a state of low-speed development. At this stage, urbanization first started from the middle reaches of the watershed, first from the middle reaches to the lower reaches, and second from the west to the east of the middle reaches. After steady and low-speed development at this stage, the urbanization rate of the Chebei Creek watershed reached 7.94% in 2000, which was at a moderate level. In this stage, the urbanization rate is relatively slow, and the spatial distribution of each land use type changes little.

From 2000 to 2020, the Chebei Creek watershed experienced a rapid urbanization process. The spatial and temporal distribution of various land use types in the watershed began to change dramatically. The area of urban land increased rapidly, and the corresponding areas of other land use types decreased. This dramatic change trend will continue until 2020. By 2020, the urban land will increase significantly compared to 1978, and the urban land will be widely distributed in the central and southern parts of the watershed. Among them, the southern part of the watershed is almost entirely covered by urban land, and the area of urban land at this stage accounts for 33.78% of the total area of the watershed, which is at a relatively high level of urbanization, and the urbanization cycle is basically over. By 2020, most of the farm land has been converted to urban land.

The Chebei Creek watershed completed an urbanization cycle in about 40 years and developed into a highly urbanized watershed.

5   Conclusions

Based on the Landsat satellite remote sensing images from 1978 to 2020, this paper takes the typical watershed in Guangzhou, the Chebei Creek watershed as the study area, using the Gradient Boosting Regression Tree (GBRT) automatic classification method to quantitatively estimate the land use/cover change in the Chebei Creek watershed. We analyzes and finds the spatio-temporal change characteristics of land use types in Chebei Creek watershed, and finds the urbanization development model of the watershed.

From 1978 to 2020, the land use types of Chebei Creek watershed mainly include forestry land, grass land, farm land, urban land, water body and bare land. In about 40 years, land use/cover types have changed dramatically. The proportion of urban land has increased from 0.35% in 1978 to 33.78% in 2020, the proportion of forestry land has changed from 41.90% to 36.08%, and the proportion of farm land has changed from 44.42% to 12.28%. Grass land, bare land and water body have always been smaller and generally less fluctuating, and most areas of the watershed have been urbanized. The Chebei Creek watershed completed an urbanization cycle in about 40 years and developed into a highly urbanized watershed. There are many highly urbanized watershed in the Pearl River Delta region, and Chebei Creek watershed is a typical case.

Acknowledgements. This study was supported by the Science and Technology Project of China Southern Power Grid Company Ltd. (nos.080037KK52190037 and GZHKJXM20190108).

References

1. United Nations, Department of Economic and Social Affairs, Population Division: World population prospects: the 2010 revision. New York, USA, (2010)
2. Mda, R., Reginster, I., Araujo, M.B., et al.: A coherent set of future land use change scenarios for Europe. Agr. Ecosyst. Environ. **114**(1), 57–68 (2006)

3. Fang, C., et al.: Comprehensive speed measurement and improvement path of China's urbanization development quality. Geograph. Res. **30**(11), 1931–1945 (2011)
4. Shi, Z., Ma, P., Wang, H., et al.: Research progress on land use/cover classification methods in remote sensing images. China Agricult. Sci. Bulletin **28**(12), 273–278 (2012)
5. Treitz, P.M., et al.: Application of satellite and GIS technologies for land-cover and land-use mapping at the rural-urban fringe—a case study. Photogramm. Eng. Remote. Sens. **58**(58), 439–448 (1992)
6. Xie, Y., Li, D., et al.: Research and application of numerical model for urban storm urban flooding. Adv. Water Sci. **16**(3), 384–390 (2005)
7. Zhou, H., Wang, C.: Guangdong Dongguan city urban flooding causes analysis and prevention measures. China Flood Drought Manag. **23**(4), 70–71 (2013)
8. Friedman, J.H.: Greedy function approximation: a gradient boosting machine. Annals Statis., 1189–1232 (2001)
9. Friedman, J.H.: Stochastic gradient boosting. Comput. Stat. Data Anal. **38**(4), 367–378 (2002)
10. Zhang, J., Xu, L., Lu, H.: Estimation of aboveground biomass of alpine pine using Landsat8 OLI and GBRT. J. Northeast Forestry Univ. **46**(8), 25–30 (2018)

Image Fusion Method Based on Improved Framelet Transform

Weiwei Kong[1,2,3(✉)], Yang Lei[4], and Chi Li[1,2,3]

[1] School of Computer Science and Technology, Xi'an University of Posts and Telecommunications, Xi'an 710121, China
{kongweiwei,lichi}@xupt.edu.cn

[2] Shaanxi Key Laboratory of Network Data Analysis and Intelligent Processing, Xi'an 710121, China

[3] Xi'an Key Laboratory of Big Data and Intelligent Computing, Xi'an 710121, China

[4] Engineering University of People's Armed Police Force, Xi'an 710086, China

Abstract. Recently, more and more images with different imaging sensors on the same scene can be obtained. To increase the storage efficiency and the overall visual performance, the image fusion technology has been becoming a research hot topic in the field of image processing. As a typical image fusion framework, framelet transform (FT) not only has much better ability to capture and describe the information of the source image, but also consumes much lower computational costs than its counterparts such as non-subsampled contourlet transform and non-subsampled shearlet transform. In this paper, for sake of further improving the fusion performance, the classical framelet transform (FT) is modified to be an improved version called improved FT (IFT), which is adept of expressing the main features and details of the source image. A series of simulation experiments have been conducted to verify the superiorities of IFT. The corresponding results demonstrate that IFT outperforms the current representative ones.

Keywords: Image fusion · Framelet transform · Filter kernel · Transformed domain

1 Introduction

With the emergence of a large number of imaging sensors, we can obtain many images towards the same scene. Often, owing to the inherent different imaging mechanism, these images also have different visual effects. For example, the computed tomography (CT) image and the magnetic resonance imaging (MRI) image mainly express the information of the bones or tissues, while the positron emission tomography (PET) image and the single-photon emission CT (SPECT) image place much emphasis on reflecting the biological activity of cells and metabolic activity of tissues or organs. Obviously, there are lots of complementary information between the above CT/MRI and PET/SPECT images. Therefore, the fusion of them may produce a single image with more reliable and more comprehensive descriptions.

N. Xiong et al. (Eds.): ICNC-FSKD 2022, LNDECT 153, pp. 656–662, 2023.
https://doi.org/10.1007/978-3-031-20738-9_75

With respect to the fusion methods, a lot of ones was presented in recent years. Overall, the mainstream fusion methods can be classified into several categories, including spatial domain-based methods, transformed domain-based methods [1–9], and deep learning-based methods [10–12]. As for the spatial domain-based methods, it mainly falls into two types, namely the weighting-based method and the neural networks-based one. The weights of the pixel pair require to be set in advance in the former one, but it often suffers from contrast reduction and spectrum distortion. With regard to the neural networks-based method, a number of parameters need to be precisely determined. However, it is a tricky task to acquire a set of proper parameters. What is the worse, different fusion applications often correspond to different parameters. Besides, deep learning-based methods often require a mass of training datasets and enormous computational costs, which has been becoming the bottleneck that limits the further development. Under the background, the transformed domain-based method has been widely applied in the field of medical image fusion. The core process of the transformed domain-based method can be mainly divided into three steps. Firstly, decomposition is conducted towards the source image to obtain many sub-images. Then, the fused sub-images can be obtained via several fusion rules. Thirdly, the final fused image can be reconstructed. Throughout the development of the transformed domain-based method, it can fall into two types namely subsampled transform methods and non-subsampled transform methods. The representatives of the former, such as the discrete wavelet transform (DWT) [1], the contourlet transform [2], the shearlet transform [3, 4], all involve down sampling, directly leading to the emergence of the Gibbs phenomenon. In comparison, the non-subsampled ones are shift-invariant due to the absence of the down sampling. As the representatives of the non-subsampled methods, NSCT [5, 6] and NSST [7–9] have been widely applied during the past decade, but they often suffer from enormous computational costs. Compared with NSCT and NSST, framelet transform (FT) [13] owns much better ability to capture the inherent information of the source image, and lower computational cost.

Under the background, to improve the performance of the traditional FT, a modified version called improved FT (IFT) is investigated and proposed here.

The remainder structure is arranged as follows. FT is reviewed in Sect. 2. Section 3 presents IFT in detail. Simulation experiments and results are reported in Sect. 4. The conclusion is drawn in the end.

2   Framelet Transform

The large computational resources costs of NSCT and NSST have increasingly been becoming a bottleneck restricting its applications in the occasions of highly demand of real-time performance. In comparison to its previous classical version named DWT, FT successfully overcomes the Gibbs effect owing to the property of shift-invariance. In addition, compared with NSCT and NSST, the information ability of FT is superior and the required costs are low.

Structurally, FT is similar to DWT but still has its inherent characteristics. Concretely speaking, FT was designed via tight frame filter banks which enable FT to own the properties of symmetry and shift-invariance.

Here, let the low-pass and high-pass filters be $h_0(n)$, $h_1(n)$ and $h_2(n)$ respectively. Thus, the mathematical expressions of $\varphi(t)$, $\psi_1(t)$ and $\psi_2(t)$ are as follows.

$$\varphi(t) = \sqrt{2} \sum_n h_0(n)\varphi(2t - n) \tag{1}$$

$$\psi_i(t) = \sqrt{2} \sum_n h_i(n)\varphi(2t - n), \quad i = 1, 2 \tag{2}$$

Based on Eqs. (1 and 2), any function named $f(t)$ is

$$f(t) = \sum_{k=-\infty}^{\infty} c(k)\varphi_k(t) + \sum_{i=1}^{2}\sum_{j=0}^{\infty}\sum_{k=-\infty}^{\infty} d_i(j, k)\psi_{i,j,k}(t) \tag{3}$$

where $c(k)$ and $d_i(j, k)$ are the scaling function and wavelet function related to $\varphi(t)$ and $\psi_i(t)$, respectively. The mathematical expressions of $c(k)$ and $d_i(j, k)$ are

$$c(k) = \int_{-\infty}^{\infty} f(t)\varphi_k(t)d(t) \tag{4}$$

$$d_i(j, k) = \int_{-\infty}^{\infty} f(t)\psi_{i,j,k}(t)d(t) \tag{5}$$

The two summations in Eq. (3) indicate the low-frequency function and two high-frequency ones of $f(t)$, respectively. Here, the combination of the above two summations is described as one-level FT of $f(t)$. The structure diagram of one-level FT is shown in illustrated in Fig. 1.

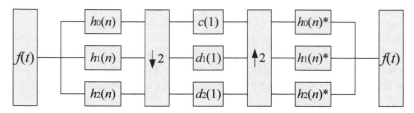

Fig. 1. Structure diagram of one-level FT

Similar to DWT, we can easily extend FT to a high dimensional version via using 1D FT along different directions. Firstly, 1D row transform is to get the data denoted by DATA. Then, 1D column transform is done on DATA. Suppose $\{\varphi(x), \psi_1(x), \psi_2(x)\}$ and $\{\varphi(y), \psi_1(y), \psi_2(y)\}$ are the basic functions of 1D FT along x and y directions, respectively. Then, the scaling and wavelet functions related to 2D FT can be obtained via 1D filters convolving.

3   Improved Framelet Transform

In the traditional FT, the commonly involved filter during the decomposition and reconstruction courses is the piecewise cubic framelet filter, whose concrete coefficients are as follows.

$$
\begin{cases}
D\{1\} = [1 \ \ 4 \ \ 6 \ \ 4 \ \ 1]/16 \\
D\{2\} = [1 \ \ 2 \ \ 0 \ \ -2 \ \ -1]/8 \\
D\{3\} = [-1 \ \ 0 \ \ 2 \ \ 0 \ \ -1]/16 * \text{sqrt}(6) \\
D\{4\} = [-1 \ \ 2 \ \ 0 \ \ -2 \ \ 1]/8 \\
D\{5\} = [1 \ \ -4 \ \ 6 \ \ -4 \ \ 1]/16
\end{cases}
\tag{6}
$$

$$
\begin{cases}
R\{1\} = [1 \ \ 4 \ \ 6 \ \ 4 \ \ 1]/16 \\
R\{2\} = [-1 \ \ -2 \ \ 0 \ \ 2 \ \ 1]/8 \\
R\{3\} = [-1 \ \ 0 \ \ 2 \ \ 0 \ \ -1]/16 * \text{sqrt}(6) \\
R\{4\} = [1 \ \ -2 \ \ 0 \ \ 2 \ \ -1]/8 \\
R\{5\} = [1 \ \ -4 \ \ 6 \ \ -4 \ \ 1]/16
\end{cases}
\tag{7}
$$

where "D" and "R" are the decomposition and reconstruction coefficients of the framelet filters, respectively.

It can be easily found that the original filter given in Eq. (7) includes both symmetric and anti-symmetric coefficients. For example, $D\{1\}, D\{3\}, D\{5\}, R\{1\}, R\{3\}$ and $R\{5\}$ are all symmetric kernels, while $D\{2\}, D\{4\}, R\{2\}$ and $R\{4\}$ are anti symmetric ones. However, twenty-five sub-images can be obtained once per decomposition, resulting in large computational costs. What is the worse, some information of the source image cannot be sufficiently extracted and captured via the filter. To this end, a modified FT called IFT is constructed, in which a simplified filter is devised. For sake of declining the number of sub-images produced and enhancing the information capturing ability, the coefficients of the new filter is given as

$$
\begin{cases}
D\{1\} = [4 \ \ 8 \ \ 4]/16 \\
D\{2\} = [4 \ \ 0 \ \ -4]/16 * \text{sqrt}(2) \\
D\{3\} = [-4 \ \ 8 \ \ -4]/16
\end{cases}
\tag{8}
$$

$$
\begin{cases}
R\{1\} = [4 \ \ 8 \ \ 4]/16 \\
R\{2\} = [-4 \ \ 0 \ \ 4]/16 * \text{sqrt}(2) \\
R\{3\} = [-4 \ \ 8 \ \ -4]/16
\end{cases}
\tag{9}
$$

Compared with the original filter, the kernel in the new filter is fewer. Thus, the number of the sub-images obtained will be less also, which is only nine. The next section will report the simulation experimental results based on different filters.

4 Experimental Results

4.1 Experimental Settings

For objectively evaluate the fusion performance of different methods, two pairs of pre-registered multimodal medical source images have been selected, which can be accessed on the website of Harvard medical school. All images all share the size of 256×256.

The simulation software is MATLAB R2019a. In addition, as two representative non-subsampled transform methods, NSCT and NSST are selected to be compared with IFT. As for the objective metrics, spatial frequency (SF) [14], edge information retention ($Q^{AB/F}$) [15], and sum of the correlations of differences (SCD) [16] are adopted in this paper. The larger the values of the above three metrics, the better the fusion performance. The same fusion rule namely "Ave-Max" is adopted in all of the methods for comparison.

4.2 Fusion Performance

Two pairs of source images are fused here. Note that the common medical images mainly involve four categories including the CT image, the MRI image, the PET image and the SPECT image. The first two types belong to the area of anatomical images, while the last two ones belong to the field of functional images. Therefore, the CT image and the MRI image are gray ones, but the PET image and the SPECT image are pseudo color ones.

In the first experiment, a pair of CT/MRI medical source images are fused by four methods, respectively. The corresponding performance is shown in Fig. 2.

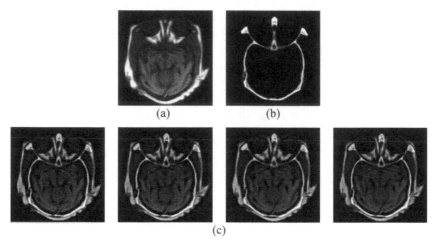

Fig. 2. Fusion results on the first experiment: (a) CT image. (b) MRI image. (c) Fused images based on four methods, including NSCT, NSST, FT and IFT.

In the second experiment, a pair of MRI/PET medical source images are fused by four methods, respectively. Note that the PET image is in pseudo color, and it should be converted into the gray version before fusion. Here, the YUV color space is used. The corresponding performance is shown in Fig. 3.

Due to the same fusion rule, the subjective visual performance of the four methods is approximate. For sake of carefully comparing the fusion effects, the objective evaluation is conducted, and the corresponding results is given in Table 1. The bold indicates the best result, and the italic denotes the sub-optimal result. It can be found that IFT generally outperforms the other three ones.

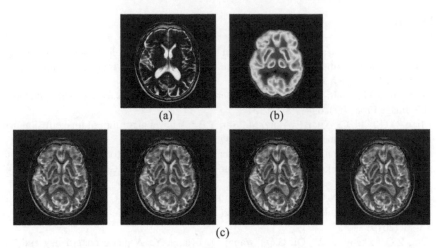

Fig. 3. Fusion results on the second experiment: (a) MRI image. (b) PET image. (c) Fused images based on four methods, including NSCT, NSST, FT and IFT.

Table 1. Objective evaluation results on two experiments

	CT/MRI			MRI/PET		
	SF	$Q^{AB/F}$	SCD	SF	$Q^{AB/F}$	SCD
NSCT	19.5219	0.6827	1.6200	29.6596	0.5700	1.3697
NSST	*19.7244*	0.6787	1.6007	**30.4444**	0.5668	1.3631
FT	18.1321	*0.7105*	**1.6594**	30.0273	*0.5837*	**1.4146**
IFT	**19.7996**	**0.7135**	*1.6213*	*30.3508*	**0.5987**	*1.3757*

The bold and italics indicates the best result among the methods.

5   Conclusions

In this paper, the improved version named IFT is proposed. Compared with its predecessors such as NSCT, NSST and classical FT, IFT is equipped with a new filter kernel to enhance the information capturing ability. Two pairs of source images are selected, and the fusion experiments are conducted on four different methods. The results demonstrate that IFT owns better performance compared with the current representative non-subsampled transform methods. It is noteworthy that, assuming the same fusion rule is adopted by all involved methods, only the fusion scheme is considered during the simulation experiments. How to devise more effective fusion rules and combine them with IFT are the core research direction in the future.

Acknowledgements. The work was supported by the Natural Science Basic Research Program of Shannxi Province of China under Grant 2022JM-369.

References

1. Vijayarajan, R., Muttan, S.: Discrete wavelet transform based principal component averaging fusion for medical images. AEU: Archiv fur Elektronik und Ubertragungstechnik: Electron Commun **69**(6), 896–902 (2015)
2. Darwish, S.M.: Multi-level fuzzy contourlet-based image fusion for medical applications. IET Image Proc. **7**(7), 694–700 (2013)
3. Miao, Q.G., Shi, C., Xu, P.F., Yang, M., Shi, Y.B.: A novel algorithm of image fusion using shearlets. Opt. Commun. **284**(6), 1540–1547 (2011)
4. Liu, X., Zhou, Y., Wang, J.J.: Image fusion based on shearlet transform and regional features. AEU: Archiv fur Elektronik und Ubertragungstechnik: Electron. Commun. **68**(6), 471–477 (2014)
5. Chang, L.H., Feng, X.C., Zhu, X.L., Zhang, R., He, R.Q., Xu, C.: CT and MRI image fusion based on multiscale decomposition method and hybrid approach. IET Image Proc. **13**(1), 83–88 (2019)
6. Zhu, Z.Q., Zheng, M.Y., Qi, G.Q., Wang, D., Xiang, Y.: A phase congruency and local Laplacian energy based multi-modality medical image fusion method in NSCT domain. IEEE Access **7**, 20811–20824 (2019)
7. Panigrahy, C., Seal, A., Mahato, N.K.: MRI and SPECT image fusion using a weighted parameter adaptive dual channel PCNN. IEEE Signal Process. Lett. **27**, 690–694 (2020)
8. Gai, D., Shen, X.J., Chen, H.P., Xie, Z.Y., Su, P.X.: Medical image fusion using the PCNN based on IQPSO in NSST domain. IET Image Proc. **14**(9), 1870–1880 (2020)
9. Ullah, H., Ullah, B., Wu, L.W., Abdalla, F.Y.O., Ren, G.H., Zhao, Y.Q.: Multi-modality medical images fusion based on local-features fuzzy sets and novel sum-modified-Laplacian in non-subsampled shearlet transform domain. Biomed. Signal Process. Control **57**(3), 101724 (2020)
10. Xiang, L., et al.: Deep-learning-based multi-modal fusion for fast MR reconstruction. IEEE Trans. Biomed. Eng. **66**(7), 2105–2114 (2019)
11. Liang, X.C., Hu, P.Y., Zhang, L.G., Sun, J.G., Yin, G.S.: MCFNet: Multi-layer concatenation fusion network for medical images fusion. IEEE Sens. J. **19**(16), 7107–7119 (2019)
12. Liu, Y., Chen, X., Wang, Z.F., Wang, Z.J., Ward, R.K., Wang, X.S.: Deep learning for pixel-level image fusion: Recent advances and future prospects. Inf. Fusion **42**(7), 158–173 (2018)
13. Cai, J.F., Ji, H., Liu, C.Q., Shen, Z.W.: Framelet-based blind motion deblurring from a single image. IEEE Trans. Image Process. **21**(2), 562–572 (2012)
14. Zheng, Y., Essock, E.A., Hansen, B.C., Haun, A.M.: A new metric based on extended spatial frequency and its application to DWT based fusion algorithm. Inf. Fusion **8**(2), 177–192 (2007)
15. Piella, G., Heijmans, H.: A new quality metric for image fusion. In: Proceedings of International Conference on Image Processing, 1, 173–176 (2003)
16. Aslantas, V., Bendes, E.: A new image quality metric for image fusion: the sum of the correlations of differences. AEU: Archiv fur Elektronik und Ubertragungstechnik: Electron. Commun. **69**(12), 1890–1896 (2015)

Video Stream Forwarding Algorithm Based on Multi-channel Ring Buffer Technology

Yingbao Cui[✉], Xiaoguang Huang, Jing Liu, Linyu Zhang, and Xiaodong Wang

State Grid Information and Telecommunication Group, Beijing, China
{cuiyingbao,huangxiaoguang,liujing,zhanglinyu,
wangxiaodong}@sgitg.sgcc.com.cn, cyb_happy@163.com

Abstract. With the development of the country and the progress of the society, people's daily life is more and more inseparable from electricity. With the rapid development of the speed and scale of my country's power grid construction, the scale of the power grid has jumped to the first place in the world. Guaranteeing the stability, continuity and quality of power supply has become an important task of the State Grid Corporation, which requires daily power inspections. However, at present, most of my country's electric power inspections rely on manual labor. On the one hand, due to the influence of geographical space, complex terrain, and changing weather, manual inspections have many limitations and dangers. On the other hand, due to the rising labor cost, the cost of power inspection is becoming more and more heavy. If power companies want to improve the efficiency and accuracy of inspections, they must use emerging technology push streaming live platforms to help inspectors in their daily inspections. In order to improve the inspection stability and inspection efficiency, this paper proposes and implements a multi-channel ring buffer video stream forwarding algorithm. This algorithm solves the problem of long delay in video stream forwarding on traditional inspection live broadcast platforms, so that inspectors can receive video streams in a timely and stable manner and play them smoothly. Based on the video fluency experiment in the production environment, the effectiveness of this algorithm is verified.

Keywords: Grid stability · Inspection video stream · Multi-channel · Ring buffer · High performance

1 Introduction

As an important application form of streaming media technology, live video has developed rapidly in recent years. Compared with traditional streaming media on-demand services, live video streaming is more challenging. The live video inspection service requires higher real-time performance. Therefore, in order to improve the inspection quality of electric power inspectors, it is necessary not only to consider factors such as video quality, smoothness, and freeze, but also to minimize the inspection delay. Streaming media refers to the video data that is streamed during the live broadcast of the inspection. The transmitted video is encoded and compressed to reduce the amount of transmitted data.

© The Author(s), under exclusive license to Springer Nature Switzerland AG 2023
N. Xiong et al. (Eds.): ICNC-FSKD 2022, LNDECT 153, pp. 663–672, 2023.
https://doi.org/10.1007/978-3-031-20738-9_76

Currently, it is compressed in the H.264 encoding format. The inspection client can pass the encoding device. The video can be played after decoding the compressed data stream.[1, 2].

Real-time streaming is a transmission method in streaming media, mainly to ensure that the data on the transmitted streaming media matches the network transmission performance, so that inspectors can obtain video streams in real time. Real-time streaming requires dedicated streaming media and transmission protocols. Dedicated streaming media servers include QuickTime Streaming Server, Windows Media Server, etc., and dedicated transmission protocols include RTSP (real-time Streaming Protocol) or MMS (Microsoft Media Server).

Streaming media is intermittently transmitted on the network by means of packet transmission. The TCP protocol is a reliable transmission that can ensure the sequence of packets. For continuous streaming media video data streams, the TCP protocol can be used to ensure its sequence. Since the network changes dynamically, and the video stream data collected is obtained continuously and at a constant speed, it is necessary to ensure that the speed of the two is matched to achieve a real-time preview effect. To control the video sampling device, the user needs to feed back the video image after the control in real time, so it also requires the control of real-time performance.

Hikvision's front-end camera equipment uses the H.264 encoding standard to compress the captured images. The H.264 encoding and compression standard is widely used in the field of video compression due to its advantages of high compression and high quality and support for multiple network streaming media. At the same time, three formats of video frames are defined in the H.264 protocol. Among them, the video frame with the complete encoding format is called I frame, and the frame that only includes the difference part is generated by referring to the previous I frame, which is called P frame. There is also a reference A frame encoded with a frame before and after is called a B frame. Included in the I frame is a timed decoded refresh frame called an IDR frame. The purpose of defining IDR is to allow the video to be refreshed immediately during the playback process, playing the newly generated picture, and then discarding all historical frames generated between the IDR frames. This encoding method of H.264 provides a lot of convenience in the design and implementation of the video stream forwarding mechanism in the future, and also greatly helps in ensuring the real-time performance of video stream playback [3].

This paper studies the video stream forwarding technology. Aiming at the shortcomings of its inability to play smoothly in real time, this paper proposes a real-time stream multi-channel high-performance forwarding algorithm based on ring buffer. Improve the real-time performance of forwarding video, and improve the receiving and sending efficiency of the server and client.

2   Related Work

As a video-centric IoT solution provider, Hikvision will provide users with corresponding SDKs for secondary development and use when products are sold, and provide a set of algorithms for video stream collection and forwarding for users' reference. Hikvision's original forwarding algorithm is that the client directly connects to the Hikvision camera through the network protocol to obtain the video stream and play it in real time [4]. On the basis of the existing cameras in the laboratory, the video stream forwarding algorithm provided by the manufacturer is realized by using the VS2010 development environment. Read the video signal of the front-end camera and decode and display it in the front-end interface. It mainly includes three aspects: user registration module, preview module, capture module [5].

The user preview module is the core of the video stream forwarding mechanism provided by Hikvision. Before previewing, the user needs to register on the client. After the registration is successful, log in to the camera according to the IP of the IP camera, and complete the preparations before obtaining the video stream. The client calls the NET_DVR_RealPlay_V30 interface to obtain the video stream from the logged-in camera for playback. Playing images meets real-time requirements, but this forwarding mechanism only satisfies the real-time acquisition and playback of video streams in the case of a single user and single channel, and cannot meet the requests of multiple users to view video streams from cameras with different channels at different locations [6].

Based on the above research results, combined with the characteristics of live broadcast inspection in the power field, this paper proposes and implements a multi-channel ring buffer video stream forwarding algorithm. This algorithm solves the problem of long delay of video stream forwarding on traditional live broadcast platforms, so that the client can receive the video stream from the server in time and play it smoothly [7, 8].

3   Video Stream Forwarding Algorithm

3.1   Basic Process

In this part, a real-time stream multiplexing forwarding algorithm based on ring buffer is designed. The server and client respectively adopt multi-thread concurrency and buffering technology to improve the real-time performance of forwarding video and improve the receiving and sending efficiency of server and client. Mainly include:

(1)   The framework structure of ring buffer and multi-thread concurrency is established on the server side and the client side respectively;

(2)   When the user of the client side requests to view the video of a certain video sampling device ID, the client side sends the request message to the server side;

(3)   When the server receives the video request message, it requests the video from the NVR, and the main thread receives the video stream sent by the NVR, hooks it to the ring buffer, and the sub-thread corresponding to the video stream channel sends the video stream to the client;

(4)   After the main thread of the client receives the video stream, it mounts it on the ring buffer, and the sub-thread corresponding to the video channel plays it to the corresponding window. By improving the algorithm, the system can be optimized, the waste of system resources can be reduced, the comprehensive performance of the system can be improved, and the real-time user experience can be improved.

3.2 Algorithm Implementation Process

In the process of technical implementation, the video server regularly stores the video stream data by establishing a ring buffer, the video stream obtained from the NVR, and then sends the video stream in an orderly manner through the applied sub-thread and the main thread to operate the buffer asynchronously to the client. The specific implementation process is shown in the Fig. 1.

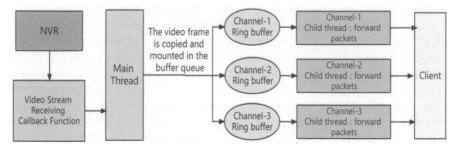

Fig. 1. Algorithm structure of multi-channel multi-buffer

The video server and client respectively establish a multi-thread concurrent frame structure based on ring buffer. Taking the server side as an example, the server creates a buffer and a sub-thread for each channel, each channel corresponds to a video sampling device, and the ring buffer consists of a ring array and several video data buffers. It points to the data buffer (Video Data Buffer referred to as VDB) received from the NVR.

The ring buffer includes but is not limited to the following members, as shown in the Table 1.

Table 1. Ring buffer member table

Name	Short name	Initial value
Video stream head pointer	Head	−1
Video stream end pointer	Tail	−1
Forwarded IDR frame	FuteureIDR	−1
Forwarded IDR frame	LastIDR	−1
Currently forwarding video frames	nCurP	−1
Channel number	nID	Null
Keyframe	nIDR	0

The main thread on the server side is responsible for placing the VDB received from the NVR on the ring buffer corresponding to the channel, and the sub-thread corresponding to each channel is responsible for extracting video data from the ring buffer and sending the video data to the client. The following is the process of the main thread and child thread manipulating the buffer.

Each buffer will have two threads to operate and access. The main thread is responsible for fetching the video stream into the buffer, the operation object is Tail, and the forwarding sub-thread is responsible for fetching the video stream from the buffer, and the operation object is Head.

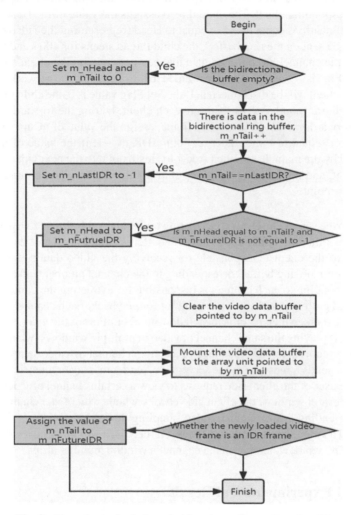

Fig. 2. Flow chart of sub-thread video stream forwarding algorithm

(1) The main thread operation process, as shown in Fig. 2: the buffer is empty in the initial stage, when video data is received for the first time, both Head and Tail are set to 0, and the video data is linked to the array unit pointed to by Tail; When the main thread continuously receives video data packets and the buffer is not full,

Tail continues to increase by 1; and when the data buffer is full, if Tail is equal to LastIDR, LastIDR is −1; if Tail is equal to Head, and FutureIDR is not −1, then set Head to FutureIDR; and clear the video data buffer pointed to by Tail, and hang the newly received video data buffer on the array unit pointed to by Tail. And if the newly added data is a key frame, assign the value of Tail to FutureIDR;

(2) Sub-thread operation process: as shown in Fig. 3, The sub-thread of each channel executes an eternal loop, fetches video data from the ring buffer, and sends it to the corresponding client. The child thread assigns the value of the Head pointer to nCurp. If nCurp is equal to −1 or equal to Head (representing the video data buffer that can be sent on the ring buffer), the child thread sleeps for 100 s and restarts the loop. If the channel has buffered data that can be sent, and the client's forwarding flag SocFlag is −1, it will not be forwarded to the client, and the network connection with the client will be disconnected. If the SocFlag value is 0, the child thread sends the data buffer pointed to by nCurp to each client. During the forwarding process, if the forwarded video data is a key frame, assign the value of nCurp to LastIDR; if nCurp is equal to FutureIDR, set FutureIDR to −1; if the value of head is not changed by the main thread, then Add 1 to Head, point to the next video data buffer to be sent, and then return to the loop start point; otherwise, return directly to the loop start point.

Similar to the server, the client creates a ring buffer and a sub-thread for each channel. The structure of the ring buffer and video data buffer is similar to that of the server. The main thread of the client is responsible for receiving the video data stream from the server. Put it into the ring buffer corresponding to the channel number, and then the sub-thread corresponding to each channel is responsible for extracting the content from the ring buffer and putting it in the video data buffer for playback. Each channel has an array of window numbers, which records the window number nPort of the video to be played, so as to support playing the same channel of video in multiple windows; when the user of the client side requests to view the video of a certain channel of video sampling device, the client side sends a message to the server side Send a message requesting a video.

When the user of the client side requests to view a certain channel of video sampling device on a certain window, the client side checks whether there is a channel and ring buffer corresponding to the required video sampling device locally. If not, it wants the server to send a video stream request message., if there is, it will not send a video request, but add the new window number to the channel's window number array.

4   Related Experiments and Results

4.1   Experimental Process

In order to evaluate the performance and efficiency of the forwarding mechanism, this paper conducts an experimental analysis, and proposes two indicators to analyze the advantages and disadvantages of the forwarding mechanism, namely the number of data packets to be forwarded on the server side and the forwarding delay time of key frames on the server side. The laboratory environment is Windows 8.1, based on the VS2010

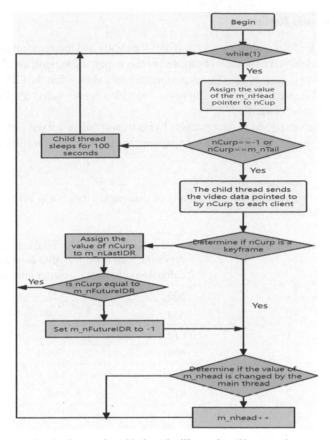

Fig. 3. A figure caption is always placed below the illustration. Short captions are centered, while long ones are justified. The macro button chooses the correct format automatically.

IDE platform, and the interior scene of the laboratory is shot. There are pedestrians passing by in this area. The fluency of these images can be used as the main observation object of the experiment.

The list of equipment used in the experiment is shown in Table 2:

Table 2. Experimental equipment table.

Name	Device model	Quantity
Hikvision Network DVR	DSA-7300	1
Video sampling equipment	Yuanxin gun and ball machine, Hikvision gun machine, Hikvision dome machine	4
Server	Windows Server 2012, Access2007	1
Client	Windows 8.1, Windows7	2

4.2 Experimental Results

For the ring buffer forwarding algorithm (CBF for short) and the basic streaming media forwarding algorithm (BT for short) proposed in this paper, the experimental test of video stream forwarding is carried out. The experimental data shows that the CBF mechanism proposed in this paper significantly improves the video server video. Performance and efficiency of stream forwarding.

The total amount of video data packets to be forwarded in a fixed period of time is recorded on the server side. Through comparison, the experimental results are shown in Table 3:

Table 3. Comparison table of the total amount of data packet forwarding before and after the algorithm improvement.

Number of video channels	Number of clients	Total amount of video data to be forwarded before improvement	Total amount of video data to be forwarded after improvement
1	1	503	22
1	2	553	26
2	1	612	38
2	2	589	35
3	1	734	53
3	2	812	64
4	1	923	72
4	2	898	78

As shown in the figure above, the number of data to be forwarded before and after the improvement of the forwarding mechanism is recorded under different video channels and different numbers of clients. The changes in different situations are shown in Fig. 4.

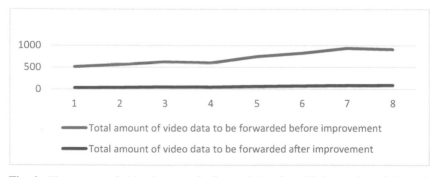

Fig. 4. The amount of video frames to be forwarded varies with the number of channels

After the implementation of different streaming media forwarding mechanisms, if you want to observe the effect of real-time forwarding, the delay time of forwarding response is also an important reference indicator. By writing logs on the server side, the time point when the client starts to request video data and the time point when the server starts to return video data is recorded, and the response delay time of the two different forwarding mechanisms can be calculated. The response delay time of different forwarding mechanisms is shown in Table 4.

Table 4. Comparison of data forwarding delay time before and after algorithm improvement

Number of video channels	Number of clients	Delay time before improvement (s)	Delay time after improvement (s)
1	1	2.91	2.46
1	2	3.12	2.57
2	1	4.09	2.69
2	2	4.23	2.76
3	1	6.24	2.71
3	2	6.93	2.82
4	1	8.27	2.78
4	2	8.91	2.86

As shown in the figure above: when the number of video channels remains the same, with the increase of clients, the forwarding delay time before the improvement will increase significantly, while the delay time after the improvement will not change significantly; the same before the improvement The forwarding delay time will increase with the increase of video channels when the number of clients remains the same, and the delay event will not change significantly after the improvement.

5 Conclusion

This chapter elaborates the process of inspection live video stream forwarding in detail, and proposes a forwarding algorithm for multi-channel ring buffering, which solves various problems in the forwarding process of streaming media. Through the comparison and analysis of the experiment and the existing forwarding algorithms, the experimental results show that the forwarding algorithm based on the multi-channel ring buffer proposed in this paper has significant performance advantages.

Acknowledgment. This work was supported by State Grid Information and Communication Industry Group Bipolar Collaborative R&D Project: Research and development of power AIOT basic capability platform (K102000071).

References

1. Sha, S.: Design and implementation of cross-platform video surveillance Client Based on Qt . Xidian University (2013)
2. Yongbo, D.: Research on remote video surveillance system based on H.264 algorithm. Wuhan University of Technology (2014)
3. Yingda, J.: Design of intelligent video monitoring system for substation. Jinan University (2016)
4. Chu, J.: Development of remote update system for field controller program based on TCP/IP. Dalian University of Technology (2009)
5. Gao, H.: Research on intelligent analysis technology of surveillance video and its application in building energy conservation. Shandong Jianzhu University (2017)
6. Migliore, D.A., Matteucci, M., Naccari, M.A.: Revaluation of frame difference in fast and robust motion detection. In: Proceedings of the 4th ACM international workshop on Video surveillance and sensor networks. ACM, 215–218 (2006)
7. Hua-Yu, J., Xiao-Yong, W., Zhi-long, Z., Dan-Pu, L., Fang-Fang, Y.: Rate adaptive algorithm for live video streaming media research. J. Chinese Media University (2022)
8. Jiali, Y., Yimin, L.: Research on overseas communication of Chinese long-term video streaming platforms in the post-epidemic era. Southeast Communication (2021)

An Intelligent Annotation Platform for Transmission Line Inspection Images

Jing Liu[1]($\boxtimes$), Xiaodong Wang[1], Yingbao Cui[1], Xiaoguang Huang[1], Linyu Zhang[1], and Yang Zhang[2]

[1] State Grid Information and Telecommunication Group, Beijing, China
{liujing2,wangxiaodong,cuiyingbao,huangxiaoguang,
zhanglinyu}@sgitg.sgcc.com.cn, 460887059@qq.com
[2] State Grid Corporation of China, Beijing, China
yang-zhang@sgcc.com.cn

Abstract. With the development of digitalization, the traditional industries represented by the power industry have also undergone earth-shaking changes. In the electricity industry, the combination of artificial intelligence technology and transmission line inspection has greatly promoted the advancement of digital technology and the changes of the electricity industry. However, compared with traditional machine learning algorithms, AI technology today requires a large amount of data. But the uneven quality of samples in the power sector, low sample labeling efficiency, and lack of unified data management limit the use of AI technology in transmission line inspections wide application. Therefore, a unified intelligent annotation platform for transmission line inspection images is established to provide service capabilities such as sample annotation and sample management. In addition, considering the cost of labeling, the use of deep learning methods to pre-label the data reduces the workload of manual labeling. Ultimately, the construction of large-scale datasets supports the application of artificial intelligence models in power business scenarios.

Keywords: Transmission line inspection · Platform · Image annotation

1 Introduction

With the rapid development of artificial intelligence technology in recent years, how to combine power grid business scenarios with artificial intelligence technology is a very important issue. In order to response to the national new digital infrastructure construction strategy [1] and promote the rapid implementation of artificial intelligence-related products, State Grid Corporation of China (hereinafter referred to as the "Company") issued the "State Grid Corporation Artificial Intelligence Special Plan", and compiled the "State Grid Corporation of China Co., Ltd [2]" "Two Repositories and One Platform" Architecture Design Plan" and "State Grid Co., Ltd. Artificial Intelligence Technology Application Guide", accelerate the integration of artificial intelligence technology and power business scenarios, and lay the foundation for the era of intelligence for power.

© The Author(s), under exclusive license to Springer Nature Switzerland AG 2023
N. Xiong et al. (Eds.): ICNC-FSKD 2022, LNDECT 153, pp. 673–681, 2023.
https://doi.org/10.1007/978-3-031-20738-9_77

At present, all industries are actively applying artificial intelligence technology and all majors of State Grid Corporation are also actively exploring the landing of artificial intelligence [3]. In the field of equipment operation and maintenance, applications such as inspection robots, power transmission channel monitoring, and unmanned aerial vehicle (UAV) intelligent inspection have been carried out. Deep learning has shown great capabilities in computer vision field [4]. Using deep learning for image detection and recognition of transmission lines can achieve better processing results than traditional methods. However, deep learning relies on a large amount of data with annotation information, and there are the following problems in the field of transmission line inspection:

- There are many types of defect samples in transmission line inspection, and there is no corresponding sample management mechanism.
- In the process of labeling the transmission line inspection samples, there is a lack of labeling review mechanism, and the overall sample labeling quality is low.
- The cost of sample labeling is high, and there is a lack of professional inspectors, resulting in low sample labeling efficiency.

Aiming at the above problems, this paper researches and designs an intelligent labeling system for transmission line inspection, provides sample management and multi-person labeling services, and supports large-scale labeling of transmission line data [5]. At the same time, an intelligent labeling module is designed to pre-label the original dataset using deep learning methods, thereby improving the efficiency and quality of manual labeling. Through this system, a large-scale labeled data set can be constructed to provide large-scale data support for artificial intelligence technology, which is of great significance to the digital process of transmission line inspection.

2   Method

2.1   Overall Design of Intelligent Labeling System

The transmission line image intelligent annotation system can be divided into application layer, service layer and data layer as a whole, as shown in Fig. 1.

The bottom layer is the data layer, which can be divided into structured data storage and unstructured data storage. Among them, unstructured data such as transmission line image data is stored in a distributed file management system, image annotation information is stored in a distributed full-text search engine, and system-related form data is stored in a relational database.

The service layer builds a sample management module, a multi-person labeling management module and an intelligent labeling module based on the data layer. Provide data management, data labeling management and intelligent labeling algorithm services respectively.

The application layer is the functions provided by the annotation system, including applications such as annotation modules, administrator modules, and algorithm applications.

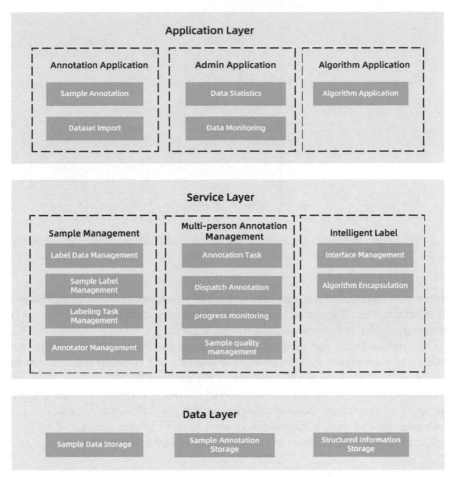

Fig. 1. The overall architecture of the intelligent annotation platform for transmission line inspection images

2.2   Technical Architecture

The overall technical architecture of the transmission line image intelligent annotation system is shown in Fig. 2. The annotation system is mainly divided into four logical parts, including client, logical background, database buffer part and database storage part. The specific technical implementation of each part is as follows:

The client uses Google Chrome browser to access web pages. Modern browsers have a good interactive experience. Web development based on the VUE framework allows you to create data-driven UI components through simple and flexible APIs, providing users with a good experience. And easy access.

The logical backend can be divided into algorithm service and labeling system backend service. The model in the algorithm service adopts the python language, the deep learning framework adopts pytorch, and the Flask web framework is used to encapsulate

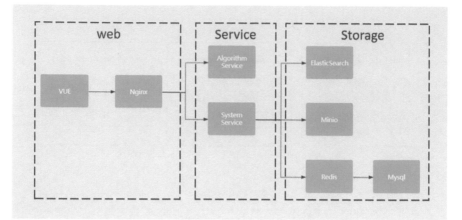

Fig. 2. The overall technical architecture of the transmission line image intelligent annotation system

it into a web service, and the algorithm service is provided externally by exposing the API interface. The background service of the labeling system adopts a single service architecture, which is convenient and quick to deploy.

The database buffer part uses the Redis database as the cache server, as the buffer between the background application and the database [6], to avoid frequent database access which will cause excessive pressure on the database IO.

Database storage is divided into three blocks. Minio distributed storage is used to store massive image data. Because Elasticsearch is good at searching, analyzing and calculating massive data, Elasticsearch is used to store labeled data [7]. Finally, use MySQL as the backend database to provide structured data storage and reading [8].

2.3   Intelligent Annotate Detection Algorithm

The intelligent labeling module adopts Faster-RCNN algorithm [9]. The Faster-RCNN network framework is mainly composed of two sub-networks, namely the region proposal network (RPN) and the Fast-RCNN detection network [10], as shown in Fig. 3.

RPN mainly generates high-quality region proposal candidate boxes, judges and corrects the candidate boxes through the classification function and the bounding box regression function, and initially locates the target. The input image is first extracted with features by a convolutional neural network, which is shared by the RPN and Fast-RCNN detection networks. After entering the two sub-networks, the RPN is first convolved with a 3 × 3 convolution kernel, and then the convolution result is input into two 1 × 1 convolution kernels for operation. One of the convolution kernels converts the feature map into a format after reshape processing, and the softmax function determines whether the obtained anchor frame has a target object; the other convolution kernel is followed by a regression function to determine the object coordinates.

Fast-RCNN detects the region of interest (ROI) in the network, integrates the convolutional features with the proposed candidate frame information, and extracts the

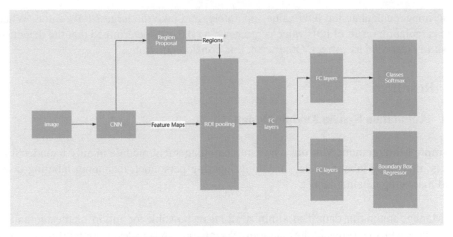

Fig. 3. Faster-RCNN network framework

proposed feature blocks of uniform size and inputs them to the fully connected layer [7]. Again, the classification function is used to calculate the category of the proposed feature block, and the bounding box regression function is used to precisely locate the position of the detection box.

2.4 Intelligent Annotate Evaluation Standard

The performance evaluation indicators of the detection model in this study are precision and recall.

The intersection-over-union (IOU) is an important parameter in the calculation of precision and recall.

$$IOU = \frac{A \cap B}{A \cup B} \tag{1}$$

where: A is the area of the artificially labeled frame; B is the area of the target frame detected by the model.

The precision and recall are calculated as:

$$precsion = \frac{TP}{TP + FP} \tag{2}$$

$$recall = \frac{TP}{TP + FN} \tag{3}$$

In the formula (2) and (3): TP stands for the boxes that the detection model has detected and the result is consider correct; FP stands for the boxes that the model has detected but is wrong; FN stands for the boxes that the model has not detected but is correct. The sum of TP and FN is the number of manual annotation boxes. For each frame detected by the model, traverse the manually marked frames in the same transmission line image, calculate the IOU value separately, and take the largest IOU value. When the calculation result of IOU value is greater than 0.5, it is determined that the detection result is recorded as correct. Otherwise, it is considered wrong.

3   Result

3.1   Annotation System Module Design

Sample Management Module. The sample management module mainly includes four parts: managing labeling data, managing labeling personnel, managing labeling tasks and managing labeling labels.

- Manage annotation data: The administrator is responsible for importing the annotation source data, the format of the data source is jpg, png and the description information of the data can be added. After the data is imported, the administrator can choose to open the data for intelligent processing, and can perform data sampling, data clustering, or data pre-classification on the data.
- Manage labelers: The administrator is responsible for managing labeler information, adding labeler information, and can view, edit, and delete labeler information; set labeling groups, and add, edit, and delete labeling team members.
- Manage labeling tasks: Manage label labels: Administrators can create new labels, view the label list, edit and delete labels.
- Administrators can manage labeling tasks, create labeling tasks and assign labeling tasks to specific labeling groups.

Multi-person Annotation Module. Multi-person labeling module, the team leader needs to manage the labeling tasks of the group, mainly including issuing labeling tasks, monitoring task progress and starting quality inspection.

- Issue the labeling task: After receiving the labeling task assigned by the administrator, the team leader divides the task into several small parts, selects the corresponding labeling personnel for each part of the task, and sends the labeling task to a specific labeling team in the hands of members. Finally, the annotation team members complete the annotation task and store the annotations in coco and voc formats.
- Monitor the progress of members' marking: The team leader can check the completion progress of the team members' marking tasks through the team members' individual task progress bars.
- Start quality inspection: when all the labeling tasks are completed, the team leader starts quality inspection to check the quality of the labelled data.

Intelligent Label Module. The smart labeling module provides algorithm services through api methods. It uses the python flask framework for web application development, designs API interfaces that conform to the rest specification based on the http protocol, and the underlying algorithm framework uses pytorch.

3.2 Dateset

This paper builds a data set based on transmission line inspection, with a total of 55,729 pictures. The inspection fields mainly include: towers, insulators, large-sized fittings, small-sized fittings and conductors, as shown in Fig. 4.

Fig. 4. Image display of inspection defects in some transmission lines

Pole tower categories include: foreign objects on the tower body (bird's nest, honeycomb, etc.), tower body corrosion (obvious yellowing, redness, peeling and rusting, and no obvious marking), lack of bolts.

Insulator categories include: contamination of insulators, self-explosion of glass insulators, damage to composite insulator sheds, damage to porcelain insulators, burns on the glaze surface of porcelain insulators, corrosion of steel feet and steel caps, displacement of equalizing rings, falling off of equalizing rings, damage to equalizing rings, Pressure equalizing ring burns. Large-size fittings include: suspension clamp offset, anti-vibration hammer slip, anti-vibration hammer falling off, and anti-vibration hammer deflection. Small-sized fittings include: missing pins/pins/pins, pins not installed properly, nuts installed irregularly, and nuts missing. Ground wire categories include: broken strands, loose strands/loose strands, foreign bodies, damage, and corrosion.

3.3 Analysis of Intelligent Labeling Results

In this paper, the 5-fold cross-validation method is used. In the experiments, we use 80% of the data as the training set and 20% of the data as the test set. During the training phase, set the learning rate to 0.001 and the number of iterations to 10000. The detection algorithm adopts Faster-RCNN, and the detection results are shown in Table1.

Table 1. Intelligent annotation detection results

Detection type	Detection subtype	Accuracy
Towers	Tower body foreign body	95
	Corrosion of tower body	92
	Missing bolt	82
Insulators	Dirty insulators	82
	Glass insulators explode	85
	Broken porcelain insulator	83
	Porcelain insulator glaze surface burns	84
	Corrosion of steel feet and steel caps	72
	Equalizing ring shift	80
	The pressure equalizing ring falls off	82
	Pressure equalizing ring damage	77
	Pressure equalizing ring burns	81
Large size fittings	Suspended Clip Offset	91
	Anti-vibration hammer slip	85
	Anti-vibration hammer falls off	87
	Anti-vibration hammer deflection	84
Small size fittings	Out of stock	75
	The pin is not installed properly	77
	Improper nut installation	81
	Missing nut	74
Ground wire	Broken stock	88
	Rust	76

Among them, the recognition rate of foreign objects in the tower body can reach up to 95%. In the insulator category, the corrosion of steel feet and steel caps is at least 72%. By analyzing Table 1, it can be seen that the algorithm can accurately detect large targets such as insulators and grading rings, and the recognition ratio is more than 90%. For small targets such as small-sized fittings and conductors, the average accuracy rate is above 75%. On the whole, the recognition results achieve a high recognition accuracy, which proves that the intelligent labeling module has the ability to initially label sample defects, and lays the foundation for subsequent manual labeling adjustments.

4 Summarize

Broken strands and foreign object hanging faults of power lines have a serious impact on the safe operation of power lines. It is of great significance to identify faulty images from UAV aerial power line images with complex backgrounds. However, compared with traditional machine learning methods, deep learning relies on massive data, and the lack of transmission line inspection data sets limits the research of deep learning-based algorithms.

This paper designs and implements a transmission line intelligent labeling platform, which mainly includes three modules: sample management, multi-person annotation and intelligent annotation. Sample management includes management of sample sets, multi-person labeling includes management of sample allocation, and intelligent labeling includes preliminary labeling of defect data. While realizing the basic annotation function, it can also significantly improve the efficiency of data annotation and the quality of data annotation. The high-quality sample input training platform outputs high-quality models, so as to better serve the intelligent development of the power industry.

Acknowledgement. The authors would like to thank State Grid Corporation of China Science and Technology Project "Research on Key Technologies of Autonomous and Controllable Power Artificial Intelligence Open Platform" (5700-202155260A-0-0-00) for the support.

References

1. Ming, L.: Study on monitor system of power transmission line based on unmanned aerial vehicle. Beijing Jiaotong University, Beijing (2017)
2. Alhassan, A.B., et al.: Power transmission line inspection robots: A review, trends and challenges for future research. Inter. J. Elect. Pow. Energy Syst. **118**, 105862 (2020)
3. Ross, G., Jeff, D., et al.: Rich feature hierarchies for accurate object detection and semantic segmentation. In: Proceedings of the IEEE Conference on Computer Vision and Pattern Recognition. Boston, MA, USA: IEEE, 705–713 (2015)
4. Evermann, J., Wand, Y.: Ontological modeling rules for UML: an empirical assessment. J. Comp. Inform. Syst. **46**(5), 56–59 (2016)
5. Ögren, I.: Possible tailoring of the UML for systems engineering purposes. Syst. Eng. **3**(4), 212–224 (2015)
6. Guan, F., Zhao, D., Zhang, X., Shan, B., Liu, Z.: Study on the intelligent decision support system for power grid dispatching. (2009)
7. Sainath, T.N., et al.: Convolutional, long short-term memory, fully connected deep neural networks. In: 2015 IEEE international conference on acoustics, speech and signal processing (ICASSP). IEEE (2015)
8. Bin, C., Hui, C., Kangli, Z.: Design of power intelligent safety supervision system based on deep learning. In: 2018 IEEE International Conference on Automation, Electronics and Electrical Engineering (AUTEEE). IEEE, 154–157 (2018)
9. Ren, S., He, K., Shicker, G., et al: Faster R-CNN: towards real- time object detection with region proposal networks. IEEE Trans Pattern Analysis Machine Intelligence **39**(6), 1137–1149 (2017)
10. Ross, G.: Fast R-CNN. In: Proceedings of the IEEE Conference on Computer Vision and Pattern Recognition. Honolulu, HI, USA, 6517–6525 (2017)

Image Interpolation Algorithm Based on Texture Complexity and Gradient Optimization

Yinbo Wang[✉] and Huimin Du

School of Electronic Engineering, Xi'an University of Posts and Telecommunications, Xi'an, China

2002210080@stu.xupt.edu.cn, 1534961911@qq.com, fv@xupt.edu.cn

Abstract. A novel image interpolation algorithm based on texture complexity and edge structure enhancement is proposed to solve the problems of blurring and edge jagging caused by traditional image enlargement. This interpolation method is useful for magnifying a single image. Firstly, polynomial model is used in complex areas, and then the horizontal, vertical and diagonal gradients of other regions are extracted. According to the gradient, the image is divided into large gradient pixels, medium gradient pixels and flat areas, and the corresponding interpolation method is adopted for the pixels with different features. This method only relies on a single original image to achieve the amplification effect, the interpolation amplification effect of low bit rate video decoded image is better, and it is convenient for real-time image amplification, so it has practical significance. Compared with other algorithms, the algorithm has great advantages in subjective and objective effects.

Keywords: Image interpolation · Texture complexity · Gradient projection

1 Introduction

Image super-resolution technology [1] has been applied in many fields, such as medical image analysis [2], video monitoring industry, satellite image processing and multimedia communication [3]. The interpolation method [4] is fast, easy to reconstruct the image quickly, and can be realized with only one image information. The reconstruction method [5] uses mathematical model to establish the function relation of high and low resolution image, so as to get the enlarged image. Therefore, the establishment of model is the key of image reconstruction. Method based on learning need to be trained in the mapping relationship between high and low resolution image, and rebuild the large image, but this method needs lots of images as the training set, after a long time of training to get the mapping relation, and the types of training set of image may also affect the mapping relationship.

This paper aims at real-time interpolation amplification of video decoding images. In addition to interpolation effect, the execution speed of the algorithm is also a key factor to be considered. So this paper mainly studies reconstruction super-resolution technology of image. Linear interpolation [6] has a simple algorithm and a fast implementation

© The Author(s), under exclusive license to Springer Nature Switzerland AG 2023
N. Xiong et al. (Eds.): ICNC-FSKD 2022, LNDECT 153, pp. 682–692, 2023.
https://doi.org/10.1007/978-3-031-20738-9_78

speed. However, regardless of the specific image content, the same convolution kernel is used to process any image. After interpolation, the image will appear fuzzy and the edge will have obvious jagged marks. Image texture feature [7] can reflect the content of the image information, texture characteristics generally includes texture complexity, the gradient information, edge direction, such as nonlinear interpolation for image texture characteristics were analyzed, and different texture characteristics using different interpolation algorithm, can better interpolation image edges, improve the subjective effect. Aiming at the problem of high complexity of the current interpolation algorithm, if you want to enlarge the image quickly, the texture complexity and gradient information of the image [8] are used in the interpolation process, which can achieve a balance between the interpolation effect and implementation speed.

2 Algorithm of This Paper

The image interpolation algorithm based on gradient [9] guidance can effectively maintain the texture details of the image. On this basis, this paper improves the gradient guidance interpolation algorithm. The algorithm flow is shown in Fig. 1. The gradient reflects the intensity of the gray change of the image, effectively extracts the multi-directional gradient of the image, and then determines the edge and texture information of the image. According to these prior information interpolation, we can get a high-resolution image with rich texture details and edge structure. In the process of edge detection, the pixel gradient in the complex texture block will also be large, which will affect the judgment of the edge. 2 therefore, this paper first uses the variance of Hadamard transform [10] to judge the texture complexity of the image block, interpolates the image block with high complexity with polynomial model [11], and then processes the rest of the image with improved edge detection algorithm, The rest areas are divided into strong edge, weak edge and flat area. The strong edge is interpolated by one-dimensional directional interpolation, the weak edge is interpolated by improved one-dimensional directional interpolation, and the flat area is interpolated by polynomial model.

2.1 Judgment of Texture Complex Blocks

As shown in Fig. 2, the smaller the spatial correlation of pixels, that is, the greater the change of neighborhood pixel values, the higher the texture complexity, and the sum of absolute values of Hadamard transformation (SAH) will be greater than the set threshold. This algorithm needs judgment 4×4 texture complexity of image block. Images of 4×4 block can be divided into four 2×2 smaller pieces. For four 2×2 small pieces separately the absolute value of the sum of the hadamard transform transform SAH_1, SAH_2, SAH_3 and SAH_4. According to the variance of the four number judgment texture complexity, variance is greater than the threshold, the 4×4 block for complex texture block. Hadamard transformation matrix of Second order: $H = \frac{1}{\sqrt{2}}\begin{bmatrix} 1 & 1 \\ 1 & -1 \end{bmatrix}$. Assuming that the 4×4 image block is denoted as X and the 2×2 small blocks are denoted as

X1, X2, X3 and X4 respectively, the calculation formula of SAH1, SAH2, SAH3 and SAH4 is shown in Eq. (1):

$$SAHm = \sum_{i=1}^{2} \sum_{j=1}^{2} |HX_m H|, m = 1, 2, 3, 4 \tag{1}$$

Is the current block texture complexity of 4×4 formula such as (2). The are averages of SAH_i.

$$complexity_{4 \times 4} = \frac{1}{4} \sum_{i=1}^{4} (SAH_i - \overline{SAH})^2 \tag{2}$$

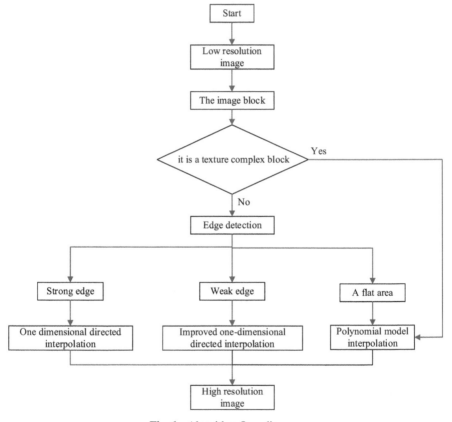

Fig. 1. Algorithm flow diagram

2.2   Edge Detection

Edges of the block of texture complexity, for the convenience of calculating the gradient and the need of subsequent interpolation, still need to be on the LR picture processing.

The gradient of LR picture is calculated by Sobel operator, and then the initial gradients G_x and G_y of high-resolution (HR) image are calculated by bicubic interpolation. The gradient angle of HR image is calculated by formula (3):

$$\theta = \arctan \frac{G_y}{G_x} \tag{3}$$

The algorithm fully considers the relationship between the spatial correlation of pixels and the gradient distribution θ Adjust the initial gradient of high-resolution image. Figure 3 is the schematic diagram of slope adjustment, and the formula is displayed as (4):

$$G'_{x_{i,j}} = \frac{\sum\limits_{\beta<45°} G_{x_{x,y}}}{m}, \quad G'_{y_{i,j}} = \frac{\sum\limits_{\beta<45°} G_{x_{x,y}}}{m} \tag{4}$$

The $G'_{x_{i,j}}$ and $G'_{y_{i,j}}$ respectively adjusted HR image as a bit of x and y direction of the gradient. $Gx_{i,j}$ and $Gx_{i,j}$ are gradients before adjustment. Sitting a little (i,j) searched within a radius of 5 x5 neighborhood and the center of the gradient Angle of less than 45° Angle, a total of m. Use formula (4) to calculate and replace the gradient of the center point to be adjusted until the gradient adjustment of the whole HR image is completed, and use the adjusted gradient components G'_x and G'_y to synthesize the gradient G of the HR image.

Divide the edge direction of the image into 0°, 90°, 45° and 135°, use the adjusted HR image gradient to project in these four directions, and the gradient component is calculated as shown in formula (5–8):

$$G_{0°} = |G| \cos \theta \tag{5}$$

$$G_{45°} = |G| \left(\frac{\sqrt{2}}{2} \cos \theta + \frac{\sqrt{2}}{2} \sin \theta \right) \tag{6}$$

$$G_{90°} = |G| \sin \theta \tag{7}$$

$$G_{135°} = |G| \left(\frac{\sqrt{2}}{2} \sin \theta - \frac{\sqrt{2}}{2} \cos \theta \right) \tag{8}$$

where G is the amplitude of the adjusted HR image gradient, θ is the gradient angle.

Figure 4 shows the schematic diagram of horizontal, vertical and diagonal pixels, and black dots represent known pixels, white dots are pixels to be interpolated, b, d, j and i are horizontal pixels, e, g, m and o are vertical pixels, and f, h, n and p are diagonal pixels.

For horizontal and vertical pixels: if $|G_0(i, j) - G_{90}(i, j)| > \lambda$, Then the pixels to strong edge pixels, and if $G_0(i, j) \geq G_{90}(i, j)$, then edge direction of 90°,if $G_0(i.j) < G_{90}(i, j)$, then edge direction of 0°.

For diagonally class pixels: if $|G_{45}(i,j) - G_{135}(i, j)| > \lambda$, Then the pixels to strong edge pixels, and if $G_{45}(i, j) \geq G_{135}(i, j)$, then edge direction of 135°,if $G_{45}(i, j) < G_{135}(i, j)$, then edge direction of 45°.The above $\lambda = 0.010$.

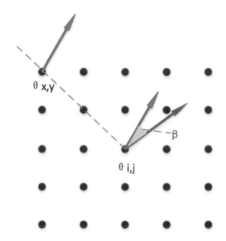

Fig. 2. Texture complex block

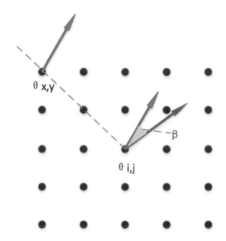

Fig. 3. Gradient adjustment schematic diagram

Rough edges of the image to grayscale change larger point, is a collection of all possible edge pixel image, use the formula (9) for a maximum of four gradient component:

$$G_{\max} = \max(G_{0°}, G_{45°}, G_{90°}, G_{135°}) \tag{9}$$

Set a low threshold value $\sigma = 75$ for $G_{\max}$, find all possible edge pixels, and then remove the strong edge points to obtain the position of weak edge pixels, and the acquisition process of weak edge direction is the same as that of strong edge.

3 Image Interpolation

For strong edge pixels, adopt one-dimensional directed interpolation, diagonal pixel interpolation diagram as shown in Fig. 5, black spots to known pixels, remember for

$a(i, j)$, The red dot is the diagonal pixel to be interpolated, which is recorded as $D_s(i, j)$. The calculation formula is shown in Eq. (10):

$$D_s(i,j) = \begin{cases} \omega(a_{i-1,j+1} + a_{i+1,j-1}) + (1-\omega)(a_{i-3,j+3} + a_{i+3,j-3}), & \theta = 45° \\ \omega(a_{i-1,j-1} + a_{i+1,j+1}) + (1-\omega)(a_{i-3,j-3} + a_{i+3,j+3}), & \theta = 135° \end{cases} \quad (10)$$

Blue dots represent unknown pixels, as shown in Fig. 6, which is recorded as $H_s(i, j)$. The calculation formula is Eq. (11):

$$H_s(i,j) = \begin{cases} \omega(a_{i-1,j-1} + a_{i-1,j+1}) + (1-\omega)(a_{i-1,j-3} + a_{i-1,j+3}), & \theta = 0° \\ \omega(a_{i-2,j} + a_{i,j}) + (1-\omega)(a_{i-4,j} + a_{i+2,j}), & \theta = 90° \end{cases} \quad (11)$$

Figure 7 is a schematic diagram of vertical pixel interpolation. The yellow dot indicates the point to be interpolated, which is recorded as $V_s(i, j)$, and the calculation formula is Eq. (12):

$$V_s(i,j) \begin{cases} \omega(a_{i,j} + a_{i,j-2}) + (1-\omega)(a_{i,j-4} + a_{i,j+2}), & \theta = 0° \\ \omega(a_{i-1,j-1} + a_{i+1,j-1}) + (1-\omega)(a_{i-3,j-1} + a_{i+3,j-1}), & \theta = 90° \end{cases} \quad (12)$$

In the Formulas (10), (11) and (12), $\omega = \frac{1}{3}$.

For weak edge pixels, diagonal, horizontal and vertical pixels are respectively recorded as $D_w(i, j)$, $H_w(i, j)$ and $V_w(i, j)$, and the calculation formula is formula (13–15):

$$D_w(i,j) = \begin{cases} \gamma_1(a_{i-1,j+1} + a_{i+1,j-1}) + \gamma_2(a_{i-3,j+3} + a_{i+3,j-3}) \\ +\gamma_1(a_{i-1,j-1} + a_{i+1,j+1}), & \theta = 45° \\ \gamma_1(a_{i-1,j-1} + a_{i+1,j+1}) + \gamma_2(a_{i-3,j-3} + a_{i+3,j+3}) \\ +\gamma_3(a_{i+1,j-1} + a_{i-1,j+1}), & \theta = 135° \end{cases} \quad (13)$$

$$H_w(i,j) \begin{cases} \gamma_1(a_{i-1,j-1} + a_{i-1,j+1}) + \gamma_2(a_{i-1,j-3} + a_{i-1,j+3}) + \gamma_3(a_{i-2,j} + a_{i,j}), & \theta = 0° \\ \gamma_1(a_{i-2,j} + a_{i,j}) + \gamma_2(a_{i-4,j} + a_{i+2,j}) + \gamma_3(a_{i-1,j-1} + a_{i-1,j+1}), & \theta = 90° \end{cases} \quad (14)$$

$$V_w(i,j) \begin{cases} \gamma_1(a_{i,j} + a_{i,j-2}) + \gamma_2(a_{i,j-4} + a_{i,j+2}) + \gamma_3(a_{i-1,j-1} + a_{i+1,j-1}), & \theta = 0° \\ \gamma_1(a_{i-1,j-1} + a_{i+1,j-1}) + \gamma_2(a_{i-3,j-1} + a_{i+3,j-1}) + \gamma_3(a_{i,j-2} + a_{i,j}), & \theta = 90° \end{cases} \quad (15)$$

Among them: $\gamma_1 = 0.286$, $\gamma_2 = 0.143$, $\gamma_3 = 0.071$.

4 Experimental Results and Analysis

This article selects 8 standard test image as shown in Fig. 8, and the algorithm compared with several representative algorithms, they respectively are: Bicubic interpolation [12], edge oriented image interpolation algorithm (NEDI) [13], image interpolation method based on multi-directional filtering and data fusion (DFDF) [14], image interpolation technology based on improved soft decision 7 strategy (RSAI) [15], image interpolation based on nonlocal autoregression (NARM) [16], image interpolation algorithm based on the combination of constructed surface and gradient guidance [17].

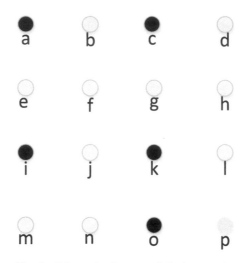

Fig. 4. Schematic diagram of pixel categories

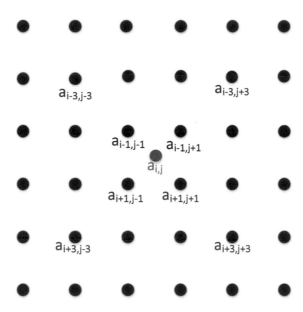

Fig. 5. Diagonal-class pixel interpolation diagram

The comparison results of PSNR and SSIM of different algorithms are shown in Table 1.The interpolation effect and local magnification effect of different algorithms

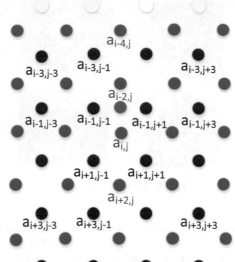

Fig. 6. Schematic diagram of horizontal class pixel interpolation

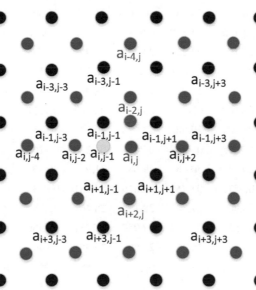

Fig. 7. Schematic diagram of vertical

on baboon image are shown in Figs. 9 and 10. The results show that the effect of this algorithm is better than other algorithms, especially for edges and textures.

Fig. 8. Standard test image

Table 1. Data comparison of different interpolation algorithms

The test image		Bicubic 1981	NEDI 2001	DFDF 2006	RSAI 2012	NARM 2013	Wu's 2017	This paper 2021
Baboon	PSNR	22.67	22.57	22.81	23.18	22.64	22.76	23.25
	SSIM	0.8413	0.8726	0.8637	0.8753	0.8624	0.8659	0.8971
Lenna	PSNR	30.18	30.58	30.50	31.35	31.39	31.05	31.35
	SSIM	0.9115	0.9129	0.9116	0.9242	0.9248	0.9209	0.9245
Barbara	PSNR	23.19	22.21	24.45	22.99	23.36	24.18	24.52
	SSIM	0.7955	0.8429	0.8714	0.8549	0.8588	0.8544	0.8662
Dollar	PSNR	18.07	18.92	19.06	18.96	18.77	18.93	19.03
	SSIM	0.7169	0.8031	0.8023	0.8022	0.7921	0.7784	0.8053
Boats	PSNR	29.68	29.58	29.67	30.02	30.25	29.27	29.89
	SSIM	0.7475	0.7358	0.7054	0.7276	0.7207	0.7606	0.7656
Pepper	PSNR	29.74	30.68	29.68	31.94	30.86	31.97	32.15
	SSIM	0.9572	0.9756	0.8819	0.9781	0.9632	0.9653	0.9784
Goldhill	PSNR	25.96	27.34	28.13	28.46	28.98	30.12	30.45
	SSIM	0.6526	0.6554	0.6942	0.7751	0.7966	0.7919	0.8064
Monarch	PSNR	26.14	26.45	26.89	26.56	27.49	27.82	27.95
	SSIM	0.9513	0.9549	0.9546	0.9583	0.9612	0.9737	0.9786

5   Conclusion

The key of image interpolation lay in the processing of edge texture details. The accuracy of edge detection determined the effect of image interpolation. Therefore, An image enlargement method based on image complexity and gradient is proposed, the blocks with complex texture were selected by Hadamard transform variance and interpolated by

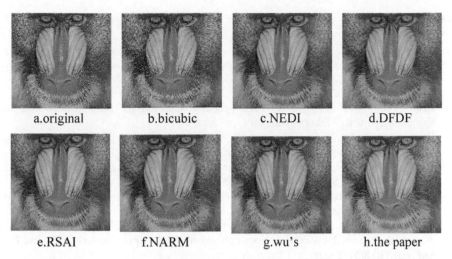

Fig. 9. Interpolation renderings of Baboon images by different algorithms

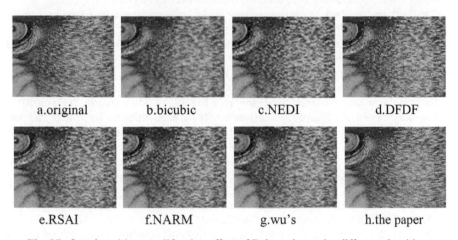

Fig. 10. Local position amplification effect of Baboon image by different algorithms

polynomial model, which effectively avoided the influence of complex blocks on edge detection and improved the accuracy of subsequent edge detection; Secondly, the Sobel operator was used to project in the four directions of $0°$, $45°$, $90°$ and $135°$, the gradient intensity variation in multiple directions was fully analyzed, the relationship between spatial pixel correlation and gradient distribution was considered, the edge detection was optimized, and the edge of the image was effectively extracted; The edge was divided into strong and weak, and calculated with different interpolation methods to improve the accuracy of interpolation. The results showed that the interpolation algorithm can enlarge all kinds of images, especially the interpolation at the edge structure.

References

1. Wang, Y., Wu, J., Wang, H.: Residual networks with channel attention for single image super-resolution. In: Conference proceedings of 2021 4th International Conference on Algorithms, Computing and Artificial Intelligence (ACAi 2021), 683–687 (2021). https://doi.org/10.26914/c.cnkihy.2021.055306
2. Pawar, M., Marab, S.: A novel edge boosting approach for image super-resolution. Evolutionary Intell (2021) (prepublish)
3. Hardiansyah, B., Lu, Y.: Single image super-resolution via multiple linear mapping anchored neighborhood regression. Multimedia Tools Appl. (2021)
4. Jiahao, L.: Analysis and comparison of three classical color image interpolation algorithms. J Phys.: Conference Series. **1802**(3) (2021)
5. Du, H., Zhang, Y., Bao, F., Wang, P., Zhang, C.: A texture preserving image interpolation algorithm based on rational function. Inter. J. Multi. Data Eng. Manag. (IJMDEM) **9**(2) (2018)
6. Hyma Lakshmi, T.V., Sri Kavya, K.C., Madhu, T., Kotamraju, S.K.: Satellite image enhancement using optimized wavelet decomposition and bicubic interpolation. Inter. J. Recent Tech Eng. (IJRTE) **8**(5) (2020)
7. Ying, W., Jikun, L.: Research on college gymnastics teaching model based on multimedia image and image texture feature analysis. Discover Internet of Things. **1**(1) (2021)
8. Cao, J., Zhang, A., Shi, L.: Orthogonal sparse fractal coding algorithm based on image texture feature. IET Image Process. **13**(11) (2019)
9. Tian, C., Chen, L.: An image dehazing method using image gradient distortion prior. Wireless Net. 2021 (prepublish)
10. Sarycheva, A., Adamov, A., Poteshin, S.S., Lagunov, S.S., Sysoev, A.A.: Influence of multiplexing conditions on artefact signal and the signal-to-noise ratio in the decoded data in Hadamard transform ion mobility spectrometry. European J. Mass Spectro. **26**(3) (2020)
11. Pengbin, F., Huijie, T., Huirong, Y.: Texture details and edge structures keep image interpolation algorithm. Comp. Appl. Res. **38**(4), 1203–1218 (2021). https://doi.org/10.19734/j.iSSN.1001-3695.2020.01.0075
12. Yang, Z.X., Lu, F., Guan, L.T.: Image enlargement and reduction with arbitrary accuracy through scaling relation of B-spline. J. Comp.-Aid. Design & Comp. Graph. **13**(9), 824–827 (2001)
13. Li, X., Orchard, M.T.: New edge-directed interpolation. IEEE Trans. Image Process. **10**(10), 1521–1527 (2001). https://doi.org/10.1109/83.951537
14. Zhang, L., Wu, X.L.: An edge-guided image interpolation algorithm via directional filtering and data fusion. IEEE Trans. Image Process. **15**(8), 2226–2238 (2006). https://doi.org/10.1109/TIP.2006.877407
15. Hung, K.W., Siu, W.C.: Robust soft-decision interpolation using weighted least square [J]. IEEE Trans. Image Process. **21**(3), 1061–1069 (2012). https://doi.org/10.1109/TIP.2011.2168416
16. Lee, S.J., Kang, M.C., Uhm, K.H., et al.: An edge-guided image interpolation method using Taylor series approximation. IEEE Trans. Cons. Elect. **62**(2), 159–165 (2016). https://doi.org/10.1109/TCE.2016.7514715
17. Liqiong, W.: Research on image interpolation and amplification algorithm based on gradient . Shandong University (2017)

A Defect Detection Method for Semiconductor Lead Frame Based on Gate Limite Convolution

Wanyu Deng, Jiahao Jie[✉], Dunhai Wu, and Wei Wang

Xi'an University of Posts and Telecommunications, Xi'an 710121, China
dengwanyu@xupt.edu.cn, {china_jjh_1998,wudunhai,
wangwei}@stu.xupt.edu.cn, 1062999098@qq.com

Abstract. Aiming at the problems of high labor intensity, high missed detection rate, and low accuracy in artificial examination of semiconductor lead frame defects, A Defect Detection Method for Semiconductor Lead Frame based on Gate Limite Convolution is proposed. The method uses a U-Net architecture and uses a training method based on small data set. Firstly, the data set required for training is generated based on the characteristics of the image itself. Furthermore, the Gate Limite Convolution Layer is used to replace the original convolution layer to train the model. The trained network strengthens the local perception of the overall image, makes full use of the overall image information and the network self-learning ability, avoids the problems of artifact, color blur and gray shadow, and thus reconstructs the flawless image. Finally, the location information of the defect is obtained by the residual method. Through the quantitative evaluation of the peak signal-to-noise ratio, structural similarity and the accuracy of the detection results of the reconstructed image. The results show that the defect detection accuracy of semiconductor lead frame by this method is as high as 98%, which meets the actual detection requirements.

Keywords: Defect detection · Small data set · Gate limite convolution

1 Introduction

As the foundation of electronic information industry, semiconductor components are of great importance to the fields of communication and chip. Nowadays, the consumer electronics market, solar energy and the manufacture of new energy vehicles, intelligent industries and other fields have promoted China has become the world's largest semiconductor market, and the demand for semiconductor industry is increasing day by day.

In the current Semiconductor Lead Frame production, more than 80% of the manufacturers still use traditional manual inspection. First, the Limited by the human eye resolution, it is largely unable to meet the demand of high precision. Secondly, for inspection accuracy, due to Due to the increase in demand, front-line workers are working at

This work is supported by Science Research Plan of Shaanxi Provincial Department of Education under Grant No. 19JC036.

high intensity for a long time and are easily affected by subjective emotions, which makes the accuracy of inspection deteriorate.

On the other hand, some Lead Frames also need to be coated with a more conductive silver due to the different needs of the actual use. If the location of the silver plating deviates, it will directly lead to the reduction of the conductivity of other components encapsulated in this area, the package is not stable and so on, thus causing a significant loss of product life. Coupled with the fact that its production is in bulk, this will cause unnecessary losses.

Eman Hussein Saleh et al. extracted the approximate sub-image of the first layer decomposition by means of à trous wavelet. In which, the defect energy is enhanced and the background energy is energy was attenuated. n turn, the standard deviation of BBs is estimated to detect defective areas of the fabric [1].

By constructing a generative adversarial network model, Wang Ming et al. put the sample images into the model for restoration, and then compared the input images with the restored images using image differencing to determine the exact defect areas [2].

Liu Kun et al. used generative adversarial networks to fit their normal surfaces. The region of interest is obtained by using a pre-trained generator for test image reconstruction and performing a difference operation between the test image and the reconstructed image; after that, the defect location is determined by binarizing the difference image and performing a connected domain size filter [3].

Yajiao Liu et al. enhanced the learning ability of the model for defect features by variable convolution, improved the characterization ability of defects by increasing the prediction scale, and finally determined the defect location by assigning to the prediction scale. The method is conceivably computationally intensive, and it is difficult to meet the real-time demand for mass-produced semiconductor lead frames, and secondly, the 90% accuracy is difficult to meet the demand for accuracy [4].

Liu et al. used the Gabor wavelet transform to obtain wafer surface texture features for defect detection at macroscopic dimensions. The generated high-dimensional features are randomly projected and downscaled, and finally, the binarization algorithm is able to detect the defect location quickly and very accurately. For microscopic defects, they are processed using a regional convolutional neural network approach to detect the defect location [5].

Zhisan Chen et al. designed a contour feature-based wafer localization method based on the characteristics of wafers. Then, to solve the problem of how to separate the wafer from the background pattern with too much similarity, a geometric pattern contour affine transformation and sub-region detection method are designed using the affine transformation principle. The defects are extracted by separating the internal and external patterns, and then the threshold segmentation and morphological processing are performed separately, and then the intersection of the defects is done to get the total defects [6].

Fityanul akhyar et al. Converted the training image from BGR to HSV channel and rearranged the color differences. The method of bilinear interpolation is used to carry out up sampling processing to achieve the purpose of maintaining small resolution. Through data enhancement, initialization operations are performed using the high resolution network (hrnet). Then, in the second transformation, the output of the trunk is taken as the input. The results show that the accuracy of surface defect detection of grade 1 Severstal steel is significantly improved compared with previous methods [7].

Zhu et al. by adding the overall cross-layer transmission structure of network results to the traditional convolutional neural network.By linking the encoding and decoding of the image to be restored, we realize the whole process from the input of the image with masking to the output of the restored image [8].

Since the defective part usually affects the network parameters when training, this paper adopts a Gate Limite Convolution-based network training method. This method introduces a mask with the same size as the training image. This mask invalidates the defective part of the data, verifies the normal part, and updates the mask with the training. The results show that this method has good detection effect and is also reliable for the detection of small defects.

Figure 1 shows a subgraph S of a model of semiconductor lead frame collected with a size of 9344 × 7000 and an actual size of about 1.96 × 1.49 cm. The red part of it is a base lead unit block U.

Fig. 1. Sub-diagram of the lead frame

When a new model of Semiconductor Lead Frame is detected, the user needs to input information including the width, height, and number of base conduction blocks of the complete leadframe. Firstly, after finding the size of the base guide unit according to the user input, the base guide unit block U is randomly cropped in the image S with a size slightly larger than the actual size of the base guide unit block, and then a 512 × 512 size image is randomly intercepted inside U as the data set of the training.

2 Algorithm Flow

The flow chart of this method is shown in Fig. 2. Firstly, according to the model data entered by the user and the planned motion matrix, the image of the sample to be tested is collected through the image acquisition platform. Then enter the detection part. According to the data entered by the user, in the original figure s, randomly cut a base guide unit block u whose size is slightly larger than the actual size of the base guide unit block, and Randomly intercept 512 × 512 size images inside the U as the training data set. At the same time, the training is carried out by manually adding defects. Then the image to be detected is put into the trained network model to reconstruct a new image and then processed by binarization to classify the defects of different materials. Finally, the final defects are obtained and displayed by filtering the residual score and defect size.

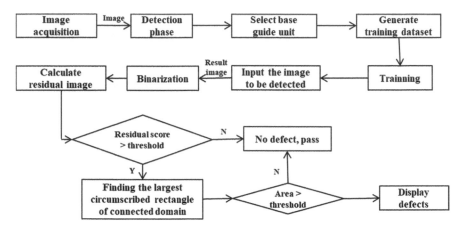

Fig. 2. Algorithm processing flow chart

3 Small Data Set

Traditionally, the training of the model uses a large number of datasets, thus training the network. Since in actual production, usually a new Lead Framework often does not have a relevant dataset when it is first detected, if the model of collecting data first and then training is still used, it is often found cumbersome in practical applications. Therefore, this method uses a small data set for training.

Lead Frames is shown in Fig. 3, with a size of about 25.8 × 7.8 cm. It is made up of several Base Guide Units arranged and combined. The top, bottom, left, and right boundaries are the protective edges, also called arguments, whose purpose is to protect the Base Conductor Units on the Lead Frame.

Obviously, a single image cannot be captured in its entirety, so it is first divided according to a planning matrix to obtain several slightly overlapping but non-repetitive sub-images. The actual size of the acquired image data is 1.96 × 1.49 cm (9344 ×

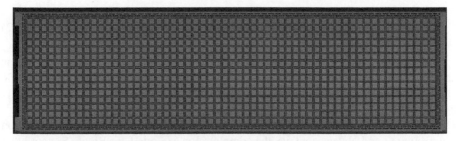

Fig. 3. Plan view of lead frame

7000 pixels). Several common Lead Frame images were captured as shown in Fig. 4. Obviously, it also contains one or more blocks of base conductor units.

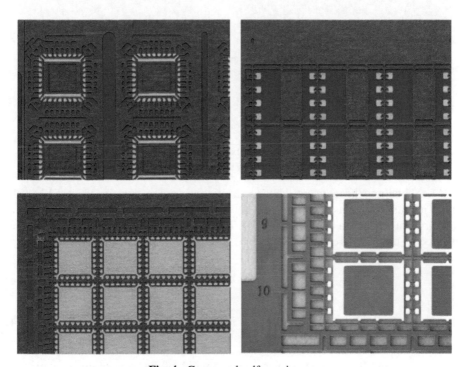

Fig. 4. Common leadframe images

The input parameters contain the width W_{block} of the Lead Frame, the height H_{block} of the Lead Frame, and the number N_{Unit} of the Base Guide cell. Therefore, the width W_{Unit} and height H_{Unit} of the Base Guide Unit are calculated as shown in Eqs. (1) and (2).

$$W_{Unit} = \frac{W_{Block}}{N_{Unit}} \tag{1}$$

$$H_{Unit} = \frac{H_{Block}}{N_{Unit}} \tag{2}$$

In the actual calculation process, it calculates the size of the cut Base Guide Unit is larger than the real size, which can ensure the interception of a complete Base Guide Unit. Although its intercepted Base Guide Unit block may not be shown in the red box in Fig. 1, it can be reassembled by cutting and translating to obtain the same standard Base Guide Unit Block as in Fig. 1. Therefore, the starting position of the interception is not required and always contains the complete Base Guide Unit Block. Suppose that the image of the intercepted Base Guide Unit Block is shown in Fig. 5.

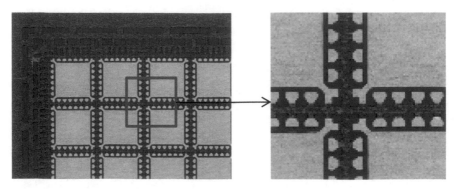

Fig. 5. Intercepted base guide unit block

After obtaining the intercepted images, a randomly cropped image of size 512×512 within it is used as a dataset to train the network. The x, y coordinates of the cropping starting point are shown in Eqs. (3) and (4). The width and height size of the training set obtained by cropping are 512.

$$x = rand\% \left(\frac{W_{Unit}}{W_{VieW}} \times W_{PixCameraW} - W_{Train} \right) \tag{3}$$

$$y = rand\% \left(\frac{H_{Unit}}{H_{VieH}} \times H_{PixCameraH} - H_{Train} \right) \tag{4}$$

where: H_{Train} and W_{Train} denotes the height and width of the cropped image in pixel.

W_{Unit} Indicates the width of the base guide unit in centimeters.

W_{View} denotes the width of camera field of view in centimeters.

$W_{PixCameraH}$ denotes the width of camera field of view in pixel.

H_{Unit} Indicates the height of the base guide unit in centimeters.

H_{View} denotes the height of camera field of view in centimeters.

$H_{PixCameraH}$ denotes the height of camera field of view in pixel.

4   Image Reconstruction Based on Gate Limite Convolution

Since the traditional pixel-based reconstruction methods are mostly based on the pixel information around the current pixel point to perform restoration, ignoring the overall structure of the image, this will lead to the initial position and size of the difference window, etc. which will cause some interference with the reconstruction effect. Therefore, this paper proposes a reconstruction method with Gate Limite Convolution, which makes full use of the overall image information and the network's autonomous learning ability to finally output the restored image.

In this paper, we use the U-Net network model, whose model structure is not too much described here, and the network hierarchy is shown in Fig. 6.

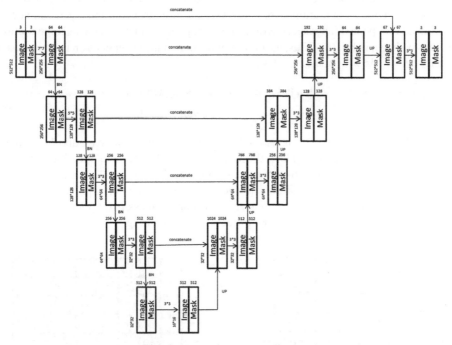

Fig. 6. Network structure diagram

The traditional deep learning convolution is shown in Eq. (5), where X denotes the input matrix, W^T denotes the filter weight, and b denotes the convolution layer filter offset.

$$X' = W^T \times X + b \tag{5}$$

Gated Limite Convolution is equivalent to a switch, the valid region corresponding to the "open" door is involved in the calculation, the invalid region corresponding to the "closed" door is not involved in the calculation. Therefore, the input X of Eq. (5) is Gate Limite on the basis of the original. The non-defective part is set to 1 to open the door, and the defective part is set to 0 to close the door. Therefore, the input X of the original Eq. (5) is updated to that shown in Eq. (6). $\odot$ denotes the dot product operation. Since the defective part results in small dot product results for individual nearby defective pixels, a scaling factor μ is introduced whose definition is shown in Eq. (7).

$$X = X \odot M \tag{6}$$

$$\mu = \frac{Sum(1)}{Sum(M)} \tag{7}$$

Equation (7), 1 indicates that all the elements of the value of 1, and the size of the same filter size matrix. M represents the mask matrix with the same size as the filter centered on the current pixel. The amount of variation of the input is adjusted by μ. We replace the original convolution layer with a Gated Limite Convolution layer, which will be masked for update after each convolution is completed. Thus the updated convolution process is shown in Eq. (8). Thus, the output of the convolution depends on the unmasked inputs, i.e., pixels with non-zero Mask.

$$x' = \begin{cases} \mu W^T (x \odot M) + b, & Sum(M) > 0 \\ 0, & Sum(M) \le 0 \end{cases} \tag{8}$$

Traditional local convolution also only uses Mask in the first layer, and Mask does not get updated. This method updates the Mask after each local convolution is completed. Assuming the current convolution using a sliding window of size 3×3, if the current pixel point corresponds to the sliding window, the sum of the corresponding values at the current mask position is greater than zero, the current mask pixel point is updated to 1, and vice versa to 0. Then the update process is shown in Fig. 7.

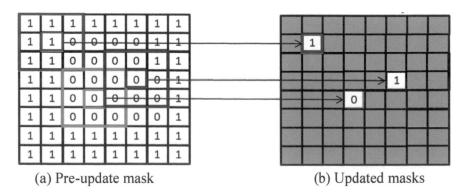

(a) Pre-update mask   (b) Updated masks

Fig. 7. Schematic diagram of mask update

As the number of layers increases, the pixels in Mask's output M' that are 0 gradually become smaller, i.e., the effective area becomes larger and larger, and Mask's influence on the whole becomes smaller and smaller. Compared with the pixel point-based pass reconstruction complex method, the local convolutional neural network-based reconstruction method can solve the problems of initial point selection and interpolation window size setting. Compared to traditional convolutional neural networks, local convolutional neural networks insert the results of previous convolutional layers by cross-layer. Its automatic mask update mechanism can solve the common convolutional network practice of using unrealistic substitution values in the mask during training to avoid artifacts, blurred colors, gray shadows, etc.

Figure 8 shows the details of the results of several different types of Lead Frame reconfigurations. These include non-silver-plated Lead Frames and silver-plated Lead Frames. Although the defects are randomly generated, they contain real defect types and are more complex than real defects, so the experimental results are equally convincing. The first image of each group represents the original image with simulated defects, where black is the simulated defect. Its size is used to simulate defects of different sizes. The second image of each group represents the predicted result image. The third image represents the real defect-free image.

Six different lead frames are now randomly repaired, which are sampled and repaired separately. Table 1 shows the results of its repair. From Table 1, for the models with and without added Gate Limite Convolution, both have the same detection accuracy, and both have good repair results for repairing broken pins and hole defect repair.

Peak Signal-to-Noise Ratio (PSNR) and Structural Similarity (SSIM) are commonly used to evaluate the effect of image reconstruction. The larger the values of PSNR and SSIM, the better the image reconstruction effect is. The quantitative evaluation of the restoration effect of six different Lead Frames is now made as shown in Table 2.

In Tables 1 and 2, A indicates the relevant data with gated convolution added, and B indicates the relevant data without Gate Limite Convolution added. Both meet the practical detection requirements with equal detection accuracy. However, the Gate Limite Convolution-based model is greater than the conventional model in both PSNR and SSMI, and the reconstruction is better.

Analysis of the above table shows that the SSMI were above 90%, but their PSNR values were low. The reason for this problem is that when the image is captured, only the skeletonized part is provided with color by a constant backlight, so its pixel values do not vary relatively much. Since the production of Lead Frames is ultimately a process, it is difficult to achieve a uniform distribution in the production of silver and copper plating. Even if it is uniformly distributed, in its copper production and silver plating process by the influence of material traces there must be color differences, the copper and silver material trace texture as shown in Fig. 9.

Although the colors of copper and silver are slightly different from the real picture after reconstruction, the slight difference in their repaired colors does not affect the inspection results because the binary map is used for defect localization when determining defects. The reconstructed image is now denoted as B and the original image is denoted as A. Firstly, according to the set threshold value, B and A are segmented according to Eq. (9) respectively, where $S(x, y)$ denotes the pixel value of the input image at (x, y), O denotes the output of that pixel point, and value denotes the threshold value.

$$O(x, y) = \begin{cases} 255 & S(x, y) > value \\ 0 & S(x, y) \leq value \end{cases} \tag{9}$$

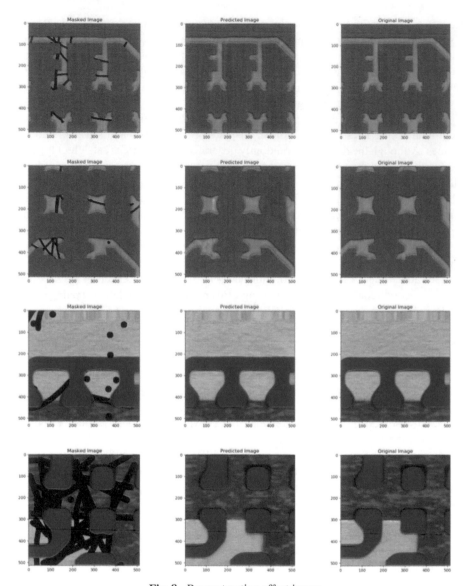

Fig. 8. Reconstruction effect image

The outputs of the original and reconstructed images are now noted as O_A and O_B. The residual plots are obtained by performing with or operations on the two. The ratio of the number of residual pixels to the number of pixels with an O_B pixel value of 255 is the residual fraction, and if the residual fraction is greater than a predetermined threshold, it is considered defective, otherwise it is considered defect-free.

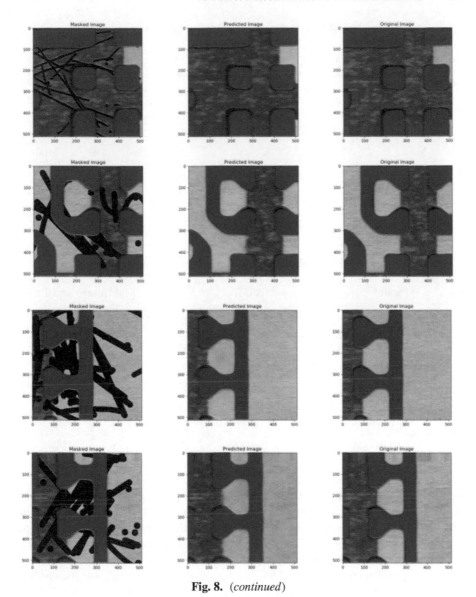

Fig. 8. (*continued*)

Figure 10 shows a demonstration of the detection results in a practical application using this method.

5 Conclusion

For the traditional manual inspection, leakage and low accuracy are the main problems in the inspection, and this paper proposes a local convolution-based lead frame defect detection method for this problem. A subgraph of size 512 × 512 is first generated as

Table 1. Different Lead Frame repair results

No	Silver plated or not		Number of images		Total number of defects		Total number of defects		Accuracy	
	A	B	A	B	A	B	A	B	A (%)	B(%)
1	N		10		17		17		100	100
2	N		10		21		21		100	100
3	Y		10		8		8		100	100
4	Y		10		13		13		100	100
5	Y		10		19		18		94.74	94.74
6	Y		10		22		22		95.45	95.45

Table 2. Quantitatively describe the reconstruction effect of different Lead Frames

No	PSNR		SSMI		Meet the actual testing needs	
	A	B	A (%)	B (%)	A	B
1	31.747	27.233	92.315	89.88	Y	Y
2	31.587	27.572	92.357	90.12	Y	Y
3	31.541	27.355	90.107	87.58	Y	Y
4	31.307	27.120	93.823	90.89	Y	Y
5	31.924	27.572	90.114	88.61	Y	Y
6	30.750	27.113	93.202	89.52	Y	Y

the training dataset based on the base guide units within the Lead Frame. Incorporating the idea of local convolution, which enables the network to learn the overall structure better. It also avoids the appearance of artifacts, blurred colors, gray shadows and other problems. After obtaining the repaired image, the location of the defect is obtained by the residual method. From the results, it is not hard to see that the method detection results meet the actual detection needs and has been applied on the ground, and its feasibility has been verified. This method can be applied not only to the semiconductor field, but also to other similar defect detection methods and image restoration problems to provide feasible solutions for reference and ideas. The method can be subsequently optimized to further improve the speed and accuracy as well as the image restoration quality.

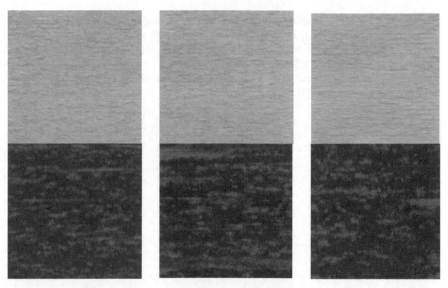

Fig. 9. Lead frame material trace image

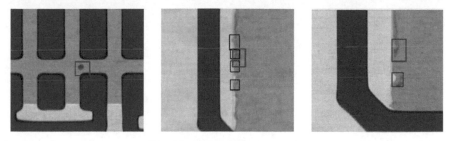

Fig. 10. Defect detection results

References

1. Saleh, E.H., Fouad, M.M., Sayed, M.S., Badawy, W., Abd El-Samie, F.E.: Automated fabric defect detection using à trous wavelet transform and bollinger band (BB). In: Proceedings of the 2020 9th International Conference on Software and Information Engineering (ICSIE) (ICSIE 2020). Association for Computing Machinery, New York, NY, USA, 131–134
2. Wang, Q., Yang, R., Wu, C., Liu, Y.: Surface defect recognition of cylinder head based on generative adversarial networks. Manuf. Auto. **42**(11):7 (2020)
3. Hebei University of Technology, 28 August (2020). CN111598877A
4. Liu, Y., Yu, H., Wang, J., Yu, L., Zhang, C.: Surface detection of multi-shape small defects for section steel based on deep learning. J Comp Appl, 1–8 (2021)
5. Liu, M.: Research on multiscale defect detection technology of water surface based on Gabor and RCNN. Zhejiang University, MA thesis
6. Zhishan, C.: Analysis and design of wafer defect detection system based on machine vision. Guizhou University, MA thesis

7. Akhyar, F., Lin, C.Y., Kathiresan, G.S.: A beneficial dual transformation approach for deep learning networks used in steel surface defect detection. In: Proceedings of the 2021 International Conference on Multimedia Retrieval (ICMR '21). Association for Computing Machinery, New York, NY, USA, 619–622
8. Zhu, T., GUO, Y., LI, Y.: Inpainting of ground-based on partial convolutional network. China Sciencepaper **17**(3), 269–273(2022)

YOLOx-M: Road Small Object Detection Algorithm Based on Improved YOLOx

Jiaze Sun[1,2,3] and Di Luo[1(✉)]

[1] Xi'an University of Posts & Telecommunications, Xi'an 710121, Shaanxi, China
sunjiaze@xupt.edu.cn, luodi4517@163.com,
2003200056@stu.xupt.edu.cn
[2] Shaanxi Provincial Key Laboratory of Network Data Analysis and Intelligent Processing,
Xi'an 710121, Shaanxi, China
[3] Xi'an Key Laboratory of Big Data and Intelligent Computing, Xi'an 710121, Shaanxi, China

Abstract. In order to improve the detection accuracy of small objects in road scenarios, a detection algorithm, YOLOx-M is proposed. First, the last layer of the backbone uses the MobileVit Block network, which learns a stronger global representation. Moreover, we use the CIOU loss function because it is more sensitive to location information. Finally, a moderate increase for the input scale to improve road small object detection accuracy. Experiments conducted on BDD100k and self-built datasets show that YOLOx-M has a mAP@0.5 of 0.5860 for six categories in the dataset improves 0.0481 over YOLOx, while the detection speed is 31.69 fps, which satisfies the real-time requirements. YOLOx-M has significantly improved the detection accuracy for road small objects, providing a better algorithm for detecting road small objects in autonomous driving environments.

Keywords: Small object · BDD100k · CIOU loss function

1 Introduction

With the continuous development of computer vision-related technologies, object detection algorithms are becoming more and more widely used for pedestrian and vehicle detection in the field of autonomous driving. However, small object detection in autonomous driving scenarios is often less effective when porting traditional object detection algorithms to autonomous driving platforms due to the complexity of the scenario and the real-time requirements. Single stage road object detection algorithms can use lightweight networks as the main structure of the algorithm framework, reducing the complexity of the model and meeting the real-time requirements of autonomous driving for object detection.

Common deep learning object detection algorithms fall into two categories. One is based on a two-stage object detection algorithm, where the first stage first extracts the object location region to ensure sufficient accuracy and recall. The second stage classifies the proposed box to seek a more accurate location. Usually, the type of object detection

© The Author(s), under exclusive license to Springer Nature Switzerland AG 2023
N. Xiong et al. (Eds.): ICNC-FSKD 2022, LNDECT 153, pp. 707–716, 2023.
https://doi.org/10.1007/978-3-031-20738-9_80

algorithm has higher accuracy but a slower detection speed. Common two stage detection algorithms include RCNN, Fast R-CNN, and so on. Another type of algorithm is the single stage object detection algorithm. The type of algorithm usually puts together location regression and category classification. The parameters are relatively more difficult to learn without doing detection and classification specifically. Usually, these algorithms are faster but less accurate. Common single stage object detection algorithms include OverFeat [2], SSD [3], and YOLO [4] series of algorithms.

YOLOx is an Anchor Free object detection framework introduced by MEGVII in 2021. That performs optimization of a sample allocation strategy and decoupled head part. YOLOx is a single stage object detection algorithm. The single stage object detection algorithm is divided into Backbone, Neck and Head. The current optimization methods for single stage object detection networks are divided into three main categories: modification of the backbone, modification of the enhanced feature extraction network and modification of the loss function.

Modifying the backbone means during the network's feature extraction of an image. The network has to extract as many features from the image as possible during the feature extraction process. Yu Qiang [5] proposed a multiscale YOLOv3 algorithm for road scene object detection. Two new additional scales were used to obtain more small object feature information so that while obtaining more small object information the network parameters become more complex and the training overhead is higher. In terms of input scale improvement, Qian Wu [6] improved the YOLOv5 algorithm for the real-time detection of traffic lights. The accuracy of YOLOv5 for traffic light detection was improved by a memory feature fusion network and by increasing the input scale, and the network did not experiment with other road objects.

In different scenarios, modify regression or classification losses to improve problem like masking, positive and negative sample imbalance, etc. YUE Xiaoxin [7] detects road small objects by introducing a Focal Loss function that solves the imbalance between positive and negative samples. This is more complex for the model and will increase the training overhead.

In summary, feature extraction for small objects in autonomous driving scenarios is difficult and prone to large overheads. To address these issues, this paper proposes a road small object detection algorithm, YOLOx-M. First, MobileVit Block is used to obtain global dependencies when the feature layer passes through the backbone. Second, in the prediction box for regression, the loss function uses the CIOU loss function that considers the location relationship. The CIOU loss function increases the model complexity relatively low, and the model training cost is low. Finally, the input scale of the network is upgraded so that the network can obtain as much feature information about small objects as possible to improve the detection accuracy of small objects.

2   YOLOx Algorithm

YOLOx is divided into three main parts, the backbone, path aggregation feature pyramid network (PAFPN) and decoupled head. The backbone uses the CSPDarkNet structure to extract the feature layers from the input images. PAFPN uses PANet, an FPN that upsampling and downsampling the features for feature fusion. The decoupled head is the classifier and regressor of YOLOx, it can prediction analysis on the prediction box.

3 YOLOx-M Algorithm

In this paper, a new algorithm YOLOx-M is proposed to improve the backbone and loss function from three perspectives respectively for the feature extraction of road small objects with a few features: (1) By using a larger scale of input, the network can extract feature information of small objects in pictures to a greater extent. (2) In the process of prediction box localization, for the situation that the prediction box and the ground truth box do not overlap. In this paper, we use the CIOU loss function to obtain the location information of the two boxes. (3) YOLOx-M using MobileViT Block structure in the backbone, we get better generalization ability by obtaining global dependency through the self-attentive based visual transformer (Fig. 1).

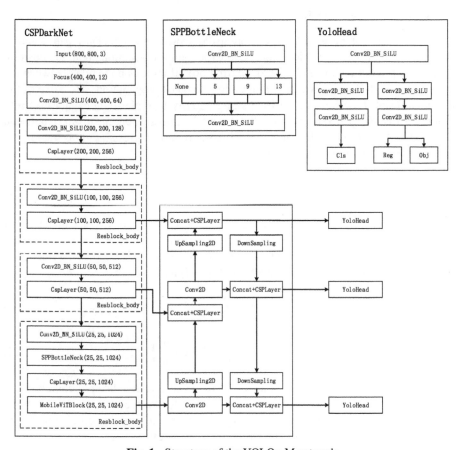

Fig. 1. Structure of the YOLOx-M network

3.1 Backbone Input Scale

In the process of feature extraction by the backbone feature network, the network down-sampling the object by 8x, 16x and 32x. During which the features of small objects are gradually compressed to retain more detailed features. For feature extraction, the three active feature layers are output from the backbone to the augmented feature extraction network. In the autonomous driving scenario, the localization information fades out in the higher order feature maps due to the small size and low resolution of the road small objects. By increasing the input scale while acquiring more small object detail information in the shallow convolutional layer can detect small objects on the image more accurately. However, during the experiments, it was found that there is a certain range for increasing the input scale of images. And if it exceeds a certain range, it will produce an inverse effect and the ability of the model to extract feature information will become weaker.

3.2 Regression Loss Function

YOLOx uses IOU for the regression positioning of the prediction box. During the filtering of the prediction boxes, IOU does not take into account information about the position between the prediction box and the ground truth box. If there is no overlap between the two boxes, then IOU $= 0$, which will not allow for backpropagation. Also, one of the more significant drawbacks of IOU is its insensitivity to scale. IOU cannot take into account the distance between objects. CIOU can take into account the distance between the object and the Anchor. This makes the prediction box regression more stable. The specific formula of CIOU is shown in Eq. (1)

$$CIOU = IOU - \frac{\rho^2\left(b, b^{gt}\right)}{c^2}\alpha_v \tag{1}$$

where the α is the weight function. The v is used to measure the consistency of the aspect ratio. The c denotes the diagonal distance between the prediction box and the minimum external matrix of the ground truth box. The $d = \rho^2\left(b, b^{gt}\right)$ denotes the Euclidean distance. The α and the v are calculated as in Eqs. (2) and (3)

$$\alpha = \frac{v}{1 - IOU + v} \tag{2}$$

$$v = \frac{4}{\pi^2}\left(\arctan\frac{w^{gt}}{h^{gt}} - \arctan\frac{w}{h}\right)^2 \tag{3}$$

where the CIOU is a direct minimization of the distance between the prediction box and the ground truth box after normalization. So that the loss can converge faster than the IOU loss, shorten the training cycle, and improve the model training efficiency. At the same time can get a prediction model with higher accuracy.

3.3 Backbone Global Dependency Acquisition

The MobileViT is a lightweight general purpose vision Transformer for mobile devices and lightweight neural network models are useful in vision tasks. The road object detection algorithm as an important part of autonomous driving can also be combined with a self-attentive mechanism to extract some hard-to-capture features in the network through

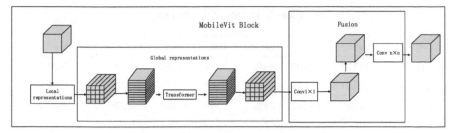

Fig. 2. MobileVit Block network structure

excellent generalization capabilities. In order to get more global reliance on the network, this paper adds the MobileVit Block module to the last layer of the backbone. The overall structure of the network is shown in Fig. 2.

The MobileVit Block is shown in Fig. 2, it consists of three modules: the local information encoding module, the global information encoding module, and the feature fusion module. The functions of these three modules are to extract the local feature information, extract the global feature information, and fuse the feature information. The MobileVit Block uses standard convolution and Transformer to effectively combine local and global visual representation information. This gives the network a much better performance.

4 Experiment

4.1 Experimental Environment

In this paper, 7000 images were extracted from the BDD100k dataset as the training set, 1000 images were extracted as the validation set, and 1734 images were extracted as the test set. The number of ground truth boxes in the training set, validation set, and test set are shown in Table 1. In order to test the generalization ability of the YOLOx-M, the self-built dataset was also tested. There are 10 categories in the BDD100k, and since there are fewer objects in the four categories of Bike, Motor, Train, and Rider, only Car, Person, Truck, Bus, Traffic_light, Traffic_ sign were taken to balance the number of objects. The pre-training weights for YOLOx are from YOLOx-s, and the model was trained on an Nvidia Tesla K80 12G GPU and tested on an NVIDIA GeForce GTX 2060 GPU. To verify the effectiveness of improving the accuracy of road small object detection in autonomous driving scenarios in this paper, the following questions need to be answered: (1) Whether the prediction box of road small objects can be more accurately regressed by location information. (2) How more feature information can be obtained for road small objects.

4.2 Experimental Design and Procedure

To address the problem (1), experiments were designed to obtain the relative positions of the prediction box and the ground truth box using the CIOU loss function in the

Table 1. Details of the data set

Category	Number	Car	Person	Truck	Bus	Traffic_light	Traffic_sign
Training set	128552	72109	8939	3021	1157	19062	24264
Validation set	18191	9845	1467	495	172	2737	3475
Test set	31885	17287	2578	749	295	4897	6079

prediction box regression process. With the inclusion of scaling and penalty terms, the model can regress the prediction box more consistently.

To address the problem (2), in order to extract more effective information during feature extraction. The network increases the image input scale so that the output to the shallow effective feature layer of the enhanced feature extraction network obtains more detailed information. Afterwards, the MobileVit Block network structure is used in the last layer of the backbone to obtain global dependencies.

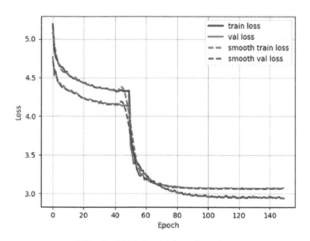

Fig. 3. YOLOx-M loss function

As shown in Fig. 3, during the training process, the model backbone is first frozen in the first 50 epochs, the feature extraction network does not change, and the network is only fine-tuned. During 50 to 150 epochs, the freezing of the model backbone ends, the feature network and all parameters will be changed. Around 100 epochs, the losses start to converge. After 120 epochs, the model shows a more obvious overfitting phenomenon, which is specifically represented in the form of reduced losses in the training set and unchanged or larger losses in the validation set. At the same time, mAP@0.5 is lower than the calculated mAP@0.5 of the previous model.

4.3 Assessment Indicators

In this paper, mAP@0.5 is used as the evaluation benchmark of the detection model. Specifically, m is the average value of AP0.5 for different categories, and 0.5 in AP0.5

means that the threshold value of IOU is 0.5. Where AP needs to be calculated for Precision and Recall of the model, as shown in the Eq. (4–6)

$$Precision = \frac{TP}{TP + FP} \tag{4}$$

$$Recall = \frac{TP}{TP + FN} \tag{5}$$

$$AP = \frac{1}{101} \sum_{i=0}^{100} Precision(Recall = \frac{i}{100}) \tag{6}$$

where the metric of Precision in the formula measures the accuracy of the model classification, which means, the proportion of positive samples in the total sample. The metric of Recall refers to the proportion of positive model predictions in the total positive sample. The metric of TP is the number of accurately predicted labels. The metric of FP is the number of false detections of non-existent objects or false detections of already existing objects. The metric of FN is the number of missed detections of objects.

4.4 Experimental Results

Experiment (1) used the CIOU loss function, under the same road object detection dataset detection accuracy, as shown in Table 2. The average increase of mAP@0.5 in six categories is 0.0053. For small objects, mAP@0.5 of Traffic_light and Traffic_sign improves by 0.0104 and 0.0078 respectively, which is higher than the average. This suggests in favor of the location information of the CIOU loss function, the overlap rate and the penalty term, more small objects allow the model to get a more accurate approximation of the prediction box to the ground truth box through. The results of the experiment were as expected.

Table 2. Comparison of detection models mAP@0.5

Class name	IOU	CIOU
Car	0.7160	0.7175
Person	0.5349	0.5335
Truck	0.4735	0.4755
Bus	0.4379	0.4495
Traffic_light	0.5265	0.5369
Traffic_sign	0.5387	0.5465
average value	0.5379	0.5432

Experiment (2) took the measure of changing the input scale of the backbone and the use of MobileVit Block to improve the network. Under the same training weights, the six object types and detection accuracies based on the BDD100k dataset are shown in Table 3. The average increase of mAP@0.5 in 6 categories is 0.0481. For small objects, Traffic_light and Traffic_sign improved by 0.0630 and 0.0667 respectively, which are higher than the average values. This indicates that more small objects acquired more features through the backbone. The results of the experiment are as expected.

Table 3. Comparison of detection model mAP@0.5

Class name	YOLOv5	YOLOx	Paper 1[5]	YOLOx-M
Car	0.6277	0.7160	0.7398	0.7566
Person	0.4613	0.5349	0.5651	0.5876
Truck	0.3836	0.4735	0.5733	0.5035
Bus	0.3023	0.4379	0.5354	0.4736
Traffic_light	0.3463	0.5265	0.4947	0.5895
Traffic_sign	0.3859	0.5387	0.5770	0.6054
average value	0.4178	0.5379	0.5809	0.5860

YOLOx-M was also compared with other detection algorithms in experiments. The average values of mAP@0.5 in YOLOv5 and paper 1[5] are 0.4178 and 0.5809, respectively. The average mAP@0.5 of the algorithm YOLOx-M proposed in our paper is 0.5860, which is higher than other comparison algorithms. YOLOx-M provides a new choice for road small objects in autonomous driving scenarios.

4.5 Comparative Analysis of Experimental Effects

In order to test the generalization ability of our model, the comparison experiment uses a self-built dataset. Which contains two scenes, the Scene 1 is a road scene inside a tunnel and the other is a highway scene at night. The images inside the tunnel are generally dim and poorly lit, which makes detection more difficult. The experiment results show that the object confidence of YOLOx-M is higher than that of YOLOx, and the detection in the images is generally better (Fig. 4).

There are four cars and one truck in this scenario. Furthermore, two cars and one truck were detected with a higher confidence level using YOLOx-M than that derived directly from YOLOx. And the overall results were better than the YOLOx detection results. In the nighttime highway scenario, the YOLOx algorithm detected five cars and one bus, while the algorithm in our paper detected one more car, with a higher confidence level than YOLOx. In summary, the accuracy of YOLOx-M is significantly improved over YOLOx in both BDD100k and self-built datasets.

5 Conclusion

Aiming at the problem that single stage object detection algorithms are difficult to detect small objects and have insufficient feature extraction in autonomous driving scenarios,

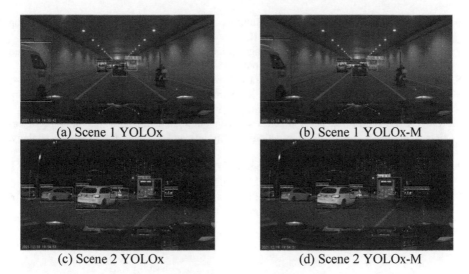

(a) Scene 1 YOLOx

(b) Scene 1 YOLOx-M

(c) Scene 2 YOLOx

(d) Scene 2 YOLOx-M

Fig. 4. YOLOx-M detection results

this paper proposes a detection algorithm YOLOx-M for small objects in road scenarios. YOLOx-M performs well in the BDD100k dataset as well as in self-built datasets. YOLOx-M is 0.481 higher than YOLOx's mAP@0.5, reaching 0.5860 with relatively high detection accuracy. At the same time, the speed of detecting each image through the system reached 31.69 fps, which exceeds 30 fps. This satisfies the requirement of 30 fps for object detection by autonomous driving. The algorithm proposed in this paper has been improved for the backbone and loss function. The authors' subsequent research will do more in-depth research on the feature fusion of road small objects for FPN.

Acknowledgement. The work is supported by the Key Industrial Chain Core Technology Research Project of Xi'an (2022JH-RGZN-0028) and the Special Fund for Key Discipline Construction of General Institutions of Higher Learning from Shaanxi Province.

References

1. Lin, T.Y., Dollár, P., Girshick, R.: Feature pyramid networks for object detection. In: Proceedings of the IEEE conference on computer vision and pattern recognition, pp. 2117–2125 (2017)
2. Sermanet, P., Eigen, D., Zhang, X.: Overfeat: integrated recognition, localization and detection using convolutional networks. arXiv:1312.6229 (2013)
3. Liu, W., Anguelov, D., Erhan, D.: Ssd: single shot multibox detector. In: European conference on computer vision, pp. 21–37. Springer, Cham (2016)
4. Bochkovskiy, A., Wang, C.Y., Liao, H.Y.M.: Yolov4: optimal speed and accuracy of object detection. arXiv:2004.10934 (2020)
5. Qiang, Y., Kuan, W., Hai, W.: A multi-scale YOLOv3 detection algorithm of road scene object. J. Jiangsu Univ. (Natural Science Edition), 628–633 (2021)

6. Wu, Q., Guozhong, W., Guoping, L.: Improved YOLOV5 traffic light real-time detection robust algorithm. Comp. Sci. Exp., 231 (2022)
7. Xiaoxin, Y., Junxia, J., Xidong, C., Guangan, L.: Road small target detection algorithm based on improved YOLO V3. Comp. Eng. Appl., 218–223 (2020)
8. Viola, P., Jones, M.J.: Robust real-time face detection. Int. J. Comp. Vision, 137–154 (2004)
9. Hariharan, B., Arbeláez, P., Girshick, R.: Simultaneous detection and segmentation. In: European conference on computer vision, pp. 297–312. Springer, Cham (2014)
10. Kang, K., Li, H., Yan, J.: T-cnn: tubelets with convolutional neural networks for object detection from videos. In: IEEE Transactions on Circuits and Systems for Video Technology, pp. 2896–2907 (2017)
11. Everingham, M., Eslami, S.M., Van Gool, L.: The pascal visual object classes challenge: a retrospective. Int. J. Comp. Vision, 98–136 (2015)
12. Lin, T.Y., Goyal, P., Girshick, R.: Focal loss for dense object detection. In: Proceedings of the IEEE international conference on computer vision, pp. 2980–2988 (2017)

Printed by Printforce, the Netherlands